GAIAM REAL GOODS

SOLAR LIVING
SOURCEBOOK

ELEVENTH EDITION

Robert C. Davis 5/04

THE REAL GOODS SOLAR LIVING BOOK SERIES

The Real Goods Solar Living Sourcebook,
Eleventh Edition
edited by John Schaeffer

The Independent Home: Living Well with Power
from the Sun, Wind, and Water
by Michael Potts

The Real Goods Independent Builder: Designing &
Building a House Your Own Way
by Sam Clark

The Straw Bale House
by Athena Swentzell Steen, Bill Steen, David
Bainbridge, with David Eisenberg

Serious Straw Bale: A Home Construction Guide
for all Climates
by Paul Lacinski & Michel Bergeron

The Beauty of Straw Bale Homes
by Athena & Bill Steen

The Passive Solar House: Using Solar Design
to Heat and Cool Your Home
by James Kachadorian

The Rammed Earth House
by David Easton

Building With Earth
by Paulina Wojciechowska

The Earth-Sheltered House: An Architect's Sketchbook
by Malcolm Wells

Treehouses
by David Pearson

Circle Houses
by David Pearson

The Natural House
by Daniel D. Chiras

Mortgage-Free!
Radical Strategies for Home Ownership
by Rob Roy

This Organic Life
by Joan Gussow

Wind Energy Basics: A Guide to Small and
Micro Wind Systems
by Paul Gipe

Wind Power for Home & Business:
Renewable Energy for 1990s and Beyond
by Paul Gipe

Hemp Horizons: The Comeback of the
World's Most Promising Plant
by John W. Roulac

A Place in the Sun:
The Evolution of the Real Goods Solar Living Center
by John Schaeffer &
the Design/Construction Team

Real Goods Trading Company in Ukiah, California, was founded in 1978 to make available new tools to help people live self-sufficiently and sustainably. In 2001 Real Goods merged with Gaiam, Inc., to become Gaiam Real Goods. Through seasonal catalogs, a biannual *Solar Living Sourcebook*, a website (www.realgoods.com), and retail outlets, Gaiam Real Goods provides a broad range of tools for independent living.

"Knowledge is our most important product" is the Real Goods motto. To further its mission, Gaiam Real Goods has joined with Chelsea Green Publishing Company to co-create and co-publish the Real Goods Solar Living Book series. Many of these books are quoted extensively throughout the *Sourcebook*. The titles in this series are written by pioneering individuals who have firsthand experience in using innovative technology to live lightly on the planet. Chelsea Green books are both practical and inspirational, and they enlarge our view of what is possible as we enter the next millennium.

Stephen Morris
President, Chelsea Green

John Schaeffer
President & Founder, Real Goods

ELEVENTH EDITION

GAIAM REAL GOODS

SOLAR LIVING SOURCEBOOK

The Complete Guide to Renewable Energy Technologies
and Sustainable Living

EXECUTIVE EDITOR John Schaeffer

TECHNICAL EDITOR Doug Pratt

with Contributions from the Real Goods Staff and Friends

A REAL GOODS SOLAR LIVING BOOK

Distributed by:
Chelsea Green Publishing Company
White River Junction, Vermont

Printed in the United States of America with soy-based inks on recycled paper
1 2 3 4 5 6 7 8 9 10

ISBN 0–916571-04-1

The Real Goods Solar Living Sourcebook is the eleventh edition of the book originally published as the *Alternative Energy Sourcebook*, with over 400,000 in print, distributed in 44 English speaking countries.

DISTRIBUTED BY:
CHELSEA GREEN PUBLISHING COMPANY
P.O. Box 428, White River Junction, Vermont 05001

Gaiam Real Goods
Real Goods Solar Living Center
P. O. Box 836
13771 South Highway 101, Hopland, California 95449
Business office: 303-464-3600
To order: 1-800-762-7325 or fax 800-508-2342
For technical information: 707-468-9292 ext. 8700
For renewable energy orders: 800-919-2400 or fax 707-462-4807
Renewable energy email: techs@realgoods.com
Email: realgood@realgoods.com
Home page: http://www.realgoods.com

FOR THE EARTH

May we preserve and nurture it in our every action.

CONTENTS

ACKNOWLEDGMENTS

It is impossible to thank everyone who has been responsible for all eleven editions of our *Sourcebook* since its inception in 1982. The fact that we've sold over 400,000 copies in 44 English-speaking countries and that the *Sourcebook* now sits on all reputable renewable energy bookshelves is a testament to the success of our many contributors. Doug Pratt (our own Dr. Doug) has done a tremendous job of being a prolific writer for this edition and has managed to stay on schedule while continuing his day job of selling and supporting renewable energy systems at Real Goods Renewables. Doug's work on both the 10th and the 11th edition *Sourcebooks* has been thoroughly invaluable. Claudia Kaufman, our primary graphic artist on this edition, has been a joy to work with, has performed an amazing amount of work flawlessly, and has been diligent beyond compare in keeping us all on schedule. Claudia has been aided behind the scenes by Creative Director Marcie Long.

Many thanks to Steve Rogers, Real Goods Renewables Manager; Jeff Oldham, Real Goods Design and Consulting Manager; and all of our RG Renewables technicians who have supported and found the products featured within and shared their research and inspiration with us. We are also very grateful to our new partners at Gaiam, including Gaiam CEO Jirka Rysavy, President Lynn Powers, and Creative VP Howard Ronder, who have continued to support Real Goods, the *Solar Living Sourcebook* and all of our educational activities. Thanks to Chelsea Green President Stephen Morris, a colleague and friend for over 20 years, for helping me to understand the book trade and how the *Sourcebook* fits in. And, finally, for their financial support, I'd like to thank all 10,000 former Real Goods shareowners, who made this exhilarating adventure possible.

—John Schaeffer

The Real Goods Nonprofit Institute for Solar Living

The Real Goods Institute for Solar Living, operating as a 501(C)3 nonprofit organization, educates and promotes the knowledge of renewable energy and other sustainable living practices through its hands-on workshops; through SolFest, an annual renewable energy educational event happening in August each year; and through displays and demonstrations at the solar and wind-powered Solar Living Center in Hopland, California. There is more information about the Solar Living Center and Institute workshops in the Appendix or on our website at www.realgoods.com. Call 800-762-7325 for the current schedule, or contact Institute personnel directly at 707-744-2017 for information about tax deductible donations and other Institute activities. The Institute encourages you to become a Partner for $35 per year and support its inspirational educational activities. To become a Partner call 707-744-2012.

INTRODUCTION

MARKET CONDITIONS FOR THE CONVERSION from our fossil fuel–based economy to a solar economy have never been better. In the two years since we published our *10th Edition Sourcebook*, an energy meltdown has occurred in California and the Western United States. On the heels of California's ill-conceived deregulation scheme, out-of-state utilities have cashed in on an energy scarcity opportunity, increasing prices for electricity and natural gas as much as sixfold.

From the peoples' standpoint, the stage is set for a rapid conversion from a fossil fuel–based energy economy to one based on renewable energy.

From the Gaiam Real Goods perspective, this is not a new or unexpected turn of events. Before long the entire country will be coping with the same forces that are shaping California's energy crisis. This is good news for solar energy.

When electricity prices climb, the whole world wants to go solar. We continue to add staff as demand spirals upward, but the issues are more endemic than can be handled by just one company. The world needs to come to its senses and eliminate its dependence upon fossil fuels. We see lots of positive movement in this direction.

The era of oil and fossil fuels is coming to an end as we move through the early days of the third millennium. Behind us (we hope) is the wanton devastation and destruction of natural habit, while before us is the bountiful opportunity for a fruitful and fulfilling future. We have to play our cards correctly, and this means understanding our finite energy resources.

The era of oil and fossil fuels is coming to an end as we move through the early days of the third millennium.

From the people's standpoint, the stage is set for a rapid conversion from a fossil fuel–based energy economy to one based on renewable energy. Obstructionist policies by entrenched political and business interests, however, threaten to engulf our planet in a deathly haze of greenhouse gases. Whether we have the foresight to take heed of our natural limits, or whether we will continue our unfettered consumption becomes the ultimate question for our species. Our fate is in our own hands.

Ours is the first generation in history to know that we are in danger of self-destruction. Signposts are everywhere:

- Fresh water is scarce. Worldwide water use has tripled since 1950, resulting in huge water deficits in key river basins in China and India. In India, with its one billion population, the extraction of water from aquifers is twice its annual recharge. What will happen when the population grows by an expected 600 million by 2050.

- We have exceeded the sustainable yield of oceanic fisheries. Eleven of the world's 15 most important fishing areas and 70% of the major fish species are either depleted or overexploited.

- The world's forests face similar devastation. Nearly half of the forests that once covered the Earth are gone. In just 15 years, between 1980 and 1995, more than 400 million acres—an area larger than all of Mexico—were lost. And the rate of forest destruction is accelerating. The number one indicator of the Earth's failing health is the shrinking number of plant and animal species. Of 242,000 plant species on the planet, 14%—or 33,000—are threatened because of habitat destruction. Of 9,600 bird species in existence, 67% are in decline and 11% are facing extinction. The outlook for fish and mammals is equally bleak.

The U.S. Congress' own Office of Technology Assessment estimates that all known oil reserves will have been depleted by 2037.

Our legacy from the era of oil and fossil fuels is global warming. Carbon emissions exceed the capacity of the Earth's natural systems to "fix" carbon dioxide. Since scientists began recording average annual Earth temperatures in 1866, the sixteen warmest years on record all occurred since 1980. 1998 was the warmest year on record, and also represented the largest single year increase ever. The impact of unchecked global warming is the stuff of a Hollywood disaster movie, with oceans rising an estimated 3″–10″ in the next century. Moreover, the impact on other species will likely be more devastating than on humans.

The U.S. Congress' own Office of Technology Assessment estimates that all known oil reserves will have been depleted by 2037. And yet, the playing field remains tilted in favor of the entrenched oil interests. Over $100 billion in subsidies are available to the fossil fuel industry, while incentives for renewables are minimal. One hopeful sign is the renewable energy buydown programs administered on a state level.

The answer is so obvious—conservation!

There are strong factions—oil companies and utilities to name two—who have vested interests in maintaining the status quo. In the 1996 election, oil and gas companies gave $11.8 million to congressional candidates in order to protect tax breaks worth at least $3 billion. George W. Bush makes no secret of his ties to big oil, nor the extent of contributions received from it during his campaign. The powers of entrenchment are indeed formidable but there is also lots of good news.

The good news is that the revolution is occurring at the grassroots level. There has been a blossoming of many "culturally creative" pursuits. Sales of organic food grew 19-fold from $180 million in 1980 to $3.5 billion in 1996. The same kind of "paradigm shift" can be seen with the current energy "crisis," which we prefer to call a giant energy "opportunity." With rolling blackouts accelerating in the summer of 2001, energy consciousness is unquestionably rising. The only true "blackouts" are the intelligence blackouts on the part of politicians. The answer is so obvious—conservation! Do the math!

Fact: There are 1,000 power plants in California putting out 53 MW (megawatts) on average for a total combined output of 53,204 MW. Thus with 24 hours in a day, the average California power plant can put out 1.3 gWh (one gigawatt is 1000 MW) per day. There are 34 million people in California and 15 million households.

Simple conservation would enable the state of California to shut down 100 average-sized power plants, or better yet, leave them running and eliminate any need to construct new ones for years. For a mere cost of $135 million, the state could solve this "crisis."

- If every household in California replaced 4 (average 100 watt) incandescent light bulbs with 4 (equivalent 27 watt) compact fluorescent light bulbs, burning on average 5 hours per day, we would save 22 gWh per day—or enough energy saved to shut down 17 power plants.

If the state bought these lamps for every household at $2 each, total cost would be $120 million.

- If every household in California replaced one average-flow showerhead with an energy saving showerhead we would save 1.3 kWh per day per household or 19.2 gWh per day—or enough energy saved to shut down another 15 power plants.

If the state bought these low flow showerheads for every household at $1 each, total cost would be $15 million.

If every household in California installed a solar hot water heater, which saves 5.8 kWh/day, we would cumulatively save 87 gWh/day—or enough energy saved to shut down another 67 power plants.

Conclusion: Simple conservation would enable the state of California to shut down 100 average sized power plants, or better yet, leave them running and eliminate any need to construct new ones for years. For a mere cost of $135 million, the state could solve this "crisis."

(Energy facts were derived from California Energy Commission's website: www.energy.ca.gov)

By spending just two days of utility company bailout money, the crisis would be over.

CONSIDER THIS: Since the first rolling blackouts occurred in California in January 2001, the governor has spent nearly $8 billion bailing out the utilities. He continues to spend $73 million per day or $3 million per hour—this is money that will never be recouped and will go down in history as one of the greatest public ripoffs ever. By spending just two days of utility company bailout money, the crisis would be over.

Solar energy is the other piece to the equation. Solar provides maximum power just when it's needed—in the middle of the summer afternoon, when air conditioning loads are heaviest and the huge bottleneck occurs in transmission lines. There are signs everywhere that solar is gaining and moving mainstream. Federal, state, and local governments are now subsidizing renewables through manufacturer and end-user rebates. The California Energy Commission (CEC) enables us to sell renewable power in California for 50% off or give a rebate for

While the use of coal, oil, and nuclear power during the 1990s expanded by only slightly more than 1% each annually, the use of photovoltaics (solar electric modules) has grown by 17% annually and wind-generated electricity by 26%.

$4.50 per watt. The cities of San Francisco, Alameda County and the Sacramento Municipal Utilities District are strongly considering huge PV arrays to keep their municipalities running during summer blackouts.

Prices keep declining. With the price of solar now approaching $3–$4 per watt (after a 50% state rebate) on installed PV, the payback period is 11 years at a price of $0.14/kWh and only 5½ years at $0.28/kWh. With utilities now paying up to $0.40/kWh, the return on investment for the homeowner is now approaching what it was for dot.com stocks in the halcyon days of the NASDAQ. All around the world, there are signs that the stage is set for positive change. While the use of coal, oil, and nuclear power during the 1990s expanded by only slightly more than 1% each annually, the use of photovoltaics (solar electric modules) has grown by 17% annually and wind-generated electricity by 26%. While there are still two billion people in the world without access to electricity (and another billion who have it less than ten hours per day), there are now 500,000 homes worldwide (mostly in third-world villages) powered by photovoltaics. Maybe they know something that we don't know—that power from the sun is the best power you can get.

We opened our first Real Goods store in Willits, California in June, 1978. Our mission was, and still is, to demonstrate and provide renewable energy alternatives. Real Goods is better positioned than ever to help you realize your solar dreams. In January 2001 we merged with Gaiam Inc., a like-minded company that promotes healthy and sustainable lifestyles through five catalog titles, a website (www.gaiam.com), and a large wholesale division. Now we are Gaiam Real Goods—the same people but with deeper resources. Over the years, we've gathered an unbeatable team of renewable energy experts, perhaps the best-informed and most experienced group of specialists on the planet. Our commercial consulting division (Real Goods Design and Consulting Group) works with eco-tourism resorts, developers, architects, builders, and green businesses to design, procure, install, and maintain renewable energy systems. Real Goods Renewables specializes in residential renewable energy design and sends out its own catalog twice per year.

We are on the brink of a major change in consciousness. Whether the scales tip toward exceeding our natural limits or proactively managing our fate is unknown.

We are on the brink of a major change in consciousness. Whether the scales tip toward exceeding our natural limits or proactively managing our fate is unknown. We know which side we're on! As we go to press with this 11th edition of the *Sourcebook*, the buzz about energy is everywhere, providing us with an unprecedented opportunity to educate others about self-reliance, renewable energy, and sustainable living.

My efforts are now focused on our Solar Living Center, our Renewable Energy Division and our nonprofit Institute for Solar Living. Our 12-acre demonstration site in Hopland, California, which is fully powered by wind and

It is not too late to build a society that is environmentally sustainable; where water is safe to drink, air is safe to breathe, and communities—even countries—share resources equitably.

solar energy and demonstrates permaculture, sustainability, and ecological building practices. The Institute provides education and inspirational solutions to our "beyond-the-limits" predicament.

We host more than 160,000 visitors every year and our annual SolFest celebration, held each August, presents the seminal thinkers of our time to thousands of kindred spirits. It's an occasion as inspirational as it is fun. If you'd like to support our nonprofit Institute, please consider a donation or becoming a partner (707-744-2017).

We know our species can't survive the continued loss of biodiversity, the decimation of forests, the shrinking of ocean fisheries, and the fouling of our atmosphere. We simply have to make things right. It is not too late to build a society that is environmentally sustainable; where water is safe to drink, air is safe to breathe, and communities—even countries—share resources equitably. Let us work toward a future where our great-grandchildren can look back and say, "Thank goodness they finally came to their senses. . . ."

For The Earth,

John Schaeffer

John Schaeffer
Founder & President

CHAPTER 1

LAND & SHELTER

"Buy land, son; the Lord isn't making more of it."

—Attributed by Mark Twain to his father

THERE IS NOTHING MORE REAL IN THIS LIFE THAN LAND. Earth gives us each thing we have, and at the end we return those gifts to it. "Owning land" is a relatively modern notion and by no means a universal practice. In fact, to many of the world's peoples who live intimately with the land, the idea of ownership is incomprehensible. We who seek an alternative way of living also find dissonance between the accident of "ownership" and our visceral fealty to the land, which owns us all. We resolve this contradiction by becoming the land's stewards, assuming responsibility to and for the land that gives us life. By looking about ourselves at a globe marred by exploitation, overcrowding, and shortsightedness, we have discovered that only through honoring the land, with all its creatures and qualities, can we truly honor ourselves.

The search for good land is like the search for a mate. We hope to find sustenance, partnership, comfort, and stability.

The search for good land is like the search for a mate. We hope to find sustenance, partnership, comfort, and stability. In the end, if we truly wish to settle peacefully and productively, we must find love and passion for the land as well. This nation began in a wave of western migration motivated by a hunger for land and independence, but somehow the habit of migration stuck, while the regard for land went astray. Our national failure to settle well has led to our tragic pattern of exploitation, insensitivity, restlessness, and estrangement. Real estate sales and the vertiginous overvaluation of land depends on the predictability of this pattern: the average American family uproots and moves every four to seven years, and our whole culture has come to thrive on this lonely rootlessness. The "back to the land" movement, which began in the late 1960s and which fostered the renewable energy industry during a prolonged petroleum winter, differs from prevailing American culture precisely in its love for the land.

Most of us thrill to the magnificence of mountain vistas or the glittering promise of tropical islands in the sun. Such prospects, even as photographs in magazines or on the flickering television screen, make our spirits soar. But those images depict land on a grand scale; for most of us, the land we can own is a much smaller plot, and the emotions we feel about it are subtler. Coming as we do from a time when holdings grew ever smaller and more urban, and where mates and property were described as chattels, a word derived from the same root as cattle, it is hardly surprising that we in the back-to-the-land generation have had to look deeply into our hearts and our past to rediscover love, whether for another human or for a piece of land. Even our words undercut the strong but subtle ties that grow between ourselves, our human partners, and our land.

Almost a century of easy energy has led to habits of use and patterns of settlement completely divorced from the inherent qualities of the land. As we begin to rediscover our connection with the earth, it is only to be expected that we should have lost the art of choosing land well. A conventional house is situated not to take best advantage of the

Almost a century of easy energy has led to habits of use and patterns of settlement completely divorced from the inherent qualities of the land.

sun and the special features of a site, but to fit within arbitrary boundaries and to accommodate the whims of the planner and catskinner who bring in the road and powerlines; such persons are not responsible to the land and the families that will live on it, but to the developers who expect a quick profit. With horrible uniformity, buildings accumulate in dark gulches, cheek-by-jowl along narrow streets, or like pimples on a smooth mountain brow, without regard for what is best for land, dwelling, or family. Occasionally, more sensitive developers proudly leave some of the more imposing trees, or employ building designs and materials evolved over centuries, but seldom as well as possible. We who are third- and fourth-generation inhabitants of these cold boxes have neither the tools nor the tradition to evaluate, with meaning, the suitability of these structures for the lives we wish to lead.

In a triumphant throwing-out of the baby with the bath water, many back-to-the-land pioneers sold their luxuries, packed up their necessities, and headed for the boonies. We admire their spirit, and acknowledge that there was wisdom as well as desperation in their movement away from the urban center. Since the beginning of land stewardship, back when agriculture was being invented and humans took responsibility for the abiding fertility of the land, it has been wise counsel to seek land that no one else wants, and make it home. In a few brutal years of inadequate shelter—flapping plastic windows, leaking roofs, flickering lamplight, harsh mornings on splintery floors—we regained the immediate sense of the elements from which our energy-rich tradition had sheltered us, and a renewed understanding of the value of a well-conceived, well-built, and well-provisioned home. For many of us now, a pilgrimage to the denatured houses of our past, of our parents and our unawakened peers, is fraught with discomforts and puzzlements we are loathe to share with them: how could we have lived this way, so heedlessly, so wastefully, so uncomfortably? What great energy, what denial, is invested in ignoring the unsuitability of these cheerless domiciles, whose inhabitants are usually awaiting a chance to move to newer, larger, better-situated discomfort!

A new tradition, more than a quarter of a century old, has grown up now, and a new wave of settlers is moving back to the land without suffering the privations of those early pioneers. These lucky souls have as their guide and inspiration all of the attempts, failures, renewed trials, and successes of the original back-to-the-landers. Today our vision of what works, and what does not, is clearing, and the tools and techniques found in this Sourcebook are better than ever before. By contrast, the crowded suburban single-family dwellings distill the worst of transient Americana, wind whistling coldly between ticky-tacky unshared walls through a space too small to garden. Co-housing, be it intentional or from necessity, where families huddle warmly together within shared walls, is a better way to crowd people. Most of us spend much of our lives crowded thus, and many prefer it; the point is, it should be done well, so families may thrive. A hardy few will wish to live stoutly independent beyond the end of pavement and powerlines, and for them the solutions of factory housing—stud and mud walls, aluminum sash windows, all-electric heating—are also unworkable. For them, the stuff of survival is found in simple solutions: indigenous materials, inspirations from the region's original inhabitants, and self-built homes that look to elemental forces (the sun, wind, falling water, and unfailing bounty of the soil) for sustenance.

The sun works. It is the original source of all earthly energy. All land bends first to the sun, and to its minions, wind and weather. Even the flattest plain and craggiest hill slopes have clement spots sheltered from the tempest, often protected by no more than a slight fold in the earth, or a line of trees. Geomancers (diviners of earth signs) claim to be able to find these spots, as dowsers find buried water, by sensing the feng shui, the flow of earth, air, fire, and water. Much is made, in Asia, of the meeting of feng shui master and architect when a skyscraper is planned. On a homelier scale but with equal solemnity, in the woods and mesas, hill slopes, prairies, and rich bottomlands of this continent, a generation of conscious homebuilders deliberates the best accommodation to seasonal sunlight, prevailing wind and water flows, available local materials, and appropriate technologies for making a home.

It takes time to learn a site. Too often, we come as urbanized new settlers from afar, and the secrets of storms, heat and cold, and successful regional designs are mysterious to us. Under time pressure, we act hastily, when the sense of the land abides and so must be divined. A site cannot be known in less than a full cycle of seasons—a year— or with less effort than by living with it for every hour of every day through every season. Perhaps where a longer cycle, like the proverbial hundred-year winter pertains, we may not learn the land for seven years, and may still be surprised after seven times seven. Traditionally, at the end of the road, old-time residents born of native parents hold that only by growing up on the land can it be known. To them, a half century is far too short a perspective. Newcomers are wise to solicit and hearken to their quiet wisdom.

The sun works. It is the original source of all earthly energy.

In our view, the oldest and most honorable inhabitants are the aboriginal trees, grasses, and the whole community of native biota. Because this community has endured, many of its lessons can be learned in an afternoon, if we are still and pay attention: those wind-sculpted trees bending southeast have taken root on the south slope below the rim of a little ridge. In their lee on the western side, moss and an orchid grow. From this modest observation we can see how to build a home that will be proof against a chill prevailing northwest wind, and below a berm facing just east of south. These gross considerations can (and should) be incorporated in our plans. When there is an acre of land or more, we must devote many pleasant afternoons to judging the many sites and orientations. Stories abound of how, finally, in conducting this gentle homage of learning the land, the proper site calls out to the settler.

Wise are the settlers who first build modestly and live with the land for the statutory year or even longer. By building once in best accord with our precepts of a site and the simple principles of sun and experience, we will not go far wrong, and will learn much. Gradually, as we learn the rhythms and particularities of our land, unexpected blessings and hazards become apparent. When we come at last to build our Home, if we do so as passionate advocates of the land, aware of all its favors, we are most likely to settle well within our community and endure.

By cunning use of weight-bearing, sheathing, sealing, retaining, and insulating members, with mass and glass, rock, wood, metal, and plastic, we fashion a living machine. A dwelling's first responsibility, and where the most money is spent, is in sheltering comfortable space. Cheap energy led to a building style that ignored natural forces like sun, wind, and water and achieved comfort through powerful technology: heating, ventilating, and air conditioning, or HVAC, as it is called by the building industry. This machinery is rated in tons and consumes prodigious amounts of energy. As energy becomes more precious, we find that appropriate building techniques and clever adaptations to our site can nearly always accomplish the same effect while consuming much less energy. To heat our homes passively, we employ the "greenhouse effect" of mass and glass, whereby sunlight is converted to warmth and stored in masonry or earthen masses behind glass, which hold the heat within the envelope. To cool our homes, we encourage natural ventilation and prevailing winds to pass through a shaded interior. In both cases, we use as much insulation as we can. Whenever possible, we use thermal mass, a substantial pile of material with a high specific heat, like rock or water, which tends to maintain a constant, preferably comfortable, temperature. When passive means are insufficient, we seek nonpolluting, sustainable, and generally indigenous sources of energy, such as biomass or surplus wind or hydroelectricity.

Most of us live, and will continue to live, in existing houses. By making a strong commitment to the land, even to a tiny city lot, apartment roof, or balcony, we find that economies and unexpected opportunities abound. By bringing house and site into accord, we may discover (for example) that a disused guest room or study is a "morning room," one blessed with warming morning sun, and this space may become our home's solar "furnace room" by improving the exposure, adding thermal glass, a greenhouse, or a Trombe wall. At the very least, we should commit to the planet, and resolve to be as efficient in our dwelling as we may be.

In this chapter we have assembled the finest array available of tools for finding, restoring, and assuming stewardship; locating and orienting a home; and beginning the process of homebuilding.

Learning the Land

David and Mary Val Palumbo live in a large independent home on a forested site in upstate Vermont. Their super-insulated wood frame home uses biomass for heat and hot water, and a hybrid solar, micro-hydro, and propane gen-set system. David and his family practiced the classic two-dwelling approach to their land, building an efficient solar cabin, living on and learning the land, then building the big house.

Homecoming: David Palumbo's Story

We knew we were looking for a good sized piece, with woods; I guess we just assumed we'd be connected to utilities, but when it became clear that the properties that were right for us didn't have power, I set about getting myself educated. We were glad not to add to the need for more nuclear power, and we're individualists who pride ourselves on adventure, so relying on ourselves for our own power fit right in.

We looked at a lot of properties. I often have to hold people back when they move away from the city; few people have the necessary discipline. You don't want to buy land without studying it carefully: you want to make sure you understand about rights of way for utilities, about neighbors, about how the seasons and weather extremes treat your property before you commit. We went slowly, and ended up with a better site than we knew.

This land needed some bushwhacking before we could appreciate it. We knew from checking the soil and drainage that there were two good house sites. We lived in a tent on the best site while we built a small, efficient, quick house, 24´ x 24´, facing exactly south on the second best site, so we could start living on the land. We did a lot right with it, and it's still a very workable place. The PVs are within easy broom reach for clearing off the snow. We learned what we needed to know to start planning this house.

This is a cloudy place, and PV is not enough by itself this far north. Also, we use some big power tools, so we knew we would need a good-sized generator for back-up. Our original concept was to use as much PV as we could, and make up the difference with a propane generator. There are two houses and a big shop on our grid, so we manage it like a small utility.

We got the idea for hydro while prospecting for sites for a pond. We found three streams on the land, and one of them runs year-round. Getting the hydro system working was more hit-and-miss than it had to be; I know how to do it now with much less

fuss, and I'm looking forward to doing this one over, but it works well, so we only use the propane generator when we run the bigger tools, like the planer.

We didn't know how many children we would have when we built this place, so we made it big. All the framing and trim lumber came from the land, so we used as much indigenous material as we could. We used a lot of wire, because we wired for 12, 24, and 110 volts. During the winter, it takes some time to manage all the systems, the electricity, the gasifier, the children. The kids go to school and childcare, and Mary Val commutes to her work. During the week I run the house and do my work helping people put their technology together properly.

The big Essex wood gasifier runs all year long. It burns at about 1800°–2000°F; when it's cold out, I fill the firebox two or three times a day, but when it's warm, I fill it once every two or three days. We thought a lot about the way the systems would interact as we planned and built the house, so there's a wood chute in the garage for getting the wood to the basement and the gasifier. And my workshop, which is above the garage, has a trap door and stair so we can move equipment up and down easily.

I studied the way Native Americans, especially the ones who lived around here, treated the land. Flatlanders get emotional about cutting trees, but the regenerative capacity of the northern woods is staggering. You don't need to replant, and you can't keep the woods back. There are 250 wild apple trees on our land, planted, I suppose, by the original farmer who cleared it. The forest is crowding them out, but they are important for the wildlife, for bear, deer, and grouse. We've worked with the Department of Agriculture to release those trees from competition to the south, and now we're investigating some edge and patch cutting. Patch cutting (clear-cutting a small area, no more than a half-acre, defined by the forester) can make flatlanders really howl, but the edge of the regenerating forest, where the poplars start, is a crucial part of the woods habitat, where most of the animals thrive. The young poplars, for instance, are necessary for grouse reproduction. We're doing what the original inhabitants did, bringing sunlight into the forest.

Hyde Park, Vermont is one of the coldest spots in the lower states, and it takes some serious heat to keep a house warm. The large, comfortable Palumbo home relies on wood, a renewable resource in good supply locally, which fuels a gasifier. A gasifier is an efficient, high-tech woodstove which burns so hot that the fuel is literally turned into a gas before it burns, so combustion is efficient and complete and flue gases are clear and nonpolluting. The gasifier runs year-round, and produces domestic hot water in abundance as well as hydronic heating. This strategy would be problematic in an area where ice storms and other causes take the utility power down frequently, but the Palumbos are their own power company, and their energy flow is trouble-free.

The Palumbo Power Company began as a necessity, but has clearly become a work of conscience and vocation as the house has grown. The hybrid supply side of the Palumbo electrical system employs three sources: solar, hydro, and fossil fuel backup. Most of the electricity comes from a micro-hydroelectric generator and from a sizeable array of photovoltaic modules on the shop roof. The hydro resource was not part of the original plan, but was discovered in the process of developing a pond. Domestic water and water for the micro-hydro come from sources up the hill on the Palumbos' land. In times of extra need or low productivity from the renewable sources, a propane generator can be turned on to recharge the batteries. Propane was selected as the generator fuel because it is relatively abundant, nonpolluting, and runs the generator's engine efficiently. Energy is stored in a large bank of industrial batteries kept in a fireproof, vented battery vault in the basement between the root cellar and the furnace room. Generally house current (110-volt AC) is supplied by an inverter. A transfer switch directs electrical traffic when the generator cuts in. David Palumbo urges caution when switching on loads after the generator starts, because the generator produces off-spec electricity until it has warmed up and settled down. He has solved the problem by installing a two-minute time-delay relay.

Flatlanders get emotional about cutting trees, but the regenerative capacity of the northern woods is staggering.

Reclaiming Spoiled Land

Most of the best land on our continent has already been used, and often we must start our tenure and stewardship by healing its prior injuries. Many of the original tenants on this land religiously avoided excesses against the land, but European settlers over the past four centuries have done untold harm. The wounds are various, ranging from the inadvertent introduction of unexpectedly vigorous transplants, which kill native species and dominate the landscape (gourse and kudzu), to the deliberate and perhaps even criminal contempt for the land that lays acreage to perpetual waste in poison dumps, slag heaps, and mine tailings. We must believe that, with few exceptions (nuclear contamination comes to mind, along with some particularly nasty aromatic chemical dumps like Love Canal) all land is salvageable. New mycological discoveries promise to soon make this a reality.

We will never be able to restore the ecosystem to its pristine, presettlement state. Some of our species' earlier and more careless land-grabs, perpetrated very recently by biological standards, will diminish us forever if not promptly redressed. The most complex systems—the American wetland, now filled in for parking lots and flatlander development, and the global rainforest, still being sacrificed for cellulose and beef—are now all but lost. These great habitats, wetlands and rainforest, together with great plains and our ancestral forests, take millennia to heal, and so we must start their convalescence prayerfully.

Fertile and arable lands that we have claimed for homes and farms, as well as the low hills, plains, and bottomlands that have made this continent such an idyllic home and bountiful breadbasket, recover more quickly, and we can hope to heal them through sensitive stewardship within our lifetimes. Independent agriculture, like independent living, finds the same fault with factory farms as we found with factory housing: the site is battered to maximize immediate product, and succeeds only temporarily or through heroic infusions of imported energy in the form of pesticides, fertilizers, and intensive fossil-fueled tilth. Many of us are coming to a prefer a sustainable, Earth-attuned and life-enhancing, organic agriculture. A first step, then, is to return to natural measures. We discover, after more than a century of tinkering, that many exotics, plants and animals imported from other continents, carry with them "hitchhikers" and lack the biological or environmental controls that keep them in balance in their native ecosystems. Again and again we find that kudzu, telapia, mango, eucalyptus, mongeese, and feral cats, goats, and pigs are scourges on a fragile land where they were never meant to range. While abating past errors, we must be careful not to commit new ones in a rush for quick miracles. Monocropping—covering an expanse with a single life-form—is abhorrent to nature, and succeeds only if nature is overwhelmed. Nature's backlash often comes with surprising vigor and direction. Only by restoring land to its happy, indigenous state, in which the aboriginal biota regain their original diversity, integrity, and interrelatedness, may we prevent environmental catastrophe.

Shelter and Passive Solar Design

Building and living in houses is one of the most energy-intensive activities that humans do.

Sustainability and restoration are the keys to the future. To live sustainably, humans must find ways to live in harmony with other species, indefinitely, within the limits of the earth's capacity to support us all. To accomplish this goal, we must savor every drop of our resources, so they can be used to their maximum potential. We must use energy and materials as efficiently as possible, and slow our consumption of nonrenewable resources. We must curb the rate at which we put pollution into the atmosphere, land-fills, and surface water, before we reach the point at which our planet's natural systems can no longer handle the contamination and before we get to sustainability, we must restore and regenerate what we have already destroyed.

All too often, we rely on scientists and technology to pull us out of our crises. "Oh, they'll figure out a way to fix it before it gets too bad," we say. But many of our most prominent scientists are the ones telling us that we have to change to survive; they have no tricks up their sleeves. Now is the time to make these necessary changes.

Shelter is one of the necessities of life. Building and living in houses is one of the most energy-intensive activities that humans do. We hope to find sustenance, partnership, comfort, and stability. We should strive to minimize construction waste, avoid the use of nonrenewable materials, and, most importantly, design structures that require minimal energy input to maintain indoor comfort. Today's common building and design practices are far from sustainable, as forests and fossil fuels dwindle and global warming becomes more evident. Most houses are built with little regard for their natural sites and microclimates, so that we have to outfit them with huge air conditioners and heaters. Then we pour thousands of dollars and tons of fossil fuels into these machines, at untold cost to the environment and our heirs. This practice clearly is not sustainable, and the only question is whether the inevitable changes will occur sooner or later. If we keep waiting for later, we risk serious ecological and social chaos. Now is the time to investigate and utilize sustainable building practices. Many of the best methods are thousands of years old, and by combining these traditional techniques with modern technology, we can help open the door to a truly sustainable future. This chapter on shelter provides information on some of the best and most innovative design and construction methods, ones that will carry us through the 21st century and beyond.

When it comes to shelter, Real Goods will serve you better by providing ideas, concepts, and how-to information than by providing you with major hardware.

When it comes to shelter, Real Goods will serve you better by providing ideas, concepts, and how-to information than by providing you with major hardware. For the most part, this is a simple case of bulky materials making the shortest possible journey. You'll find Real Goods to be an incredibly rich source of information on all matter of energy-saving and energy-efficient technologies. After we've made you the local expert on passive solar and straw bale or rammed earth, buy your building materials locally as much as possible. There's a well-established industry out there waiting to provide all your basic bulky supplies, and a good relationship with your local lumberyard will be mutually beneficial while you build your energy-efficient home.

With the accent on ideas more than hardware, we'll begin with a discussion of sustainable design principles and effective planning strategies. Then we'll move on to various building materials and styles, including a primer on passive solar design. If you pay attention to energy and resources at each step, you will build a house that enables you to live lightly on the earth.

Sustainable Design

The principles of ecological or sustainable design are holistic. They ask you to consider efficient energy and resource use at every stage of the home-building process: choosing a site, planning and design, choice of materials, construction, setting up home systems, landscaping, and even the end of a structure's life cycle. Almost a century of easy energy has led to habits of use and patterns of settlement completely divorced from the inherent qualities of the land.

Ecological Design

The following is adapted from The Rammed Earth House *by David Easton, a Real Goods Solar Living Book.*

Design criteria for an integrated project must grow out of the site. It goes without saying that architecture should be responsive to the climate of its region . . . but working with specific site conditions is equally important to successful design. Where does the sunlight enter the house? When and from where do the winter storms come and the prevailing breezes blow? Does the topography or a great view dictate a preferred spot for the garden, the kitchen, or the living room? How about privacy, road noise, or potential future development?

Above all, we want to look for property with good "character traits." These include: abundant solar access, good drainage, a deep and fertile soil, a reasonable (for you) distance from work and shops, and the right "aspect." This has to do with you and how well the site conforms with your sense of place. Wherever you choose to live, each building site will offer unique situations for the enhancement of a building's performance. Resources that are available on site—sunlight, water, natural shading, prevailing winds, topography, soil— become key factors in the decision of what kind of property to buy, and what kind of house to build.

In addition to "fitting" the site, a dwelling must also fit the lifestyle of the occupants. Rather than designing a building for its outward appearance, consider how to create spaces that bring comfort, security, happiness, efficiency, and functionality. Allow the architectural plan to evolve from the site and from the needs of the residents.

Sustainable Building

The following is adapted from The Real Goods Independent Builder, *by Sam Clark, a Real Goods Solar Living Book.*

The idea of sustainable building is simple. Building, as practiced today, accounts for a large proportion of the energy use, resource depletion, and pollution in our world. It's not just the materials that end up in the building. The energy consumed in their manufacture, transport, and installation must be factored in. Of course, the fuels used to heat and operate the house over time also play a large part in its overall ecological impact.

Sustainable building minimizes these effects through energy-efficient building design, recycling, use of local or native materials, and other methods. A related goal is to create a "healthy house," whose interior space contains clean, nontoxic air. This is usually attained through the use of natural or unprocessed building materials.

An ideal sustainable building is one that sits lightly on the land, requires little fuel to heat or cool, operates on little or no outside electricity, and is healthy to live in. Some of the materials will have come right from the site or from nearby sources, such as stone, earth, and wood. Some of its fabric (doors, windows, or lumber) will have come from other buildings. Other parts of the house will have been processed or reprocessed from low-grade or waste materials.

If you pay attention to energy and resources at each step, you will build a house that enables you to live lightly on the earth.

An ideal sustainable building is one that sits lightly on the land, requires little fuel to heat or cool, operates on little or no outside electricity, and is healthy to live in.

Off-the-Grid Design

Living off-the-grid, disconnected from the conventional systems in our society that produce and deliver energy and other utilities, is the ultimate frontier of sustainable design.

The following is adapted from The Independent Home *by Michael Potts, a Real Goods Solar Living Book.*

Today, many of us who live in independent homes, or are moving toward independence in our homesteading, do so because we have concluded that the schemes by which we bring commodities and fuel from far away to satisfy our most basic needs are mindless, abusive, and likely to crash. We seek a happier, sustainable balance between dependence and independence. Energy is essential, and falls freely from the sky wherever the sun shines. Through careful reduction and management of my needs, and a comprehensive energy-harvesting strategy, I can produce enough power for all of my household uses—heating, cooling, lighting, pumping, washing, working, playing music—with energy generated in my own yard.

To take responsibility for our own power, we must manage and maintain energy generation equipment, and the hardware that makes up a system: transmission, storage, distribution, and metering. To do a good job, we must have an intelligent load management plan. The starting point for such a plan is a reasonable estimate of how much power your household requires, and when it is needed. If we take energy for granted, we think nothing of leaving a light on so that a room will be "friendlier" when we enter it; appliance engineers have taken the cue, and designed our appliances so they are never really off. Off-the-grid-ders ask: Do you really need a clock in your coffeepot?

The secret of an independent home's success is the integration of separate systems into a responsive machine. Conventional systems are meant to be stuffed inside walls and forgotten, but the independent homeowner expects to be involved in the day-to-day management of the home's systems. The power-gathering, heating and cooling, lighting and control systems work best when they are steadily harmonized and adapted. Once these systems are in place, the designer must pass along the knowledge essential to their management. Three conditions contribute to "bulletproof" residential systems: the system must be robust and clearly informative; adjustments and periodic servicing requirements must be manageable and self-explanatory; and the operators must know that competent help is available. In many locales, pioneers have taken their expertise into the marketplace and made it available to new settlers, resulting in happy and thriving energy self-sufficient communities.

Energy is essential, and falls freely from the sky wherever the sun shines.

Planning, Design, and Collaboration

If sustainability is the key to the future, planning and collaboration are the keys to successfully building a sustainable home. Whatever techniques, styles, materials, and systems you choose, careful planning throughout the process, and effective collaboration among all the various participants, will enhance your ability to obtain local and recycled products, minimize waste, reduce costs, and end up with the home you really want. It's well worth the time and effort spent before you start to build.

The following is adapted from The Real Goods Independent Builder, *by Sam Clark, a Real Goods Solar Living Book.*

To be in control of your building project, and have fun doing it, you need realistic expectations of what your house will cost and how much work it will take to make it livable. You will especially make your home-building experience more efficient, sane, and pleasurable if you follow a logical procedure for turning your ideas into designs and plans. Start with the information you have about your land, your needs, and your budget. Draw a site plan, a map that shows the characteristics of the land and locates important features such as the well, septic system, gardens, and roads. Make overlapping lists of the general features, qualities, functions, and activities you want in your house; then see how they can be combined in different patterns. Include everything. This is not the time to be realistic: things can be eliminated later if necessary. But underline those activities and features that are central. To me, the layout or plan, is the heart of a house design. It should largely dictate the structural system, the heating design, and other features.

It's possible to build a house with few if any drawings. But designing your house will be easier and more thorough if you learn to make and use scale floor plans, elevations, cross sections, and details. Making drawings is not just a way to put ideas on paper. It is a way to develop ideas and make them work for you. Basically, you make a scale drawing, such as a floor plan, and then systematically ask questions about it. Is the sun orientation good? Is the circulation efficient? Does the layout give enough privacy? Then you revise the drawing, until you have solved as many of the problems as possible. This labor may seem excessive when your real interest is not drawing but building, yet every hour you spend drawing will save you five hours of building. You can solve problems in advance on paper and avoid big and costly mistakes later.

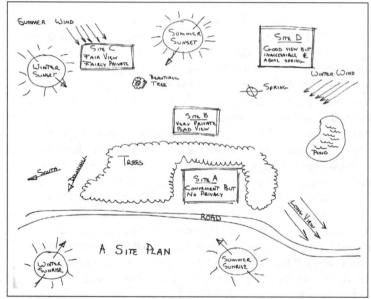

Beginning Site Plan

The builder and designer should both be hired early. They should be asked to collaborate from start to finish. My suggested team model is this: During the design phase, the builder is supporting and consulting with the designer, checking costs, and suggesting those details that he or she prefers to execute. Later, the roles reverse: During construction, the designer supports and consults with the builder on detailing and other problems that arise. Other team members, such as an energy consultant, plumber, or electrician, should also be involved early in the process, so they can collaborate with the designer and the contractor. And, of course, as the owner, you are a key team member, whether or not you do any building. Most likely, you will be coordinating the process.

Architecture and Building Materials

Depending upon where we are in the world we will find different climates, as well as different resources and methods of construction. Working with the tools, materials, and techniques most appropriate to the local area can save both money and energy. In most cases there may be a variety of appropriate responses to environmental and personal economic situations, so it is helpful to consider, research, and prioritize your options. Let's look at the advantages and disadvantages of several alternatives, to help you make an appropriate choice for your particular endeavor.

As we consider alternatives, remember that the specific nature of a region and site will exert a great deal of influence in your choice. What materials exist on site, or are available locally? How much sunshine does the site receive, what spots are shaded, where do the prevailing winds come from? Also important are issues of timing, financial constraints, and your own personal capabilities as a builder or those of builders in your area. Before you select any building material, familiarize yourself with the indigenous homes in your area. Such homes best exemplify the traditional and successful uses of materials and building strategies.

Wood

The following is adapted from The Real Goods Independent Builder, *by Sam Clark, a Real Goods Solar Living Book.*

Most people build houses that consist largely of wood. But even a stone house, or a straw bale house, will be partly wood. Wood is an easy and forgiving material—but it is also a precious, often mismanaged resource. If you want to use wood responsibly, consider cutting wood from your own land and hiring a portable sawmill, see if native timber is available at local sawmills, and plan to incorporate recycled lumber in your structure.

The most obvious way to conserve land, energy, and building resources, especially wood, is to build smaller houses. A smaller house uses less materials, and causes less pollution, at every stage. Its economies are more than proportional to size, because a small house needs not just fewer timbers, but smaller ones, since the spans are shorter. This is the point most often ignored by people who talk sustainability.

The most obvious way to conserve land, energy, and building resources, especially wood, is to build smaller houses.

Stick-Building

Change and innovation have become the constants in our building environment. Managing them is a critical skill. Over the years, I have learned that traditional approaches to building were often smarter than they appeared.

My starting point is a respect for the way most American houses have been built since 1860, which is often called stick-building, or light timber framing. In a stick-built house, the load-carrying frame consists of small, closely spaced framing members concealed inside floors, walls, and roof. Though it is often criticized, standard stick-building systems are particularly durable and adaptable. No system is as easy to build, fix, and change as the ordinary stick-built system: it is a people's technology. Unless it burns down, a house built this way will last forever if the roof is maintained and the basement is ventilated. It is every bit as strong as a timber building, and can be changed more easily.

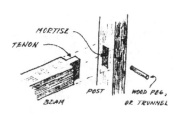

Mortise and Tenon Joint
From A Shelter Sketchbook. *Used with permission.*

Post and Beam

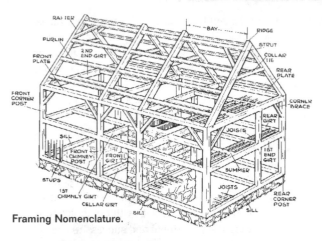

Framing Nomenclature.

Post-and-beam framing has considerable aesthetic appeal. However, this time-honored method is wood-intensive, and utilizes mostly high-quality, large, and sometimes long timbers, which are increasingly hard to find and expensive. If you have a good source of indigenous wood that fits these criteria, post and beam can be an excellent and attractive way to build.

Post-and-beam structures are prevalent in the historic communities of the eastern United States. The frame of the building is typically prefabricated and can be worked on during the non-building season, then brought to the site and erected efficiently.

As with any wood structure, the frame must be detailed properly with rigid geometry, lateral knee-bracing, and integration of frame and exterior sheathing to make a strong, resilient, and secure unit. Because wood is susceptible to rot, similar care must be taken in detailing the vapor barriers and insulation. Heavy timbers require heavy lifting and teamwork to erect, and skill in constructing joints that fit. Special tools and huge saws are required for production work, but these structures can also be built using only hand tools.

The infill walls can be conventional balloon framing, or more commonly now, stress-skin panels, which are a sheet of solid foam insulation with waferboard skins or drywall on the interior side. Interior finish can be paneling, drywall, or plaster. If wood timbers are planed, sanded, and finished with a pickling stain, the wood tends to stay lighter in color, and brighten the interior of the home. Chamfering the edges of the timbers softens the corners and is friendlier to the touch and eye. It is tempting to hurry, but taking care to finish exposed timbers before they go into place is well worth the time and effort.

Laid Stone Construction
From Step-by-Step Outdoor Stonework. *Used with permission.*

Formed Wall Stone Construction
From The Independent Builder. *Used with permission.*

Stone

Many people are attracted to the beauty and durability of stone, an ancient building material. Stone is fire-resistant, and a well-engineered stone home can be made earthquake-resistant too. Stonework buildings take a long time to construct, and they demand a lot of physical labor, but the work is rewarding and these homes can be very cost-effective if owner-built. Because of its large thermal mass, a well-designed stone structure insulated with sprayed foam or batts in a frame can be very energy efficient.

There are two common ways to work with stone: formed wall construction, in which stones and mortar are built up in courses within reusable slipforms, and laid stone, where each stone is nestled and mortared atop the stones below. Laid stone requires great skill and much time, which usually means a costly finished product.

A stone building requires particular attention to wall footings and reinforcement, and careful choice and sizing of the stones, especially at the corners and around openings. These issues are more significant for laid stone than formed stone work. Pointing and fastidious cleanup can produce a beautiful, almost ageless structure. For guidance and inspiration, look to indigenous buildings in your area and also research the stone-building techniques of Helen and Scott Nearing, Frank Lloyd Wright, Ernest Flagg, and Ken Kern.

This is one of the more unusual domes ever constructed.

Domes

Domes continue to exercise a strong appeal over our cultural imagination. However, while domes do have some unique advantages, these must be balanced against some equally unique disadvantages.

The great thing about domes is that, as Buckminster Fuller observed, the dome uses a minimum of materials to enclose a maximum of space. Also, domes are easy to erect, because all the pieces are light and the structure can go up quickly. Heavy lifting is not required, material use can be efficient (although waste is unavoidable because you'll be turning square materials into triangles), and materials are affordable. The structure of a dome is composed of a few standardized and repeated strut and hub sizes of small dimension. These components can be fabricated off-site, indoors, and without a great deal of space. Because of the unusual shape and the angles, fabrication requires extreme accuracy, which can be achieved by using jigs. Sheathing the dome can be fast as well, since standardization is the name of the game. These structures are extremely strong and can withstand almost any force acting upon them if they are firmly anchored to a strong foundation.

Domes enclose maximum space with minimal material, but are very difficult to finish internally.

However, once the shelter goes up and is sheathed, work proceeds very slowly. There is no free dome lunch: the equalizer comes in the detailing of windows and doors, roofing, interior finishing, and cabinetry. Interiors are slow to build and tedious to finish, as all shapes are unusual and either round or triangular. Placement of windows and doors can compromise the integrity of the dome shape, resulting in slow sagging, fatigue of the adjacent members, and, ultimately, collapse. The removal of any struts is discouraged, but if that's necessary addi-

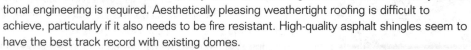

tional engineering is required. Aesthetically pleasing weathertight roofing is difficult to achieve, particularly if it also needs to be fire resistant. High-quality asphalt shingles seem to have the best track record with existing domes.

Perhaps more important, it is difficult to create an energy-efficient dome. Domes have low thermal mass, and they are difficult to insulate well because they have such a thin frame (four to six inches) and usually no attic space. Loose foam insulation could be used but I suspect that infiltration losses would be quite high. Nevertheless, the efficient shape can be easily vented and heated.

Domes are probably best employed as workshops, barns, and garages. In other words, buildings that require a minimum of interior finishing, minimal heating, and don't mind a drip or two when it rains. For more information, take a look at Buckminster Fuller's works and *The Dome Book.*

While not particularly resource or energy efficient, log cabins offer a homespun charm many folks find irresistible.

From How to Build Your Own Log Home for Less than $15,000. *Used with permission.*

Log Cabins

Although log construction is neither resource efficient nor particularly energy efficient, it has its place in areas with a good source of logs, such as the Northwest and Southeast United States. Many people are drawn to the beauty of a log home, including the scale, proportion, and elegance of the corners and openings, and the use of small logs for railings, floor beams, etc. This method of construction lends itself to prefabrication and off-site construction. Work can be done during the off-season and prepared so that, when the weather is right, the logs, already cut, notched, kerfed, and fitted, can be assembled, usually with a crane. As with timber frames, it is important to choose logs free of wood-eating insects and creatures. Log homes seem to work best in areas with cold, dry, snowy winters and mild, wet summers.

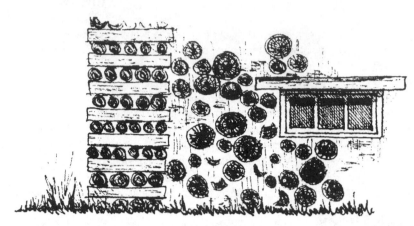

Cordwood Construction

From A Shelter Sketchbook. *Used with permission.*

A log cabin requires a foundation adequate to support the enormous weight of the structure. Room sizes and proportions need to be oversized due to the often massive thickness of the logs. The sealant used between the logs and the detailing at corners, windows, and doors are of great importance in providing a tightly sealed interior, to prevent cold air infiltration and heat loss. Detailing at door and window openings also needs to allow for settling, since the logs shrink as they dry. Increasing passive solar gain, by using lots of south-facing windows, requires special engineering to ensure the continuity of the structure's lateral resistance.

Prefabricated log structures—made with smaller dimension softwood that has been totally milled and prepackaged—seldom make satisfactory homes. Their walls are too thin, the wood quality is often poor, and their thermal properties are marginal; these countrified, cookie-cutter structures are very expensive to heat and resource-inefficient.

Cordwood construction, an unusual but promising log-building technique, uses smaller dimensioned wood to good effect. It involves laying-up short pieces of logs perpendicular to the wall like bricks. For more information, see Rob Roy's book *The Sauna*.

Adobe

Adobe is one of the oldest and most widely used building methods. It is most economical in regions with prevalent adobe (clay) soils and a relatively dry climate, with bricks made and cured on-site. Building with adobe is laborious. However, if the house is owner-built, using materials found on site, it can be inexpensive.

Traditional Laid Adobe Block

From Build With Adobe. *Used with permission.*

In the U.S. adobe is primarily found in the Southwest and California. In regions that have hot, dry summers and cold winters, adobe can be utilized without insulation, since its mass provides a "thermal flywheel" that moderates temperature swings and ensures a comfortable home. In less sunny areas, it is advisable to provide insulation, such as foamboard applied to the exterior of the walls, then plastered over with stucco cement or adobe mud plaster.

The major concern with adobe is its seismic resistance. Classic adobe buildings have withstood centuries of temblors because their buttressed walls are relatively thick for their height, and are topped with sturdy wood bondbeams. Modern, thin-walled adobe structures have proven to be unsafe in an earthquake, unless carefully reinforced with steel and concrete, braced frequently, tied together with a bond-beam, and securely linked to their foundations. The mix for the bricks (sand/clay content and waterproofing compounds used to keep the blocks from deteriorating in the rain), the detailing at the foundation and roof, and the integration of reinforcing frames are all quite important to the shelter's solidity and longevity.

Adobe lends itself to fantastic and wonderful building shapes.

From Build With Adobe. *Used with permission.*

Rammed Earth

The following is adapted from The Rammed Earth House *by David Easton, a Real Goods Solar Living Book.*

Rammed earth produces walls with beautiful color variations.

From The Rammed Earth House.
Used with permission.

Rammed earth requires serious form work during construction.

From The Rammed Earth House.
Used with permission.

Rammed earth construction combines many of adobe's best qualities with the advantages of slipformed walls. Soil from the site (some soil types are more suitable than others) is mixed with proper proportions of clay, sand, water, and cement. The earth is then tamped into reusable forms to infill walls between concrete frames, foundations, and bond-beams, which tie the structure into an earthquake-resistant frame. If attention is paid to detailing at window and door openings, the rammed earth dwelling can have the same gracious and solid feeling as adobe, for less labor. Rammed earth walls contain tremendous thermal mass, so rammed earth homes are highly energy efficient. Further insulation may not be necessary, depending on your climate and site orientation.

It is possible to apply this material using a gunnite spray rig (Impacted Stabilized Earth or PISE), as Real Goods has done to cover the straw bale walls in our Solar Living Center. For a solid PISE wall the material is sprayed against a plywood form from the outside to a thickness of 18 to 24 inches. In mild climates, such as parts of California, Arizona, and New Mexico, this thickness may be sufficient insulation. With this technique, the walls need to be reinforced with steel to make a seismically safe home.

Think of rammed earth as a sort of "instant rock." The earth rammer plies his trade in an environment filled with the dust of soil and cement, the roar of diesel engines, and the staccato thump thump thump of backfill tampers. To me, this is magic . . . watching soil become stone beneath your feet, and knowing that, when the forms are removed, a well-built wall will be here that will survive the test of centuries.

To recombine into a strong and durable rammed earth wall, a soil should be a well-graded blend of different-sized particles. Large particles provide the bulk of the matrix, while the smaller particles fill in the spaces. With ever-decreasing particle sizes, virtually all of the air space within the wall matrix can be eliminated, resulting in the densest wall possible. Density is one of the contributors to ultimate wall strength.

Earthships

Earthships give new meaning to "sweat equity." Where do you suppose all the cans came from?

From The Independent Home.
Used with permission.

Earthships—a sort of cross between rammed earth and adobe—employ opportunistic resources (used tires, aluminum cans) in a clever passive-solar strategy, often sunk into a hillside, or "earth-integrated." This innovative refuse-disposal and home-building concept was created by Michael Reynolds, a Taos architect.

A hole is excavated into the slope, then the tires are laid in a bricklike pattern and laboriously filled with soil and compacted. The tires swell and interlock under the pressure of manually rammed earth, becoming very thick and resilient. Chinks between tires are stuffed with used and partially crushed aluminum cans. Like an adobe wall, integrity is further secured by a bond-beam atop the wall. Roofing consists of the classic vegas (large wooden girders) and latillas, or modern laminated beams, along with plywood and foam sheathings. A sloping glass wall along the front, oriented generally to the south, exposes the thermal mass of the tire-and-earth frame to direct solar gain. Exterior walls and rounded, sculpted interior surfaces are plastered and painted to look like adobe and rammed earth homes.

Earthships are often designed to be completely self-sufficient: water from roof catchment, photovoltaic electricity, and innovative indoor waste disposal are all common features. Effective passive solar design can keep a well-balanced earthship hovering around 65°F with no expenditure of energy, winter and summer.

We can judge earthship longevity only by their short (ten-year) history, but they incorporate the benefits and share the risks of rammed earth and adobe construction: they are fireproof, earthquake-resistant, thermally massive, made of appropriate materials, inexpensive, and indigenous. Earthships can be built for very little money by owner-builders, although building one will give you an intimate understanding of the term "sweat equity." Along with straw bale and modern rammed earth techniques, they represent a refreshing and optimistically innovative approach to shelter.

Straw Bale

The idea of building anything permanent out of straw may seem laughable. After all, we all grew up with the tale of the Three Little Pigs. Yet traditional cultures throughout the world have used straw, grasses, and reeds as building materials, usually in combination with earth and timber, to create shelter for thousands of years. It must work!

Straw bales are made from the leftover stems of harvested grasses after the seed heads have been removed. Hay bales are made from grasses that are harvested green with the seed heads included. Both can be used for building, but because hay is valued as an animal feed, it is much more expensive than straw. The embedded seed heads in hay may also attract vermin. Straw, with no nutritional value, is quite unattractive to mice and rats. Straw bales are super energy-efficient, environmentally safe, inexpensive, and easy to work with. So long as they are protected from moisture, and their structural characteristics are taken into account, straw bales can be used to build structures that are attractive, safe, and durable.

Modern straw bale buildings first appeared shortly after the invention of mechanical baling equipment in the late 1800s. Pioneers in the timber-poor regions of Nebraska started using bales as expedient short-term shelters. When these temporary structures turned out to be both durable and comfortable in the extremes of Nebraska weather, they were soon plastered and adopted as permanent housing. A number of these 80- to 90-year-old buildings are still standing and in good repair.

A straw bale foundation and door frame ready to begin stacking.

From The Straw Bale House.
Used with permission.

Rediscovery of this century-old building technique may prove to be the answer to our current search for low-cost, energy-efficient, sustainable housing. Inexpensive, owner-builder friendly, fire-resistant, highly insulative, and ecologically benign, straw bales are revolutionizing home construction methods. As lumber prices rise and quality declines, environmentally conscious builders are looking to less expensive alternative building materials. At the same time, plagued by poor air quality, California and other states are banning agricultural burning of straw because it creates a major air pollution problem. If straw is baled rather than burned, a waste product can be turned into one of the most promising alternative building materials: straw bales. Since straw matures in a matter of months, millions of tons of these sustainable building materials are produced annually, compared to the decades it takes for trees to grow large enough for the sawmill. Straw bale construction is a win/win solution for farmers and builders.

A straw bale wall built with three-string bales (tight construction-grade bales are typically 16″ x 23″ x 46″) will produce a super-insulated wall of R-50. That's two to three times more insulation than most kinds of new construction, without the environmental hazards of formaldehyde-laced fiberglass. Straw bale walls are highly fire-resistant because, with stucco on each side of the wall, there is too little oxygen inside to support combustion. And even without stucco protection, burn-through times greatly exceed building code requirements. Two-string bales can be used for construction if they are sufficiently compacted. But they are smaller in size, have a lower insulation rating, and do not have the structural strength of three-string bales. Good, tight, three-string bales are really preferable.

Stacking walls is the easy part! Stucco application takes a bit of elbow grease.

From The Straw Bale House.
Used with permission.

Straw bale construction is quick, inexpensive, and results in a highly energy-efficient structure. The technique is easy to learn, and the wall system for a whole building can

Straw bale interior finished in traditional Southwestern style.

From The Straw Bale House. *Used with permission.*

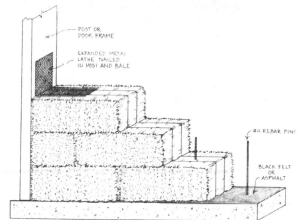

Typical straw bale construction. Like bricks, only bigger.

be stacked in just a day or two. This method also allows for some interesting building shapes. To change the shape of a wall, or put a friendly looking radius on a sharp corner, use a weed whacker to trim the bales to the desired shape. Bales can also be bent in the center to form curved walls, and interesting insets and artwork can easily be carved into the straw before the stucco is applied. The 2-foot-thick walls leave room for attractive window seats like those found in old stone castles in Europe.

Modern load-bearing, plastered straw bale wall construction goes like this: a concrete slab is poured with short pieces of rebar (metal concrete-reinforcement rods) sticking out of the concrete about 12 inches high on 2-foot centers around the slab perimeter. After waterproofing the top of the foundation, the bottom bales of the wall are wrapped with a polyethylene sheet (to prevent moisture migration from below) and impaled on the rebar. Subsequent bales are stacked on this bottom layer like bricks, with each new layer of bales offset by half over the bales below. Rebar pieces are driven into selected bale layers pinning them to the bales below. Plumbing is generally placed in interior frame walls. Electrical wiring is done on (or recessed into) the bales after the wall is up. Electrical outlet boxes are attached to wooden stakes driven into the bales. Pre-assembled window and door frames are set in place as the walls are constructed. The frames are then pinned to the surrounding bales with dowels to hold them in place before the stucco is applied.

When the walls are finished and have completely compressed under the weight of the finished roof and ceiling for six weeks or so, the bales are wrapped with stucco mesh or chicken wire. Stucco is then troweled onto the wire, coating the walls inside and out. Plaster can be used on interior walls if desired. This forms a bug-proof, fire-resistant envelope around the straw, and securely attaches the electrical fixtures to the wall.

A large roof overhang and gutters help keep rain off the exterior walls and lessen the possibility of moisture migration. Straw bales can take some moisture on the exposed bale-ends, since the gaps between the straws are too big for capillary action to carry moisture into the bale. Of course, painting the stucco also goes a long way toward keeping the bales dry. If the bales are kept dry the building will last for many decades.

If the current trend continues, it won't be long before construction-grade straw bales appear at your local lumberyard. If you want to save money, reduce your heating and cooling costs, and help prevent the current overcutting that is destroying our nation's forests and watersheds, straw bale construction is definitely the way to go. See our product section for books, hardware, and specialized tools for straw bale construction.

A Passive Solar Design Primer

Warm in the Winter—Cool in the Summer

That's how we want our homes, right? And if it takes a minimum amount of heating or cooling energy to keep them that way everybody wins. Passive solar design is the way to get there. Using exactly the same pile of building materials and labor costs, you can have an energy-efficient, sunny, easy-to-maintain house, or an energy-sucking, expensive, cave-like house. Obviously the warm, sunny, low-maintenance house is going to be a lot nicer to live in, and it will be worth far more if and when you decide to sell. The following is an excerpt from *The Passive Solar House*, a book in our Solar Living series.

Passive Solar Design

The following is adapted from The Passive Solar House *by James Kachadorian, a Real Goods Solar Living Book.*

As a society, we had managed to "forget" certain fundamental ways of building in harmony with the sun, the seasons, and the landscape—ways that have formed the basis for many architectural traditions for thousands of years.

During the summer of 1973, the U.S. economy was booming. We were all whizzing down the highway at 70 miles per hour, the legal speed limit. Gasoline was about 39 cents per gallon. That year, my wife Lea and I purchased a lovely old Vermont farmhouse, heated by a coal-stoking boiler that had been converted to oil. The base of this monster boiler was about three feet by six feet, and when it fired, it literally shook the house. We tapped our domestic hot water directly off the boiler, so we had to run the unit all four seasons: Every time we needed hot water, the boiler in the basement fired up. We were burning about 2,500 gallons of fuel oil each year, and in the coldest winter months, it was not unusual to get an oil delivery every two weeks.

Since we had no other way to heat our home, we were entirely dependent on the oil-gobbling monster and those biweekly oil deliveries to survive the Vermont winter. Our only alternative source of heat was an open fireplace. Though aesthetically pleasing, the fireplace actually took more heat out of the house than it gave off.

At that time, I was the vice president and general manager of a prefabricated post-and-beam home operation. Like most builders, I shared the industry opinion that the heating contractor's job was to install the heating system that the homeowner wanted. As designers and building contractors, we were not responsible for that aspect of a new home's construction. Moreover, in those days house plans typically were insensitive to the home's relationship to a specific environment, meaning its site and the shifting seasonal position of the sun. Our prefabricated post-and-beam packages were simply labeled "front, back, right side, left side," not "south, east, west, north." We offered little or no advice on siting, except that we needed enough room to get a tractor-trailer to the job site. Our design discussions were rarely concerned with energy efficiency, but rather with how the house would look, for instance whether it would have vaulted or standard-height ceilings.

The point is, we were not yet approaching design and construction in an integrated, comprehensive way. We had not yet recognized that all aspects of a design must be coordinated, and that every member of the design team, including the future resident, needs to be thinking about how the home will be heated and cooled from the first moment they step onto the site. As a society, we had managed to "forget" certain fundamental ways of building in harmony with the sun, the seasons, and the landscape—ways that have formed the basis for many architectural traditions for thousands of years.

There Had to Be a Better Way

In the fall of 1973, an international crisis forever changed the way Americans thought about home heating costs. In response to Israel's victory in the "Six Day War," the Arab oil-producing nations raised oil prices and then initiated an embargo and stopped shipping oil to the United States. These actions ushered in what became known as "the energy crisis."

...it has never been more practical and economical for homebuilders to use the ancient art of passive solar design.

The years immediately following that energy crisis saw a remarkable emergence of new ideas about solar energy. Solar conferences were held, and the public was treated to frequent articles describing new solar home designs in popular magazines. The results of this collective effort were largely positive. Many new ideas were tested, some successfully. Building specifications for energy efficiency developed during that time have now become standard practice. For example, double-pane high-performance glass is now used almost universally in new windows and patio doors. Standard wall insulation is now R-20, which was previously the roof standard; standard roof insulation is now R-32 ("R-value" is a measure of thermal resistance). New and improved, highly effective vapor barriers are also now standard. Exterior house wraps, such as Typar and Tyvek, are applied on most new construction to tighten up air leaks. Appliances are now more energy efficient. Heating systems have undergone major improvements. In sum, we are now building better energy-efficient houses, in large part due to the wake-up call we got in the winter of 1973–1974.

As a result of these improved "standards," it has never been more practical and economical for homebuilders to use the ancient art of passive solar design. In fact, by siting a home to maximize the benefits it receives from the sun, from local breezes, and from existing features of the landscape such as shade trees and natural windbreaks, a contemporary homesteader can build a house that heats and cools itself for most of the year, and at no additional expense over the cost of constructing a comparably sized non-solar house. And despite the equivalence of their initial construction costs, the non-solar house will cost a great deal more over time, year in and year out, to supply and maintain a fossil-fuel-guzzling or electric furnace and air-conditioner.

Why Fear Solar?

As we begin a new century, it seems that we might still be suffering from collective amnesia. We continue to import more than half of our oil from foreign sources. Speed limits are rising. A Vermont utility recently considered a plan to reward consumers who use more electricity this year than last year (fortunately it was withdrawn). Are we headed toward another energy crisis?

From my work building solar homes over the past 20 years, I've found that people resist solar for four main reasons:

- *They are afraid that the house will get too hot.*
- *They are afraid that the house will be too cold.*
- *They are afraid that a solar house by definition will be ugly and futuristic looking.*
- *And they are afraid that a solar house will require expensive, fickle gadgetry and materials, with entire walls of glass, or black-box collectors hanging from every rooftop and wall.*

None of these fears are well-founded. The design and building strategies that work best are carefully engineered for constructing solar homes with traditional features, incurring no added expenses in the process. The solar approach is really a rearrangement of materials you would otherwise need to build any home.

Far more important than buying special materials or incorporating add-on features such as sunspaces, is using common sense and sensitivity to the home's specific environment. Where does the sun rise and set at your location at various times of year? Where do the dominant breezes come from, and how can the house be effectively sheltered from harsh weather while utilizing cooling air patterns?

Let's consider ten basic principles of solar design, none of which will seem controversial. What's striking about solar design is how logical and even "obvious" its tenets really are.

Solar Principle #1

Orient the house properly with respect to the sun's relationship to the site.

Use a compass to find true south, and then by careful observation site the house so that it can utilize the sun's rays from the east, south, and west during as much of the heating season as possible. Take into account features of the landscape, including trees and natural land forms, that can buffer the house against harsher weather or winds from the north in winter, and shade the house from too much sun in the summer.

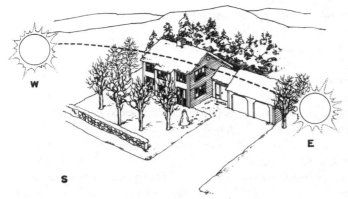

Solar Principle #2

Design on a 12-month basis.

A home must be comfortable in summer as well as winter. When designing a solar home, carefully plan to accommodate and benefit from the sun's shifting patterns and other natural, seasonal cycles. Before finalizing a building plan, spend time at the site at different times of the day and year, and pay close attention to the sun, wind, and weather.

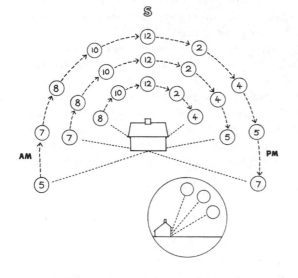

Solar Principle #3

Provide effective thermal mass to store free solar heat in the daytime for nighttime use.

Design the home's thermal mass to effectively absorb the sun's free energy as it enters the building in winter, thereby avoiding overheating. Achieve thermal balance by sizing the storage capacity of the thermal mass to provide for the heating needs of the building through the night. In summer, properly sized thermal mass will serve to cool the building by providing "thermal lag"— that is, excess heat will be absorbed during the daylight hours; by the time the mass has heated up, the day is over and that stored heat can be discharged by opening windows and increasing circulation during the night.

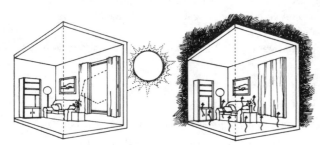

Solar Principle #4

Insulate thoroughly and use well-sealed vapor barriers.

Build tightly constructed, properly insulated walls and roofs. Contemporary standards for wall and roof insulation are very compatible with solar design. Carefully install and seal discrete (or "positive") vapor barriers on the living-space side of walls, ceilings, and/or roofs to prevent moisture from migrating into the insulation along with heat, which tends to travel outward towards the cooler exterior. Incorporate an air-lock entrance for primary doors.

Solar Principle #5

Utilize windows as solar collectors and cooling devices.

This idea sounds obvious, but many people overlook the obvious and spend large amounts of money purchasing, fueling, and maintaining furnaces and air conditioners to address needs that high-quality, operable windows can also address. Vertical, south-facing glass is especially effective for collecting solar heat in the winter, and these windows will let in much less heat in summer, since the sun's angle is more horizontal in winter and steeper in summer. This difference in seasonal angle can be exploited in very sunny locales by using awnings or overhangs that shade windows from the steep sun in summer, yet not from winter sun, which will penetrate further into rooms, supplying solar heat when it's most needed. Provide insulated window and patio door coverings to decrease nighttime heat loss in winter, and to control solar gain in spring, summer, and fall. Operable windows can be used to release excess heat and direct cooling breezes into the house.

Solar Principle #6

Do not overglaze.

Incorporate enough windows to provide plenty of daylight and to permit access to cooling breezes for cross-ventilation, but do not make the common mistake of assuming that solar design requires extraordinary allocations of wall space to glass. An overglazed building will overheat. As emphasized in Principle #5, locate your windows primarily on the home's south side, with fewer windows on the east- and west-facing sides, and only enough windows on the dark north side to let in daylight and fresh air.

Solar Principle #7

Avoid oversizing the backup heating system or air conditioner

Size the conventional backup systems to suit the small, day-to-day heating and cooling needs of the home. A well-insulated house, with appropriately sized and located thermal mass and windows, will need a backup heating system much smaller than conventional wisdom might dictate. Take into account the contribution of solar energy, breezes, and shade to the heating and cooling of the home. Do not oversize backup oil, gas, or electric furnaces, as these units are inefficient, cycling on and off when not supplying heat at their full potential. Do not oversize air conditioners, as they are likewise expensive and wasteful when operated inefficiently. Remember that a heating and cooling contractor will tend to approach the problem of sizing from a "worst-case" perspective, and may not have the skills or experience to factor in the contributions of solar energy and other natural forces.

Solar Principle #8

Provide fresh air to the home without compromising thermal integrity.

To maintain high-quality indoor air, a well-insulated and tightly con-structed home needs a continual supply of fresh air, equivalent to replacing no less than two-thirds of the building's total volume of air every hour. This air exchange should occur through intended openings (such as exterior-wall fans) in both the kitchen and bathroom, rather than through leakage around poorly sealed doors and windows.

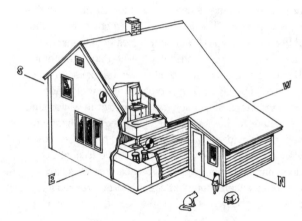

Solar Principle #9

Use the same materials you would use for a conven-tional home, but in ways that maximize energy efficiency and solar gain.

With exactly the same construction materials, it is possible to build either an energy-efficient, sunny, and easy-to-maintain solar house, or an energy-gluttonous, dark, and costly-to-maintain house. When designing a solar home, rearrange and reallocate materials to serve dual functions—adding solar benefits as well as addressing architec-tural or aesthetic goals. Placing a majority of the home's windows on the south side is an example. The carefully designed and constructed solar home need not cost any more to build than a comparably sized non-solar conventional home.

Solar Principle #10

Remember that the principles of solar design are compatible with diverse styles of architecture and building techniques.

Solar homes need not look weird, nor do they require complicated, expensive, and hard-to-maintain gadgetry to function well and be comfortable year-round. In solar design, good planning and sensitivity to the surrounding environment are far more crucial than special technologies or equipment. Over thousands of years, many traditional cultures have used the sun and other natural forces to facilitate year-round heating and cooling: ancient examples come readily to mind, including the pueblos of the American Southwest and the rammed earth structures of North Africa and the Middle East. Build your home in the style you like, but use the solar principles for siting, glazing, thermal mass, and insulation to get the greatest possible benefit from free solar energy.

The Passive Solar Concept

A French engineer named Felix Trombe is credited with the simple idea of building a solar collector comprising a south-facing glass wall with an air space between it and a blackened concrete wall. The sun's energy passes through the glass, and is absorbed by the blackened wall. As the concrete warms, air rises in the space between the glass and the concrete. Rectangular openings at the bottom and top of the Trombe wall allow this warm air to flow to and from the living space. This movement of air is called thermosiphoning. At night the blackened concrete wall will radiate, or release, its heat.

Unfortunately the process can reverse at night—pulling warm air from the living space over to the cold glass. As this warmer air is cooled by the glass, it drops to the floor, which in turn pulls more warm air from the living space. The colder it is outside, the more energetically the Trombe wall will reverse thermosiphon. One way to control this heat loss is to mechanically close the rectangular openings at night, then reopen them when the sun comes out.

The Trombe wall is the "Model A" of passive solar design; that is, it is elegant in its simplicity and dependability, but has been largely supplanted by improved modern techniques. But even if Trombe walls are rarely used in contemporary buildings, they illustrate some essential characteristics of applied passive solar design: the system requires no moving parts, no witches to turn motors on or off, and no control systems, yet it will collect and store solar energy when it is functioning properly, and then radiate heat back into the living space, even after the sun has gone down.

By contrast, an active solar collector is an ancillary system; instead of incorporating heat collection, storage, and release into the structure of the building, active systems are made up of devices

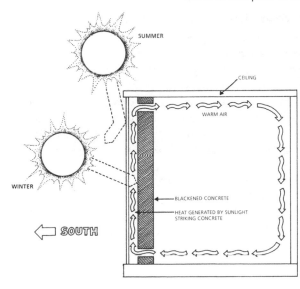

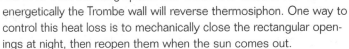

attached to the structure. Active systems also represent "add-on" expenses for a home—features that are additional to those that you would normally purchase. Active systems will not work without a pump or blower operating. Typically, solar collectors are placed on the roof. Pipes deliver water heated by the collectors to a storage tank, and heated water is pumped out of the tank as needed. These systems will not work by themselves: they need to have sensors "tell" switches to turn on pumps or blowers in order to mechanically activate the circulation of water.

The "passive" Trombe wall and the active solar collector system represent the technological range of solar heating systems, from most basic to most complicated.

Keep It Simple and Let Nature Help You

The most widely applicable system for utilizing solar energy is simple, passive, and does not add cost to construction of the home. Consider the materials that homebuilders are generally committed to purchasing, whatever kind of house they're building. We need concrete to build the base of the house, and we all like windows and patio doors. Used properly, these same materials become the building blocks of the naturally heated and cooled home. Also, let's remember to make a careful study of the prospective building site before settling on final plans and layouts, because much can be done with orientation and landscaping to aid in heating and cooling over the lifetime of the house.

Let's start by finding a south-facing house site. For the sake of discussion, let's locate this house in Hartford, Connecticut, which is at north latitude 40 degrees. If the home faces true south, you will get the maximum solar benefit, but as you rotate your home off true south, the solar benefit will be reduced accordingly. For example, at solar noon in February in Hartford, the cost of being oriented to the sun at an angle other than true south is indicated at left.

As you can see, the solar benefit decreases exponentially as you rotate the home's orientation away from true south. Within roughly 20 degrees of true south, the reduced solar benefit is minimal, which allows some latitude in placing the house on a site that presents obstacles like slopes and outcroppings.

Ideally, the north side of the site will provide a windbreak, with evergreen trees and a protective hillside. These natural features will protect the home from the harsher northerly winds and weather. Deciduous trees on the east, south, and west will shade the home in summer, yet drop their leaves in winter, allowing sunlight to reach the home.

Know Your Site

Spend some time on your proposed homesite. Camp there to learn about its sun conditions in different seasons. Make a point of being on the site at sunrise and sunset at different times of the year. Develop a sense for which direction the prevailing wind comes from. Mark the footprint of your new home on the ground using stakes and string, and use your imagination to picture the view from each room. In addition to solar orientation, consider access, view, weather patterns, snow removal, power, septic, and of course, water.

The long axis of a solar home should run east to west, presenting as much surface area to the sun as possible. If your new

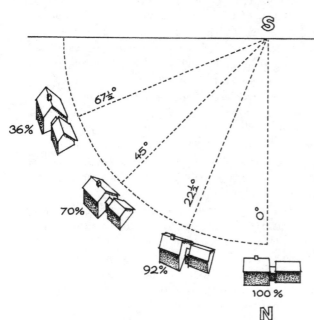

home measures 24 by 48 feet, maximize the amount of surface that the sun will strike by siting your home with the 48-foot dimension running east-west.

Use Windows as Solar Collectors

If you locate the majority of the windows and patio doors on the east, south, and west walls of your home, they can function as solar collectors as well as ventilators, gathering warm solar energy when you need heat, and letting in breezes when you need fresh and cool air. One often sees pictures of solar homes with huge expanses of south-facing glass, tilted to be perpendicular to the sun's rays. Let's remember that you want your home to be comfortable all year round. Tilted glass, though technically more efficient at heat-gathering during winter months, is very detrimental in summer, and will result in overheated living spaces. It's very important to design on a 12-month basis, and understand where the sun is in each season.

DECEMBER 21
NORTH LATITUDE 40°

MARCH 21
NORTH LATITUDE 40°

JUNE 21
NORTH LATITUDE 40°

SUN'S ORIENTATION AT SOLAR NOON

The figure shows the sun's angles at three different times of the year at north latitude 40 degrees—December 21, March 21, and June 21. In December the sun's low altitude almost directly strikes the south-facing vertical glass, and we can see again the importance of facing a home true south.

The March 21 and June 21 illustrations show that as the days grow longer, the breadth of solar aperture widens. Meanwhile, the altitude of solar noon rises to 50.0 degrees on March 21 and 73.5 degrees on June 21.

Design on a 12-Month Basis

In a northern state such as Vermont at the winter solstice (December 21), the sunlight shining through a south-facing patio door will penetrate 22 feet into the home. On the summer solstice (June 21), the sun will only enter the building a few inches.

A dentist in New Hampshire placed a small round dental mirror flat on the sill of his south-facing patio door, and each day at noon he made a mark on the ceiling where the reflected sunlight hit. In 12 month's time, can you guess what kind of geometrical pattern was on his ceiling? An elongated figure-8. The mark closest to the south wall was made at the summer solstice, and the mark farthest from the wall was made at the winter solstice.

Let's imagine south-facing glass at solar noon. If you plant a deciduous tree on the south side of your home, the sun will shine through the canopy in winter when the leaves are gone. Yet in summer, the tree's canopy will absorb almost all of the sun's heat. Plant deciduous trees at a distance from the home, based on the height to which the tree is expected to grow and the size of the anticipated canopy. If deciduous trees already exist on your site, cut down only those that directly obstruct the clearing needed to build the home. Thin adjacent trees' branches after you have gained experience with their shading patterns in both winter and summer.

Remember that because of the high arc of the summer sun, its heat will mostly bounce off vertical south-facing glass, unlike the almost direct horizontal hit your solar collectors will get in winter. This "device" called a solar home will "automatically" turn itself on during the coldest

months and shut itself off during the summer months, so that solar collection is maximized for heat gain when you need the extra heat, and minimized when heat would be uncomfortable. If you can grasp these basic dynamics, you have started to let nature work for you.

Table 1, below, shows the amount of energy received by south, north, and east/west facing vertical glass, plus horizontal glass at solar noon at 40 degrees north latitude (measured in Btus, or British Thermal Units, the standard unit of measurement for the rate of transfer of heat energy). As you can see, the amount of energy received by vertical south-facing glass in December or January is over double the amount received in June. Note how the south-facing glass graph lines run counter to all other directions.

What about east- and west-facing glass? We've been singing the praises of south-facing glass, but at the beginning and end of the heating season, east- and west-facing glass can be effective solar collectors too. However, because the angle of the sun is perpendicular to east-facing glass as the sun rises, and perpendicular to west-facing glass as the sun sets, east- and west-facing glass do not "turn off" as solar collectors in summer. For this reason, location is a critical factor in deciding how to allocate east- and west-facing glass. For example, a solar home located in central California, which has a summer cooling load, should have less east- and west-facing glass than a home located in northern Washington state.

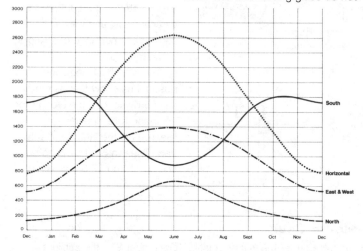

Table 1
Insolation Year-Round for Latitude 40 Degrees North

You may want to provide retractable night time insulation for at least some of the windows, since the same glass that admits heat during the daytime will steadily lose heat during the cooler night time hours. Window insulation can also keep heat out at times of overly intense sun. Another technique for reducing the amount of sunlight that reaches the windows during the hottest months of year is to design overhanging eaves (or movable awnings, which can be added later) to shade your glazing.

Now that you understand how effectively windows and patio doors function as solar collectors, you will see why I emphasize that you should use the components that you are already committed to purchase not only to enhance the livability of your home, but also to serve as an automatic solar collection system.

Passive vs. Active

It's Hard to Get a Drink in a Drizzle

The other important material we always need for a new home is concrete and/or concrete blocks. Heat storage requires thermal mass, or a body of material that can hold heat. Water is the storage medium of choice in active solar systems. Concrete has only about half the thermal storage capacity per unit of volume as water, but it has an advantage: when building a new home, we have already committed to buying tons of it. Used properly, concrete becomes another integral component in a household solar heating and cooling system.

Heating system designers think in terms of heat transfer from warmer to cooler. The typical home furnace warms air to 140°F, and the warm air is delivered to the rooms in the

Passive Solar Checklist

✔ *Small is beautiful*

✔ *East-West axis*

✔ *South-facing glazing*

✔ *Overhangs*

✔ *North side earth berming*

✔ *Thermal mass inside building envelope*

✔ *High insulation levels*

✔ *Radiant barriers in roof*

✔ *Open airways to promote internal circulation*

✔ *Tight construction to reduce air infiltration*

✔ *Air to air heat exchanger*

✔ *Best high-tech windows available*

✔ *Reduced glazing on north and west sides*

✔ *Day lighting*

✔ *Invest in any energy-saving features possible*

✔ *Pay attention to the little details*

✔ *K.I.S.S. (keep it simple, stupid)*

home via ducts. When the thermostat reads 72°F or another desired setting, the furnace shuts off. Heat has been transferred from the warmer body (the furnace at 140°F) to the cooler body (the house at 72°F). The design of a conventional heating system represents a straightforward problem which has a direct solution: determine the heat loss of the building, then size the furnace and duct work to provide a continual, or "on-demand" supply of replacement heat.

Active systems are easy to visualize (boilers, ductwork, pipes, and radiators), whereas the elements of a passive heat collection and storage system may be almost "invisible." Early solar designers copied the elements of an active, furnace-based system. Exterior solar collectors build up high temperatures, and ducts or pipes transport the heat to the interior, where it is stored in beds of rocks or tanks of water. Such active systems are complicated, and often plagued with mechanical problems. The equipment can be costly, and the system requires continuous oversight and maintenance. Active solar heating systems are sometimes difficult to justify financially.

"Solar gain" is the free heat derived from the sun. Sunlight is ubiquitous, but diffuse. Systems that involve rock beds and hot-water storage tanks attempt to concentrate a diffuse form of energy, which is both difficult and expensive. Solar energy can be likened to a drizzle: there are tons of water in the air but it's very difficult to get a cupful to drink. Passive systems represent transient engineering problems; many elements of the calculations necessary to design these systems occur simultaneously, making the processes they involve difficult to analyze. Most heating designers don't like this kind of "fuzzy" problem, and often they are not given all the site information necessary to design on anything more subtle than a worst-case or generic basis.

It's a straightforward calculation to size a furnace on a worst-case basis and to provide a system of ducts to carry heat from a 140°F furnace to areas of 72°F. But to calculate exactly what is going on when heat is entering the home from the sun is much more complicated—some heat is being used directly to heat the home, some is being stored, and some is being lost back to the outside. Moreover, each of these events influences each of the others.

The Solar Slab

The Trombe wall described previously is elegant in its simplicity but aesthetically crude. Pictures of a blackened concrete wall along the south side of a home certainly would not survive among the glossy photo spreads of Better Homes and Gardens. In addition, the Trombe wall blocks out a good portion of the cheery southerly sun. From a technical standpoint, the movement of warm air over the surface of a smooth vertical wall causes laminar flow; that is, a thin boundary layer of air will build up and the warm air passing over this boundary layer will not readily give up its heat to the concrete. An airplane wing is an example of a surface that produces laminar flow. Very little heat is transferred to the wing as it slips smoothly through the air.

On the other hand, a rough surface interrupts the flow of air, causing turbulence, which in turn causes greater heat transfer. Picture the fins in a baseboard radiator versus a smooth pipe along the baseboard. The fins provide much more surface area per running foot than smooth pipe would provide. This increase in surface area allows the heated water inside the pipe to give up its heat to the air. This concept will be crucial as we discuss the construction of the Solar Slab.

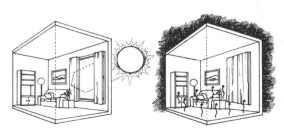

Remember Solar Principle #9: utilize materials you are already committed to purchasing for the construction of a home, but rearrange them in a different configuration in order to collect and store solar energy. Consider what you would have to buy for a full basement. The cellar floor will require a 4-inch concrete slab, and you will need a poured or concrete block cellar wall. That gives us tons of material with which to work. Let's see how we can rearrange these materials.

Start by moving the 4-inch concrete slab from the cellar floor to the

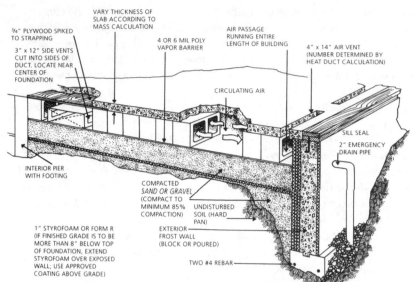

VARY THICKNESS OF
SLAB ACCORDING TO
MASS CALCULATION

¾" PLYWOOD SPIKED
TO STRAPPING

4 OR 6 MIL POLY
VAPOR BARRIER

AIR PASSAGE
RUNNING ENTIRE
LENGTH OF BUILDING

3" x 12" SIDE VENTS
CUT INTO SIDES OF
DUCT. LOCATE NEAR
CENTER OF
FOUNDATION

4" x 14" AIR VENT
(NUMBER DETERMINED BY
HEAT DUCT CALCULATION)

CIRCULATING AIR

SILL SEAL

2" EMERGENCY
DRAIN PIPE

INTERIOR PIER
WITH FOOTING

COMPACTED
SAND OR GRAVEL
(COMPACT TO
MINIMUM 85%
COMPACTION)

UNDISTURBED
SOIL (HARD
PAN)

1" STYROFOAM OR FORM R
(IF FINISHED GRADE IS TO BE
MORE THAN 8" BELOW TOP
OF FOUNDATION, EXTEND
STYROFOAM OVER EXPOSED
WALL; USE APPROVED
COATING ABOVE GRADE)

EXTERIOR
FROST WALL
(BLOCK OR POURED)

TWO #4 REBAR

first floor, eliminating the basement. This is the equivalent of placing the concrete Trombe wall on its side. Next, let's take some of the concrete blocks that we would have used to build the cellar wall, and place them under the concrete slab. Instead of arranging them with their holes aligned vertically, let's lay them on their sides with the holes lining up horizontally to form air passages running north to south. When the concrete is poured over these blocks, it will bond to the blocks and make a huge concrete "radiator"—the radiator's "fins" are the ribs in the concrete blocks. See the drawing to the left.

If this combination of poured concrete slab over horizontally laid blocks is ventilated by air holes along the north and south walls, air will naturally circulate through this concrete radiator when the sun is out. According to Solar Principle #1, we oriented your home with the long axis of the building running east to west. When the sun is out in winter, the south wall will be warmer than the north wall. As heat is transferred into the home by the south glass or by the heat transfer through the wall, air inside the south wall will rise. Warmed air will then be pulled out of the ventilated slab, and the cooler air along the north wall will drop into the holes along the north wall. This thermosiphoning effect will continue to pull air naturally through the Solar Slab.

Storage of Trapped Solar Heat

Heat from the sun comes to us as light energy. Since glass is transparent to light, sunlight passes through glass and strikes objects within the interior of the home. As soon as sunlight strikes an object, for instance the floor covering above the slab, light changes form into heat. In a highly insulated solar home, this heat will now be trapped. The temperature of the ventilated slab will rise as the trapped heat is absorbed by the concrete. Since concrete has no R-value, it has no resistance or ability to stop the transfer of heat. Any heat given up to the ventilated slab anywhere in the building will migrate evenly throughout the array of concrete blocks and poured slab.

The solar home, properly designed, can achieve thermal balance every day. The energy produced by the east-, south-, and west-facing glass will either be consumed directly by the heat demand of the home, or absorbed by the first floor heat sink. If the heat comes in too fast to be absorbed by the mass, the home overheats. In many respects, passive solar design presents a significant cooling challenge.

The surface area inside the blocks of the Solar Slab described in *The Passive Solar House* calculates to be 366 square inches, while the top surface is 119 square inches (7⅝ inches x 15⅝ inches). The ratio of square feet of floor to square feet of surface area within the blocks below the floor surface is 366÷119 = 3. This means that air passing through the blocks is exposed to three times more surface area than if the air had simply passed over a flat surface. This ratio and the roughness of the surface inside the blocks make the Solar Slab an effective heat exchanger.

Surely you have experienced sitting on a sun-warmed rock after sundown. It's nice and warm, and takes a long time to cool off. The Solar Slab produces this effect. Remember, the design goal is keeping the furnace off, or requiring it to do very little work. The heat stored in the first floor of the living space, and dispersed evenly throughout the first floor, has to be beneficial to the heating and comfort of the home.

Keep the Home Comfortable All Day

A properly sited passive solar home with properly sized glazing will hold steady at a temperature between 68°F and 70°F, and will not overheat. The home must have the correct balance between the square footage of its glass solar collectors and the dimensions of its effective thermal-storage mass. A prevalent mistake in solar design is using too much glass. People seem to think that if some south glazing is good, a lot more is better. Yet overglazing will cause overheating and large temperature drops at night. In fact, in some cases, the cost to heat the overglazed home at night will exceed the benefit derived on sunny days. Greenhouses are examples of spaces that overheat during the day and get very cold at night. On the other hand, too much thermal mass, and not enough glass to collect heat, will result in a chilly, cave-like space that will never come up to the comfortable temperature.

These considerations are especially important in the northeast, for instance, where we have about 50% sunshine in the winter and long, cold winter nights. The good news in the northeast is that our heating season is so long and so severe that almost any measure we take to utilize the sun's free heat can result in significant cost and energy savings.

Let's revisit our main objective of keeping the furnace off. The furnace was off all day as the solar home collected and stored heat. During the evening, the occupants will need very little supplemental heat to maintain 68°F to 70°F until 10:00 PM (bedtime), because the entire first floor of the house was 68°F at 5:00 PM. The backup heat is only heating the difference between the Solar Slab temperature and the desired room temperature. If 68°F feels comfortable, then no backup heat is needed. As the Solar Slab gives up its heat to the first floor living space, the Solar Slab temperature will start to decline. The first floor room temperature at 7:00 AM will be the same as the Solar Slab temperature. Stored heat has been given up to the house through the night, and the Solar Slab is now ready to absorb the next day's free solar heat. This solar home will stay ready to instantaneously accept any solar heat available. Even if the sun comes out for just a few minutes between clouds, that heat will be collected.

A particularly nice way to heat a solar home is to throw a party and invite lots of people over on a cold winter day!

In addition, this solar home will absorb excess heat from cooking, lights, and yes, even the heat given off by human bodies. A particularly nice way to heat a solar home is to throw a party and invite lots of people over on a cold winter day! Remember, heat travels from warm bodies to cold bodies. We are each a small furnace, running at 98.6°F.

Do you remember the old John Deere tractors that had an external flywheel? The tractor's small engine slowly got the huge flywheel spinning. Once up to speed, very little energy was needed to keep the tractor moving. That is called mechanical inertia. A body in motion doesn't want to stop. Likewise, the Solar Slab provides thermal inertia to the home so that the home "wants" or tends to stay at a steady temperature, using very little purchased fuel in the process. With this kind of thermal inertia built into the solar home, we can undersize the backup heater, instead sizing equipment for those rare occasions when the tremendous thermal stability of the house isn't sufficient to cope with unusually hot or cold conditions outside.

Every Location Is a Potential Solar Home Site

All too frequently we hear someone say, "Solar won't work here." How can solar energy not work? Although in some locales, as gardeners know, more sunlight is available for greater portions of the year, we all live in solar locations. Does our emphasis on ideal orientation and siting of a home mean that only if the solar conditions are ideal should you plan on building a passive solar house? No. Any contribution made by the sun and natural breezes to heating and cooling a home will be money not spent on fossil fuels, wood, and electricity.

The basic premises for a good solar home are simply the premises of good home design:

- *Make the most of what's available to you, in terms of both your environment and the materials that you are planning to use in your home construction.*

- *Let the tendencies of nature work for you and not against you.*

- *Work toward the goal of keeping the conventional furnace and air conditioner switched off, and also try to minimize your reliance upon alternative backup fuels such as wood. Only sunlight is free.*

A 2,000-square-foot home burning 431 gallons of oil per year in Ann Arbor, Michigan, may not sound as successful as the same size home burning 246 gallons of oil per year in Cheyenne, Wyoming, which is a colder but sunnier place. And yet, if you live in Michigan, then you did the best you could with the solar energy available to you. That 431 gallons of oil a year bought in the summer as an advance, one-time purchase at $1.00 a gallon will mean only about $431.00 per year for heating. And your solar home will be bright and cheerful.

Too often people wishing to heat with alternative fuels spend spring, summer, and fall getting ready for winter. Remember, cutting and stacking two cords of wood is a lot easier than cutting and stacking eight. And in addition to reducing your annual heating load, a highly insulated home with proper vapor barriers and stained natural sidings will minimize the need for periodic summertime exterior painting, staining, and weather-stripping. Furthermore, when it snows, a properly designed roof will not cause ice jams and water dams.

A passive solar home, properly designed, sited, and built, will make life a lot easier by working for you, day in and day out, instead of requiring you to be constantly working for it.

The previous text (pages 24–36) was adapted from The Passive Solar House *by James Kachadorian, a Real Goods Solar Living Book.*

Build Your Own

Building your own house is a courageous undertaking, but it can also be extremely gratifying, and can save you a lot of money. Those of you who have experience in carpentry, construction, electrical work, or plumbing may be able to do some, most, or all of the work yourselves, with a little help from your friends.

Whatever kind of structure you are going to build, there are certain things you will have to deal with. Many of us here at Real Goods have built our own houses, and earned some good down-to-earth advice from the school of hard knocks in the process. The following tips are a distillation of lessons we learned along the way. Hopefully some of the mistakes we have made can help you not make the same ones. (New mistakes are always more interesting.)

On Where to Build

"Stay away from north-facing slopes. They're very cold."

—**Terry Hamor**

"I would have chosen a building site nearer to the main road, and with better gravity-flow for water instead of building a water tower."

—**Debbie Robertson**

"Don't buy too far out in the boonies. Elbow room is great, but there are practical limits. My property was five miles out a steep, rough dirt road. That's fine if you're independently wealthy, and don't have kids or friends. The difficult commute was my primary reason for eventually selling the property."

—**Doug Pratt**

On Being Realistic with Time and Money Plans

"Learn patience. There is always so much to do. Be patient and it all gets done eventually."

—**Doug Pratt**

"If I had to do it over again, I'd realize that it takes three times as long and costs three times as much as expected. I'd have my finances together so it could be built within a year's time instead of… 15 years!"

—**Debbie Robertson**

On When to Move In

"Get a cheap portable living space initially. I had a refurbished school bus that allowed me to move onto the property with a minimum of development. This saved rent and allowed me to check out solar access and weather patterns before choosing a building site and designing a house. A house trailer can do the same thing.

Don't move in until it's finished. It's real tempting to move in once the walls and roof are up. Resist if at all possible."

—**Doug Pratt**

"I wouldn't have moved into an unfinished structure."

—**Debbie Robertson**

"I lived in a school bus until it was finished. [It was good to be able to] move into the shade in the summer and into the sun in the winter. Buses are cheaper than trailers, and come with a motor and charging system. One of my biggest mistakes was moving into the house before it was 100% completed. Most plugs and switches are still not done 13 years later! Finish it before you move in, or you never will."

—**Jeff Oldham**

"Good advice above, if you're that kind of person. I've got to live with something awhile before I see the best solution, so I say, identify essential living segments—a dry bed and a functional kitchen at least—and move right in. Keep projects contained with dropcloths and clean up after, and you end up with a functional, well-tuned house."

—**Michael Potts**

On General Building Plans

"Use passive solar design. This was one of the things I did right. My house was cool in the summer and warm in the winter, and, with an intelligent design, yours can be too. Avoid the temptation to overglaze on the south side. You'll end up with a house that's too warm in the winter and cools off too quickly at night.

Buy quality stuff. Cheap equipment will drive you crazy with frustration and eat up your valuable time."

—Doug Pratt

"Hire a carpenter and become his apprentice."

—Terry Hamor

"Take an honest assessment of your skills. Decide up front which tasks you can accomplish through your own efforts and which will require assistance. Assign a dollar amount and time required for each. Establish a working budget, which you should update throughout the project."

—Robert Klayman

"Money is no object when it comes to insulation! Don't use aluminum door or window frames. Aluminum is too good a conductor of heat. Use wood, vinyl, or fiberglass instead.

Pay attention to little details. It's not a matter of which roofing or siding system you choose; it's the small details like flashing, corner joints, drainage runoffs that really count."

—Jeff Oldham

"Design your wood storage with the ability to load from the outside and retrieve from the inside. Build a laundry chute and a dumbwaiter. Have a small utility bathroom accessible from a back door. Plan for, or install, a dishwasher."

—Debbie Robertson

"If you're hiring a contractor, get him or her involved from the beginning of the design phase. Get his/her buy-in for the project before you start. Always find a contractor you can trust and pay him/her time and materials instead of doing a contract. With a contract one of you loses every time—with time and materials, it's a win-win."

—John Schaeffer

www.realgoods.com

Books

Gaviotas

A Village to Reinvent the World

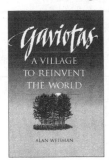

By Alan Weisman. 25 years ago, the village of Gaviotas was established in one of the most brutal environments on earth, the eastern savannas of war-torn, drug-ravaged Colombia. Read the gripping, heroic story of how Gaviotas came to sustain itself, agriculturally, economically, and artistically, eventually becoming the closest thing to a model human habitat since Eden. These incredibly resourceful villagers invented super-efficient pumps to tap previously inaccessible sources of water, solar kettles to sterilize drinking water, wind turbines to harness the mildest breezes—they even invented a rain forest! To call this book inspiring is akin to calling the Amazon a winter creek. You won't be able to put it down. 240 pages, hardcover. USA.

80-023 Gaviotas $22⁹⁵

Mortgage Free!

By Rob Roy. This book tells you how to live without being enslaved to financial institutions. A complete guide to strategies that allow you to own your land and home—free and clear—without the bank, this book will help you redefine what is "necessary," and teach you how to find land that you love and can afford. Author Rob Roy, director of Earthwood Building School, has taught the arts of natural building, renewable energy, and homesteading for years—distilled now into one comprehensive volume that will teach you to live mortgage free. 356 pages, paperback. USA.

80-382 Mortgage Free! $24⁹⁵

The New Cottage Home

By Jim Tolpin. Ask children to draw a picture of a house and they invariably draw a cottage. That's because the cottage form evokes the ancient storybook dreams of home that we all share. Journey with Jim Tolpin as he visits some of the most charming new cottages in North America—homes where quality of place is set well above quantity of space. By the water, on a mountain, or in a forest, field, or town, the new American cottage has a unique, heartfelt appeal. 231 pages, hardcover. USA.

80-011 The New Cottage Home $29⁹⁵

Country Bound!

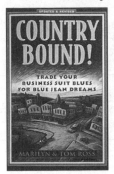

By Marilyn and Tom Ross. *Country Bound!* contains hands-on information on such topics as adapting your mind-set to life in a smaller town, cultivating a satisfying social life, finding the right rural job, and using the Internet advantageously. Authors Marilyn and Tom Ross write from personal experience. This comprehensive book will guide you through issues like finding the right rural community for you, settling into your new environment, and cultivating business prospects from your new rural setting. 341 pages, paperback. USA.

80-162 Country Bound! $19⁹⁵

The Passive Solar House

Using Solar Design to Heat & Cool Your Home

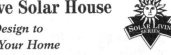

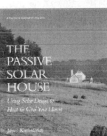

By James Kachadorian. The use of solar energy to heat and cool buildings predates recorded history, yet in recent years homes have been designed and constructed as though there were an endless and inexpensive supply of fossil fuels. James Kachadorian, a civil engineer with degrees from MIT and Worcester Polytechnical Institute, has developed successful and practical solar alternatives to building; a technique for building homes that heat and cool themselves in a wide range of different climates. Kachadorian's company has built nearly 300 solar homes with ordinary building materials and techniques familiar to all contractors and do-it-yourselfers. Included are the Solar Slab heat exchanger plans available to anyone motivated by the desire for a home that rarely, if ever, needs a backup furnace or air conditioner. Passive solar design is one of the primary keys to a sustainable society. Buildings should be warm in the winter and cool in the summer with a minimum of energy input. This is a building book for the next century. 211 pages, paperback. USA.

80-372 The Passive Solar House $24⁹⁵

Finding & Buying Your Place In The Country

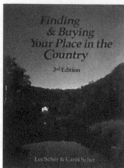

By Les and Carol Scher. Leave behind urban stress and realize your dream to "get back to the land." This informative, comprehensive guide to rural real estate provides detailed, step-by-step action plans, a checklist for property evaluation, information on financing and how to negotiate to get the best home for the money. 436 pages, paperback.

80-184 Finding & Buying Your Place in the Country $25⁹⁵

The NEW Independent Home

People and Houses that Harvest the Sun

By Michael Potts. In 1993 Michael wrote the original *Independent Home*, launching our entire Independent Living series of books, and bringing the almost unknown promise of solar energy to tens of thousands of readers. In this newly revised and expanded edition Michael again profiles the solar home-steaders whose experiments and innovation have opened the possibility of solar living for others. The focus of an independent home has evolved beyond the energy system to encompass the whole process of planning a renovation or construction of a new home. Beautiful homes are being built with age-old materials such as straw bales and rammed earth combined with state-of-the-art electronic technologies for free energy harvesting.

The New Independent Home features 50% more pages than the original, a new 16-page color insert section, and dozens of Michael's wonderful graphic illustrations that help make the technically complex clear and understandable. 350 pages, 1999, softcover, USA.

80-013 The New Independent Home $29⁹⁵

Small Houses

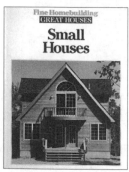

From Taunton Press. Small houses are less expensive to build, more energy efficient and easier to maintain. And as this volume proves, small houses don't have to feel small. A collection of 37 articles from *Fine Homebuilding* magazine, showcasing new houses, remodels, urban rowhouses, country retreats, guest cottages, and studios. Some are professionally designed, others are simple owner-built creations. 160 pages, paperback. USA.

80-012 Small Houses $19⁹⁵

More Small Houses

In *More Small Houses*, you'll find that smaller is beautiful—and more intimate and affordable. Twenty homes, each less that 2,000 sq. ft., each a study in craft and efficiency, are used to explore space-saving design ideas. 159 pages, hardcover.

82-449 More Small Houses $24⁹⁵

The Not So Big House

A Blueprint for the Way We Really Live

Sarah Susanka, with Kira Obolensky. Is it a surprise that people gather in the kitchen? It shouldn't be; human beings are drawn to intimacy, and this book proposes housing ideas that serve our spiritual needs while protecting our pocketbook. Commonsense building ideas combined with gorgeous home design. 199 pages, hardback.

82-375 The Not So Big House $29⁹⁵

Ecological Design

By Sim Van der Ryn and Stuart Cowan. This groundbreaking book profiles how the living world and humanity can be reunited by making ecology the basis for design. The synthesis of nature and technology, ecological design can be applied to create revolutionary forms of buildings, landscapes, and cities—as well as new technologies for a sustainable world. Some examples of innovations include sewage treatment plants that use constructed marshes to purify water and industrial "ecosystems" in which the waste from one process becomes fuel for the next. The first part of the book presents an overview of ecological design. The second part devotes chapters to five design principles fundamental to ecological design. A resource guide and annotated bibliography are included. 201 pages, paperback, 1996. USA.

80-512 Ecological Design $19⁹⁵

The Natural House Catalog

By David Pearson. This indispensable sourcebook will tell you everything you need to know to create a home that's healthy for both your family and the environment. It draws on the latest environmental technology to provide you with useful, up-to-date information. Part One features over a hundred illustrated Topic Pages, covering everything from water power to worm bins, thermal storage to feng shui. Full-color Shoppers' Pages display a wide range of environmentally friendly products, as well as where to buy them. Divided into sections for easy reference, Part Two contains over 2,000 listings to provide a comprehensive directory of suppliers, products, and resources. Resource Pages cite groups, organizations, and publications offering more information on specific topics. 287 pages, paperback. Hong Kong.

80-543 The Natural House Catalog $22⁹⁵

Eco-Renovation Revisited

Eco-Renovation, published by Chelsea Green's U.K. affiliate, Green Books, is organized under four main headings: space, energy, health, and materials. The book guides you through each of these fields, and prioritizes choices according to the basic ecological principals of recycling, self-sufficiency, renewability, conservation, and efficiency. It includes a checklist for making an ecological assessment of your home, and detailed appendices that will enable you to locate "greener" products. *Eco-Renovation* will help direct your ecological thinking to a wide range of home improvements. It will show you how to substantially reduce your heating bills, choose the most energy-efficient appliances, select building materials and finishes that are ecologically benign, organize the space in your home to the maximum effect, and protect your family from toxic substances. By following the suggestions in this book, you can cut your domestic fuel bills by over 50 percent and reduce your household's carbon dioxide emissions by as much as 75 percent. 243 pages, paperback. United Kingdom.

80-385 Eco Renovation Revisited $16⁹⁵

Architectural Resource Guide

From the Northern California Architects/Designers/Planners for Social Responsibility. Ever wondered where to find all those nontoxic, recycled, sustainable, or just plain better building materials? Well, here it is. This absolute gold mine of information has a huge wealth of products listed logically and intelligently following the standard architect's Uniform Specification System (i.e., concrete; thermal & moisture protection; windows & doors; finishes; etc.). Each section is preceded by a few pages explaining the pros and cons of particular materials, appliances, etc. Each listing says what it is, gives the brand name, a brief explanation of the product or service, and access info, both national and NorCal local, for the manufacturer. Although written for the professional architect or builder, there's plenty of info for everyone here. Updated continuously, we order small quantities to ensure shipping the latest edition. Softcover, 152 pages (approx), paperback. USA.

80-395 Architectural Resource Guide $29⁹⁵

Energy-Efficient Houses

Fine Homebuilding is renowned for both its art and authority and this book offers construction basics and a departure point for anyone contemplating building or renovating. *Energy-Efficient Houses* explores the variety of ways builders have responded to the challenge of staying cool and keeping warm using less energy. Discussions of new forms of insulation, house orientation, passive solar collection and more. 159 pages, hardcover.

82-374 Energy-Efficient Houses $24⁹⁵

The Real Goods Independent Builder

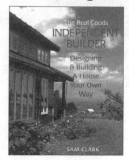

By Sam Clark. Twenty years ago author Sam Clark wrote *Designing and Building Your Own House Your Own Way,* and now he revisits the subject enhanced by time, experience, wisdom, and a battle scar or two. Widely recognized as an expert in bringing simplicity and practicality to complex building decisions, Clark details traditional, post-and-beam, rammed earth and straw bale construction. *The Real Goods Independent Builder* is essential to the home renovator as well. Topics include accessibility, ergonomics, scale, materials, "pattern languages," choosing a site, drawing layouts, making scale models, estimating costs and time, selecting materials, and establishing friendly contractual relations with building professionals. 522 pages, paperback. USA.

80-069 The Real Goods Independent Builder $29⁹⁵

Healthy By Design

By David Rousseau and James Wasley. Create a nontoxic home environment for your family, with the most comprehensive resource of its kind. Find out how to test for pollutants, which building materials pose the most risk to your health (and their alternatives), how to locate and work with a contractor who shares your environmental concerns, and how to deal with the permit process. Includes concrete plans and resources for 79 problem areas, including plumbing, insulation, ventilation, materials, foundations, and more. 290 pages, paperback. USA.

80-910 Healthy By Design $24⁹⁵

Build A Fantasy House In the Trees

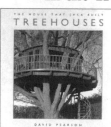

This eclectic and delightful collection of tree-dwellers and their dwellings will inspire even the diehard couch potato to grab a hammer and head for the backyard. Featuring real projects from all around the globe, this book depicts twenty distinctive tree dwellings, each one a masterwork of inventiveness. Lavishly illustrated with color photography, this book portrays not only the finished project, but also the construction process, giving enough information to complete a simple project. The extensive resource section offers places to visit, both real and virtual. David Pearson, noted author of *The Natural House Book,* has captured the combined spirits of innocence and inspiration in this awesome little book. Hardcover, 8″ x 8″, 120 pages.

82-614 Treehouses $16⁹⁵

The Straw Bale House

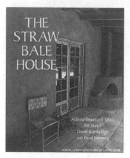

By Athena Swentzell Steen, Bill Steen, and David Bainbridge. This fascinating and useful book describes the exceptionally durable and inexpensive method of plastered straw bale construction. Whether building an entire house or home office, building with straw bales is easy to learn and can be more time, cost, and energy efficient than traditional construction methods. Benefits of straw bale construction include: 1) superinsulation, with R-values as high as R-57; 2) good indoor air quality and noise reduction; 3) speedy construction processes (walls can be erected in a single weekend); 4) low construction costs, as little as $10 per square foot; 5) use of a natural and abundant renewable resource that can be grown sustainably in one season; and 6) reduction of air pollutants created by burning agricultural waste straw. This valuable book describes these benefits in an understandable and interesting way and is a beneficial resource, whether you're ready to build a quiet, comfortable space with straw or just exploring the idea. 336 pages. USA.

80-248 The Straw Bale House $29⁹⁵

Serious Straw Bale

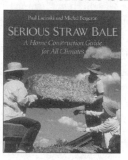

Straw bale construction is a delightful form of natural, ecological building. But until the publication of *Serious Straw Bale: A Home Construction Guide for All Climates,* it was assumed appropriate only for dry, arid climates. This book does not advocate universal use of straw bale, but provides practical solutions to the challenges brought on by harsh climates. This is the most comprehensive, up-to-date guide to building with bales ever published. Builders, designers, regulators, architects, and owner/builders will benefit from learning how bale building technique now includes climates and locales once thought impossible. 371 pages, paperback.

82-485 Serious Straw Bale $29⁹⁵

The Beauty Of Straw Bale Homes

By Athena and Bill Steen. This book is a visual treat, a gorgeous pictorial celebration of the tactile, timeless charm of straw bale dwellings. A diverse selection of building types is showcased—from personable and inviting smaller homes to elegant large homes and contemporary institutional buildings. The lavish photographs are accompanied by a brief description of each building's unique features. Interspersed throughout the book are insightful essays on key lessons the authors have learned in their many years as straw bale pioneers. 128 pages, paperback.

82-484 The Beauty of Straw Bale Homes $22⁹⁵

The Resourceful Renovator

A Gallery of Ideas for Reusing Building Materials

Jennifer Corson combines ecological awareness with frugal creativity to produce this treasure trove of inspiration for home-owners, decorators, forward-thinking architects, and do-it-yourselfers. The book is filled with nuts-and-bolts techniques and imaginative tips for giving new life to scavenged goods. Readers will be surprised and delighted with what they can build using a few materials, a small amount of money, and a living-lightly consciousness. 250 pages, paperback.

82-483 The Resourceful Renovator $24⁹⁵

A Guide To "Dirt Cheap" And "Earth Beautiful"

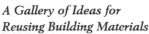

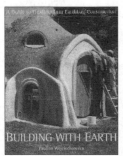

Building with Earth is the first comprehensive guide to the re-emergence of earthen architecture in North America. Even inexperienced builders can construct an essentially tree-free building, from foundation to curved roof, using recycled tubing or textile grain sacks. Featuring beautifully textured earth- and lime-based finish plasters for weather protection, "earthbag" buildings are being used for retreats, studios, and full-time homes in a wide variety of climates and conditions. This book tells (and shows in breathtaking detail) how to plan and build beautiful, energy-efficient earthen structures. Author Paulina Wojciechowska is founder of "Earth, Hands & Houses," a nonprofit that supports building projects that empower indigenous people to build shelters from locally available materials. Paperback, 8˝ x 10˝, 200 pages, 180 illustrations and photos.

82-613 Building with Earth $24⁹⁵

The Rammed Earth House

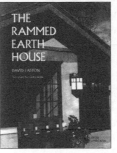

By David Easton. For more than ten thousand years humans have been using earth as a primary building material. Rammed earth, as practiced today, involves mixing basic materials on-site—earth, water, and a little cement—then tamping the mixture into wooden forms to create thick, sturdy masonry walls. Earth-built homes have a permanence and solidity altogether lacking in so many of today's modular, pre-fab, stick-built houses. *The Rammed Earth House* offers clear and complete information for owner-builders, contractors, architects, and anyone interested in housing innovations that put into practice the philosophy of sustainability. Author David Easton is an experienced designer and builder of homes and commercial buildings, and has long been the leading authority on rammed earth construction. Complementing and enriching the book's text, Cynthia Wright's photographs capture the sensuous character and abiding appeal of this beautiful and appropriate style of building. 224 pages, paperback. USA.

80-063 The Rammed Earth House $29⁹⁵

A Shelter Sketchbook

By John Taylor. This book reminds us that most of what we need to know in the 21st century is simply what we've already forgotten. Rather than technologically manufactured synthetics, people are now incorporating the most basic natural materials—earth, mud, straw—into even their high-end environmentally responsive dream houses. For many of those building such projects, the inspiration does not come from *Architectural Digest* but rather from the lessons of ancient or Third World cultures. John Taylor's trained eye and skilled hand celebrate the details of 1000-year-old earth-sheltered houses in China, passive solar heating designs of the Pueblo Indians, natural air conditioning systems built in the Middle East in the 13th century, modular building techniques used in Japan 500 years ago, and more. This is a book for builders, students, and anyone seeking stimulation for the imagination. 160 pages, paperback. USA.

80-402 A Shelter Sketchbook $18⁹⁵

Build With Adobe

By Marcie Southwick. This practical guide was written from the author's many years of experience with adobe. This revised and enlarged third edition has a lifetime of experience behind it. A refreshingly no-nonsense treatment answers many common questions: What can you spend? How do you get your money's worth? How can you do it yourself? And finally, how does your garden grow? Includes hundreds of pictures, drawings, and eight complete floor plans. Fully covers all aspects of adobe construction, plus information on passive solar heating and cooling. Not a flashy coffee-table book, but a practical and easily readable guide that will actually help you build a sturdy, beautiful adobe home. 229 pages with index. USA.

80-392 Build With Adobe $14⁹⁵

Buildings Of Earth And Straw

Structural Design for Rammed Earth and Straw Bale

By Bruce King. Straw bale and rammed earth construction are enjoying a fantastic growth spurt in the United States and abroad. When interest turns to action, however, builders can encounter resistance from mainstream construction and lending communities unfamiliar with these materials. *Buildings of Earth and Straw* is written by structural engineer Bruce King, and provides technical data from an engineer's perspective. Information includes special construction requirements of earth and straw; design capabilities and limitations of these materials; and most importantly, the documentation of testing that building officials often require.

This book will be an invaluable design aid for structural engineers, and a source of insight and understanding for builders of straw bale or rammed earth. 190 pages, paperback with photos, illustrations, and appendices. USA.

80-041 Buildings Of Earth And Straw $24⁹⁵

Build it with Bales—Version II

A Step-by-Step Guide to Straw-Bale Construction

By Matts Myhrman and Steve McDonald. When you're ready to stop reading about straw bale and start doing it, this is the book you want! It's like having an experienced and funny friend on-site with you; someone who knows straw bale construction inside out, and is delighted to share that knowledge with you. Using plain language and a great wealth of detailed drawings, this new improved 2nd edition will help you from design through completion. Presented in logical progression from planning, through foundation, walls, roof, and finishes. Written by Matts Myhrman and Steve MacDonald, a couple of the earliest straw bale pioneers. Paperback, 143 pages. USA.

80-710   Build it with Bales II $29⁹⁵

How To Build Your Elegant Home with Straw Bales

Video & Manual Package

The 90-minute video in this how-to package is professionally done, informative and entertaining. A single building is carefully documented from site layout through move-in and many innovative and labor-saving building tricks are revealed. The producers of this package have built 20+ straw bale structures over the past seven years, and have distilled much of their hard-won knowledge into a product anyone can enjoy. All types of bale construction are covered, as well as intelligent site orientation, climate-specific design considerations, flooring options, utilities, and even a bit on solar systems, composting toilets, and graywater use. The 62-page companion manual covers everything on the tape and is valuable as a quick on-site review. For simple, inexpensive, elegant, do-it-yourself building technology you can't beat straw bale, or the presentation of this inspired package. 90 minutes. USA.

80-095 How-To Video/Manual $59⁹⁵

Building With Straw—Volume I

A Straw Bale Workshop

BUILDING WITH STRAW
— VOLUME I —
A STRAW BALE WORKSHOP

LEARN ALL THE BASICS OF STRAW BALE CONSTRUCTION during this weekend workshop with a group of novice volunteers, as they help to build a two-story greenhouse addition onto a century-old Lodge.

Learn all the basics of straw bale construction in this video of a weekend workshop. Follow along as a group of novices gains hands-on experience constructing a two-story post-and-beam and bale addition to a century-old lodge. As the weekend unfolds you'll see and hear how to design to maximize the solar aspects of your site, build a straw bale wall from the foundation up, then finish off your walls with a simple stucco coating. Includes slide presentation showcasing a variety of straw bale structures, and a printed insert detailing cost factors. 73 minutes. USA.

80-237 Building with Straw—Volume 1 $29⁹⁵

Building With Straw—Volume II

A Straw Bale Home Tour

BUILDING WITH STRAW
VOLUME 2
A STRAW BALE HOME TOUR

WHAT DOES A STRAW BALE HOUSE LOOK LIKE? See for yourself in this video which takes you on a tour of ten straw bale structures.

What does a straw bale house look like? See for yourself in this video, which takes you on a tour of ten straw bale structures, ranging from a simple owner-built home costing $7.50 per square foot to a custom bank-financed home costing $100.00 per square foot. Straw bale construction, a technique developed nearly a century ago in Nebraska, is currently experiencing a worldwide renaissance centered in the American Southwest. In this video you'll hear the personal insights and hindsight of ten modern-day pioneers in New Mexico and Arizona who have built their dream homes out of straw. Learn from their experiences and get ideas for your own straw bale home. 60 minutes. USA.

80-238 Building with Straw—Volume 2 $29⁹⁵

Straw Bale Building

Straw Bale Building
How to plan, design & build with straw
Chris Magwood & Peter Mack

Special focus on building-code compliance and northern climates

By Chris Magwood and Peter Mack. Both of the authors have extensive hands-on knowledge of building with straw bales, and they offer you a warehouse of ideas and a workshop filled with tips for practice projects and dwellings. From planning to finishing, they address costs, technique, building code compliance and energy efficiency. 234 pages, paperback.

82-479 Straw Bale Building $24⁹⁵

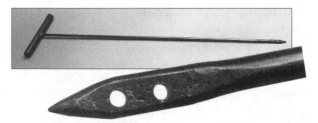

Straw Bale Needle

This is the tool you need to make your own nonstandard bales. Use it to make half bales at the ends of every other row or those odd-length bales needed when you set a window into your straw wall. Since the needle is 3 feet long, it can be used to penetrate bales horizontally for pulling ties through to anchor chicken wire to each side of the wall. USA.

54-316 Straw Bale Needle $29⁹⁵

The Alternative Building Sourcebook

The Alternative Building
SOURCEBOOK

Traditional, Natural and Sustainable Building Products and Services

The first comprehensive resource guide for professional builders, owner-builders, and homeowners. This single-source guide will put you in touch with the builders, suppliers, and manufacturers who can make your next project, or your first home, a breeze. It contains an introduction to many traditional and alternative building styles such as timber framing, log building, straw bale, cob, and bamboo, followed by supplier and information listings. Also covers roofing systems, sawmills, heating, interior finishes, restoration, and education. Over 900 listings. 144 pages, paperback. USA.

82-177 The Alternative Building Sourcebook $19⁹⁵

Step-by-Step Outdoor Stonework

Over Twenty Easy-to-Build Projects for Your Patio or Garden

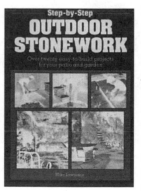

Step-by-Step
OUTDOOR STONEWORK
Over twenty easy-to-build projects for your patio and garden

Mike Lawrence

By Mike Lawrence. Designed for do-it-yourselfers of all levels of expertise, this practical book will help you create a variety of beautiful stonework projects—patios, walls, arches, paths, furniture, fountains, rock gardens, and more. Each project is accompanied by clear text, step-by-step illustrations, and inspiring photographs. In this definitive guide to garden stonework, you'll learn to estimate costs and materials, prepare ground sites, and use the ideal tools and techniques to build stone creations of exceptional strength and timeless beauty. 96 pages, paperback. USA.

80-533 Step-by-Step Outdoor Stonework $18⁹⁵

Building Tipis & Yurts

Authentic Designs for Circular Shelters

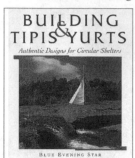

By Blue Evening Star. Circular shelters have a special resonance in the landscape; their construction is uniquely graceful and practical, and their portability promotes an extraordinary sense of personal freedom. The clear instructions and easy-to-follow diagrams here make simple shelters even simpler. History, lore, tools, and tips. 128 pages, paperback.

82-428 Building Tipis & Yurts $16⁹⁵

The Natural House

A Complete Guide to Healthy, Energy-Efficient Environmental Homes

By Daniel D. Chiras. Aimed at both builders and dreamers, this extraordinary book explores environmentally sound, natural, and sustainable building methods including rammed earth, straw bale, earthships, adobe, cob, and stone. Besides sanctuary to shelter our families, today we want something more—we want to live free of toxic materials. Chiras is an unabashed advocate of these new technologies, yet he openly addresses the pitfalls of each, while exposing the principles that make them really work for natural, humane living. This book is devoted to analysis of the sustainable systems that make these new homes habitable. This is the short course in passive solar technology, green building material, and site selection, addressing more practical matters—such as scaling your dreams to fit your resources—as well. 469 pages, paperback.

82-474 The Natural House $34⁹⁵

The Slate Roof Bible

By Joseph Jenkins. Slate roofs are rare, but books about them are even more rare. If you're fortunate enough to own one of these extremely long-lived treasures count yourself lucky. This detailed, richly illustrated, and authoritative work is complete with historical data, anecdotes, and how-to information. Written without jargon, the *Slate Roof Bible* reveals every aspect of slate roofs, from their historical beginnings to modern slate roof restoration secrets.

A two-part, indexed and referenced book which includes a detailed step-by-step, illustrated repair manual, followed by sources of tools, equipment, materials, and new or used slates. Paperback, 287 pages. USA.

80-025 The Slate Roof Bible $34⁹⁵

The Sauna

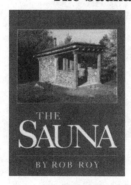

By Rob Roy. For well over a thousand years, saunas have been used to promote health, fitness, and peace of mind. In this informative book, Rob Roy offers a fascinating history of this ancient rite. He explains how and why our ancestors used saunas to instill deep relaxation, prepare for marriage and childbirth, and relieve such maladies as colds, arthritis, rheumatism, and asthma. You'll learn the best way to use a sauna, from seasoning the storeroom to cooling down after an invigorating session. Inviting you to experience the pleasures of a handcrafted sauna, this illustrated guide provides step-by-step instructions for building different types of saunas using traditional techniques. You'll also find complete advice on heating alternatives, a glossary of sauna terms and a comprehensive list of sauna resources, including equipment, manufacturers, and suppliers. 120 pages, paperback. USA.

80-371 The Sauna $19⁹⁵

Build Your Dream House For A Song

By David Cook. A different breed of self-help book. In the belief that everyone should be able to live in a dream home they own free and clear, the author supplies a diverse wealth of tactics to obtain land and building materials for a fraction of retail value. These aren't bizarre, once-in-a-blue-moon lucky breaks that you need to make a career of. Most ideas are followed by examples of how the author arranged or found this deal with a minimum of time and fuss. Learn how to buy at auctions and tax sales, design for ease of construction, avoid paying rent while building, and get the best help for the lowest cost.

What would your life be like without mortgage or rent payments every month? Want to find out? (Sorry but you'll have to pay retail just this one last time. . . .) 151 pages, paperback. USA.

80-709 Your Dream Home for a Song $20⁰⁰

A Sampler of Alternative Homes Video

Approaching Sustainable Architecture

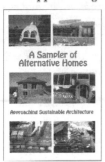

Enjoy a fascinating variety of homes and the creative people who built them. Discover how passive solar design and environmentally forward materials can be used to create comfortable and economical homes. Adobe, rammed earth, straw bale, hermitage, earthships, papercrete, sand bags, and recycled materials are covered in this informative 1 hour, 55 minute video. USA.

80-904 A Sampler of Alternative Homes Video $29⁹⁵

The Kitchen Idea Book

By Joanne Kellar Bouknight. The kitchen is the most important room in the house: It must accommodate intimate family conversations and meal preparation, and still stand up to casual entertaining. A former editor of Fine Homebuilding gives you a dream book with floor plans, design tips, material evaluations and photos galore. 201 pages, hardcover.

82-466 The Kitchen Idea Book $29⁹⁵

Earthship Books

By Michael Reynolds. A builder's guide for an innovative construction technique integrating recycled waste with conventional building materials. The primary construction material is used tires, a post-consumer disposal nightmare, which can be recycled into a building, free. A single tire filled with tamped earth weighs 300 pounds and is stable, immovable, and cheap: the houses in the book cost from $20 per square foot. Interior walls made with adobe-covered cans are strong, light, and can be sculpted to any shape—an exciting concept. *Earthship*, Volume 1 (230 pages), Volume 2 (260 pages), and Volume 3 (257 pages), are comprehensive how-to manuals showing how to use these new building materials. They provide detailed information from design to completion. Vol. 1 shows how to build an Earthship, and is an excellent introduction. Vol. 2 explains interior details and integrates water, waste, lighting, heating and electrical systems, and includes a chapter about getting permits. Vol. 3 adds structural and mechanical details, and provides a vision of community and urban planning concepts. USA.

80-118 Earthship, Volume 1 $27⁰⁰

80-131 Earthship, Volume 2 $27⁰⁰

80-245 Earthship, Volume 3 $29⁹⁵

The Cobber's Companion

How to Build Your Own Earthen Home

By Michael Smith. What is a cob? Cob is a traditional building technique originating in Western Europe, using handformed lumps of earth mixed with straw. In England alone, you can find over 10,000 cozy cob homes, many of which are still warm and tidy after 500 years of continuous occupation. Cobbing surpasses other earthen construction techniques (adobe, rammed earth, pise) both in ease of construction and freedom of design. This book is a practical guide to cob construction, outlining ideas and techniques developed and tested during the construction of over 50 buildings in North America, including the rainy Pacific Northwest, where earthen building was previously uncommon. No exploration of alternative construction is complete without a peek at this book. 117 pages, spiralbound. USA.

80-558 The Cobber's Companion $21⁹⁵

The Cob Builders Handbook

By Becky Bee. One-third of the world's population lives in homes made of unbaked earth. Fifty thousand cob buildings are still in use in England today, most of them built in the 18th and 19th centuries. This is not a fad or craze, but rather a building technique that has stood the test of time. *The Cob Builders Handbook* contains all the necessary basics: design, site selection, materials, foundations, floors, windows and doors, roofs, finishes, and of course, making and using cob—a mixture of sand, straw, and clay formed into load-bearing walls by hand. 173 pages, paperback. USA.

82-106 The Cob Builders Handbook $23⁹⁵

Pocket Reference To The Physical World

Compiled by Thomas J. Glover. This amazing book, measuring just 3.2 by 5.14 inches, is like a set of encyclopedias or internet search engine in your shirt pocket! It's a comprehensive resource of everything from the physical properties of air to pipe-sizing information, and from calculating joist spacing to converting a broad range of weights and measures. A big chunk of knowledge in a very small package. No desktop is complete without one. 544 pages, paperback.

80-506 Real Goods Pocket Ref $9⁹⁵

Pocket Handyman

Expanded from the already incredibly handy *Pocket Ref*, *Handyman In-Your-Pocket* includes extensive additional info on carpentry, construction, electrical wiring, sheet metal, wire and cable, plumbing for water and compressed air, welding, paints, finishes, and solvents, screws, bolts, and anchors, weights and properties of materials. Anyone doing building or maintenance will love this jam-packed little book. 768 pages. USA.

82-615 Handyman In-Your-Pocket $12⁹⁵

Code Check—A Field Guide To Building A Safe House

By Redwood Kardon. An astoundingly simple flip-book reference to the impenetrable morass of codes that govern homebuilding. The author, a building inspector, provides tips and a comprehensive checklist for each inspection that reflects Uniform Building Code requirements. Completely updated. Get it right the first time. 28 pages, paperback.

82-448 Code Check $14⁹⁵

82-447 Code Check Electrical $16⁹⁵

Pacific Yurts

The yurt is an architectural wonder that has been in continuous use for centuries. Invented by the nomadic Mongols of Siberia, this circular, domed abode is as near-perfect a blend of beauty, simplicity, and functionality as any human habitation in history. Established in 1978, Pacific Yurts is the original manufacturer of the modern lattice wall yurt. The high-quality Pacific yurt can be either a recreation retreat, or a year-round residence. It can be insulated and equipped with a heater or woodstove. The low cost per square foot and ease of transport make the yurt an outstanding value. Easily transported in a pickup, even the largest yurt can be erected by two people in less than a day. Materials are of the highest quality, with select #1 quality fir, durable electronically bonded top cover and 100% woven polyester side cover, with large clear vinyl windows and screens. Many custom options are available including heavy duty top cover with 15-year material warranty, insulation, door awnings, water catchment, French doors, engineered high wind and snow package, ceiling fan support, screen door curtain, screened ring insert, custom colors, and more. Prices listed are representative for basic yurts.

Dimensions				
Diameter	Sq. Ft.	Height	Ship Weight	Price
12 feet	115	8′	700 lbs.	$2,680⁰⁰
14 feet	155	8′ 9″	850 lbs.	$3,590⁰⁰
16 feet	200	9′ 3″	950 lbs.	$4,050⁰⁰
20 feet	314	10′	1,350 lbs.	$5,190⁰⁰
24 feet	452	1′ 6″	1,700 lbs.	$6,110⁰⁰
30 feet	706	13′	2,200 lbs.	$7,790⁰⁰

Do not order from Real Goods; call us and we will send you a free literature package and refer you directly to the manufacturer.

Pacific Domes

An Affordable, Portable Shelter for All Seasons,
Pacific Domes create an ideal environment for many uses in the art of living, such as . . .

- Family shelter
- Retreat space
- Guest housing
- Workshops
- Artist studio
- Dance and yoga studio

The frame is sturdy ¾″ EMT galvanized steel tubing that has been proven to handle hurricanes and heavy snow loads. The waterproof, heavy-duty 100% cotton canvas skin has a life expectancy of 5 to 10 years with simple routine care, although with the use of special coatings it will last indefinitely. All models include the large vinyl bay window as pictured, with additional 2-foot round vinyl windows with interchangeable screens. Pacific Domes also feature a zip-off roof and base screens for summer ventilation. A setup for a wood stove is provided for winter use. Easy to transport and set up, with portable floor plans available for each size.

Model Approx.	Floor Area	Height	# of Windows	Total Weight	Price
16 ft.	200 sq. ft.	8.5 ft.	6	200 lb.	$2,400
20 ft.	300 sq. ft.	12 ft.	8	420 lb.	$3,900
24 ft.	425 sq. ft.	14 ft.	10	490 lb.	$4,900
30 ft.	700 sq. ft.	15 ft.	12	700 lb.	$6,200
36 ft.	1000 sq. ft.	20 ft.	14	1000 lb.	$8,700
44 ft.	1500 sq. ft.	22 ft.	16	2000 lb.	$12,300

To order, contact Pacific Domes directly at 888-488-8127
For a free brochure, call 541-488-8127
www.PacificDomes.com   247 Granite St., Ashland, OR 97520

Nesting Bird Yurts

Yurts are great as year-round homes, vacation homes, studios, offices, retreats, kids' rooms, or guest houses. The Nesting Bird is the highest-quality, lowest-impact yurt kit available. Nesting Bird carefully evaluates all their materials based on quality, longevity, environmental friendliness, and aesthetics. Natural, renewable, nontoxic, and recycled materials are used whenever possible, as well as sustainably harvested woods. The sides are clean, breathable poly-cotton, and the roofs are fabricated from a special low-outgassing vinyl that is German-made. The 19-oz. standard roof has a 7-year prorated warranty, or the optional 28-oz. heavy-duty roof has a 15-year prorated warranty. All textiles including optional insulation and cotton liner are fire retardant treated.

The Nesting Bird yurt incorporates a number of innovative features that make it the easiest yurt to set up, take down, and transport. The no-lacing, clip-on wall system, simple doorframe connections, and easily managed lattice wall sections are some examples. Your yurt can be purchased with or without insulation, and you can add insulation at any time.

Nesting Bird has a 5-year warranty on their entire yurt, with a 7- or 15-year prorated warranty on the roof. This company has shown their commitment to using environmentally responsible materials, and promoting sustainable building practices. These are nice folks, building the best yurt known.

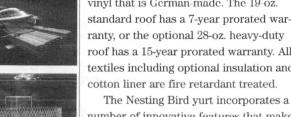

Dimensions				
Diameter	Sq. Ft.	Height	Ship Weight	Price
16 feet	198	10′7″	1,000 lbs.	$5,195
18 feet	263	11′3″	1,100 lbs.	$6,275
21 feet	337	11′9″	1,400 lbs.	$7,225
24 feet	468	12′10″	1,800 lbs.	$8,275
24 feet	566	13′3″	2,200 lbs.	$9,875
30 feet	731	14′4″	2,500 lbs.	$10,975

Do not order from Real Goods; call us and we will send you a free literature package and refer you directly to the manufacturer.

CHAPTER 2

SUNSHINE TO ELECTRICITY

"The glorious lamp of Heav'n, the radiant sun is Nature's eye."

Dryden 1631-1700; The Fable of Acis

ALL THE EARTH'S ENERGY COMES FROM THE SUN (with the possible exception of geo-thermal and nuclear power). Solar panels receive this energy directly. Both wind and hydro power sources use solar energy indirectly. The coal and petroleum resources that we're so busy burning up now represent stored solar energy from the distant past, yet every single day, enough free sunlight energy falls on the Earth to supply our energy needs for four to five years at our present rate of consumption. Best of all, with this energy source, there are no hidden costs, and no borrowing or dumping on our children's future. The amount of solar energy we take today in no way diminishes or reduces the amount we can take tomorrow and tomorrow and tomorrow.

Solar energy can be directly harnessed in a variety of ways. One of the oldest uses is heating domestic water for showering, dishwashing, or space heating. At the turn of the century solar hot water panels were an expected part of as many as 80% of homes in Southern California and Florida until gas companies, sensing a serious low-cost threat to their business, started offering free water heaters and installation. What we're going to cover in this chapter is one of the more recent, cleanest, and most direct ways to harvest the sun's energy: photovoltaics or PV.

What are Photovoltaic Cells?

In 1954 Bell Telephone Systems announced the invention of the Bell Solar Battery, a "forward step in putting the energy of the sun to practical use."

Photovoltaic cells were developed at Bell Laboratories in the early 1950s as a spin-off of transistor technology. Very thin layers of pure silicon are impregnated with tiny amounts of other elements. When exposed to sunlight, small amounts of electricity are produced. They were mainly a laboratory curiosity until the advent of spaceflight in the 1960s, when they were found to be an efficient, long-life, although staggeringly expensive, power source for satellites. Since the early '60s PV cells have slowly but steadily come down from prices of over $40,000 per watt to current prices of around $6 per watt, or below $3 per watt in large quantities with state rebates. Using the technology available today we could equal the entire electric production of the United States with photovoltaic power plants using about 12,000 sq. miles, or less than 12% of the state of Nevada. See map on page 62 for details. As the true environmental and societal costs of coal and petroleum become more apparent, PVs promise to be a major power resource in the future. Who says that space programs have no benefits for society at large?

A Brief Technical Explanation

A single PV *cell* is a thin semiconductor sandwich, with two layers of highly purified silicon. The layers have been slightly doped with boron on one side, and with phosphorous on the other side. Doping produces either a surplus or a deficit of electrons depending on which side we're looking at. Electronics-savvy folks will recognize these as p- and n-layers. When our sandwich is bombarded by sunlight, photons knock off some of the excess electrons. This creates a voltage difference between the two sides of the wafer, as the excess electrons try to migrate to the deficit side. In silicon this voltage difference is just under half a volt. Metallic contacts are made to both sides of the wafer. If an external circuit is attached to the contacts, the electrons have a complete circuit and a current flows. The PV cell acts like an electron pump. There is no storage capacity in a PV cell, it's simply an electron pump. Each cell makes just under half a volt, regardless of size. The amount of current is determined by the number of electrons that the solar photons knock off. Bigger cells will deliver more electrons, but so do more efficient cells, or cells exposed to more intense sunlight. There are practical limits however to size, efficiency, or how much sunlight the cell can tolerate.

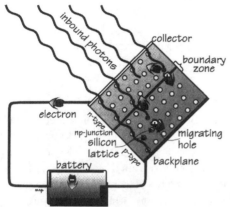

Graphics from
The Independent Home
by Michael Potts

Since 0.5 volt solar panels won't often do us much good, we need to assemble a number of PV cells. A PV *module* consists of many cells wired in series to produce a higher voltage. Modules consisting of about 36 cells in series have become the industry standard for large power production. This makes a module that delivers power at 17 to 18 volts, a very handy level for 12-volt battery charging. The module is encapsulated with tempered glass (or some other transparent material) on the front surface, and with a protective and waterproof material on the back surface. The edges are sealed for weatherproofing, and there is often an aluminum frame holding everything together in a mountable unit. A junction box, or wire leads, providing electrical connections is usually found on the module's back. Although truly weatherproof encapsulation was a problem with the early modules assembled 20 years ago, we have not seen any encapsulation problems with glass-faced modules in many years.

A PV *array* consists of a number of individual PV modules that have been wired together in series and/or parallel to deliver the voltage and amperage a particular system requires. An array can be as small as a single pair of modules, or large enough to cover acres.

PV costs are down to a level that makes them the clear choice for most remote, and many not so remote, power applications. They are routinely used for roadside emergency phones and most temporary construction signs, where the cost and trouble of bringing in utility power outweighs the higher initial expense of PV, and where mobile generator sets present more fueling and maintenance trouble. More than 100,000 homes in the United States, largely in rural sites, depend on PVs as a primary power source, and this figure is growing rapidly as people begin to understand how clean and reliable this power source is, and how deeply our current energy practices are borrowing from our children. Because they don't rely on miles of exposed wires, residential PV systems are more reliable than utilities, particularly when the weather gets nasty. PV modules have no moving parts, degrade very, very slowly, and boast a lifespan that isn't fully known yet, but will be measured in multiple decades. Standard factory PV warranties are 10 to 25 years. Compare this to any other power generation technology or consumer goods.

Astropower AP7105

Construction Types

There are currently four commercial production technologies for PV cells.

Single Crystalline

Astropower Single Crystal Module

This is the oldest and most expensive production technique, but it's also the most efficient sunlight conversion technology commercially available. Complete modules have sunlight to wire output efficiency averages of about 10% to 12%. Efficiencies up to 20% have been achieved in the lab, but these are single cells using highly exotic components, like unobtainium, that couldn't economically be used in commercial production.

Boules (large cylindrical loafs) of pure single-crystal silicon are grown in an oven, then sliced into wafers, doped, and assembled. This is the same process used in manufacturing transistors and integrated circuits, so it is very well-developed, efficient, and clean. Degradation is very slow with this technology, typically 0.25% to 0.5% or less degradation per year. Silicon crystals are characteristically blue, and single crystalline cells look like deep blue glass. Examples are Siemens and Astropower single crystalline products.

Polycrystalline or Multicrystalline

Solarex Multicrystalline Module

In this technique, pure, molten silicon is cast into loaves, then sliced into wafers off the large block of multicrystalline silicon. It is just slightly lower in conversion efficiency compared to single crystal, but the process is less exacting, so manufacturing costs are a bit lower. Module efficiency averages about 10% to 11%, sunlight to wire. Degradation is very slow and gradual, similar to single crystal discussed above. Crystals measure approximately a centimeter (two-fifths of an inch) and the multicrystal patterns can be clearly seen in the cell's deep blue surface. Doping and assembly are the same as for single-crystal modules. Examples are Solarex and Kyocera polycrystalline products.

String Ribbon

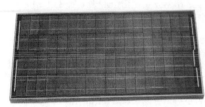

Evergreen String Ribbon Module

This clever technique is a refinement of polycrystalline production. A pair of strings are drawn up through molten silicon pulling up a thin film of silicon like a soap bubble. It cools, crystallizes, and you've got ready-to-dope wafers. The ribbon width can be controlled by the string spacing, and the thickness is controlled by the string diameter. There's far less slicing, dicing, or waste, so production costs are lower. Sunlight to wire conversion efficiency is about 7% to 8%. Degradation will be similar to ordinary slice and dice polycrystal. Examples are Evergreen modules and the APX line from Astropower.

Amorphous or Thin-Film

Uni-Solar Amorphous Module

In this technique, silicon material is vaporized and deposited on glass or stainless steel. This production technology costs less than any other method, but the cells are less efficient, so more surface area is required. Early production methods in the 1980s produced a product that faded up to 50% in output over the first three to five years. Present day thin-film technology has dramatically reduced power fading, although it's still a long-term uncertainty. Uni-Solar has a "within 20% of rated power at 20 years" warranty, which relieves much nervousness, but we honestly don't know how these cells will fare with time. Sunlight to wire efficiency averages about 5% to 7%. These cells are often almost black in color. Unlike other modules, if glass is used on amorphous modules it is not tempered, so breakage is more of a problem. Tempered glass can't be used with this high-temperature deposition process. If the deposition layer is stainless steel and a flexible glazing is used, the resulting modules will be somewhat flexible. These are often used as marine or RV modules. Uni-Solar makes examples of flexible, unbreakable modules. Solarex's Millennia series of modules are examples of thin-film on glass.

New Technology

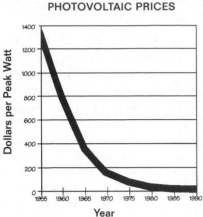

PHOTOVOLTAIC PRICES

Photovoltaic prices have decreased dramatically since 1955. Prices continue to drop 10%–15% per year as demand and production increase. We have seen prices as low as $3 per watt for very large systems in mid 2001.

We're going out on a limb here to talk about emerging technology in a book with a multi-year shelf life, so bear with us. We first wrote this New Technology blurb in early 1999 for the *10th Edition Sourcebook*, now, two years later, there has been surprisingly little substantial change. Let's see what's new below. Of course the latest industry catalogs will showcase the newest commercial technology, but gazing now into our solar-powered silicon-crystal ball in early 2001 reveals:

Thin-film technology continues to offer the best hope for PV price reductions in the future. We have seen the commercial introduction of several thin-film offerings, but are still awaiting significant progress on the thin-film promise of improved efficiency (less surface area required) and lower costs. On the down side, thin-film modules have had problems with long-term performance degradation. The challenge is to produce a low-cost thin-film module that doesn't degrade unacceptably when exposed to sunlight.

The first commercial introduction of newer thin-film technology was Uni-Solar's Triple-Junction modules, which became available in 1997. Instead of the usual single junction of p- and n-layers, the Uni-Solar modules feature three light spectrum-sensitive junction layers. This boosts amorphous cell output slightly. Early experiences are very encouraging, as these modules come out of the box with outputs substantially over specs. Long-term durability is still an open question, although Uni-Solar is offering a good "within 20% of rated output at 20 years" warranty.

In 1998 Astropower and Evergreen Solar both introduced versions of a continuously produced polycrystalline silicon module. The production technologies of these two companies differ, but the end result is lower production costs, because far less slicing and dicing is required if the silicon rolls off the line in precisely the thickness required. As a polycrystalline product, these don't have the long-term degradation problem that dogs amorphous thin-film. At this point, only the Evergreen product has hit the U.S. market, as nearly all of Astropower's production has been shipping overseas.

Several major companies stand poised to introduce new thin-film technologies in the foreseeable future. These technologies are identified by the new, and as yet unfamiliar, trace elements used in doping the cells. These include Siemens Solar, with copper indium gallium diselenide (CIGS cells), BP Solar, Golden Photon; and Solar Cells Inc. with cadmium telluride (CdTe cells); and several other international corporations sporting a bewildering variety of other unpronounceable labels. Many of these thin-film technologies have been under testing at the National Renewable Energy Labs since 1988, so we should have a good understanding of durability and degradation rates when these products finally hit the market.

Putting It All Together

The PV industry has standardized on a nominal 12-volt module output for battery charging. This means a module usually produces 14 to 18 volts output, as the source voltage must be higher for effective battery charging. Twelve volts is a relatively safe voltage, and is familiar as the standard voltage for automotive electrical systems. Dr. Doug advises, "While not impossible, it's pretty difficult to hurt yourself on such a low voltage. Keep your tongue out of the sockets and you'll be okay." Batteries however, where we store large quantities of accumulated energy, can be very dangerous if mishandled. Please see the Panel to Plug chapter/booklet, which includes batteries and safety equipment, for more information.

Multiple modules can be wired in parallel or series to achieve any desired output. As systems get bigger, we usually run collection and storage at higher DC voltages. Small

A Mercifully Brief Glossary of PV System Terminology

AC—Alternating Current. This refers to the standard utility-supplied power, which alternates its direction of flow 60 times per second, and for normal household use has a voltage of approximately 120 or 240 (in the USA). AC is easy to transmit over long distances, but it is impossible to store. Most household appliances require this kind of electricity.

DC—Direct Current. This is electricity that flows in one direction only. PV modules, small wind turbines, and small hydroelectric turbines produce this type, and batteries of all kinds store it. Appliances that operate on DC very rarely will operate directly on AC and vice versa. Conversion devices are necessary.

Inverter—An electronic device that converts the low voltage DC power we can store in batteries, to conventional 120-volt AC power as needed by lights and appliances. This makes it possible to utilize the lower cost (and often higher quality) mass-produced appliances made for the conventional grid-supplied market. Inverters are available in a wide range of wattage capabilities. We commonly deal with inverters from 50- thru 10,000-watts capacity.

— continued on next page

systems processing up to about 2,000 watt-hours are fine at 12 volts. Systems processing 2,000 to 7,000 watt-hours will function better at 24 volts, and systems running more than 7,000 watt-hours should probably be running at 48 volts. These are guidelines, not hard and fast rules! The modular design of PV panels allows systems to grow and change as system needs change. Modules of different manufacture and age can be intermixed with no problems, so long as all modules have rated voltage output within about 1.0 volt of each other. Buy what you can afford now, then add to it in a few years when you can afford to expand.

Efficiency

On average, the sun delivers 1000 watts (1 kilowatt) per square meter at noon on a clear day at sea level. This is defined as a "full sun" and is the benchmark by which modules are rated and compared. That is certainly a nice round figure, but it is not what most of us actually see. Dust, water vapor, air pollution, seasonal variations, altitude, and temperature all affect how much power your modules actually receive. For instance, the 1991 eruption of Mt. Pinatubo in the Philippines reduced available sunlight worldwide by 10% to 20% for a couple of years. It is reasonable to assume that most sites will actually average about 85% of full sun, unless they are over 7,000 feet in elevation, in which case they'll probably receive more than 100% of full sun.

PV modules do not convert 100% of the energy that strikes them into electricity (we wish!). Current commercial technology averages about 10% to 12% conversion efficiency for single- and multicrystalline modules, and 5% to 7% for amorphous modules. Conversion rates slightly over 20% have been achieved in the laboratory by using experimental cells made with esoteric and rare elements. But these elements are far too expensive to ever see commercial production. Conversion efficiency for commercial single and multicrystalline modules is not expected to improve, this is a mature technology. There's better hope for increased efficiency with amorphous technology, and much research is currently underway.

How Long Do PV Modules Last?

PV modules last a long, long time. How long we honestly don't yet know, as the oldest terrestrial modules are barely 30 years old, and still going strong. In decades-long tests the fully-developed technology of single- and poly-crystal modules has shown to degrade at fairly steady rates of 0.25% to 0.5% per year. First-generation amorphous modules degraded faster, but there are so many new wrinkles and improvements in amorphous production that we can't draw any blanket generalizations for this module type. The best amorphous products now seem to closely match the degradation of single-crystal products, but there is little long-term data. All full-size modules carry 10- to 25-year warranties, reflecting their manufacturers' faith in the durability of these products. PV technology is closely related to transistor technology. Based on our experience with transistors, which just fade away after 20 years of constant use, most manufacturers have been confidently predicting 20-year or longer lifespans. However, keep in mind that PV modules are only seeing six to eight hours of active use per day, so we may find that lifespans of 60 to 80 years are normal. Cells that were put into the truly nasty environment of space in the late 1960s are still functioning well. The bottom line? We're going to measure the life expectancy of PV modules in many decades—how many, we don't know yet.

— continued from previous page

PV Module—*A "solar panel" that makes electricity when exposed to direct sunlight. PV is shorthand for photovoltaic. We call them PV modules to differentiate from solar hot water panels, which are a completely different technology, and are often what folks think of when we say "solar panel." PV modules do not make hot water.*

Payback Time for Photovoltaic Manufacturing Energy Investment

In the early years of the PV industry there was a nasty rumor circulating that said PV modules would never produce as much power over their lifetimes as it took to manufacture them. During the very early years of development, when transistors were a novelty, and handmade PV modules costing as much as $40,000 per watt were being used exclusively for spacecraft, this was true. The truth is that PV modules pay back their manufacturing energy investment in 1.4 to 10 years time (less than half of the typical warranty period), depending on module type, installation climate, and other conditions. Now, in all honesty, this information comes to us courtesy of the module manufacturers. The National Renewable Energy Labs have done some impartial studies on payback time (see the results at www.nrel.gov/ncpv/pdfs/24596.pdf). They conclude that modules installed under average U.S. conditions reach energy payback in three to four years, depending on construction type. The aluminum frame all by itself can account for six months to one year of that time. Quicker energy paybacks, down to one to two years, are expected in the future as more "solar grade" silicon feedstock becomes available, and simpler standardized mounting frames are developed.

Maintenance

It's almost laughable how easy maintenance is for PV modules. Having no moving parts makes them practically maintenance-free. Basically, you keep them clean. If it rains irregularly or if the birds leave their calling cards, you would be wise to hose the modules down. Do not hose them off when they're hot, since uneven thermal shock could theoretically break the glass. Wash them in the morning or evening. For PV maintenance, that's it.

Control Systems

Controls for PV systems are usually simple. When the battery reaches a full-charge voltage, the charging current can be directed elsewhere, or more commonly, the charging circuit is simply opened. Open-circuited, PV module voltage rises 5 to 10 volts and stabilizes harmlessly. It does no harm to the modules to sit at open-circuit voltages, but they aren't doing any work for you either. When the battery voltage drops to a certain setpoint, the charging circuit is closed and the modules go back to charging. With the newer solid-state controllers, this opening and closing of the circuit happens so rapidly that you'll simply see a stable voltage. Most controllers offer a few other whistles and bells, like nighttime disconnect, LED indicator lights, etc. See the Controls and Monitors section in the Panel to Plug chapter (page 125) for a complete discussion of controllers.

Powering Down

The downside to all this good news is that the initial cost of a PV system is still high. After decades of cheap, plentiful utility power, we've turned into a nation of power hogs. The typical American home consumes 10 to 20 kilowatt hours daily. Supplying this demand with PV-generated electricity can be costly; however, it makes perfect economic sense as a long-term investment. Fortunately, at the same time that PV-generated power started becoming affordable and useful, conservation technologies for electricity started becoming popular, and given the steadily rising cost of utility power, even necessary. The two emerging technologies dovetail together beautifully. Every watt-hour you can trim off your projected use in a stand-alone PV-based system will reduce your initial setup cost by $3 to $4. Using a bit of intelligence and care in your lighting and appliance selection will allow you all the conveniences of the typical 15 kWh per day American home, while consuming less than 4 kWh per day. At $4 per installed watt-hour that's

$44,000 we just shaved off the initial system cost! With this kind of careful analysis applied to electrical use, most of the full-size home electrical systems we design come in between $10,000 to $20,000, depending on the number of people and intended lifestyle. Simple weekend cabins with a couple of lights and a boom box can be set up for $1,000, sometimes less. With up to 50% renewable energy buydowns in many states, PV can be very cost-effective.

Other chapters in the Sourcebook present an extensive discussion of electrical conservation, for both off and on grid (utility power), and offer many of the lights and appliances discussed. We strongly recommend reading these sections before beginning system sizing. We are not proposing any substantial lifestyle changes, just the application of appropriate technology and common sense. Stay away from 240-volt watt-hogs, cordless electric appliances, standard incandescent light bulbs, instant-on TVs, and monster side-by-side refrigerators, and our friendly technical staff can work out the rest with you.

PV Performance in the Real World

Okay, here's the dirt under the rug. Skeptics and pessimists knew it all along: PV modules could not possibly be all that perfect and simple. Even the most elegant technology is never quite perfect. There are a few things to watch out for, beginning with . . . Wattage ratings on PV modules are given under *ideal* laboratory conditions. As most of us figured out a long time ago, the world isn't a perfect place. Assuming you can avoid or eliminate shadows, the two most important factors that affect module performance out in the real world are percentage of full sun and operating temperature.

Shadows

Short of outright physical destruction, hard shadows are the worst possible thing you can do to a PV module. Even a tiny amount of shading dramatically affects module output. Electron flow is like water flow. It flows from high voltage to low voltage. Normally the module is high and the battery or load is lower. A shaded portion of the module drops to very low voltage. Electrons from other portions of the module and even from other modules in the array will find it easier to flow into the low-voltage shaded area than into the battery. These electrons just end up making heat, and are lost to us. This is why bird droppings are a bad thing on your PV module. A fist-sized shadow will effectively shut off a PV module. Don't intentionally install your modules where they will get shadows during midday, 10 a.m. to 3 p.m., prime generating time. Early or late in the day when the sun is at extreme angles there's little power, so don't sweat shadows then. Sailors may find shadows unavoidable at times, just keep them clear as much as practical.

Full Sun

As previously mentioned in the Efficiency paragraph, most of us seldom see 100% of full sun conditions. If you are not getting full, bright, shadow-free sun, then your PV output will be reduced. If you are not getting bright enough sun to cast fairly sharp-edged shadows, then you do not have enough sun to harvest much useful electricity. 80% to 85% of a "full sun" (defined as 1000 watts per square meter) is what most of us actually receive on a clear sunny day. High altitude and desert locations will do better on sunlight availability. On the high desert plateaus, 105% to 110% of full sun is normal. They don't call it the "sunbelt" for nothing.

Temperature

The power output from all PV module types fades somewhat at higher temperatures. This is not a serious consideration until ambient temperatures climb above 80°F, but that's not uncommon in full sun. The backs of modules should be as well-ventilated as practical. Always allow some airspace behind the modules if you want decent output in hot weather. On the positive side of this same issue, all modules increase output at lower temperatures, as in the wintertime, when most residential applications can use a boost. We have seen cases when the modules were producing 30% to 40% over specs on a clear, cold winter morning with a fresh reflective snow cover and hungry batteries.

As a general rule of thumb, we usually derate official PV module output by about 15% for the real world. For panel-direct systems (where the modules are connected directly to the pump without any batteries), derate by 20%, or even by 30% for really hot summer climates if you want to make sure the pump will run strongly in hot weather.

Photovoltaic Summary

Advantages	**Disadvantages**
1. No moving parts	1. High initial cost
2. Ultra-low maintenance	2. Only works in direct sunlight
3. Extremely long life	3. Sensitive to shading
4. Noncorroding parts	4. Lowest output during shortest days
5. Easy installation	5. Low-voltage output difficult to transmit
6. Modular design	
7. Universal application	
8. Safe low-voltage output	
9. Simple controls	
10. Long-term economic payback.	

Module Mounting

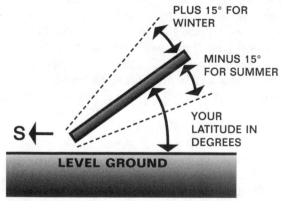

Proper PV Mounting Angle

Modules will catch the maximum sunlight, and therefore have the maximum output, when they are perpendicular (at right angles) to the sun. This means that tracking the sun across the sky from east to west will give you more power output. But tracking mounts are expensive and prone to mechanical and/or electrical problems. Due to economies of scale, they're usually only worthwhile on larger PV systems, generally ones with eight or more modules.

All systems are most productive if the modules are approximately perpendicular to the sun at solar noon, the most energy-rich time of the day for a PV module. The best year-round angle for your modules is approximately equal to your latitude. Because the angle of the sun changes seasonally, you may want to adjust the angle of your mounting rack seasonally. In the winter, modules should be at the angle of your latitude plus approximately 15 degrees; in the summer, your latitude minus a 15-degree angle is ideal. On a practical level, many residential systems will have power to burn in the summer, and seasonal adjustment may be unnecessary.

Generally speaking, PV arrays that consist of eight or more modules are more cost-effective on a tracking mount, while smaller arrays are usually more cost-effective on fixed mounts. This rule of thumb is far from ironclad; there are many good reasons to use either kind of mounting. For a more thorough examination see the PV Mounting/ Trackers Section which includes a large selection of mounting technologies.

Naming of Parts

Following are several examples of photovoltaic-based electrical systems, starting from very simple and working up to very complex. All the systems that use batteries can also accept power input from wind or hydro sources as a supplement, or as the primary power source. PV-based systems constitute better than 90% of our renewable energy sales, so the focus here will be mostly on them.

A Simple Solar Pumping System

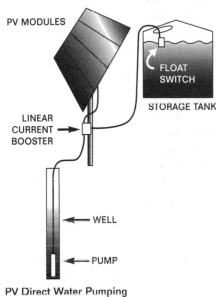

PV MODULES

FLOAT SWITCH

STORAGE TANK

LINEAR CURRENT BOOSTER

WELL

PUMP

PV Direct Water Pumping

In this simple system all energy produced by the PV module goes directly to the water pump. There's no storage of the electrical energy; it's used immediately. Water delivered to the raised storage tank is our stored energy. The brighter the sun, the faster the pump will run. This kind of system (without battery storage) is called PV-direct, and is the most efficient way to utilize PV energy. Eliminating the electrochemical conversion of the battery saves about 20%–25% of our energy, a very significant chunk! However, PV-direct systems only work with DC motors that can use the variable power output of the PV module, and of course this simple system only works when the sun shines.

There is one component of a PV-direct system we won't find in other systems. The PV direct controller, or Linear Current Booster (LCB) is unique to systems without batteries. This solid-state device will down-convert excess voltage into amperage that will keep the pump running under low light conditions when it would otherwise stall. An LCB can boost pump output by as much as 40%, depending on climate and load conditions. We usually recommend them for PV-direct pumping systems.

For more information about solar pumping see the extensive Water chapter in this Sourcebook.

A Utility Intertie System Without Batteries

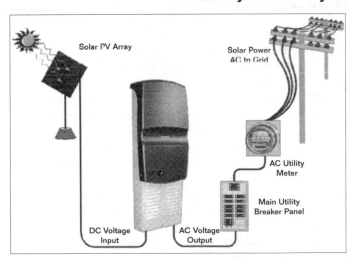

Solar PV Array

Solar Power AC to Grid

AC Utility Meter

Main Utility Breaker Panel

DC Voltage Input

AC Voltage Output

This is the simplest and most cost-effective way to connect PV modules to regular utility power. All incoming PV-generated electrons are converted to household AC power by the intertie inverter, and delivered to the main household circuit breaker panel, where they displace an equal number of utility-generated electrons. That's power you didn't have to buy from the utility company. If the incoming PV power exceeds what your house can use at the moment, the excess electrons will be forced backwards through your electric meter, turning it backwards. If the PV power is insufficient, that shortfall is automatically and seamlessly made up by utility power. It's like water seeking its own level (except it's really fast water!). When your intertie system is pushing excess power backwards through the meter, the utility is paying you regular electric rates for your excess power. You sell power to the utility during the daytime, it sells it back to you at night. This treats the utility grid like a big 100% efficient battery. But if utility power fails with one of these systems, even if it's sunny, the PV system will be off for the safety of utility workers.

There is a minimum of hardware for these intertie systems; a power source (the PV modules), an intertie inverter, a circuit breaker, and some wiring to connect everything. See the separate section specifically on Utility Intertie for more information.

A Small Cabin Solar Electric System

Most PV systems are designed to store some of the collected energy for later use. This allows you to run lights and entertainment equipment at night, or to temporarily run an appliance that takes more energy than the PV system is delivering. Batteries are the most cost-effective energy storage technology available so far, but batteries are a mixed blessing. The electrical/chemical conversion process isn't 100% efficient, so you have to put back about 20% more energy than you took out of the battery, and the supply is finite. Batteries are like buckets, they can only get so full, and can only empty just so far. A charge controller becomes a necessary part of your system to prevent over- (and sometimes under-) charging. Batteries can also be dangerous. Although the lower battery voltage is generally safer to work around than conventional AC house current, it is capable of truly awe-inspiring power discharges if accidentally short-circuited. So fuses and safety equipment also become necessary whenever one adds batteries to a system. Fusing ensures that no youngster, probing with a screwdriver into some unfortunate place, can burn the house down. And finally, a monitoring system that displays the battery's approximate state of charge is essential for reliable

The Real Goods exclusive SolProvider is a good example of a small cabin electrical system.

performance and long battery life. Monitoring could be done without, just as you could drive a car without any gauges, warning lights, or speedometer, but this doesn't encourage system reliability or longevity.

The Real Goods exclusive SolProvider Kit is an example of a small cabin or weekend retreat system. It has all the basic components of a residential power system. A *power source* (the PV modules), a *storage system* (the deep-cycle batteries), a *controller* to prevent overcharging, *safety equipment* (fuses), and *monitoring equipment* (the red-green battery charge indicator lights).

The SolProvider is supplied as a simple DC-only system. It will run 12-volt DC equipment, such as RV lights and appliances. An optional inverter can be added at any time to provide conventional AC power, and that takes us to our next example.

A Full-Size Household System

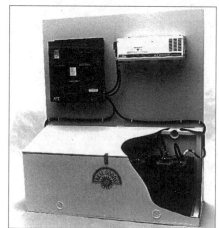

The typical Real Goods Fulltimer system takes only 2´ x 5´ inside your utility room.

Let's look at an example of a full-size residential system to support a family of three or more. The *power source, storage, control, monitoring,* and *safety* components have all been increased in size from the small cabin system, but most importantly, we've added an *inverter* for conventional AC power output. The majority of household electrical needs are run by the inverter, allowing conventional household wiring, and a greater selection in lighting and appliance choices. We've found that when the number of lights grows above five, AC power is much easier to wire, plus fixtures and lamps cost significantly less due to mass production, so the inverter pays for itself in appliance savings.

Often, with larger systems like this, we combine and preassemble all the safety, control, and inverter functions using an engineered, UL-approved Trace Power Panel. This isn't a necessity, but we've found that most folks appreciate the tidy appearance, fast installation, UL-approval, and ease of future upgrades that Power Panels bring to the system.

Because family sizes, lifestyles, local climates, and available budgets vary widely, the size and components that make up a larger residential system like this are customized for each individual application. System sizing is based on the customer's estimate of needs, and an interview with one of our friendly technical staff. See the System Sizing hints and worksheets at the end of Chapter 3.

A Utility Intertie System With Batteries

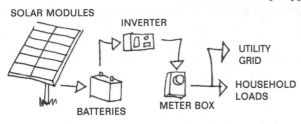

SOLAR MODULES
INVERTER
UTILITY GRID
HOUSEHOLD LOADS
BATTERIES
METER BOX

In this type of intertie system, the customer has both a renewable energy system and conventional utility-supplied grid power. Any renewable energy beyond what is needed to run the household and maintain full charge on an emergency back-up battery bank is fed back into the utility grid earning money for the system owner. If household power requirements exceed the PV input, e.g., at night or on a cloudy day, the shortfall is automatically and seamlessly made up by the grid. If the grid power fails, power will be drawn instantly from the backup batteries to support the household.

Switching time in case of grid failure is so fast that only your home computer may notice. This is the primary difference between intertie systems with and without batteries. Batteries will allow continued operation if the utility fails. They'll provide emergency power backup, and will allow you to store and use any incoming PV energy.

While intertie systems are federally mandated—i.e., the utility must buy your renewable-generated power—they don't have to make it easy or profitable. Because the concept and possibility of small-scale residential intertie systems is very new, many utilities haven't formulated their policies concerning buyback rates, insurance requirements, and required safety equipment. Our technical staff can only supply limited assistance in dealing with your local utility company. This is still a new and emerging technology.

In order to hasten this emerging technology, a number of federal and state sponsored programs exist, and some of them have real dollars to spend! These dollars usually appear as refunds or buydowns to the consumer. That's you. Programs and available funds vary with time and state. For the latest information, call your State Energy Office, listed on pages 132–137.

System Sizing

We've found from experience that there's no such thing as "one size fits all" when it comes to energy systems. Everyone's needs, expectations, site, and climate are individual, and your power system, in order to function reliably, must be designed with these individual factors accounted for. Our friendly and helpful technical staff, with over 25,000 solar systems under their collective belts, has become rather good at this. We don't charge for this personal service, so long as you purchase your system components from us. We do need to know what makes your house, site, and lifestyle unique. So filling out the household electrical demands portion of our sizing worksheets is the first, and very necessary step, usually followed by a phone call (whenever possible), and a customized system quote. Worksheets, wattage charts, and other helpful information for system sizing, are included at the end of Chapter 3, our Panel to Plug chapter—which also just happens to cover batteries, safety equipment, controls, monitors, and all the other bits and pieces you need to know a little about to assemble a safe, reliable energy system.

—Doug Pratt

Residential Power System Basics

Every residential system, from a small weekend cabin up through the largest multifamily system, consists of the following basic components:

*A **Power Source** can be PV, hydro, wind, or fossil-fueled generator.*

*A **Storage System** is usually a lead-acid battery bank.*

***Safety Equipment** includes properly sized fusing for any wire connected to a battery.*

*A **Controller** prevents battery overcharging, and sometimes over-discharging.*

*A **Monitor** displays battery state of charge as a minimum, and more as needed.*

How Much PV Area to Equal
U.S. Electric Production?

According to the Energy Information Administration of the U.S. Department of Energy, www.eia.doe.gov, the U.S. produced 3,678 billion kilowatt-hours of electricity in 1999. Let's round that up and aim for an even 4,000 billion kWh, which is a production figure we'll hit sometime between the years 2005 and 2010. Note that this is "production," not "use." Transmission inefficiencies and other losses are covered. This is the amount of power we need to stuff into the pipeline inlet.

We'll want our PV modules in a good sunny area to make the best of our investment, so looking at the National Solar Radiation Data Base (NREL document # TP-463-5607)

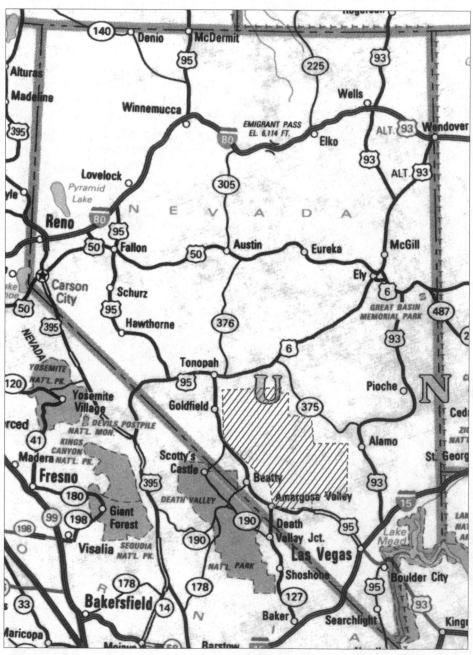

This area—the Nevada Test Site and Nellis Air Force Base—could deliver enough energy to meet the entire U.S. needs, using existing photovoltaic technology.

for Tonopah, Nevada, we see that a flat-plate collector on a fixed-mount facing south at a fixed-tilt equal to the latitude, 38.07° in this case, saw a yearly average of 6.1 hours of "full-sun" per day in the years 1961 through 1990. A "full-sun" is defined as 1,000 watts per square meter.

For PV modules we'll use the large Astropower 120-watt module, which the California Energy Commission, www.energy.ca.gov/greengrid/certified_pv_modules.html, rates at 107 watts output, based on observed, real-world performance. 107 watts times 6.1 hours equals 652.7 watt-hours or 0.6527 kilowatt-hours per day per module at our Tonopah site. At 26″ x 58.1″ this module presents 10.5 square feet of surface area. We'll allow some space between rows of modules for maintenance access, and for sloping wintertime sun, so let's say that each module will need 15 square feet.

Conversion from PV module DC output to conventional AC power isn't perfectly efficient. Looking at the real-world performance figures from the California Energy Commission again, www.energy.ca.gov/greengrid/certified_inverters.html, we see that the Trace Engineering 20kW model PV-20208 is rated at 96% efficiency. We'll probably be using larger inverters, but this is a typical efficiency for large intertie inverters. So our 0.6527 kWh per module per day becomes 0.6266 kWh by the time it hits the AC grid.

A square mile, 5,280 feet times 5,280 feet equals 27,878,400 square feet. Divided by 15 sq.ft. per module, we can fit 1,858,560 modules per square mile. At 0.6266 kilowatt-hours per module per day, our square mile will deliver 1,164,574 kWh per day on average, or 425,069,510 kWh per year. Back to our goal of 4,000,000,000,000 kWh, divided by 425,069,510 kWh per year per square mile, it looks like we need about 9,410 square miles of surface to meet the electrical needs of the U.S. That's a square area a bit less than 100 miles on a side. This is a bit over half of the approximate 16,000 square miles currently occupied by the Nevada Test Site and the surrounding Nellis Air Force Range (www.nv.doe.gov/nts/ and www.nellis.af.mil/environmental/default.htm).

As a practical measure, we need to point out that PV power production happens during the daytime, and so long as we persist in turning the lights on at night, there will continue to be substantial power use at night. Also, so long as we're out in the desert, solar thermal collection might be a more efficient power generation technology. But, however you run the energy collection system, large solar-electric farms on what is otherwise fairly useless desert land could add substantially to the electrical independence and security of any country. The existing infrastructure of coal, nuclear, and hydro power plants could continue to provide reliable power at night, but non-renewable resource use and carbon dioxide production would be greatly reduced.

—**Doug Pratt**

www.realgoods.com

Photovoltaic Products

System Design

New Solar Electric Home

Joel Davidson. Gives you all the information you need to set up a first-time PV system, whether it's a remote site, grid connect, marine or mobile, stand-alone, or auxiliary. Good photos, charts, graphics, and tables. Written by one of the pioneers. A good all-around book for getting started with alternative energy. 408 pages, paperback. USA.

80-101 The New Solar Electric Home $18⁹⁵

The Solar Electric House

Steven Strong. The author has designed more than 75 PV systems. This fine book covers all aspects of PV, from the history and economics of solar power to the nuts and bolts of panels, balance of systems equipment, system sizing, installation, utility intertie, stand-alone PV systems, and wiring instruction. A great book for the beginner. 276 pages, paperback. USA.

80-800 The Solar Electric House $21⁹⁵

The Solar Electric Independent Home Book

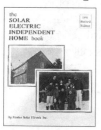

Paul Jeffrey Fowler. A good, basic primer for getting started with PV, written for the lay person. Lots of good charts, clear graphics, a solid glossary, and appendix. Includes 75 detailed diagrams, and the text includes recent changes in PV technology. This is one of the best all-round books on wiring your PV system. 174 pages, paperback. USA.

80-102 The Solar Electric Independent Home Book $16⁹⁵

Wiring 12 Volts For Ample Power

David Smead and Ruth Ishihara. The most comprehensive book on DC wiring to date, written by the authors of the popular book *Living on 12 Volts with Ample Power*. This book presents system schematics, wiring details, and troubleshooting information not found in other publications. Leans slightly toward marine applications. Chapters cover the history of electricity, DC electricity, AC electricity, electric loads, electric sources, wiring practices, system components, tools, and troubleshooting. 240 pages, paperback. USA.

80-111 Wiring 12 Volts For Ample Power $19⁹⁵

RVers Guide To Solar Battery Charging

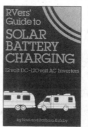

Noel and Barbara Kirkby. The authors have been RVing for over 20 years and have applied photovoltaics to their independent lifestyle. This book includes numerous example systems, illustrations, and easy-to-understand instructions. *The Whole Earth Catalog* calls it, "A finely detailed guide to installing PV systems in your motorhome, trailer, boat, or cabin." 176 pages, paperback. USA.

80-105 RVers Guide To Solar Battery Charging $12⁹⁵

Who Owns the Sun?

The Most Important Book on Solar Energy in Three Decades!

Solar technologies—including photovoltaics, wind and hydroelectric turbines, and solar water and space-heating—are more cost-effective than ever before, and they offer an inexhaustible source of power for the entire planet.

You can be certain that there will be a global shift to a solar-based economy; the question is when and how such a transformation will occur. Can this be the beginning of the long-awaited solar revolution that Real Goods has been trying to instigate for the last ten years?

Authors Dan Berman and John O'Connor ask the hard-hitting questions about who will control the next generation of utilities. Will it be the the same giant corporations that dominate the market today? Or will citizens seize this once-in-a-lifetime opportunity, and reassert responsibility for the sources and supplies of their energy?

Who Owns the Sun? argues that democratic control of solar energy is the key to revitalizing America—putting power back into the hands of local people. 356 pages, tables, index, bibliography. USA.

80-034 Who Owns the Sun? (hardbound edition) $25⁰⁰

CREST SolarSizer™ Software
A Photovoltaic System Design Tool

This new software from the non-profit Center for Renewable Energy and Sustainable Technology is without a doubt the best PV sizing program available. Initially developed as a professional design tool, we found it incredibly easy to navigate and a blessing for anyone looking into the possibility of a PV installation. It uses a simple graphic interface to choose and customize both household loads and potential system components. Once selected, you may "install" appliances and components in the virtual house, and access detailed reports on initial, annual, and lifetime costs. Keep working until your virtual PV system meshes with power and budget needs. Features extensive Help files, and yearly sunlight data for 250 U.S. sites, and 46 sites in 27 foreign countries. USA.

System Requirements: Supplied on two 3.5" disks Requires Windows 3.1 or later (Windows 95, 98), an Intel 80486/Pentium processor, 4MB of RAM, 4MB hard drive space, and a graphics card set to display 256 colors.

80 712 CREST SolarSizer™—Windows version $125⁰⁰

Editor's Note

For a great website to calculate solar needs and payback times, check out www.energy.ca.gov/cleanpower/

Renewable Energy With The Experts

A brand new, professionally produced video series that covers basic technology introductions as well as hands-on applications. Each tape is presented by the leading expert in that field, and tackles a single subject in detail. Includes classroom explanations of how the technology works with performance and safety issues to be aware of. Then action moves into the field to review actual working systems, and finally assemble a real system on camera as much as possible. Other tapes in this series cover storage batteries with system sizing, and water pumping. You'll find these tapes in appropriate chapters. This is the closest to hands-on you can get without getting dirt under your fingernails. All tapes are standard VHS format. USA.

80-360 Set of Three Renewable Experts Tapes $100⁰⁰

Johnny Weiss has been designing and installing PV systems for over 15 years. He teaches hands-on classes at Solar Energy International in Colorado. 48 minutes.

80-361 Renewable Experts Tape—PV $39⁰⁰

Don Harris, the owner of Harris Turbines, practically invented the remote microturbine industry, and has manufactured over 1,000 in the past 15 years. 42 minutes.

80-362 Renewable Experts Tape—Hydro $39⁰⁰

Mick Sagrillo, the owner of Lake Michigan Wind and Sun, has been in the wind industry for over 16 years with over 1,000 turbine installation/repairs under his belt. 63 minutes.

80-363 Renewable Experts Tape—Wind $39⁰⁰

AstroPower Photovoltaic Modules

Based in Newark, Delaware, AstroPower is the only independent, U.S.-owned company in the business of solar module production, and they're one of the fastest-growing PV companies in the world. Cells and modules are entirely made in the U.S. The single crystal 55-, 60-, 75-, and 120-watt modules we're offering are conventional construction with 36 series-connected cells, tempered-glass glazing, Tedlar backsheet, anodized aluminum frame, a large weather-tight junction box, and are UL-listed. Astropower's single crystal modules use silicon wafer stock that is recycled from the computer chip industry. Now featuring one of the industry's strongest warranties at 20 years!

AstroPower AP120 120-Watt Module

Our Best-Selling PV by Wattage

There's only one word to describe this low-cost, single-crystal, 120-watt module. Stunning! The 36 cells are the largest we've seen at slightly over six inches. Higher wattage allows fewer modules to get the same job done. This means less time bolting things down and interconnecting. Thirty-six series-connected cells assure good performance in hot weather, and this module is intermixable with other 36-cell modules, regardless of manufacturer. The weather-tight junction box is large, has three standard ½″ knockouts, features two large screw terminals for each polarity, plus a pair of spare terminals, all clearly labeled. Will accept up to 8-gauge wire. A pair of bypass diodes are built into the junction box. 20-year mfr.'s warranty. UL-listed. USA.

Rated Watts: 120.0 watts @ 25°C
Rated Power: 16.9 volts @ 7.1 amps
Open Circuit Volts: 21.0 volts @ 25°C
Short Circuit Amps: 7.7 amps @ 25°C
L x W x D: 58.1″ x 26.0″ x 1.4″
Construction: single crystal, tempered glass.
Warranty: within 20% of rated output for 20 yrs.
Weight: 26.1 lb/11.9 kg

11-166 AstroPower AP120 120-Watt Module $649⁰⁰

AP-120 modules are in short supply! Check before ordering, please.

AstroPower AP75 75-Watt Module

Our Best-Selling PV Ever

This state-of-the-art 75-watt module has 36 series-connected cells that assure excellent performance in hot weather, and this module is intermixable with other 36-cell modules (like the Siemens SP75), regardless of manufacturer. The weather-tight junction box is large, has three standard ½″ knockouts, terminals for each polarity, plus a pair of spare terminals, all clearly labeled. Will accept up to 8-gauge wire. A pair of bypass diodes are built into the J-box. Twenty-year warranty. USA.

Rated Watts: 75.0 watts @ 25°C
Rated Power: 17.0 volts @ 4.4 amps
Open Circuit Volts: 21.0 volts @ 25°C
Short Circuit Amps: 4.8 amps @ 25°C
L x W x D: 47.2″ x 20.7″ x 1.4″
Construction: single crystal, tempered glass
Warranty: within 20% of rated output for 20 yrs.
Weight: 18.1 lb/8.2 kg

11-165 AstroPower AP75 75-Watt Module $419⁰⁰

Siemens Photovoltaic Modules

Siemens is a German multinational company with photovoltaic production facilities in Vancouver, WA, and Camarillo, CA. Siemens purchased Arco Solar several years ago. Same American employees, same facilities, same high quality. Siemens modules use the most-efficient single crystal silicon, low-iron tempered glass, EVA encapsulation, Tedlar back sheets, anodized aluminum frame, and weather-tight junction boxes with built-in bypass diodes. All larger Siemens modules feature an industry-leading 25-year warranty.

Siemens SP-Series

The SP series is based on a 5-inch cell that's grown in an octagonal shape that requires less trimming and waste. All SP series modules benefit from Siemens TOPS (Textured Optimized Pyramidal Surface) coating, which reduces the amount of light lost to reflection. Modules feature 36 or 72 cells wired in series for good nonfading performance in hotter climates. Other features include an anodized aluminum frame, tempered glass covers, EVA encapsulation, polymer backsheets, built-in bypass diodes, and a roomy, conduit-ready J-box with easy terminal-strip connections. All SP modules have an industry-leading 25-year mfr.'s warranty. USA.

Siemens SP75

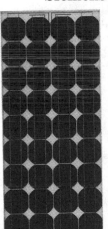

The SP75 is a 75-watt, 12-volt, single crystal power production module. The SP75 is the base from which all other modules in the SP-series have grown. This 12-volt nominal module has 36 cells in series, and is interchangeable with the Astropower AP75 75-watt module. USA.

Rated Watts: 75 watts @ 25°C
Rated Power: 17.0 volts @ 4.4 amps
Open Circuit Volts: 22.0 volts @ 25°C
Short Circuit Amps: 4.8 amps @ 25°C
L x W x D: 47.3" x 20.8" x 2.5"
Construction: single crystal, tempered glass
Warranty: within 20% of rated output for 25 yrs.
Weight: 16.7 lb/7.6 kg

11-106 Siemens SP75 Module $449⁰⁰

Siemens SP130

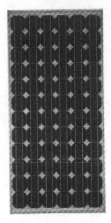

Using 72 of the SP-series cells, this module is wired from the factory for 24-volt nominal output. Since so many PV arrays and applications are running at higher voltages, we're starting to see higher voltage modules directly from manufacturers. This saves mounting and wiring effort on site. This larger module requires truck or FedEx shipping. USA.

Rated Watts: 130 watts @ 25°C
Rated Power: 33.0 volts @ 3.95 amps
Open Circuit Volts: 42.8 volts @ 25°C
Short Circuit Amps: 4.5 amps @ 25°C
L x W x D: 63.76" x 32.05" x 1.57"
Construction: single crystal, tempered glass
Warranty: within 20% of rated output for 25 yrs.
Weight: 32.6 lb/14.8 kg

11-575 Siemens SP130 Module $729⁰⁰

Siemens SP150

The most robust of the SP-series, these modules pass the flash test at the end of the production line with the highest marks. They aren't bigger, they just deliver a little better. These are 24-volt modules to reduce wiring and mounting effort on site. This larger module requires truck or FedEx shipping. USA.

Rated Watts: 150 watts @ 25°C
Rated Power: 34.0 volts @ 4.4 amps
Open Circuit Volts: 43.4 volts @ 25°C
Short Circuit Amps: 4.8 amps @ 25°C
L x W x D: 63.76" x 32.05" x 1.57"
Construction: single crystal, tempered glass
Warranty: within 20% of rated output for 25 yrs.
Weight: 32.6 lb/14.8 kg

11-576 Siemens SP150 Module $849⁰⁰

Siemens SM110 Modules

Based on the SM-series PV cells that have been in proven production for over a decade, this high-wattage module has 72 cells, and is available in either a 12-volt or a 24-volt output configuration. The 12-volt model has a pair of watertight electrical junction boxes; the 24-volt model has a single watertight junction box. Both models feature all the usual SM-series performance enhancements, including the new TOPS anti-reflective coating, a multi-layer backing that protects from cuts and scratching while reflecting any light energy back into the cells, and multiple contacts on cell front and back for better long-term reliability. Large, weather-tight junction boxes with built-in bypass diodes provide multiple terminals. Full-perimeter anodized aluminum frame, tempered ultra-clear glass, and a 25-year mfr.'s warranty complete the package. USA.

Specifications at 24-volt. 12-volt models will double the amps and halve the voltage.

Rated Watts: 110 watts @ 25°C
Rated Power: 35.0 volts @ 3.15 amps
Open Circuit Volts: 43.5 volts @ 25°C
Short Circuit Amps: 3.45 amps @ 25°C
L x W x D: 51.8″ x 25.98″ x 1.6″
Construction: single crystal, tempered glass
Warranty: within 20% of rated output for 25 yrs.
Weight: 25.1 lb/11.5 kg

11-577 Siemens SM110-24v PV Module \$629⁰⁰

11-578 Siemens SM110-12v PV Module \$639⁰⁰

Siemens SM55

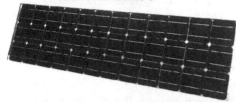

The SM55 is a standard module, consisting of 36 cells in series. It is ideal for water pumping applications, or rooftop RV applications. Like all 36-cell modules, it works well in hot climates as high-temperature voltage drop is kept tolerable, and is intermixable with other 36-cell modules, regardless of manufacturer. UL-listed. USA.

Rated Watts: 55 watts @ 25°C
Rated Power: 17.4 volts @ 3.15 amps
Open Circuit Volts: 21.7 volts @ 25°C
Short Circuit Amps: 3.45 amps @ 25°C
L x W x D: 50.9″ x 13″ x 1.4″
Construction: single crystal, tempered glass
Warranty: within 20% of rated output for 25 yrs.
Weight: 12.0 lb/5.5 kg

11-105 Siemens SM55 Module \$399⁰⁰

Siemens SM50-H

The Siemens SM50-H is efficient, attractive, easy to install, and comes with a wired-in bypass diode in the junction box. The SM50-H consists of 33 cells in series. This is the module that used to be sold as the M75. Not for hot climates (above 80°F). UL-Listed. USA.

Rated Watts: 50 watts @ 25°C
Rated Power: 15.9 volts @ 3.15 amps
Open Circuit Volts: 19.8 volts @ 25°C
Short Circuit Amps: 3.4 amps @ 25°C
L x W x D: 48″ x 13″ x 1.4″
Construction: single crystal, tempered glass
Warranty: within 20% of rated output for 25 yrs.
Weight: 11.5 lb/5.2 kg

11-101 Siemens SM50-H Module \$359⁰⁰

Siemens SR50 Module

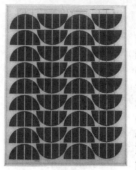

The SR50 module uses 36 half-round cells, a design that reduces silicon trimming and waste. These modules also feature the new TOPS anti-reflective coating, a multi-layer backing that protects from cuts and scratching while reflecting any light energy back into the cells, and multiple contacts on cell front and back for better long-term reliability. Large, weather-tight junction boxes with built-in bypass diodes provide multiple terminals. Full-perimeter anodized aluminum frame, tempered ultra-clear glass, UL-listing, and a 25-year mfr.'s warranty complete the package. USA.

Rated Watts: 50 watts @ 25°C
Rated Power: 17.0 volts @ 2.95 amps
Open Circuit Volts: 21.6 volts @ 25°C
Short Circuit Amps: 3.2 amps @ 25°C
L x W x D: 30.8″ x 23.4″ x 1.34″
Construction: single crystal, tempered glass
Warranty: within 20% of rated output for 25 yrs.
Weight: 13.0 lb/5.9 kg

11-236 Siemens SR50 Module \$329⁰⁰

Siemens SM20

The SM20 is a compact, self-regulating module ideal for RVs, boats, and remote homes where needs are minimal, use is intermittent, or space is limited. As the battery approaches full charge, the SM20 decreases the current output from 1.37 amps to less than 0.25 amp, eliminating the need for a charge controller. Siemens recommends at least 70Ah of battery storage for each SM20 module. Not recommended for hot climates. USA.

Rated Watts: 22 watts @ 25°C
Rated Power: 14.5 volts @ 1.38 amps
Open Circuit Volts: 18.0 volts @ 25°C
Short Circuit Amps: 1.60 amps @ 25°C
L x W x D: 22.3″ x 12.9″ x 1.3″
Construction: single crystal, tempered glass
Warranty: within 10% of rated output for 10 yrs.
Weight: 5.6 lb/2.5 kg

11-107 Siemens SM20 Module \$259⁰⁰

Solarex Photovoltaic Modules

Solarex is an American-owned company with photovoltaic assisted production facilities in Maryland. The Solarex line offers exceptional performance in high-temperature climates, as all the modules offer 16.8 volts or higher output. Solarex uses polycrystalline cells coated with a titanium dioxide antireflective material, low-iron tempered glass, EVA encapsulation, Tedlar back sheets, bronzed aluminum framing, and large weather-tight junction boxes for all their modules, unless noted otherwise.

Solarex MSX-60

The MSX-60 is Solarex's production solar panel. Now with a 20-year warranty on the 60-watt modules! USA.
Rated Watts: 60 watts @ 25°C
Rated Power: 17.1 volts @ 3.5 amps
Open Circuit Volts: 21.1 volts @ 25°C
Short Circuit Amps: 3.8 amps @ 25°C
L x W x D: 43.63" x 19.75" x 1.97"
Construction: polycrystalline, tempered glass
Warranty: within 20% of rated output for 20 yrs.
Weight: 15.9 lb/7.2 kg

11-501 Solarex MSX-60 Module $359⁰⁰

The Solarex Lite Series

The Lite series of Solarex modules are made without glass and are very thin, lightweight, and highly portable. Typically they're used for boating, camping, RVs, scientific expeditions, or mobile communications. This series of panels is great for recharging laptop computers and video cameras in remote locations. Consists of a layer of aluminum bonded to fiberglass mat with the silicon cells on top. Environmentally sealed with Tedlar (a non-yellowing plastic glazing material). These modules will flex slightly, but are extremely tough. Has 3-meter lead wire attached. USA.

Solarex MSX30L Unbreakable Lite

This module is a good choice for sailboats or scientific expeditions. Will bend to mount on boat cabin tops, can be stepped on or have things dropped on it without damage. USA.
Rated Watts: 30 watts @ 25°C
Rated Power: 17.8 volts @ 1.68 amps
Open Circuit Volts: 21.3 volts @ 25°C
Short Circuit Amps: 1.82 amps @ 25°C
L x W x D: 24.25" x 19.5" x 0.38"
Construction: polycrystalline, Tedlar coating
Warranty: one year
Weight: 4.84 lb/2.2 kg

11-511 Solarex MSX30L Module $284⁰⁰

Solarex MSX20L Unbreakable Lite

The 20-watt module in this series is a good choice for recharging video camera batteries, some small water pumps, or heavy laptop computer users. USA.
Rated Watts: 18.5 watts @ 25°C
Rated Power: 17.8 volts @ 1.12-amps
Open Circuit Volts: 21.0 volts @ 25°C
Short Circuit Amps: 1.16 amps @ 25°C
L x W x D: 17.5" x 19.5" x 0.38"
Construction: polycrystalline, Tedlar coating
Warranty: one year
Weight: 3.28 lb/1.49 kg

11-517 Solarex MSX20L Module $226⁰⁰

Solarex MSX10L Unbreakable Lite

The 10-watt module is an excellent choice for backpackers, charging laptop computers, or recharging nicad batteries. USA.
Rated Watts: 10 watts @ 25°C
Rated Power: 17.5 volts @ 0.57 amps
Open Circuit Volts: 21.0 volts @ 25°C
Short Circuit Amps: 0.6 amps @ 25°C
L x W x D: 17.5" x 10.5" x 0.38"
Construction: polycrystalline, Tedlar coating
Warranty: one year
Weight: 1.8 lb/0.82 kg

11-513 Solarex MSX10L Module $135⁰⁰

Solarex Millennia Series

Millennia amorphous modules use a 3rd generation thin-film process. They offer greatly improved thin-film stability and efficiency. Their tandem-junction construction puts two silicon junction layers on the panel face, with each junction tuned for optimum conversion of different sunlight spectral segments. These are some of the most resource-efficient, and least expensive modules available.

Solarex makes several frame and wiring versions of the Millennia. We keep life simple by offering the conventional full-perimeter, anodized aluminum universal frame, with standard watertight junction box, EVA encapsulation, and glass backsheets. The larger size and universal frames of Millennia modules lend themselves to do-it-yourself mounting structures. The cost of ready-made factory mounts can negate any savings on these lower-cost, but larger modules.

Millennia modules are available in two voltage ranges. The LV series are conventional 12-volt nominal modules with a 16.7 volt maximum power point for battery charging. The MV series are nominal 48-volt output and can be used for direct utility intertie, or for battery charging applications.

Millennia modules are made in Solarex's new automated thin-film manufacturing facility. A patented laser-scribing procedure forms the individual solar cells, with all inter-cell electrical connections internal to the module. This monolithic construction is innately more reliable, and is unique among large power modules.

Note: Like all amorphous silicon photovoltaics, Millenia module' power output decreases during the first few months of exposure. Initial power will be as much as 18% higher than rated, with voltage 12% higher and current 6% higher. This power bonus must be considered when designing the power, control, and safety systems. Listed module specifications are post-attenuation, at standard test conditions.

Solarex Millennia MST-43LV

The nominally 12-volt LV modules are designed for use in traditional battery-equipped applications. They can be wired in series for higher voltage, and/or in parallel for higher amperage, just like any conventional 36-cell modules, with which they can also be intermixed. UL-listed. USA.
Rated Watts: 43.0 watts @ 25°C
Rated Power: 16.5 volts @ 2.6 amps
Open Circuit Volts: 22.7 volts @ 25°C
Short Circuit Current: 3.3 amps
Construction: amorphous, plate glass
Warranty: within 20% of rated output for 10 yrs.
Weight: 34.8 lb/15.8 kg
L x W x D: 48.4" x 26.3" x 2.0"

11-506 Solarex Millennia MST-43LV $199⁰⁰

Solarex Millennia MST-43MV

The MV modules are nominally 48 volts, and are intended for use in utility intertie or other higher-voltage arrays. This high-voltage module allows intertie systems to start smaller, and grow in less-expensive chunks. UL-listed. USA.

Rated Watts: 43.0 watts @ 25°C
Rated Power: 72 volts @ 0.6 amps
Open Circuit Volts: 98 volts @ 25°C
Short Circuit Current: 0.8 amps
Construction: amorphous, plate glass
Warranty: within 20% of rated output for 10 yrs.
Weight: 34.8 lb/15.8 kg
L x W x D: 48.4" x 26.3" x 2.0"

11-502 Solarex Millennia MST-43MV $199⁰⁰

Kyocera Photovoltaic Modules

Kyocera is a Japanese-based company with many, many years of experience in polycrystalline solar cells. Their cells have some of the highest efficiency found in commercial production. All Kyocera modules feature standard 36-cell construction, tempered, low-glare glass, EVA encapsulation, polyvinyl backsheets, anodized and clear-coated aluminum frames, with large, excellent, weatherproof junction boxes. Bypass diodes are installed halfway through the 36-cell series to reduce the effects of partial shading. The KC-series modules we're offering here are all UL, and cUL listed, and CE certified. A 25-year limited warranty is standard.

Kyocera KC120

Bigger wattage modules have some real advantages. You buy fewer modules, which means less bolting down and wiring up; thus, a faster, easier installation. Larger modules usually offer some economies of scale; it costs less to build one big module than two smaller ones. Kyocera's top-of-the-line module, the KC120, delivers a powerful 120 watts. No other module we offer delivers as much power in as small a footprint. The multi-crystal KC120 features standard 36-cell construction, tempered, low-glare glass, EVA encapsulation, polyvinyl backsheet, anodized and clear-coated aluminum frame, with a large, excellent, weatherproof junction box. The J-box has four standard ½" knockouts. Bypass diodes are installed halfway through the 36-cell series to reduce the effects of partial shading. UL and cUL listed. CE certified. Twenty-five-year warranty. Japan.

The Kyocera J-box is large, watertight, and simple, but requires wire terminals.

Rated Power: 120 watts
Voltage @ max. power: 16.9 volts
Current @ max. power: 7.1 amps
Open Circuit Voltage: 21.5 volts
Short Circuit Current: 7.45 amps
L x W x D: 56.1" x 25.7" x 2.0"
Construction: polycrystalline, tempered glass
Weight: 26.3 lb/11.9 kg
Warranty: within 20% of rated output for 25 yrs.

11-568 Kyocera KC120 Module $685⁰⁰

Kyocera KC80

Sporting the same features and construction as the KC120 above, the KC80 is assembled with ⅔-size cells. Features include standard 36-cell construction, tempered, low-glare glass, EVA encapsulation, polyvinyl backsheet, anodized and clear-coated aluminum frame, with a large, excellent, weather-proof junction box. The J-box has four standard ½″ knockouts. Bypass diodes are installed halfway through the 36-cell series to reduce the effects of partial shading. UL and cUL listed. CE certified. Twenty-five-year warranty. Japan.

Rated Power: 80 watts
Voltage @ max. power: 16.9 volts
Current @ max. power: 4.73 amps
Open Circuit Voltage: 21.5 volts
Short Circuit Current: 4.97 amps
L x D x W: 38.4″ x 25.7″ x 2.0″
Construction: polycrystalline, tempered glass
Weight: 17.7 lb/8.0 kg
Warranty: within 20% of rated output for 25 yrs.

11-567 Kyocera KC80 Module $465⁰⁰

Uni-Solar Triple Junction PV Technology

Uni-Solar delivers a nice breakthrough in lower cost, higher output amorphous film technology. For higher efficiency and lower production costs, each cell is composed of three spectrum-sensitive semiconductor junctions stacked on top of each other: blue light sensitive on top, green in the middle, red on the bottom. This gives 30% more output per square foot than previous Uni-Solar products. Like all previous Uni-Solar products these flexible, unbreakable modules are deposited on a stainless steel backing with DuPont's Tefzel glazing and, due to unique construction, have the ability to continue useful output during partial shading. The unbreakable construction and good resistance to heat-fading makes these modules an excellent choice for RVs, or any site where vandalism is a possibility. The new triple-junction technology brings the price down to a level equal to or better than single crystal technology.

Uni-Solar has crafted their basic flexible amorphous PV material into several product lines, including; conventional aluminum-framed modules; marine modules with flexible frames; and several roofing products.

Like all amorphous modules, power output will fade approximately 15% over the first month or two. Initial power will be as much as 18% higher than rated. This power bonus must be considered when designing the power, control, and safety systems. Rated power is post fade, and is warranted for 3 to 20 years, depending on the model.

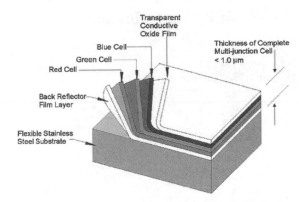

Conventional Uni-Solar Modules

Frames are anodized aluminum. Large weatherproof junction boxes are supplied on 64-, 42- and 32-watt modules with a 20-year warranty. Wire pigtails are supplied on 21-, 11-, and 5-watt sizes with a 10-year warranty. These are all excellent choices for RVs, vandalism-prone sites, or systems that need to move easily without threat of breakage. UL and CUL listed. USA.

Uni-Solar 5-Watt Module

Rated Watts: 5.0 watts @ 25°C
Rated Power: 16.5 volts @ 300 milliamps
Open Circuit Volts: 23.8 volts @ 25°C
Construction: triple junction, Tefzel glazing
Warranty: within 20% of rated output for 10 yrs.
Weight: 2.5 lb/1.13 kg
L x W x D: 19.3″ x 8.5″ x 0.9″

11-230 Uni-Solar US5 Module $89⁰⁰

Uni-Solar 11-Watt Module

Rated Watts: 10.3 watts @ 25°C
Rated Power: 16.5 volts @ 620 milliamps
Open Circuit Volts: 23.8 volts @ 25°C
Construction: triple junction, Tefzel glazing
Warranty: within 20% of rated output for 10 yrs.
Weight: 3.6 lb/1.63 kg
L x W x D: 19.3″ x 15.1″ x 0.9″

11-231 Uni-Solar US11 Module $149⁰⁰

Uni-Solar 21-Watt Module

Rated Watts: 21.0 watts @ 25°C
Rated Power: 16.5 volts @ 1.27 amps
Open Circuit Volts: 23.8 volts @ 25°C
Construction: triple junction, Tefzel glazing
Warranty: within 20% of rated output for 10 yrs.
Weight: 6.6 lb/3.0 kg
L x W x D: 36.5″ x 15.1″ x 1.25″

11-232 Uni-Solar US21 Module $195⁰⁰

Uni-Solar 32-Watt Module

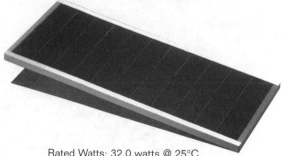

Rated Watts: 32.0 watts @ 25°C
Rated Power: 16.5 volts @ 1.94 amps
Open Circuit Volts: 23.8 volts @ 25°C
Construction: triple junction, Tefzel glazing
Warranty: within 20% of rated output for 20 yrs.
Weight: 10.6 lb/4.8 kg
L x W x D: 53.8″ x 15.1″ x 1.25″

11-227 Uni-Solar US32 $239⁰⁰

Uni-Solar 42-Watt Module

Rated Watts: 42.0 watts @ 25°C
Rated Power: 16.5 volts @ 2.54 amps
Open Circuit Volts: 23.8 volts @ 25°C
Construction: triple junction, Tefzel glazing
Warranty: within 20% of rated output for 20 yrs.
Weight: 13.8 lb/6.3 kg
L x W x D: 36.5″ x 29.1″ x 1.25″

11-233 Uni-Solar US42 $285⁰⁰

Uni-Solar 64-Watt Module

Rated Watts: 64.0 watts @ 25°C
Rated Power: 16.5 volts @ 3.88 amps
Open Circuit Volts: 23.8 volts @ 25°C
Construction: triple junction, Tefzel glazing
Warranty: within 20% of rated output for 20 yrs.
Weight: 20.2 lb/9.2 kg
L x W x D: 53.8″ x 29.1″ x 1.25″

11-228 Uni-Solar US64 $419⁰⁰

Flexible Marine Modules

With a flexible, floating, marine module design that Uni-Solar has been offering for years, this series uses triple-junction film for greater efficiency in tight spaces. Can be rolled down to an 8-inch diameter without damage. Features blocking diodes within the potted junction block to prevent battery discharge and nickel-plated brass grommets at corners. Includes eight feet of cable with built-in fuse holder and a 2-pin SAE connector. We've seen these flexible modules cleverly velcroed to the bottom of cockpit cushions. Just flip the cushions over for charging! UL-listed. USA.

Uni-Solar Marine Flex USF-5

A 5-watt module that's ideal for battery maintenance charging on boats or RVs. Can be used without charge control on all but the smallest batteries. USA.
Rated Watts: 5.0 watts @ 25°C
Rated Power: 16.5 volts @ 0.30 amps
Open Circuit Volts: 23.8 volts @ 25°C
Short Circuit Amps: 0.37 amps @ 25°C
L x W x D: 21.8″ x 9.7″ x 0.3″
Construction: triple-junction, Tefzel glazing, flex frame
Warranty: within 20% of rated output for 3 yrs.
Weight: 1.18 lb/0.54 kg

11-234 Uni-Solar Flex 5-Watt Module $109⁰⁰

Uni-Solar Marine Flex USF-11

An 11-watt module for modest battery charging needs. Good for small boats, laptop computers, or other mobile uses. USA.
Rated Watts: 10.3 watts @ 25°C
Rated Power: 16.5 volts @ 0.62 amps
Open Circuit Volts: 23.8 volts @ 25°C
Short Circuit Amps: 0.78 amps @ 25°C
L x W x D: 21.8″ x 16.7″ x 0.3″
Construction: triple-junction, Tefzel glazing, flex frame
Warranty: within 20% of rated output for 3 yrs.
Weight: 2.0 lb/0.91 kg

11-221 Uni-Solar Flex 11-Watt Module $179⁰⁰

Uni-Solar Marine Flex USF-32

The largest flexible module available, this full-power 32-watt model is the ticket for extended travel, cruising, or independent living. USA.
Rated Watts: 32.0 watts @ 25°C
Rated Power: 16.5 volts @ 1.94 amps
Open Circuit Volts: 23.8 volts @ 25°C
Short Circuit Amps: 2.40 amps @ 25°C
L x W x D: 56.3" x 16.7" x 0.3"
Construction: triple-junction, Tefzel glazing, flex frame
Warranty: within 20% of rated output for 3 yrs.
Weight: 4.70 lb/2.14 kg

11-235 Uni-Solar Flex 32-Watt Module $395⁰⁰

Evergreen Cedar Series PV Modules

Innovative, Low-Cost,
String Ribbon Silicon Production

EC-51

Evergreen Solar is a rapidly growing U.S. company with an innovative String Ribbon™ polycrystalline production technology. String Ribbon produces a silicon wafer by drawing up a thin film of molten silicon between two strings, like a soap bubble. Production is a simple, nearly continuous process. With none of the usual slicing or dicing required to get thin blanks of silicon to build PV cells on, there's very little waste, just fast, efficient cell production. That means PV modules that use less energy and waste less raw materials in the process. Like all polycrystalline modules, this is a very stable power production material with a life expectancy measured in decades.

As they've gained experience with String Ribbon production, Evergreen has been producing larger cells, resulting in higher wattage modules. Their latest offerings, the Cedar Series, produces 51 watts, or in the double-size configuration, 102 watts. Features include standard 36 cells in series configuration, tempered-glass covers, clear anodized aluminum frames with standardized 24" mounting width, Tedlar backsheets, and roomy watertight junction boxes.

The 102-watt module features dual-voltage output. With 72-cell construction, it can be wired at the junction box to deliver either 12- or 24-volt output. This makes for less wiring and interconnection requirements on 24-volt systems, which are increasingly the standard voltage for household and pumping systems.

EC-102

All Evergreen modules are UL-listed and feature a great 20-year warranty. USA.

EC-102 specs below are listed at 12-volt output. 24-volt output is the same wattage, but double the voltage, and half the amperage.
Rated Power: 51 watts
Voltage @ max. power: 16.9 volts
Current @ max. power: 3.02 amps
Open Circuit Voltage: 20.0 volts
Short Circuit Current: 3.54 amps
L x W x D: 32.1" x 25.7" x 2.2"
Construction: polycrystalline, tempered glass
Weight: 15.0 lb/6.8 kg
Warranty: within 20% of rated output for 20 yrs.

11-002 Evergreen EC-51 Module $269⁰⁰
Rated Power: 102 watts
Voltage @ max. power: 16.9 volts
Current @ max. power: 6.04 amps
Open Circuit Voltage: 20.0 volts
Short Circuit Current: 7.09 amps
L x W x D: 62.4" x 25.7" x 2.2"
Construction: polycrystalline, tempered glass
Weight: 27.0 lb/12.2 kg
Warranty: within 20% of rated output for 20 yrs.

11-003 Evergreen EC-102 Module $499⁰⁰

Evergreen 94- and 47-Watt Modules

The Cedar Series modules also come in slightly lower wattages. They have all of the other specifications of size, warranty, and construction. USA.

11-570 Evergreen EC-47 Module $249⁰⁰
11-571 Evergreen EC-94 Module $459⁰⁰

Please check before ordering these lower wattage modules; they may not always be available.

Evergreen 32-Watt PV Module

String ribbon polycrystalline technology allows continuous production of smaller cells at a lower cost. Since it is crystalline, it is more efficient than any amorphous module and thus takes up less space per watt. At this price for this size PV, you can't beat it. Industry standard 20-year warranty. USA.
Rated Watts: 32 watts @ 25°C
Rated Power: 15.8 volts, 2.0 amps
Open Circuit Volts: 20.1 volts @ 25°C
Short Circuit Amps: 2.4 @ 25°C
L x W x D: 45.0" x 13.9" x 1.38"
Construction: polycrystalline, tempered glass
Warranty: within 20% of rated output for 20 yrs.
Weight: 11.3 lb/5.1 kg

11-204 Evergreen 32-Watt Module $219⁰⁰

New Evergreen 64-Watt Dual Voltage Module

The smaller cells allow construction of a 64-watt module that is essentially two 32-watt, 12-volt modules. This has two advantages. First, it is less affected by hard shadows than traditional 36-cell modules. Second, it may be configured either as a 12-volt module *or* a 24-volt module. This makes it particularly cost-effective for small 24-volt PV-direct pumping systems. Compare its cost with the cost of two separate 32-watt PVs that would otherwise be required! Twenty-year warranty. USA.

Rated Power: 64 watts
Voltage @ max. power: 15.5 volts*
Current @ max. power: 4.1 amps*
Open Circuit Voltage: 20.3 volts*
Short Circuit Current: 4.9 amps*
L x W x D: 45.0″ x 26.0″ x 1.4″
Construction: polycrystalline, tempered glass
Weight: 20.5 lb/9.3 kg
Warranty: within 20% of rated output for 20 yrs.

This data is for 12-volt configuration (for 24-volt configuration, wattage is the same, voltages are almost exactly double, and the currents are halved).

11-205 Evergreen 12/24-Volt 64-Watt Module $369⁰⁰

13-Watt General Purpose Module

Delivering a peak 13.5 watts, this single-crystal, tempered glass glazed module is useful for any 12-volt charging applications. Features include weather-tight construction, a sturdy anodized aluminum frame with four adjustable mounting tabs, a standard 36 cells in series, built-in reverse-current diode, and 7-foot wire pigtail output with color-coded clips. Good for trickle battery recharging on RVs, farm equipment, boats, and a perfect match with the Aquasolar 700 pump for backyard fountains. China.
Rated Watts: 13 watts @ 25°C
Rated Power: 17.5 volts, 710 mA
Open Circuit Volts: 18.0 volts @ 25°C
Short Circuit Amps: 750 mA @ 25°C
L x W x D: 15.1″ x 14.4″ x 1.0″
Construction: single crystal, tempered glass
Warranty: 90 days
Weight: 3.5 lb/1.6 kg

11-562 13-Watt Module With Clips $139⁰⁰

Photowatt PW 1000-100 100-Watt 12- or 24-Volt

Produced by one of Europe's largest PV manufacturers, the Photowatt PW 1000 modules feature polycrystalline cells, EVA encapsulant, tempered glass covers, Tedlar backsheets, a watertight J-box, and a full-perimeter aluminum frame. PV cells are produced and assembled in France. With 72 cells, these modules can be configured for either 12-volt or 24-volt output, and will intermix with conventional 36-cell modules. Has a prorated 25-year land-based warranty, or a prorated 5-year warranty in marine environments. UL-listed. CEC-approved. France.

Rated Power: 100 watts
Voltage @ max. power: 17.2 volts*
Current @ max. power: 5.8 amps*
Open Circuit Voltage: 21.6 volts*
Short Circuit Current: 6.0 amps*
L x W x D: 52.6″ x 26.5″ x 1.8″
Construction: polycrystalline, tempered glass
Weight: 21.1 lb/10.5 kg
Warranty: within 20% of rated output for 25 yrs.

This data is for 12-volt output. 24-volt output is the same wattage, half the amperage, double the voltage.

11-572 Photowatt PW1000-100w Module $520⁰⁰

Photowatt PW 750-80 80-Watt 12-Volt

Also produced by Europe's largest PV manufacturer, the PW 750 modules feature polycrystalline cells, EVA encapsulant, tempered glass covers, Tedlar backsheets, a watertight J-box, and a full-perimeter aluminum frame. PV cells are produced and assembled in France. This is a standard 36-cell module, and will intermix with any other 36-cell modules. Has a prorated 25-year land-based warranty, or a prorated 5-year warranty in marine environments. UL-listed. CEC-approved. France.

Rated Power: 80 watts
Voltage @ max. power: 17.3 volts
Current @ max. power: 4.6 amps
Open Circuit Voltage: 21.9 volts
Short Circuit Current: 5.0 amps
L x W x D: 48.7″ x 21.9″ x 1.8″
Construction: polycrystalline, tempered glass
Weight: 17.2 lb/7.8 kg
Warranty: within 20% of rated output for 25 yrs.

11-573 Photowatt PW750-80w Module $370⁰⁰

PV Accessories

PV Battery Chargers

Prevent Dead Car Batteries the Solar Way

To keep your battery fully charged, even after long-term parking, place our solar-powered, 12-volt trickle charger on your dashboard in the sun, and plug it into your cigarette lighter. If your car's lighter doesn't work when the ignition is off, Real Goods includes an exclusive adaptor and inline fuse holder to connect directly to your fuse box, or battery. Output of this charger is a powerful 100 mA. (Most dash chargers are only 40-50 mA.) The 14.5″ x 4.8″ module is weather/rust/shock/UV resistant, and stores a generous 8-foot cord on the back. Not recommended for small motorcycle batteries. Made in Canada.

11-563 Battery Saver Plus $29⁹⁵

Improved Solar Super Charger

EXCLUSIVE *A Real Goods Exclusive for 20 Years!*

Our new high-powered Solar Super Charger will charge any two nicad batteries at a time. The holder fits AAA, AA, C, or D sizes. The improved single-crystal solar module measures 3.5″ x 10″, is waterproof, nearly unbreakable, and has a 16″ wire lead. Battery box is not waterproof. Output is 4.5 volts at 250 mA. Will easily charge a pair of our high-capacity AA nicads in 6 hours or less. Taiwan.

50-212 Improved Solar Super Charger $29⁹⁵

Your Vehicle's Battery Will Always Be Ready To Go

There's nothing quite like the frustration felt when your vehicle won't start. Don't let a weak battery ruin your fun or keep you from missing your next appointment. This Deluxe Clip-on Solar Charger is a 250-milliamp, 12-volt solar module that keeps your boat, RV, tractor, golf cart, or motorcycle battery topped off, even when it's been sitting unused for extended periods of time. With five times the current output of our old model, this newer version is also water, weather, and rustproof, corrosion-resistant, and extremely durable. Its built-in, adjustable stand allows the panel to receive the maximum sunlight. Measures 10″ W x 7.25″ H x .5″ D, and it comes with a 6-foot connecting cable and alligator clips.

11-184 Deluxe Solar Charger With Clips $69⁹⁵

Direct Solar Charging For Battery-Powered Appliances

The SolarVerter line of small PV modules cuts out the middle man, allowing you to recharge any battery-operated appliance directly from the sun. The systems recharge lights, games, audio equipment, cell phones, and radios without hassling with a separate battery charging unit. All SolarVerter models include an innovative DC jack adaptor that allows solar charging even without an external DC jack. Also included are a complete assortment of DC plugs with polarity adjustment tips, a well-illustrated, clearly written instruction sheet, and a generous 12-foot cord, so you and your appliance can stay in the shade while the PV module is hard at work in the sun. One-year mfr.'s warranty. Hong Kong.

SolarVerter 4150

Supplies 4 volts @ 150 mA. Good for any device that uses just two batteries, C size or smaller. Good for most portable radios, Walkmans®, or flashlights. Measures 3.2″ x 5.9″ x 0.3″. Weighs 6 oz.

11-172 SolarVerter 4150 Module $35⁹⁵

SolarVerter 6220

Supplies 6 volts @ 220 mA. A higher-powered model for devices that use two to four batteries. Good for cell phones, two-way radios and cassette players. Measures 6.4″ x 4.5″ x 1.4″. Weighs 11 oz.

11-173 SolarVerter 6220 Module $59⁹⁵

SolarVerter 9150

Supplies 9 volts @ 150 mA. A higher-voltage model for 6- to 7.5-volt devices. Good for some games, toys, and boom boxes. Measures 6.4″ x 4.5″ x 0.4″. Weighs 11 oz.

11-174 SolarVerter 9150 Module $62⁹⁵

Solar Traveler RV & Marine Kits

Reliable Solar Electric Power

Noticed how many RV-makers are jumping on the solar bandwagon, marketing solar electric options in every RV supply center? Real Goods sold the first retail PV modules in 1979. With nearly 20 years of solid experience, we are still your best source for solar electric products and information. Our Solar Traveler Kits demonstrate our commitment to affordability and ease of use. The kits include absolutely everything you need to add solar electric charging to your RV or boat. No missing cables and plugs, no unnecessary bells and whistles, no engineering degree required. Basic kits include a Uni-Solar PV module, wiring, complete mounting hardware, a fully compatible charge controller, and easy-to-follow instructions. The triple-junction Uni-Solar modules are unbreakable, shade and heat tolerant, and the first choice for RVing and boating.

The Basic and Expansion kits are available in both 32- and 64-watt sizes. Installation of the Solar Traveler Kit will give you a reliable off-the-grid power source wherever you travel, and the peace of mind that comes from choosing a company you can trust.

The Solar Traveler Kit 64

For full-time use. The Basic 64 Kit includes a 64-watt Uni-Solar module with flush mounting hardware, 25 feet of lead-in wire, the Solar Commander 15-amp panel-mounted controller (with volt and amp meters), and 15 feet of battery wire. Accommodates up to two Expansion Kits. Expansion 64 Kits include another Uni-Solar module with mounting hardware and interconnect wiring. Six-month warranty on Kits, 20-year warranty on modules. USA.

12-202 Solar Traveler Basic Kit 64 $629⁰⁰

12-203 Solar Traveler Expansion Kit 64 (each)
$529⁰⁰

The Solar Traveler Kit 32

For occasional or weekend use. The Basic 32 Kit includes a 32-watt Uni-Solar module with flush mounting hardware, 25 feet of lead-in wire, the SunSaver 6.5-amp panel-mounted controller, and 15 feet of battery wire. Accommodates up to two Expansion Kits. Expansion 32 Kits include another Uni-Solar module with mounting hardware and interconnect wiring. Six-month warranty on Kits, 20-year warranty on modules. USA.

12-204 Solar Traveler Basic Kit 32 $399⁰⁰

12-205 Solar Traveler Expansion Kit 32 (each)
$329⁰⁰

Small Pre-Assembled Systems

The Real Goods SolProvider Energy Center

 Your Pre-Wired Solar Electric System

Our techs have already done the hard part: selecting fully compatible components and state-of-the-art safety equipment, completing the most difficult wiring, clearly labeling all landing points and circuit breakers, and pulling it all together in a comprehensive and genuinely useful manual that details every aspect of installation, use, and maintenance. Pre-wired, code-compliant, and easy to install—just add battery(ies), PV module(s), and a wall for mounting.

The system is a great jumping off point, with plenty of room for expansion and customizing. The SolProvider's 6-circuit load center delivers reliable 12-volt DC power for lighting, fans, low-voltage water pump, or entertainment equipment. Add an inverter to run your personal computer, printer, TV, or even a modest household (call our technical staff for help selecting the inverter specific to your intended use). The SolProvider/PV Package adds a Uni-Solar US64 module and rooftop RV mount to the basic package. Don't know how much PV you need? Call our techs.

SolProvider includes the Trace C-12 charge controller, charge control for up to 12 amps of PV input (up to 3 US64s), and automatic disconnect of selected loads to prevent overly discharging the battery, plus two DC lighter plug sockets. Landing points for field-installed wiring from battery. Everything is mounted on a 16″ x 18″ Meadowood™ board, a chipboard-like material made from straw. With the SolProvider and a little solar savvy, you'll get the energy self-sufficiency you've been looking for. Five-year warranty on Power Board and 20-year warranty on PV. USA.

27-180 The SolProvider $395⁰⁰

27-183 SolProvider/PV Package $795⁰⁰

Plug-And-Go Power For People On The Move

Weighing in at 15.5 lb., with a molded carry handle and nylon shoulder strap, this infinitely portable 12-volt power source tucks into your camper, cabin, or trunk for reliable or emergency power in any setting. It has multiple mini-pin, cigarette lighter, and large terminal outputs, and will accommodate any of the lamp, fan, or voltage supply options listed below. Can recharge from a standard AC outlet, or from an automotive cigarette lighter plug. Add a PV module for solar recharging. The built-in charge controller will handle up to 50 watts, and protect the 17Ah sealed battery from overcharging. Multi-step LED gauges monitor charging and battery capacity, and a 25-amp fuse protects everything (spare fuse included). A low-voltage circuit will shut off all outputs if the battery gets too low, and play an annoying electronic tune (sorry no choice of tune). The built-in, high-powered flashlight runs over 20 hours on a full charge. Battery is designed for replacement with minor technical skill. 13″ x 8.6″ x 4.0″. Taiwan.

27-182 Double-Size Multi Power Source $149⁰⁰

15-209 Repl. 17Ah Sealed Battery $49⁰⁰

11-252 10W Solar Panel Option $129⁰⁰

Multi Power Supply

All the features, plugs, switches, lights, straps, handles, and fuses, of our Double-Size Multi Power Supply above, but with half the battery capacity and weight. (Has a single 9Ah sealed battery.) It works with all Multi Power Supply accessories, and will fit inside the Wattz-In-A-Box package. 10.5″ x 7.5″ x 4.6″. 9.75 lbs. Taiwan.

27-195 Multi Power Supply $119⁰⁰

15-200 Repl. Sealed Battery $27⁹⁵

Oscillating Fan Option

An 8″ personal fan that mounts securely to the top of the Multi Power Supply unit. 180° oscillation. Simple on/off operation. Offers multi-speeds when used with the optional 8-step voltage supply. Taiwan.

64-210 Oscillating Fan $24⁹⁵

13-Watt PL Lamp Option

Included with our Wattz-In-A-Box this energy-efficient, compact fluorescent, gooseneck lamp option gives the equivalent of a 60 watt incandescent light for car repairs, base-camp card games, or emergency first aid. It mounts securely to the top of either Power Supply unit. Taiwan.

25-425 13W PL Lamp $29⁹⁵

8-Step Voltage Supply Adds Flexibility

Included with our Wattz-In-A-Box, this reliable DC regulator clips to the side of either Power Supply. It can directly run almost any battery-operated appliance with a power input plug. Includes a four-way adapter cord. It provides stable voltage at any power level, and cannot be damaged by short circuit. Solid-state regulated voltage at 1.5V, 2V, 3V, 4.5V, 6V, 7.5V, 9V, and 12V, with a 3-amp output limit. Can *only* be used with one of Multi Power Supply boxes. Taiwan.

25-366 8-Step Voltage Supply $29⁹⁵

High Intensity Spotlight Option

Keep the hyenas at bay, set up your field camp after dark, or light your way to a midnight swimming hole. This powerful beam gives you brilliant high-intensity illumination for ¼ mile. The handheld spotlight plugs directly into either Multi Power Supply unit. Taiwan.

25-426 High Intensity Spotlight $19⁹⁵

Portable Power
Any Place In The World

Rafting the Colorado? Hiking the Himalayas? Braving the wilds of Central Park? Take along our deluxe Wattz-In-A-Box portable power station, a practically indestructible travel case outfitted with a reliable solid state DC power supply (up to 12 volts), 13W PL fluorescent gooseneck camp light, a ProWatt 150-watt inverter for AC applications, and plenty of room for your laptop or video camera. The case itself is airtight, watertight, and unbreakable, with a foam-padded, custom-fit interior. The Multi Power Supply unit contains a sealed 9Ah battery with multiple mini-pin, cigarette lighter, and large terminal outputs. It will power up your laptop, cell phone, video camera, or camp lights on demand. So how do you recharge your power supply in the Serengeti? Use the cigarette lighter in your Land Rover, the AC outlet at the nearest hotel, or add an optional PV module, up to 50-watts, for solar charging. Multi-step LED indicators monitor battery and charge levels, and the unit shuts off automatically when full. The Multi Power Supply has a built-in high-powered flashlight that will run for over 12 hours on a full charge. PV module sold separately. Taiwan and USA.

27-200 Wattz-In-A-Box $395⁰⁰

11-252 10W Charger Option $129⁰⁰

15-200 Repl. Sealed Battery $27⁹⁵

A Higher Capacity
Portable Power System

Looking for something bigger than Wattz-In-A-Box? The Sunwize Portable Power Generator boasts ten times the capacity of the Multi Power Supplies on the two previous pages. The Portable Power Generator can supply lighting, or run entertainment equipment at your occasional-use cabin, or on your boat. Also ideal for emergency use and disaster relief, the PPG is housed in a rugged weather-tight Rubbermaid box (21˝ x 15˝ x 12˝). Already prewired are a 15-amp PV controller, a 150-watt Statpower Prowatt inverter for AC output, a pair of lighter plugs for DC output, an externally mounted DC voltmeter "fuel gauge," an input plug for PV module(s), and all the fusing to keep everything safe. All you need to add is the PV module(s) or other charging source of choice, and a 98Ah (or smaller) battery. All components except the battery and PV module(s) are prewired and ready to rock. The Sunwize PPG weighs 9 lb. as delivered, or 81 lb. with the maximum-capacity sealed battery listed below. One-year mfr.'s warranty. USA.
*PV Module & Battery **not** included.*

12-107 Sunwize Portable Power Generator $599⁰⁰

15-214 98AH/12V Sealed Battery $169⁰⁰

PV-Powered Automatic
Sign Lighting Kit

Our premium sign lighting kit will provide a full 12 hours of dusk-to-dawn runtime for a 13-watt floodlight, based on a wintertime low of 3.0 hours sun. The sealed, maintenance-free battery will provide five days of backup power. The controller provides daytime charge control, turns on the light at dusk for a selected time period, and shuts it off if the battery gets too low. This is truly a maintenance-free system. Turn it on, and it will take care of itself. Add a second floodlight if wanted and either reduce runtime, or add a second PV module. The controller can handle the extra wattage, and it's a poletop rack for two modules already.

Kit provides: one 13-watt outdoor floodlight; one 75-watt Astropower PV module; one pole top mount; one Morningstar Sunlight 10 controller; one 98-AH sealed battery; and a 30A fuseholder with fuse and spare to keep everything safe. You provide 2˝ steel pipe for the mount, a box for battery and control, and a mount for the light. Kit pricing saves about $40 over individual parts. All USA components.

12-238 Solar Sign Lighting Kit $829⁰⁰

Solar Power For Computers

Rugged Power For Laptops And Portable Electronics

The Sunwize Portable Energy System can deliver charging power for almost any laptop computer, or any other portable electronic device (cell phones, GPS units, portable audio units, camcorders). Features efficient, space-saving, single-crystal solar cells, unbreakable construction, a regulated Multi-Volt Controller,

The Opti-Meter shows percentage of full sun, making aiming and power output estimates easy.

(delivers 3 to 18 volts), an adjustable fold-out stand, 9 feet of power extension cord on a wind-up spool in the back, a zippered nylon storage bag, and the patented Opti-Meter, which displays an easily understood percentage of full sun output to make aiming the module easy. Two computer cables and a multi-plug pinwheel cable are included for connecting to electronic devices.

A pre-addressed padded mailer is supplied in case the supplied cables don't fit. Tell the manufacturer what you have, they'll get you a cable to match, no charge. Peak power is 9.9 watts. Two systems can be doubled into a single output. One-year mfr.'s warranty. USA.

11-164 Portable Energy System $449⁰⁰

Laptop Power Adapters
Run Your Laptop from Direct Solar Power, Cigarette Socket or Airline Power

Laptops come in a huge variety of different input voltage and power plug configurations. Even within a single brand there is rarely a standard. Here is the gizmo to standardize them. On the input side, our adapter either plugs into a standard lighter socket or, with a separate cord, into an airline power socket. It accepts DC input over a wide voltage range, including PV modules. On the output side, the adapter provides a matching power plug and carefully emulates your laptop's standard charger, while providing additional high and low voltage protection,

and spike/noise protection. Output power will always be clean and safe. The adapter is in a rugged housing, and sealed from moisture.

These adapters will accept direct PV module input. Some laptops can recharge with as little as 10 watts. Most require 20 to 50 watts. Too big a solar panel causes no problem; too small means your laptop won't charge. Call our tech staff for help. The module needs a female lighter socket for output.

Due to the wide variety of output voltages and power plug types, adapters are drop shipped directly from the manufacturer. We can supply adapters for most current-production laptops. Call if there's any doubt. Delivery time is two weeks or less. Size and weight may vary slightly with laptop model, but averages 4″ x 2.5″ x 1.25″, weighs 11 oz., supplied with nylon bag. Three-year mfr.'s warranty. USA.

**◈ 50-251 Laptop Power Adapter
(specify laptop model) $99⁹⁵**

◈ Means shipped directly from manufacturer.

10-Watt PV With Cigarette Lighter
Ready to Go!

We have found numerous uses for 10-watt PVs prewired with a cigarette lighter receptacle. This is one of the best. Many laptops will require a higher output PV module. Please call for technical support. The fiberglass-backed lightweight PV measures 17.75″ x 9.5″ and weighs 1.1 lb. Nominal output is 0.6 amps at 16.5 volts. China.

**11-252 Lightweight 10-Watt PV
with Cigarette Lighter Output $129⁰⁰**

Photovoltaic Mounting and Tracking Equipment

To Track or Not to Track, That Is the Question

Photovoltaic modules produce the most energy when situated perpendicular to the sun. A tracker is a mounting device that follows the sun from east to west and keeps the modules in the optimum position for maximum power output. At the right time of year, and in the right location, tracking can increase daily output by as much as 50%. But beware of the qualifiers: not every site or every system is a good candidate for tracking mounts.

Tracking is an option only if there is clear access to the sun from early in the morning until late in the afternoon. A solar window from 9AM to 4PM is workable; if you have greater access, more power to you (literally). Tracking will add the most power in summertime when the sun is making a high arc overhead. Trackers will add 35% to 50% to your incoming power during the summer. In wintertime, when the sun is making a low arc for most of us, tracking will only add 10% to 20%. But this

— continued on next page

Larger PV modules are usually supplied with aluminum frames, but some kind of mounting structure is required to hold several modules together and to get the modules off the roof and allow some cooling air behind them. PV systems are most productive if the modules are approximately perpendicular to the sun at solar noon, the most energy-rich time of the day for a PV module. The best year-round angle for your modules is approximately equal to your latitude. Because the angle of the sun changes seasonally, you may want to adjust the angle of your mounting rack seasonally. In the winter, modules should be at the angle of your latitude plus approximately 15 degrees; in the summer, your latitude minus a 15-degree angle is ideal.

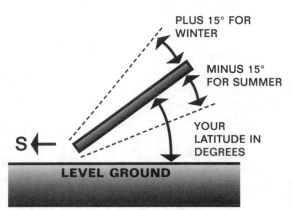

PLUS 15° FOR WINTER

MINUS 15° FOR SUMMER

YOUR LATITUDE IN DEGREES

S ←

LEVEL GROUND

Proper PV Mounting Angle

On a practical level, many residential systems will have power to burn in the summer, and seasonal adjustment may be unnecessary.

The perfect mounting structure would follow the sun across the sky every day. Tracking mounts do this, but their mechanical and/or electrical complexity drives the cost up significantly, and it's one more item that needs maintenance. For smaller systems it's usually more cost-effective to use fixed mounts.

In the following pages you'll find numerous mounting structures, each with its own particular place in an independent energy system. We'll try to explain the advantages and disadvantages of each style to help you decide if a particular mount belongs in your system.

In ascending order of complexity, your choices are:

- *RV Mounts*
- *Home-Built Mounts*
- *Fixed Mounts (with or without adjustable tilt)*
- *Top-of-Pole Fixed Mounts*
- *Passive Trackers*
- *Active Trackers*

RV Mounts

Because of wind resistance and never knowing which direction the RV will be facing next, most RV owners simply attach the module(s) flat on the roof. RV mounts raise the module an inch or two off the roof for cooling. They can be used for small, simple home systems as well. Simple and inexpensive, most of them are made of aluminum for corrosion resistance. Disadvantages are their fixed tilt and orientation.

— continued from previous page

depends heavily on your site and most importantly, your latitude. If you live north of 45° latitude, winter tracking isn't going to help much; the sun makes such a low arc across the sky that fixed mounts could catch it just as well. If you live south of 35° latitude, then winter tracking holds some promise for you. In general, the closer to the equator you are, the more sense wintertime tracking makes.

Tracking mounts are expensive. If you've got inexpensive modules, like the used Quad-Lams that flooded the market a few years ago, you're better off just buying more modules and putting them on fixed mounts. Economies of scale make larger trackers less costly per module. The only smaller systems that routinely use trackers are water pumping systems. (It's the only way to get more gallons per day out of a pumping system.) Residential systems usually need to be larger than six modules to make tracking cost-effective.

Tracking is most effective at temperate latitudes if your power needs peak in the summer. Pumping and cooling applications are the most common uses. Residential power needs usually peak in the winter, in which case tracking is only cost-effective on systems of six modules or larger. In tropical latitudes, below 30°, tracking makes sense for almost any system of four modules or larger.

Home-Built Mounts

Want to do it yourself? No problem. Fixed racks are pretty easy to put together. Anodized aluminum or galvanized steel are the preferred materials due to corrosion resistance, but mild steel can be used just as well, so long as you're willing to touch up the paint occasionally. Slotted steel angle stock is available in galvanized form at most hardware and home-supply stores, and is exceptionally easy to work with. Wood is not recommended because your PV modules will last longer than any exposed wood. Even treated wood won't hold up well when exposed to the weather for over 40 years. Make sure that no mounting parts will cast shadows on the modules. Adjustable tilt is nice for seasonal angle adjustments, but most residential systems have power to spare in the summer, and seasonal adjustments are usually abandoned after a few years.

Fixed Mounts

This is probably the most common mounting structure style we sell. The ground or roof mounts, or the side-of-pole mounts are examples. The ground/roof racks have telescoping back legs for seasonal tilt and variable roof pitches. This mounting style can be used for roof or ground, or even flipped over and used on south-facing walls. Use concrete footings for ground mounts. The racks are designed to withstand wind speeds of up to 120 mph. They don't track the sun though, and getting snow off of them is sometimes troublesome. Ground mounting leaves the modules vulnerable to grass growing in front, or to rocks kicked up by mowers. Roof, wall, or pole mounting is preferred.

Top-of-Pole Fixed Mounts

A very popular and cost-effective choice, pole top mounts are designed to withstand winds up to 80 mph; in some cases up to 120 mph. This mounting style is the best choice for snowy climates. With nothing underneath it, snow tends to slide right off. For small or remote systems, pole-top mounts are the least expensive and simplest choice. We almost always use these for one- or two-module pumping systems. Tilt and direction can be easily adjusted; in fact you can track the sun manually with this mounting style. The pole is standard steel pipe and is not included (pick it up locally to save on freight). Make sure that your pole is tall enough to allow about 25% to 33% burial depth, and still clear livestock, snow, weeds, etc. Ten feet total pole length is usually sufficient. Taller poles are sometimes used for theft deterrence. Pole diameter depends on the specific mount.

Passive Trackers

Tracking mounts will follow the sun from east to west across the sky, significantly increasing the daily power output of the modules. See the "To Track or Not to Track" sidebar for a discussion on when tracking mounts are appropriate.

Made by Zomeworks, the Track Rack follows the sun from east to west using just the heat of the sun for energy. No extra source of electricity is needed at all; a simple, effective, and brilliant design solution. The north-south tilt axis is seasonally adjustable manually. Maintenance consists of two squirts with a grease gun once every year.

Tracking will help substantially in the summer and somewhat in the winter. The two major problems with passive technology are wind disturbances, and slow "wake-up" when cold. The tracker will go to "sleep" facing west. On a cold morning it may take more than an hour for the tracker to warm up and roll over toward the east. In winds over 25 mph, the passive tracker may be blown off course. This tracker can withstand winds of up to 85 mph (provided you follow the manufacturer's recommendations for burying the pipe mount), but will not track at high wind speeds. If you have routine high winds you should have a wind turbine to take advantage of those times, but that's a different subject.

Active Trackers

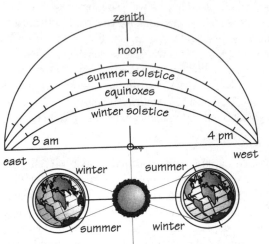

Active trackers use photocells, electronics, and linear actuators like those on satellite TV dishes to track the sun very accurately. A small controller bolted to the array is programmed to keep equal illumination on the photocells at the base of an obelisk. Power usually comes from tapping off the PV array. Power use averages a minuscule 0.5 watt per hour. Wattsun is the primary active tracker manufacturer. All their trackers feature electronic east-west tracking with manual north-south adjustment. Wattsun has a dual-axis option, which automatically adjusts north-south tilt. This option is highly recommended with their azimuth-style racks.

Active trackers average slightly more energy collection per day than a passive tracker in the same location, but historically they have also averaged more mechanical and electrical problems too. Active mounts usually cost more initially. Due to fixed costs for controllers and linear actuators, active trackers are more economical for larger PV arrays. Other disadvantages are reliance on electronic gizmos and mechanical parts. Wattsun's warranty is good, but you can count on having to do some tinkering occasionally.

PV Mounting Products

Wattsun Active Trackers

Wattsun has upgraded all their mid-sized trackers to a new-style azimuth actuation design. This is an improvement from the old style "tilt and roll" to a new stronger style of azimuth tracker that rotates around the pole mount first, and then tilts (if the Dual Axis option is used) for elevation position. This tracking style requires less ground clearance, as the array is always parallel to the ground or rooftop. There are no large array corners pointing down toward the ground, or up into the air to catch the wind. This new azimuth tracker provides a full 180° of east to west rotation, and up to 80° of tilt. The tilt angle is adjusted automatically with the Dual Axis option, which is strongly recommended for all applications. The tilt angle on the standard single axis tracker is manually adjustable.

The controller will accept power directly from the PV array, up to 55 volts open circuit. Approximate power use for the tracker is a minuscule 12 watt-hours per day. Outputs to the tracking motors are short circuit protected. The electronics are completely potted, waterproof, and over-temperature protected.

Trackers, depending on model, mount on 2″, 3″, 4″, or 6″ Schedule 40 steel pipe (not included). They can withstand winds up to 100 mph. Because they break down into smaller component parts, most Wattsuns can be shipped by UPS. Stainless steel hardware is available for severe marine environments at additional cost. All necessary hardware, including stainless module mounting bolts, is included with the standard package. Mounting pole size is in parentheses following price. Mfr.'s warranty is 10 years on controller and frame, 2 years on actuators. USA.

◈ 13-476   Wattsun TR 2 Siemens SM-series $470⁰⁰ (2″)
◈ 13-250   Wattsun TR 4 Siemens SM-series $845⁰⁰ (3″)
◈ 13-251   Wattsun TR 6 Siemens SM series $875⁰⁰ (3″)
◈ 13-252   Wattsun AZ 8 Siemens SM-series $1,325⁰⁰ (3″)
◈ 13-160   Wattsun AZ 10 Siemens SM-series $1,375⁰⁰ (4″)
◈ 13-244   Wattsun AZ 12 Siemens SM-series $1,395⁰⁰ (4″)
◈ 13-253   Wattsun AZ 16 Siemens SM-series $1,495⁰⁰ (4″)
◈ 13-254*   Wattsun AZ 18 Siemens SM-series $1,795⁰⁰ (6″)
◈ 13-255*   Wattsun AZ 24 Siemens SM-series $2,095⁰⁰ (6″)
◈ 13-477   Wattsun TR 2 SP75 or AP75 $470⁰⁰ (2″)
◈ 13-285   Wattsun TR 4 SP75 or AP75 $895⁰⁰ (3″)
◈ 13-284   Wattsun AZ 6 SP75 or AP75 $1,345⁰⁰ (3″)
◈ 13-283   Wattsun AZ 8 SP75 or AP75 $1,395⁰⁰ (4″)
◈ 13-282*   Wattsun AZ 10 SP75 or AP75 $1,495⁰⁰ (4″)
◈ 13-281*   Wattsun AZ 12 SP75 or AP75 $1,775⁰⁰ (6″)
◈ 13-286*   Wattsun AZ 18 SP75 or AP75 $2,195⁰⁰ (6″)
◈ 13-478   Wattsun TR 2 Siemens SR90/100 $895⁰⁰ (2″)
◈ 13-479   Wattsun TR 4 Siemens SR90/100 $1,325⁰⁰ (3″)
◈ 13-480   Wattsun AZ 6 Siemens SR90/100 $1,345⁰⁰ (4″)
◈ 13-481*   Wattsun AZ 8 Siemens SR90/100 $1,495⁰⁰ (4″)
◈ 13-482*   Wattsun AZ 12 Siemens SR90/100 $1,995⁰⁰ (6″)
◈ 13-154   Wattsun TR 2 Astropower 120 $845⁰⁰ (3″)
◈ 13-155   Wattsun AZ 4 Astropower 120 $1,345⁰⁰ (3″)
◈ 13-156   Wattsun AZ 6 Astropower 120 $1,395⁰⁰ (4″)

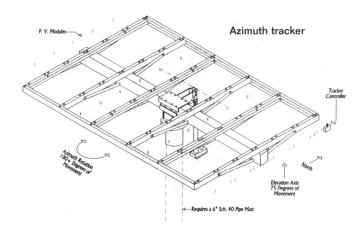

Azimuth tracker

◈ 13-157*   Wattsun AZ 8 Astropower 120 $1,895⁰⁰ (4″)
◈ 13-158*   Wattsun AZ 12 Astropower 120 $2,295⁰⁰ (6″)
◈ 13-256   Wattsun TR 4 Solarex MSX-60 $895⁰⁰ (3″)
◈ 13-257   Wattsun AZ 6 Solarex MSX-60 $1,325⁰⁰ (3″)
◈ 13-258   Wattsun AZ 8 Solarex MSX-60 $1,345⁰⁰ (4″)
◈ 13-164   Wattsun AZ 10 Solarex MSX-60 $1,395⁰⁰ (4″)
◈ 13-259*   Wattsun AZ 12 Solarex MSX-60 $1,495⁰⁰ (4″)
◈ 13-261*   Wattsun AZ 18 Solarex MSX-60 $2,195⁰⁰ (6″)

Note: TR denotes a Tilt & Roll tracker, AZ denotes an Azimuth tracker

Must be shipped by common carrier; too large for UPS. Call our technicians for prices on larger quantities, or for more information.

Wattsun TR trackers for south of the equator require special ordering due to mirror-image operation

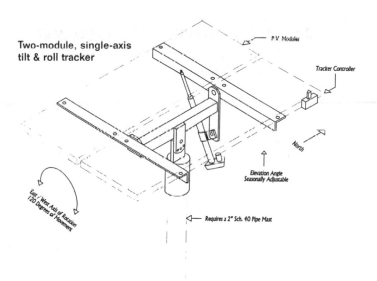

Two-module, single-axis
tilt & roll tracker

Tilt and Roll Tracker

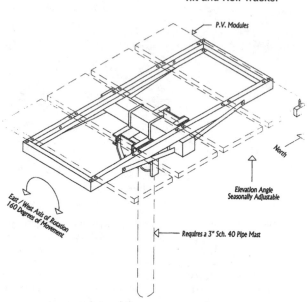

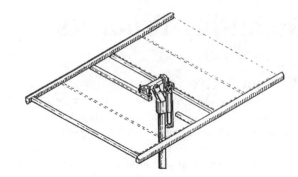

Wattsun Dual Axis Options

With the new style of azimuth tracker, the Dual Axis option is strongly recommended for all trackers. The tilt range is approximately 80°.

◈ **13-208 Wattsun Dual Axis Option for 6 Modules & Larger $395⁰⁰**

Optional All Stainless Steel Hardware

◈ **13-274 Wattsun Hardware, 4 Modules $40⁰⁰**

◈ **13-275 Wattsun Hardware, 6 Modules $50⁰⁰**

◈ **13-276 Wattsun Hardware, 8 Modules $60⁰⁰**

◈ **13-277 Wattsun Hardware, 12 Modules $80⁰⁰**

◈ **13-278 Wattsun Hardware, 16 Modules $90⁰⁰**

◈ **13-279 Wattsun Hardware, 18 Modules $110⁰⁰**

◈ **13-280 Wattsun Hardware, 24 Modules $120⁰⁰**

Zomeworks Universal Track Racks
Passive Solar Trackers

How the Zomeworks Track Rack Follows the Sun . . .

1. *The Track Rack begins the day facing west. As the sun rises in the east, it heats the unshaded west-side canister, forcing liquid into the shaded east-side canister. As liquid moves through a copper tube to the east-side canister, the tracker rotates so that it faces east.*

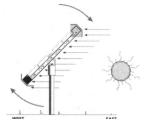

2. *The heating of the liquid is controlled by the aluminum shadow plates. When one canister is exposed to the sun more than the other, its vapor pressure increases, forcing liquid to the cooler, shaded side. The shifting weight of the liquid causes the rack to rotate until the canisters are equally shaded.*

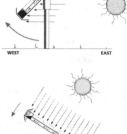

3. *As the sun moves, the rack follows (at approximately 15 degrees per hour), continually seeking equilibrium as liquid moves from one side of the tracker to the other.*

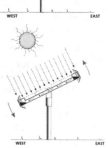

4. *The rack completes its daily cycle facing west. It remains in this position overnight until it is "awakened" by the rising sun the following morning.*

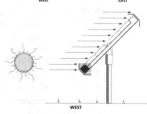

Zomeworks new universal racks feature adjustable cross pieces and stainless J-clips that allow six standardized racks to fit any type of PV module; no holes are drilled in the frames. Cross members slide to fit length or width and the J-clips slide to fit the module. This means lower costs, faster shipping, and greater adaptability. The passive tracking mechanism is the same reliable, heat-driven, simple drive that Zomeworks has used for decades. The rack will go to sleep facing west in the evening. The warmth of the rising sun will flip it over to the east in the morning. Wake-up can take a couple of hours in cold weather, in warm weather less than an hour. Painted mild steel construction will withstand wind speeds up to 85 mph (except a few of the very largest module configurations), although tracking may be iffy at high wind speeds. East-west tilt range is approximately 110° with a shock absorber to dampen motion in windy conditions. North-south tilt is manually adjustable over a 40° range. Easy pole-top installation, with Schedule 40 steel mounting pipe supplied locally to save on shipping. 10-year manufacturer's warranty. USA.

All models except smallest shipped by common carrier.

See Application Chart for rack sizing and pricing.

Four J-clips are required per module. An average number of J-clips is supplied with each rack, as indicated in our chart below. For some configurations extra clips may need to be purchased.

Cross-members and clips slide to fit

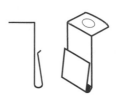

J-clips are stainless steel

ZOMEWORKS UNIVERSAL TRACK RACK APPLICATION CHART

Manufacturer	Module	Maximum Number Of Modules					
Astropower	AP-1206	2	4	6	8	10	12
	AP-7105	2	4	8	10	12	14
	APX-90	2	4	6	8	10	12
BP	BP00 & 75	2	4	8	10	12	14
Evergreen	E32						
	E64						
Kyocera	KC-120	2	4	6	8	10	12
	KC-80	2	4	6	8	10	12
	KC-60	2	4	6	8	10	12
	KC-51	3	6	10	12	14	18
Siemens	SR100 & SR90	2	4	6	8	10	12
	SP75	2	5	8	10	12	14
	SM55 & 65	4	8	12	16	20	24
	SM50-H	4	8	12	16	20	24
	SR50	2	4	6	8	10	12
Solarex	MSX120	1	2	4	-	6	8
	MSX83 & 77	2	4	6	8	10	12
	MSX64 & 60	3	6	8	10	12	16
Uni-Solar	US-64	2	4	6	-	8	10
	US-42	2	4	6	-	8	10
	US-32	3	6	10	14	18	-
Universal Track Rack Model		020	040	055	070	085	100
Nominal Pipe Sz (Sch. 40 steel)		2.5″	3″	4″	6″	6″	6″
Recommended Min/Max Pole Height		30″/75″	50″/62″	56″/82″	60″/120″	60″/120″	76″/120″
Recommended Hole Depth		36″	36″	48″	60″	60″	60″
Recommended hole diameter filled with concrete is 3 times the pipe diameter – minimum.							
Ship Weight, lbs per box		45/50	25/140	55/180	65/250	67/265	85/190/250
J-clips supplied for # of modules		2	4	7	8	12	14
Item Order Number		**13-950**	**13-951***	**13-952***	**13-953***	**13-954***	**13-955***
Price		**$489⁰⁰**	**$859⁰⁰**	**$1,235⁰⁰**	**$1,359⁰⁰**	**$1,599⁰⁰**	**$1,849⁰⁰**

*Too large for UPS, ships freight collect. All Track Racks drop shipped directly from New Mexico.

13-956 Extra set of 4 J-clips $10⁰⁰

See product description for use of this chart.

Uni-Rac Fixed Racks

Uni-Rac makes several mounting styles; all feature the highest quality, non-corrosive components. Constructed of aluminum and galvanized steel with stainless steel fasteners, there are no painted or zinc-plated pieces. Durability is excellent, assembly is fast, and instructions are clear and complete. Uni-Racs ship directly from the manufacturer via UPS, usually within 24 hours, thanks to standardized construction. There's no waiting for racks, and no exorbitant shipping charges. Engineered to sustain wind loads of 50 lbs. per sq.ft. (about 120 mph). USA.

Uni-Rac Ground/Roof Mounts

These conventional mounts have adjustable length rear legs for easy site adjustments and seasonal tilt angle adjustments. Multiple modules are mounted ladder style. Can be used as roof, ground, or wall mounts.

GROUND/ROOF MOUNTS

SKU #	UniRac #	Bolts for # modules	AP/SP 75	AP120	Evergreen E64, EC94/102	KC 60/80/120	Siemens SM	SM110	SP130/150	US42/64	Price
13-558	GR-24	1	1	-	-	-	1	-	-	-	$90.00
13-559	GR-28	1	-	1	1	1	2a	1	-	-	$95.00
13-560	GR-36	2	-	-	-	-	-	-	1	1	$110.00
13-561	GR-44	2	2	-	-	-	3a	-	-	-	$130.00
13-562	GR-52	2	-	2	2	2	-	2	-	-	$150.00
13-563	GR-60	3	-	-	-	-	4a	-	-	-	$160.00
13-564	GR-68	3	3	-	-	-	-	-	2	2	$170.00
13-565	GR-72	4	-	-	-	-	-	-	-	-	$185.00
13-566	GR-80	4	-	3	3	3	-	3	-	-	$210.00
13-567	GR-88	4	4	-	-	-	6a	-	-	-	$220.00
13-568	GR-96	4	-	-	-	-	-	-	3	3	$255.00
13-569	GR-104	4	-	4	4	4	7a	4	-	-	$265.00
13-570	GR-112	6	5	-	-	-	8a	-	-	-	$325.00
13-571	GR-120	6	-	-	-	-	-	-	-	4	$340.00
13-572	GR-136	6	6	5	5	5	9a	5	4	-	$370.00
13-573	GR-144	8	-	-	-	-	10a	-	-	-	$385.00
13-574	GR-160	8	7	6	6	6	11a	6	5	5	$415.00

a – Extra Module Bolt Set(s) required: **13-640 Uni-Rac Module Bolt Set $3.50**

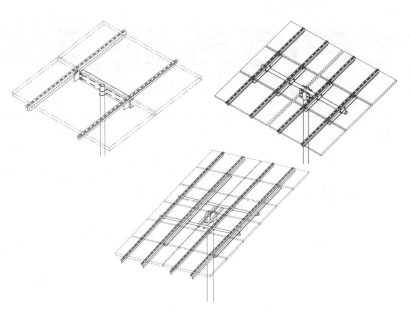

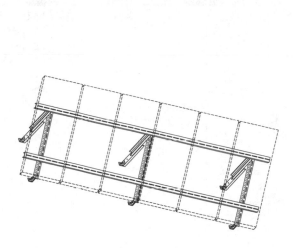

Uni-Rac Low-Profile Ground/Roof Mounts

Many consider the low-profile style more aesthetically pleasing when you have a larger number of modules for roof or ground mounting. Has adjustable length rear legs. Leg assemblies can slide back and forth during assembly, so you'll always hit framing members with mounting screws. Multiple racks can line up without visible joints.

Uni-Rac Pole-Top Mounts

Pole-Top mounts will shed snow better than any other mounting style. Tilt and direction adjust easily. You supply the Schedule 40 steel pipe locally. Approximately ⅓ of the total pipe length should be buried and backfilled with concrete. Pole heights more than about 10 feet above ground are okay, but should have guy wires or be firmly attached to the side of a building. Listed on next page.

LOW PROFILE GROUND/ROOF MOUNTS

SKU #	UniRac #	Bolts for # modules	AP/SP 75	AP120	Evergreen E64, EC94/102	KC 60/80/120	SM	Siemens SM110	SP130/150	US42/64	Price
13-575	LP-80	4	-	3	3	3	-	3	-	-	$245⁰⁰
13-576	LP-88	4	4	-	-	-	6a	-	-	-	$255⁰⁰
13-577	LP-96	4	-	-	-	-	-	-	3	3	$265⁰⁰
13-578	LP-104	4	-	4	4	4	7a	4	-	-	$275⁰⁰
13-579	LP-112	6	5	-	-	-	-	-	-	-	$335⁰⁰
13-580	LP-120	6	-	-	-	-	8a	-	-	4	$345⁰⁰
13-581	LP-136	6	6	5	5	5	9a	5	4	-	$365⁰⁰
13-582	LP-144	8	-	-	-	-	10a	-	-	-	$390⁰⁰
13-583	LP-160	8	7	6	6	6	11a	6	5	5	$405⁰⁰
13-584	LP-176	8	8	-	-	-	12a	-	-	-	$415⁰⁰
13-986	LP-192	10	9	-	-	-	-	-	6	-	$435⁰⁰

a – Extra Module Bolt Set(s) required: **13-640 Uni-Rac Module Bolt Set** $3⁵⁰

POLE-TOP MOUNTS

SKU #	UniRac #	Pipe Size	Bolts for # modules	Astropower		Evergreen	Kyocera		Siemens			Siemens	Uni-Solar	Price
				AP75	AP120	E64, EC94/102	60/80	120	SM	SP75	SM110	SP130/150	US42/64	
13-585	22-20M	2.5	1						1					$103⁰⁰
13-586	22-24M	2.5	1	1						1				$106⁰⁰
13-587	22-24L	2.5	1											$113⁰⁰
13-588	22-28M	2.5	1		1		1		2a		1			$109⁰⁰
13-589	22-32L	2.5	1			1		1b				1	1	$119⁰⁰
13-591	22-44M	2.5	2	2						2				$127⁰⁰
13-593	22-52M	3	2		2		2		4a		2			$139⁰⁰
13-594	22-52L	3	2			2		2b						$146⁰⁰
13-596	PT-30M	4	3										2	$210⁰⁰
13-597	PT-32S	4	3	3						3		2		$205⁰⁰
13-599	PT-40S	4	4		3		3		5a		3			$225⁰⁰
13-601	PT-40L	4	4					3						$245⁰⁰
13-602	PT-40M	4	4			3			6a					$235⁰⁰
13-603	PT-44S	4	4	4						4				$230⁰⁰
13-604	PT-48S	4	4						7a			3		$245⁰⁰
13-606	PT-52S	4	4			4	4							$285⁰⁰
13-607	PT-52M	4	4		4				8a		4			$295⁰⁰
13-608	PT-52L	4	4					4						$320⁰⁰
13-610	PT-60M	4	6											$320⁰⁰
13-611	PT-60L	4	6										4	$345⁰⁰
13-613	PT-64M	4	6	6						6				$335⁰⁰
13-614	PT-68M	4	6						10a			4		$350⁰⁰
13-616	PT-72L	4	6											$390⁰⁰
13-617	PT-80S	4	6			6	6							$385⁰⁰
13-618	PT-80M	4	6		6				12a		6			$395⁰⁰
13-619	PT-80L	4	8					6						$420⁰⁰
13-621	PT-84M	4	8	8						8				$410⁰⁰
13-625	PT-104S	4	8			8	8							$475⁰⁰
13-626	PT-104M	6	10	10	8				16a	10	8	6		$595⁰⁰
13-627	PT-104L	6	10					8						$620⁰⁰
13-631	PT-120L	6	12										8	$685⁰⁰
13-632	PT-136S	6	10			10	10					8		$705⁰⁰
13-633	PT-136M	6	12	12	10				20a	12	10			$730⁰⁰
13-634	PT-136L	6	10					10						$755⁰⁰
13-636	PT-144M	6	12											$760⁰⁰
13-637	PT-144L	6	14											$785⁰⁰
13-638	PT-160S	6	12			12	12							$795⁰⁰
13-639	PT-152M	6	14	14					22a	14				$790⁰⁰

a – Extra Module Bolt Set(s) required: **13-640 Uni-Rac Module Bolt Set $3⁵⁰**

b – Extra Long Channel Required. Specify Rack & Modules: **13-641 Uni-Rac Long Channel $10⁰⁰**

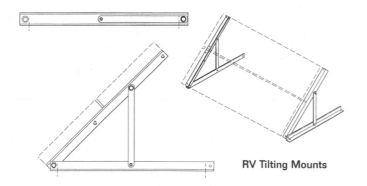

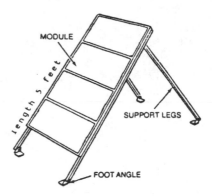

RV Tilting Mounts

RV Mounting Structures

Direct Power offers two lines of RV rooftop mounting structures: simple flush mounts or more efficient tilt-up mounts. You may not always park in a direction to make use of the tilt-up feature, but when you can, it increases output by 30% to 40%. Always drive with panels in the flat position. All units are constructed of aluminum and available for one- or two-panel systems. If you have more than two panels, use more than one mount. USA.

Tilting Mounts

◈ **13-915 RV Tilt Mount 1 Siemens SM-series $50⁰⁰**

◈ **13-918 RV Tilt Mount 2 Siemens SM-series $70⁰⁰**

◈ **13-917 RV Tilt Mount 1 MSX60, SP75 or AP75 $60⁰⁰**

◈ **13-920 RV Tilt Mount 2 MSX60, SP75 or AP75 $84⁰⁰**

◈ **13-152 RV Tilt Mount 1 AP120 $65⁰⁰**

◈ **13-386 RV Tilt Mount 1 US64 $80⁰⁰**

Flush Mounts

◈ **13-909 RV Flush Mount 1 Siemens SM-series $26⁰⁰**

◈ **13-912 RV Flush Mount 2 Siemens SM-series $40⁰⁰**

◈ **13-911 RV Flush Mount 1 MSX60, SP75 or AP75 $32⁰⁰**

◈ **13-914 RV Flush Mount 2 MSX60, SP75 or AP75 $45⁰⁰**

◈ **13-153 RV Flush Mount 1 AP120 $35⁰⁰**

◈ **13-907 RV Flush Mount 1 US64 $45⁰⁰**

◈*Means shipped directly from manufacturer.*

Aluminum Universal Solar Mounts

For simple installation it's hard to beat the economy of our aluminum PV mounting structure. Each mount will hold up to four Siemens SM-Series modules, three Siemens SP75 modules, three Kyocera modules, or three Solarex modules (MSX-53, 56, 60). Aluminum angle rails adjust to three seasonal tilt angles. It can mount on a deck, wall, or roof. All stainless-steel hardware is included to secure your modules to the mounts, but anchor bolts must be provided by the installer. Instructions are included. Specify module type. USA.

13-702 Solar Mount, 3 or 4 PV Modules $109⁰⁰

Roof Standoff Kits

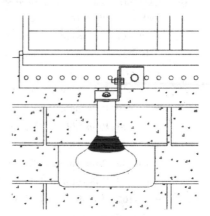

Here's the clean, secure, leakproof way to attach rooftop photovoltaic mounts. These 6-inch standoffs penetrate the roof like a plumbing vent, and use a conventional off-the-shelf vent flashing for 1.25″ pipe that every roofer knows how to deal with. Use a flashing type that's appropriate for your roof surface. Our line drawings show asphalt shingles. Comes with a pair of stainless ⁵⁄₁₆″ x 3½″ lag bolts to secure the standoff to a framing member, and a stainless ⅜″ bolt to secure the rack to the standoff. Flashing not included. Uni-Rac ground/roof mounts up to /104 size require 4 standoffs. Sizes /112 to /136 require 6. Larger racks require 8. Made in USA.

13-984 Roof Standoff Kit $17⁰⁰

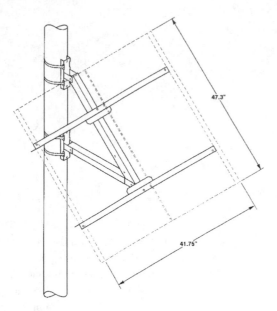

Side-Of-Pole Racks

These racks are adjustable from 0° to 90°. Two or four hose clamps secure the rack to the side of a mounting pole (pole not included). Mounting poles, typically 1½- to 3-inch OD pipe, should be purchased locally. Racks have mill finish aluminum module mounting rails, and milled steel structural parts. The item numbers below specify which modules each rack will hold. Only the most common modules are listed; others are available. Call our technical department. USA.

❖ **13-311 Pole-Side 1 Siemens SM-series $85⁰⁰**

❖ **13-321 Pole-Side 2 Siemens SM-series $150⁰⁰**

❖ **13-331 Pole-Side 3 Siemens SM-series $170⁰⁰**

❖ **13-341 Pole-Side 4 Siemens SM-series $200⁰⁰**

❖ **13-316 Pole-Side 1 MSX-60 $95⁰⁰**

❖ **13-323 Pole-Side 2 MSX-60 $170⁰⁰**

❖ **13-464 Pole-Side 1 SP75 or AP75 $90⁰⁰**

❖ **13-465 Pole-Side 2 SP75 or AP75 $170⁰⁰**

❖ **13-466 Pole-Side 3 SP75 or AP75 $210⁰⁰**

❖ **13-467 Pole-Side 4 SP75 or AP75 $260⁰⁰**

❖ *Means shipped directly from manufacturer.*

Wind and Hydro Power Sources

Wind and hydro energy sources are most often developed as a booster or bad weather helper for a solar-based system.

Sunlight, wind, and falling water are the renewable energy big three. These are energy sources that are commonly available at a reasonable cost. Solar or sunlight, the single most common and most accessible renewable energy source, is well-covered at the beginning of this chapter. We've found in our years of experience that wind and hydro energy sources are most often developed as a booster or bad weather helper for a solar-based system. These hybrid systems have the advantage of being better able to cover power needs throughout the year, and are less expensive than a similar capacity system using only one power source. When a storm blows through, the solar input is lost but a wind generator more than makes up for it. The short, rainy days of winter may limit solar gain, but the hydro system picks up from the rain and delivers steady power 24 hours a day. This is not to say that you shouldn't develop an excellent single-source power system if you've got it; like a year-round stream dropping 200 feet across your property, for instance. But for most of us, we'll be further ahead if we don't put all our eggs in one basket. Diversify!

Our experienced technical staff is well-versed in supplying the energy needs of anything from a small weekend getaway cabin, through an upscale, but energy-conserving resort. We'll be glad to help put a system together for you. There is usually no charge for our personalized services.

Wind Systems

We generally advise that a good, year-round wind turbine site isn't a place that you'd want to live. It takes average wind speeds of 8 to 9 mph and up, to make a really good site. That's honestly more wind than most folks are comfortable living with. But this is where the beauty of hybrid systems comes in. Many, very happily livable sites do produce 8+ mph during certain times of the year, or when storms are passing through. Tower height and location also make a big difference. Wind speeds average 50% to 60% higher at 100 feet compared to ground level (see chart in the wind section). Wind systems these days are almost always designed as wind/solar hybrids for year-round reliability. The only common exceptions are systems designed for utility intertie; they feed excess power back into the utility, and turn the meter backwards.

Hydro Systems

For those who are lucky enough to have a good site, hydro is really the renewable energy of choice. System component costs are lower, and watts per dollar return is higher for hydro than for any other renewable source. The key element for a good site is drop. A small amount of water dropping a large distance will produce the same quantity of energy as a large amount of water dropping a small distance. The turbine for the small amount of water is smaller, lighter, easier to install, and vastly cheaper. We offer several turbine styles for differing resources. The small Pelton wheel Harris systems are well-suited for mountainous territory that can deliver some drop and high pressure to the turbine. The propeller-driven Jack Rabbit is for flatter sites with less drop, but more volume, and the Stream Engine, with a turgo-type runner, falls in-between. It can handle larger water volumes and make useful power from less drop.

Read on for detailed explanations of wind generators and hydro turbines. If you need a little help and guidance putting a system together or simply upgrading, our technical staff, with over a hundred years of combined experience in renewable energy, will be glad to help. Call us toll-free at: 800-919-2400.

Hydroelectricity

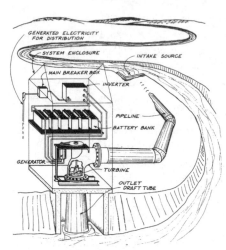

Low Head Installation

If you could choose any renewable energy source to use, hydro is the one. If you don't want to worry about a conservation-based lifestyle—always nagging your kids to turn off the lights, watching the voltmeter, basing every appliance decision on energy efficiency—then you had better settle next to a nice year-round mountain stream! Hydropower, given the right site, can cost as little as a tenth of a PV system of comparable output. Hydropower users are often able to run energy-consumptive appliances that would bankrupt a PV system owner, like large side-by-side refrigerators and electric space heaters. Hydropower will probably require more effort on-site to install, but even a modest hydro output over 24 hours a day, rain or shine, will add up to a large cumulative total. Hydro systems get by with smaller battery banks because they only need to cover the occasional heavy power surge rather than four days of cloudy weather.

Hydro turbines can be used in conjunction with any other renewable energy source, such as PV or wind, to charge a common battery bank. This is especially true in the West, where seasonal creeks with substantial drops only flow in the winter. This is when power needs are at their highest and PV input is at its lowest. Small hydro systems are well worth developing, even if used only a few months out of the year, if those months coincide with your highest power needs. So, what makes a good hydro site and what else do you need to know?

What is a "Good" Hydro Site?

High Head Installation

The Columbia River in the Pacific Northwest has some really great hydro sites, but they're not exactly homestead-scale (or low-cost). Within the hydro industry the kind of home-scale sites and systems we deal with are called micro-hydro. The most cost-effective hydro sites are located in the mountains. Hydropower output is determined by water's volume times its fall (jargon for the fall is "head"). We can get approximately the same power output by running 1,000 gallons per minute through a 2-foot drop as by running 2 gallon per minute through a 1,000-foot drop. In the former scenario, where lots of water flows over a little drop, we are dealing with a low-head/high-flow situation, which is not truly a micro-hydro site. Turbines that can efficiently handle thousands of gallons are usually large, bulky, expensive, and site-specific. But if you don't need to squeeze every last available watt out of your low-head source, the propeller-driven Jack Rabbit generator will produce very useful amounts of power from low-head/high-flow sites, or the Turgo runner used on the Stream Engine is good at higher volume, lower head sites.

For hilly sites that can deliver a minimum of 50 feet of head, Pelton wheel turbines offer the lowest-cost generating solution. The Pelton-equipped Harris turbine is perfect for low-flow/high-head systems. It can handle a maximum of 120 gallons per minute, and requires a minimum 40-foot fall in order to make any useful amount of power. In general, any site with more than 100 feet of fall will make an excellent micro-hydro site, but many sites with less fall can be very productive also. The more head, the less volume will be necessary to produce a given amount of power. Check the output charts in the products section for a rough estimate of what your site can deliver.

Our Hydro Site Evaluation service will estimate output for any site, plus it will size the piping and wiring, and factor in any losses from pipe friction and wire resistance. See the example at the end of this editorial section.

A hydro system's fall doesn't need to happen all in one place. You can build a small collection dam at one end of your property and pipe the water to a lower point, collecting fall as you go. It's not unusual to use several thousand feet of pipe to collect a hundred feet of head (vertical fall).

What If I Have a High-Volume, Low-Head Site or Want AC-Output?

The Jack Rabbit is a drop-in-the-creek generator

Typically, high-volume, low-head, or AC-output hydro sites will involve engineering, custom metalwork, formed concrete, permits, and a fair amount of initial investment cash. None of this is meant to imply that there won't be a good payback, but it isn't an undertaking for the faint-of-heart, or thin-of-wallet. AC generators are typically used on larger commercial systems, or on utility intertie systems. DC generators are typically used on smaller residential systems. Real Goods can provide engineering, specification, and procurement services for larger hydro systems up to 8 megawatts (8 million watts). Contact our technical staff directly for assistance.

For the homestead with a good creek, but no significant fall, the Jack Rabbit turbine is a far less costly alternative. With just a foot or two of fall, and a narrow spot in the creek to speed things up, this simple drop-in-the-creek turbine can provide for modest power needs. A simple rock or timber crib can be constructed to channel and speed the flow.

The Jack Rabbit was originally designed for towing behind seismic research ocean barges. So it's salt water tolerant, and will generate whether spun clockwise or counterclockwise. This could be useful for tidal flows. Output is a maximum of 100 watts, so daily output averages 1.5 to 2.4 kWh depending on the site. Of course, if you'd rather look into the typical low-head scenario, contact the DOE's Renewable Energy Clearinghouse at 800-363-3732, or use Internet access at www.eren.doe.gov for more free information on low-head hydro than you ever thought was possible.

How Do Micro-Hydro Systems Work?

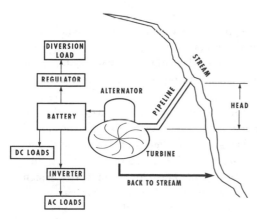

Typical Micro-Hydro System

The basic parts of micro-hydro systems are: the *pipeline* (called the penstock), which delivers the water; the *turbine*, which transforms the energy of the flowing water into rotational energy; the *alternator* or *generator*, which transforms the rotational energy into electricity; the *regulator*, which controls the generator or dumps excess energy, depending on regulator style; and the *wiring*, which delivers the electricity. Our micro-hydro systems also use batteries, which store the low voltage DC electricity, and usually an inverter, which converts the low voltage DC electricity into 120 volts or 240 volts AC electricity.

Most micro-turbine systems use a small DC alternator or generator to deliver a small but steady energy flow that accumulates in a battery bank. This provides a few important advantages. First, the battery system allows the user to store energy and expend it, if needed, in short powerful bursts (like a washing machine starting the spin cycle). The batteries will allow substantially more energy for short periods than a turbine is producing, as long as the battery and inverter are designed to handle the load. Second, DC charging means that precise control of alternator speed is not needed, as is required for 60 Hz AC output. This saves thousands of dollars on control equipment.

DC Turbines

Several micro-hydro turbines are available with simple DC output. We currently offer the Harris, the Stream Engine, and the Jack Rabbit.

Harris Turbine

The Harris turbine uses a hardened cast silicone bronze Pelton turbine wheel mated with a low-voltage DC alternator. Pelton wheel turbines work best at higher pressures and lower volumes. Life begins at about 50 feet of head for these turbines, and has no practical upper limit. The alternators used for the Harris turbine are the world's most common models: vintage 1960s and 1970s Delco or Motorcraft automotive units with windings that are customized for each individual application. Bearings and brushes will require replacement at intervals from one to five years, depending on how hard the unit is working. These parts are commonly available at any auto parts store.

Depending on the volume and fall supplied, Harris turbines can produce from 1 kWh (1000 watt-hours) to 35 kWh per day. Maximum alternator instantaneous output is about 1500 watts with cooling options required. The typical American home consumes 10 to 15 kWh per day with no particular energy conservation, so with a good hydro site, it is fairly easy to live a conventional lifestyle.

The Harris turbine can be supplied with one, two, or four nozzles. The maximum flow rate for any single nozzle is approximately 30 gallons per minute (gpm). The turbine can handle flow rates to about 100 gpm before the sheer volume of water starts getting in its own way. Many users buy two- or four-nozzle turbines, so that individual nozzles can be turned on and off to meet variable power needs or water availability. The brass nozzles are easily replaceable because they eventually wear out, especially if there is grit in the water. They are available in 1/16″ increments, from 1/16″ thru 1/2″. The first nozzle doesn't have a shutoff valve, while all nozzles beyond the first one are supplied with ball valves for easy, visible operation.

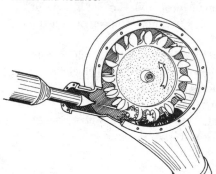

Two- and four-nozzle Harris turbines. The four-nozzle is upside down to show the Pelton wheel and nozzles.

An 1880s vintage single-nozzle Pelton wheel. Only the generator technology has changed.

Stream Engine

The Stream Engine uses a cast bronze turgo-type runner wheel, which can handle a bit more volume—up to about 150 gpm before it starts choking—and starts to deliver useful output at 15–20 foot head. Nozzles are cone-shaped plastic casings; you cut them at the size desired, from 1/8″ up to 1″. Two- or four-nozzle turbines are available, with replacement nozzles being a readily available bolt-in.

The Stream Engine uses a unique brushless permanent-magnet alternator with three large, sealed shaft bearings. Permanent magnets mean there are no field brushes to wear out. Magnetic field strength is adjusted by varying the air gap between the magnet disk and the stationary alternator windings. Once the unit is set up there is almost no routine maintenance required. Setup does require some time with this universal design however, and involves trial and adjustment: selecting one of four alternator wiring setups, and then adjusting the permanent-magnet rotor air gap until peak output is achieved. A precision shunt and digital multi-meter are supplied to expedite this setup process. Output voltage can be user-selected at 12, 24, or 48 volts. Maximum instantaneous output is 800 watts for this alternator.

Jack Rabbit

The Jack Rabbit is a 100-watt, low-speed alternator that was originally designed for towing behind seismic sleds for ocean-floor mapping. This drop-in-the-stream propeller-driven charger reaches maximum output at 9 to 10 mph stream speed. This is a relatively low-cost, low-impact hydro solution for folks in the flatlands. But it's also relatively low-output. It isn't worth installing unless you've got a choke point in your stream that delivers 9 mph or higher flow rates. The 24-volt version in particular seems to require higher flow rates to produce any useful power output. It is available in 12- or 24-volt versions, and is not field adjustable or switchable.

Power Transmission

Transmission distances of more than 500 feet often require expensive large-gauge wire or technical gimmickry.

One disadvantage of hydro systems is the difficulty of transmitting power from the turbine to the batteries, particularly with high-output sites. A typical installation places the batteries at the house on top of the hill, where the good view is, and the turbine at the bottom of the hill, where the water ends its maximum drop. Low-voltage power is difficult to transmit if large quantities or long distances are involved. The batteries should be as close to the turbine as is practical, but if there's more than 100 feet of distance involved, things will work better if the system voltage is 24 or even 48 volts. Transmission distances of more than 500 feet often require expensive large-gauge wire or technical gimmickry. With really long distances we sometimes recommend installing the batteries and inverter at the turbine site, which allows you to transmit 120-volt AC power. The higher voltage eases transmission problems, but introduces other potential problems. When the inverter trips off due to accidental overload, you won't be pleased about having to hike down to the turbine site to reset it—particularly if it's dark and raining, which Murphy's Law can practically guarantee. Please consult with the Real Goods Technical Staff about this, or other transmission options.

Controllers

Controllers designed for photovoltaics may damage the hydro generator, and will very likely become crispy-critters themselves if used with a hydro generator.

Hydro generators require special controllers or regulators. Controllers designed for photovoltaics may damage the hydro generator, and will very likely become crispy-critters themselves if used with a hydro generator. You can't simply open the circuit when the batteries get full like you can with PV. So long as the generator is spinning, there needs to be a place for the energy to go. Controllers for hydro systems take any power beyond what is needed to keep the batteries charged, and divert it to a secondary load, usually a water- or space-heating element. So extra energy heats either domestic hot water or the house itself. These diversion controllers are also used with some wind generators, and can be used for PV control as well if this is a hybrid system.

The Jack Rabbit turbine is a special case. This low-speed turbine breaks the usual control rules, and can be used with an inexpensive PV control. Because of the Jack Rabbit's slow speed, it won't generate high voltage if disconnected from the battery.

Site Evaluation

Okay, you have a fair amount of drop across your property, and/or enough water flow for the Jack Rabbit, so you think micro-hydro is a definite possibility. What happens next? Time to go outside and take some measurements, then fill in the necessary information on the Hydro Site Evaluation form. By looking at your completed form, the Real Goods technicians can calculate which turbine and options will best fill your needs, as well as what size pipe and wire and which balance-of-system components you require. Then we can quote specific power output and system costs so that you can decide if it is worth the installation. The Jack Rabbit doesn't need the fall or flow measurements; it just requires a minimum of 13 inches of water depth and a flow that's moving at 6 miles per hour (a slow jog), or up.

Distance Measurements

Keep the turbine and the batteries as close together as practical. As discussed earlier, longer transmission distances will get expensive. The more power you are trying to move, the more important distance becomes.

You'll need to know the distance from the proposed turbine site to the batteries (how many feet of wire), and the distance from the turbine site to the water collection point (how many feet of pipe). These distances are fairly easy to determine, just pace or tape them off.

Fall Measurement

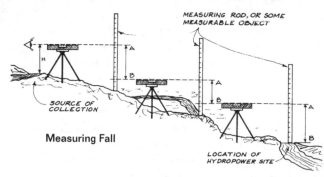

Measuring Fall

Next, you'll need to know the fall from the collection point to the turbine site. This measurement is a little tougher. If there is a pipeline in place already, or if you can run one temporarily and fill it with water, this part is easy. Simply install a pressure gauge at the turbine site, make sure the pipe is full of water, and turn off the water. Read the static pressure (which means no water movement in the pipe), and multiply your reading in pounds per square inch (psi) by 2.31 to obtain the drop in feet. If the water pipe method isn't practical, you'll have to survey the drop or use a fairly accurate altimeter. A number of relatively inexpensive sports watches come with a built-in altimeter now. If the altimeter can read ±10 feet, that's close enough. Strap it on, take a hike, and record the difference.

The following instructions represent the classic method of surveying. You've seen survey parties doing this and if you've always wanted to attend a survey party, this is your big chance to get in on the action. You'll need a carpenter's level, a straight sturdy stick about six feet tall, a brightly colored target that you will be able to see a few hundred feet away, and a friend to carry the target and to make the procedure go faster and more accurately. (It's hard to party alone.)

Stand the stick upright and mark it at eye level. (Five feet even is a handy mark that simplifies the mathematics, if that's close to eye level for you.) Measure and note the length of your stick from ground level to your mark. Starting at the turbine site, stand the stick upright, hold the carpenter's level at your mark, make sure it is level, then sight across it uphill toward the water source. With hand motions and body English, guide your friend until the target is placed on the ground at the same level as your sightline, then have your friend wait for you to catch up. Repeat the process, carefully keeping track of how many times you repeat. It is a good idea to draw a map to remind you of landmarks and important details along the way. If you have a target and your friend has a stick (marked at the same height, please) and level, you can leapfrog each other, which makes for a shorter party. Multiply the number of repeats between the turbine site and the water source by the length of your stick(s) and you have the vertical fall. People actually get paid to have this much fun!

Flow Measurement

Finally, you'll need to know the flow rate. If you can, block the stream and use a length of pipe to collect all the flow. Time how long it takes to fill a 5-gallon bucket. Divide 5 gallons by your fill time in seconds. Multiply by 60 to get gallons per minute. Example: The 5-gallon takes 20 seconds to fill. 5 divided by 20 = 0.25 times 60 = 15 gpm. If the flow is more than you can dam up or get into a 4-inch pipe, or if the force of the water sweeps the bucket out of your hands, forget measuring: you've got plenty!

Conclusion

Now you have all the information needed to guesstimate how much electricity your proposed hydro system will generate based on the manufacturer's Output Charts on the next pages. This will give you an indication whether or not your hydro site is worth developing, and if so, which turbine option is best. If you think you have a real site, fill out the Hydro Site Evaluation form, and send it along with $10 to the Technical Dept. at Real Goods. We will run your figures through our computer sizing program, which allows us to size plumbing and wiring for the least power loss at the lowest cost, and a myriad of other calculations necessary to design a working system. You'll find an example of our return form on the next page, followed by the form for the info *we* need from you.

—Doug Pratt

CALCULATION OF HYDROELECTRIC POWER POTENTIAL

ENTER HYDRO SYSTEM DATA HERE: **Customer: Meg A. Power**

PIPELINE LENGTH:	1300.00 FEET
PIPE DIAMETER:	4.00 INCHES
AVAILABLE WATER FLOW:	100.00 G.P.M.
VERTICAL FALL:	200.00 FEET
HYDRO TO BATTERY DISTANCE:	50.00 FEET (1 way)
TRANSMISSION WIRE SIZE:	2.00 AWG #
HOUSE BATTERY VOLTAGE:	24.00 VOLTS
HYDRO GENERATION VOLTAGE:	29.00 VOLTS

Power produced at hydro: *Power delivered to house:*

49.78 AMPS	49.78 AMPS
29.00 VOLTS	28.20 VOLTS
1443.53 WATTS	1403.59 WATTS

Four nozzle, 24V, high output w/cooling turbine required

PIPE CALCULATIONS

HEAD LOST TO PIPE FRICTION:	7.61 FEET
PRESSURE LOST TO PIPE FRICTION:	3.29 PSI
STATIC WATER PRESSURE:	86.62 PSI
DYNAMIC WATER PRESSURE:	83.33 PSI
STATIC HEAD:	200.01 FEET
DYNAMIC HEAD:	192.40 FEET

HYDRO POWER CALCULATIONS

OPERATING PRESSURE:	83.33 PSI
AVAILABLE FLOW:	100.00 GPM
WATTS PRODUCED:	1443.53 WATTS
AMPERAGE PRODUCED:	49.78 AMPS
AMP-HOURS PER DAY:	1194.65 AMP-HOURS
WATT-HOURS PER DAY:	34644.83 WATT-HOURS
WATTS PER YEAR:	12645362.71 WATT-HOURS

LINE LOSS (USING COPPER)

TRANS. LINE ONE-WAY LENGTH:	50.00 FEET
VOLTAGE:	29.00 VOLTS
AMPERAGE:	49.78 AMPS
WIRE SIZE #:	2.00 AMERICAN WIRE GAUGE
VOLTAGE DROP:	0.80 VOLTS
POWER LOST:	39.95 WATTS
TRANSMISSION EFFICIENCY:	97.23 PERCENT
PELTON WHEEL RPM WILL BE:	2969.85 AT OPTIMUM WHEEL EFFICIENCY

This is an estimate only! Due to factors beyond our control (construction, installation, incorrect data, etc.) we cannot guarantee that your output will match this estimate. We have been conservative with the formulas used here and most customers call to report more output than estimated. However, be forewarned! We've done our best to estimate conservatively and accurately, but there is no guarantee that your unit will actually produce as estimated.

Real Goods
Hydroelectric Site Evaluation Form

Name: _____

Address: _____

Phone #: _____ Date: _____

Pipe Length:_____ (from water intake to turbine site)

Pipe Diameter:_____(only if using existing pipe)

Available Water Flow:_____(in gallons per minute)

Fall:_____(from water intake to turbine site)

Turbine to Battery Distance:_____(one way, in feet)

Transmission Wire Size:_____(only if existing wire)

House Battery Voltage:_____(12, 24, ??)

Alternate estimate (if you want to try different variables)

Pipe Length:_____ (from water intake to turbine site)

Pipe Diameter:_____(only if using existing pipe)

Available Water Flow:_____(in gallons per minute)

Fall:_____(from water intake to turbine site)

Turbine to Battery Distance:_____(one way, in feet)

Transmission Wire Size:_____(only if existing wire)

House Battery Voltage:_____(12, 24, ??)

For a complete computer printout of your hydroelectric potential, including sizing for wiring and piping, please fill in the above information and send to Real Goods along with $10.00.

17-001 HydroElectric Evaluation $10⁰⁰

Hydroelectric Products

Hydro Site Evaluation

In order to accurately size your hydroelectric system, some specific site information is needed. Once received, your site information will be fed into our computer for our custom hydro-sizing analysis. Please fill out the Hydroelectric Site Evaluation form on the preceding page, send it in with $10 and we'll evaluate your potential site and recommend a system for you. Allow two weeks processing time.

17-001 Hydro Site Evaluation $10⁰⁰

Harris Hydroelectric Turbines

Solar only generates power when the sun is shining; hydro generates power 24 hours a day. It only takes a small input to cumulatively add up to a large daily total. The generating component of the Harris turbines is an automotive alternator equipped with custom-wound coils appropriate for each installation. Field strength, and therefore output, is site-adjustable with a large rheostat. The built-in ammeter shows the immediate effects of tweaking the field. The rugged turbine wheel is a one-piece Harris casting made of tough silicon bronze. There are hundreds of these wheels in service, with no failures to date. The cast aluminum wheel housing serves as a mounting for the alternator and up to four nozzle holders. It also acts as a spray shield, redirecting the expelled water down into the collection/drain box. Expelled water depends on gravity to drain.

Harris Hydroelectric turbines are available with one, two or four nozzles to maximize the output of the unit. Having an extra nozzle or two beyond what is absolutely required makes it easy to change flow rates or power output for seasonal or household variations. Just turn different size nozzles on and off. The individual brass nozzles are screwed in from the underside of the turbine. They do wear with time, and will need eventual replacement. The Low Head Option is equipped with larger ball valves to reduce friction loss. Any turbine delivering 30 amps or more (at whatever the system voltage is), needs the cooling fan option. USA.

FLOW RATES AND NOZZLE SELECTION

Gallons Per Minute	Number Of Nozzles
5 to 30	1
30 to 60	2
60 to 120	4

All turbines have custom windings; we need the Hydro Site Evaluation information to properly assemble a turbine for your site!

OUTPUT LIMITS FOR HARRIS TURBINES

	12-Volt	24-Volt
High Output:	750 watts	1500 watts

The Harris Standard Output Turbine is no longer available; all turbines are the High Output model now.

◈**17-101 1-Nozzle Turbine $950⁰⁰**
◈**17-102 2-Nozzle Turbine $1,050⁰⁰**
◈**17-103 4-Nozzle Turbine $1,249⁰⁰**
◈**17-132 24-Volt Option $49⁹⁵**
◈**17-136 48-Volt Option $149⁰⁰**
◈**17-133 Low Head Option (less than 60´) $39⁹⁵**
◈**17-134 Extra Nozzles $4⁹⁵**
◈**17-135 Optional Cooling Fan Kit $69⁹⁵**

◈*Means shipped directly from manufacturer.*

HARRIS HIGH OUTPUT HYDRO TURBINE (IN WATTS)

GPM	Feet Of Net Head						
	25	50	75	100	200	300	600
3	-	-	-	-	30	70	150
6	-	-	25	35	100	150	300
10	-	35	60	80	180	275	550
15	20	60	95	130	260	400	800
20	30	80	130	200	400	550	1100
30	50	125	210	290	580	850	1500
50	115	230	350	500	950	1400	-
100	200	425	625	850	1500	-	-

Wattage output figures based on actual measurement with High Output alternator option. A fan option must be used on all systems producing over 30 amps. (420 watts @ 12V, 840 watts @24V.)

Harris Pelton Wheels

For do-it-yourselfers interested in a small hydroelectric system, we offer the same tough and reliable Pelton wheel used on the complete turbines above. Harris silicon bronze Pelton wheels resist abrasion and corrosion far longer than polyurethane or cast aluminum wheels. These are 5″ diameter quality castings that can accommodate nozzle sizes of ¹⁄₁₆″ through ½″. Designed with threads for Delco or Motorcraft alternators. USA.

◈**17-202 Silicon Bronze Pelton Wheel $269⁰⁰**

The Stream Engine

The highly adjustable Stream Engine uses a turgo-type runner wheel that can produce useful power from lower heads. The permanent-magnet generator has no brushes to wear out, and uses a clever combination of alternator wiring changes and adjustable air gap to compensate for differing flow rates. Setup does require some time with this universal design, and involves trial and adjustment, selecting one of four alternator wiring setups, and then adjusting the permanent-magnet rotor air gap until peak output is achieved. A precision shunt and digital multimeter are supplied to expedite this setup process. Output

voltage is user-settable from 12 to 48 volts. Available in two- or four-nozzle turbines, the Stream Engine is supplied with universal cut-to-size nozzles of ⅛″ to 1″. See the Nozzle Chart to determine flow rates at various heads. Head can be from 5 feet to 400 feet, and flows of 5 gpm to 300 gpm can be used (note that flow rates above 150 gpm have diminishing returns, as the sheer volume of water starts to get in its own way). Maximum output is approximately 800 watts. See performance chart. Canada.

◈**17-109 Stream Engine, 2-nozzle $1,695⁰⁰**

◈**17-110 Stream Engine, 4-nozzle $1,885⁰⁰**

◈**17-111 Turgo Wheel only (bronze) $550⁰⁰**

◈**17-112 Extra Universal Nozzle $39⁹⁵**

STREAM ENGINE NOZZLE FLOW CHART (IN GPM)

Head Pressure Feet	PSI	⅛	3/16	¼	5/16	⅜	7/16	½	⅝	¾	⅞	Turbine 1.0	RPM
5	2.2					6.18	8.40	11.0	17.1	24.7	33.6	43.9	460
10	4.3			3.88	6.05	8.75	11.0	15.0	24.2	35.0	47.0	62.1	650
15	6.5		2.68	4.76	7.40	10.7	14.6	19.0	29.7	42.8	58.2	76.0*	800
20	8.7	1.37	3.09	5.49	8.56	12.4	16.8	22.0	34.3	49.4	67.3	87.8*	925
30	13.0	1.68	3.78	6.72	10.5	15.1	20.6	26.9	42.0	60.5	82.4*	107*	1140
40	17.3	1.94	4.37	7.76	12.1	17.5	23.8	31.1	48.5	69.9	95.1*	124*	1310
50	21.7	2.17	4.88	8.68	13.6	19.5	26.6	34.7	54.3	78.1*	106*	139*	1470
60	26.0	2.38	5.35	9.51	14.8	21.4	29.1	38.0	59.4	85.6*	117*	152*	1600
80	34.6	2.75	6.18	11.0	17.1	24.7	33.6	43.9.	68.6	98.8*	135*	176*	1850
100	43.3	3.07	6.91	12.3	19.2	27.6	37.6	49.1	76.7	111*	150*	196 *	2070
120	52.0	3.36	7.56	13.4	21.0	30.3	41.2	53.8	84.1*	121*	165*	215*	2270
150	65.0	3.76	8.95	15.0	23.5	33.8	46.0	60.1	93.9*	135*	184*	241*	2540
200	86.6	4.34	9.77	17.4	27.1	39.1	53.2	69.4	109*	156*	213*	278*	2930
250	106	4.86	10.9	19.9	30.3	43.6	59.4	77.6*	121*	175*	238*	311*	3270
300	130	5.32	12.0	21.3	33.2	47.8	65.1	85.1*	133*	191*	261*	340*	3590
400	173	6.14	13.8	24.5	38.3	55.2	75.2	98.2*	154*	221*	301*	393*	4140

STREAM ENGINE POWER OUTPUT (IN WATTS)

Net Head Feet	5	10	15	20	30	40	50	75	100	150	200*	300*
5			5	8	10	15	20	30	40			
10		7	12	18	23	30	45	60	80	100		
15	5	10	15	20	30	40	50	75	100	125	150	200
20	8	16	25	32	50	65	85	125	170	210	275	350
30	12	30	45	60	90	120	150	225	300	400	500	700
40	16	40	60	80	120	160	200	300	400	500	600	
50	20	50	75	100	150	200	250	375	500	600		
75	30	75	110	150	225	300	375	560	700			
100	40	100	150	200	300	400	500	650				
150	60	150	225	300	400	550	650					
200	80	200	300	400	550	700						
300	120	240	360	480	720							
400	160	320	480	640								

Flow rates above 150 gpm are possible, but will suffer from diminishing returns.

Jack Rabbit Submersible Hydro Generator

No pipes or dams! Power from any fast-running stream!

The Jack Rabbit is a special low-speed alternator mounted in a heavy-duty, oil-filled cast aluminum housing with triple shaft seals. Originally designed for towing behind seismic sleds for oil exploration, this marine-duty unit is ideal for home power generation near a reasonably fast-moving stream. In a 9-mph stream (slow jog) the Jack Rabbit produces about 2,400 watt-hours daily. In a 6-mph stream (brisk walk) it produces over 1,500 watt-hours. See the output chart. The 12.5″ propeller requires 13″ of water depth. A rock or timber venturi can often be constructed to increase stream speed and power output.

The Jack Rabbit turbine can be mounted in a variety of ways—on bridging logs, on a counter-weighted cantilever, or on an adjustable frame. For streams that are subject to flooding violent enough to dislodge logs or boulders, you should have a method of raising the generator.

For best output the prop should face upstream, but in tidal flow applications output happens in either direction due to the rectified alternator output. Comes with three meters of output wire and a separate rectifier pack.

The Jack Rabbit is a very rugged machine which requires no maintenance. Small debris that would clog or slow a Pelton turbine passes through harmlessly. The prop is top-grade ductile aluminum, which can be hammered back into shape in case it gets really walloped by a log. Props are available as replacement parts.

The Jack Rabbit rarely requires a diversion controller and dummy load to prevent overcharging. In all but the highest output sites, it can simply be shut off with a PV controller. The slow speed prevents any high-voltage buildup.

The Jack Rabbit weighs 22 pounds, is 14.4″ long, 8.4″ diameter, has a 12.5″ prop, and carries a 3-year mfr.'s warranty. Voltage not field switchable. Great Britain.

◈**17-104 Jack Rabbit Hydro Generator 12V $1,195⁰⁰**

◈**17-105 Jack Rabbit Hydro Generator 24V $1,195⁰⁰**

◈**17-108 Jack Rabbit Spare Prop $169⁰⁰**

◈*Means shipped directly from manufacturer.*

Experience has shown the Jack Rabbit isn't worth installing unless you've got a choke point in your stream that delivers 9-mph or higher flow rates. The 24-volt version in particular seems to require higher flow rates to produce any useful power output.

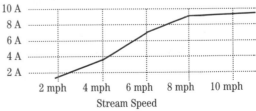

JACK RABBIT POWER CURVE @12V

Hydro System Controls

Trace C-Series Multi-Function DC Controllers

Trace C40
Controller

The new C-series of solid-state controllers has grown. Now featuring models at 35, 40, and 60 amps, these multi-controllers can be used for PV charge control, DC load control, or DC diversion. They will operate in only one mode however, so PV charge control and DC load control would require two controllers. The 35- and 40-amp controls can be manually selected for 12, 24, or 48 volts. The 60-amp control is available in 12 and 24 volts. All setpoints are field adjustable with removable knobs to prevent tampering. They are electronically protected against short-circuit, overload, overtemp, and reverse polarity with auto-reset. No fuses to blow! Two-stage lightning and surge protection are included. The three-stage (bulk, absorb, float) PV charge control uses solid-state pulse width modulation control for the most effective battery charging. Has an automatic "equalize" mode every 30 days, which can be turned off for sealed batteries, and a manually selected "nicad" mode is included. A battery temperature sensor is optional, and recommended if your batteries will see temps below 40°F or above 90°F. The diversion control mode will divert excess power to a dummy heater load, and offers the same adjustments and features as the PV control. The DC load controller has adjustable disconnect and auto-reconnect voltages with a time delay for heavy surge loads. Electronic short-circuit protected. 9″ x 5″ x 2″, weighs 4 lbs. Certified by ETL to UL standards. Two-year mfr.'s warranty. USA.

25-027 Trace C-35 Multi-Controller $119⁰⁰

25-017 Trace C-40 Multi-Controller $159⁰⁰

25-028 Trace C-60 Multi-Controller $199⁰⁰

When used as a diversion control, the diversion load must have a smaller amperage rating than the Trace Controller to prevent problems. For instance, 40 amps of heater element will trigger the electronic overcurrent protection on a C-35 Controller, but a 30-amp heater would be okay.

Optional C-Series LCD Display

C40 with Digital Display

The LCD display with backlighting continuously displays battery voltage, DC amperage, cumulative amp-hours, and a separate resettable "trip" amp-hour meter (it only displays current flows passing through the controller). It can be mounted on the front of the C-Series Controller, or remotely using simple phone cable and plugs. USA.

25-018 Trace C-Series LCD Display $99⁰⁰

Universal Air Heating Element

A simple mount-it-yourself 12-volt resistor. Use two or more elements in series for higher voltage systems. Use more elements in parallel for higher amperage. These resistors can get very hot during regulation and must be mounted a safe distance from flammable surfaces. USA.

25-109 12V/20A Air Heating Element $24⁹⁵

Air Heater Diversion Loads

Unit comprises air heating element(s) mounted in an aluminum box with a small fan to disperse the heat. Order one or more to match the total amperage of your charging inputs. Terminal block accepts up to 1/0 wire. Do not exceed the amperage rating of your diversion regulator. 18″ x 7″ x 7″. Two-year mfg. warranty. USA.

❖ **25-750 12V/30A Diversion Load $160⁰⁰**

❖ **25-751 12V/60A Diversion Load $187⁰⁰**

❖ **25-752 12V/120A Diversion Load $192⁰⁰**

❖ **25-753 24V/30A Diversion Load $162⁰⁰**

❖ **25-754 24V60A Diversion Load $220⁰⁰**

❖ **25-755 48V/30A Diversion Load $220⁰⁰**

Water Heating Elements

Industrial-grade elements designed for years of trouble-free service. Screws into standard 1″ water heater NPT (National Pipe Thread) fittings. Order one element for each external load circuit. 12/24-volt element rated at 600 watts (40 amps at 12 volts, 20 amps at 24 volts), 48-volt element rated at 30 amps. The 12/24-volt element has a pair of 12-volt elements and four terminals. Wire in parallel for 12V, or in series for 24V operation.

❖ **25-078 12/24V Water Heating Element $89⁹⁵**

❖ **25-155 48V/30A Water Heating Element $116⁰⁰**

❖ *Means shipped directly from manufacturer.*

Enermaxer III Universal Battery Voltage Controller

The Enermaxer III Universal Battery Voltage Controller is an excellent voltage controller that has proven extremely reliable for hydro and wind systems. It is now available in a new re-engineered 120-amp version. This is the controller of choice for hydro systems, but will work with multiple charging sources at once (wind, hydro, PV). It is a parallel shunt regulator that diverts any excess charging power to a dummy heating load, typically a water- or air-heating element. It is made of all solid-state components so that there are no relay contacts to wear out. The desired float voltage is selected with the front-mounted rheostat, and diversion activity is indicated by the LED light. During charging, as battery voltage rises, just enough power will be diverted to maintain the float voltage. The battery will float at any user-selected voltage and any excess power, up to the maximum amperage rating for the Enermaxer, will be diverted. Water heating is the most popular diversion.

When ordering the Enermaxer, keep in mind that you must have an external load at least equal to your charging capacity. This is usually performed with either water heating elements or air-heating elements. We offer elements above. Two Enermaxer models are offered, one for up to 120 amps at 12 to 36 volts, or a separate 48-volt unit that handles up to 60 amps. The lower voltage unit is set for 24V operation from the factory, but is easily adjusted for 12, 24, 32, or 36 volts on site. Two-year manufacturer's warranty. USA.

25-153 Enermax 120A Control 12V, 24V 32V, 36V
349^{00}

25-220 Enermax 60A Control 48V 395^{00}

120-AMP ENERMAX SPECIFICATIONS

Battery Voltage	Adjustment Range	Maximum Amperage	Idle Amperage
12	13.3–17.0	120	0.01
24	26.6–34.0	120	0.01
32	35.5–45.0	120	0.01
36	39.9–51.0	120	0.01

Wind Energy

This section was adapted from Wind Energy Basics, *a Real Goods Solar Living Book by Paul Gipe. He is also the author of* Wind Power for Home & Business *(1993),* Wind Energy Comes of Age *(1995), and* Energía Eólica Práctica *(2000). Gipe has written and lectured extensively about wind energy for more than two decades.*

Small Wind Turbines Come of Age

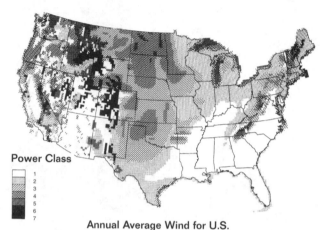

Power Class

1
2
3
4
5
6
7

Annual Average Wind for U.S.

The debut of micro wind turbines has revolutionized living off-the-grid. These inexpensive machines have brought wind technology within reach of almost everyone. And their increasing popularity has opened up new applications for wind energy previously considered off-limits, such as electric fence charging and powering remote telephone call boxes, once the sole domain of photovoltaics.

Micro wind turbines have been around for decades for use on sailboats, but they gained prominence in the 1990s as their broader potential for off-the-grid applications on land became more widely known. While micro wind turbines have yet to reach the status of widely available consumer commodities such as personal computers, the day may not be far off.

The use of wind power is "exploding," say Karen and Richard Perez, the editors of *Home Power* magazine. "There are currently over 150,000 small-scale RE (renewable energy) systems in America and they are growing by 30% yearly. [But] [t]he small-scale use of wind power is growing at twice that amount—over 60% per year." And a large part of that growth is due to Southwest Windpower of Flagstaff, Arizona.

Southwest Windpower awakened latent consumer interest in micro wind turbines with the introduction of its sleek Air series. Since launching the line in 1995, they've shipped more than 30,000 of the popular machines.

"What Americans, and folks all over the world, are finding out," the Perezes say, "is that wind power is an excellent and cost-effective alternative" to extending electric utility lines, and fossil-fueled backup generators.

Hybrid Wind and Solar Systems

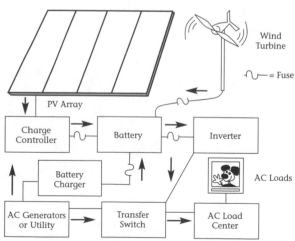

Both wind and PV can happily feed a common battery.

You might say that joining wind and solar together is a marriage made in heaven. The two resources are complementary: in many areas wind is abundant in the winter when photovoltaics are least productive, and sunshine is abundant in the summer when winds are often weakest. The sun and wind together not only improve the reliability of an off-the-grid system, but are also more cost-effective than using either source alone.

Hybrid systems include a DC source center (for DC circuit breakers), batteries, inverters, and often an AC load center. These components are necessary whether you're using just wind or solar. So, it's best to spread the fixed cost of these components over more kilowatt-hours by using photovoltaic panels in addition to a wind turbine.

Installing a 1,500-watt wind turbine on a tilt-up tower using an electric winch.
©Paul Gipe

Engineers have found that these hybrids perform even better when coupled with small backup generators to reduce the battery storage needed. Many of those living off-the-grid reach the same conclusion by trial and error.

Typically a micro turbine, such as Southwest Windpower's Windseeker, or a small wind turbine, such as the Bergey 1500, a modest array of PV panels, batteries, and a small backup generator will suffice for most domestic uses. Though Pacific Gas & Electric Co. found that most Californians living off-the-grid had backup generators, they seldom used them. In a well-designed hybrid system, the backup generator provides peace of mind, but little electricity.

Size Matters

In wind energy, size, especially rotor diameter, matters. Nothing tells you more about a wind turbine's potential than its diameter—the shorthand for the area swept by the rotor. The wind turbine with the bigger rotor will intercept more of the wind stream and will almost invariably generate more electricity than a turbine with a smaller rotor, regardless of their generator ratings.

The area swept by a wind turbine rotor is equivalent to the surface area of a photovoltaic array. When you need more power in a PV array, you increase the surface area of the panels exposed to the sun by adding more panels. In the case of wind, you find a wind turbine that sweeps more area of the wind—a turbine with a greater rotor diameter.

Micro turbines range in size from 3 to 5 feet in diameter. The lower end of the range is represented by the Southwest Windpower's Air series, the upper end by their Windseeker brand. These machines are suitable for recreational vehicles, sailboats, fence-charging, and other low-power uses. Micro turbines will generate about 300 kilowatt-hours (kWh) per year at sites with average wind speeds typical of the Great Plains (about 12 mph at the height of the turbine).

Mini wind turbines are slightly larger than micro turbines and are well suited for vacation cabins. They span the range from 5 to 9 feet in diameter and include World Power Technologies' Whisper H40 as well as Bergey Windpower's new XL1. Wind turbines in this class can produce 1,000 to 2,000 kWh per year at sites with an average annual wind speed of 12 mph.

Household-size turbines, as the name implies, are suitable for homes, farms, ranches, small businesses, and telecommunications. They span an even broader range than the other size classes, and encompass turbines from 10 to 23 feet in diameter. This class includes World Power Technologies' 175, Bergey's 1500 as well as their Excel model. Household-size wind turbines can generate from 2,000 kWh to 20,000 kWh per year at 12 mph sites.

Generators

Most small wind turbines use permanent-magnet alternators. This is the simplest generator configuration and is nearly ideal for micro and mini wind turbines. There is more diversity in household-size turbines, but again nearly all use permanent-magnet alternators. Some manufacturers use Ferrite magnets, others use rare-earth magnets. The latter have a higher flux density than ferrite magnets. Both types do the job.

Power Curves

The power curve indicates how much power (in watts or kilowatts) a wind turbine's generator will produce at any given wind speed. Power is presented on the vertical axis; wind speed on the horizontal axis. In the advertising wars between wind turbine manufacturers, often the focus is the point at which the wind turbine reaches it's "rated" or nominal power. Though rated power is just one point on a wind turbine's power curve,

Bob & Ginger Morgan's Bergey Excel being installed by a crane near Tehachapi, California (see page 115).
©Paul Gipe

The sleek Air 403 Industrial Turbine

many consumers mistakenly rely on it when comparison shopping. But not all power curves are created equal. Some power curves are, to be diplomatic, more "aggressive" than others.

Manufacturers may pick any speed they choose to "rate" their turbine. In the 1970s it was easy for unscrupulous manufacturers to manipulate this system to make it appear that their turbines were a better buy than competing products. By pushing "rated power" higher they were able to show lower relative costs (turbine cost/rated power) or they were able to increase their price—and profits—proportionally.

Unlike wind farm turbines, the performance of many small wind turbines have not been tested to international standards. As a rule, don't place much faith in power curves, unless the manufacturer has clearly stated the conditions under which the curve was measured (usually in a detailed footnote), or in "rated power." Some power curves can be off by 40% in low winds, and as much as 20% at "rated" power. And it's the performance in lower winds that matters most.

Most homeowners seldom see their turbines operating in winds at "rated" speeds of, say, 28 mph. Small wind turbines operate most of the time in winds at much lower speeds, often from 10 to 20 mph. For size or price comparisons, stick with rotor diameter or swept area. Both are more reliable indicators of performance than power curves.

Robustness

Wind turbines work in a far more rugged environment than photovoltaic panels that sit quietly on your roof. You quickly appreciate this when you watch a small wind turbine struggling through a gale. There's no foolproof way to evaluate the robustness of small wind turbine designs.

In general, heavier small wind turbines have proven more rugged and dependable than lightweight machines. Wisconsin's wind guru Mick Sagrillo is a proponent of what he calls the "heavy metal school" of small wind turbine design. Heavier, more massive turbines, he says, typically last longer. Heavier in this sense is the weight or mass of the turbine relative to the area swept by the rotor. By this criteria, a turbine that has a relative mass 10 kg/m2 may be more robust—and rugged—than a turbine with a specific mass of 5 kg/m2.

The Bergey Excel Turbine

Siting

To get the most out of your investment, site your wind turbine to best advantage: well away from buildings, trees, and other wind obstructions. Install the turbine on as tall a tower as you're comfortable working with. Jason Edworthy's experience at Nor'wester Energy Systems in Canada convinces him that the old 30-foot rule still applies. This classic rule from the 1930s dictates that for best performance, your wind turbine should be at least 30 feet above any obstruction within 200 feet of the tower. Under the best of conditions, a tower height of 30 feet is the absolute minimum, says Edworthy.

Putting a turbine on the roof is no alternative. Seldom can you get the turbine high enough to clear the turbulence caused by the building itself. Imagine trying to mount a 30-foot tall tower on a steeply pitched roof. It's a recipe for disaster. Even if you could, turbine induced vibrations will quickly convince you otherwise. (It's like putting a noisy lawn mower on your roof.) While many small wind turbines are relatively quiet, some are not—another good reason to put them out in the open, well away from buildings.

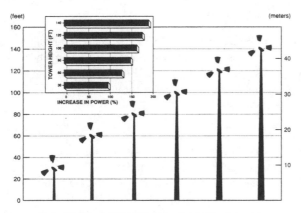

Increase in Power with Height Above 30 ft (10m)

Chart adapted from Wind Power for Home & Business.

Towers

Most small wind turbines are installed on guyed, tubular masts that are hinged at the base. With an accompanying gin pole, these towers can be raised and lowered, simplifying installation and service. Some tilt-up towers use thin-walled steel tubing, others use thick-walled steel pipe. Household-size turbines also use guyed masts of steel lattice, as well as free-standing truss towers. With the advent of pre-engineered, tubular mast kits, there's now less excuse than ever for installing micro and mini wind turbines on inadequate towers.

Installation

Those with good tool skills—who work safely—can install a micro turbine themselves using a pre-engineered, tilt-up tower kit. Installing mini wind turbines, because of the greater forces involved, requires considerably more skill. Household-size turbines should be left only to professionals—and even the pros have made tragic mistakes. If you like doing the work yourself, start with a micro turbine and a tilt-up tower. Once you're satisfied you know what you're doing and you're comfortable with the technology, you can try your hand with a larger turbine.

ESTIMATED ANNUAL ENERGY OUTPUT AT HUB HEIGHT IN THOUSAND KWH/Y

Avg Wind Speed	Rotor Diameter, m (ft)					
	1	1.5	3	7	18	40
	(3.3)	(4.9)	(9.8)	(23)	(60)	(130)
(mph)	thousands of kWh per year					
9	0.15	0.33	1.3	7	40	210
10	0.20	0.45	1.8	10	60	290
11	0.24	0.54	2.2	13	90	450

Typical Costs

The cost of a wind system includes the cost of the turbine, tower, ancillary equipment (disconnect switches, cabling, etc.), and installation. The total cost of a micro turbine can be as little as $1,000, depending upon the tower used and its height, while that of household-size machines can exceed $30,000. When comparing prices remember that bigger turbines cost more but often are more cost-effective. For an off-the-grid power system, the addition of a wind turbine will almost always make economic sense by reducing the number of photovoltaic panels or batteries needed.

For grid-intertie systems, the economics depend upon the winds at your site and a host of other factors, including the average wind speed, the cost of the wind system, the cost of utility power, and whether your utility provides net billing.

Some Do's and Don'ts

Bergey Excel and photovoltaic panels at home of Dave Blittersdorf, founder of NRG Systems.
©Paul Gipe

Do plenty of research. It can save a lot of trouble—and expense—later.

Do visit the library. Books remain amazing repositories of information. (I do have a bias about books, since I write them. But I've always been a firm believer in "you get what your pay for" when it comes to free information, whether it's from the Web, manufacturers, or their trade associations.)

Do talk to others who use wind energy. They've been there. You can learn from them what they did right, and what they'd never do again.

Do read and, equally important, follow directions.

Do ask for help when you're not sure about something. The folks at Real Goods Renewables are there to help.

Do build to code. In the end it makes for a tidier, safer, and easier-to-service system.

Do take your time. Remember, there's no rush. The wind will always be there.

Do be careful. Small wind turbines may look harmless, but they're not.

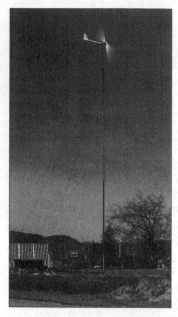

Real Goods Solar Living Center's Whisper 3000 wind turbine atop a hinged, tilt-up tower in Hopland, California.
©Paul Gipe

Don't skimp and don't cut corners. Taking short cuts is always a surefire way to ruin an otherwise good installation.

Don't design your own tower—unless you're a licensed mechanical engineer.

Don't install your turbine on the roof despite what some manufacturers may say!

And, of course, **don't** believe everything you read in sales brochures.

In general, doing it right the first time may take longer and cost slightly more, but you'll be a lot happier in the long run.

Sources of Information:

For more on micro turbines see *Wind Energy Basics*, White River Junction, VT: Chelsea Green Publishing, 1999. 222 pp., ISBN 1-890132-07-1, available from Real Goods (see page 110). Describes a new class of small wind turbines, dubbed *micro turbines*. These inexpensive machines, when coupled with readily available photovoltaic panels, have revolutionized living off-the-grid.

For more on small wind turbine technology see *Wind Power for Home & Business*, White River Junction, VT: Chelsea Green Publishing, 1993. 414 pp., ISBN 0-930031-64-4, also available from Real Goods (see page 110). Explains how modern, integrated wind turbines work, and how to use them most effectively.

For information about the commercial wind power industry see *Wind Energy Comes of Age*, New York: John Wiley & Sons, 1995. 613 pages. ISBN 0-471-10924-X. A chronicle of wind energy's progress from its rebirth during the oil crises of the 1970s to its maturation on the plains of northern Europe in the 1990s. Selected as one of the outstanding academic books published in 1995.

To determine your local wind resources visit http://rredc.nrel.gov/wind/pubs/atlas/ or search for the National Renewable Energy Laboratory's *Wind Energy Resource Atlas of the United States*.

For tips on installing micro and mini wind turbines using a Griphoist brand winch, see "Get a Grip!" in *Homepower Magazine* #68, December 1998/January 1999 or visit www.chelseagreen.com/Wind/articles/Getagrip.pdf.

The Gipe Family Do-It-Yourself Wind Generator Slide Show

1. Taking delivery of a new Air turbine and 45-foot tower.

2. Checking the packing list against parts delivered.

3. Securing the tower's base plate.

4. Aligning the guy anchors.

5. Driving the screw anchors.

6. Unspooling the guy cable.

7. Assembling the mast.

8. Clevis and gated fitting hook.

9. Gin pole and lifting cables.

10. Strain relief for supporting power cables.

11. Final assembly of the turbine.

12. Disconnect switch and junction box.

13. Slowly raising the turbine with a Griphoist-brand hand winch.

14. Air turbine safely installed with a Griphoist-brand hand winch.

All photos ©Paul Gipe

Wind Products

Wind Power For Home & Business

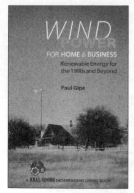

Paul Gipe. A Real Goods Independent Living Book. Real Goods and Chelsea Green joined together to publish the most complete reference on all aspects of modern (post-1970s) wind energy machines for homes and businesses. Wind energy technology has changed a lot in the last 20 years. No longer characterized by a do-it-yourself contraption in the backyard that could barely power a water pump, today's state-of-the-art wind generators are efficient, powerful and inexpensive in the long run. New turbine designs can now generate all the power you need for home use if you live in a good wind area. Whether you're committed to using a wind power system or just want to learn about it, this book will tell you all you need to know. Author Paul Gipe has had a part in nearly every aspect of wind energy's development since the mid-1970s—from measuring wind regimes, to siting and installing wind turbines. Gipe is currently president of the American Wind Energy Association, and works at Tehachapi Wind Farm, the world's largest producer of wind-generated electricity. He has written and lectured extensively about wind power. Extensive appendixes and references. 414 pages, paperback. USA.

80-192 Wind Power for Home & Business $34⁹⁵

Wind Power Diskette

The wind industry has been evolving more rapidly than we can update books. This companion 3.5″ diskette contains the most up-to-date and comprehensive information on wind power available anywhere. There are also extensive worksheets and output programs that are not found in the original book. This disk is the least expensive and most adaptable way to present all the newest models and output specs. Works with any Windows-based computer system. USA.

80-035 Wind Power Diskette $15⁰⁰

Complete set of Wind Power Book and Diskette

80-036 Wind Power Book and Diskette $44⁹⁵

Wind Energy Basics
A Guide to Small and Micro Wind Systems

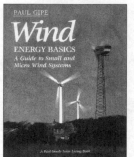

By Paul Gipe. Step-by-step manual for reaping the advantages of a windy ridgetop as painlessly as possible. Chapters include wind energy fundamentals, estimating performance, understanding turbine technology, off-the-grid applications, utility intertie, siting, buying, and installing, plus a thorough resource list with manufacturer contacts, and international wind energy associations. A Real Goods Solar Living Book. 222 pages, paperback. USA.

80-928 Wind Energy Basics $19⁹⁵

Residential Wind Power With Mick Sagrillo

Mick Sagrillo, the owner of Lake Michigan Wind and Sun, has been in the wind industry for over 16 years, with over 1,000 turbine installation/repairs under his belt.

This professionally produced video includes basic technology introductions, hands-on applications, and performance and safety issues. The action moves into the field to review actual working systems, including assembling a real system. Tape runs 63 minutes. Standard VHS format. USA.

80-363 Renewable Experts Tape—Wind $39⁰⁰

Wind Generators

Windseeker 500-Watt Turbines
Wind Generators Made Simple

Windseeker uses a brushless design with high-powered neodymium permanent magnets. A dependable solid state voltage regulator is mounted on the generator. Start-up speed is about 6 mph, with peak output at 27 mph. No routine maintenance is required. The 20-lb. lightweight design is largely cast aluminum, with stainless steel hardware for corrosion-free longevity. All Windseeker models mount on standard 2″ steel pipe. An elegantly simple tilt-back governor is used to prevent overspeed in high winds, yet allows useful charging to continue. Wind speeds up to 120 mph can be survived unattended.

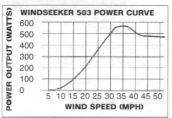

Windseeker features quieter, smoother, longer-lasting 3 basswood blades with a tough urethane two-coat finish, and polyurethane UV tape to protect the leading edge. Magnets are nickel-plated for corrosion resistance, and the entire assembly is powder-coated white. Voltage must be specified at time of order, 12V and 24V are standard. See Wind Accessories for tower and anchor options. Two year mfr.'s warranty. USA.

Rated Wattage Output: 500 @ 27 mph Wind Speed
Available Voltages: 12 or 24 volts regulated DC
Max. Design Wind Speed: 120 mph
Cut-in Wind Speed: 6 mph
Number of Blades: 3
Rotor Diameter: 5.0 ft.
Tower Top Weight: 20 lb
Generator Type: Brushless Permanent Magnet Alternator
Tower Type: 2″ steel pipe

◈**16-139 Windseeker 503, 3-blade $1,075⁰⁰**

◈*Means shipped directly from manufacturer.*

Southwest Windpower Air 403
Our Favorite Small Turbine Gets Upgraded

In early '99 Southwest Windpower released the improved second-generation Air 403. The non-corrosive swoopy cast aluminum body remains the same. Refinements include a 1.5″ tower socket that tilts upward at 4.5° for increased blade-to-tower clearance and all new die castings for improved fit and finish. Blade airfoil shape, neodymium magnet alternator, and charge controller have all been upgraded. The new alternator boosts output by 30% to 400 watts, and blade tip speed has been reduced for quieter operation. Blades are flexible, presenting the most aggressive airfoil shape at lower speeds, then gradually twisting as speed increases. At wind speeds above 35 mph the blades will act as brakes, maintaining a steady speed (although things get increasingly noisy at higher speeds in our own tests). The built-in charge controller, which can stand alone and be adjusted for any battery type or be run through a conventional PV controller, has always been unique. Now it is more robust, with protection against static, line spikes, and EMI. It will sense battery voltage, and electromagnetically slow the blades when batteries approach full charge. This makes for less noise and less bearing wear.

Standard Air 403 models are unpainted, the marine versions are white powder-coated, with special bearing and seal upgrades for protection in corrosive environments, both models feature a red LED under the belly to indicate output. The stop switch shorts turbine output to ground, to stop or greatly slow turbine in extreme high winds. This option is particularly recommended for marine installations if you plan on sleeping aboard. Surface-mounted circuit breakers provide basic circuit protection. 12-volt systems need 75A breaker, 24- and 48-volt systems can use 50A breaker. See Wind Accessories for tower kits, anchor augers, and marine mounting kits.

All Air 403 turbines have a 3-year mfr.'s warranty. USA.

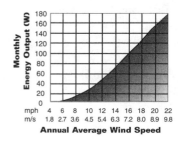
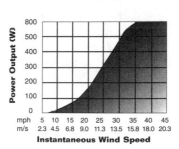

(See following page for Air 403 specifications and prices.)

Rated Wattage Output: 400 @ 28 mph wind speed
Available Voltages: 12, 24, or 48 volts regulated DC
Max. Design Wind Speed: 100 mph
Cut-in Wind Speed: 7 mph
Number of Blades: 3
Rotor Diameter: 46 inches
Tower Top Weight: 13 lb
Generator Type: Brushless Permanent Magnet Alternator
Tower Type: 1½" steel pipe

Specify voltage (12, 24, or 48) at time of order.

16-151 Air 403 Turbine Std. $595⁰⁰

16-152 Air 403 Turbine Marine $845⁰⁰

16-153 Air 403 50-Amp Stop Switch $19⁹⁵

25-167 Blue Seas 50-Amp Circuit Breaker $34⁹⁵

25-168 Blue Seas 75-Amp Circuit Breaker $34⁹⁵

Air 403 Industrial

The Industrial model is specially designed for survival in the most extreme environments; on remote mountain-

tops and off-shore platforms. It will survive winds in excess of 100 mph and temperatures down to –60°F for extended periods without attention. Features of this fully powder-coated model include anodized cooling fins for continuous high-power output, larger internal wiring, and special carbon fiber blades which are twice the strength of standard Air blades. It's designed for hybrid solar/wind systems in telecommunication stations, offshore platforms, monitoring stations, lighthouses, and cathodic protection systems. Three-year mfr.'s warranty. Industrial models are made to order. Allow 6 to 8 weeks delivery time. USA.

Specify voltage (12, 24, or 48) at time of order.

❖16-150 Air 303 Industrial $995⁰⁰

16-153 Air 403 50-Amp Stop Switch $19⁹⁵

25-168 Blue Seas 75-Amp Circuit Breaker $34⁹⁵

❖*Means shipped directly from manufacturer.*

Whisper Wind Turbines
Now with Complete Wiring Center and Diversion Control!

Whisper H40 Turbine   **Whisper H80 Turbine**

Whisper currently offers three turbine models that are designated by the square footage of blade sweep. Nothing measures output potential better than blade size. Big generators can't deliver when connected to a small blade. Whisper is trying to start an honest industry trend here.

The H40 model sweeps 40 square feet, and delivers a peak of 900 watts.

The H80 model sweeps 80 square feet, and delivers a peak of 1000 watts.

The 175 model sweeps 175 square feet, and delivers a peak of 3200 watts.

All Whisper models use modern, lightweight, brushless, direct-drive, low-speed alternators with start-up speeds of 7 to 7.5 mph. The 175 model provides over-speed protection by tipping the generator and blades back on their mount; the H models twist to the side. Both schemes simply move the blades at right angles to the incoming wind, return automatically when wind speed drops, and rely on nothing more complex than wind pressure and gravity. The fiber composite blades will last a lifetime without attention or repainting. The 3-blade design of the H-models offers better balance and less machine strain in shifting winds.

All Whisper turbines are delivered with the capable and good-looking E-Z wiring center, which includes a complete monitoring package with easy-to-read LED display, circuit breaker input for wind, a brake switch for the wind plant, and a built-in diversion controller with sufficient wattage to handle the turbine input.

H40 Turbine

The H40 model has a 7-foot (2.1 meter) blade diameter, and is designed for medium to higher wind speed sites. If you have 12 mph (5.4 m/s) or higher average winds, this turbine is for you. Got lower wind speeds? Check the H80 below.

The H40 features a neodymium permanent magnet alternator combined with the high-efficiency composite airfoil 3-blade design. The H40 is available in 12-, 24-, or 48-volt output, preset from the factory. Voltage is not field switchable on this turbine.

H80 Turbine

The H80 model has a 10-foot (3.0 meter) blade diameter, and is designed for low to moderate wind speed sites. Got wind speeds averaging 12 mph (5.4 m/s) or less? This 1000-watt turbine is for you. The large swept area does better at harvesting useful energy from low winds. The advanced 3-blade airfoil design is optimized for the lower wind speeds that most sites see daily. The H80 is available in 24- or 48-volt output, preset from the factory. Voltage is not field switchable on this turbine.

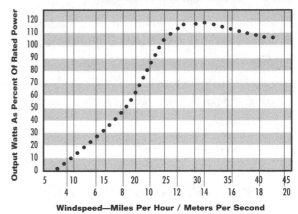

WHISPER 175 POWER CURVE

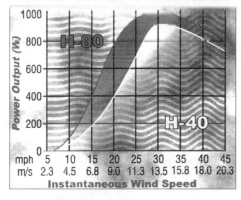

Tower kits and anchors are offered below in Wind Accessories. All Whisper turbines feature a two-year manufacturer's warranty. USA.

	Model H40	Model H80	Model 175
Available Voltage:	12-48 VDC	12-48 VDC	24-48 VDC
Rotor Diameter:	7 feet	10 feet	14.8 feet
Tower Top Weight:	47 lb	65 lb	155 lb
Tower Type:	2.5" steel pipe	2.5" steel pipe	5" steel pipe
Max. Design Wind Speed:	120 mph	120 mph	120 mph
Cut-in Wind Speed:	7.5 mph	7.0 mph	7.1 mph
Number of Blades:	3	3	2
Rated Wattage (@mph):	900@28	1000@24	3200@27
Generator Type:	Brushless Permanent Magnet Alternator		

◈**16-239** Whisper H40 w/wiring center, 24 or 48V **$1,495⁰⁰** *(specify voltage)*

◈**16-240** Whisper H40 w/wiring center, 12V **$1,795⁰⁰**

◈**16-241** Whisper H80 w/wiring center, 24 or 48V **$1,995⁰⁰** *(specify voltage)*

◈**16-242** Whisper 175 w/wiring center, 24 or 48V **$5,455⁰⁰**

◈*Means shipped directly from manufacturer.*

175 Turbine

The 175 is Whisper's most powerful turbine. The 14-foot (4.26 meter) 2-blade rotor features a hand-laid fiberglass construction with foam core for light weight and easy startup. This 3000-watt turbine has been upgraded with a stronger, larger diameter yaw shaft, new blade stabilizer straps, and a third spindle bearing for longer life. The 175 is shipped as 48-volt. It can be field switched to 24-volt.

EZ-Wire Universal Charge Controller

The Whisper 175 Turbine that's been cranking out power at the Solar Living Center for the past 6 years.

I'd like to tell you that your company was instrumental in my getting started in this industry. When I was in school, I used to get the old RG "Hard Corps" catalog and would memorize the thing front-to-back. After college, I knew exactly what I wanted to do. So—my thanks to Real Goods.

—Steve Wilke
National Sales Manager
Bergey Windpower Company

Bergey Wind Generators

Bergey is one of the oldest and most experienced American windplant manufacturers. Today, they are still thriving, growing, and delivering new models. Bergey turbines feature simple direct drive, passive controls, extensive corrosion protection, and freedom from routine maintenance.

Bergey wind generators are wonderfully simple: no brakes, pitch-changing mechanism, gearbox, or brushes. An automatic furling design forces the generator and blades partially out of the wind at high wind speeds, but still maintains maximum rotor speed. AUTOFURL uses only aerodynamics and gravity, there are no brakes, springs, or electromechanical devices to reduce reliability. These turbines are designed to operate unattended at wind speeds up to 120 mph.

No part of wind turbine gets worked harder than the blades. Bergey Windpower pioneered pultruded fiberglass blades in 1980. This award-winning process results in blades with a tensile strength exceeding 100,000 psi, twice as strong as steel. The patented POWERFLEX blades combine torsion flexibility with precisely located pitch weights near the tips. Aerodynamic and centrifugal forces act together to twist the blades toward the best angle for each wind speed.

Power is transmitted from the turbine to controller as 3-phase AC, making for easy transmission and much smaller wiring requirements. Bergey models include the most comprehensive Owner's Manuals of any wind turbine on the market, with site selection, anchoring, erecting, and tuning help. Recommended maintenance is this: once a year on a windy day, walk out to the tower and look up. If the blades are turning, everything is okay. The Bergey Windpower line features models at 1,000 watts, 1,500 watts, and the Excel at 7,500 watts when battery charging, or 10,000 watts when utility intertied. All Bergey turbines and controls carry a 5-year manufacturer's warranty, the best in the wind business.

Bergey XL.1 Wind Generator

The 1000-watt XL.1 is Bergey's newest turbine. With 24-volt output, this wind charger is intended for off-the-grid or remote-location battery charging. This 2.5-meter (8.2-foot) turbine is designed for high reliability, low maintenance, and automatic operation under adverse conditions. The XL.1 features Bergey's proven AUTOFURL storm protection, a direct-drive permanent-magnet alternator, excellent low-wind performance, and nearly silent operation thanks to the state-of-the-art, patent-pending, blade airfoils combined with the oversized, low-speed alternator. Low-wind-speed performance is enhanced by low-end boost circuitry that increases output at wind speeds down to 5.6 mph (2.5 m/s).

The BWC PowerCenter provides battery-friendly constant voltage charging, and will control up to 30 amps of PV input also. The simple controller shows battery voltage and which charging sources are active with at-a-glance LED indicators.

A direct-intertie, no-battery version of the XL.1 is under development and should be available for utility intertie by late 2001. Call us for details.

See the Wind Accessories section for complete XL.1 tower kits that include tubing, anchors, and will ship by UPS. Excellent 5-year mfg. warranty. China & USA.

Rated Wattage Output: 1000 @ 24.6 mph (11 m/s) Wind Speed
Available Voltage: 24 volts DC
Cut-in Wind Speed: 5.6 mph (2.5 m/s)
Max. Design Wind Speed: 120 mph (54 m/s)
Rotor Diameter: 8.2 feet (2.5 m)
Number of Blades: 3
Tower Top Weight: 75 lb (34 kg)
Tower Type: Conventional guyed section or freestanding
Generator Type: Brushless Permanent Magnet Alternator

❖**16-243 Bergey XL.1-24 Turbine w/controller $1,695⁰⁰**

❖*Means shipped directly from manufacturer.*

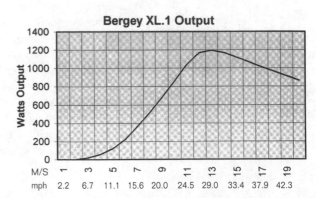

Bergey XL.1 Output

Bergey 1500-Watt Wind Generator

This modern 10-foot (3.0 m) wind turbine delivers 1500 watts for battery charging or water pumping. Uses the POWER-FLEX® non-weathering fiberglass blades discussed above, and Bergey's AUTO-FURL system that allows continued unattended output at speeds up to 120 mph. Battery charging units can be supplied at 12, 24, 48, or 120VDC. Voltage output and use must be specified when ordering. The required control unit is included and features a solid-state regulator that tapers the final charge and provides transient (lightning) protection between the turbine and batteries. Output is 3-phase AC for ease of transmission and smaller cable sizes. Special high-voltage output models with step-down transformers are available for 24-volt sites with exceptionally long transmission distances. See Wind Accessories for Bergey 1500 towers. Five-year mfr.'s warranty. USA.

Rated Wattage Output: 1500 @ 28 mph (12.5 m/s) Wind Speed
Available Voltages: 12 to 24 volts regulated DC
Cut in Wind Speed: 8 mph (3.6 m/s)
Max. Design Wind Speed: 120 mph (54 m/s)
Rotor Diameter: 10 feet (3.0 m)
Number of Blades: 3
Tower Top Weight: 168 lb (76 kg)
Tower Type: Conventional guyed section or freestanding
Generator Type: Brushless Permanent Magnet Alternator

◈ **16-201 Bergey 1500-xx Turbine w/controller**
$4,700⁰⁰ (specify voltage: 24, 48, or 120)

◈ **16-202 Bergey 1500-12 Turbine w/controller**
(12V) $4,900⁰⁰

◈ **16-244 Bergey 1500-120-24 Turbine w/controller & transformer (24V) $5,100⁰⁰**
Includes 4:1 step-down transformer for long wire runs, 24V output only

◈ **16-245 Bergey 1500-240-24 Turbine w/controller & transformer (24V) $5,400⁰⁰**
Includes 8:1 step-down transformer for very long wire runs, 24V output only

◈ *Means shipped directly from manufacturer.*

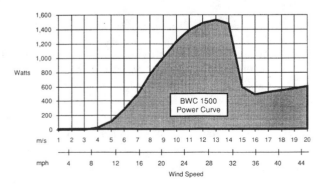

Bergey Excel 10,000-Watt Wind Generator

This modern 7-meter (23-foot) diameter, 10,000-watt wind turbine is designed for high reliability, low maintenance, and automatic operation under severe weather conditions. The Excel can be used for utility intertie, battery charging, or water pumping. Uses the extraordinarily strong and flexible POWERFLEX® non-weathering fiberglass blades discussed above, and Bergey's unique AUTOFURL system that allows continued unattended output at speeds up to 120 mph. The Excel was introduced in 1983, and has been installed at over 600 sites worldwide.

As a battery charger, the Excel is rated at 7,500 watts, and can be supplied with outputs at 24, 48, 120, or 240 volts DC. They are well suited for large rural homesteads, remote villages, eco-tourism resorts, and larger telecommunications sites.

Connected to the grid without batteries via the unique, UL-certified GridTek 10 inverter, the Excel is rated at 10,000 watts. All incoming wind energy is converted to 240 vac, and will displace utility power. Any extra energy beyond what the house can use at the moment is sold back to the utility company. Anyone living in a "net-metering" state can turn their meter backwards.

The water pumping model delivers 3-phase variable voltage, variable frequency output that will directly run a 240-volt, 3-phase AC pump without batteries. Wind-electric pumping systems allow the turbine to be some distance from the actual well, and perform well over a wide variety of wind speeds. Wind-electric pumping systems require less maintenance than old-style water pumpers.

Output voltage must be specified when ordering. Charge control or Gridtek inverter units are included in price. Output is 3-phase AC for ease of transmission and smaller cable sizes. See Wind Accessories for tower options on this turbine. Five-year mfr.'s warranty. USA.

Rated Wattage Output: 10,000 or 7,500 @ 31 mph (13.8 m/s)
Available Voltages: 24, 48, 120, or 240VDC, 240VAC/60Hz (or 220VAC/50Hz)
Cut-in Wind Speed: 7 mph (3.1 m/s)
Max. Design Wind Speed: 120 mph (54 m/s)
Governing System: Self Furling, auto return
Rotor Diameter: 7 meter (23 feet)
Number of Blades: 3
Tower Top Weight: 1050 lb (477 kg)
Tower Type: Conventional guyed section or freestanding
Generator Type: Brushless Permanent Magnet Alternator

(See following page for prices)

◈16-200 Bergey Excel-R/48 Turbine, 48VDC $19,400⁰⁰

◈16-246 Bergey Excel-R/24 Turbine, 24VDC $20,500⁰⁰

◈16-247 Bergey Excel-R/120 Turbine, 120VDC $18,400⁰⁰

◈16-248 Bergey Excel-R/240 Turbine, 240VDC $18,400⁰⁰

◈16-203 Bergey Excel-S/60 240VAC/60Hz $20,900⁰⁰

◈16-203 Bergey Excel-S/50 220VAC/50Hz $20,900⁰⁰

◈16-204 Bergey Excel-PD, for water pumping $18,400⁰⁰

◈Means shipped directly from manufacturer.

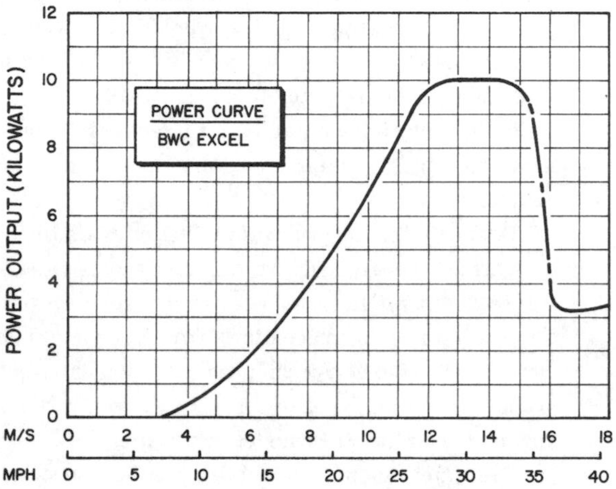

Wind Accessories

Tower Kits for Wind Turbines

The basic rule of thumb when installing any wind turbine: *Be 30-feet higher than any trees, buildings, or other obstructions within a 200-foot radius.* Anything that ruffles up the wind before it hits your turbine is going to reduce the power available, and it's going to beat up your turbine. You really want to be up in the clean, smooth airflow. Plus, there's more power available as you go up. At 100 feet there's about 40% more energy than at ground level. "The taller the tower, the greater the power."

Towers come in a variety of styles. The most basic division is between self-supporting towers, like the old water pumpers used, and guy-wire supported towers. Guyed towers are significantly less expensive, and are usually the tower of choice for smaller homestead turbines. Tubular guyed towers that use steel pipe or lighter weight steel tubing for the actual tower, with guy wire supports, have become the most common do-it-yourself tower type. This style completely assembles the tower, guy wire system, and turbine while safely on the ground, then stands it all up. Rigging and setup takes awhile during initial assembly, but standing up or folding down the tower only takes a few minutes, and everything happens with your feet firmly planted on the ground. Most of the tower kits offered below are stand-up tubular guyed towers.

Roof Mounting Kit For Air 403

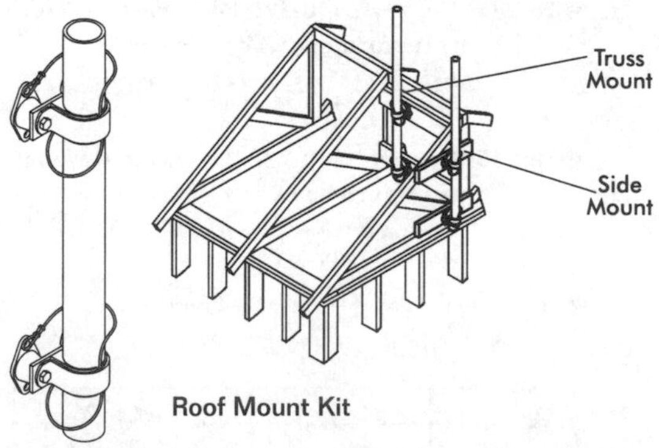

Roof Mount Kit

We don't recommend this mounting option for any building that people plan to sleep (or lose sleep) in. Wind turbines make noise, and it transmits down the tower very nicely. This kit includes vibration isolators, which helps, but won't entirely cure the noise problem. This kit also includes clamps, straps, safety leashes, bolts, and roof seal. Mounting pipe and lag bolts are not included. USA.

16-235 Air 403 Roof Mount w/Seal $99⁰⁰

Tower Kits For Air 403

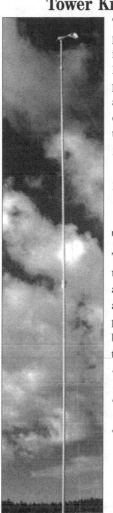

These low-cost, guyed, tubular tower kits provide all the hardware, cabling, and instructions to assemble a stand-up tower. Pipe and anchors are not included. Steel pipe can be purchased locally; guy anchors are offered below. 36″ anchors recommended with 25′ tower, 48″ anchors with 47′ tower. USA.

16-233 Air 403 25′ Guyed Tower Kit
$169⁰⁰

16-234 Air 403 47′ Guyed Tower Kit
$259⁰⁰

Tower Kits For Windseeker

These guyed, tubular tower kits provide all the hardware, cabling, and instructions to assemble a stand-up tower. Pipe and anchors are not included. Steel pipe can be purchased locally; guy anchors are offered below. 36″ anchors recommended with 25′ tower; 48″ anchors with 47′ tower. USA.

❖**16-125 Windseeker 25′ Guyed Tower Kit** $219⁰⁰

❖**16-126 Windseeker 47′ Guyed Tower Kit** $279⁰⁰

❖*Means shipped directly from manufacturer.*

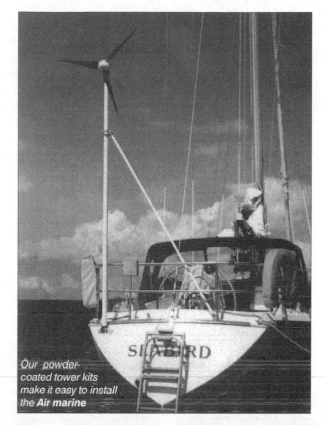

Our powder-coated tower kits make it easy to install the Air marine

Saddle Clamps   **Tilting Mast Base**   **Stay Base**

Mounting Kits For Air 403 Marine

A quality 9-foot mast mount that's built to withstand hurricane-force winds, and looks good too. Includes white powder-coated aluminum poles, all stainless steel hardware, self-locking nuts, and vibration dampening mounts. Main mast is 1.9″ od (nominal 1.5″ pipe), stays are 1″ od. Masts and hardware kits are sold separately for those who want to supply their own masts. USA.

❖**16-148 Air Marine Tower Hardware Kit** $169⁰⁰

❖**16-149 Air Marine Tower Mast Kit** $179⁰⁰

❖*Means shipped directly from manufacturer.*

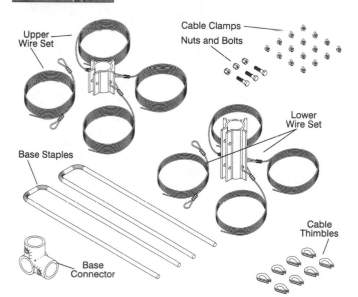

Upper Wire Set · Cable Clamps · Nuts and Bolts · Lower Wire Set · Base Staples · Base Connector · Cable Thimbles

Tower Kits include these pieces.

Tower Kits For Whisper H40 Or H80 Turbines

These guyed, tubular tower kits provide all the hardware, hinged base plate, cabling, and instructions to assemble a stand-up tower. Pipe and anchors are not included. Steel pipe can be purchased locally; guy anchors are offered below. See chart below for anchor recommendations. See Tower Sizing graphic for layout and amount of ground space required. All these towers will fit either the H40 or the H80 turbine, and require nominal 2.5″ pipe (which is actually 2.875″ od). USA.

All tower kits drop shipped from manufacturer.

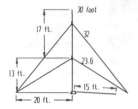

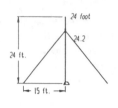

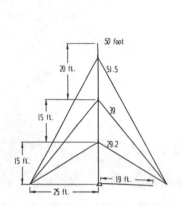

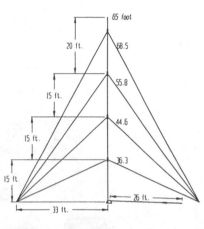

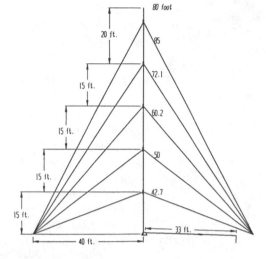

WHISPER H40 OR H80 TOWER KITS

Item #	Tower Height	Anchor Hole Dimensions	Guy Wire Radius	Auger Size	Price
◈16-256	24′	1.5′ dia., 3′ deep, 12″ concrete fill	15′	36″	$290⁰⁰
◈16-257	30′	2′ dia., 3′ deep, 9″ concrete fill	20′	48″	$480⁰⁰
◈16-258	50′	2′ dia., 3′ deep, 10″ concrete fill	25′	48″	$590⁰⁰
◈16-259	65′	2′ dia., 3′ deep, 12″ concrete fill	33′	60″	$760⁰⁰
◈16-260	80′	2′ dia., 4′ deep, 12″ concrete fill	40′	60″	$910⁰⁰

◈*Means shipped directly from manufacturer.*

Anchors For Wind Towers

These anchor augers are sold in sets of four. With smaller turbines like the Air 403 or the Windseeker you can simply screw them into place. No digging! With larger turbines you need to dig anchor holes, per the recommendations above, and fill the bottom with concrete. USA.

16-127 36″ Anchor Augers (4) $59⁹⁵

◈**16-253 36″ Galvanized Anchor Augers (4)** $175⁰⁰

◈**16-128 48″ Anchor Augers (4)** $69⁹⁵

◈**16-254 48″ Galvanized Anchor Augers (4)** $199⁰⁰

◈**16-255 60″ Galvanized Anchor Augers (4)** $270⁰⁰

◈*Means shipped directly from manufacturer.*

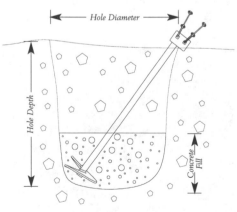

36-Inch Auger = 6-Inch Disk

48-Inch Auger = 8-Inch Disk

60-Inch Auger = 10-Inch Disk

Tower Kits For Whisper 175 Turbines

These guyed, tubular tower kits provide all the hardware, hinged base plate, cabling, and instructions to assemble a stand-up tower. Pipe and anchors are not included. Steel pipe can be purchased locally; guy anchors are offered below. See chart below for anchor recommendations. See Tower Sizing graphic for layout and amount of ground space required. These towers will fit only the 175 turbine, and require nominal 5˝ pipe (which is actually 5.563˝ od). USA.

All tower kits drop shipped from manufacturer.

WHISPER 175 TOWER KITS

Item #	Tower Height	Anchor Hole Dimensions	Guy Wire Radius	Auger Size	Price
◈16-249	30´	2´ dia., 3´ deep, 12˝ concrete fill	20´	48˝	$980⁰⁰
◈16-250	42´	2.5´ dia., 4´ deep, 12˝ concrete fill	25´	48˝	$1,060⁰⁰
◈16-251	70´	2.5´ dia., 5´ deep, 16˝ concrete fill	35´	60˝	$1,455⁰⁰
◈16-252	84´	3´ dia., 5´ deep, 12˝ concrete fill	45´	60˝	$1,700⁰⁰

Towers For Bergey XL.1 Turbines

These are complete tilt-up, guyed tower kits that include galvanized tubular tower sections with associated hardware, guy anchors, guy wires, and gin pole. Only the tower electrical wiring is needed to complete your installation.

Nine- and 13-meter towers have 2-meter long, 3.5-inch diameter tower sections. These tower kits are shipped via UPS. 19-, 25-, and 32-meter towers have 3-meter long, 4.5-inch diameter tower sections, and are shipped via common carrier (semi truck).

◈16-261 Bergey XL.1 Tower Kit, 9m (30´) $490⁰⁰

◈16-262 Bergey XL.1 Tower Kit, 13m (42´) $690⁰⁰

🚚 16-263 Bergey XL.1 Tower Kit, 19m (64´) $990⁰⁰

🚚 16-264 Bergey XL.1 Tower Kit, 25m (84´) $1,190⁰⁰

🚚 16-265 Bergey XL.1 Tower Kit, 32m (104´) $1,490⁰⁰

◈Means shipped directly from manufacturer.

🚚 Means shipped freight collect.

Guyed Lattice Towers For Bergey 1500 & Excel

These standard guyed towers are intended for conventional gin pole construction, or for crane installation. They cannot be tilted up or down with the weight of the turbine attached. Kits include galvanized lattice-type tower sections with associated hardware, guy anchors, guy wires, and associated hardware. Also includes tower grounding hardware, ground rods, and furling cable with furling winch. Tower wiring not included. USA.

🚚 16-218 1500 Tower Kit, 12m (40´) $1,500⁰⁰

🚚 16-219 1500 Tower Kit, 18m (60´) $1,900⁰⁰

🚚 16-220 1500 Tower Kit, 24m (80´) $2,500⁰⁰

🚚 16-221 1500 Tower Kit, 30m (100´) $3,000⁰⁰

🚚 16-274 1500 Tower Kit, 37m (120´) $3,900⁰⁰

🚚 16-211 Excel Tower Kit, 18m (60´) $5,400⁰⁰

🚚 16-212 Excel Tower Kit, 24m (80´) $6,000⁰⁰

🚚 16-213 Excel Tower Kit, 30m (100´) $6,900⁰⁰

🚚 16-214 Excel Tower Kit, 37m (120´) $8,100⁰⁰

◈Means shipped directly from manufacturer.

🚚 Shipped freight collect from Oklahoma.

Tilt-Up Guyed Towers For Bergey 1500 & Excel

These stronger guyed towers can be tilted up or down with the weight of the turbine attached. A little more expensive, but easier to work with if the turbine ever needs to come down for maintenance or an approaching hurricane. 1500 kits use galvanized 3-meter-long tubular tower sections. Excel kits use 6-meter-long lattice type tower sections. Both come with all guy anchors, guy wires, and associated hardware. Also includes tower grounding hardware, ground rods, and furling cable with furling winch. Tower wiring not included. USA.

🚚 16-266 1500 Tilt-Up Tower Kit, 16m (54´) $1,800⁰⁰

🚚 16-267 1500 Tilt-Up Tower Kit, 24m (80´) $2,800⁰⁰

🚚 16-268 1500 Tilt-Up Tower Kit, 31m (102´) $3,600⁰⁰

🚚 16-269 Excel Tilt-Up Tower Kit, 18m (60´) $7,700⁰⁰

🚚 16-270 Excel Tilt-Up Tower Kit, 24m (80´) $8,700⁰⁰

🚚 16-271 Excel Tilt-Up Tower Kit, 30m (100´) $10,000⁰⁰

🚚 16-272 Tilt-Up Tower Jackstand $270⁰⁰

🚚 16-273 Tilt-Up Tower Raising Kit $1,800⁰⁰

🚚 Shipped freight collect from Oklahoma.

Lightning Protection

Do you need lightning protection on a wind system? Is the Pope Catholic? You need serious industrial-grade protection here, and it's surprisingly simple and inexpensive.

The first level of protection depends on you to ground the tower, the guy wires, and make sure *All* the grounds in your wind and household system are connected together. Lightning wants to go to ground. Make it easy, and give it a path outside your system wiring. The second level of protection is needed when lightning decides to take a trip through your system wiring. Delta arrestors will clamp any high voltage spike, and direct most of the energy to ground. They will withstand multiple strikes until capacity is reached and then rupture, indicating the need for replacement. Delta arrestors can be installed at any convenient point, or points, between the turbine and the battery. Multiple arrestors are a good idea in lightning-prone areas. Lightning *does* strike twice!

The 3-wire DC model is for turbines like the Air and Windseeker with 2-wire DC output. Connect one wire to positive, one to negative, and one to ground. It can take up to 60,000 amps current, or 2000 joules per pole. Response time to clamp is 5 to 25 nanoseconds depending on amp load.

The 4-wire model is for turbines like the Whisper and Bergey with 3-wire wild AC output. Connect one wire to each of the output wires and one to ground. It can take up to 100,000 amps current, or 3000 joules per pole. Same clamp time as above. USA.

25-194 DC 3-wire Lightning Protector $44⁹⁵

25-749 AC 4-wire Lightning Protector $59⁹⁵

2100 Totalizer Anemometer

The 2100 Totalizer is a moderately-priced instrument that accurately determines your average windspeed. The Totalizer is an odometer that counts the amount of wind passing through the anemometer, and you find the average windspeed in miles per hour by dividing miles of wind over elapsed time. An excellent choice for determining the average wind speed at your site. Read it once a day or once a month; so long as you note when the last reading was taken, you can determine the average. A 9-volt alkaline battery (included) provides one year of operation. The readout can be mounted in any protected environment. Includes a Maximum #40 anemometer, 60 feet of sensor cable, battery, stub mast for mounting, and instructions.

63-354 The 2100 Totalizer $299⁰⁰

Handheld Windspeed Meter

This is an inexpensive and accurate windspeed indicator. It features two ranges, 2 to 10 mph and 4 to 66 mph. A chart makes easy conversions to knots. Speed is indicated by a floating ball viewed through a clear tube. Many of our customers are curious about the wind potential of various locations and elevations on their property, but are reluctant to spend big money for an anemometer just to find out. Here is an economical solution. Try taping this meter to the side of a long pipe, (use a piece of tape over the finger hole for high range reading) and have a friend hold it up while you stand back and read the meter with binoculars. This is very helpful in determining wind speeds at elevations above ground level. Incudes protective carrying case and cleaning kit.

63-205 Windspeed Meter $20⁰⁰

Fuel Cells

Power Source of Our Bright and Shining Future?

This is still a very young, very rapidly developing technology. The first commercial residential units probably won't show up till about 2003 or 2005, and they're liable to be fairly expensive.

Fuel cells seem to be the darlings of the energy world lately. If we believe all the hype, they'll bring clean, quiet, reliable, cheap energy to the masses, allowing us to continue an energy-intensive lifestyle with no penalties or roadblocks. Are fuel cells going to save our energy-hog butts? Maybe. They sure will help. Can you buy one yet? Maybe. It helps if you're the Department of Defense, or have a fat government grant or contract in your back pocket. This is still a very young, very rapidly developing technology. The first commercial residential units probably won't show up till about 2003 or 2005, and they're liable to be fairly expensive. Want a bit more background info before you order one? Good plan, read on.

Traditional energy production, for either electricity or heat, depends on burning a fuel source like gasoline, fuel oil, natural gas, or coal to either spin an internal combustion engine, or to heat water, then piping the resulting steam or hot water to warm our buildings, or run the turbines to make electricity. Burning fuels produce some byproducts we'd be better off not letting loose into the environment, and every energy conversion costs some efficiency. The increasing industrialization of the world is requiring more and more fuel, creating our current unsustainable run on the world's energy reserves. The ultimate effect of all this fossil fuel burning is to create global climate change and global warming. Most scientists predict that if we fail to curb our fossil fuel consumption, the average world temperature will rise 4° to 10°C in the next 50–100 years, and the oceans will rise three to eight feet in the same time.

Fuel cells are chemical devices that go from a source, like natural gas, straight to heat and electrical output, without the combustion step in the middle. Fuel cells increase efficiency by two- or three-fold, and dramatically reduce the unintended byproducts. A fuel cell is an electrochemical device, similar to a battery, except fuel cells operate like a continuous battery that never needs recharging. So long as fuel is supplied to the negative electrode (anode), and oxygen or free air is supplied to the positive electrode (cathode), they will provide continuous electrical power and heat. Fuel cells can reach 80% efficiency when both the heat and electric power are utilized.

How Fuel Cells Operate

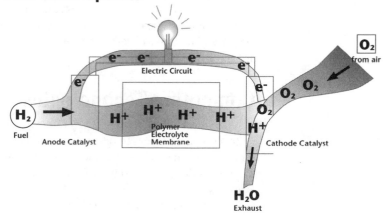

A fuel cell is composed of two electrodes, sandwiched around an electrolyte material. Hydrogen fuel is fed into the anode. Oxygen (or free air) is fed into the cathode. Encouraged by a catalyst, the hydrogen atom splits into a proton and an electron. The proton passes through the electrolyte to reach the cathode. The electron takes a separate, outside path to reach the cathode. Since electrons flowing through a wire is commonly known as "electricity," we'll make those free electrons do some work for us on the way. At the cathode, the electrons, protons, and oxygen all meet, react and form water. Fuel cells are actually built up in "stacks" with multiple layers of electrodes and electrolyte. Depending on the cell type, electrolyte material may be either liquid or solid.

Like other electrochemical devices such as batteries, fuel cells eventually wear out and the stacks have to be replaced. Stack replacement cost is typically 20% to 25% of the initial fuel cell cost. Current designs have run times of 40,000 to 60,000 hours. Under continuous operation that's 4.5 to 6.8 years.

What's "Fuel" to a Fuel Cell?

In purest form, a fuel cell takes hydrogen, the most abundant element in the universe, and combines it with oxygen. The output is electricity, pure water (H_2O), and a bit of heat. Period. Very clean technology! No nasty byproducts and no waste products left over. There are some real obvious advantages to using the most abundant element in the universe. We've got plenty of it on hand, and, as most anyone can clearly see, a hydrogen-based economy is far more clean and sustainable than a petroleum-based economy.

On the downside, hydrogen doesn't usually exist in pure form on earth. It's bound up with oxygen to make water, or with other fuels like natural gas or petroleum. If you run down to your friendly local gas supply to buy a tank of hydrogen, what you get will be a byproduct of petroleum refining.

Since hydrogen probably isn't going to be supplied in pure form, most commercial fuel cells have a fuel-processing component as part of the package. Fuel processors, or "reformers" do a bit of chemical reformulation to boost the hydrogen content of the fuel. This makes a fuel cell that can run on any hydrocarbon fuel. Hydrocarbon fuels include natural gas, propane, gasoline, fuel oils, diesel oil, methane, ethanol, methanol, and a number of others. Natural gas or propane are favored for stationary generators, and methanol or even gasoline have been used in experimental automotive fuel cells. But if you feed the fuel cell something other than pure hydrogen, you're going to get something more than pure water in the output, carbon dioxide usually being the biggest component. Fuel cells emit 40% to 60% less carbon dioxide than conventional power generation systems using the same hydrocarbon fuel. Other air pollutants such as sulfur oxides, nitrogen oxides, carbon monoxide, and unburned hydrocarbons are nearly absent, although you'll still get some trace byproducts we'd be better off without.

The dream is to build a fuel cell that accepts water input. But it takes more energy to split the water into hydrogen and oxygen than we'll get back from the fuel cell. One potential future scenario has large banks of PV cells splitting water. The hydrogen is collected, and used to run cars and light trucks. Efficiency isn't great, but nasty byproducts are zero.

Fuel cells emit 40% to 60% less carbon dioxide than conventional power generation systems using the same hydrocarbon fuel.

Competing Technologies

Just like there's more than one "right" chemistry with which to build a battery, there are quite a number of ways to put together a functional fuel cell. As an emerging technology there are several fuel cell chemical combinations receiving experimental interest, substantial funding, and absolutely furious development. At the risk of boring you, or scaring you all away, we're going to list the major chemical contenders for fuel cell power, and their strong or weak points.

Phosphoric Acid

The typical phosphoric acid fuel cell is about the size of a freight car. This one's in New York's Central Park.

This is the most mature fuel cell technology. Quite a number of 200-kilowatt demonstration/experimental units are in everyday operation at hospitals, nursing homes, hotels, schools, office buildings, and utility power plants. The Department of Defense currently runs about 30 of these with 150- to 250-kilowatt output. If you simply *must* have a fuel cell in your life right now, this is your baby. They are available for sale . . . for a price. Currently that price is about $3,000 per kilowatt. These are large, stationary power plants, usually running on natural gas. Phosphoric acid cells do not lend themselves well to small-scale generators. Locomotives and buses are probably about as small as they will go. Operating temperatures are 375°–400°F, so they need thermal shielding. Efficiency for electric production alone is about 37%; if you can utilize both heat and power, efficiency hits about 73%. Stack life expectancy is about 60,000 hours.

Proton Exchange Membrane

PEM cells are smaller, run at lower temperatures, and are more consumer-friendly.

PEM is the fuel cell technology that is probably seeing the most intense development, and is the type of fuel cell you are most likely to see in your lifetime. Because it offers more power output from a smaller package, low operating temperatures, and fast output response, this is the favored cell technology for automotive and residential power use. Thanks to their low noise level, low weight, quick start-up, and simple support systems, experimental PEM fuel cells have even been produced for cell phones and video cameras. Recent advances in performance and design raise the tantalizing promise of the lowest cost of any fuel cell system. Current prices are about $3,000 per kilowatt, but they will drop as mass-production economies begin to kick in. Operating temperature is a very low 175°–200°F, so minimal thermal shielding is needed. Efficiency for electric production alone is about 36%, if you can utilize both heat and power, efficiency hits about 70%. PEM cells can start delivering up to 50% of their rated power at room temperature, so start-up time is minimal. Stack life expectancy is about 40,000 hours.

There are some small demonstration PEM fuel cells available for sale currently, but for something that's going to run your house you'll need some patience. Several manufacturers have residential units under development. We may see the first units available by 2001 or 2002, although this technology isn't expected to fully mature till sometime after 2005. For current information visit the nonprofit Fuel Cells 2000 website at www.fuelcells.org.

Molten Carbonate and Solid Oxide

These two fuel cell technologies aren't related, except that they are both large baseload type cells that utility companies can use to supply grid power. These technologies have barely delivered their first prototypes, so operational plants are still far in the future. Very high operating temperatures, 1,200° to 1,800°F, are the norm for these cells. Electrical generation efficiencies are over 50%, and pollution output is way, way down, so these technologies are going to be very attractive to utility companies. The high operating temperatures mean these cells don't turn on or off casually. Start-up can take over 10 hours. These aren't fuel cells for off-the-grid homes.

Cruisin' the City on Fuel Cells?

One of civilization's biggest fossil fuel uses, and the one we seem to have the hardest time weaning ourselves from, is automobiles. Do fuel cells offer any hope for hopeless driving addicts? Yes, they do. Every major auto manufacturer has a well-funded, active fuel cell development program. Several manufacturers, Daimler-Chrysler, Toyota, and Honda in particular, have several generations of prototypes behind them already. The stakes are high. The company that develops a workable fuel cell-powered auto holds the keys to the future. Several manufacturers are aiming for the first commercial models to be available by 2003 or 2004, with full-scale production a couple years later. These will probably be methanol-fueled PEM cells with a modest battery or capacitor pack to help meet sudden acceleration surges.

One of the primary problems has been the size of the fuel cell stack required. Mercedes' first prototype was a small bus; its second prototype was a van; and its latest, fifth, prototype is in the new compact A-series sedan. Each prototype just has room for a driver and passenger. So cell stacks are getting smaller and more efficient, but we could still stand some improvement (or an expectation adjustment).

Crash-safe hydrogen storage in a vehicle also continues to be a trying problem. We demand much safer vehicles than we used to. The first generations of fuel cell vehicles will be using on-board reformers with methanol or other liquid fuels, so this isn't a problem that requires an immediate solution.

How Do Fuel Cells Compare to Solar or Wind Energy?

As an electrical supplier, a fuel cell is closer kin to an internal combustion engine than to any renewable energy source. Think of a fuel cell as a vastly improved generator. It burns less fuel, it makes less pollution, there are no moving parts, and it hardly makes any noise. But unless you have a free supply of hydrogen, it will still use nonrenewable fossil fuels for power, which will cost something.

Photovoltaic modules, wind generators, hydro generators, and solar hot water panels all use free, renewable energy sources. No matter how much you extract today, it doesn't impact how much you can extract tomorrow, the next day, and forever. This is the major difference between technologies that harvest renewable energy, and fuel cells, which are primarily going to continue using nonrenewable energy sources.

In the long run, we think that fast-starting PEM fuel cells will take the place of backup generators in residential energy systems. They'll be in place to pick up the occasional shortfall in renewable-powered off-the-grid systems, or to provide backup AC power for grid-supplied homes when the utility fails. If your home is off-the-grid, then solar, wind, or hydro is still going to be the most cost-effective, and environmentally responsible primary power source. In an ideal world, every backup power fuel cell would be supplied by a small PV-powered hydrogen extractor that cracks water all day and stores the resulting hydrogen for later use. Guess we'd better get busy designing hydrogen extractors. . . .

—Doug Pratt

CHAPTER 3

FROM PANEL TO PLUG
or How to Get Your Renewable Energy Source Powering Your Appliances

A Renewable Energy System Primer

How All the Pieces Fit Together

IN OTHER CHAPTERS AND SECTIONS OF THE SOURCEBOOK we give you details of how individual bits like PV modules or inverters work, and why they're needed. This is the section where all the bits and pieces come together. Read this section first to get an overall idea of how everything works, and then move on to details as needed.

Stand-Alone Systems (Battery Back-up)

All explanations in this section will revolve around systems that use batteries and are capable of stand-alone operation. If your interest lies in simple utility intertie systems, with no battery backup, please see the separate section dealing with utility intertie, which is immediately following this primer.

It All Revolves Around the Battery

The battery, or battery pack, is the center of every stand-alone system. Everything comes and goes from the battery, and much of the safety and control equipment is designed to either protect the battery, or protect other equipment from the battery. The battery is your system reservoir, it stores up electrons until they're needed to run something. The size of your reservoir determines how many total electrons you can store, but has minimal effect on how fast electrons can be poured in or out, at least in the short term. A small battery will limit how many hours or days your system will run without recharging. A larger battery will allow more days of run time without recharging, but there are practical limits in each direction. Batteries are like the muscles of your body, they need some exercise to stay healthy, but not too much exercise or they get depleted. We usually aim for three to five days worth of battery storage. If you have less than three days worth of battery storage, your system will be cycling the battery deeply on a daily basis. With more than five days worth of storage, your battery starts getting more expensive than a backup generator, or other recharging source.

Battery Bank

Once you start removing electrons from your battery, you'd better have a plan for replacing them, or you'll soon have an empty reservoir—what we call a dead battery—which brings us to . . .

The Charging Source

Batteries aren't the least bit fussy about where their recharging electrons come from. So long as the voltage from the charging source is a bit higher than the battery voltage, the electrons will flow "downhill." So you can haul your battery into town to get it charged, or you can connect it to your vehicle charging system and charge it on the way to town and back. But batteries are heavy and the sulfuric acid in them just loves to eat your clothes, so hauling batteries around gets old, real fast. You'll want an on-site charging source. This can be a fossil-fueled (gasoline or propane) generator, a wind turbine, a hydroelectric generator, solar modules (PVs), a stationary bicycle generator, or combinations of the above.

So long as incoming energy keeps up with outgoing energy, everybody's happy. Up to about 80% of full charge, a battery will accept very large current inputs with no problem. As the battery gets closer to full, we have to start applying some charge control to prevent overcharging. Batteries will be damaged if they are severely or regularly overcharged—which brings us to . . .

Keeping Your Batteries Under Control

Someplace in your system there needs to be a charge controller. Charge controllers use various strategies, but they all have a common goal—to keep the battery from overcharging and toasting. A battery that's regularly overcharged will use excessive amounts of water, may run dry, and sheds flakes of active lead from the battery plates, reducing future capacity. A charge controller watches the battery voltage, and if it gets too high, it will either open the charging circuit, or dump excess electrons into a heater element. We use different strategies depending on what the charging source is. Solar modules can use simple controls between the solar array and the battery that simply open the circuit if voltage gets too high. Rotating generators like wind and hydroelectric turbines use a controller that diverts excess energy into a heater element. (Really bad things happen if you open the circuit between a generator and the battery while the generator is spinning!)

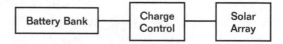

Now you have a battery, a way to recharge it, a controller to prevent overcharging, but you don't know how full or empty it is. You need a system monitor.

Monitoring Your System

System monitors all have a common goal: to keep you informed about your system performance, and prevent the exceptionally deep discharges that sap the life out of your batteries. There's quite a lot of variety in how they try to accomplish this. They can be as simple as a red or green light, or as complicated as a PC-linked system that logs dozens of system readings every few minutes. At the simplest level, displaying the battery voltage gives an approximate indication of how full or empty your battery is, and what's going on at the moment. More complete monitors may show how many amps are flowing in or out at the moment, and the best monitors keep tabs on all this information, then give it to you as a simple "% of full charge" reading. Many charge controllers offer some simple monitoring abilities, but the best monitors stand alone, and get mounted in the kitchen or living room, where they're more likely to get some regular attention.

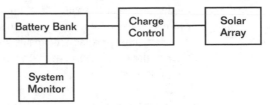

Can you run a system without a monitor? Sure, but it's not a smart idea. You could run your car without any gauges or meters too, but you'd be much more likely to exceed speed limits, run out of gas, or do accidental damage. System monitors are a real bargain, compared to batteries.

Now you have a battery, a way to recharge it, a controller to prevent overcharging, and a system monitor to tell how full or empty the battery is. Now you need a way to use some of your stored energy.

Using Your Power

Batteries deliver direct current, or DC, when you ask them for power. You can tap this energy to run any DC lights or appliances directly. This is fine if you're running an RV or a boat; vehicles pretty much run on DC, thanks to their automotive heritage.

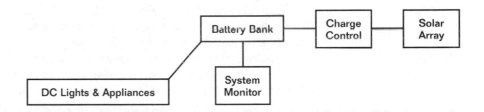

But what if you want to run a household, or a microwave in your RV? Household appliances run on 120 volt AC, or alternating current. AC power is easier to transmit over long distances, so it has become the world standard. Your AC appliances won't run on DC from your battery. You could convert all your household lights and appliances to DC, but there are some definite conveniences and savings to be had by joining the mainstream. AC appliances are mass-produced, making them less expensive, more universally available, and generally better quality, when compared to "RV appliances." But AC power can't be stored, it has to be generated as demanded. To meet this highly variable demand, you'll need a device called an *inverter*. The inverter converts low-voltage DC from the battery storage, into high-voltage AC as your household lights and appliances demand it.

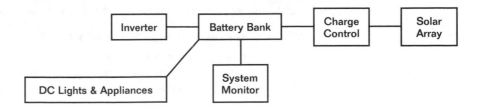

Inverters come in a variety of wattage outputs from 50 watts up through hundreds of thousands of watts, and range in cost from under $50 to more than you want to know. You want one that'll deliver enough wattage to start and run all the AC appliances you might turn on at the same time, but no bigger than necessary. Bigger inverters cost more, and use a bit more power on standby, waiting for something to turn on. With inverters, cheap usually means leaving out some protection equipment, which means a shorter life expectancy. (We've seen really cheap inverters blow up the instant a compact fluorescent lamp was turned on.)

Many of the larger, household-size inverters come with built-in battery chargers that come on automatically anytime outside AC power becomes available by starting a generator or plugging into an RV park outlet. This makes a simple backup charging source for those times when the primary solar or wind charger just isn't keeping up.

Now you have a battery, a way to recharge it, a controller to prevent overcharging, a system monitor to tell how full or empty it is, and a way to use your stored energy. There's only one thing missing, and it's the most important part, safety!

How to Avoid Burning Down Your House

For the most part, low-voltage systems and batteries are a lot safer than jittery high-voltage AC systems. Just keep your tongue out of the DC sockets and you'll be fine. However, if something does short-circuit a battery, look out! Because they're right close by, batteries can deliver more amps into a short-circuit than AC systems, strung over many miles, can usually manage. So common sense, and the Electric Code, says that you really must put some overcurrent protection on any wire attached to a battery. "Overcurrent protection" is code-speak for a fuse or circuit breaker. These fuses or circuit breakers must be rated for DC use. A DC short-circuit is harder to interrupt than AC shorts, so DC-rated equipment usually has some extra stuff inside to snuff any DC arc. It isn't just an excuse to raise the price. DC-rated fuses and circuit breakers are tougher, and you really, really want them that way. You only need to see wires glowing red with burning insulation dripping off them once, to appreciate how appropriate fuses are.

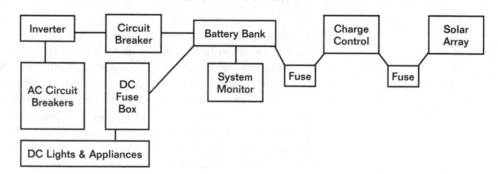

Now you have a battery, a way to recharge it, a controller to prevent overcharging, a system monitor to tell how full or empty it is, a way to use your stored energy, and the safety equipment to prevent burning down your house. That's a complete system!

What System Voltage Should I Use?

Renewable energy systems can be put together with the battery bank running at 12, 24, or 48 volts. The more energy your system processes per day, the higher your system voltage is likely to be. As systems get bigger, it's easier to push the power around on the DC side if the voltage is higher. If you double the voltage, you only need to push half the amperage through the wires to perform a given job. There are no hard and fast rules here, but generally systems processing up to 2,000 watt-hours per day are fine at 12 volts. From 2,000 to about 7,000 watt-hours per day things will work easier at 24 volts, and from 7,000 watt-hours on up we prefer to run at 48 volts. These are guidelines, not rules. We can make a system run at any of these three common voltages, but you'll buy less hardware and big wires if system voltage goes up as the watt-hours increase.

Ready . . . Set . . . Build!

Are you ready to put a renewable energy system together? Of course not. You've probably still got a lot of questions about how big, how many, what brand, and more. Each of the individual *Sourcebook* sections takes on a particular part of your RE system. We have sections on Batteries, Controllers, Monitors, Safety & Fusing, Inverters, Photovoltaics, Wind, Hydro, Wiring, and a system sizing worksheet. Each of these sections explains a particular RE component in more detail, and helps you make the best choice for your individual needs.

And finally, if you aren't the technical type, and don't take discovery joy from figuring this stuff out yourself—or even if you do, but would appreciate a little guidance—there's our highly experienced technical staff, which we think is the best in the business. We're available for consultation, system design, and troubleshooting Monday through Saturday, 7:30 a.m. to 6:00 p.m.

—Doug Pratt

Utility Intertie Systems

How to Sell Power to Your Electric Utility Company (*and get yours for free!*)

Utility electric costs are rising, blackouts are increasing, and the power grid we once took for granted is becoming increasingly more unreliable. Many of our customers are turning to the reliability and consistency of solar power and utility intertie systems. Solar is clean, reliable, nonpolluting energy that is quickly becoming cost-effective with payback periods approaching six years where utility rates are high and rebate programs are in force. State and government rebate programs may reduce initial costs by as much as 50%. The time for solar is *now*.

A utility intertie system makes it possible to generate your own solar power and sell it back to your utility company. Your utility electric meter will run backwards anytime your renewable energy (RE) system is making more power than your house needs at the moment. If the meter runs backwards, your utility company is buying power from you, and at the same retail cost that you pay them. It feels great! Even if your RE system isn't making more energy than your house needs at the moment, any watt-hours your RE system delivers will displace an equal number of utility watt-hours, directly reducing your utility bill, and reducing your exposure to the fluctuating and often increasing utility rates that are becoming the norm as utility deregulation proceeds.

The Legalities of Electricity and Net Metering

The 1978 federal Public Utilities Regulatory Policies Act (PURPA) states that any private renewable energy producer in the USA has a right to sell power to their local utility company. The federal law doesn't state that utilities have to make this easy, or profitable however. Under PURPA, the utility will usually pay their "avoided cost," otherwise known as wholesale rates. This law was an outgrowth of the Arab oil embargo in the early '70s and was intended to encourage renewable energy producers. Two to four cents per kilowatt-hour was not sufficiently encouraging to have much of an effect on home utility intertie installations. The next wave of encouragement started arriving in the mid-'80s with *net metering laws*. There is no blanket federal net metering law (yet).

As of early 2001, 33 states have net metering laws that allow small-scale renewable energy producers to sell excess power to their utility company, through the existing electric meter, for the same price that the utility company charges the homeowner for power. Details of these state laws vary widely. For an up-to-date summary see the Department of Energy website at www.eren.doe.gov/greenpower/netmetering/nmtable.shtml, or call them at 800-363-3732. We have also listed a directory of State Energy Offices at the end of this section.

How Utility Intertie Systems Work . . .

An intertie system uses an inverter that takes any available solar and wind power, turns it into conventional AC, and feeds it into your electrical circuit breaker panel. Electricity flow is much like water flow; it runs from higher to lower voltage levels. When you turn on an electrical appliance it slightly lowers the level, letting current flow in from the utility. As long as the electricity level in your house is being maintained by the intertie inverter, no current flows in from the utility. You are displacing utility power with renewable power (in most cases solar power). If your house demands more current than the intertie system is delivering at the moment, utility power makes it up. If your house is using less current than the intertie system is delivering, the excess flows out through your meter, turning it backwards, back into the grid. This treats the utility like a big 100% efficient battery. You sell it power during the day, it sells power back to you at night for the same price. If you live in one of the majority of States that allow net metering, then everything goes in and out through a single residential meter. Very simple, and extremely encouraging to renewable energy producers.

Keep It Safe!

Utility companies are naturally concerned about any source that might be feeding power into their network. Is this power clean enough to sell to your neighbors? Is the frequency, voltage, and waveform within acceptable specifications? And most importantly, what happens if utility power goes off? Utility companies take a very dim view of any power source that might send power back into the grid when there's a utility power failure, as this could be a serious threat to powerline repair workers.

All present-day intertie inverters have enough precise output specifications, voltage limits, and automatic shut-off protection features to allow even the most paranoid utility engineer to enjoy a good night's sleep. If the utility fails, the inverter will disconnect in no more than 30 milliseconds! Most utility companies agree this is a faster response than their repair crews could perform. Even an "island" situation will be detected and shutoff within two minutes. An island happens (theoretically) when your little neighborhood is cut off, isolated, and just happens to require the amount of energy that your intertie inverter is delivering. Utility engineers wake up in cold sweats worrying about things like this.

Because small-scale intertie is a new and evolving development, some smaller utility companies may not yet have worked out their official policies and procedures. Most simply follow the lead of Pacific Gas & Electric, the largest utility company in the U.S. (now in Chapter 11 bankruptcy). If PG&E approves a particular product, it's probably safe enough for the rest of us.

Still, some folks have run into stone walls when they try to get approval for small-scale residential intertie systems with their friendly local utility. If you've already tried to work through the proper channels and have run into brick walls, then be assured that neither you nor the local utility has anything to fear from connecting that intertie system and turning it on. Safety has been the top concern during development of intertie technology. These inverters just can't cause any problems for the utility. The worst problem they could cause the homeowner is to shut off the intertie system because utility power drifts outside specifications.

A tacit unspoken approval has also evolved for really small intertie systems of 250 watts or smaller that will rarely be turning the meter backwards. Most utilities would rather not be bothered with such small projects. So long as the inverter is intended for intertie and absolutely, positively, shuts off with the utility, and has UL approval, you can feel safe and secure simply turning it on.

Be aware that a small number of electric meters will not support net metering. Some meters are equipped with a ratchet that will only allow them to turn forward. Some of the newer wireless remote-reading data collection meters simply count how many times the hash mark on the wheel goes by; they just assume it's moving forward. Also remember that not all states have net metering laws for small-scale renewable energy producers.

Do I Need Batteries or Not?

Utility intertie inverters come in two basic configurations: those that support backup batteries, and those that don't. This is a major fork in the road of system design, as the two system types operate very differently, and have big price differences. To help you choose wisely, read on . . .

Battery-based systems are basically a large stand-alone renewable energy system, using a sophisticated inverter that can also intertie with the utility. These are the highly capable Trace SW-series inverters. These inverters do not come in small or medium sizes. They will allow selected circuits in your house to continue running on battery or solar power if utility power fails. With inverter, safety, control, battery, and a modest PV array, these systems start out at about $7,500.

Direct intertie systems are much simpler, usually consisting of PV module(s), mounts, inverter, a circuit breaker, and wire to connect everything. Without batteries, there's less safety and control equipment required, and more of your investment tends to end up buying PV power. But if utility power goes off, so does your intertie system. Direct intertie inverters come in medium and large sizes from 1000 watts up to 2,500 watts. Direct intertie systems become meaningful at $5,000 and up. Direct intertie inverters are produced by Trace, ISMA, and Advanced Energy Systems (see pages 138-141).

Which type of intertie system is right for you? That primarily depends on how reliable utility power is for you. If you live in an area where storms, poor maintenance, or mangled deregulation schemes tend to knock out power regularly, then a battery-based system is going to greatly increase your security and comfort level. Looking out over the darkened neighborhood from inside your comfortable, well-lit home does wonders for your sense of well-being. On the downside, adding batteries will raise your system cost by $3,000 to $5,000, and adds system components that will need some on-going maintenance and eventual replacement. If utility power is reliable and well maintained in your area, then there is little incentive for a battery-based system. Those dollars could be better spent on PV wattage. You'll see more benefits and get to enjoy almost zero maintenance. Just be aware that if utility power does fail, your solar system will automatically shut off too for safety reasons.

Rebates Can Reduce Your Total System Cost by Up to 50%!

There's money available out there for doing the right thing. A number of states have programs that will rebate/buydown/grant—whatever you want to call it—up to 50% of your initial renewable energy system cost. It is difficult to keep tabs on all of the rapidly changing programs in all of the states anymore. There's a list of State Energy Offices below. Give yours a call and ask about net metering and money available for renewable energy projects. If you're in California, there's a buydown program for intertie systems that we are fully versed on, and will be happy to help you with. Our fully trained technicians have years of experience sizing these systems (1-800-919-2400).

Do It for Your Children and the Planet . . .

Don't go into utility intertie expecting to cash in immediately. You won't. Most intertie systems will pay for themselves in saved electric bills, but the payback point is usually 6–10 or more years. Also, most state intertie regulations won't let residential customers deliver more than their average electric use. The details vary from state to state, but if you make more power than you use, you'll either give it away, or sell it at wholesale rates. Don't convert your backyard into a PV farm just yet. Intertie is something you do because it feels good to be independent of the utility and to cover your own electric power needs directly with a clean, nonpolluting, renewable power source. Solar modules last for decades, require almost no maintenance, and don't borrow our kids' future, which is probably the best possible reason to invest in renewable energy.

State Energy Offices Access Info

Here is all the straight, up-to-date information on utility intertie, and on any state programs that might help pay for a renewable energy system. All states have provided mail, phone, and fax access. Many also provide web and email access. We've listed everything available as of press time. The **n** denotes a state that allows net-metering (as of early 2001).

—Doug Pratt

Alabama Energy Office
Department of Economic and Community
Affairs
401 Adams Avenue, P.O. Box 5690
Montgomery, Alabama 36103-5690
Tel: (334) 242-5292
Fax: (334) 242-0552
Web: adeca.state.al.us/science/index-sep.html

Alaska Energy Office
Alaska Housing Finance Corporation
Research and Rural Development Division
P.O. Box 101020
4300 Boniface
Anchorage, Alaska 99510-1020
Tel: (907) 338-8115
Fax: (907) 330-8183
Email: ahfcric@corecom.net
Web: www.ahfc.state.ak.us/

American Samoa Energy Office
Territorial Energy Office
American Samoa Government
Samoa Energy House, Tauna
Pago Pago, American Samoa 96799
Tel: (684) 699-1101
Fax: (684) 699-2835

n Arizona Energy Office
Department of Commerce
3800 North Central Avenue, Suite 1200
Phoenix, Arizona 85012
Tel: (602) 280-1402
Fax: (602) 280-1445
Email: energy@ep.state.az.us
Web: www.azcommerce.com/energy.htm

Arkansas Energy Office
Arkansas Industrial Development Commission
One Capitol Mall
Little Rock, Arkansas 72201
Tel: (501) 682-7377
Fax: (501) 682-2703
Web: www.aedc.state.ar.us/energy/

n California Energy Office
California Energy Commission
1516 Ninth Street, MS #32
Sacramento, California 95814
Tel: (916) 654-5000
Fax: (916) 654-4420
Email: energia@energy.ca.gov
Web: www.energy.ca.gov

◼ Colorado Energy Office
Governor's Office of Energy Management
and Conservation
225 East 16th Avenue, Suite 650
Denver, Colorado 80203
Tel: 303-894-2383
Fax: 303-894-2388
Email: oemc@state.co.us
Web: www.state.co.us/oemc

◼ Connecticut Energy Office
Policy Development and Planning—Energy
Connecticut Office of Policy and Management
450 Capitol Ave.
PO Box 341441
Hartford, Connecticut 06134-1441
Tel: (860) 418-6297
Fax: (860) 418-6495

◼ Delaware Energy Office
Division of Facilities Management
149 Transportation Circle
Dover, Delaware 19001
Tel: (302) 739-5644
Fax: (302) 739-6148

District of Columbia Energy Office
2000 14th Street, NW, Suite 300E
Washington, DC 20009
Tel: (202) 673-6700
Fax: (202) 673-6725
Email: clintondc@aol.com
Web: www.dcenergy.org

Florida Energy Office
Florida Department of Community Affairs
2555 Shumard Oak Boulevard
Tallahassee, Florida 32399-2100
Tel: (850) 488-2475
Fax: (850) 488-7688

Georgia Energy Office
Division of Energy Resources
Georgia Environmental Facilities Authority
100 Peachtree Street, NW, Suite 2090
Atlanta, Georgia 30303-1911
Tel: (404) 656-5176
Fax: (404) 656-7970
Web: www.gefa.org/gefa/energy.html

Guam Energy Office
Pacific Energy Resources Center
548 N. Marine Drive
Tamuning, Guam 96911
Tel: (671) 646-4361 / 647-1403 / 647-1407 / 649-4362
Fax: (671) 649-1215
Email: energy@ns.gov.gu

◼ Hawaii Energy Office
Energy Resources and Technology Division
Department of Business, Economic
Development and Tourism
253 South Beretania, 5th Floor
P.O. Box 2359
Honolulu, Hawaii 96804-2359
Tel: (808) 587-3807
Fax: (808) 586-2536
Email: gmishina@dbedt.hawaii.gov
Web: www.hawaii.gov/dbedt/ert/energy.html

◼ Idaho Energy Office
Energy Division
Idaho Department of Water Resources
1301 N. Orchard Street
Boise, Idaho 83706
Tel: (208) 327-7900
Fax: (208) 327-7866
Email: kbaker@idwr.state.id.us
Web: www.idwr.state.id.us/energy/

◼ Illinois Energy Office
Bureau of Energy and Recycling
Illinois Department of Commerce and
Community Affairs
620 East Adams
Springfield, Illinois 62701
Tel: (217) 785-2009
Fax: (217) 785-2618
Web: www.commerce.state.il.us/resource_
efficiency/

◼ Indiana Energy Office
Energy Policy Division
Indiana Department of Commerce
One North Capitol, Suite 700
Indianapolis, Indiana 46204-2288
Tel: (317) 232-8939
Fax: (317) 232-8995

🔳 Iowa Energy Office
Energy Bureau
Energy and Geological Resources Division
Iowa Department of Natural Resources
Wallace State Office Building
East 9th & Grand Avenue
Des Moines, Iowa 50319
Tel: (515) 281-8681
Fax: (515) 281-6794
Web: www.state.ia.us/dnr/energy

Kansas Energy Office
Energy Programs Section
Kansas Corporation Commission
1500 SW Arrowhead Road
Topeka, Kansas 66604-4027
Tel: (785) 271-3170
Fax: (785) 271-3354
Email: j.ploger@kcc.state.ks.us
Web: www.kcc.state.ks.us/energy/energy.htm

Kentucky Energy Office
Kentucky Division of Energy
663 Teton Trail
Frankfort, Kentucky 40601
Tel: (502) 564-7192
Fax: (502) 564-7484
Web: www.nr.state.ky.us/nrepc/dnr/energy/
dnrdoe.html

Louisiana Energy Office
Energy Section
Technology Assessment Division
Department of Natural Resources
P.O. Box 44156
625 North 4th Street, Room 234
Baton Rouge, Louisiana 70804-4156
Tel: (225) 342-1399
Fax: (225) 342-1397
Email: tangular@dnr.state.la.us
Web: www.dnr.state.la.us/SEC/EXECDIV/
TECHASMT/ENERGY/

🔳 Maine Energy Office
Energy Conservation Division
Department of Economic & Community
Development
#59 State House Station
Augusta, Maine 04333-0059
Tel: (207) 297-2656
Fax: (207) 287-5701

🔳 Maryland Energy Office
Maryland Energy Administration
1623 Forest Drive, Suite 300
Annapolis, Maryland 21403
Tel: (410) 260-7511
Fax: (410) 974-2875
Email: mea@energy.state.md.us
Web: www.energy.state.md.us

🔳 Massachusetts Energy Office
Division of Energy Resources
Department of Economic Development
70 Franklin Street, Seventh Floor
Boston, Massachusetts 02202-1313
Tel: (617) 727-4732
Fax: (617) 727-0030
Email: energy@state.ma.us
Web: www.magnet.state.ma.us/doer

Michigan Energy Office
Energy Division
Michigan Department of Consumer and
Industry Services
P.O. Box 30221
Lansing, Michigan 48909
Tel: (517) 241-6228
Fax: (517) 241-6229
Web: www.cis.state.mi.us/opla/erd

🔳 Minnesota Energy Office
Minnesota Energy Division
Minnesota Department of Commerce
121 7th Place East, Suite 200
St. Paul, Minnesota 55101-2145
Tel: (651) 297-2545
Fax: (651) 282-2568

Mississippi Energy Office
Energy Division
Mississippi Department of Economic and
Community Development
P.O. Box 850
Jackson, Mississippi 39205-0850
Tel: (601) 359-6600
Fax: (601) 359-6642
Web: www.decd.state.ms.us/

Missouri Energy Office
Energy Center
Department of Natural Resources
P.O. Box 176
169A E. Elm Street
Jefferson City, Missouri 65102
Tel: (573) 751-4000
Fax: (573) 751-6860
Web: www.dnr.state.mo.us/de/homede.htm

n **Montana Energy Office**
Planning, Prevention, and Assistance Division
Department of Environmental Quality
P.O. Box 200901
1520 East 6th Avenue
Helena, Montana 59620-0901
Tel: (406) 444-6754
Fax: (406) 444-6836

Nebraska Energy Office
State Capitol Building, 9th Floor
P.O. Box 95085
1111 "O" Street, Suite 223
Lincoln, Nebraska 68509-5085
Tel: (402) 471-2867
Fax: (402) 471-3064
Email: energy@mail.state.ne.us
Web: www.nol.org/home/NEO/

n **Nevada Energy Office**
Department of Business and Industry
727 Fairview Drive, Suite F
Carson City, Nevada 89701
Tel: (775) 687-5975
Fax: (775) 687-4914
Email: dparsons@govmail.state.nv.us
Web: www.state.nv.us/b&i/eo/

n **New Hampshire Energy Office**
Governor's Office of Energy & Community
Services
State of New Hampshire
57 Regional Drive
Concord, New Hampshire 03301
Tel: (603 271-2611
Fax: (603) 271-2615
Web: www.state.nh.us/governor/energycomm/
index.html

n **New Jersey Energy Office**
Division of Energy
New Jersey Board of Public Utilities
2 Gateway Center
Newark, New Jersey 07102
Tel: (973) 648-3717
Fax: (973) 648-7420
Email: energy@bpu.state.nj.us
Web: www.bpu.state.nj.us/wwwroot/energy/
energy.htm

n **New Mexico Energy Office**
Energy Conservation & Management Division
NM Energy, Minerals and Natural Resources
Department
408 Galisteo Street
Santa Fe, New Mexico 87501
Tel: (505) 827-1373
Fax: (505) 827-8177
Web: www.emnrd.state.nm.us/ecmd/

n **New York Energy Office**
New York State Energy Research and
Development Authority
Corporate Plaza West
286 Washington Avenue Extension
Albany, New York 12203-6399
Tel: (518) 862-1090
Fax: (518) 862-1091
Web: www.nyserda.org

North Carolina Energy Office
(net-metering is pending in NC)
Energy Division
North Carolina Department of Commerce
1830A Tillery Place
Raleigh, North Carolina 27604
Tel: (919) 733-2230
Fax: (919) 733-2953

n **North Dakota Energy Office**
Energy Programs
North Dakota Division of Community Services
Office of Management and Budget
600 East Boulevard Avenue, 14th Floor
Bismarck, North Dakota 58505-0170
Tel: (701) 328-2094
Fax: (701) 328-2308
Web: www.state.nd.us/dcs/Energy/default.html

North Mariana Islands Energy Office
Commonwealth of the Northern Mariana
Islands
P.O. Box 340
Saipan, MP 96950
Tel: (670) 322-9229
Fax: (670) 322-9237

𝕟 Ohio Energy Office
Office of Energy Efficiency
Ohio Department of Development
77 South High Street, 26th Floor
Columbus, Ohio 43215-6108
Tel: (614) 466-6797
Fax: (614) 466-1864
Web: www.odod.ohio.gov/cdd/oee/

𝕟 Oklahoma Energy Office
Division of Community Affairs and
Development
Oklahoma Department of Commerce
P.O. Box 26980
6601 North Broadway
Oklahoma City, Oklahoma 73126
Tel: (405) 815-6552
Fax: (405) 815-5344
Web: www.odoc.state.ok.us/

𝕟 Oregon Energy Office
625 Marion Street, N.E.
Salem, Oregon 97310
Tel: (503) 378-4131
Fax: (503) 373-7806
Email: energy.in.internet@state.or.us
Web: www.energy.state.or.us

𝕟 Pennsylvania Energy Office
Office of Pollution and Compliance Assistance
Department of Environmental Protection
P.O. Box 2063
400 Market Street, RCSOB
Harrisburg, Pennsylvania 17105
Tel: (717) 783-9981
Fax: (717) 783-8926
Web: www.dep.state.pa.us/dep/deputate/
pollprev/pollution_prevention.html

Puerto Rico Energy Office
Energy Affairs Administration
Department of Natural and Environmental
Resources
P.O. Box 9066600, Puerta de Tierra
San Juan, Puerto Rico 00906-6600
Tel: (787) 724-8774
Fax: (787) 721-3089

𝕟 Rhode Island Energy Office
One Capital Hill, 2nd Floor
Providence, Rhode Island 02908
Tel: (401) 222-3370
Fax: (401) 222-1260
Email: riseo@ids.net

South Carolina Energy Office
1201 Main Street, Suite 820
Columbia, South Carolina 29201
Tel: (803) 737-8030
Fax: (803) 737-9846
Web: www.state.sc.us/energy/

South Dakota Energy Office
Governor's Office of Economic Development
711 E. Wells Avenue
Pierre, South Dakota 57501-3369
Tel: (605) 773-5032
Fax: (605) 773-3256
Web: www.state.sd.us/state/executive/oed/

Tennessee Energy Office
Energy Division
Department of Economics & Community
Development
Tennessee Tower
312 8th Avenue, North, 9th Floor
Nashville, Tennessee 37243
Tel: (615) 741-2994
Fax: (615) 741-5070
Web: www.state.tn.us/ecd/energy.htm

𝕟 Texas Energy Office
State Energy Conservation Office
Texas Comptroller of Public Accounts
111 E. 17th Street- 11th Floor
Austin, TX 78701
Tel: (512) 463-1931
Fax: (512) 475-2569
Web: www.seco.cpa.state.tx.us/

Utah Energy Office
Utah Office of Energy Services
Department of Community & Economic
Development
324 South State Street, Suite 500
Salt Lake City, Utah 84111
Tel: (801) 538-8690
Fax: (801) 538-8660
Email: dbeaudoi@dced.state.ut.us
Web: www.dced.state.ut.us/energy

𝑛 Vermont Energy Office
Energy Efficiency Division
Vermont Department of Public Service
112 State Street, Drawer 20
Montpelier, Vermont 05620-2601
Tel: (802) 828-2811
Fax: (802) 828-2342
Email: vtdps@psd.state.vt.us
Web: www.state.vt.us/psd/ee/ee.htm

𝑛 Virginia Energy Office
Division of Energy
Virginia Department of Mines, Minerals &
Energy
202 North Ninth Street, 8th Floor
Richmond, Virginia 23219
Tel: (804) 692-3200
Fax: (804) 692-3238
Web: www.mme.state.va.us/de

Virgin Islands Energy Office
Oscar E. Henry Customs House
200 Strand Street
Frederiksted, USVI 00840
Tel: (340) 772-2616
Fax: (340) 772-2133
Email: USVIEOOH@VIACCESS.Net

𝑛 Washington Energy Office
Washington Energy Policy Group
Department of Community, Trade and
Economic Development
P.O. Box 43173
Olympia, Washington 98504-3173
Tel: (360) 956-2098
Fax: (360) 956-2180

West Virginia Energy Office
Energy Efficiency Program
West Virginia Development Office
Building 6, Room 645
State Capitol Complex
Charleston, West Virginia 25305
Tel: (304) 558-0350
Fax: (304) 558-0362
Web: www.wvdo.org/community/eep.htm

𝑛 Wisconsin Energy Office
Energy Bureau
Division of Energy and Public Benefits
Department of Administration
101 E. Wilson Street, 6th Floor
P.O. Box 7868
Madison, Wisconsin 53707-7868
Tel: (608) 266-8234
Fax: (608) 267-6931
Web: www.doa.state.wi.us/depb/boe/index.asp

Wyoming Energy Office
Energy Program
Wyoming Business Council
214 West 15th Street
Cheyenne, Wyoming 82002
Tel: (307) 777-2800
Fax: (307) 777-2837
Email: jnunle@missc.state.wy.us
Web: www.wyomingbusiness.org/energy.htm

Utility Intertie Inverters

Straight from the Sun to Your Utility Meter

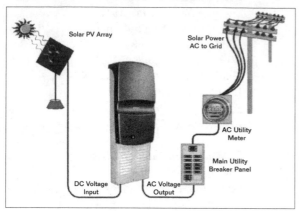

PV modules that produce conventional AC power have long been a dream of the renewable energy industry. We can finally offer several cost-effective ways to connect PV modules directly to your household AC system.

Trace provided the first generation of utility intertie inverters with their large multi-tasking sine wave series in the mid-'90s. The original full-featured SW-series can intertie and support emergency backup batteries. The ability to intertie and provide battery backup remains unique to this inverter line. They are the best choice if you have a really large PV array to intertie, or absolutely must have backup batteries due to unreliable utility power, but they're expensive for smaller arrays. The SW-series products are listed below, and in the regular inverter product section, because they are, after all, still regular inverters.

Direct intertie inverters are the best choice for systems that don't need backup batteries. Trace, Advanced Energy Systems, and SMA make direct intertie inverters. Any incoming PV power is converted to conventional AC power, displacing utility use. There are no batteries. The utility grid is treated as a large battery. When there's excess power, your meter runs backwards, when there's a deficit, power is bought back.

An additional advantage of direct intertie inverters is the high energy efficiency of the system. PV modules are controlled with power point tracking, which allows the modules to run at their maximum power output, rather than a lower battery charging voltage. Batteries are eliminated, which typically costs 20% of charging power; and all transmission is done at high voltage, allowing smaller wire sizes, and less loss.

A multiple number of small intertie inverters can be paralleled for system expansion. Direct intertie inverters will not operate without an AC power source to feed back into, and will not charge batteries. All models are UL-listed.

Some states and utility companies are offering rebate or buy-down programs for intertie systems. Please call your local utility or our technical staff for details.

Utility Intertie Inverters Without Batteries

Advanced Energy Systems Intertie Inverters

AES has gained a good name for itself engineering and manufacturing large stand-alone power systems. This New Hampshire company is heavy on engineering capability. Among many large projects, they provided the custom 120 kW Static Power Pack inverter and controller for the largest solar installation in Latin America at the Essene Way in Belize, which was a Real Goods project. When small intertie systems became possible, AES was the first manufacturer with a UL-listed mid-sized (1,000-watt) direct intertie inverter.

Advanced Energy Systems GC-1000 Inverter

The first mid-sized, UL-listed intertie inverter available on the market. The wall-mounted GC-1000 is a cost-effective solution for connecting up to 1,000 watts of PV modules directly to the grid without the use of batteries. The GC-1000 offers economies of scale and allows easy system expansion in the future. Features include nominal 48-volt input, an outdoor rated enclosure, innovative thermal design that requires no fans, peak efficiency of 93%, maximum power point tracking for best performance under any condition, rugged, industrial-rated components with a low parts count, and no night time power losses. Like other direct utility intertie inverters, no batteries are used. 19″ x 8″ x 6.5″, weighs 41 lbs. Five-year mfr.'s warranty.

The GC-1000 is available as a single unit, or may be purchased as a complete system, which includes an outdoor-rated, fused, string combiner, GFI protection, and DC disconnects. The complete system is highly recommended. With the string combiner, as shown, height increases to 28.5″.

A simple blinking LED shows status, activity, and any problems. An optional complete Data Monitoring System can be mounted on the face of the inverter, or remotely. Features of the Data Monitoring System include a 4-line x 40 character LCD display and keypad; monitors AC volts, amps, watts, frequency, with daily and cumulative kWh; also monitors DC volts, amps, watts; inverter temperature and run time; monitors up to 10 GC-1000 units; can be accessed with modem or RS232 interface; compatible with Campbell Data Logger software. USA.

27-157 GC-1000 Utility Intertie Inverter
      Complete Package $1,695⁰⁰

◆**27-158 GC-1000 Utility Intertie Inverter $1,375⁰⁰**

◆**27-159 AES Internal Monitor Option $449⁰⁰**

◆**27-160 AES External Monitor Option $539⁰⁰**

◆*Means shipped directly from manufacturer.*

Trace Sun Tie
XR-Series Inverters

Connect the Sun to Your Utility Meter

Trace Engineering's second-generation utility intertie ST-XR-series inverters deliver PV power directly to your home or business AC electric system without the use of batteries. Any available PV energy is instantly and efficiently converted to 240VAC, offsetting utility power use. If there is an excess of PV-generated power, it's sold to the utility company. Those with "net-metering" laws can actually make their utility meter turn backwards during the daytime. The all-in-one ST-XR inverter design includes NEC-required disconnects, DC and AC circuit breakers, lightning arrestor, and a built-in LCD display that shows system status, current performance, and daily cumulative power production. Maximum power point tracking technology is used to gain the best output from your PV array under any conditions. The solid-state inverter delivers 90% to 94% efficiency.

The ST-XR-series is available in four models at 1000, 1500, 2000, and 2500 watts. The 1500 and 2500 models include PV ground fault interrupters and combiner boards that are required if your PV array is mounted on a residence. The protective rain shield, as pictured, is an option for all models, and is a requirement for outdoor installations. Multiple inverters of different capacities can be wired in parallel. Each operates independently.

The optional remote meter connects with standard phone cable and will display daily, weekly, monthly, or cumulative energy production, as well as current production data.

All ST-XR models are 48-volt nominal, so PV modules are usually added in groups of four. They operate with any PV technology, either crystalline or thin film. The wall-mounted, powder-coated aluminum enclosure is fully insect screened, weighs 35 pounds, and measures 13.25″ W x 33.25″ H x 5.3″ D. UL-listed to UL 1741–1999. Made in USA. Two-year mfr.'s warranty with five-year option for $225. Homeowner must purchase the extended warranty directly from Trace, following installation. A picture and a signed-off final building permit are required.

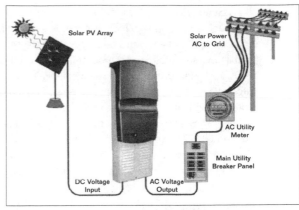

27-844 Trace ST-XR1000 Intertie Inverter $1,849⁰⁰

27-845 Trace ST-XR1500 Intertie Inverter
      w/PVGFP & combiner $2,239⁰⁰

27-846 Trace ST-XR2000 Intertie Inverter $2,259⁰⁰

27-847 Trace ST-XR2500 Intertie Inverter
      w/PVGFP & combiner $2,759⁰⁰

27-848 ST-XR Rain Shield Option $125⁰⁰

27-864 ST-XR Remote Meter $179⁰⁰

SMA Inverters

The World's Most Experienced
Utility-Intertie Inverter Manufacturer

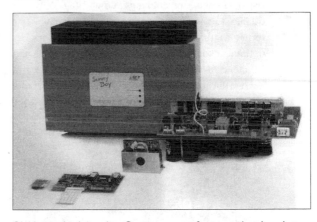

SMA is a high-quality German manufacturer that has been making photovoltaic-powered utility intertie inverters since 1985. SMA has put more than 50,000 units into intertie service, primarily in the green German and European markets. The direct-intertie Sunny Boy series is their third-generation design, and has benefited greatly from all the experience. These are modular string inverters using 6 to 22 modules in a high-voltage string wired in series, which eliminates the need for combiner boxes and large wire. Maximum power point tracking delivers the highest possible wattage under all conditions. Maximum inverter efficiency and safety have been the primary design goals during development of all SMA products.

All Sunny Boy inverters feature powder-coated, stainless steel, outdoor-rated, NEMA 4 cases that can be wall mounted right outside with the array. They are designed for full-power operation at temperatures up to 60°C (140°F). Higher temperatures will cause limited output or automatic shut-off to protect the inverter. Ground-fault protection is built into each inverter.

The first Sunny Boy model to gain the U.S. intertie-required UL 1741 listing is the 2500-watt unit. 1600-, 1100- and 700-watt units are in the works, and will be available sometime in 2002; please call for details.

Sunny Boy 2500 Direct-Intertie Inverter

The Sunny Boy 2500 inverter features the best of high-quality German engineering. High efficiency, redundant processors, easy data communications, maximum power point tracking, an excellent reliability record, and powder-coated stainless steel outdoor-rated boxes are only a few of the design features. Sunny Boy inverters use a high-voltage DC input with maximum power point tracking from 275 to 550 volts. Output is 240 volt AC. Using conventional single-crystal, 12v nominal modules, the Sunny Boy 2500 can accept series strings of 17 to 24 modules, with 20 to 22 modules in series being about perfect. Shorter series strings may drop under voltage on hot days, longer series strings may rise over voltage on cold days. Low voltage only causes shut off, like sunset. High voltage may cause damage. Internal terminal strips accept up to three strings per inverter. The optimal maximum PV array size is 2750 to 3450 watts. Efficiency is an excellent 94%, with less than 7 watts internal use during operation, or less than 0.1 watt during standby. At night the inverter disconnects from the utility to eliminate any phantom load.

A series of Sunny Boy inverters on a large array.

Several monitoring options are offered. The on-board 2-line LCD display will satisfy most folks. It's mounted on the face of the inverter and scrolls through Instantaneous AC Watts; DC Volts; Total Energy Delivered in kWh; Daily Energy Delivered in kWh;

Operating Hours; and Inverter Status. The backlit display is activated by knocking on the lid. (When German engineers start questing for high efficiency, there are very few limits.) The on-board display must be ordered initially.

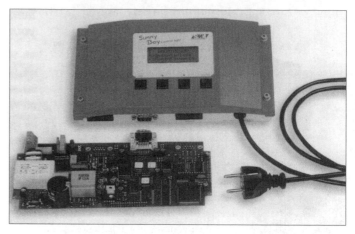

The Sunny Boy Control (Light) can monitor up to 20 inverters through the existing AC wiring. Just plug it in. Each inverter must be equipped with the optional plug-in Power Line Carrier Module. Data is stored for the last 200 days, and is displayed on the 4-line LCD, or can be exported to a computer with the RS232 plug and Sunny Boy software installed. Other Sunny Boy Controls are available for larger arrays; please call for details.

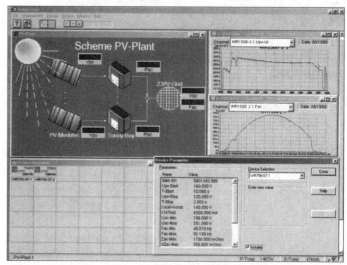

Data from the Sunny Boy Control is easily imported to a PC.

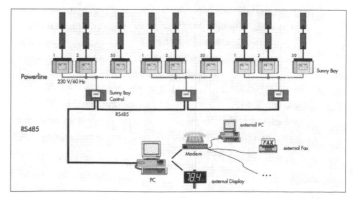

Sunny Boy 2500 size is 17.1″ x 11.6″ x 8.4″, weight is 15.5 lbs. Control (Light) is 9" x 5" x 3" and weighs 3.5 lb. Made in Germany. Five-year mfr.'s warranty (No extended warranty required for California buy-down programs!)

27-856 **Sunny Boy 2500 Standard** $2,495⁰⁰
27-857 **Sunny Boy 2500 w/Display** $2,595⁰⁰
27-858 **Sunny Boy Control (Light)** $549⁰⁰
27-859 **Power Line Carrier Module** $139⁰⁰
25-775 **Hi-Voltage Lightning Protection** $39⁹⁵

PV Power Packages For Sunny Boy 2500

We're offering two pre-designed power packages for the 2500. These provide PV modules and Uni-Rac low-profile roof or ground mounts with adjustable rear legs. We can supply pole top mounts for about the same price, but you'll need to call us to put together a customized package. Site-specific items such as wire, conduit, and small parts are supplied by the customer. The inverter is not included with these packages.

Our 1.5 kilowatt package provides 20 Astropower 75-watt modules, and takes the 2500 to about half of its potential, so you've got room for future expansion. Based on the North American yearly average of about 5.5 hours peak equivalent sunlight per day, this system will deliver approximately 7 kWh/day. Combined with the Sunny Boy 2500, it's eligible for a $5,752 rebate from the California Energy Commission on completion.

Our 2.8 kilowatt package provides 22 Siemens SP130 modules, and powers the Sunny Boy 2500 fully. Based on the North American yearly average of about 5.5 hours peak equivalent sunlight per day, this system will deliver approximately 12.7 kWh/day. Combined with the Sunny Boy 2500, it's eligible for a $10,860 rebate from the California Energy Commission on completion.

12-242 **1.5 kW Power Package for SB2500** $9,575⁰⁰
       [consists of: 20 Astropower AP75 modules
       (11-165), plus 4 Uni-Rac LP-112 (13-579)]

12-243 **2.8 kW Power Package for SB2500** $17,700⁰⁰
       [consists of 22 Siemens SP130 modules
       (11-575), plus 2 Uni-Rac LP-192 (13-986),
       and 2 Uni-Rac LP-160 (13-583)]

Utility Intertie Inverters With Batteries

Trace Sine Wave Inverters

This series of sine wave inverter/chargers does it all. Conventional remote inverter, uninterruptible power supply, generator management, or utility intertie, they can do any, or sometimes all, of these functions simultaneously!

These microprocessor controlled inverters produce a multistepped approximation of a sine wave that meets all utility requirements for line intertie and will run any AC appliance within its wattage capabilities. And they do it at up to 94% efficiency. All models in the SW series are Electrical Testing Lab approved to UL specifications.

There are three primary modes of operation for the SW-series.

Inverter Mode

In this mode the SW series acts like a typical remote power inverter with a battery charger. It has adjustable threshold watts to trigger turn-on, and adjustable search spacing to save power. It will produce rated surge power for two minutes. These inverters will automatically protect themselves from overload, high temperature, and high or low battery voltage. Fan cooling is temperature regulated; low voltage battery disconnect is adjustable.

The sophisticated generator management system, which works in all operation modes, provides a wide range of features, including auto start by battery voltage or time, number of start attempts, cranking time for each attempt, warm-up time before loading, automatic battery charging, maximum generator load management, auto low-voltage disconnect, auto shut off, and more.

Standby Mode

Grid power is fed through and partially conditioned by the SW series inverters. In case of utility failure, the Trace will pick up instantly and support the load. Transfer time is 16 to 30 milliseconds. The SW series inverter will support the grid if it "browns out." If the amount of AC power demanded is greater than the breaker size you've programmed into the SW inverter, it will contribute power to the system. If the grid fails and the SW inverter is running all the loads, then all the automatic generator management and battery charging discussed above will occur. Once utility power is restored the inverter will match phases and shift the AC loads back to the utility, then shut off the generator if it is running.

Utility Interactive Mode

This mode will allow you to sell excess power back to the utility. The setup is simple, and the unit was designed to meet all utility requirements, but regulations will vary from one utility to another, and may not be formalized at all utilities. We may only be able to offer limited assistance obtaining approval from your local utility. Talk to your local utility first; possession of the equipment and the ability to do so does not constitute a right to sell them power.

A small battery bank is still required in the Utility Interactive Mode. A larger battery bank will protect you from blackouts for as long as it lasts, or until the generator starts and takes over. The inverter will float the batteries at a programmed voltage; any additional power input will be converted to AC and fed back into the grid. If the utility fails in the Utility Interactive Mode, the SW inverter will go into the normal Inverter Mode to run your house without a hiccup.

All the SW inverters are stackable in series for 240 volts AC, three wire splittable output at double the rated wattage continuously. Stacking only requires plugging in an inexpensive stacking cord. Designed for wall mounting with conventional 16-inch stud spacing. All Trace SW series inverters carry a two-year manufacturer's warranty, with a five-year option. Available directly to the homeowner for $350 with a signed-off final building permit and a picture of the installation. USA.

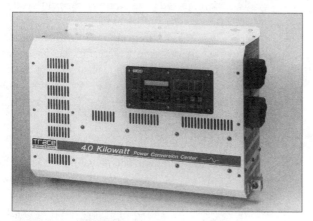

Trace SW-Series, 24-Volt Models

SW4024 Sine Wave Inverter/Charger

Input Voltage: 20 to 34 volts DC
Output Voltage: 120 volts AC true RMS ±2%
Continuous Output: 4,000 watts
Surge Capacity: 10,000 watts for 2 minutes
Standby Power: adjustable from 1 to 16.8 watts
Charging Rate: 0 to 120 amps
Transfer Amps: 60 amps
World Voltages Available: yes, call for details
Average Efficiency: 94% peak, 85% to 90% average

Recommended Fusing: 400A Class T or Trace DC-250
Dimensions: 22.5″ x 15.25″ x 9.0″
Weight: 105 lb

27-236 SW4024 Trace Sine Wave, 24V, 4 kW $3,495⁰⁰

◆**27-240 SW3024E Export Model, various volts/Hz $3,495⁰⁰**

◆*Means shipped directly from manufacturer.*

Trace SW Series, 48-Volt Models

SW4048 Sine Wave Inverter/Charger

Input Voltage: 40 to 68 volts DC
Output Voltage: 120 volts AC true RMS ±2%
Continuous Output: 4,000 watts
Surge Capacity: 10,000 watts for 2 minutes
Standby Power: adjustable from 1 to 16.8 watts
Charging Rate: 0 to 60 amps
Transfer Amps: 60 amps
World Voltages Available: yes, call for details
Average Efficiency: 94% peak, 85% to 90% average
Recommended Fusing: 200A Class T or Trace DC-175
Dimensions: 22.5″ x 15.25″ x 9.0″
Weight: 105 lb

27-241 SW4048 Trace Sine Wave, 48V, 4 kW $3,495⁰⁰

◆**27-248 SW3048E Export Model, various volts/Hz $3,495⁰⁰**

◆*Means shipped directly from manufacturer.*

SW5548 Sine Wave Inverter/Charger

Input Voltage: 40 to 68 volts DC
Output Voltage: 120 volts AC true RMS ±2%
Continuous Output: 5,500 watts
Surge Capacity: 10,000 watts for 2 minutes
Standby Power: Adjustable, 1 to 20 watts
Charging Rate: 0 to 75 amps
Transfer Amps: 60 amps
World Voltages Available: yes, call for details
Average Efficiency: 96% peak, 85% to 90% average
Recommended Fusing: 200A Class T or Trace DC-250
Dimensions: 22.5″ x 15.25″ x 9.0″
Weight: 130 lb

27-249 SW5548 Trace Sine Wave, 48V, 5.5kW $3,995⁰⁰

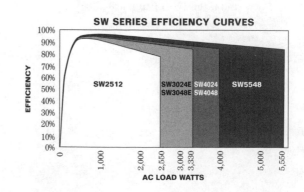

Trace Inverter Options

SW Remote Control Panel Option

Offers a complete duplicate control panel with all controls, indicator lights, and LCD display. Includes 50 feet of connection cable. (Uses standard 25-pin computer cable and connectors.) USA.

27-210 Remote Control Panel for SW series
        (specify voltage) **$329⁰⁰**

SW Communications Adapter

This option allows PC connection and monitoring of one SW-series inverter with Rev. 4.01 or later software. Includes adapter, DOS-based software, 50-foot cable, and DB9 connector. USA.

27-305 SW Comm. Adapter $175⁰⁰

SW Stacking Interface

To stack a pair of SW inverters, this 42″ comm-port cable is all you need. Simple phone-type plug-in connectors. 240V output. USA.

27-314 SW Stacking Intertie Cable $39⁹⁵

SW Paralleling Kit

A new item from Trace, connects two SW inverters in parallel to double the wattage output at 120V output. (The less expensive SW Stacking Interface doubles the wattage, but at 240V output.) For U.S. voltage inverters only. USA.

27-415 SW Paralleling Kit $345⁰⁰

Conduit Box Options

The optional Conduit Box allows the very heavy input wiring from the batteries to be run inside conduit to comply with NEC. Has three ½″, ¾″, and 2″ knockouts per side, and plenty of elbow room inside for bending large wire. USA.

27-209 Conduit Box option-SW Series $89⁹⁵

Trace T240 Autotransformer

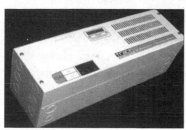

A very handy 2:1 bi-directional transformer with a 4,500 watt continuous rating. It can be used to step 120-volt inverter output up to run the occasional 240-volt appliance, like a water pump. It can be used to balance the load on a 240-volt generator, by delivering 120-volts to an inverter/charger while evenly loading the generator. Or it can be used as a step-down transformer, allowing a stacked pair of inverters to share the start-up and running of a large 120-volt appliance. Supplied in a

powder-coated enclosure with multiple ¾″ and 1″ knockouts and a 2-pole 25-amp circuit breaker. Weighs 39.4 lb. Measures 6.3″ x 7″ x 21″. Accepts 14 to 4 AWG wire sizes. This unit replaces the Trace T220 at the same price. USA.

27-326 T240 Autotransformer $349⁰⁰

Battery and Inverter Cables

The wire sizing between the battery and the inverter is of critical importance. Small cables limit inverter performance. We offer seriously beefy 2/0 and 4/0 cables using super flexible welding-type cable. Highly recommended to complete your inverter installation. They're available in 3-foot, 5-foot or 10-foot pairs. One-lug cables have bare wire on one end to accommodate set-screw lugs, with a crimped, soldered, and color-coded shrink-wrapped ⅜″ lug terminal on the other end. Two-lug cables have the same high quality terminals installed at both ends. See chart below for recommended cable sizing. Extra lugs offered for the occasional installation that needs an extra lug end. USA.

15-770 2/0 3-ft. Cables, 1 lug, pair $29⁹⁵
15-771 2/0 5-ft. Cables, 1 lug, pair $49⁹⁵
15-772 2/0 10-ft. Cables, 1 lug, pair $84⁹⁵

15-773 4/0 3-ft. Cables, 1 lug, pair $39⁹⁵
15-774 4/0 5-ft. Cables, 1-lug, pair $69⁹⁵
15-775 4/0 10-ft. Cables, 1 lug, pair $115⁰⁰

27-311 4/0 5-ft. Cables, 2 lug, pair $69⁹⁵
27-312 4/0 10-ft. Cables, 2 lug, pair $125⁰⁰

27-099 2/0 5-ft. Cables, 2 lug, pair $49⁹⁵
27-098 2/0 10-ft. Cables, 2 lug pair $89⁹⁵

26-605 Copper Lug, solder-type 2/0 $4⁹⁵
26-606 Copper Lug, solder-type 4/0 $13⁹⁵
26-623 Solderless Lug, #2 to 4/0 $7⁹⁵

DC DISCONNECT SIZING FOR TRACE INVERTERS

Disconnect Size	Recommended With These Inverters	Minimum Cable Size
DC-175	DR-1512, 2412, 1524, & 2424. SW-3048E, & 4048	2/0
DC-250	DR-3624, SW-2512, 4024, & 5548	4/0

Complete Intertie Systems

Direct Intertie Systems

Rolling Blackouts and Skyrocketing Electricity Prices Got You Worried? Make Your Own Solar Power!

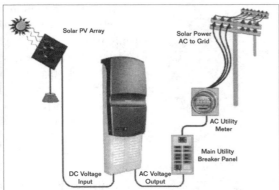

Trace Engineering's second-generation utility intertie ST-XR-series inverters deliver PV power directly to your home or business AC electric system without the use of batteries. Any available PV energy is instantly and efficiently converted to 240 VAC and fed to your circuit breaker panel, displacing utility power use. If there is an excess of PV-generated power it's sold to the utility company. The all-in-one ST-XR inverter design includes NEC-required disconnects, DC and AC circuit breakers, lightning arrestor, and a built-in LCD display that shows system status, current performance, and daily cumulative power production. Maximum power point tracking technology is used to gain the best output from your PV array under any conditions. The solid-state inverter delivers 90% to 94% efficiency.

The ST-XR-series is available in four models at 1000, 1500, 2000, and 2500 watts. The 1500- and 2500-watt models include PV ground fault interrupters and combiner boards that are required if your PV array is mounted on a residence. The protective rain shield, as pictured, is an option for all models, and is a requirement for outdoor installations. Multiple inverters can be wired in parallel. Each operates independently.

All ST-XR models are 48-volt nominal; they operate with any PV technology, either crystalline or thin film. The wall-mounted, powder-coated aluminum enclosure is fully insect screened, weighs 35 pounds, and measures 13.25″ w x 33.25″ h x 5.3″ d. UL-listed. Made in USA. Two-year mfr. warranty.

SYSTEM PAYBACK ANALYSIS

Your Cost After Rebate	.$8,020
Approximate Installation Cost	.1,500
Total System Cost	**.$9,520**

If you pay $0.20/kWh, payback = 9.8 years.
If you pay $0.30/kWh, payback = 6.5 years.

The ST-XR1000 Utility Intertie Power System

This system includes the Trace ST-XR1000 intertie inverter, eight 120-watt PV modules, and a poletop mount. Based on the North American yearly average of 5.5 hours sunlight, this system will deliver approximately 4.3 kWh/day (about 30% of the usage of the average American home). The PV array must be mounted separately from the house with this package. Wiring, as required by your installation site, and 6″ steel pipe for the poletop mount are supplied locally. System pricing saves $280 over individual pricing. Qualifies for a California Energy Commission rebate of $3,500.

12-234 ST-XR1000 Complete Intertie System $7,440⁰⁰

The ST-XR1500 Utility Intertie Power System

This system includes the Trace ST-XR1500 intertie inverter, twelve 120-watt PV modules, and a pair of low-profile rooftop mounts. Based on the North American yearly average of 5.5 hours sunlight per day, this system will deliver approximately 6.5 kWh/day (about 40% of the usage of the average American home). This package allows mounting on a residence. Wiring is supplied locally, as required by your installation site. System saves $327 over individual pricing. Qualifies for a California Energy Commission rebate of $5,251.

12-235 ST-XR1500 Complete Intertie System $10,630⁰⁰

The ST-XR2000 Utility Intertie Power System

This system includes the Trace ST-XR2000 intertie inverter, sixteen 120-watt PV modules, and a pair of poletop mounts. Based on the North American yearly average of 5.5 hours sunlight per day, this system will deliver approximately 9.4 kWh/day (about 60% of the usage of the average American home). The PV arrays must be mounted separately from the house with this package. Wiring, as required by your installation site, and a pair of 6″ steel pipes for the poletop mounts are supplied locally. System pricing saves $951 over individual pricing. Qualifies for a California Energy Commission rebate of $7,153.

12-236 ST-XR2000 Complete Intertie System $13,050⁰⁰

The ST-XR2500 Utility Intertie Power System

This system includes the Trace ST-XR2500 intertie inverter, twenty-four 120-watt PV modules, and three low-profile rooftop mounts. Based on the North American yearly average of 5.5 hours sunlight per day, this system will deliver approximately 13.3 kWh/day (about 85% of the usage of the average American home). This package allows mounting on a residence. Wiring is supplied locally, as required by your installation site. System saves $1,202 over individual pricing. Qualifies for a California Energy Commission rebate of $10,730.

12-237 ST-XR2500 Complete Intertie System $18,750⁰⁰

Intertie Systems with Batteries

Because battery-based intertie systems have more variables and consumer choices, we often end up designing a custom package for each particular customer (and we're happy to do that). But let's give you some idea about what an intertie system with batteries is made of, and what it costs.

All battery-based intertie systems are built around one or more of the Trace SW-series inverters. These are the only inverters on the market, so far, that will support utility intertie and batteries. SW inverters are available in 4,000- or 5,500-watt sizes. That's large or larger.

In "sell mode" operation, the SW inverter will allow battery voltage to rise to a set float voltage point. Any incoming renewable energy power beyond this point is converted into standard AC power and delivered to the household. The SW delivers AC power at a volt or two higher than utility power. Like water seeking its own level, this will seamlessly displace utility power. If utility power fails, the SW will instantly switch to battery power for continued operation of any AC appliances connected to the SW's output terminals. This is usually a separate electric subpanel with all the circuits the homeowner wants to continue running during a power outage. For utility lineworker safety, the SW cannot "sell" any power if it can't sense utility power on its AC input terminals.

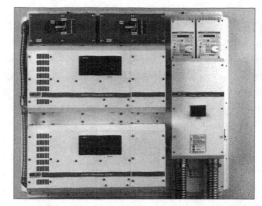

Trace SW-Series Power Panel with Dual Inverters

There is a second set of AC input terminals for a backup generator, along with a sophisticated generator management system that can automatically start and stop any remote-start capable backup generator.

In the kits below we are supplying the SW inverter pre-assembled on a Power Panel. This provides all the safety and control equipment assembled and wired on a 43" x 40" powder coated base plate. A Power Panel makes for a faster, cleaner-looking installation.

Batteries are not included in these kits, although we make minimum recommendations, and include interconnect cabling. Battery capacity is probably something you want to talk with one of our techs about. Sealed batteries are usually the best choice for systems that will float on utility power most of the time.

The Mendocino—Basic Starter Kit—2.3 kWh/Day

Here's the minimum you need for an intertie system with battery backup. It consists of the Trace SW4048 inverter on a Power Panel with a C40 controller option. Code-required ground-fault and combiner boxes are included, plus a common sense-required lightning protector, and heavy-duty battery cabling. It employs four 120-watt PV modules on a low-profile roof or ground mount, with capacity for about 36 more similar PV modules in the future.

Based upon the North American yearly average of 5.5 hours sunlight per day, this system will deliver approximately 2.3kWh/day (with lots of future expansion capacity). Wiring is supplied locally, as required by your installation site. It qualifies for a California Energy Commission rebate of $1,807 on completion.

12-239 The Mendocino Intertie Kit $8,939⁰⁰

Item #	Qty.	Description	Unit Price	Total Price
27-293	1	PP-SW4048 w/C40 option	$ 5,253⁰⁰	$ 5,253⁰⁰
27-312	1	4/0, 10-ft. cable pair	$ 125⁰⁰	$ 125⁰⁰
15-785	4	4/0, 18″ interconnect cable	$ 19⁹⁵	$ 79⁸⁰
24-104	1	100A ground fault protection	$ 275⁰⁰	$ 275⁰⁰
24-214	1	12″ x 10″ x 4″ raintight j-box	$ 50⁰⁰	$ 50⁰⁰
25-194	1	DC lightning protector	$ 44⁹⁵	$ 44⁹⁵
24-228	1	Fused PV combiner box	$ 229⁰⁰	$ 229⁰⁰
24-643	1	15A fuse pack for combiner box (10 fuses)	$ 11⁹⁵	$ 11⁹⁵
11-568	4	Kyocera KC120 PV module	$ 649⁰⁰	$ 2,596⁰⁰
13-578	1	Uni-Rac low profile for 4 modules (LP-104)	$ 275⁰⁰	$ 275⁰⁰
		Total:		$ 8,939⁷⁰
		MEMBER PRICE:		$ 8,492⁷²
Recommended Minimum Battery Pack:				
15-214	4	12v/98Ah gel battery	$ 169⁰⁰	$ 676⁰⁰

The Palo Alto—Comfort Level Kit—11.6 kWh/Day

Our mid-size system uses the same basic hardware as the previous system, but with a fuller complement of twenty 120-watt PV modules on low-profile mounts, to seriously shave kilowatts off the top of your electric bill.

Based on the North American yearly average of 5.5 hours sunlight per day, this system will deliver approximately 11.6kWh/day (still with plenty of future expansion capacity). Wiring is supplied locally, as required by your installation site. Qualifies for a California Energy Commission rebate of $9,037 on completion.

**12-240 The Palo Alto Intertie Kit
$20,543⁰⁰**

Item #	Qty.	Description	Unit Price	Total Price
27-293	1	PP-SW4048 w/C40 option	$5,253⁰⁰	$ 5,253⁰⁰
27-312	1	4/0, 10-ft. cable pair	$ 125⁰⁰	$ 125⁰⁰
15-785	10	4/0, 18″ interconnect cable	$ 19⁹⁵	$ 199⁵⁰
24-104	1	100A ground fault protection	$ 275⁰⁰	$ 275⁰⁰
24-214	1	12″ x 10″ x 4″ raintight j-box	$ 50⁰⁰	$ 50⁰⁰
25-194	1	DC lightning protector	$ 44⁹⁵	$ 44⁹⁵
24-228	1	Fused PV combiner box	$ 229⁰⁰	$ 229⁰⁰
24-643	1	15A fuse pack for combiner box (10 fuses)	$ 11⁹⁵	$ 11⁹⁵
11-568	20	Kyocera KC120 PV module	$ 649⁰⁰	$12,980⁰⁰
13-578	5	Uni-Rac low profile for 4 modules (LP-104)	$ 275⁰⁰	$ 1,375⁰⁰
		Total:		$20,543⁴⁰
		MEMBER PRICE:		$19,516²³
Recommended Minimum Battery Pack:				
15-214	8	12v/98Ah gel battery	$ 169⁰⁰	$ 1,352⁰⁰

The Topanga—Deluxe—23 kWh/Day

Using the same basic hardware package we started with, this is what an SW4048 inverter can do fully employed. Want more? The SW units are stackable. There's no limit. This system has forty 120-watt PV modules on low-profile mounts.

Based on the North American yearly average of 5.5 hours sunlight per day, this system will deliver approximately 23kWh/day. Wiring is supplied locally, as required by your installation site. Qualifies for a California Energy Commission rebate of $18,074 on completion.

**12-241 The Topanga Intertie Kit
$35,480⁰⁰**

Item #	Qty.	Description	Unit Price	Total Price
27-294	1	PP-SW4048 w/2 C40 options	$5,511⁰⁰	$ 5,511⁰⁰
27-312	1	4/0, 10-ft. cable pair	$ 125⁰⁰	$ 125⁰⁰
15-785	10	4/0, 18″ interconnect cable	$ 19⁹⁵	$ 199⁵⁰
24-105	1	200A ground fault protection	$ 325⁰⁰	$ 325⁰⁰
24-214	1	12″ x 10″ x 4″ raintight j-box	$ 50⁰⁰	$ 50⁰⁰
25-194	2	DC lightning protector	$ 44⁹⁵	$ 89⁹⁰
24-228	2	Fused PV combiner box	$ 229⁰⁰	$ 458⁰⁰
24-643	1	15A fuse pack for combiner box (10 fuses)	$ 11⁹⁵	$ 11⁹⁵
11-568	40	Kyocera KC120 PV module	$ 649⁰⁰	$25,960⁰⁰
13-578	10	Uni-Rac low profile for 4 modules (LP-104)	$ 275⁰⁰	$ 2,750⁰⁰
		Total:		$35,480³⁵
		MEMBER PRICE:		$33,706³³
Recommended Minimum Battery Pack:				
15-214	8	12v/98Ah gel battery	$ 169⁰⁰	$ 1,352⁰⁰

TYPICAL PAYBACK CALCULATION

Your Cost After Rebate $17,406
Approximate Installation Cost 2,000
Total System Cost $19,406
If you pay $0.20/kWh, payback = 11.6 years.
If you pay $0.30/kWh, payback = 7.7 years.

Intertie System Accessories

Outdoor Rated Fused PV Combiner Box

At Last an Affordable Code-Compliant Combiner Box!

This new PV combiner box from Trace Engineering provides a UL-listed solution for National Electric Code requirements on PV wiring. When wiring a PV array, many modules or substrings of modules have to be wired together. This box provides fusing on each substring of up to 15 amps per string and up to 10 substrings (total amperage is limited to 80 amps). Includes grounding bus bar. Fuses sold separately. USA.

24-228 Trace Fused PV Combiner Box $229⁰⁰

24-643 10 Pack of 15-Amp Fuses $11⁹⁵

DC Ground Fault Protection

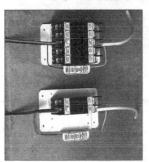

Trace is offering a Ground Fault Protection Device to meet NEC requirements for PV arrays that are mounted on a residential building. Available in one to four poles, with each pole capable of handling 100 amps of input. Must be installed in a customer-supplied electrical box. Our raintight box below is acceptable. Can be used with any system voltage. One- and four-pole versions shown. UL-listed. USA.

24-104 Trace 100A Ground Fault Protect $275⁰⁰

24-105 Trace 200A Ground Fault Protect $325⁰⁰

24-106 Trace 300A Ground Fault Protect $375⁰⁰

24-107 Trace 400A Ground Fault Protect $425⁰⁰

24-214 Raintight J-Box 12 x 10 x 4 $50⁰⁰

Lightning Protector

Simple, effective protection for any PV, wind, or hydro system up to 300 volts DC. These high-capacity silicone oxide arrestors will absorb multiple strikes until their capacity is reached and then rupture, indicating need for replacement. Three-wire connection, positive, negative, and ground, can be done at any convenient point in the system. For any AC lines that go outside, use the AC version. USA.

25-194 DC Lightning Protector $44⁹⁵

25-277 AC Lightning Protector $44⁹⁵

Safety Disconnect For Utility Intertie

Some utilities require a lockable disconnect between your intertie inverter and the main panel. This allows your friendly local lineman to indulge his paranoia by positively locking your intertie off. This one is a raintight, 60-amp, 3-pole, unfused model with knockouts from ½″ up to 1¼″. 13.5″ H x 8.5″ W x 4″ D. USA.

24-229 Safety Disconnect 60A 3-Pole Unfused $159⁰⁰

Fully Integrated Controls

Trace Power Panel

Fully integrated controls give you everything you need in one compact, well engineered package. We're presenting them first because for all but the most simple cabin systems, they make life much simpler and easier. Most of the remote home power systems we've sold in the past few years have been built around one of these integrated control packages.

Some intelligent manufacturers have taken the safety, charge control, and monitoring functions and combined them onto one tidy, preassembled, prewired, UL-approved board, or enclosed box. The technical staff at Real Goods endorses this concept wholeheartedly, as it generally results in safer, faster, better-looking installations that rarely, if ever, cause problems with building inspectors. Most integrated controls are easily adaptable for diverse system demands, and expandable in the future. So now, instead of all these bits and pieces that you (or your paid electrician) have to wire together on site, you can have it all delivered prewired.

The Ananda Power Technologies (APT for short) and Pulse Powercenters were the forerunners of this concept. Trace Engineering has also taken an interest in delivering integrated controls. The excellent Trace Power Panel offerings are the functional core of the majority of remote power systems being delivered now. The Power Panels also include an inverter or inverters in the preassembled package.

Integrated control packages are custom assembled to match the particular needs of a renewable energy system. You must communicate with one of our technical staff to order an integrated control. If there's anything better, cheaper, or safer for your application, we'll let you know during this interview period.

Now that we've got you all fired up about the simplicity of an integrated control system, please read on to understand the reasons for, and functions of, the charge controller, safety, and monitoring equipment that will be part of the integrated control system if you decide to go that route.

Trace Power Panel Systems

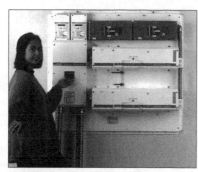

Trace DR Power Panel

Trace SW Power Panel

The Trace Power Panel is a pre-assembled, integrated "balance of systems" panel that provides charge control, system safety, inverter(s), and circuit breakers, all ready to hang on the wall. Power panels are available with a range of capacities, DC inputs, and AC output configurations, and are certified to UL standards as a complete system. Power Panels include all NEC required components, wiring, safety circuitry, and are NEMA 1 rated for indoor mounting. Power Panels offer the proven reliability and features of Trace inverters, the industry residential standard. Each Power Panel is shipped by truck in a wooden crate. USA.

Power Panels come ready for installation, pre-wired with the following:

- Inverter(s) of choice with adjustable three-stage battery charger
- 175 or 250-amp DC disconnect breaker(s)
- 60-amp AC disconnect/bypass breaker(s)
- Powder-coated steel backplate (40″ H x 43″ W)
- Pre-wired 500A/50mV shunt for remote monitoring
- Battery temperature sensor(s)
- Inverter conduit box(es)
- Negative/Ground bonding block

Inverter and Charge Control Options

Two inverter types are available. The DR-series inverters are lower-cost, modified sine wave units with sophisticated three-stage battery charging ability, but without automatic generator management. The SW-series inverters offer cleaner sine wave power, three-stage battery charging, automatic generator management, and the ability to intertie with utility power. Details of each inverter type are covered in the inverter section of this chapter. Each inverter model is available assembled with either single or dual inverters, signified at the end of the Power Panel model # as either /S or /D.

PV charge controller(s), extra DC circuit breakers, and battery cables are optional. See listings below. Please consult with one of our technical staff before placing an order. We'll be happy to help you put together the proper components.

DR-Series Power Panel Systems

Item Number	Model Number	AC Output	DC Input	Price
🚚 27-258	PP-DR1512/S	1.5kW	12VDC	$1,979⁰⁰
🚚 27-259	PP-DR1512/D	3.0kW	12VDC	$3,637⁰⁰
🚚 27-260	PP-DR2412/S	2.4kW	12VDC	$2,295⁰⁰
🚚 27-261	PP-DR2412/D	4.8kW	12VDC	$4,337⁰⁰
🚚 27-262	PP-DR1524/S	1.5kW	24VDC	$1,879⁰⁰
🚚 27-263	PP-DR1524/D	3.0kW	24VDC	$3,529⁰⁰
🚚 27-264	PP-DR2424/S	2.4kW	24VDC	$2,295⁰⁰
🚚 27-265	PP-DR2424/D	4.8kW	24VDC	$4,337⁰⁰
🚚 27-266	PP-DR3624/S	3.6kW	24VDC	$2,529⁰⁰
🚚 27-268	PP-DR3624/D	7.2kW	24VDC	$4,795⁰⁰

Voltage and frequency configurations are also available for international use. Please call.

🚚 *Means shipped freight collect.*

SW-Series Power Panel Systems

Item Number	Model Number	AC Output	DC Input	Price
🚚 27-271	PP-SW4024/S	4.0kW	24VDC	$4,995⁰⁰
🚚 27-272	PP-SW4024/D	8.0kW	24VDC	$8,995⁰⁰
🚚 27-273	PP-SW4048/S	4.0kW	48VDC	$4,995⁰⁰
🚚 27-274	PP-SW4048/D	8.0kW	48VDC	$9,197⁰⁰
🚚 27-275	PP-SW5548/S	5.5kW	48VDC	$5,695⁰⁰
🚚 27-276	PP-SW5548/D	11.0kW	48VDC	$9,995⁰⁰

Voltage and frequency configurations are also available for international use. Please call.

🚚 *Means shipped freight collect.*

Charge Control Options For Power Panels

The Trace C40 charge controller is available as an option prewired into any Trace Power Panel. It can also be supplied with the Digital Volt Meter option for the C40. The DVM can be installed on the C40, or with a 50′ cable for remote installation.

🚚 **27-277 PPO-C40 C40 DC Charge or Load Controller installed w/60A breaker $258⁰⁰**

🚚 **27-278 PPO-DVM/C40 Digital Meter Display installed on Controller $99⁰⁰**

🚚 **25-025 DVM Display w/ 50' cable $115⁰⁰**

Circuit Breaker Options For Power Panels

The DC disconnect box will accept up to four auxiliary breakers. Note that each charge controller option above comes with an installed 60A breaker.

🚚 **27-279 PPO-CD15 15-amp DC Load Breaker installed in DC Disconnect $59⁰⁰**

🚚 **27-280 PPO-CD20 20-amp DC Load Breaker installed in DC Disconnect $59⁰⁰**

🚚 **27-281 PPO-CD60 60-amp DC Load Breaker installed in DC Disconnect $59⁰⁰**

Optional Shipping Box
Becomes Battery Enclosure

This heavy-duty plastic shipping enclosure for the Power
Panel becomes a great battery box on site. ID measures
42″ x 46″ x 24″ tall. Fits 20 golf cart or L-16 batteries
comfortably.

15-749 Optional shipping/battery box $359⁰⁰

Battery And Inverter Cables

Provided in pairs using super flexible
welding-type cable, with color-coded
heat shrink on both ends. These
cables have crimped and soldered ⅜″
lug terminals on both ends. Use with
Pulse System Controls and Trace
Disconnects. Power Panels will
require one lug terminal to be cut off
the positive cable. USA.

See chart below for recommended cable size based
on inverter use.

27-311 4/0, 5-ft. Battery Cables, 2-lug pair $69⁹⁵

27-312 4/0, 10-ft. Battery Cables, 2-lug pair $125⁰⁰

15-099 2/0, 5-ft. Battery Cables, 2-lug pair $49⁹⁵

15-098 2/0, 10-ft. Battery Cables, 2-lug pair $89⁹⁵

CABLE SIZING FOR TRACE INVERTERS

Disconnect Size	Recommended with These Inverters	Minimum Cable Size
DC-175	DR-1512, 2412, 1524, & 2424. SW-3048E, & 4048	2/0
DC-250	DR-3624, SW-2512, 4024, & 5548	4/0

Safety and Fusing

Trace DC Disconnect

Any renewable energy system that stores energy for future use will contain a battery, or a number of batteries grouped together in a battery bank. The battery chapter will go into a full explanation of the hows and whys. We're here to talk about battery safety.

In many ways batteries are safer than traditional AC power. It's pretty difficult to shock yourself with the low voltage direct current (DC) in the 12- or 24-volt battery systems we commonly use. On the other hand, batteries, because they are right there locally, can deliver many more amperes into an accidental short circuit than a long-distance AC transmission system is capable of. These high amperage flows from battery systems can turn wires red-hot almost instantly, burning off the insulation and easily starting a fire. So Dr. Doug's #1 safety rule is: **Any wire that attaches to a battery must be fused!** (The National Electrical Code says the same thing, although they spend a lot more words saying it.) As if red-hot wires and melting insulation weren't enough, it's sometimes difficult to stop a DC short circuit. In a really severe short circuit, popping open a circuit breaker or fuse may not stop the electrical flow. The current will simply arc across the gap and keep on cooking! Once a DC arc is struck, it has much less tendency to self-snuff than an AC arc. Arc-welders greatly prefer using DC for this very reason, but impromptu welding is not the sort of thing you want to introduce into your house. Fuses that are rated for DC use a special snuffing powder inside to prevent an arc after the fuse blows. AC rated fuses rarely use this extra level of protection. Many DC circuit breakers require an upstream main fuse that's designed to stop very large (20,000-amp) current flows. All the Class T fuses we offer have this ability and rating.

The National Electrical Code wisely requires a fuse and a safety disconnect between any power source and any appliance. Our renewable energy systems have two or more power sources, the RE generator(s) (PV, wind, hydro) and the battery bank.

Inverter safety requires some specialized equipment. When the 1990 National Electrical Code, the first to address low voltage systems, came out there was no product to meet the needs of full-size inverters. These large inverters can safely handle up to 800 amps at times. Existing AC equipment either produced unacceptable voltage drop at high amperage flows, or was prohibitively expensive. Pulse and Trace Engineering have provided several solutions to our problem. The integrated full-system solution is to use one of the Pulse System Controls which provides all the fusing, safety disconnect, monitoring, and controls for the entire system in one compact, pre-wired, engineered, UL-approved package.

For systems that run most electrical appliances at 120 volts AC, and therefore run most power through an inverter, the new Trace DC Disconnect may be the most cost-effective way to fuse your system. This product can provide fusing and disconnect means for one or two inverters, a charge controller, and up to three additional DC loads.

For mid-size, 400- to 1000-watt inverters a 110-amp class T fuse assembly with an appropriate Blue Sea circuit breaker makes a cost-effective and legal safety disconnect. The small inverters of 100 to 300 watts simply plug or hardwire into a fused outlet.

Fuses are sized to protect the wires in a circuit. See the chart near the fuse listings for standard sizing/ampacity ratings. These ratings are for the most common wire types. Some types of wire can handle slightly more current, and all types of wires can safely handle much more current for a few seconds. If a particular appliance, rather than the wire, wants protection at a lower amperage, then usually a separate or inline fuse is used just for that appliance.

Safety Equipment

Automatic Transfer Switch

Transfer switches are designed as safety devices to prevent two different sources of voltage from traveling down the same line to the same appliances. This transfer switch made by Todd Engineering will safely connect an inverter and an AC generator to the same AC house wiring. If the generator is not running, the inverter is connected to the house wiring. When the generator is started, the house wiring is automatically disconnected from the inverter and connected to the generator. A time-delay feature allows the generator to start under a no-load condition and warm up for approximately one minute. Available in both 110-volt and 240-volt models. Each will handle up to a maximum of 30 amps. These switches are great for applications where utility power may be available only a few hours per day, or if frequent power outages are experienced. Housed in a metal junction box with hinged cover. Wires are clearly marked and installation schematic is included inside cover. Two-year warranty. USA.

23-121 Transfer Switch, 30A, 110V $55⁰⁰

23-122 Transfer Switch, 30A, 240V $99⁰⁰

50-Amp/3-Pole Transfer Switch

For those who need to transfer more than 30 amps, we offer this larger model with three heavy-duty DPDT 50-amp contactors. This contactor can be used with either 120-volt or 240-volt AC. Has a 15-second time delay to allow generator start-up under no load. Clearly marked, easy terminal strip wiring. Schematic included inside cover. 12″ x 10″ x 4″ enclosure with multiple knockouts provided. Two-year warranty. USA.

23-118 Transfer Switch, 50A, 3 pole $170⁰⁰

Blue Sea DC Circuit Breakers—High Amperage Protection, Small, Inexpensive Package

Here is a compact DC circuit breaker at an affordable price. These sealed, surface-mounting breakers are water- and vaporproof, they are thermally activated, and manually resettable, with the reset handle providing a positive trip identification. The red trip button can also be manually activated, allowing the breaker to function as a switch. They are "trip-free," and cannot be held closed against an overload current. Thermal tripping allows high starting loads, without tripping. It has ¼″ studs for easy electrical connection.

These simple to install, surface-mount breakers are an excellent low-cost choice for small to midsized inverters, wind generators, control and power diversion equipment, or any other application with current flows of 50 to 150 amps. For use with DC power systems up to 30 volts. USA.

Rated at 3,000-amps interrupting capacity. Requires a class T fuse inline before the circuit breaker for NEC compliance. See below.

25-167 Blue Sea 50A Circuit Breaker $34⁹⁵

25-168 Blue Sea 75A Circuit Breaker $34⁹⁵

25-169 Blue Sea 100A Circuit Breaker $34⁹⁵

25-183 Blue Sea 125A Circuit Breaker $34⁹⁵

25-184 Blue Sea 150A Circuit Breaker $34⁹⁵

Class T DC Fuse Blocks With Covers

For those who want to protect their midsized inverter or other DC systems we also offer the excellent DC-rated, slow-blowing Class T fuses with block, cover, and #2/0 lugs. Dimensions are 2″ x 2.5″ x 5.5″, including the cover. Rated for DC voltage up to 125 volts. USA.

Does not provide a disconnect means, and so will not satisfy NEC code unless used with the Blue Sea circuit breakers above.

24-518 Fuse Block, 300A Fuse & Cover $75⁰⁰

24-212 Fuse Block, 200A Fuse & Cover $52⁹⁵

24-507 Fuse Block, 110A Fuse & Cover $52⁹⁵

24-217 300-Amp Class T Replacement Fuse $38⁰⁰

24-211 200-Amp Class T Replacement Fuse $17⁹⁵

24-508 110 Amp Class T Replacement Fuse $17⁹⁵

Safety Disconnects

Every PV system should employ a fused disconnect for safety. The function of this component is to disconnect all power-generating sources and all loads from the battery so that the system can be safely maintained and disconnected in emergency situations with the flick of a switch. We carry several different safety disconnects for different sized applications. Fuses should be sized according to the wire size you are using.

We recommend using a 2-pole disconnect for the charging side of your batteries. One pole is used for the line between the panels and charge controller, and the other between the controller and the batteries. Make sure this fused disconnect can handle the maximum amperage of your array output. Another fuse or fused switch can be used to handle the discharge of your battery. Raintight boxes should be used if the box will be exposed to the weather.

The 60-amp 3-pole fused disconnect should be employed when the Enermaxer voltage regulator is used, which requires the third pole for disconnection. USA.

24-201 30A 2-Pole Safety Disconnect $34⁹⁵

24-644 30A 2-Pole Safety Disconnect, Raintight $54⁹⁵

24-404 30A 3-Pole Safety Disconnect $99⁰⁰

24-202 60A 2-Pole Safety Disconnect $54⁹⁵

24-645 60A 2-Pole Safety Disconnect, Raintight $95⁹⁵

24-203 60A 3-Pole Safety Disconnect $99⁰⁰

◈ 24-204 100A 2-Pole Safety Disconnect $144⁹⁵

Fuses sold separately at right.

Minimum Protection Fuseblocks

For simple, inexpensive protection we offer larger diameter, lower resistance, fuseblocks that offer more contact surface area than automotive-type fuses. Single fuseblocks with spring clips. 30A block takes up to 10 gauge, 60A block takes up to 2 gauge. Order fuses separately, below. USA.

24-401 30A Fuseblock $9⁹⁵

24-402 60A Fuseblock $12⁹⁵

DC-Rated Fuses

These Class RK5 fuses are rated for DC service up to 125 volts. They are time delay types that will allow motor starting or other surges. The 45A fits a 60A holder. Always buy extras to have a spare on hand! USA.

24-501 30A DC Fuse $3⁹⁵

24-502 45A DC Fuse $7⁹⁵

24-503 60A DC Fuse $7⁹⁵

24-505 100A DC Fuse $14⁹⁵

WIRE AMPACITY CHART

Wire Gauge	Maximum Ampacity
14	15
12	20
10	30
8	45
6	65
4	85
2	115
0	150
2/0	175
4/0	250

DC Ground Fault Protection

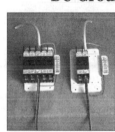

Trace is offering a Ground Fault Protection Device to meet NEC requirements for PV arrays that are mounted on a residential building. Available in one to four poles, with each pole capable of handling 100 amps of input. Must be installed in a customer-supplied electrical box. Our raintight box below is acceptable. Can be used with any system voltage. One- and four-pole versions shown. UL-listed. USA.

Provided without enclosure.

24-104 Trace 100A Ground Fault Protect $275⁰⁰

24-105 Trace 200A Ground Fault Protect $325⁰⁰

24-106 Trace 300A Ground Fault Protect $375⁰⁰

24-107 Trace 400A Ground Fault Protect $425⁰⁰

24-214 Raintight J-Box 12 x 10 x 4 $50⁰⁰

Trace DC Disconnects

To provide DC system overload/disconnect equipment for everything passing in or out of the battery bank, here's an alternative that meets all the code requirements, and costs substantially less. Best used with systems that only use DC power for one or two loads, the Trace DC Disconnects are available in two basic versions, 175 amps and 250 amps with a variety of options.

No fumbling for expensive replacement fuses in the dark, these units feature easy to reset circuit breakers. Breakers are rated for 25,000-amps interrupting capacity at 65 volts, and are UL-listed for DC systems up to 125 volts. Main breaker lugs accept up to #4/0 AWG fine strand cable, no ring terminals required. .75″, 2″, and 2.5″ knockouts are provided. Mates up with the conduit box option for DR and SW series Trace inverters. (Matching color scheme, too.) Space is available inside the disconnect for the DC current shunt(s) required by most monitors.

Options, all of which can be installed in the field, include a second main breaker for dual inverter systems, a DC bonding block for negative and ground wires, and up to four 15- or 60-amp circuit breakers for PV or other charging input, or DC power output to loads.

The Negative/Ground bonding block provides lugs for up to four 4/0, two #1/0, two #2, and four #4 cables.

A properly sized DC Disconnect with a 60-amp auxiliary breaker for the PV input gives all the protection an inverter-based system requires. Add another 15-amp auxiliary breaker if you've got a DC water pump or fridge to run, and you're covered. USA.

25-010 Trace DC175-amp Disconnect $329⁰⁰
25-011 Trace DC250-amp Disconnect $329⁰⁰
25-012 Trace 2nd 175-amp Breaker $195⁰⁰
25-013 Trace 2nd 250-amp Breaker $195⁰⁰
25-014 Trace 15-amp Auxiliary Breaker $25⁰⁰
25-008 Trace 20-amp Auxiliary Breaker $29⁰⁰
25-015 Trace 60-amp Auxiliary Breaker $40⁰⁰
25-016 Trace Neg/Grd Bonding Block $50⁰⁰

DC DISCONNECT SIZING FOR TRACE INVERTERS		
Disconnect Size	Recommended With These Inverters	Minimum Cable Size
DC-175	DR-1512, 2412, 1524, & 2424. SW-3048E, & 4048	2/0
DC-250	DR-3624, SW-2512, 4024, & 5548	4/0

Code Approved DC Load Centers

The continuing move toward inverters running more and more of the household loads has reduced our selection of DC load centers. We find that a simple six-circuit center will handle the DC fusing needs for 99% of the systems. The six-circuit center meets all safety standards and utilizes only UL-listed components. All current-handling devices have been UL-approved for 12-volt through 48-volt DC applications.

These DC circuit breakers require a main fuse inline before the breakers to provide catastrophic overload protection. If you already have a large Class T fuse for your inverter it can be used for this protection. If not, you must use the 110-Amp Class T Fuse Block on the input. USA.

Breakers must be ordered separately.

Load center does not include the 110-amp main fuse.

23-119 6-Circuit Load Center $49⁹⁵
24-507 Fuse Block, 110A Fuse and Cover $52⁹⁵
23-131 10A Circuit Breaker $15⁹⁵
23-132 15A Circuit Breaker $15⁹⁵
23-133 20A Circuit Breaker $15⁹⁵
23-134 30A Circuit Breaker $15⁹⁵
23-136 40A Circuit Breaker $15⁹⁵
23-135 50A Circuit Breaker $15⁹⁵
23-140 60A Circuit Breaker $19⁹⁵
23-137 70A Circuit Breaker $29⁹⁵

DC Lightning Protector

Here is simple, effective protection that can be used on any PV, wind, or hydro system up to 300 volts DC. These high-capacity silicon oxide arrestors will absorb multiple strikes until their capacity is reached and then rupture, indicating need for replacement. Three-wire connection, positive, negative, and ground, can be done at any convenient point in the system. The AC unit is identical operation and capacity; only the wire color codes vary. The 4-wire unit is for wind turbines with 3-phase output (Whisper and Bergey turbines). USA.

25-194 DC Lightning Protector LA302 DC $44⁹⁵

25-277 AC Lightning Protector LA302 R $44⁹⁵

25-749 4-Wire Lightning Protector $59⁹⁵

ATC-Type DC Fusebox

Using the newer, safer, ATC fuse style, this covered plastic box has wire entry ports on sides. 12 individually fused circuits with a pair of main lugs for battery positive, and a negative bus bar. Will accept up to 10-gauge wire. Requires a Torx T-15 screwdriver. 8.75″ x 5.5″ x 2″. USA.

24-218 ATC-Type Fusebox $39⁹⁵

Inline ATC Fuse Holder

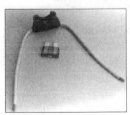

A simple inline holder for any ATC-type fuse. With 14-gauge pigtails. Not for exposed outdoor use.

25-540 ATC Inline Fuse Holder $2⁹⁵

ATC-Type DC Fuses

These are the newer plastic body DC fuses that most automotive manufacturers have been using lately. Compared to old-style round glass fuses, they're safer, as energized metal parts aren't exposed, they have more metal to metal surface contact, the amp rating is easier to read and color-coded, and with the different colors they look cool! These come in little boxes of 5 fuses. USA.

24-219 ATC 2A Fuse $2⁹⁵

24-220 ATC 3A Fuse $2⁹⁵

24-221 ATC 5A Fuse $2⁹⁵

24-222 ATC 10A Fuse $2⁹⁵

24-223 ATC 15A Fuse $2⁹⁵

24-224 ATC 20A Fuse $2⁹⁵

24-225 ATC 25A Fuse $2⁹⁵

24-226 ATC 30A Fuse $2⁹⁵

15-Position Bus Bar

You can use this 15-position bus bar as a common negative for any fuse block.

24-431 15-Position Bus Bar $11⁹⁵

Rain-Tight Junction Boxes & Power Distribution Block

To make your wiring at the PV array easier, mount our 2-pole power distribution block inside a rain-tight junction box. This is the easy way to join your large lead-in cables to the smaller interconnect wires from the PV array. Insert cables and tighten set screws. Primary side accepts one large cable, #6 to 350 MCM, secondary side accepts six smaller cables, #14 to #4. For use with copper or aluminum conductors.

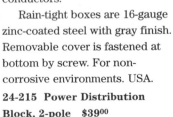

Rain-tight boxes are 16-gauge zinc-coated steel with gray finish. Removable cover is fastened at bottom by screw. For non-corrosive environments. USA.

24-215 Power Distribution Block, 2-pole $39⁰⁰

24-213 Junction Box 10″ x 8″ x 4″ $40⁰⁰

24-214 Junction Box 12″ x 10″ x 4″ $50⁰⁰

Controllers

Whenever batteries are used in a renewable energy system, we need a controller to prolong battery life. The most basic function of a controller is to prevent battery overcharging. If batteries are allowed to routinely overcharge, their life expectancy will be dramatically reduced. A controller will sense the battery voltage, and reduce or stop the charging current when the voltage gets high enough. This is especially important with sealed batteries where we can't replace the water that is lost during overcharging.

The only exception to controller need is when the charging source is very small and the battery is very large in comparison. If a PV module produces 1.5% of the battery's ampacity or less, then no charge control will be needed. For instance, a PV module that produces 1.5 amps charging into a battery of 100 amp-hours capacity won't require a controller as the module will never have enough power to push the battery into overcharge.

PV systems generally use a different type of controller than a wind or hydro system requires. PV controllers can simply open the circuit when the batteries are full without any harm to the modules. Do this with a rapidly spinning wind or hydro generator and you will quickly have a toasted controller, and possibly a damaged generator. Rotating generators make electricity whenever they are turning. With no place to go, the voltage will escalate rapidly until it can jump the gap to some lower voltage point. These mini lightning bolts can do damage. With rotating generators we generally use diversion controllers that take some power and divert it to other uses. Both controller types are explained below.

PV Controllers

SunSaver Charge Controller

Most PV controllers simply open or restrict the circuit between the battery and PV array when the voltage rises to a set point. Then, as the battery absorbs the excess electrons and voltage begins dropping, the controller will turn back on. With some controllers these voltage points are factory preset and nonadjustable, while others can be adjusted in the field. Earlier PV controllers used a relay, a mechanically controlled set of contacts, to accomplish this. Newer solid state controllers use power transistors and PWM (Pulse Width Modulation) technology to rapidly turn the circuit on and off, effectively floating the battery at a set voltage. PWM controllers have the advantage of no mechanical contacts to burn or corrode, but have the disadvantage of greater electronic complexity. Both types are in common use, with most new designs favoring PWM, as electronics have been gaining greater reliability with experience.

Controllers are rated by how much amperage they can handle. National Electric Code regulations require controllers to be capable of withstanding 25% over amperage for a limited time. This allows your controller to survive the occasional edge-of-cloud effect, when sunlight availability can increase dramatically. Regularly or intentionally exceeding the amperage ratings of your controller is the surest way to turn your controller into a crispy critter. It's perfectly okay to use a controller with more amperage capacity than you are generating. In fact, this is a piece of hardware where buying larger to allow future expansion is often smart planning, and usually doesn't cost much.

A PV controller usually has the additional job of preventing reverse current flow at night. Reverse current flow is the tiny amount of electricity that can flow backwards through PV modules at night, discharging the battery. With smaller one or two module systems the amount of power lost to reverse current is really negligible. A dirty battery top will cost you far more power loss. Only with larger systems does it become anything to be concerned about. Much has been made of reverse current loss in the past, and almost all charge controllers now deal with it automatically. Most of them do this by sensing that voltage is no longer available from the modules, when the sun has set, and then opening the relay or power transistor. A few older or simpler designs still use diodes—a one-way valve for electricity—to accomplish this, but the relay, or the power transistor, has become the preferred method. (See sidebar on Dinosaur Diodes.)

Hydroelectric & Wind Controllers

Trace C-40 Diversion Controller

Hydroelectric and wind controllers have to use a different strategy to control battery voltage. A PV controller can simply open the circuit to stop the charging and no harm will come to the modules. If a hydro or wind turbine is disconnected from the battery while still spinning it will continue to generate power, but with no place to go, the voltage rises dramatically until something gives. With these types of rotary generators we usually use a diverting charge controller. Examples of this technology are the Trace C35, C40, or C60 series of controllers, or the Enermaxer controllers. A diverter-type control will monitor battery voltage, and when it reaches the adjustable set point, will dump excess power into some kind of dummy load. A heater element is the most common dummy load, but DC incandescent lights can be used also. (Incandescent lights are nothing but heater elements that give off a little incidental light, anyway.) Both air and water heater elements are commonly used, with water heating being the most popular. It's nice to know that any power beyond what is needed to keep your batteries fully charged is going to heat your household water or hot tub. Diversion controllers can be used for PV regulation as well, in addition to hydro or wind duties in a hybrid system with multiple charging sources. This controller type is rarely used for PV regulation alone, due to its higher cost, and higher internal power consumption.

Controllers for PV-Direct Systems

Dinosaur Diodes!?

When PV modules first came on the market a number of years ago it was common practice to use a diode, preferably a special ultra-low forward-resistance Schottky diode, to prevent the dreaded reverse current flow at night. Early primitive charge controllers didn't deal with the problem, so installers did, and gradually diodes achieved a mystical must-have status. Over the years the equipment has improved and so has our understanding of PV operation. Even the best Schottky diodes have 0.5 volt to 0.75 volt forward voltage drop. This means the

— continued on next page

A PV-direct system connects the PV module directly to the appliance we wish to run. They are most commonly used for water pumping and ventilation. By taking the battery out of the system, initial costs are lower, control is simpler, and maintenance is virtually eliminated.

Although they don't use batteries, and therefore don't need a control to prevent overcharging, PV-direct systems often do use a device to boost pump output in low light conditions. While not actually a controller, these booster devices are closer kin to controllers than anything else, so we'll cover them here. The common name for these booster devices has become LCB; short for Linear Current Booster. Technically (and legally) the term LCB belongs to Bobier Electronics who pioneered the technology, and whose products we're proud to offer.

LCBs or Linear Current Boosters

LCBs are usually only used on simple PV-direct systems where there are no batteries. They are a solid state device that helps motors start and keep running under low light conditions. The LCB accomplishes this by taking advantage of some PV module and DC motor operating characteristics. When a PV module is exposed to light, even very low levels, the voltage jumps way up immediately, though the amperage produced at low light is very low. A DC motor on the other hand, wants just the opposite conditions to start. It wants lots of amps, but doesn't care much if the voltage is low. The LCB provides what the motor wants by down-converting some of the high voltage into amperage. Once the motor starts, the LCB will automatically raise the voltage back up as much as power production conditions allow without stalling the motor. Meanwhile, on the PV module side of the LCB, the module is being allowed to operate at its maximum power point, which is usually a higher voltage point than the motor is operating at. It's a constant balancing act until the PV module gets up closer to full output, when, if everything has been sized correctly, the LCB will check out of the circuit, and the module will be connected directly to the motor. LCBs cost us something in efficiency, which is why we like to get them out of the circuit as the modules approach full power. All that fancy conversion comes at a price in power loss. LCBs usually suffer about a 10% to 20%

— continued from previous page

module is operating at a slightly higher voltage than the battery. The higher the module voltage, the more electrons that can leak through the boundary layer between the positive and negative silicon layers in the module. These are electrons that are lost to us; they'll never come down the wire to charge the battery.

The module's 0.5 volt higher operating voltage usually results in more power being lost during the day to leakage than the minuscule reverse current flow at night we're trying to cure. Modern charge controllers use a relay or power transistor, which has virtually zero voltage drop, to connect the module and battery. The relay opens at night to prevent reverse flow. Diodes have thus become dinosaurs in the PV industry. Larger, multimodule systems may still need blocking diodes if partial shading is possible. Call our tech staff for help.

conversion loss, but the gains in system performance more than make up for this. An LCB can boost pump output as much as 40% by allowing the pump to start earlier in the day and run later. Under partly cloudy conditions an LCB can mean the difference between running or not. We usually recommend them strongly with most PV-direct pumping systems. With PV-direct fan systems, where start-up isn't such a bear, LCBs are less important.

Ask Dr. Doug:

To Track or to Boost?
Which delivers more power?

Until recently, tracking mounts have been the only way to increase the daily output from your PV array. By following the sun from east to west, trackers can increase output as much as 30%. Tracking works best in the summer, when the sun makes a high arc across the sky. Tracking is the best choice for power needs that peak in the summer, like water pumping and cooling equipment. In the winter, tracking will increase output by only 10%.

What if you're looking for more winter output? The new booster charge controls can increase your cold-weather output up to 30%. Boost controls are a refinement of linear current boosting which allows the PV module to run at its maximum power point, while downconverting the excess voltage into amps. Boost controls perform best with cold modules and hungry batteries, a wintertime natural! Boost controllers also allow wiring and transmitting at higher voltage. A 48-volt array can charge a 24-volt battery, for instance. This saves further in wire costs and reduces transmission losses.

Bottom line: For most residential power systems a booster control will do more for your power well-being than a tracker.

PV, Hydro, & Wind Charge Controllers

Morningstar SunGuard PV Charge Controller

A Small Controller at the Right Price!

For systems with a single module of 75 watts or smaller, you can't beat this controller. Featuring a five-year warranty, temperature compensation, 25% overload rating, extreme heat (and cold) tolerance, reverse current leakage less than 0.01 mA, solid-state series type 0 to 100% duty cycle PWM, epoxy encapsulated circuitry, 1500W transorb lightning protection and more! For 12-volt systems only, it is rated for up to 4.5 amp PV and regulates at 14.1 volts. Okay for sealed batteries. USA.

**25-739 Morningstar SunGuard
4.5-Amp 12-Volt Controller $29⁹⁵**

8-Amp PWM PV Charge Controller

The 8-amp Mini is a series-type PWM solid-state controller with no relays to wear out. Two LEDs indicate the operation of the controller and relative battery charge condition. Featuring an adjustable set point (for sealed or vented batteries), five-year warranty, reverse current leakage protection, reverse polarity and short circuit protection, and full circuit encapsulation for outdoor weather resistance. One unit is for small 12-volt applications and the other is for 36-volt applications. USA.

**25-233 Mini 12/8A Adjustable 8-Amp 12-Volt PV
Charge Controller $44⁹⁵**

**25-726 Mini 36/8A Adjustable 8-Amp 36-Volt PV
Charge Controller $54⁹⁵**

Marine/RV Charge Controllers

Sun Selector produces a special pair of controllers for marine or RV applications with two separate battery banks. These controllers will charge both battery banks, with the bulk of charging going to the lowest battery, but will not allow one battery to discharge into the other. Features include automatic night-time disconnect; automatic reduction of voltage cut-off point as the battery approaches full charge to minimize gassing; four LED indicators to show charger condition; and full watertight encapsulation for outdoor or marine use. Both controllers measure only 2″ x 2″ x 1.25″ and are simple to install, with just two input wires and a pair of output wires for each battery bank. Inline fuses are included on the battery plus wires. The M8 will handle up to 8 amps of charging current; the M16 up to 16 amps. These units are set up for conventional wet-cell lead-acid batteries; gel-cell units are available by special order. Five-year warranty. USA.

25-127 M8M Dual Battery Controller, 12V $99⁹⁵
25-129 M16M Dual Battery Controller, 12V $139⁹⁵

SunSaver 6-Amp And 10-Amp Controllers

The SunSaver 6-amp and 10-amp controllers provide high-reliability, low-cost PWM charge control for small 12-volt systems. Featuring an easy-to-access, graphic terminal strip for connections, a weather-resistant anodized aluminum case, fully encapsulated potting, reverse current protection, and field selection of battery type at either 14.4V for flooded, or 14.1V for sealed, batteries. Has an LED to show charging activity, and features automatic temperature compensation, based on the temperature of the controller. Accepts up to #10 gauge wire. 6″ x 2.2″ x 1.3″, 8 oz. Five-year mfr's. warranty. USA.

25-762 SunSaver 6A/12V Controller $47⁹⁵
25-763 SunSaver 10A/12V Controller $54⁹⁵
25-776 SunSaver 10A/24V Controller $59⁹⁵

Sunlight 10-Amp And 20-Amp Controllers

A Simple PV and Light Control

The Sunlight 10-amp and 20-amp are simple, 12-volt solid-state PWM charge controllers with automatic light control, and low voltage disconnect for battery protection. Connect a battery, a PV panel, and a 12V gizmo, and you've got a system that automatically turns on your sign lighting, stairway light, or other PV-powered task at dusk. The Sunlight 10 will handle up to 10 amps input or output; the Sunlight 20, up to 20 amps. If voltage falls below 11.7V, a low voltage disconnect occurs, and will not let the light run until battery voltage recovers to 12.8V. A "test" button overrides LVD for five minutes. Load control has 10 settings from off through dusk to dawn, including three unique settings that run a few hours after dusk, turn off, then run an hour or two before dawn. Charge control can be set for either sealed batteries at 14.1V or wet-cell batteries at 14.4V with a simple jumper, and provides reverse current protection at night. Automatic charge voltage temperature compensation is built in. Reverse polarity protected. Five-year mfr.'s warranty. USA.

25-742 Sunlight 10 PV/Load control $109⁰⁰

25-743 Sunlight 20 PV/Load control $139⁰⁰

Trace C-12 Charge/Load/Lighting Controller

This controller can do multiple functions. For starters it's an electronic, 3-stage PV charge controller with easily adjustable bulk and float voltages, nighttime disconnect, and automatic monthly equalization cycles.

It's also a 12-amp automatic load controller with 15-amp surge and electronic overcurrent protection, automatically reconnecting at 12-second and then 1-minute intervals. Voltage set points for connect and disconnect are also easily adjustable. Loads will blink 5 minutes before disconnect, allowing the user to reduce power use. Even after disconnect it allows one 10-minute grace period by pushing an override button.

Plus, it's an automatic lighting controller that will turn on at dusk for sign or road lighting. Run time is user adjustable from two to eight hours, or dusk to dawn. If voltage gets too low, low voltage disconnect will override run time to protect the battery.

The LED mode indicator will show the approximate state of charge and indicate if low voltage disconnect, overload, or equalization has occurred.

Controller is compatible with any 12V battery type including sealed or nicads. Interior terminal strip will accept up to #10 AWG wire. The box is rain-tight, and the electronics are conformal coated. Certified by ETL to UL specs. Mfr.'s warranty is two years. USA.

27-315 Trace C-12 Multi Controller $109⁰⁰

27-319 Battery Temperature Probe $29⁰⁰

BZ Products Model 12 Controller And Monitor

The Model 12 is a modern 12-volt, 12-amp, solid-state PWM controller with a large LCD digital meter displaying either battery voltage or PV charging current. For small systems, this is all the control and monitoring your system needs. Has an accurate three-digit display, including tenths of a volt and tenths of an amp. Features include easily user adjustable float voltage, built-in temperature compensation, internal PV and battery fusing, a manual equalization switch, lightning protection, and LED indicators for float voltage or equalizing. Has two 1⁄2″ knock-out fittings in base, wire terminals will accept up to #6 gauge, and has reverse polarity protection. Measures 6.4″H x 4.7″W x 1.5″D, weighs 1.0 lb. Five-year mfr's warranty. USA.

25-764 BZ Model 12 Controller/Monitor $99⁹⁵

ProStar PV Charge Controllers

Control and Monitoring in a Single Package

MorningStar is one of the largest and most reliable PV controller manufacturers in the world. We are proud to bring their best charge controls into the Real Goods fold.

The ProStar line, with models at 15 or 30 amps, features solid-state, constant-voltage PWM control, automatic selection of 12- or 24-volt operation, simple manual selection of sealed or flooded battery charging profile, night time disconnect, built-in temperature compensation, automatic equalize charging, reverse-polarity protection, electronic short circuit protection, moderate lightning protection, and roomy terminal strips that accept up to #6 AWG wire sizes. A built-in LCD display scrolls through battery voltage, charging current, and load current for any loads connected directly to the controller. An automatic low-voltage disconnect will shut off connected loads if batteries get dangerously low for more than 55 seconds. Maximum internal power consumption is a minimal 22 milliamps. Dimensions are 6.01″ x 4.14″ x 2.17″. Five-year mfr.'s warranty. USA.

25-729 ProStar 30-amp PV Controller w/ meter $219⁰⁰

25-771 ProStar 15-amp PV Controller w/ meter $179⁰⁰

Trace C-Series
Multi-Function DC Controllers

The new C-series of solid-state controllers has grown. Now featuring models at 35, 40, and 60 amps, these multi-controllers can be used for PV charge control, DC load control, or DC diversion. They only operate in one mode, however, so PV charge control and DC load control requires two controllers. The 35- and 60-amp controllers can be manually selected for 12 or 24 volts. The 40-amp controller will do 12, 24, or 48 volts. All setpoints are field adjustable with removeable knobs to prevent tampering. They are electronically protected against short-circuit, overload, overtemp, and reverse polarity with auto-reset. No fuses to blow! Two-stage lightning and surge protection is included. The three-stage (bulk, absorb, float) PV charge control uses solid-state pulse width modulation control for the most effective battery charging. Has an automatic "equalize" mode every 30 days, which can be turned off for sealed batteries, and a manually selected "nicad" mode is included. A battery temperature sensor is optional, and recommended if your batteries will see temps below 40°F or above 90°F. The diversion control mode will divert excess power to a dummy heater load, and offers the same adjustments and features as the PV control. Your dummy load must have a smaller amperage capability than the controller. The DC load controller has adjustable disconnect and auto-reconnect voltages with a time delay for heavy surge loads. Electronic short-circuit protected. 9″ x 5″ x 2″ and weighs 4 lb. Certified by ETL to UL standards. Two-year mfr.'s warranty. USA.

25-027 **Trace C-35 Multi-Controller** $119⁰⁰

25-017 **Trace C-40 Multi-Controller** $159⁰⁰

25-028 **Trace C-60 Multi-Controller** $199⁰⁰

Optional LCD Display

The optional LCD display with backlighting continuously displays battery voltage, DC amperage, cumulative amp-hours, and a separate resettable "trip" amp-hour meter (only displays current flows passing through charge controller). Can be mounted on the front of any C-series charge controller, or used remotely with simple phone cable and plugs. USA.

25-018 **Trace C-Series LCD Display** $89⁹⁵

Solar Boost Charge Controllers

Solar Boost 2000

Solar Boost 50 & Optional Remote Display

PV modules put out higher voltage than your batteries can accept; that's lost power. Solar Boost controllers run modules at their maximum power point, and use linear current booster technology to downconvert excess voltage into amperage your batteries can accept. With low-voltage hungry batteries and high-voltage cold modules (wintertime, in other words), the amperage boost will realistically reach 30%. The LCD display shows amps in and amps out to easily gauge actual performance.

There are several Solar Boost models featuring solid-state PWM 3-stage charge control, with adjustable charge voltage and electronic overcurrent protection. A battery temperature option is available for batteries exposed to temperatures below 40°F or over 80°F. Amperage rating is on *output*. Reduce by 15% for input amp rating.

The smaller Solar Boost 2000E is open-frame construction for panel mounting. This 12-volt, 25-amp control comes with a built-in LCD display for battery voltage, amps in, and amps out.

The larger Solar Boost is available in either a 50-amp, 12/24 volt unit, or a 30-amp, 24/48 volt unit, both fully enclosed with conduit knockouts. The optional LCD display is either built in, or a remote display. These larger models allow running the PV array at the higher voltage, while charging the battery at the lower voltage. All Solar Boost products are made in the USA, have a three-year mfr's warranty and now feature ETL listing to UL 1741 specs.

25-744 **Solar Boost 2000E 12V/25A Controller** $225⁰⁰

25-745 **Solar Boost 50L 12/24V Controller No LCD** $419⁰⁰

25-746 **Solar Boost 50L 12/24V Controller w/LCD** $499⁰⁰

25-765 **Solar Boost 30L 24/48V Controller No LCD** $479⁰⁰

25-001 **Solar Boost 30L 24/48V Controller w/LCD** $559⁰⁰

25-747 **Solar Boost LCD Remote Display** $115⁰⁰

25-748 **Solar Boost Battery Temp Sensor** $29⁹⁵

Dr. Doug recommends: Boost controllers are the best way to increase PV power output in the winter.

Enermaxer III Universal Battery Voltage Controller

The Enermaxer III Universal Battery Voltage Controller is an excellent voltage controller that has proven extremely reliable for hydro and wind systems. It is now available in a 120-amp version for 12 through 36 volts. This is the controller of choice for hydro systems, but will work with multiple charging sources at once (wind, hydro, PV). It is a parallel shunt regulator that diverts any excess charging power to a dummy heating load, typically a water- or air-heating element. It is made of all solid-state components so that there are no relay contacts to wear out. The desired float voltage is selected with the front-mounted rheostat. During charging, as battery voltage rises, just enough power will be diverted to maintain the float voltage. The battery will float at any user-selected voltage and any excess power, up to the maximum amperage rating for the Enermaxer, will be diverted.

When using the Enermaxer, you must have an external load at least equal to your charging capacity. This load is usually either a water-heating element or air-heating element. We offer elements below. Enermaxer is set for 24V operation from the factory, but is easily adjusted for 12, 32, or 36 volts on site. Two-year mfr.'s warranty. USA.

25-153 Enermaxer III 120A Controller 12V, 24V, 32V, 36V $349⁰⁰

120-AMP ENERMAX SPECIFICATIONS

Battery Voltage	Adjustment Range	Maximum Amperage	Idle Amperage
12	13.3–17.0	120	0.01
24	26.6–34.0	120	0.01
32	35.5–45.0	120	0.01
36	39.9–51.0	120	0.01

60-AMP ENERMAX SPECIFICATIONS

48	53.2–68.0	60	0.01

48V Enermax model not currently available. May be in production again late 2001; please call for details.

Water Heating Elements

Industrial-grade elements designed for years of trouble-free service. Screws into standard 1″ water heater NPT (National Pipe Thread) fittings. Order one element for each external load circuit. 12/24-volt element rated at 600 watts (40 amps at 12 volts, 20 amps at 24 volts), 48-volt element rated at 30 amps. The 12/24-volt element has a pair of 12-volt elements and four terminals. Wire in parallel for 12V, or in series for 24V operation.

◈ **25-078 12/24V Water Heating Element $89⁹⁵**

◈ **25-155 48V/30A Water Heating Element $116⁰⁰**

◈*Means shipped directly from manufacturer.*

Air Heating Element For Diversion Loads

These naked 12V resistors can get very hot during regulation and must be mounted a safe distance from flammable surfaces. Use two or more elements in series for higher-voltage systems. Air elements tend to "sing" during regulation when used with the Enermax control. USA.

25-109 12V/20A Air Heating Element $24⁹⁵

Air Heater Diversion Loads

These feature air heating element(s) mounted in an aluminum box with a small fan to disperse the heat. Order one or more to match the total amperage of your charging inputs. Terminal block accepts up to 1/0 wire. Do not exceed the amperage rating of your diversion regulator. 18″ x 7″ x 7″. Two-year mfg. warranty. USA.

◈ **25-750 12V/30A Diversion Load $160⁰⁰**

◈ **25-751 12V/60A Diversion Load $187⁰⁰**

◈ **25-752 12V/120A Diversion Load $192⁰⁰**

◈ **25-753 24V/30A Diversion Load $162⁰⁰**

◈ **25-754 24V60A Diversion Load $220⁰⁰**

◈ **25-755 48V/30A Diversion Load $220⁰⁰**

◈ *Means shipped from manufacturer.*

About Diodes

Diodes are one-way valves for electricity. In PV systems they help the electrons get where they're supposed to go, and stay out of places they aren't supposed to go. But even the best valves have some restriction to flow in the forward direction, so diodes are used for strictly limited jobs.

The most common diode is a blocking diode. These are usually already installed inside the J-box of most larger modules. Installed in the middle of the series string of PV cells that make up the individual module, a blocking diode prevents a shaded module from stealing power from its neighbors. Smaller modules can have blocking diodes added inline. But unless you have a multiple-module array that is so poorly positioned that it will get hard shade during prime midday sun hours, don't worry about blocking diodes. Large RV or sailboat arrays need blocking diodes (if they aren't already installed by the module manufacturer).

Another common use is the bypass diode, used in series strings. If one module in the string has limited output due to shading or other problems, a bypass diode installed across the output terminals will allow the other modules in the string to continue output. They are usually only necessary on long, high-voltage series strings. You must have an amperage rating higher than the module string.

Isolation diodes are used to prevent backflow from batteries, or when combining parallel strings. Reverse current flow from batteries is usually better handled by the charge controller (see the sidebar on Dinosaur Diodes, pages 157–158), but a few controllers don't do night time disconnect.

Blocking or Bypass Diode

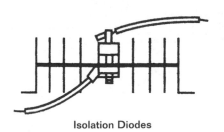

Isolation Diodes

60-Amp Isolation Diode

If PV modules are used with the Enermaxer, then you need this 60-amp Schottky diode kit installed between the PV modules and batteries to prevent night time reverse current flow. USA.

25-757 60-amp Diode with Heat Sink $49⁹⁵

Blocking Or Bypass Diodes

These are flat pack diodes with operating current derated 50% from manufacturer's specs for reliable operation and low voltage drop. 45-volt maximum. Use for 12- or 24-volt systems. May require the addition of shrink tubing or fork terminals for installation in module J-boxes. USA.

25-758 5-amp Diode (set of 3) $9⁹⁵

25-759 10-amp Diode (set of 2) $9⁹⁵

Specialty Controllers

12-Volt Digital Timer

This accurate quartz clock will control lights, pumps, fans, or other appliances from any 12VDC power source. The switched voltage can be up to 36 volts DC or 240 volts AC, and up to 8 amps. Normally open and normally closed contacts are provided, as well as a manual on/off/auto button. The 7-day timer can accept up to 8 on/off events per day. The internal battery will retain programming for one year after loss of 12V power (but won't switch on/off events). 4″ x 4″ x 2.5″. One-year mfr.'s warranty. USA.

25-064 12V Digital Timer $89⁹⁵

Battery Selector Switch

Few renewable energy homes use dual battery banks, but most boats and RVs do. This high-current switch permits selection between Battery 1 or Battery 2, or both in parallel. The off position acts as a battery disconnect. Wires connect to 3/8″ lugs. Capacity is 250 amps continuous, 360 amps intermittent. Okay for 12- or 24-volt systems. 5.25″ diameter x 2.6″ deep. Marine UL-listed. USA.

25-715 Battery Switch $34⁹⁵

Dual Battery Isolator

For RVs or boats, this heavy-duty relay automatically disconnects the household battery from the starting battery when the engine is shut off. This prevents inadvertently running down the starting battery. When the engine is started, the isolator connects the batteries, so both will be recharged by the alternator. The relay is large enough to handle heavy, sustained charging amperage. Easy hookup: Connect a positive wire from each battery to the large terminals. Connect ignition power to one small terminal, ground the other small terminal. Also see our dual battery PV charge controllers on page 159. USA.

25-716 Dual Battery Isolator $19⁹⁵

GenMate Auto-Start Controller

GenMate Controller (shown without case or cover)

GenMate is a super-versatile computerized generator controller that can be customized to automatically start and stop nearly any electric start generator.

Flipping tiny switches programs the GenMate, a one-time operation. This controller can be set to automatically start a generator at a user-selected voltage, run until another selected voltage is reached, and then turn off. Besides monitoring battery voltage, there are also two additional inputs: one line that is normally open, and one that is normally closed. If either change state, then the generator will start. These can be used for manual starting, automatic pumping on a low water signal, automatic generator-driven fire pumps, a timed generator exercise cycle, or practically anything you can imagine. You can also select a pre-start warning beeper, and the number of start-cycle retries to attempt before sounding a start-failure alarm. To use this controller, be sure that your generator is equipped with a low oil cutoff switch and automatic choke. 12- or 24-volt operation. Full five-year warranty. Specify generator make and model when ordering! Hondas and some others require customized controllers. USA.

❖ **25-170 GenMate Auto-Start Controller $449⁰⁰**

❖ *Means shipped directly from manufacturer.*

Spring Wound Timer

Our timers are the ultimate in energy conservation; they use absolutely no electricity to operate. Turning the knob to the desired timing interval winds the timer. Timing duration can be from 1 to 12 hours (or 1 to 15 minutes), with a hold feature that allows for continuous operation. This is the perfect solution for automatic shut-off of fans, lights, pumps, stereos, VCRs, and Saturday morning cartoons! It's a single pole timer, good for 10 amps at any voltage, and it mounts in a standard single gang switch box. A brushed aluminum faceplate is included. One-year warranty.

25-400 Spring-Wound Timer, 12 hour $29⁰⁰

25-399 Spring-Wound Timer, 15 minute, no hold $29⁰⁰

30-Amp DC-Controlled Relays

These general purpose DPDT (double-pole double-throw) power relays have either a 12-volt/169-milliamp or 24-volt/85-milliamp pull-in coil, depending on model chosen. DC timer, load disconnect, float switch, etc., can be limited in their amperage capability. These relays can be used remotely to switch larger loads, up to 30 amps per pole. They can also be used to switch AC circuits from DC-powered timers or sensors. Because these are highly adaptable DPDT relays, they can control multiple switching tasks simultaneously. Choose your system voltage. USA.

25-112 12V/30A Relay $39⁹⁵

25-151 24V/30A Relay $39⁹⁵

DC-To-DC Converters

Got a DC appliance you want to run directly, but your battery is the wrong voltage? Here's your answer from Solar Converters. These bi-directional DC-to-DC converters work at 96% efficiency in either direction. For instance, the 12/24 unit will provide 12V @ 20A from a 24V input, or can be connected backwards to provide 24V @ 10A from a 12V input. Current is electronically limited and has fuse back-up. Three bi-directional voltages are offered: The 12/24 unit described above; a 24/48 unit that delivers 5A or 10A depending on direction, and a 12/48 unit that delivers 2.5A or 10A depending on direction. The housing is a weatherproof NEMA 4 enclosure measuring 4.5″ x 2.5″ x 2″, good for interior or exterior installation (keep out of direct sunlight to keep the electronics happy). Rated for –40°F to 140°F (–40°C to 60°C). One-year mfr.'s warranty. Canada.

25-709 12/24 DC-to-DC Converter, 20A $159⁰⁰

25-710 24/48 DC-to-DC Converter, 10A $179⁰⁰

25-711 12/48 DC-to-DC Converter, 10A $179⁰⁰

12 VDC Isolator For 12-volt Answering Machines

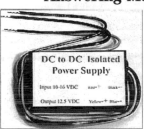

This power supply allows using a 12VDC answering machine on a 12-volt home energy system without phone interference. The telephone company grounds their system on the positive side. When connected to your home system through the answering machine, the result is often phones that buzz loudly, or don't work at all. This isolated power supply solves the problem. Works with 10–16VDC input. Runs any answering machine labeled 9–12VDC input. 300mA output. USA.

25-714 12 VDC Isolator $39⁹⁵

A Fax Timer For Sleepy Inverters

On the first ring this clever timer wakes up, turns on your inverter and your fax machine. It stays on for 10 minutes, or any time length you adjust, then turns everything off and lets the inverter go back to sleep. There's a plug for a small-wattage lamp to make sure your inverter wakes up. If the phone rings a second time before time out, it will reset for another full time period. The included manual override switch will keep things on until switched off for outgoing faxes.

This timer will work with any fax machine that can be turned on and off without requiring settings input, and is ready to receive immediately when turned on. Most simple jet-type (plain paper) faxes will work. Simple plug-in installation. Timer plugs into phone jack and standard AC outlet, fax and lamp plug into timer. USA.

◈ **25-712 Fax Timer $159⁰⁰**

◈ *Means shipped directly from manufacturer.*

PV Direct Pump Controllers

These current boosters from Solar Converters will start your pump earlier in the morning, keep it going longer in the afternoon, and give you pumping under lower light conditions when the pump would otherwise stall. Features include Maximum Power Point Tracking, to pull the maximum wattage from your modules; a switchable 12- or 24-volt design in a single package; a float or remote switch input; a user-replaceable ATC-type fuse for protection; and a weatherproof box. Like all pump controllers, a closed float switch will turn the pump off, an open float switch turns the pump on.

Four amperage sizes are available in switchable 12/24 volt. Input and output voltages are selected by simply connecting or not connecting a pair of wires. Amperage ratings are surge power. Don't exceed 70% of amp rating under normal operation. The 7-amp model is the right choice for most pumping systems, and is the controller included in our submersible and surface pumping kits. 7- and 10-amp models are in a 4.5″ L x 2.25″ W x 2″ D plastic box. 15-amp model is a 5.5″ L x 2.9″ W x 2.9″ D metal box. 30-amp model is supplied in a NEMA 3R rain-tight 10″ x 8″ x 4″ metal box. One-year mfr.'s warranty. Canada.

25-002 7A, 12/24V Pump Controller $89⁹⁵

◈ **25-003 10A, 12/24V Pump Controller $125⁰⁰**

◈ **25-004 15A, 12/24V Pump Controller $199⁰⁰**

◈ **25-005 30A, 12/24V Pump Controller $349⁰⁰**

◈ *Means shipped from manufacturer.*

Monitors

If you're going to own and operate a renewable energy system, you might as well get good at it. The reward will be increased system reliability, longer component life, and lower operating costs. The system monitor is the gizmo that allows you to peer into the electrical workings of your system and keep track of what's going on.

The most basic, indispensable, minimal piece of monitoring gear is the voltmeter. A voltmeter measures the battery voltage, which in a lead-acid battery system can be used as a rough indicator of system activity and battery state of charge. By monitoring battery voltage, you can avoid the battery-killing over and under voltages that come naturally with ignorance of system voltage. No battery, not even the special deep-discharge types we use in RE systems, likes to be discharged beyond 80% of its capacity. It drastically reduces the life expectancy of the battery. Since a decent voltmeter is a tiny percentage of battery cost, there is no excuse for not equipping your system with this minimal monitoring capability.

Beyond the basic voltmeter, the next most common monitoring device we use is the ammeter. The ammeter measures current flow (amps) in a circuit. They can tell us how much energy is flowing in or out of our system. Some ammeters use multiple sensors, called shunts, so that we can see how much energy is flowing in from our energy source(s), and how much is flowing out to our loads. Others simply use a single shunt that gives us the net amperage in or out for the entire system.

The most popular monitoring device is the accumulating amp-hour meter. This is an ammeter with a built-in clock and simple computing ability that will give a running cumulative total of amp-hours in or out of the system. This allows the system owner to easily and accurately monitor how much energy has been taken from the battery bank. Both of our most popular meters, the Tri-Metric and the E-Meter can display remaining energy as an easily understood percentage. When the meter says your battery system is 80% full, that's a number anyone can understand. Not every system or every person needs this kind of high-powered monitor/controller, but for those with an in-depth interest or an aversion to technology, it's available at modest cost.

Install Your Monitor Where It Will Get Noticed

The best, most feature-laden monitor won't do you a bit of good if nobody ever looks at it. We usually recommend installing your monitor in the kitchen or living room so it's easy for all family members to notice and learn what it's telling us. It's surprising how fast kids can learn given the proper incentive. A *No computer games or Saturday morning cartoons unless the batteries are charged* rule works wonders to teach the basics of battery management. Sometimes this even works on adults, too.

www.realgoods.com

System Monitors

Analog Volt And Ammeters

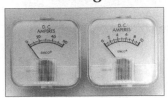

These ammeters operate without the need for external shunts. The meters measure 2.625″ x 2.375″ with a standard 2″ panel mount that has a four-bolt pattern. All voltmeters should be installed with an inline fuse. All meters can be recalibrated to assure long-term accuracy. USA.

25-304 20 to 32V Voltmeter $29⁹⁵

25-311 0 to 10A Ammeter $19⁹⁵

25-312 0 to 20A Ammeter $19⁹⁵

25-313 0 to 30A Ammeter $19⁹⁵

25-315 0 to 60A Ammeter $19⁹⁵

A Larger Voltmeter

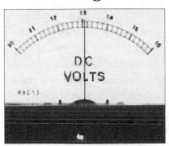

Our expanded-scale 10- to 16-volt voltmeters have a larger 2.5″ x 3″ face that makes them easier, and more accurate to read. Designed for panel mounting, it drops into a 2.25″ round hole. Accuracy is ±5%, and needle position is adjustable. Voltmeters should be installed with an inline fuse. Great for 12-volt systems. The most accurate we've found. Inexpensive, and can be user-calibrated. China.

25-301 10-16V Voltmeter $9⁹⁵

Brackets For Volt & Ammeters

For quick, easy meter mounting when you can't recess behind a panel, here's your ticket. These aluminum brackets are pre-punched to fit specific meters perfectly. Has screw mounting holes for surface mounting. USA.

25-292 Meter Bracket for 10-16V meter $8⁹⁵

25-294 Meter Bracket for one 2.5″ square meter $6⁹⁵

25-295 Meter Bracket for two 2.5″ square meters $8⁹⁵

25-296 Meter Bracket for three 2.5″ square meters $9⁹⁵

Digital Voltmeter For 12- & 24-Volt Systems

This easy-to-read surface-mount digital voltmeter monitors battery voltage up to 40 volts, and reads three digits (tenths of a volt). Accuracy is a very tight ±1%, and calibration can be field-adjusted. Draws only 8 mA to operate. 4″ x 2″ x 1.75″. One-year mfr.'s warranty. USA.

25-297 Digital Voltmeter $50⁰⁰

Digital Ammeter

Same enclosure as the Digital Voltmeter. Works on 12- or 24-volt systems. Reads four digits up to 199.9 amps. Shows reverse flow with a negative sign, and includes 100A shunt. 4″ x 2″ x 1.75″. One-year mfr.'s warranty. USA.

25-298 Digital Ammeter $75⁰⁰

Battery Capacity Meter

This is a great meter for the technically challenged, or when you want a simple supplemental meter for a remote location. With ten LED lights ranging from green to yellow to red to flashing red, this solid-state, watertight meter provides easy-to-understand basic monitoring at a glance. Each LED shows a 10% charge increment. Only one LED lights at a time, making it easy to see if the system is charging, discharging, or the approximate state of charge. Reverse polarity protected, draws only 25 mA, supplied with peel-and-stick mounting and five-foot cable. Can easily be remote mounted with cable extension. 4″H x 2″W x 0.5″D, 1.0 lb. 5-year mfr's warranty. USA.

25-760 12V Battery Capacity Meter $59⁹⁵

25-761 24V Battery Capacity Meter $69⁹⁵

Induction Ammeters

These low-cost meters read DC amps without any electrical connection. Simply place the meter over the wire. Accuracy is only fair, but will give good indication of general activity. The meter can be placed over the wire at any convenient point. Both meters are dual scale with 0 in the center. The 30A meter is 2″ diameter. The Dual Range meter has a 75A and a 600A scale, and is 2.5″ diameter. Neither meter is accurate for current flow under 10 amps. USA.

25-299 30A Induction Ammeter $24⁹⁵

25-300 Dual Range Induction Ammeter $29⁹⁵

The Ultimate System Monitor

The E-Meter from Cruising Equipment is both the simplest, and the most complete system monitor available. Simple? The accurate fuel gauge type meter gives at-a-glance basic monitoring, even changes from green to yellow to red to flashing red as charge level drops. Complete? Offers volts, amps, amp-hours (or kilowatt-hours), time (or percentage) remaining based on a rolling average of the past few minutes, historical battery info, number of recharge cycles, average cycle depth, and more. Photo sensor in front panel adjusts display brightness for night time. The meter goes into "sleep" mode with only the bar graph fuel gauge active after 10 minutes of inactivity. Draws 28 mA in sleep mode.

It can be ordered with an optional RS-232 computer plug output for logging of data, and we offer a unit without the needed shunt for use with Powercenters or Trace Power Panels, which come with the shunt pre-installed. Mounts in a 2″ round hole, is 3″ deep, and accepts 9.5 to 40 volts input. Two pre-scalers, 0 to 100V or 0 to 500V are available for higher voltage systems. Requires two inline fuses, and five wires for hookup. At least two wires, to the shunt, must be a twisted pair. Shielded meter wire sold below. Two-year mfr.'s warranty. USA.

25-346 Standard E-Meter w/500A Shunt $239⁰⁰

25-340 E-Meter w/RS-232 Plug & 500A Shunt $289⁰⁰

25-337 E-Meter without Shunt $199⁰⁰

26-524 Shielded Meter Wire, 3 twisted pair $0.70/ft (specify length)

25-338 0-500 Volt E-Meter Pre-Scaler $95⁰⁰

25-339 0-100 Volt E-Meter Pre-Scaler $95⁰⁰

TriMetric 2020 Monitor

Our Best-Selling System Monitor

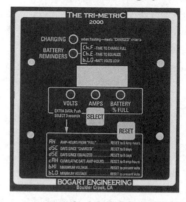

The TriMetric is a larger, user-friendly 12- to 48-volt digital monitor that displays volts, amps, or percentage of full-charge, and tracks a number of battery management items. This second-generation monitor features a 33% larger 3-digit LED display, the face and label are larger, and the deeper battery management data is easier to access and is all listed right on the label. Just hold down the "Select" button for two seconds, then push Select again to cycle thru all the extra data functions. A "Charging" light shows when positive amp flow is happening, or flashes to show that "charged" criteria was met during the last day.

A new feature is the "Battery Reminder" light, which can be programmed to flash a warning for any of three conditions: 1) low battery voltage (you set the voltage); 2) equalize batteries (you set 1 to 250 days); 3) full charge batteries (you set 1 to 60 days). Every five seconds the digital display will spell out what the Reminder light is warning you about.

Fits a standard double-gang electric box, turned sideways. Standard TriMetric works from 8–35 volts; adding the Lightning Protector/48V adaptor, the range is 12–65 volts, and it will still all fit inside a deep double-gang electric box. 100A shunt reads to 250 amps in 0.01 increments; 500A shunt reads to 1000 amps in 0.1 increments. Shunt must be purchased separately. Accuracy is ±1%. Draws 16 to 32 mA depending on display mode. Requires 4 wires for connections up to 55-feet between meter and batteries. Longer distances need 6 wires for best accuracy. One-year mfr.'s warranty. USA.

25-371 TriMetric 2020 Monitor $169⁰⁰

25-303 Lightning Protector/48V Adaptor $29⁹⁵

25-351 100-amp Shunt $19⁹⁵

25-364 500-amp Shunt $29⁹⁵

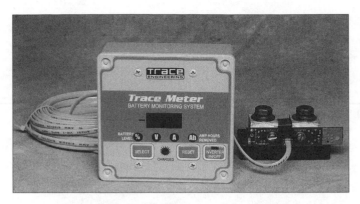

Trace TM500 System Monitor

Batteries are a significant and expensive part of any independent energy system. They are also the part of your system that is most vulnerable to mistreatment. The Trace TM500 meter keeps track of the energy your system has available as well as energy consumed, ensuring adequate reserve power and proper treatment of your batteries. It measures DC system voltage (needs the TM48 adapter for 48-volt systems), net amperage, cumulative amp-hours, days since full, peak voltage, minimum voltage, and more! Measuring 4.55 inches square by 1.725 inches deep, it may be surface or flush mounted. A 500-amp shunt is required (Trace Power Centers, Modules and Panels already have a shunt wired in). We strongly recommend purchasing the version with the shunt included. Their shunt is prewired for simply plugging in. Otherwise some tedious wiring is required. USA.

25-740 Trace TM500 Meter With Prewired Shunt $245⁰⁰

25-724 Trace TM500 Meter Without Shunt $195⁰⁰

25-725 Trace 48-Volt Adapter for TM500 $39⁹⁵

Brand Electronics ONE Meter

Multi-Channel Digital Monitoring for Utility Intertie Systems

Oak Display

Ever wonder how many watt-hours your battery-based intertie system is really contributing? Here's the ultimate meter to answer all your questions, and a few you hadn't even thought of yet. In the three-channel basic package we're offering below, the ONE Meter monitors DC power input from your PV, wind, or hydro system, plus AC power output through the inverter output terminals, plus AC power input/output through the inverter AC1 terminals. Results are displayed on the 4 x 20 backlit LCD display. Two displays are available, the flush-mounted oak-trimmed one we've shown, which mounts in a standard 3-gang electric box, or a fully enclosed stand-alone display. We'll supply the oak-trimmed flush display unless ordered otherwise. Additional displays (up to 3) can be run from a single base.

AC Current Sensor

Stand-Alone Display

The ONE Meter is computer-ready with a common RS-232 serial-port connection for downloading to a PC, and comes with a Windows-compatible BASIC terminal program for logging, compiling, and display. The ONE Meter also has stand-alone logging capability. AC current sensors are simple snap-on types, DC current sensors are toroid hall effect types, cabling is standard 8-conductor CAT-5 LAN type with RJ-45 connectors, widely used for computer networking. Cabling is supplied with the meter.

Additional data channels (up to 9) can be monitored with a single base display. Optional extra channels must be specified for AC or DC. Our basic package is set up for a single Trace SW-series inverter. Each additional SW inverter in your system will require two more AC channels for complete monitoring. Made in USA. One-year mfr.'s warranty.

25-766 Brand ONE Meter Basic Package $799⁰⁰

25-767 Brand Optional DC Channel $150⁰⁰

25-768 Brand Optional AC Channel $150⁰⁰

25-769 Brand Optional Oak Display $150⁰⁰

25-770 Brand Optional Stand-Alone Display $150⁰⁰

Homeowner's Digital Multimeter

If you're going to be the owner/operator of a renewable energy system, you need one of these. A multimeter is your first line of help when things go wrong. This is a nice one with all the features you'll ever need, and a great price. LCD 0.5˝ display reads 31⁄2 digits. Reads DC volts, AC volts, amps up to 10, plus continuity and ohms. Has good quality, flexible test leads. Powered by an included 9-volt battery. China.

25-000 Digital Multimeter $12⁹⁵

Watt's Up? An AC Watt-Hour Meter

Find the watt-suckers in your home! This inexpensive AC watt/hour meter plugs in between any standard 120VAC appliance and the outlet, and shows you exactly how much power is being used. Displays watts of usage, accumulated watt-hours, time since plug-in or reset, and even dollars. A great learning tool for any household, and a practical way for those planning a PV system in the future to monitor current and project future usage. Can be quickly set for your local electric rate from $0.05–$0.20/kWh. Shows cumulative costs up to $999, and time up to 99.9 hours. Measures up to 15 amps (1800 watts). Accurate to ±5%. One-year mfr's. warranty. USA.

25-344 Watt's Up? $119⁰⁰

No returns unless defective.

Protect Your Monitor Investment

Monitors are expensive, protection is cheap. What more do we need to say? Inline fuse holder works with all monitors. The E-Meter requires two fuse holders.

25-540 Inline ATC Fuse Holder $2⁹⁵

24-219 2A ATC Fuse 5/box $2⁹⁵

Large Storage Batteries

Why Batteries?

It's a reasonable question. Why do renewable energy systems use batteries? Why don't they simply make 120-volt AC? The answer is simple. No technology has ever been discovered that will store AC power; it must be produced on demand. For huge power companies with grids spread over half a state or more, considerable averaging of power consumption takes place. So they can (usually) produce AC power on demand. For the single remote homestead, we don't have that advantage. Batteries give us the ability to store energy when there's an excess coming in, and dole it back out when there's a deficit. Batteries are often the least durable and most maintenance demanding component of RE systems, but they're absolutely essential for any remote system.

Batteries provide a way to store energy for later use, and are required for any stand-alone renewable energy system. For those contemplating a utility intertie system with batteries, see the Sourcebook section on Utility Intertie. Batteries accumulate energy as it is generated by various renewable energy devices such as PV modules, wind, or hydro plants. This stored energy then runs the household at night or during periods when there is no energy input. Batteries can be discharged rapidly to yield more current than the charging source can produce by itself, so pumps or motors can be run intermittently. Batteries need to be treated with care, both for life expectancy and safety reasons. If commonsense caution is not used, batteries can provide enough power to cause impromptu welding and even explosions. See the Sourcebook section on Safety and Fusing (pages 151–155).

Battery Capacity

Think of a battery like a bucket. It will hold a specific amount of energy, and no amount of shoving, compressing, or wishing is going to make it hold any more. But damage, leaks, or old age might allow it to hold less.

A battery's capacity for holding energy is rated in amp-hours; one amp delivered for one hour equals one amp-hour. To know how much total energy is delivered, we need to know at what voltage the amp-hours are delivered. Battery capacity is listed in amp-hours at a given voltage. For instance a typical golf-cart battery is 220 amp-hours at 6 volts. Battery manufacturers typically rate storage batteries at a 20-hour rate. Our 220 amp-hour golf cart battery example will deliver 11 amps for 20 hours. The 20-hour rate is the standard we use for all the batteries in our publications. This rating is designed only as a means to compare different batteries to the same standard, and is not to be taken as a performance guarantee. Batteries are electrochemical devices, and are sensitive to temperature, charge/discharge cycle history, and age. The performance you will get from your batteries will vary with location, climate, and usage patterns. In the end, a battery rated at 200 amp-hours will provide you with twice the storage capability of one rated at 100 amp-hours.

Batteries are less than perfect storage containers. For every 1.0 amp-hour you remove from a battery, it is necessary to pump about 1.25 amp-hours back in, to bring the battery back to the same state of charge. This figure varies with temperature, battery type, and age, but is a good rule of thumb for approximate battery efficiency.

Batteries 101

A wide variety of differing chemicals can be combined to make a functioning battery. Some combinations are very low cost, but also very low power, others, like the lithium-ion batteries used in better laptops, can store astounding amounts of power, but have astounding costs to go with them. Lead-acid batteries offer the best balance of capacity per dollar, and are far and away the most common type of battery storage used in stand-alone power systems.

Units of Electrical Measurement

Most electrical appliances are rated with wattage, a measure of energy consumption per unit of time. One watt delivered for one hour equals one watt-hour of energy. Wattage is the product of current (amps) times voltage. This means that one amp delivered at 120 volts is the same amount of wattage as 10 amps delivered at 12 volts. Wattage is independent of voltage. A watt at 120 volts is the same amount of energy as a watt at 12 volts. To convert a battery's amp-hour capacity to watt-hours, simply multiply the amp-hours times the voltage. The product is watt-hours. To figure how much battery capacity it will require to run an appliance for a given time, multiply the appliance wattage times the number of hours it will run to yield the total watt-hours. Then divide by the battery voltage to get the amp-hours. For example, running a 100-watt lightbulb for one hour uses 100 watt-hours. If a 12-volt battery is running the light it will consume 8.33 amp-hours (100 watt-hours divided by 12 volts equals 8.33 amp-hours).

How Lead-Acid Batteries Work

The lead-acid battery cell consists of positive and negative lead plates of slightly different composition suspended in a diluted sulfuric acid solution called an *electrolyte*. This is all contained in a chemically and electrically inert case. As the cell discharges, sulfur molecules from the electrolyte bond with the lead plates, releasing excess electrons. A flow of electrons is what we call electricity. As the cell recharges, excess electrons bond with the sulfur compounds, forcing the sulfur molecules back into the sulfuric acid solution. A single lead-acid *cell* produces approximately two volts, regardless of size. Each individual cell has it's own cap. A battery is simply a collection of cells. A 12-volt automotive starting battery consists of six cells, each producing two volts, connected in series. Larger cells provide more storage capacity; we can run more amp-hours out and in, but the voltage output never exceeds the two-volt peak potential of the chemical reaction that drives the cell. Cells are connected in series and parallel to achieve the needed voltage and storage capacity.

Advantages and Disadvantages of Lead-Acid Batteries

Lead-acid batteries are the most common battery type available. They are well understood, and supplies for servicing and recycling them are widely available. Of all the energy storage mediums available to us, lead-acid batteries offer the most bang per buck by a wide margin. All the renewable energy equipment on the market is designed to work within the typical voltage range of lead-acid batteries.

That's the good part. Now the bad part: Lead-acid batteries produce hydrogen gas during charging, which poses a fire or explosion risk if allowed to accumulate. The hydrogen must be vented to the outside. See our drawings of ideal battery enclosures for both indoor and outdoor installations at the end of this section. Lead-acid batteries will sustain considerable damage if they are allowed to freeze. But this is harder to do than you may imagine. A fully charged battery can survive temperatures as low as –40°F without freezing, but as the battery is discharged, the liquid electrolyte becomes closer to plain water. Electrolyte also tends to stratify, with a lower concentration of sulfur molecules near the top. If a battery gets cold enough and/or discharged enough, it will freeze. At 50% charge level a battery will freeze at approximately 15°F. This is the lowest level you should intentionally let your batteries reach. If freezing is a possibility, and the house is heated more or less full-time, the batteries should be kept indoors, with a proper enclosure and vent. For an occasional use cabin, the batteries may be buried in the ground within an insulated box.

The active ingredients, lead and sulfuric acid, are toxins in the environment, and need to be handled with great respect. In addition to being a danger if it enters the water cycle, lead is a strategic metal, and has salvage value, so it should be recycled. The icing on the cake here is that, depending on current lead prices, you'll probably get paid for your old battery. Doing the right thing and getting paid for it. Ah, American ingenuity at work. Turn in old batteries for recycling at auto parts stores or recycling centers. A great majority of the lead, acid, and plastic in new car batteries comes from recycled batteries. Improper disposal is becoming less of a problem as the dangers of lead poisoning become well known, but is still a significant problem in developing countries.

Lead-acid batteries age in service. Once a bank of batteries has been in service for six months to one year, it generally is not a good idea to add more batteries to the bank. A battery bank performs like a team of horses, pulling only as well as the weakest. New batteries will perform no better than the oldest cell in the bank. All lead-acid batteries in a bank should be of the same capacity, age, and manufacturer as much as possible.

Monitoring Your Lead-Acid Battery Bank's State of Charge and State of Health

The battery state of charge should be monitored to prevent life-shortening deep discharges. Batteries can be monitored either the difficult but accurate way, with a hydrometer by sucking up a sample of electrolyte; the simple but easy to fool way, with a voltmeter; or with an accurate, but more expensive, amp-hour tracking monitor.

Using a hydrometer is the most accurate way to monitor the battery condition, but it's messy and potentially hard on both your clothes and the batteries. Don't use a hydrometer to check battery state of charge. Do wear old clothes, watch out for drips, and don't let anything drop into the battery cell. The best use of a hydrometer is to check the battery pack's state of health twice a year or so. See "Dr. Doug's Battery Care Class" later in this chapter for hydrometer checking details. The fewer times you open the battery tops, the better for your jeans and the batteries.

Battery voltage, as displayed on a voltmeter, can be used as an approximate indicator of the state of charge. However, a voltmeter is only accurate when the battery is in what is called its at-rest state, having been neither charged or discharged for several hours. Voltage is quite elastic and will stretch upward when a charge is applied, or stretch downward when a load is applied. The battery's internal chemistry needs time to settle down before a reading that is truly indicative of the battery's state of charge can be obtained. Since an at-rest voltage can't be obtained at a moment's notice, we often find it necessary to guesstimate the state of charge. Digital meters are highly recommended for their high degree of accuracy and their ability to read fine differences.

And finally, to make state of charge monitoring really easy: Several of the better system monitors available now will keep track of how many amp-hours have flowed in or out of the battery. Both the Tri-Metric and the E-Meter will display an easy-to-understand percent-of-battery-full number. These sophisticated monitors will automatically keep track of charge level, and even adjust for charging efficiency as the battery pack ages.

Different Strokes for Different Folks

Batteries are built and rated for the type of "cycle" service they are likely to encounter. Cycles can be "shallow," reaching 10% to 15% of the battery's total capacity, or "deep," reaching 50% to 80% of total capacity. No battery can withstand 100% cycling without damage, often severe. Batteries designed for shallow-cycle service, such as automotive starting batteries, will tolerate few, if any, deep cycles without sustaining internal damage. This makes them unsuitable for independent power systems. Batteries used in remote power systems must be capable of many repeated deep cycles without ill effects.

While both the typical automotive starting battery and the more unusual deep-cycle batteries that we use for renewable energy systems are lead-acid types, there are important construction and even materials differences. Starting batteries are designed to deliver several hundred amperes for a few seconds, and then the alternator takes over and the battery is quickly recharged. Deep-cycle batteries are designed to deliver a few amperes for hundreds of hours between recharges. Neither battery type is well suited for doing the other's job, and will usually suffer a short ugly life if forced to. Automotive engine-starting batteries are rated for how many amps they can deliver at a low temperature, or cold cranking amps (CCA). This rating is not relevant for storage batteries. Beware of any battery that claims to be a deep-cycle storage battery and has a CCA rating.

How Big a Battery Do I Need?

We usually size household battery banks to provide stored power for three to five days of autonomy during cloudy weather. For most folks, this is a comfortable compromise between cost and function. If your battery bank is sized to provide a typical three to five days of back-up power, then it will also be large enough to handle any surge loads that the inverter is called upon to start. A battery bank smaller than three days' capacity is going to get cycled fairly deeply on a regular basis. This isn't good for battery life. A larger battery bank cycled less deeply is going to cost less in the long run. Banks larger than five days' worth start getting more expensive than a back-up power source (like a modest-sized generator). However, we occasionally run into situations with ¾-horse-power or larger submersible well pumps or stationary power tools requiring a larger battery bank simply to meet the surge load when starting. Call the Real Goods technical staff for help if you are anticipating large loads of this type.

A Short Primer of Lead-Acid Battery Types

The common types of lead-acid batteries are presented here in order of worst to best. The "life expectancies" that we've listed for various battery types are the average that we've learned to expect with only reasonable care over the years. Please don't take these figures as a performance guarantee. We've dealt with novices who can destroy the best battery within six months, but we've also been blessed by meeting a few super conscientious people who can make their batteries last more than twice the average.

Car Batteries

The most common type of lead-acid battery is the automotive battery, sometimes called "starting batteries." This type of lead-acid battery has many thin lead plates and is designed to deliver hundreds of amps for a few seconds to start a car. Starting batteries are only designed to cycle about 10% to 15% of their total capacity and to recharge quickly from the alternator after discharging. They are not designed for the deep cycle service demanded by remote home power systems, and will fail fairly quickly when used in a deep-cycling application.

"RV" or "Marine" Deep-Cycle Batteries

This generic category includes most of the 12-volt batteries that Sears, Montgomery Ward, K-Mart, etc. sell as "deep cycle," "RV," or "marine" batteries. They are always 12-volt, and usually have between 80- and 160-amp-hour capacity. These batteries are a compromise between starting batteries and true deep-cycle batteries, as many of them are actually put into starting battery service by RV users who simply don't know better. They will give far better deep-cycle service than starting batteries, and may be the ideal choice for a beginning system that you plan to expand later. Life expectancy for these batteries is typically two to three years.

"Telephone Company" or Lead-Calcium Batteries

During the past 15 years telephone companies have been upgrading much of their switching equipment from the old style 48-volt relay type to newer solid-state equipment. When a telephone station is changed over, the huge battery bank that ran the old equipment is sold or recycled. Occasionally these shallow-cycle lead-calcium batteries are used in remote power systems. The typical life expectancy for these batteries is 15 to 20 years, although there are some on the market that claim 50 years or more. These batteries can be used in remote power systems, *if* you treat them carefully. These are shallow-cycle batteries that rarely experienced more than a 15% cycle in telephone

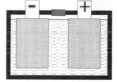

FULLY DISCHARGED

Cathode	Anode
Lead sulphate	Lead sulphate
$PbSO_4$	$PbSO_4$

Electrolyte
Water H_2O

FULLY CHARGED

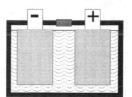

	Anode
Cathode	Lead dioxide
Lead Pb	PbO_2

Electrolyte
Suphuric acid & water
H_2SO_4 & H_2O

Fully spent, a battery's electrolyte has turned to water, all ions having been taken up by the cathode and anode. At the peak of readiness, a battery's plates have given up the ions to the electrolyte, which is now a corrosive mixture of water and sulphuric acid.

service. If you are careful never to discharge them deeply, these batteries can give years of excellent service.

While phone company batteries can sometimes be found cheap, or even free for the hauling, their sheer weight and size make them difficult to contend with. Some of these batteries weigh close to 400 pounds per 2-volt cell. Because cycle capacity is limited to 15% or 20%, you have to buy, move, and install five or six times more battery mass than is required for true deep-cycle batteries. Remember, that phone company battery may be rated at 1,680 amp-hours, but you can only use 20% of that capacity or 336 amp-hours, which isn't much by renewable energy standards. Unless you can find these batteries almost free, we don't recommend them.

Sealed Batteries

Sealed batteries have the acid either gelled or put into a sponge-like glass mat. They have the advantage/disadvantage of being completely liquid-tight. They can operate in any position, even sideways or upside down, and will not leak acid. Because the electrolyte moves more slowly, these batteries cannot tolerate high rates of charging or discharging for extended periods, although their thinner plates will allow high rates for a short time. Their sealed construction, which makes them ideal for some limited applications, makes it impossible to check individual cell conditions with a hydrometer. Although these cells are "sealed," they do have vents to prevent pressure build-up in case of gassing. Many PV charge controls will push charging voltage too high for sealed batteries. Premature failure will result due to loss of water vapor. We recommend sealed batteries only in situations where hydrogen gassing during charging cannot be tolerated, or the battery is going to be moved and handled a great deal, or in conditions where the battery needs to fit into unique, tight spaces. Boats, UPS computer power supplies, and remote expeditions are the most common uses. Special lower voltage charge controls must be used with these batteries. Life expectancy is two to five years for most AGM (absorbed glass mat) batteries, and five to ten years for the higher quality, but more difficult to manufacture, gel cell batteries. Most sealed batteries are AGM types.

True Deep-Cycle Batteries

True deep-cycle batteries are specifically designed for energy storage and deep-cycle service. They tend to have larger and thicker plates. This is the type of battery that is best suited for use with renewable energy systems. They are designed to withstand having a majority of their capacity used before being recharged. They are available in many sizes and types, the most common being 6-volt and 2-volt configurations for ease of movement. Once in place, the multiple batteries are series and/or parallel connected for your basic system voltage. These batteries are built to survive hundreds or even thousands of 80% cycles, though for best life expectancy we recommend 50% as the normal maximum discharge. This leaves you a 30% reserve for real emergencies. Never use the bottom 20% unless you like buying new batteries. The less deeply you regularly cycle your batteries, the longer they will last. The three most commonly available batteries within this group are the "golf cart" types with a three- to five-year life expectancy, the L-16 series with a seven- to ten-year life expectancy and industrial forklift-type batteries with a 15- to 20-year life expectancy. Deep-cycle batteries are usually your best battery investment.

We often recommend the golf cart types for small to medium-sized beginning systems. They make relatively inexpensive "trainer" batteries. Do your learning, make your mistakes, and in three to five years, when they wear out, you'll be in a much better position to judge your needs and what you're willing to pay for them.

New Technologies?

Compared to the electronic marvels in the typical renewable energy package, the battery is a very simple, relatively antiquated, electrochemical package. Tremendous amounts of research have been directed lately into energy storage technology. Auto manufacturers are desperately searching for a lightweight battery with high energy density and low cost—The Magic Battery. There are currently several dozen battery technologies under intense development in the laboratory. Several of these new technologies are bearing fruit now for cell phones, laptops, and hybrid vehicles, but are still far too expensive for the amount of energy storage required in a renewable energy system.

The possibilities of lead-acid technology are far from tapped out. Lead-acid batteries are also in the laboratory. This old dog is still capable of learning some new tricks. For now we must coexist with traditional battery technology, a technology that is nearly 100 years old, but is tried and true and requires surprisingly little maintenance. The care, feeding, cautions, and dangers of lead-acid batteries are well understood. Safe manufacturing, distribution, and recycling systems for this technology are in place and work well. Could we say the same for a sulfur-bromine battery?

Common Battery Questions and Answers

Ask Dr. Doug

I'm just getting started on my power system. Should I go big on the battery bank assuming I'll grow into it?

The answer is yes and no. Yes, you should start with a somewhat larger battery bank than you absolutely need. Over time most folks find more and more things to use power for once it's available. But if this is your first venture into remote power systems and battery banks, then we usually recommend that you start with some "trainer batteries." So no, don't invest too heavily in batteries your first couple of years. The golf cart type deep-cycle batteries make excellent trainers. They are modestly priced, will accept moderate abuse without harm, and are commonly available. You're bound to make some mistakes and do some horrible things with your first set of batteries. You might as well make mistakes with inexpensive batteries. In three to five years, when the golf cart type trainers wear out, you'll be much more knowledgeable about what you need and what quality you're willing to pay for.

The battery bank I started with two years ago just doesn't have enough capacity for us anymore. Is it okay to add some more batteries to the bank?

Lead-acid batteries age in service. The new batteries will be dragged down to the performance level of the worst cell in the bank. Different battery types have different life expectancies, so we really need to consider how long the bank should last. For instance, it would be acceptable to add more cells to a large set of forklift batteries at two years of age since this set is only at 10% of life expectancy. But an RV/Marine battery at two years of age is at 100% of life expectancy. Consider what you've got left, and that any new batteries will be giving up two years of life expectancy right off the bat.

I keep hearing rumors about some great new battery technology that will make lead-acid batteries obsolete in the near future. Is there any truth to this, and should I wait to invest in batteries?

Ah, pie in the sky someday. . . . The truth is there are several dozen battery technologies under intense development now. Some of them, like nickel-metal hydride, look very promising, but none of them are going to give lead-acid a run for your money within the foreseeable future. Lead-acid is also in the laboratory: this old dog can still learn some new tricks. Lead-acid technology is going to be around, and is going to continue to give the best performance per dollar for a long time to come.

Dr. Doug's Battery Care Class

What basic safety issues are important when working on my batteries?

1. Protect your eyes with goggles and hands with rubber gloves. Battery acid is a slightly dilute sulfuric acid. It will burn your skin after a few minutes of exposure, and your eyes almost immediately. Keep a box or two of baking soda and at least a quart of clean water in the battery area at all times. Flush any battery acid contact with plenty of water. If you get acid in your eyes, flush with clear water for 15 minutes and then seek medical attention.

2. Wear old clothes! No matter how careful I try to be around batteries, I always seem to end up with holes in my jeans. Wear something you can afford to lose, or at least have holes in.

3. Tape the handles of your battery tools or treat them with Plastic-Dip so that they can't possibly short out between battery terminals. Even small batteries are capable of awesome energy discharges when short circuited. The larger batteries we commonly use in RE systems can easily turn a ten-inch crescent wrench red-hot in seconds while melting the battery terminal into a useless puddle, and for the grand finale possibly explode and start a fire at the same time. This is more excitement than most of us need in our lives.

4. Now stop thinking "Oh, none of that will happen to me; I'm careful!" All of the stupid, avoidable catastrophes above have happened to me, and I'm very careful and should really know better, too. Don't take needless chances, it's too easy to be safe.

What should I do to get my batteries ready for winter?

1. Check the water level in your batteries after charging and fill to proper level with distilled water. Batteries use more water when they're being fully charged every day, like hopefully you've been doing all summer. Don't use tap water! The trace minerals can poison the battery. Don't fill before charging; the little gas bubbles will cause the level to rise and spill electrolyte.

2. Clean the battery tops. The batteries have been gassing at full charge all summer. The condensed fumes and dust on the tops of the batteries start to make a pretty fair conductor after a few months. Batteries will significantly discharge across the dirt between the terminals. That power is lost to you forever! Sponge off the tops with a baking soda/water solution, or use the battery cleaner sprays you can get at the auto parts store. Follow the baking soda/cleaner with a clear water rinse. Make sure that cell caps are tight and none of the cleaning solution gets into the battery cell! This stuff is deadly poison to the battery chemistry. Clean your battery tops once or twice a year.

3. Clean and/or tighten the battery terminals. Lead is a soft metal and will gradually "creep" away from the bolts. If you were smart and coated all the exposed metal parts around your battery terminals with grease or Vaseline when you installed them, they'll still be corrosion-free. If not, then take them apart, scrub or brush as much of the corrosion off as possible, then dip or brush with a baking soda and water solution until all

Helpful Hints for Lead-Acid Batteries

(for more detail see the Battery Care & Maintenance section)

1. All batteries will self-discharge and can sulfate when left unattended for long periods and where no charge controller or trickle charge device is employed. Disconnect them in this case. However, self-discharge is not prevented by disconnecting batteries. Clean, dry battery tops is a must as a preventative measure.

2. Keep batteries warm. At 0°F, 50% of their rated capacity is lost.

3. All wet cell battery enclosures must be vented to prevent potential explosion of hydrogen and other gases.

4. Do not, under any circumstances, locate any electrical equipment in a battery compartment.

5. Wear old clothes when working with electrolyte solution as they will soon be full of holes.

6. Be careful of metal or tools falling between battery terminals. The resulting spark can cause a battery to be destroyed or damaged.

7. Baking soda neutralizes battery acid, so keep a few boxes on hand.

8. Check the water level of your battery once a month until you know your typical usage pattern. Excessive water loss indicates gassing and the need for a charge controller or voltage regulator. Use distilled water only! The trace minerals and chemicals in tap water kill battery capacity.

9. Protect battery terminals and connectors from corrosion. Use a professional spray or any kind of grease or Vaseline. If charging has ceased or been reduced and the generator checks out, chances are corrosion has built up where the cable terminal contacts the battery post, preventing the current from entering the battery. This can occur even though the terminal and battery post look clean.

10. Never smoke or carry a lighted match near a battery, especially when charging.

11. Take frequent voltage readings. The voltmeter is the best way to accurately monitor a battery.

12. Batteries gain a memory about how they are used. Large deviations from regular use after the memory has been established can adversely affect performance. Incurred memory can be erased by discharging the battery system 95%. Recharge it to 140% of its capacity at a slow rate, then rapidly discharge the battery again completely and recharge to normal capacity. A generator/battery charger combination should be used for this procedure.

13. Use fewer, larger cells in series rather than lots of small cells paralleled. For example use two 6V batteries in series rather than two 12V batteries in parallel. This is safer in the case of a shorted cell; you have half the chance of random cell failure, half the maintenance effort. In general, larger batteries have thicker and more rugged plates, certainly more rugged than popular marine or RV deep cycle batteries.

fizzing stops and then scrape some more. Keep this up until you get all the blue/green crud off. (It's like mold, if you don't get it all, it will return). Then carefully cover all the exposed metal parts with grease when you put it back together. If your terminals are already clean, then just gently snug up the bolts.

4. Run a hydrometer test on all the cells. (Wait at least 48 hours after adding water before you run this test). Your voltmeter is great for gauging battery state of charge, the hydrometer is for gauging state of health. Use the good kind of hydrometer with a graduated float (not the cheapo floating ball type). We sell one for only $7 (#15-702). What we're looking for in this test is not the state of charge, but the difference in points between cells. In a healthy battery bank, all cells will read within 10 points of each other. Any cell that reads 20 to 25 points lower is probably starting to fail. You may have three to six months to round up a replacement set. At 50 points difference or more, the bad cell is sucking the life out of your battery bank and needs to get out now! Don't pay any attention to the color-coded good-fair-recharge markings on the float. These only pertain to automotive batteries, which use a slightly hotter acid. We're only looking for differences between cells.

What if I find a significant difference between cells on the hydrometer test?

Well, the first question is how much difference? At 10 to 20 points difference, you might just need a good equalizing charge. Run your batteries up to about 15.0 volts (that's for 12-volt systems, you folks at higher voltages can figure it out) and hold them at between 15.0 and 15.5 volts for three to five hours. This will even out any small differences between cells that have developed over time. It's a good idea to do an equalization every two to six months; it's like a minor tune-up for your batteries. Equalizing is more important in the winter when batteries tend to run at lower charge levels than in the summer.

At more than 20 points difference, you're probably looking at a dying cell. Replacing a single battery out of a larger bank usually isn't a great idea. If the batteries have seen 50% or more of their typical life expectancy, it probably isn't a good idea to simply replace the bad battery. For instance, a set of golf cart batteries (three to five years typical) with two years on them is not a good candidate for a single battery replacement. A new battery installed in an older battery bank will be dragged down to the performance level of the worst cell in the bank. So don't mix old and new batteries unless you're willing to sacrifice a considerable amount of life expectancy on the new cells. The Battery Book for Your PV Home covers in detail what I've lightly brushed over here. Item #80-104, $8.

I don't want to limp through another winter always being low on power and damaging my batteries. What's the best way to prevent damage?

BATTERY STATE-OF-CHARGE

Voltage Reading	Percent of Full Charge
12.6	100%
12.4–12.6	75%-100%
12.2–12.4	50%-75%
12.0–12.2	25%-50%
11.7–12.0	0%-25%

1. Get a decent system monitor. Nothing kills off your batteries faster than repeated excessively deep cycles. A digital voltmeter is the minimum monitoring equipment you need. The voltmeter is comparable to the gas gauge in your car: you can operate without it, but you're running blind and may do actual damage when things die at an inconvenient time. Overly discharging your batteries causes short life expectancy. Meters are far cheaper than batteries. Copy the battery state-of-charge chart at left, post it by your meter and pay attention; note that this chart is for battery voltage at rest.

Other possibilities, particularly if your system is being operated by non-technical people, is to install one of the monitors that simply give you a percent of full reading; green, yellow, or red lights; or one of the audible low battery alarms. These are monitors that almost any of us dummies can understand.

2. Add an efficient battery charger. Provided you've got a 1,000-watt or larger generator, the highly efficient Todd battery chargers put more watts per gallon of gas into your battery than any other charger on the market. For generator-based charging, you want the 15.5-volt models for the fastest bulk charging. The Todd will work just fine in conjunction with other battery chargers you might already have.

For larger systems, the battery charger capabilities of the larger Trace inverters are the simplest way to add battery charging capability, and offer more power, more control, and better reliability than the Todd chargers.

I live in a very cold climate; how do I keep my batteries from freezing, and what if they do?

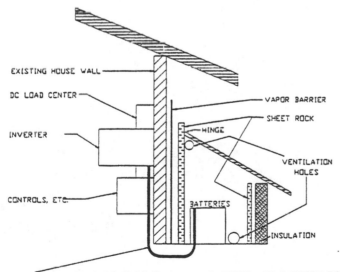

WIRES & CABLES PASS THROUGH WALL AT BOTTOM OF BATTERY BOX TO PREVENT HYDROGEN FROM ENTERING HOUSE. (HYDROGEN RISES)

Ideal Exterior Battery Enclosure

Freezing is usually fatal for lead-acid batteries. The expanding ice crystals bend internal plates, puncture separators, and push active lead material off the plates. There usually isn't much left after a hard freeze. The good news is that batteries don't freeze easily. At full charge it takes at least –40°F to freeze a battery, but as the battery discharges, the freezing point rises. At 50% depth of discharge, the lowest you ever want to take your batteries under normal conditions, they'll freeze at about 15°F.

If yours is a full-time residence, we recommend installing the batteries indoors. Build a little lean-to enclosure against an outside or basement wall, with venting from inside to the outside. It's important that the roof be sloped to help the hydrogen rise and pass outside.

You only need a little two-inch-diameter vent at the top and bottom to keep everything safe, plus you'll be pulling a little warm household air through the enclosure to keep the batteries toasty and happy.

If yours is a part-time recreational residence and wintertime temperatures can drop below –40°F, then we recommend either installing the batteries in the basement or buried in the ground with two-inch rigid foam insulation on top and sides. On the bottom just use sand or finely crushed gravel. The ground will be warmer than air temperatures and we want that ground contact.

This will prevent freezing in all but the most extreme climates. In either case, leave the PV charging system operational to ensure that the batteries stay charged! If the system is going to be unattended all winter, it's a good idea to turn down the charge voltage to reduce water usage. If you have a charge controller with adjustable voltage, then turn the charging voltage down to about 13.5–13.8 volts. Whatever kind of charge controller you have, make sure the batteries are topped off with distilled water before you leave.

With a minimum amount of care and attention you can make any set of batteries last up to twice as long as "normal." A couple of hours a year is a small investment considering what batteries cost to replace, and how much energy most folks spend worrying about them.

—Doug Pratt

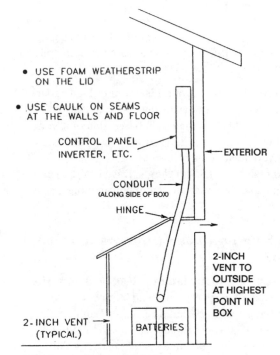

- USE FOAM WEATHERSTRIP ON THE LID
- USE CAULK ON SEAMS AT THE WALLS AND FLOOR

CONTROL PANEL INVERTER, ETC.

EXTERIOR

CONDUIT (ALONG SIDE OF BOX)

HINGE

2-INCH VENT TO OUTSIDE AT HIGHEST POINT IN BOX

2-INCH VENT (TYPICAL)

BATTERIES

Ideal Interior Battery Enclosure

Large Batteries, Chargers, and Accessories

Renewable Energy with the Experts

Batteries with Richard Perez

Richard has over 20 years experience in the renewable energy field, and wrote *The Battery Book* before he started publishing *Home Power* magazine. Richard goes into the pluses and minuses of various battery chemistries, with emphasis on safety, care, and feeding of the lead-acid deep-cycle batteries used in most residential storage systems. He also does a good job of covering energy conservation in appliance selection.

This tape was filmed at the Funky Flats Institute, which has been headquarters to *Home Power* magazine thru the testing of many, many generations of solar electric gizmos. Normal renewable energy–powered homes don't look anything like this. Funky Flats is an ongoing and valuable test bed laboratory for the industry. Don't think your house has to look this way to use renewable energy. Tape runs 57 minutes. USA.

80-369 Renewable Experts Tape—Batteries $39⁹⁵

Battery Book for Your PV Home

This booklet by Fowler Electric Inc. gives concise information on lead-acid batteries. Topics covered include battery theory, maintenance, specific gravity, voltage, wiring, and equalizing. Very easy to read and provides the essential information to understand and get the most from your batteries. Highly recommended by Dr. Doug for every battery-powered household. 22 pages, paperback. USA.

80-104 Battery Book for Your PV Home $8⁰⁰

Battery Care Kit

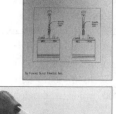

Our kit includes the informative *Battery Book for Your PV Home*, and a full-size graduated battery hydrometer. The *Battery Book* explains how batteries work and shows how to care for your batteries and get the best performance. Every battery-powered household needs a copy. The hydrometer is your most important tool for determining battery state of health. Detect and fix small problems before they require complete battery bank replacement.

Use Tip: The color-coded red, yellow, and green zones on the hydrometer are calibrated for automotive starting batteries: ignore them. Just look for differences between cells.

15-754 Hydrometer/Book Kit $12⁹⁵

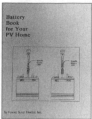

A Note on Shipping Large Batteries

Shipping of wet-cell storage batteries is very expensive. We offer free shipping on larger quantities of many batteries.

Wet Cell Batteries

"Golf Cart" Type Deep Cycle Batteries

These batteries are excellent "trainer" batteries for folks new to battery storage systems. First-time users are bound to make a few mistakes as they learn the capabilities and limitations of their systems. These are true deep cycle batteries and will tolerate many 80% cycles without suffering unduly. Even with all the millions of dollars spent on battery research, conventional wisdom still indicates that these lead-acid "golf cart" batteries are the most cost-effective battery storage solution for smaller to medium-sized systems. Typical life expectancy is approximately three to five years. Cycle life expectancy is typically about 225 cycles to 80% depth of discharge. USA.

Battery Voltage: 6 volts
Rated Capacity: 220 amp-hours
L x W x D: 10.25″ x 7.25″ x 10.25″
Weight: 63 lb/28.6 kg

15-101 Deep Cycle Battery, 6V/220Ah $79⁹⁵

Free shipping in the continental U.S. for 10 or more.

L-16 Series Deep Cycle Batteries

For larger systems or folks who are upgrading from the "golf cart" batteries, these L-16s have long been the workhorse of the alternative energy industry. The larger cell sizes offer increased ampacity and lower maintenance due to larger water reserves. L-16s typically last from seven to ten years. Cycle life expectancy is typically about 300 cycles to 80% depth of discharge. USA.

Battery Voltage: 6 volts
Rated Capacity: 370 amp-hours
L x W x D: 11.25″ x 7″ x 16″
Weight: 128 lb/58 kg

🚚 **15-102 Deep Cycle Battery, 6V/370Ah $199⁰⁰**

Free shipping in the continental U.S. for 10 or more.

Industrial-Quality IBE Batteries

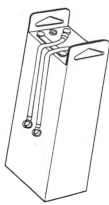

IBE industrial batteries are some of the best batteries available. With proper maintenance, they will easily last 15 to 20 years or more. Each cell is individually packaged in a steel or plastic case with two lifting handles and coated with acid-resistant paint. These individual cells can be carried by two people, making them the best choice for sites without a forklift, or with difficult access. The industrial 2-volt cells provide increased performance, greater reliability, and less frequent maintenance. Every performance and life enhancing trick known to the battery trade has been incorporated into these cells. Cycle life expectancy is 1500 cycles to 80% depth of discharge, or 5000 cycles to 20% depth of discharge.

Priced as 6-cell, 12-volt batteries. For 24-volt systems you must purchase two batteries. These batteries are totally recyclable. Interconnects and input/output leads are sold separately. USA.

🚚 **15-600 IBE Battery, 12V/1015Ah $1,850⁰⁰**

🚚 **15-601 IBE Battery, 12V/1107Ah $1,955⁰⁰**

🚚 **15-602 IBE Battery, 12V/1199Ah $2,040⁰⁰**

🚚 **15-603 IBE Battery, 12V/1292Ah $2,170⁰⁰**

🚚 **15-604 IBE Battery, 12V/1384Ah $2,340⁰⁰**

🚚 **15-605 IBE Battery, 12V/1464Ah $2,415⁰⁰**

🚚 **15-606 IBE Battery, 12V/1568Ah $2,635⁰⁰**

🚚 **15-607 IBE Battery, 12V/1673Ah $2,895⁰⁰**

Specify system voltage and plastic or steel case when ordering. Allow 4–6 weeks for delivery. Plastic cases may require an additional 3–4 weeks.

🚚 *Means shipped freight collect.*

Preassembled Industrial-Quality Batteries

The Lowest-Cost Industrial Battery Choice for Those With Forklift Access

From Yuasa-General Battery, the world's largest battery manufacturer, these consist of six propylene cells in what the industry calls a *steel tray*. Most folks would call it a steel box with a flip top. Features include heat-sealed cell covers, molded lead cell connectors, and *free freight!* Yep, free freight in the continental U.S. to a business with a forklift, or to a trucking terminal.

You must have a forklift to move these heavy, preassembled cells. Sold with six cells, or 12-volts, per tray. Purchase multiple trays for higher system voltages. With proper maintenance, these cells will last 15 to 20 years. Cycle life expectancy is 1500 cycles to 80% depth of discharge, or 5000 cycles to 20% depth of discharge. Interconnect cables are only needed between 12-volt trays; no cell interconnects required. Five-year mfr.'s warranty. USA.

GBC PREASSEMBLED 12-VOLT BATTERY PACKS

Item #	Ah @ 20 hrs	Weight	Size (inches) (L x W x H)	Price
🚚 15-799	627	498	30.75 x 7.78 x 23.25	$1,280⁰⁰
🚚 15-800	732	558	17.89 x 12.96 x 25.0	$1,430⁰⁰
🚚 15-801	836	630	20.16 x 13.0 x 25.0	$1,585⁰⁰
🚚 15-802	941	696	22.38 x 12.97 x 25.0	$1,640⁰⁰
🚚 15-803	1046	762	38.96 x 9.01 x 24.88	$1,770⁰⁰
🚚 15-804	1150	840	27.12 x 12.97 x 25.0	$1,945⁰⁰
🚚 15-805	1255	918	29.82 x 13.48 x 25.0	$2,045⁰⁰
🚚 15-806	1359	984	31.58 x 13.0 x 25.0	$2,175⁰⁰
🚚 15-807	1464	1068	33.65 x 13.0 x 25.0	$2,370⁰⁰
🚚 15-808	1568	1128	35.95 x 13.0 x 25.0	$2,470⁰⁰
🚚 15-809	1673	1260	38.96 x 13.51 x 25.0	$2,585⁰⁰

Allow 4 to 6 weeks for delivery. Must deliver to trucking terminal or forklift-equipped business.

🚚 *Means shipped by truck.*

Preassembled GBC HUP Solar Batteries

Absolutely the Best Battery Made for Renewable Energy Systems

From Yuasa-General Battery, the world's largest battery manufacturer, these consist of six propylene cells in what the industry calls a *steel tray*. Most folks would call it a steel box with a flip top. Features include heat-sealed cell covers, lead-plated copper intercell connectors, and *free freight!* Yep, free freight in the continental U.S. to a business with a forklift, or to a trucking terminal. You must have a forklift to move these heavy, preassembled cells. Sold with six cells, or 12-volts, per tray. Purchase multiple trays for higher system voltages. With proper maintenance these cells will last over 20 years. HUP stands for High Utilization Positive. These use a patented process that reduces positive plate flaking and shedding, the primary cause of eventual battery failure. No battery offers better performance, a longer life expectancy, or a better manufacturer's warranty. Cycle life expectancy is 2100 cycles to 80% depth of discharge, or 6000 cycles to 20% depth of discharge. Interconnect cables are only needed between 12-volt trays; no cell interconnects required. 10-year mfr.'s warranty. USA.

GBC PREASSEMBLED HUP SOLAR 12-VOLT BATTERY PACKS

Item #	Ah @ 20 hrs	Weight	Size (inches) (L x W x H)	Price
15-810	845	642	40.0 x 7.75 x 25.0	$1,695⁰⁰
15-811	950	708	40.0 x 8.25 x 25.0	$1,180⁰⁰
15-812	1055	780	40.0 x 8.75 x 25.0	$1,955⁰⁰
15-813	1160	859	40.0 x 9.0 x 25.0	$2,145⁰⁰
15-814	1270	936	40.0 x 10.25 x 25.0	$2,255⁰⁰
15-815	1375	1002	40.0 x 11.25 x 25.0	$2,365⁰⁰
15-816	1482	1086	40.0 x 12.0 x 25.0	$2,650⁰⁰
15-817	1585	1152	40.0 x 12.75 x 25.0	$2,725⁰⁰
15-818	1690	1236	40.0 x 13.5 x 25.0	$2,850⁰⁰

Allow 4 to 6 weeks for delivery. Must deliver to trucking terminal or forklift-equipped business.

🚚 *Means shipped by truck.*

Sealed Batteries

Sealed AGM-Type Batteries

These small, sealed batteries may be just what your next expedition needs to keep those laptops running and lights on after dark. The 6V/4Ah size is the replacement for several brands of solar sensor lights. The 12V/7Ah size is a replacement battery for our Multi Power Supply, and is a useful 12-volt storage battery for small power systems. These smaller batteries use an absorbed glass mat construction. AGM costs a bit less than gel batteries, but is slightly less durable. Usual life expectancy is 2 to 3 years. USA.

◈ **15-198 6V/4Ah Sealed Battery $11⁹⁵**
(2.8″ L x 1.9″ W x 4.25″ H 1.95 lb.)

◈ **15-200 12V/7Ah Sealed Battery $27⁹⁵**
(6″ L x 2.5″ W x 3.9″ H 5.5 lb.)

◈ **15-202 12V/12Ah Sealed Battery $49⁹⁵**
(6″ L x 3.9″ W x 3.9″ H 8.7 lb.)

◈ *Means shipped directly from manufacturer.*

Real Goods Solar Gel Batteries

These are the best sealed batteries available anywhere. These true deep cycle gel batteries come highly recommended by the National Renewable Energy Labs. They are completely maintenance-free, with no spills or fumes. Because the electrolyte is gelled, no stratification occurs, and no equalization charging is required. Self-discharge is under 2% per month. May be airline transported as approved by DOT (Department of Transportation), ICAO (International Commercial Airline Organization), and IATA (International Airline Transport Association). Life expectancy depends on battery size and use in service, but no sealed battery will perform better or longer than these. Smaller batteries will ship by UPS; larger sizes are shipped freight collect. Free shipping on any order of 10 or more batteries! Batteries will be shipped from the nearest warehouse in AZ, CA, CO, GA, IL, MD, MO, NJ, NV, TX, UT, or WA. Amp-hour rating at standard 20 hour rate. USA.

(See following page for prices)

◈ 15-210   8GU1 Solar Gel Battery 12V/32Ah, 24.2 lb
7.75″ x 5.2″ x 7.25″ $69⁰⁰

◈ 15-211   8G22NF Solar Gel Battery 12V/51Ah, 37.6
lb 9.5″ x 5.5″ x 9.25″ $99⁰⁰

◈ 15-212   8G24 Solar Gel Battery 12V/74Ah, 53.6 lb
10.8″ x 6.75″ x 9.8″ $129⁰⁰

◈ 15-213   8G27 Solar Gel Battery 12V/86Ah, 63.2 lb
12.75″ x 6.75″ x 9.8″ $145⁰⁰

◈ 15-214   8G31 Solar Gel Battery 12V/98Ah, 71.7 lb
12.9″ x 6.75″ x 9.4″ $169⁰⁰

🚚 15-215   8G4D Solar Gel Battery * 12V/183Ah,
129.8 lb 20.75″ x 8.5″ x 10″ $325⁰⁰

🚚 15-216   8G8D Solar Gel Battery * 12V/225Ah,
160.8 lb 20.75″ x 11″ x 10″ $385⁰⁰

🚚 15-217   8GGC 2 Solar Gel Battery 6V/180Ah, 68.4
lb 10.25″ x 7.2″ x 10.8″ $185⁰⁰

*Automotive posts, needs Wing Nut Battery Terminals for bolt-on cables.

◈ Means shipped directly from manufacturer.

🚚 Means shipped freight collect.

Sealed Industrial Battery Packs

**The Best Choice for
Emergency Power Backup Systems**

The GBC Renegade series offers true gel construction, not the cheaper, lower life-expectancy AGM construction. These batteries are an excellent choice for emergency power backup. No routine maintenance is required, and this battery type thrives on float service. Life expectancy is 10 to 15 years with proper care and a charger/controller that keeps voltage below 2.35v/cell. Price includes free freight in the continental U.S. to a trucking terminal or a business with a forklift. Delivered in preassembled 12- or 24-volt packs with smooth steel cases and no exposed electrical contacts. Cycle life expectancy is 1250 cycles to 80% depth of discharge. Interconnect cables are only needed between 12- or 24-volt trays; cell interconnects are soldered in place at the factory. Five-year mfr.'s warranty. USA.

GBC PREASSEMBLED SEALED 12-VOLT GEL BATTERY PACKS

Item #	Ah @ 20 hrs	Weight	Size (inches) (L x W x H)	Price
🚚 15-819	555	564	30.66 x 7.78 x 22.63	$1,945⁰⁰
🚚 15-820	925	885	31.13 x 13.0 x 22.63	$3,100⁰⁰
🚚 15-821	1110	1050	29.31 x 12.96 x 22.63	$3,665⁰⁰
🚚 15-822	1480	1375	25.69 x 19.61 x 22.63	$4,735⁰⁰

Allow 4 to 6 weeks for delivery. Must deliver to trucking terminal or forklift-equipped business.

🚚 Means shipped by truck.

GBC PREASSEMBLED SEALED 24-VOLT GEL BATTERY PACKS

Item #	Ah @ 20 hrs	Weight	Size (inches) (L x W x H)	Price
🚚 15-823	555	1128	30.76 x 12.83 x 22.63	$3,665⁰⁰
🚚 15-824	830	1596	30.06 x 19.5 x 22.63	$5,315⁰⁰
🚚 15-825	925	1770	38.5 x 16.63 x 22.63	$5,850⁰⁰
🚚 15-826	1110	2100	38.69 x 19.69 x 22.63	$6,900⁰⁰
🚚 15-827	1480	2748	38.69 x 25.63 x 22.63	$8,925⁰⁰

Allow 4 to 6 weeks for delivery. Must deliver to trucking terminal or forklift-equipped business.

🚚 Means shipped by truck .

Iota DLS-Series Battery Chargers

Iota chargers are solid-state, high-efficiency chargers, very similar to the now-out-of-business Todd charger we've offered for years. Iota features include light, compact construction; automatic fan cooling; reverse-polarity protection; easy terminal access; external, replaceable fuses; rugged reliability; and UL listing. The Iota is a constant-voltage charger, so as the battery voltage gradually rises with charging, the charger will gradually reduce amperage flow.

Chargers are available in 12-volt models at 30 amps, 55 amps, and 75 amps. A 24-volt model at 25 amps is also available. We offer high-volt and low-volt models at each amperage level. High-volt models are for wet-cell batteries, offer the fastest charging, and will help keep generator run time to a minimum. Low-volt models are for sealed batteries, or any battery pack that is constantly float charged. In addition, each charger has switchable dual voltage output, with about 0.6-volt difference. Low voltage for float charging, or high voltage for the most rapid battery charging. Voltage range is selected by inserting a plug or bridging a pair of terminals (depending on charger model); a customer-supplied switch can be easily inserted. 75-amp charger models are equipped with a 20-amp/115-volt AC plug. Two-year mfr's warranty. USA.

◈ 15-791 Iota DLS 30 Lo-Volt Charger $145⁰⁰

◈ 15-792 Iota DLS 30 Hi-Volt Charger $145⁰⁰

◈ 15-793 Iota DLS 55 Lo-Volt Charger $179⁰⁰

◈ 15-794 Iota DLS 55 Hi-Volt Charger $179⁰⁰

◈ 15-795 Iota DLS 75 Lo-Volt Charger $329⁰⁰

◈ 15-796 Iota DLS 75 Hi-Volt Charger $329⁰⁰

◈ 15-797 Iota DLS 24-25 Lo-Volt Charger $289⁰⁰

◈ Means shipped from manufacturer.

DLS CHARGER SPECS								
	DLS 30		**DLS 55**		**DLS 75**		**DLS 24-25**	
Max. Output Amps	30		55		75		25	
Max. Input Watts	500		900		1300		600	
Size: 6.75″ x 3.75″ x	7″		10″		13″		7″	
Weight	5.5 lb		7 lb		10 lb		7 lb	
DC Output Volts	Hi–	Lo–	Hi–	Lo–	Hi–	Lo–	Hi–	Lo–
	14.8–	13.6–	14.8–	13.6–	14.8–	13.6–	29.5–	27.1–
	15.4	14.2	15.4	14.2	15.4	14.2	31.0	28.4

Battery Pal Automatic Float Charger

Leaving a car, boat, motorcycle, RV or other vehicle garaged for weeks or months almost guarantees a dead battery. The process is unstoppable. A fully charged battery, even a new one, loses approximately 5% of its charge per week. Battery Pal prevents this problem. Unlike trickle chargers, it cannot damage a battery through electrolyte boilout due to overcharging. Float circuitry senses when the battery is fully charged and automatically reduces current flow to the level that maintains battery life: 25 to 60 milliamps. This is an ideal unit for garages where a solar charger won't work. For use with 5- to 100-Ah lead-acid storage batteries. Hong Kong.

Power Requirement: 100 to 130 VAC
Maximum Charging Output: 600 milliamps
Charge Voltage: 13.8±1 VDC

11-128 Battery Pal $39⁹⁵

Battery Accessories

Solartech Battery Pulse Maintenance

Applying a small pulse current to lead-acid batteries has been scientifically proven to dramatically reduce deterioration due to sulfate build-up, and to increase battery life and performance. The pulse helps dissolve sulfate crystals, and get them back into the electrolyte. This increases electrolyte strength, and frees up plate surface area for active use. Pulse technology is most effective when used on batteries that don't get routine equalization service, or have been sitting for long periods, but will increase battery life on any system regardless of treatment. Considering what battery systems cost, these are a great investment for your current and future batteries. One pulse unit can cover several large batteries within reason. Obviously, the more the pulse is dispersed, the less effect it will have. Large battery bank systems may require several pulse units for good coverage.

The solid-state Can-Pulse units from Solartech have proven to be the most effective battery pulse units available. They install directly on the battery bank, require no operator attention or service, and will only activate when the batteries are approaching full charge. This is above 14.0 volts for 12-volt batteries, or above 26.0 volts for 24-volt batteries. LED light indicates operation. Two-year mfr.'s warranty. Canada.

◈ **15-783 Can-Pulse $149⁰⁰**

Please specify 12- or 24-volt.

◈ *Means shipped directly from manufacturer.*

DC-To-DC Converters

Got a DC appliance you want to run directly, but your battery is the wrong voltage? Here's your answer from Solar Converters. These bi-directional DC-to-DC converters work at 96% efficiency in either direction. For instance, the 12/24 unit will provide 12V at 20A from a 24V input, or can be connected backwards to provide 24V at 10A from a 12V input. Current is electronically limited and has fuse backup. For loads that exceed the converter's amperage, tap directly off half the battery bank and use the converter to balance the two halves. Three bi-directional voltages are offered: the 12/24 unit described above; a 24/48 unit that delivers 5A or 10A depending on direction; and a 12/48 unit that delivers 2.5A or 10A depending on direction. The housing is a weatherproof NEMA 4 enclosure measuring 4.5″ x 2.5″ x 2″, good for interior or exterior installation (keep out of direct sunlight to keep the electronics happy). Rated for –40°F to 140°F. One-year mfr.'s warranty. Canada.

25-709 12/24 DC-to-DC Converter, 20A $159⁰⁰
25-710 24/48 DC-to-DC Converter, 10A $179⁰⁰
25-711 12/48 DC-to-DC Converter, 10A $179⁰⁰

Full-Size Battery Hydrometer

This full-size graduated-float-style specific gravity tester is accurate and easy to use in all temperatures. Specific gravity levels are printed on the tough, see-through plastic body. It has a one-piece rubber bulb with a neoprene tip. USA.

Note: Ignore the color-coded green/yellow/red zones on the float. These are calibrated for automotive starting batteries, which use a denser acid solution. Storage batteries will usually read in the red or yellow zones. Just watch for differences between individual cells.

15-702 Full-Sized Hydrometer $6⁹⁵

Wing Nut Battery Terminals

These terminals convert conventional round post-type batteries to a wing nut connection adapter with a 5/16-inch stud. Sold in pairs.

15-755 Wing Nut Terminals $4⁹⁵

Battery And Inverter Cables

The wire sizing between the battery and the inverter is of critical importance. Small cables limit inverter performance. We offer seriously beefy 2/0 and 4/0 cables using super flexible welding-type cable. Highly recommended to complete your inverter installation. They're available in 3-foot, 5-foot or 10-foot pairs. One-lug cables have bare wire on one end to accommodate set-screw lugs, with a crimped, soldered, and color-coded shrink-wrapped 3/8" lug terminal on the other end. Two-lug cables have the same high-quality terminals installed at both ends. See chart below for recommended cable sizing. Extra lugs offered for the occasional installation that needs an extra lug end. USA.

15-770 2/0 3-ft. Cables, 1 lug, pair $29⁹⁵
15-771 2/0 5-ft. Cables, 1 lug, pair $49⁹⁵
15-772 2/0 10-ft. Cables, 1 lug, pair $84⁹⁵

15-773 4/0 3-ft. Cables, 1 lug, pair $39⁹⁵
15-774 4/0 5-ft. Cables, 1-lug, pair $69⁹⁵
15-775 4/0 10-ft. Cables, 1 lug, pair $115⁰⁰

27-311 4/0 5-ft. Cables, 2 lug, pair $69⁹⁵
27-312 4/0 10-ft. Cables, 2 lug, pair $125⁰⁰

27-099 2/0 5-ft. Cables, 2 lug, pair $49⁹⁵
27-098 2/0 10-ft. Cables, 2 lug pair $89⁹⁵

26-605 Copper Lug, solder-type 2/0 $4⁹⁵
26-606 Copper Lug, solder-type 4/0 $13⁹⁵
26-623 Solderless Lug, #2 to 4/0 $7⁹⁵

RECOMMENDED SIZING FOR CIRCUIT BREAKERS & CABLES

Circuit Breaker Size	Recommended with these Inverters	Minimum Cable Size
DC-175	DR-1512, 2412, 1524, & 2424. SW-3048E, & 4048	2/0
DC-250	DR-3624, SW-2512, 4024, & 5548	4/0

Heavy-Duty Insulated Battery Cables

Serious, beefy, 2/0- or 4/0-gauge battery interconnect cables for systems using 1,000 watt and larger inverters. These single cables have 3/8" lug terminals and are shrink-wrapped for corrosion resistance. The 9" cables will fit series-connected L-16 or golf cart type batteries. Use 13" cables for parallel connections, or where batteries aren't tight together. No color choice (usually black). USA.

15-785 Battery Cable, 4/0, 18-inch $24⁹⁵
15-780 Battery Cable, 4/0, 13-inch $19⁹⁵
15-778 Battery Cable, 2/0, 13-inch $14⁹⁵
15-779 Battery Cable, 2/0, 9-inch $11⁹⁵

Light-Duty Battery Cable

Automotive-grade 4-gauge cables with 3/8" lug terminals for use on smaller systems, with inverters under 1,000 watts. No color choice. Not pictured. USA.

15-776 4-Gauge Battery Cable, 15-inch $3⁹⁵
15-777 4-Gauge Battery Cable, 24-inch $4⁹⁵

Perko Battery Selector Switch

Few renewable energy homes use dual battery banks, but most boats and RVs do. This high current switch permits selection between Battery 1, or Battery 2, or both in parallel. The off position acts as a battery disconnect. Wires connect to 3/8" lugs. Capacity is 250 amps continuous, 360 amps intermittent. Okay for 12- or 24-volt systems. 5.25" diameter x 2.6" deep. Marine UL-listed.

25-715 Perko Battery Switch $34⁹⁵

Dual Battery Isolator

For RVs or boats, this heavy-duty relay automatically disconnects the household battery from the starting battery when the engine is shut off. This prevents inadvertently running down the starting battery. When the engine is started, the isolator connects the batteries, so both will be recharged by the alternator. Relay is large enough to handle heavy, sustained charging amperage. Easy hook-up: Connect a positive wire from each battery to the large terminals. Connect ignition power to one small terminal; ground the other small terminal. USA.

25-716 Dual Battery Isolator $19⁹⁵

Small Rechargeable Consumer Batteries

Rechargeable Batteries and Chargers

If you've waded this deeply into the Sourcebook, we probably don't need to bore you with why you ought to be using rechargeable batteries as much as possible. They're far less expensive, they don't add toxics to the landfill, and they don't perpetuate a throw-away society. You know all that already, right? Well, just in case you want a review, there are a few brief paragraphs below. Already feeling secure in your knowledge and just want to choose the best rechargeable battery or charger for your purpose? Then skip ahead to "Choosing the Right Battery."

Why We Need to Use Rechargeable Batteries

Americans toss over three billion small consumer batteries into the landfill every year. The rest of the world adds another few billion to the total. The vast majority of these are alkaline batteries. Alkaline batteries are the common Duracell, Energizer, Eveready brands of batteries you find in the grocery store checkout lane. While domestic battery manufacturers have refined their formulas in the past few years to eliminate small amounts of mercury, alkaline batteries are still low-level toxic waste. Casual disposal in the landfill may be acceptable to your local health officials, but what a waste of good resources! Our grandchildren, maybe even our children, will be part of the next gold rush, when we start mining our 20th century landfills for all the refined metals and other depleted resources that are waiting there.

So throwaway batteries are wasteful and a health hazard. They're also surprisingly expensive. Alkaline cells cost around $0.90 to $2.20 per battery, depending on size and brand. Use it once, and then throw it away. In comparison, rechargeable nickel-metal hydride cells—including the initial cost of the battery, a charger, and two or three cents worth of electricity—cost $0.04 to $0.10 per cycle, assuming a very conservative lifetime of 400 cycles. Chargers outlive batteries, so these rechargeable cost figures err on the high side. Typically, throwaway batteries cost $0.10/hour to operate, while rechargeable batteries cost only $0.001/($\frac{1}{10}$ of one cent) per hour.

When *Not* to Use Rechargeable Batteries

Rechargeable batteries cost far less, they reduce the waste stream, and they usually present less disposal health hazards. What's not to love about them? It isn't a perfect world; there are downsides to rechargeable technology. Specifically, lower capacity, lower operating voltage, self-discharge, and, in the case of nicad cells only, toxic waste. Let's look at each issue individually.

Lower capacity. Rechargeable batteries typically contain about 35% to 70% of the total energy that a standard alkaline can store. So you'll change batteries more often with most appliances. High-drain appliances like digital cameras or remote control toys are an exception to this. nickel-metal hydride (NiMH) batteries will support high-drain appliances longer than any other battery chemistry, including throwaway batteries. And they will do it hundreds of times before hitting the trash.

Lower operating voltage. Nicad and nickel-metal hydride batteries operate at 1.25 volts per cell. The alkaline cells that many battery-powered appliances are expecting to use operate at 1.50 volts per cell. This quarter of a volt is sometimes a problem for voltage-sensitive appliances, or devices like large boom boxes that use many cells in series. The cumulative voltage difference over eight cells in series adds up to a problem if the device isn't designed for rechargeable battery use.

Self-discharge. All battery chemistries discharge slowly when left sitting. Some chemistries, like the alkaline and lithium cells used for throw-away batteries, have shelf lives of five to ten years. Rechargeable chemistry will self-discharge much faster, with shelf lives of two to four months. These batteries aren't the ones you want in your glovebox emergency flashlight, and probably aren't the best possible choice for battery-powered clocks. We have rechargeable alkaline batteries for those uses.

Toxic nicad cells. Nicad cells contain the element cadmium. Human beings do not fare well when ingesting even tiny amounts of cadmium. These cells absolutely must be properly recycled. We accept all nicad batteries returned to us for recycling. Radio Shack stores also accept nicads for recycling. Do *not* casually dispose of nicad cells! They are toxic waste! Now, that said, let us point out that the newer, higher-capacity nickel metal hydride batteries are not toxic at all, and will work just fine in any nicad battery charger you might already have.

Choosing the Right Battery

There are three basic rechargeable battery chemistries to choose from: rechargeable alkaline, nickel-metal hydride(NiMH), and nickel cadmium (nicad). Each has strengths and weaknesses. We've summarized everything and compared with standard throwaway alkaline batteries in the chart below.

Regardless of your chemistry choice, buy several sets of batteries, so you can always have one set in the charger while another set is working for you.

RECHARGEABLE BATTERY COMPARISONS

	Std. Alkaline (the comparison std.)	Rechargeable Alkaline	NiMH	Nicad
Volts Capacity (compared to alkaline)	1.5 100%	1.5 90% initially diminishes slowly.	1.25 75% no loss with cycles.	1.25 35% no loss with cycles.
Capacity in mAH AAA AA C D	750mAH 2000mAH 5000mAH 10000mAH	750mAH 1800mAH 3000mAH 7200mAH	600mAH 1500mAH 3300mAH 7000mAH	240mAH 700mAH 1800mAH 4000mAH
Avg. Recharge Cycles	1	12–50	400–600	400–600
Self-Discharge Rate	Negligible. 5-yr. shelf life.	Negligible. 5-yr. shelf life.	Modest. 15% loss@ 60 days.	Moderate 60% loss@ 60 days.
Strengths	1.5-volt. Long shelf life. Hi-capacity.	1.5-volt. Long shelf life. Good capacity.	Sustains high draw. Many recharges Nontoxic. No memory.	Many recharges.
Weaknesses	Highest cost. Throwaway.	Loses capacity with deep cycles. Special charger.	1.2-volt	1.2-volt. Toxic content. Self-discharges.
Best Uses	Clocks; emergency stuff.	Clocks, emergency stuff, toys, games, radios, flashlights.	Digital cameras! Remote controls, toys, flashlights, any high-drain appliance.	Toys, flashlights.

The Infinity Guarantee: Batteries That Last Forever

Our nicads and nickel-metal hydrides are so durable and reliable, we guarantee them forever. You will never have to buy batteries again. Should one fail, with proof of purchase we will replace it free of charge. Don't look for fine print, exclusions, or hidden catches. There are none. These batteries will outlast you.

More Rechargeable Battery Info For Those Who Need to Know It All

Nicad Cells: The full name of this cell type is nickel cadmium. Nicads were first produced around 1900 and came into common use in the 1950s. Nicads are made of nickel, cadmium, and potassium hydroxide in a water solution. Cadmium is a toxic material and should be recycled as such. All nicads, regardless of size, produce about 1.25 volts per cell. Voltage is stable during charging and discharging, but this stability can present problems when you want to know how much energy is left in the battery. In lead-acid battery types, checking the voltage will give you an idea of the cell's state of charge. Not so with nicads, which display the same voltage from 10% to 90% of capacity.

The capacity of nicad cells varies according to how much active material is used. Some manufacturers just take a AA cell and put it in a C or D size can, which results in very low capacity. We don't recommend (or sell) batteries made this way! Weight is a good indicator of how much material the cell really contains, if capacity isn't listed on the outside.

Nicads comfortably do better than 900 cycles under laboratory conditions. In the real world, where they'll likely suffer being overly discharged and overly charged, we give them a life expectancy of 400 to 600 cycles. Many more cycles than you're likely to use in a lifetime. What about being overly charged or discharged? Your batteries will perform better and last longer if they aren't routinely stretched to their maximum performance. When they get down to about 10% of capacity, nicads and NiMH batteries will start delivering reduced voltage. This is the time to stop and replace them. Total discharges aren't deadly, but avoid them when you can. The best chargers will shut off or go into low-current maintenance mode when the batteries are fully charged. Chargers that just keep going and going will overheat the cells and cause reduced life if you don't manually shut them off and/or remove the cells once they've been fully charged.

Nickel-Metal Hydride Cells: These are probably the best choice, and are increasingly becoming the consumer battery of choice. Compared to nicads, NiMH batteries offer about twice the capacity per cell, with no toxic chemicals. NiMH and nicads are close cousins, cadmium is replaced with metal hydrides, which are environmentally benign. In the lab, NiMH batteries have done as many as 400,000 cycles; out in the real world we can expect considerably less. Typical life expectancies are 300 to 500 cycles. The second generation of nickel-metal hydride batteries we're now offering has solved the rapid self-discharge problems of earlier NiMH technology. For heavy-discharge uses like digital cameras and remote-control toys, NiMH is clearly the best choice as there's no voltage sag until the last 10% of capacity. This means you actually get to use better than 90% of the battery's rated capacity. No other battery chemistry will deliver like this under heavy load. Like the nicad chemistry, NiMH cells produce 1.25 volts per cell, but most digital cameras and remote-control toys are designed around this voltage now. NiMH batteries present no disposal threats or problems. They contain no toxic chemicals.

Rechargeable Alkaline Batteries: This technology was popularized by Ray-O-Vac's introduction of the Renewal line in 1994, and is now available from several manufacturers. Alkaline batteries use a chemical process that isn't entirely reversible, but with a bit of chemical tweaking it has achieved limited success. The advantages of rechargeable alkalines are the same as the advantages of non-rechargeable alkalines, including highest initial capacity, long shelf life, fully charged on arrival, and higher 1.5V voltage.

Disadvantages are a bit less obvious. The biggest problem is diminished capacity with each recharge. The chemical cycle isn't fully reversible, so every cycle comes around with some capacity loss. The deeper the discharge cycle, the more capacity loss. You can lose 10% capacity per cycle if you run these batteries completely down. So there's a limited number of cycles. Alkalines aren't capable of more than 50 useful cycles even with the most careful shallow cycling. The average is about 12 to 25 cycles. All this adds up to the highest cost per cycle for rechargeable batteries. Although, of course, they are a big step up from standard throwaway alkalines.

Choosing the Right Battery Charger(s)

Now that you've seen the choices in rechargeable batteries, you'll also need to decide which charger(s) to use. The three types of chargers are AC charger (plugs into a wall socket), solar charger (charges directly from the sun), and the 12-volt charger (plugs into your car's cigarette lighter or your home's renewable energy system).

Solar Chargers

The great benefit of solar battery chargers is that you don't need an outlet to plug them into, you just need the sun! They can be used anywhere the sun shines. However, most solar chargers work more slowly than plug-in types, and it may take several days to charge up a set of batteries. This needn't be a problem. We strongly recommend you buy an extra set of batteries with your solar charger, so you can charge one set while using the other. The Accucell charger with the solar option is an exception to the solar-chargers-are-slower rule. This is a fast, 5-hour charger, but you pay for the speed. An advantage of the smaller solar chargers is that you can leave the batteries in the charger without worrying about overcharging. Given time and patience (prerequisites for a prudent, sustainable life in any case), solar chargers are the best way to go.

DC Chargers

The 12-volt charger or the Accucell charger will recharge your batteries from a car, boat, or from your home's 12-volt renewable energy system (if you're so blessed). Both are fairly fast chargers, with the Accucell providing automatic shutoff when finished. Both provide a variable charge rate depending on battery size. Most AC chargers put the same amount of power into every battery, regardless of size. Even though the 12-volt charger is a fast charger, it will not drain your car battery unless you forget and leave it charging for several days. This charger can overcharge batteries, however, so pay close attention to how long it takes to charge, and don't leave them in there much longer. Leaving them in the 12-volt charger for one-and-a-half times the recommended duration will not harm the batteries.

AC Chargers

The AC (plug in the wall) chargers are by far the most convenient for the conventional utility-powered house. Most of us are surrounded by AC outlets all day long and all we have to do is plug in the charger and let it go to work. Our favorites are the Battery Manager, the Accucell charger, and the Golden Power Fast Charger. All these chargers turn off when charging is complete, and go into a trickle charge maintenance mode. Your batteries will always be fully topped up and waiting for you. The Battery Manager is probably the best all-around charger, as it automatically figures out what chemistry and size each battery is, then charges it accordingly. But with a 24-hour charge cycle, it isn't fast. If you want speed, the Accucell for alkaline batteries, or Golden Power chargers for NiMH and nicad batteries are better choices.

Frequently Asked Questions (FAQs) on Rechargeable Batteries

Is it true that rechargeable batteries develop a "memory?"

This was a nicad problem many years ago. If you only use a little of the battery's energy every time, and then charge it up, the battery will develop a "memory" after several dozen of these short cycles. It won't be able to store as much energy as when new. This was only true with nicads, not with NiMHs or alkalines. Improved chemistry has practically eliminated any memory problems. Short cycles have never been a problem for NiMH or alkaline batteries.

How do I tell when a nicad or NiMH battery is charged?

Because these batteries always show the same voltage from 10% to 90% of capacity, this is tricky. A voltmeter tells little or nothing about the battery's state of charge. Either use a smart charger like the Battery Manager, or time the charge cycle. (Calculate the recommended charging time by dividing the battery capacity by the charger capacity and add 25% for charging inefficiency. For instance: a battery of 1100-mAh capacity in a charger that puts out 100 milliamps will need 11 hours, plus another three hours for charging inefficiency, for a total of 14 hours to completely recharge.) A cruder method to use is the touch method. When the battery gets warm, it's finished charging. (This method is only applicable with the plug-in chargers.)

My new nicad (or NiMH) batteries don't seem to take a charge in the solar-powered charger.

All batteries are chemically "stiff" when new. Nicad or NiMH batteries are shipped uncharged from the factory, and are difficult to charge initially. A small solar charger may not have enough power to overcome the battery's internal resistance; hence, no charge. The solution is to use a plug-in charger for the first few cycles to break in the battery. Another solution is to install only one battery at a time in the solar charger.

We cannot emphasize this too strongly! We get many complaints about "dead" solar chargers and nicads, and the problem is almost always overcoming the battery's shiny new stiffness. Once the batteries have been broken in, there is no problem. These batteries will give you many, many years of use for no additional cost with solar charging.

All figures are given in milliamp hours of capacity (1,000 milliamp hours = 1 amp hour). General purpose and alkaline battery capacities are average figures and are not based on any single brand name.

SMALL BATTERY CAPACITY (IN MAH)

Battery Type	Voltage	AAA	AA	C	D
General purpose	1.50	300	900	2,000	4,000
Alkaline	1.50	750	2,000	5,000	10,000
Rechargeable Alkaline	1.25	750	1,800	3,000	7,200
Golden Power (Real Goods)	1.25	240	700	1,800	4,000
Millennium	1.25	180	700	1,800	1,800
Power-Sonic	1.25	n/a	500	1,800	4,000
SAFT (no name industrial)	1.25	180	500	1,800	4,000
Nickel-Metal Hydride	1.25	600	1,500	3,300	7,000

My tape player (or other device) doesn't work when I put nicads in. What's wrong?

Some battery-powered equipment will not function on the lower-voltage nicad or NiMH batteries, which produce 1.25 volts (vs. 1.5 volts produced by fresh alkalines). Some manufacturers use a low-voltage cutoff that won't allow using rechargeables. What's worse, if the cutoff won't let you use nicads, it's also forcing you to buy new alkalines when they're only 50% depleted! Sometimes with devices that require a large number of batteries (six or more), the 1/4-volt difference between each nicad and alkaline gets magnified to a serious problem. Even a fully charged set of nicads will be seen as a depleted battery pack by the appliance.

Do I have to run the rechargeable battery completely dead every time?

No. For best life expectancy it's best to recharge nicad and NiMH batteries when you first sense the voltage is dropping. You'll be down to 10% or 15% capacity at this point. Further draining is not needed or recommended.

Alkaline rechargeable batteries really hate deep cycles. They'll do best when they're only run down about 50%.

—Doug Pratt

Rechargeable Batteries and Chargers

Batteries That Really Just Keep Going and Going!

Americans use nearly three billion disposable batteries every year, the rest of the world adds another few billion to that total. Beyond being expensive and a colossal waste of resources, this presents a huge low-level toxic disposal problem. Rechargeable batteries are the obvious, easy, money-saving solution, but buy an extra set. That way you have one set in your charger at all times and one set in the appliance. When the appliance gets weak, just swap batteries with the fresh ones. If you change batteries once a week, they will last for 10 to 15 years. With our Infinity Guarantee, your batteries will last forever.

Nickel-Metal Hydride Batteries

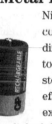

NiMH and nicad batteries are close cousins. The chemistry is similar, the difference is NiMH batteries are non-toxic, they have almost twice the storage capacity, there are no memory effects from short cycling, and they excel at hard use. This 2nd generation of NiMH has licked the rapid self-discharge problems of earlier NiMH technology. For heavy-discharge uses like digital cameras and remote-control toys, NiMH is clearly the best choice, as there's no voltage sag until the last 10% of charge. This means you actually get to use better than 90% of the battery's rated capacity. No other battery chemistry will deliver like this under heavy load. With no toxic lead, mercury, or cadmium content, and a life expectancy of 400 cycles or more, NiMH batteries are great for the environment.

Most nicad chargers will do NiMH batteries adequately, although sometimes they have trouble recognizing the greater capacity, and won't fully charge the NiMHs. Leaving the cells in the charger an extra day or two beyond the auto shutoff point often works. This is true for our Innovations Battery Manager. For best performance and fastest recharge times, pick up one of our NiMH chargers below. China.

17-9173 4 Golden Power NiMH AAA 4-pack (600mAh) $9⁰⁰

17-9173 8 Golden Power NiMH AAA 8-pack (600mAh) $16⁵⁰

17-9174 4 Golden Power NiMH AA 4-pack (1500mAh) $9⁰⁰

17-9174 8 Golden Power NiMH AA 8-pack (1500mAh) $16⁵⁰

17-9175 Golden Power NiMH C 2-pack (3300mAh) $10⁵⁰

17-9176 Golden Power NiMH D 2-pack (7000mAh) $19⁵⁰

NiMH Or Nicad Chargers

Smart Chargers for NiMH Batteries

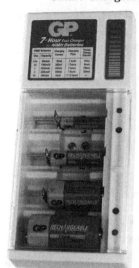

If you choose nickel-metal hydride batteries and want a charger that's both speedy and versatile, the GP Fast Charger is definitely the way to go. Designed to maximize the life of NiMH rechargeable batteries, it features an automatic current control that allows you to charge two to four AA, AAA, C, or D cells. Unit switches to trickle-charge after seven hours, so batteries are always fully charged until you're ready to use 'em. If you still have some nicad batteries on hand, try GP's other affordable charger. It handles nicads as well as NiMH batteries, and charges four AA or AAA cells in under 6 hours, then converts to trickle-charge. This charger also includes a four-pack of AA NiMH batteries. Both chargers have reverse polarity safety protections and LED indicators to show charging status. China.

50-255 GP NiMH Fast Charger (8.75" x 4" x 2.5") $29⁰⁵

57 171 GP Charger With NiMH Batteries (5" x 2.75" x 2.5") $29⁹⁵

Rechargeable Nicads

The Highest-Powered Nicads at the Lowest Prices

These are the most powerful nicads we've found on the market. The high-powered AAs have a capacity of 700 mAH (compared to 500 mAH for most others); the C is 1,800 mAH (compared to 1,200 mAH for most others); and the D is a full 4,000 mAH, more than three times greater than most competitors. We also offer a AAA with 240 mAH. These higher-capacity nicads will last much longer on a charge than other nicads and are closer in power to alkaline disposable batteries. All nicad batteries we offer are covered by our Real Goods Infinity Guarantee to last your entire lifetime. We also recycle any nicad for free. 1.25 volts. Hong Kong.

50-120 AAA Nicads (2-pack) $4⁵⁰

50-125 AA Nicads (2-pack) $4⁵⁰

50-121 AA Nicads (4-pack) $8⁵⁰

50-122 C Nicads (2-pack) $9⁵⁰

50-123 D Nicads (2-pack) $13⁵⁰

Batteries are shipped to you uncharged. You will have to charge them before first use.

Rechargeable High-Capacity Alkalines

For applications that demand long shelf life, like emergency flashlights or battery-powered wall clocks, or any device that demands 1.5-volt output, you can't do better than Accucell rechargeable alkaline batteries. These batteries arrive fully charged, have a 5-year shelf life, and feature more initial capacity than any other rechargeable battery type. Capacity will fade with use, however, and life expectancy is 12 to 50 cycles. Deeper cycles will cost capacity and life expectancy. Still, a big step up from one-use throwaways.

Accucell batteries should only be recharged in a charger designed for alkaline batteries. Our favorite chargers are the Accucell Quick Charger or the Battery Manager. Our tests show the Accucell charger delivers 10%-20% better performance throughout the Accucell battery's life. There are benefits to be had for the higher cost of pulse charging. Initial capacity listed below. Germany.

57-160 Accucell AAA, 750 mAH (2-pack) $4^{50}
57-162 Accucell AA, 1800 mAH (2-pack) $4^{50}
57-163 Accucell AA, 1800 mAH (4-pack) $8^{50}
57-164 Accucell C, 3000 mAH (2-pack) $15^{95}
57-166 Accucell D, 7200 mAH (2-pack) $16^{95}

Accucell Quick Charger

Our Fastest Charger Ever

Using revolutionary pulse charging, the Accucell Quick Charger safely puts more energy into rechargeable batteries faster than any other charger we've seen. It charges NiMH, nicad, and rechargeable alkaline batteries. It can accommodate up to 8 AA or AAA, 4 C or D, or two 9-volt batteries at a time. Batteries must all be the same type and size, with manual selection of battery type. It will switch from charging to maintenance-trickle mode automatically when the batteries are charged. Charge time runs from a minimum of 10 minutes for a AAA-cell nicad to a maximum of 9.6 hours for a D-cell NiMH. Our tests show better rechargeable battery performance with any type of battery using this charger. Batteries enjoy pulse charging, and will accept more energy from this charger.

The Accucell charger comes with both 120VAC and 12VDC charging cords. Cool! AC/DC! Add our lightweight Solar Option below and you'll have fast solar battery charging too. For rechargeable batteries only. One-year mfr.'s warranty. Germany

57-154 Accucell Quick Charger $99^{95}
11-252 Lightweight 10W Solar Option $119^{00}

A Smart Choice For Recharging

Extend the life of almost any battery using this Battery Xtender. Its "smart" technology easily accommodates alkaline, carbon-zinc, titanium, nickel-metal hydride and 1.5 volt batteries, sensing battery chemistry and automatically adapting its charging parameters for optimum recharging of each and every cell. Simultaneously accepts AAA, AA, C and D cells in any combination of sizes and chemistries. A built-in indicator gauges the status of each battery. Costs less than one cent in electricity per battery. 10″H x 6″ W x 3″D. China.

17-0181 Battery Xtender $49^{00}

120-Volt Battery Charger

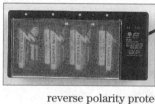

Our universal charger recharges any size NiMH or nicad from a standard 120-volt outlet in a day to a day and a half. This 4-cell model features LED lights (to indicate charging is in progress) and has reverse polarity protection, in case you accidentally put your nicads in backwards. Delivers 130 mA to any battery size. It will charge any combination of four AAA, AA, C, or D nicad batteries simultaneously or two 9-volt batteries. This low-cost charger does not shut off automatically. Okay for NiMH or nicads. China.

50-223 4-Cell AC Charger $14^{95}

Our Infinity Guarantee

Our nicads and nickel-metal hydrides are so durable and reliable, we guarantee them forever. You will never have to buy batteries again. Should one fail, with proof of purchase we will replace it free of charge. Don't look for fine print, exclusions, or hidden catches. There are none. These batteries will outlast you.

Solar Super Charger

Our high-powered Solar Super Charger will charge any two NiMH or nicad batteries at a time. The holder accommodates AAA, AA, C, or D sizes. The single-crystal solar module measures 3.5″ x 9″, is waterproof, nearly unbreakable, and has a 16″ wire lead. Battery box is not waterproof. Nominal output is 4.5 volts at 200 milliamps. Will charge a pair of our AA nicads in 7 hours of good sun. Our higher-capacity NiMH batteries will need a couple days. Mexico.

50-212 Improved Solar Super Charger $29⁹⁵

Battery Saver TravelPak

For camping, traveling abroad, or any outdoor activity where you don't have access to electricity, this travel solar charger will keep your AA nicads charged and ready for action. Delivering 140 mA output, it recharges 1 to 4 AA batteries in as little as 12 hours of sunlight exposure. It has an additional storage cavity for 4 more AA batteries. You can rotate your batteries, making sure you always have a freshly charged set. Wireless, compact and portable, this Solar Charger has tilting panel positions to maximize sunlight exposure. Great for campers,

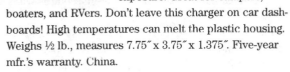

boaters, and RVers. Don't leave this charger on car dashboards! High temperatures can melt the plastic housing. Weighs ½ lb., measures 7.75″ x 3.75″ x 1.375″. Five-year mfr.'s warranty. China.

51-068 Battery Saver TravelPak $19⁹⁵

Battery Saver PowerPak

Solar or Plug-In Charging Lets You Keep AA Batteries Always Ready!

This Battery Saver PowerPak comes complete with a multi-jack for 6-volt applications and a 12-volt cigarette lighter connector. It delivers 140 mA at 6 volts or 70 mA at 12 volts with a voltage selection switch. Charges many types of cellular phone, computer, or 6-volt gel cell batteries. The PowerPak will run small portable appliances like the Sony Walkman, or fans, and you can use it to maintain 12-volt car batteries. The PowerPak charges 4 AA rechargeable nicads in less than 12 hours of direct sunlight, plus it stores an extra set of 4 AA batteries and has a LED charge light indicator. Other features include a tilting panel for maximum sunlight exposure and a carry handle for portability. Measures 9.5″ x 6.75″ x 1.75″ and weighs 1.5 lbs. Five year mfr.'s warranty. China.

51-069 BatterySaver PowerPak $39⁹⁵

Walkman® not included.

12-Volt Nicad Charger

This 12-volt charger will recharge AAA, AA, C, or D size NiMH or nicad batteries in 10 to 20 hours from your 12-volt power source. It will charge up to four batteries simultaneously. Batteries must be charged in pairs. Has a clever contact system on the positive end that will deliver different current flow depending on battery size. AAAs will receive 50 milliamps, AAs will receive 130 milliamps, Cs and Ds will receive 240 milliamps. It comes with a 12-volt cigarette lighter plug to go into any 12-volt socket. Make sure the cigarette lighter works with the key off when using it in your car. One or two charge cycles will not discharge the car's battery. China.

50-214 12-Volt Nicad Charger $24⁹⁵

Inverters

What is an inverter, and why would I want one in my renewable energy system? Reasonable question. Many small renewable energy systems don't need an inverter. The inverter is an electronic device that converts direct current (DC) into alternating current (AC). Renewable energy sources like PV modules and wind turbines make DC power, and batteries will store DC power. This is the least expensive and most universally applicable storage method available. Batteries store energy as low-voltage DC. DC is fine, in fact preferable, for some applications. A small cabin that only needs three or four lights would get by just fine running everything on 12-volt DC. But most of the world operates on higher voltage AC. AC transmits more efficiently than DC and so has become the world standard. No technology exists to store AC, however. It must be produced as needed. If you want to run conventional household appliances with your renewable energy system, you need a device to produce AC house current on demand. That device is an inverter.

Inverters are a fairly new entry on the scene. Until the early 1990s, the highly efficient, long-lived, relatively inexpensive inverters that we have now were still a pipe dream. The world of solid-state equipment has advanced by extraordinary leaps and bounds. More than 95% of the household power systems we put together now include an inverter.

Electrical Terminology & Mechanics

Electricity can be supplied in a variety of voltages and waveforms. As delivered to and stored by our renewable energy battery system, we've got low voltage DC. As supplied by the utility network, we've got house current: nominally 120 volts AC. What's the difference? DC electricity flows in one direction only, hence the name direct current. *It flows directly from one battery terminal to the other battery terminal. AC current alternates, switching the direction of flow periodically. The U.S. standard is 60 cycles per second. Other countries have settled on other standards, but it's usually either 50 or 60 cycles per second. The electrical term for the number of cycles per second is Hertz, named after an early electricity pioneer. So AC power is defined as 50 Hz or 60 Hz.*

Our countrywide standard also defines the voltage. Voltage is similar to pressure in a water line. The greater the voltage, the higher the pressure. When voltage is high it's easier to transmit a given amount of energy, but it's more difficult to contain and potentially more dangerous. House current in this country is delivered at about 120 volts for most of our household appliances, with the occasional high-consumption appliance using 240 volts. So our power is defined as 120/240 volts/60 Hz. We usually use the short form, 120 volts AC, to denote this particular voltage/cycle combination. This voltage is by no means the world standard, but is the most common one we deal with. Inverters are available in international voltages by special order, but not all models from all manufacturers.

Benefits of Inverter Use

The world runs on AC. In North America it's 120-volt, 60-cycle AC. Other countries may have slightly different standards, but it's all AC. Joining the mainstream allows the use of mass-produced components, wiring hardware, and appliances. Appliances may be chosen from a wider, cheaper, and more readily available selection. Electricity transmits more efficiently at higher voltages. Power distribution through the house can be done with conventional 12- and 14-gauge romex wiring using standard hardware, which electricians appreciate and inspectors understand. Anyone who's wrestled with the heavy 10-gauge wire that a DC system requires will see the immediate benefit.

Modern brand-name inverters are extremely reliable. The models we sell have failure rates well under 1%. Efficiency averages about 90% for most models with peaks at 95% to 98%. In short, inverters make life simpler, do a better job of running your household, and ultimately save you money on appliances and lights. But watch out for the flood of cheapo imported inverters! High failure rates and poor performance are common with these.

An inverter/battery system is the ultimate clean uninterruptible power supply for your computer, too. In fact, an expensive UPS system is simply a battery and a small inverter with an expensive enclosure and a few bells and whistles. Just don't run the power saw or washing machine off the same inverter that runs your computer. We often recommend a small secondary inverter just for the computer system. This ensures that the water pump or some other large, unexpected load can't possibly hurt you.

Reading Modified Sine Wave Output with a Conventional Voltmeter

Most voltmeters that sell for under $100 will give you weird voltage readings if you use them to check the output of a modified sine wave inverter. Readings of 80 to 105 volts are the norm. This is because when you switch to "AC Volts" the meter is expecting to see conventional utility sine wave power. What we commonly call 110- to 120-volt power actually varies from 0 to about 175 volts through the sine wave curve. 120 volts is an average called the Root Mean Square, or RMS for short, that's arrived at mathematically. More expensive meters are RMS corrected; that is, they can measure the average voltage for a complex waveform that isn't a sine wave. Less expensive meters simply assume if you switch to "AC Volts" it's going to be a sine wave. So don't panic when your new expensive inverter checks out at 85 volts: the inverter is fine, but your meter is being fooled. Modern inverters will hold their specified voltage output, usually about 117 volts, to plus or minus about 2%. Most utilities figure they're doing well by holding variation to 5%.

Description of Inverter Operation

If we took a pair of switching transistors and set them to reversing the DC polarity (direction of current flow) 60 times per second, we would have low voltage alternating power of 60 cycles or Hertz (Hz). If this power was then passed through a transformer, which can transform AC power to higher or lower voltages depending on design, we could end up with crude 120-volt/60 Hz power. In fact this was about all that early inverter designs of the 1950s and '60s did. As you might expect, the waveform was square and very crude (more about that in a moment). If the battery voltage went up or down so did the output AC voltage, only ten times as much. Inverter design has come a long, long way from the noisy, 50% efficient, crude inverters of the 1950s. Modern inverters hold a steady voltage output regardless of battery voltage fluctuations, and efficiency is typically in the 90% to 95% range. The waveform of the power delivered has also been dramatically improved.

Waveforms

The AC electricity supplied by your local utility is created by spinning a bundle of wires through a series of magnetic fields. As the wire moves into, through, and out of the magnetic field, the voltage gradually builds to a peak and then gradually diminishes. The next magnetic field the wire encounters has the opposite polarity, so current flow is induced in the opposite direction. If this alternating electrical action is plotted against time, as an oscilloscope does, we get a picture of a sine wave, as shown in our wave form gallery below. Notice how smooth the curves are. Transistors, as used in all inverters, turn on or off abruptly. They have a hard time reproducing curvy sine waves.

Early inverters produced a square-wave alternating current (see the wave form gallery again). While square-wave does alternate, it is considerably different in shape and peak voltage. This causes problems for many appliances. Heaters or incandescent lights are fine, motors usually get by with just a bit of heating and noise, but solid-state equipment has a really hard time with it, resulting in loud humming, overheating, or failure. It's hard to find new square-wave inverters anymore, and nobody misses them.

Most modern inverters produce a hybrid waveform called a quasi-, synthesized, or modified sine wave. In truth this could just as well be called a modified square wave, but manufacturers are optimists. (Is the glass half full or half empty?) Modified sine wave output cures many of the problems with square wave. Most appliances will accept it and hardly know the difference. There are some notable exceptions to this rosy picture, however, which we'll cover in detail later in the Inverter Problems section.

Full sine wave inverters have been available since the mid-'90s, but because of their higher initial cost, and lower efficiency they were only initially used for running very specific loads. This is changing rapidly. There are a variety of high-efficiency, moderate-cost sine wave inverters available now. Trace and Statpower have both unveiled a whole series of sine wave units, and it looks like the wave of the future. True sine wave inverters have very nearly become the standard for larger household power systems already. With very rare exceptions, sine wave inverters will happily run any appliance that can plug into utility power. Motors run quieter and cooler on sine wave, and solid-state equipment has no trouble. True sine wave inverters deliver top quality AC power, and are almost always a better choice for household use. For smaller systems, sine wave quality may not be worth the higher cost, however.

Inverter Output Ratings

Inverters are sized according to how many watts they can deliver. More wattage capability will cost more money initially. Asking a brand-name inverter for more power than it can deliver will result in the inverter shutting down. Asking the same of a cheapo inverter may result in a crispy critter. All modern inverters are capable of briefly sustaining much higher loads than they can run continuously, because some electric loads, like motors, require a surge to get started. This momentary overcapacity has led some manufacturers to fudge on their output ratings. A manufacturer might, for instance, call its unit a 200-watt inverter based on the instantaneous rating, when the continuous output is only 140 watts. Happily, this practice is fading into the past. Most manufacturers are taking a more honest approach as they introduce new models and are labeling inverters with their continuous wattage output rather than some fanciful number. In any case, we've been careful to list the continuous power output of all the inverters we carry. Just be aware that the manufacturer calling the inverter a 200-watt unit does not necessarily mean it will do 200 watts continuously. Read the fine print.

Wave-Form Gallery

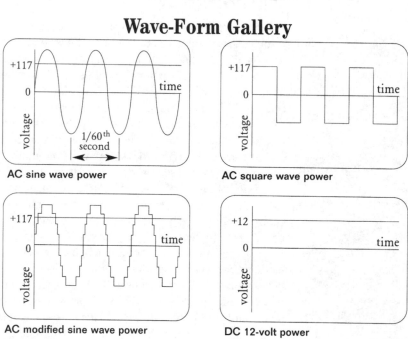

AC sine wave power

AC square wave power

AC modified sine wave power

DC 12-volt power

How to Cripple Your Inverter

Want to reduce your inverter's ability to provide maximum output? Here's how: Keep your battery in a low state of charge, use long, undersize cables to connect the battery and inverter, and keep your inverter in a small airtight enclosure at high temperatures. You'll succeed in crippling, if not outright destroying, your inverter. Low battery voltage will severely limit any inverter's ability to meet a surge load. Low voltage occurs when the battery is undercharged, or undersized. If the battery bank isn't large enough to supply the energy demand, then voltage will drop, and the inverter probably won't be able to start the load. For instance, a Trace DR2412 will easily start most any washing machine, but not if you've only got a couple of golf cart batteries to supply it with energy. The starting surge will demand more electrons than the battery bank can supply, voltage will plummet, and the inverter will shut off to protect itself (and your washing machine motor). Now suppose you have eight golf cart batteries on the DR2412, which is plenty to start the washer, but decided to save some money on the hookup cables,

Helpful Hints

1. Keep the inverter as close to the battery as possible, but not in the battery compartment. Five to ten feet and separated by a fireproof wall is optimal. The high-voltage output of the inverter is easy to transmit. The low-voltage input transmits poorly.

2. Keep the inverter dry and as cool as possible. They don't mind living outdoors if protected.

3. Don't strangle the inverter with undersize supply cables. Most manufacturers have recommendations in the owner's manual. If in doubt, give us a call.

4. Fuse your inverter cables and any other circuits that connect to a battery!

and used a set of $16.95 automotive jumper cables. The battery bank is capable, the inverter is capable, but not enough electrons can get through the undersized jumper cables. The result is the same: low voltage at the inverter, which shuts off without starting the washer. Do not skimp on inverter cables, or on battery interconnect cables. These items are inexpensive compared to the cost of a high-quality inverter and good batteries. Don't cripple your system by scrimping on the electron delivery bits.

All inverters produce a small amount of waste heat. The harder they are working, the more heat they produce. If they get too hot they will shut off or limit their output to protect themselves. Give the inverter plenty of ventilation. Treat it like a piece of stereo gear: dry, protected, and well ventilated.

How Big an Inverter Do I Need?

Inverters are rated by their continuous wattage output. The more they can put out, the more they cost initially. So you don't want to buy a bigger one than you need. On the other hand, an inverter that's too small is going to frustrate you because you'll need to limit usage.

If you have a small cabin with an appliance or two to run, it doesn't take much. For example, say you want to run a 19-inch TV, a VCR, and a light all at once. Total up all the wattages, about 80 for the TV, 25 for the VCR, and 20 for a compact fluorescent light, a total of 125 watts. Pick an inverter that can supply at least 125 watts continuously, and you have it.

To power a whole house full of appliances and lights might take more planning. Obviously not every appliance and light will be on at the same time. Mid size inverters of 600 to 1,000 watts do a good job of running lights, entertainment equipment, and small kitchen appliances, in other words, most common household loads. What a mid-size inverter will not do is run a mid- to full-size microwave, a washing machine, or larger handheld power tools. For those loads you need a full-size 2,000+ watt inverter. In truth, most households end up with one of the full-size inverters because household loads tend to grow, and larger inverters are often equipped with very powerful battery chargers. This is the most convenient way to add battery charging capability to a system (and the cheapest, too, if you are already buying the inverter).

Chargers, Lights, Bells and Whistles

As you go from small plug-in-the-lighter-socket inverters to larger household-sized units, you'll find increasing numbers of bells and whistles, most of which are actually pretty handy. At the very least, all inverters have an LED light showing it's turned on. Mid-size units may feature graphic volt and amp meters; better units have an LCD display. While these displays are handy and entertaining, household systems should still use a real system monitor, hopefully mounted in the kitchen or hallway.

Battery chargers are the most useful inverter option, and for most folks, the cheapest and easiest way to add a powerful, automatic battery charger to their system. Chargers are pretty much standard equipment on all larger household-sized inverters, and are an option on some mid-size inverters. With a built-in charger, the inverter will have a pair of "AC input" terminals. If the inverter detects AC voltage at these terminals, because you just started the backup generator for instance, it will automatically connect the AC input terminals to the AC output terminals, and then go to work charging the batteries. Inverter-based chargers tend to be extremely robust, adjustable, and will treat your batteries nicely.

Many financially challenged off-the-grid systems start out with just a generator, an inverter/charger, and some batteries. You run the generator every few days to recharge the batteries. Add PV charging as money allows, and the generator gradually has to run less and less.

How Big a Battery Do I Need?

We usually size household battery banks to provide stored power for three to five days of autonomy during cloudy weather. For most folks, this is a comfortable compromise between cost and function. If your battery bank is sized to provide a typical three to five days of back-up power, then it will also be large enough to handle any surge loads that the inverter is called upon to start. A battery bank smaller than three-days capacity is going to get cycled fairly deeply on a regular basis. This isn't good for battery life. A larger battery bank cycled less deeply is going to cost less in the long run, because it lasts longer. Banks larger than five-days capacity start getting more expensive than a back-up power source (like a modest-sized generator). However, we occasionally run into situations with ¾ horsepower or larger submersible well pumps or stationary power tools requiring a larger battery bank simply to meet the surge load when starting. Call the Gaiam Real Goods technical staff for help if you are anticipating large loads of this type.

Inverter Safety Equipment and Power Supply

In some ways batteries are safer than conventional AC power. It's fairly difficult to shock yourself at low voltage. But in other ways batteries are more dangerous: They can supply many more amps into a short circuit, and once a DC arc is struck, it has little tendency to self-snuff. One of the very sensible things the National Electrical Code requires is fusing and a safety disconnect for any appliance connected to a battery bank. Fusing is extremely important for any circuit connected to a battery! Without fusing you are risking burning your house down.

There are several products readily available to cover DC fusing and disconnect needs for inverter-based systems. For full-size inverter fusing, depending on system requirements you can either use a properly sized Powercenter, or the very popular Trace DC Disconnect. Our technical staff can answer your questions about which one is better for you. Mid-sized inverters can use a Class T fuse with a Blue Seas circuit breaker to provide safe and compliant connection. All this safety and connection equipment is covered in the Safety section of this chapter. Generally, you'll need a 200- to 400-amp fuse for full-size inverters, a 100- to 200-amp fuse for mid-size inverters, and small inverters will be plugged into a 20- or 30-amp fused outlet.

The size of the cables providing power to the inverter is as important as the fusing, as we noted earlier. Do not restrict the inverter's ability to meet surge loads by choking it down with undersize or lengthy cables. Ten feet is the longest practical run between the battery and inverter. This is true even for small 100- to 200-watt inverters. Put the extension cord on the AC side of the inverter! You'll find cable and circuit breaker requirement charts with the larger inverters. Batteries should either be the sealed type, or live in their own enclosure. Don't put your inverter in with the batteries!

Potential Problems with Inverter Use— and How to Correct Them

We have been painting a rosy picture up to this point, but let us now have a little brutal honesty to balance things out. If you know about these problems beforehand it is usually possible to work around them when selecting appliances.

All inverters have output limits. They are rated to produce a specific wattage for a specified time. Obviously a 250-watt inverter is not going to run your hairdryer or power tools. So don't pick an inverter that is too small for your needs.

Waveform Problems

This is the biggest single inverter problem! We talked about sine wave vs. modified sine wave in the technical description above. Unless noted otherwise, all inverters produce modified sine wave current. This works fine for 97% of appliances, with notable exceptions listed below. The problems discussed here are caused only by modified sine wave inverters. If you are looking at one of the new full-size sine wave units, you can skip this section. (Unless you want to read it now and feel smug.) Sometimes we employ a small sine wave inverter to power a specific appliance or two, while the rest of the household runs on the cheaper and slightly more efficient modified sine wave inverter.

Audio Equipment

Some audio gear will pick up a 60-cycle buzz through the speakers. It doesn't hurt the equipment, but it's annoying to the listener. There are too many models and brands to say specifically which are a problem and which aren't. We've had better luck with new equipment recently. Manufacturers are starting to put better power supplies into their gear. We can only recommend that you try it and see.

Some top-of-the-line audio gear is protected by SCRs or Triacs. These devices are installed to guard against power line spikes, surges, and trash (nasties which don't happen on inverter systems). However, they see the sharp corners on modified sine wave as trash and will sometimes commit electrical hara-kiri to prevent that nasty power from reaching the delicate innards. Some are even smart enough to refuse to eat any of that ill-shaped power, and will not power up. The only sure cure for this (other than more tolerant equipment) is a sine wave inverter. If you can afford top-of-the-line audio gear, chances are you've already decided that sine wave power is the way to go.

Computers

Computers run happily on modified sine wave. In fact most of the uninterruptible power supplies on the market have modified sine wave or even square wave output. The first thing the computer does with the incoming AC power is to run it through an internal power supply. We've had a few reports of the power supply being just a bit noisier on modified sine, but no real problems. Running your prize family-heirloom computer off an inverter will not be a problem. What can be a problem is large start-up power surges. If your computer is running off the same household inverter as the water pump, power tools, and microwave, you're going to have trouble. When a large motor, like a skilsaw, is starting, it will momentarily pull the AC system voltage way down. This can cause computer crashes. The fix is a small, separate inverter that only runs your computer system. It can be connected to the same household battery pack, and have a dedicated outlet or two.

Laser Printers

Many laser printers are equipped with SCRs, which cause the problems detailed above. Laser printers are a very poor choice for renewable energy systems anyway due to their high standby power use keeping the heater warm. Lower cost inkjet printers can do almost anything a laser printer can do while only using 25 to 30 watts instead of 900 to 1,500 watts.

Ceiling Fans

Most variable-speed ceiling fans will buzz on modified sine wave current. They work fine, but the noise is annoying.

Other Potential Inverter Problems

Okay, you sine wave buyers can stop smirking here, the following problems apply to all inverters.

Radio Frequency Interference

All inverters broadcast radio static when operating. Most of this interference is on the AM radio band. Do not plug your radio into the inverter and expect to listen to the ball game; you'll have to use a battery powered radio and be some distance away from the inverter. This is occasionally a problem with TV interference when inexpensive TVs and smaller inexpensive inverters are used together. Distance helps. Put the TV (and the antenna) at least 15 feet from the inverter. Twisting the inverter input cables may also limit their broadcast power (strange as it sounds, it works).

Phantom Loads and Vampires

A phantom load isn't something that lurks in your basement with a half-mask, but it's close kin. Many modern appliances remain partially on when they appear to be turned off. That's a phantom load. Any appliance that can be powered up with a button on a remote control must remain partially on and listening to receive the "on" signal. Most TVs and audio gear these days are phantom loads. Anything with a clock—VCRs, coffee makers, microwave ovens, or bedside radio-clocks—uses a small amount of power all the time.

Vampires suck the juice out of your system. Vampires are the heavy black power cube that plugs into an AC socket, and delivers lower voltage power to your answering machine, electric toothbrush, power tool charging stand, or any of the other huge variety of appliances that uses a power cube on the AC socket. These villainous wastrels usually run a horrible 60% to 80% inefficiency (which means that for every dime's worth of electricity consumed, they throw away six or eight cents worth.) Most of these nasties always draw power, even if there's no battery/toothbrush/razor/cordless phone present and charging. It would cost their manufacturers less than 25 cents per unit to build a power saving standby mode into the power cube, but since you, the consumer, are paying for the inefficiency, what do they care? The appliance might be turned off, but the vampire keeps sucking a few watts. Ever noticed that power cubes are usually warm? That's wasted power being converted to heat. By the way, cute and appropriate as it is, we can't take credit for the vampire name. That's official electric industry terminology (bless their honest souls).

So what's the big deal? It's only a few watts isn't it? The problem isn't the power consumed by vampires and phantom loads, it's what is required to deliver those few watts 24 hours per day. When there's no demand for AC power, a full-size inverter will drop into a stand-by mode. Stand-by keeps checking to see if anything is asking for power, but it takes only a tiny amount of energy, usually under 2 watts. If the phantom or vampire load is enough to awaken the inverter from stand-by, it consumes its own load plus the inverter's overhead, which is 6 to 20 watts. The inverter overhead is the real problem. An extra 10 to 25 watts might sound like grasping at straws, but over 24 hours every day this much power lost to inefficiency can easily add a couple $450 modules to the size of the PV array and maybe another battery or two to the system requirements. You want to make sure your inverter can drop into stand-by mode whenever there's no real demand for AC power.

Keeping the Vampires at Bay

With some minimal attention to details, you can keep the vampires and phantoms under tight control and save yourself hundreds or thousands on system costs.

The solution for clocks is battery power. A wall-mounted clock runs for nearly a year on a single AA rechargeable battery. We have found a wide selection of good-quality battery-powered alarm clocks, and most of the other timekeepers anyone could possibly require, just by looking around. Clocks on house current are ridiculously wasteful.

The solution for phantoms and vampires are outlets or plug strips that are switched off when not in use. The outlets only get switched on when the appliance is actually being used or recharged. As a side benefit, this cures PMS—perpetual midnight syndrome—on your VCR, too!

If you're aware of the major energy-wasting gizmos we all take for granted, it's easy to avoid or work around them. When you consider the minor inconvenience of having to flip a wall switch before turning on the TV against the $400 to $500 cost of another PV module to make up the energy waste, it all comes into perspective. A few extra switched outlets during construction looks like a very good investment. If your house is already built, then use switched plug strips.

Conclusions

Inverters allow the use of conventional, mass-produced AC appliances in renewable energy-powered homes. They have greatly simplified and improved life "off-the-grid." Modern solid-state electronics have made inverters efficient, inexpensive, and long-lived. Batteries and inverters need to live within 10 feet of each other, but not in the same enclosure. Modified sine wave inverters will not be perfectly problem-free with every appliance. Expect a few rough edges. The best inverters produce pure sine wave AC power, which is acceptable to all appliances. Sine wave units tend to cost a bit more, but are strongly recommended for all full-size household systems. Keep the power-wasting phantom loads and power cube vampires under tight control, or be prepared to spend enough to overcome them.

—**Doug Pratt**

Inverters

PowerStar 200-Watt Pocket Inverter

This compact inverter is one of the best mini-inverters in the industry. It is ideal for powering most 19-inch color TVs, personal computers, VCRs, video games, stereos, and lots of other small appliances directly from your car or 12-volt system. It's great for travelers to carry in their cars to power these appliances. Includes a cigarette lighter input plug. It will deliver 140 watts of 115 volts AC power continuously from your 12-volt battery. It will provide 400 peak watts and 200 watts for over two minutes. PowerStar has recently made an improvement to the 200-watt inverter. In case of overload, the unit now safely delivers as much power as it can into that overload. China.

Input Voltage: 10 to 15 volts DC
Continuous Output: 140 watts
Surge Capacity: 400 watts
Idle Current: 0.25 amp @ 12V (3 watts)
Recommended Fusing: 30-amp
Dimensions: 5″ x 2.6″ x 1.7″
Weight: 15 oz.
Warranty: one year

27-104 PowerStar 200 Inverter, 200W/12V $69⁹⁵

PowerStar Upgradable Inverters

PowerStar's line of upgradable inverters all use the same case, and are rated in terms of their continuous true RMS power capability. The continuous power rating on the smallest inverter is 400 watts; the first upgrade takes you up to 700 watts continuous; and the final upgrade takes you up to 1,300 watts continuous. The actual upgrade can be performed by the factory in a few days. The price for the upgrade is simply the difference between the two units. Therefore, there is no cost penalty for buying a smaller unit and later upgrading it, making these inverters extremely versatile!

All three units feature an audible low-battery warning below 10.9-volts input, a green AC power indicator lamp, one standard three-prong 115-volt outlet, optional hard-wire AC output access, and a built-in connector socket for a remote-control circuit. The PowerStar design features a dual mode limiter to enable operation of appliances rated higher in power than the inverter. A moderate overload lowers the output voltage, a severe overload causes shutdown. Reset by cycling the power switch.

The small 400-watt unit is suitable for computer systems, power tools, and small appliances. Its 3000-watt surge capacity can start and run lightly loaded ¼-hp motors.

The 700-watt unit will run a 500-watt microwave, a vacuum cleaner, a hair dryer (on medium), or a small coffee pot or hotplate. Like the smaller unit, the surge capacity is 3,000 watts.

The 1,300-watt unit will continuously operate a full-size microwave, a circular saw, or any 1,300-watt appliance. 6,000-watt surge capacity on this larger unit. Scotland.

Input Voltage: 10.5 to 16.5 volts DC
Output Voltage: 115 volts AC true RMS ± 5%
Continuous Output: 380, 700, 1,300 watts
Surge Capacity: 3,000 watts (6000 for the 1,300-watt unit)
Idle Current: 0.06-amp @ 12VDC (0.7-watt)
World Voltages Available: none
Average Efficiency: over 90% at half-rated power
Recommended Fusing: 110A Class T for 400 and 700, 200A Class T for 1,300
Dimensions: 3.15″ x 3.3″ x 12″
Weight: 15 lb. maximum (varies with model)
Warranty: two years

27-105 PowerStar Inverter, 400W/12V $339⁰⁰

27-106 PowerStar Inverter, 700W/12V $389⁰⁰

27-107 PowerStar Inverter, 1300W/12V $519⁰⁰

Statpower Inverters

Statpower Technologies Corporation has been a highly innovative and reliable inverter manufacturer since the dawn of the modern electronic inverter age. Headquartered near Vancouver, British Columbia, Statpower has pioneered sophisticated high-frequency power conversion technology since 1988. Statpower products are smaller, lighter, and less expensive than transformer-based inverters. Like all modern inverters, they will protect themselves from any conceivable problem such as overheating or overloading.

Statpower offers a full range of inverters from 50 watts for simple laptop charging on the road, to 3,000 watt pure sine wave units for full households.

Small Portable Statpower Inverters

Statpower NOTEpower 50-Watt Inverter

The NOTEpower inverter is perfect for the on the go "hacker." Notebook computers are great as long as you only want to use them for a few hours on a trip, but extended field trips can give your little brain in a pouch that run-down feeling. Plug the NOTEpower into any vehicle cigarette lighter and you can operate your computer (via its AC adapter/charger) or recharge spare computer batteries. Weighing in at just 6.5 ounces, the NOTEpower can put out up to 50 watts of power that could come in handy for cellular phones, camcorders, shavers, or other rechargeable equipment. It will shut off automatically on low voltage, leaving power to start your car. A trim 3.5″ x 2.5″ x 1.25″, the NOTEpower will fit just about anywhere! This is the easy way to adapt your laptop to 12 volts. One-year mfr.'s warranty. China.

27-134 NOTEpower Inverter $49⁹⁵

Statpower PROwatt 150

The PROwatt 150 inverter features cigarette lighter plug portability. It provides 150 watts continuously, 200 watts for five minutes, or up to 400 watts on a surge. It can run laptops, desktop computers, small TVs, and a variety of battery chargers and smaller appliances. If overloaded it will shut off without harm. Will shut off automatically before your battery gets too low for starting. Weighs 22 oz. Measures 1.5″ x 4″ x 4.9″. 10- to 15-volt DC input. One-year mfr.'s warranty. China.

27-114 PROwatt 150 Inverter $79⁹⁵

The Statpower Portawattz Series

The Portawattz is Statpower's lower priced line. Featuring reliable high-frequency switching and North American designs, they are manufactured in China to keep the price down. No battery chargers or transfer switches with these no-frills inverters. Difficult reactive loads like fluorescent lights and power tools are no problem for these quality inverters. These 12-volt inverters give an audible alarm if battery voltage falls to 10.6 volts, and shut off at 10.0 volts, or if overloaded or overheated. The Portawattz series all feature a six-month manufacturer's warranty. China.

Statpower Portawattz 300

The largest of Statpower's offerings with cigarette lighter plug portability, the Portawattz 300 can deliver 300 watts continuously, or 500 watts into a surge. Note that continuous draws over 125 watts will seriously challenge the current handling capability of the lighter plug configuration. Hard wiring is recommended, and will not void the warranty. An external ATC-type automotive fuse protects against accidental reverse polarity. The Portawattz 300 can provide reliable, portable power for camping, mobile offices, and emergency lighting. Has two 3-prong outlets. Weighs 1.8 lb. Measures 1.8″ x 4.7″ x 6″. Six-month mfr.'s warranty. China.

27-197 Portawattz 300 Inverter $49⁹⁵

Statpower Portawattz 600

The 600-watt model can comfortably run lights, entertainment equipment, computers, and small kitchen appliances. It delivers 600 watts continuously, and surges to 1,200 watts. Three 3-prong outlets are provided for power output.

Continuous Output: 600 watts
Surge Capacity: 1,200 watts
Idle Current: 0.3 A (3.6 watts)
Recommended Fusing: 110A Class T fuse
Dimensions: 11″ x 6.25″ x 2.5″
Weight: 4.5 lb (2 kg)

27-804 Portawattz 600 Inverter $109⁰⁰
24-507 110-Amp Class T Fuse w/ Cover $52⁹⁵
24-508 110-Amp Class T Repl. Fuse $17⁹⁵

Statpower Portwattz 1000

The 1000 model is a good choice for small cabins and RVs. It delivers 1,000 watts continuously and will surge to 2,000 watts. Graphic DC voltage and amperage meters are included on the front face. One 2-prong outlet, and two 3-prong outlets are provided for power output. The Class T fuse listed below is recommended for safe connection.

Continuous Output: 800 watts
Surge Capacity: 2,000 watts
Idle Current: 0.3A (3.6 watts)
Recommended Fusing: 110A Class T fuse
Dimensions: 10.25″ x 9.5″ x 3.25″
Weight: 5.2 lb (2.4 kg)

27-198 Portawattz 1000 Inverter $269⁰⁰
24-507 110 Amp Class T Fuse w/ Cover $52⁹⁵
24-508 110 Amp Class T Repl. Fuse $17⁹⁵

Statpower Portawattz 1750

The 1750 model can run most microwaves and small power tools in addition to lights, TVs and computers like the smaller models. Delivers 3,000 watts to a surge, 1,750 watts for five minutes, or 1,500 watts continuously. Has graphic DC voltage and amperage meters on the front face. One 2-prong outlet, and two 3-prong outlets are provided for power output. The Class T fuse listed below is recommended for safe connection.

Continuous Output: 1500 watts
Surge Capacity: 3,000 watts
Idle Current: 0.6A (7.2 watts)
Recommended Fusing: Trace DC-175 or 200A Class T fuse
Dimensions: 16.25″ x 9.5″ x 3.25″
Weight: 8.25 lb (3.8 kg)

27-199 Portawattz 1750 Inverter $379⁰⁰

24-212 200-Amp Class T Fuse w/ Cover $52⁹⁵

24-211 200-Amp Class T Repl. Fuse $17⁹⁵

Statpower Portawattz 3000

The beefy 3000 model is rated to deliver 5,000 watts into a surge, 3,000 watts for five minutes, or 2,500 watts continuously. This is enough power to run almost any 120VAC household appliance. Features front-face graphic DC voltage and amperage meters. Dual outlets with a 15A circuit breaker provide power output. A properly rated circuit breaker or fuse and 4/0 supply cables are needed for safe installation of this inverter.

Continuous Output: 2,500 watts
Surge Capacity: 5,000 watts
Idle Current: 0.6A (7.2 watts)
Recommended Fusing: Trace DC-250
Dimensions: 20″ x 8.5″ x 6.5″
Weight: 20 lb (9 kg)

27-805 Portawattz 3000 Inverter $599⁰⁰

Statpower PROsine Inverters & Inverter/Chargers

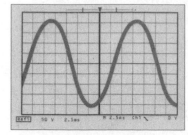

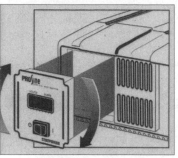

Ahh . . . the good stuff. . . . These true sine wave units produce a perfect sine wave. Using a PROsine, there will be no interference in audio, video, and electronic equipment. This is better quality power than your utility company can usually supply. Four basic PROsine models are offered at 1,000, 1,800, 2,500, and 3,000 watts. The two larger units are equipped with automatic battery chargers for when an external power source is available. All models are available in 12- or 24-volt input.

Statpower PROsine 1000 & 1800 Inverter Models

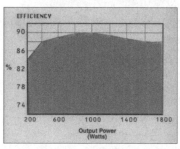

Features of these clean power models start with the intelligent, easy-to-read LCD display with backlighting that shows battery voltage, DC amperage, and a bar graph showing AC wattage output. The display can be rotated 90° to allow shelf or wall mounting and still read upright. It can also be removed and mounted remotely using standard phone plugs and extension cable. Advanced thermal design and careful internal layout allow longer run times in high ambient temperatures. The Powersave mode puts the inverter to sleep, and allows power use to drop to 1.5 watts when there is no load.

Standard PROsine 1000 & 1800 models are supplied with a duplex GFCI outlet with a 15-amp breaker. The outlet can be removed, and the inverter can be hardwired if desired. Alternatively, these models can be supplied as "xfer" versions. In this version, AC hardwire input and output terminals are provided. This turns the PROsine into an automatic back-up system, quickly transferring to inverter power if normal AC power fails. (Note: this is not intended as a computer UPS system.)

All PROsine inverters have a two-year manufacturer's warranty. China.

PROsine 1000

Continuous Output:
1,000 watts
Surge Capacity: 1,500 watts
Standby Power: 1.5 watts
Recommended Fusing: 110A
Class T fuse
Dimensions: 15.4″ x 11″ x 4.5″
Weight: 14.5 lb (6.6 kg)

27-809 PROsine 1000W/12V w/ GFCI $860⁰⁰

27-810 PROsine 1000W/12V w/ Xfer $910⁰⁰

27-811 PROsine 1000W/24V w/ GFCI $950⁰⁰

27-812 PROsine 1000W/24V w/ Xfer $1,000⁰⁰

24-507 110-Amp Class T Fuse w/ Cover $52⁹⁵

24-508 110-Amp Class T Repl. Fuse $17⁹⁵

PROsine 1800

Continuous Output:
1,800 watts
Surge Capacity: 2,900 watts
Standby Power: 1.5 watts
Recommended Fusing: 200A
Class T fuse or Trace DC-175
Dimensions: 15.4″ x 11″ x 4.5″
Weight: 16.5 lb (8.25 kg)

27-813 PROsine 1800W/12V w/ GFCI $1,250⁰⁰

27-814 PROsine 1800W/12V w/ Xfer $1,300⁰⁰

27-815 PROsine 1800W/24V w/ GFCI $1,380⁰⁰

27-816 PROsine 1800W/24V w/ Xfer $1,430⁰⁰

24-212 200-Amp Class T Fuse w/ Cover $52⁹⁵

24-211 200-Amp Class T Repl. Fuse $17⁹⁵

Statpower PROsine 2,500 & 3,000 Inverter/Charger Models

Featuring a compact profile, light weight, clean, reliable power output, good starting abilities, and a powerful built-in battery charger, these units are Statpower's flagship models.

In the inverter mode, a precise, true sine wave AC output is delivered that, in most cases, is better than what you'll receive from your local electric utility. True sine wave allows loads to operate at peak performance, smoother, cooler, and without any audible buzzing or visible distortion. Operating in a back-up mode, these inverter/charger models will switch from utility AC to inverter mode in a maximum of 20 milliseconds. (Not intended as a computer UPS supply.)

In charger mode, the powerful, precise, three-stage charger provides ripple-free DC for the fastest possible battery charging without any interference to DC equipment. It does this with great solid-state efficiency, so that you get maximum amps into the battery for every watt

taken from the generator. The PROsine 2.5 requires only 15 amps of AC to deliver 100 amps into the battery, and the PROsine 3.0 only requires 17 amps of AC to deliver 120 amps into the battery. This leaves more generator capacity to power other household loads simultaneously. Charging voltages are factory-set for wet-cell batteries, and can be adjusted for other battery types by the user.

Both of these models come with a remote control and monitoring panel that shows battery voltage and amperage, as well as easy-touch switches for status monitoring and system control. The 3.0 model is equipped with the Advanced Control System, which offers a finer degree of operating information and system control via a menu-driven LCD display. The ACS panel is an option for the 2.5 model.

Both models have hardwire output. The 3.0 model is also equipped with a dual GFCI AC outlet. This outlet is an option for the 2.5 model. A DC wiring box to allow conduit use for the battery to inverter cabling is an option for both models. A remote battery temperature sensor option is also offered to automatically adjust charging voltages if the batteries will be subjected to temperatures below 40°F or above 85°F.

Like all modern inverters, PROsine models feature automatic overload, short circuit, and overtemperature protection. Unlike most large inverters, they also feature a reverse polarity protection fuse. PROsine inverter/ chargers have a two-year manufacturer's warranty. China.

PROsine 2.5 Inverter/Charger

Input Voltage: 12 or 24 volts DC depending on model
Output Voltage: 120 volts AC RMS –10%/+4%
Continuous Output: 2500 watts
Surge Capacity: 4000 watts
Charging Rate: 0 to 100 amps (half for 24V)
Standby Power: <3 watts
World Voltages Available: not at this time
Average Efficiency: approx. 87%
Recommended Fusing: 400A Class T Fuse or Trace DC-250
Dimensions: 20″ x 15″ x 5.5″
Weight: 32 lb (14.5 kg)

27-821 PROsine 2.5, 12V $2,600⁰⁰

27-822 PROsine 2.5, 24V $2,860⁰⁰

PROsine 3.0 Inverter/Charger

Input Voltage: 12 or 24 volts DC depending on model
Output Voltage: 120 volts AC RMS –10%/+4%
Continuous Output: 3000 watts
Surge Capacity: 4000 watts
Charging Rate: 0 to 120 amps (half for 24V)
Standby Power: <3 watts
World Voltages Available: not at this time
Average Efficiency: approx. 87%
Recommended Fusing: 400A Class T Fuse or Trace DC-250
Dimensions: 20″ x 15″ x 7″
Weight: 32 lb (14.5 kg)

27-819 PROsine 3.0, 12V $3,000⁰⁰

27-820 PROsine 3.0, 24V $3,300⁰⁰

Statpower Options

Portawattz Remote On/Off Switch

Provides a plug-in remote switch for the Portawattz 1000 and 1750. Handy for RVs or other situations where the inverter is removed from the living area. Comes with simple installation instructions and 25-foot cord. Uses standard phone extension cord if you need a greater length.

27-194 Portawattz Remote Sw. $29⁹⁵

PROsine DC Wiring Box

Allows running the battery-to-inverter cabling in conduit to comply with code. Has 2.5-inch knockouts.

27-817 DC Wiring Box for PROsine 1000 & 1800 $99⁰⁰

27-825 DC Wiring Box for PROsine 2.5 & 3.0 $119⁰⁰

ACS Remote Panel For PROsine 2.5

Gives the finer management and control features of this LCD panel to the 2.5 model. This is standard equipment on the 3.0 model.

27-824 ACS Remote Panel Option $249⁰⁰

GFCI Outlet For 2.5 Model

Adds a dual GFCI protected outlet to the 2.5 model. This is standard equipment on the 3.0 model.

27-823 GFCI Outlet Option $59⁹⁵

Exeltech Sine Wave Inverters

New XP-Series, More Power, Lower Cost!

Long known for their clean, problem-free sine wave inverters, Exeltech is offering their new XP-Series of pure sine wave electronic inverters. The XP-Series features better performance and smaller size. Made in America, these lightweight, reliable units are a great choice for modest-sized clean power needs. Sine wave inverters will run any sensitive load, such as audio, video, satellite, and test equipment without buzzing or interference. XP inverters are rated at continuous power wattage, and will surge to approximately twice rated power. Output distortion is 2% maximum, and frequency is controlled to ±0.1%. Inverters automatically protect themselves from over and under voltage, overtemperature, and shorted output. Efficiency runs 85% to 90% over normal operation range. World voltages, 230V & 50Hz, and input voltages up to 108V available by special order. One-year mfr.'s warranty. USA.

ExelTech XP 250

Input Voltage: 12 or 24 volts DC
Output Voltage: 117 volts AC true RMS ±5%
Continuous Output: 250 watts
Surge Capacity: 350 watts
Standby Power: 5 watts
Average Efficiency: 87% at full-rated power
Recommended Fusing: 50-amp
Dimensions: 10.4″ x 5.3″ x 2.8″
Weight: 4.5 lb

27-143 ExelTech XP 250/12-Volt $459⁰⁰
27-144 ExelTech XP 250/24-Volt $459⁰⁰

Trace UX-Series Inverters

Trace's UX-series, replaces the old 600–800 watt units at the modest-output end of Trace's quality offerings. Models are available with 600- or 1,100-watt continuous output, and offer 3-stage automatic battery charging options. Included with the "SB" charger option is a 30-amp transfer switch to automatically switch incoming generator power to the inverter output terminals.

All UX-series inverters are 12-volt input, and feature the usual rugged Trace quality. Standard features include: powder-coated chassis, stainless fasteners, conformal-coated circuit boards, gold-plated connectors, 90% efficiency, miniscule power use on standby, automatic high- and low-battery cutout, automatic high-temp cutout, and overload shutdown. Can be wall or shelf mounted. AC power terminations are protected inside chassis. 230-volt/50Hz foreign models available by special order. Two-year mfr.'s warranty. USA.

Trace UX612 Inverter

Continuous Output: 600 watts
Surge Capacity: 2,500 watts
Standby Power: 0.4 watts
Efficiency: 92% peak
Recommended Fusing: 110-amp
Dimensions: 10.5˝ x 15.5˝ x 6.0˝
Weight: 25 lb
Battery Charger (SB option): 25 amps max.

27-192 Trace UX612 Inverter $595⁰⁰

27-193 Trace UX612SB Inverter/Charger $695⁰⁰

Trace UX1112 Inverter

Continuous Output: 1,100 watts
Surge Capacity: 3,000 watts
Standby Power: 0.5 watts
Efficiency: 90% peak
Recommended Fusing: 200 A class T or Trace DC-175
Dimensions: 10.5˝ x 15.5˝ x 6.0˝
Weight: 35 lb
Battery Charger (SB option): 50 amps max.

27-190 Trace UX1112 Inverter $795⁰⁰

27-191 Trace UX1112SB Inverter/Charger $895⁰⁰

The Trace DR-Series Inverters

The DR is the second generation series of inverters from Trace. All DR inverters come equipped with high performance battery chargers as standard equipment. They are available for 12-, 24-, or 48-volt in a variety of wattages.

DR-series inverters are conventional, high-efficiency, modified sine wave inverters. All models are Electrical Testing Lab approved to UL specifications. These inverters will automatically protect themselves from overload, high temperature, high- or low-battery voltages; all the things we've come to expect from modern inverters. You just can't hurt them externally. Fan cooling is temperature regulated, low battery cut-out is standard and is adjustable for battery bank size. All models come with a powerful three-stage adjustable battery charger and all models have a 30-amp transfer switch built in. Charge amperage is adjustable to suit your batteries or generator. Charging voltages are tailored by a front panel knob. Pick your battery type, eight common choices given plus two equalization selections, and you've automatically picked the correct bulk and float voltage for your batteries.

The optional battery temperature probe will allow the inverter to factor all voltage set points by actual battery temperature. This is a good feature to have when your batteries live outside and go through temperature extremes. Comes with a 15-foot cord. The conduit box option allows the battery cables to be run in two-inch conduit to meet NEC.

The DR-series is stackable in series (must be identical models) for three-wire splitable 240-volt output. Stacking is now easier and less expensive, too; just plug in the inexpensive stacking cord, bolt up the battery cable jumpers, and it's done.

All models are designed for wall mounting with conventional 16-inch stud spacing. All Trace inverters carry a two-year manufacturer's warranty. USA.

Trace DR-Series, 12-Volt Models

Trace DR1512 Inverter/Charger

Input Voltage: 10.8 to 15.5 volts DC
Output Voltage: 120 volts AC true RMS + 5%
Continuous Output: 1,500 watts
Surge Capacity: 3,200 watts
Idle Current: 0.045A @ 12V (0.54 watt)
Charging Rate: 0 to 70 amps
World Voltages Available: yes, call for details
Average Efficiency: 94% at half rated power
Recommended Fusing: 200A Class T or Trace DC-175
Dimensions: 8.5˝ x 7.25˝ x 21˝
Weight: 35 lb

27-237 Trace DR1512 Inverter/Charger, 12V $1,029⁰⁰

Trace DR2412 Inverter/Charger

Input Voltage: 10.8 to 15.5 volts DC
Output Voltage: 120 volts AC true RMS ± 5%
Continuous Output: 2,400 watts
Surge Capacity: 6,000 watts
Idle Current: 0.045A @ 12V (0.54 watt)
Charging Rate: 0 to 120 amps
World Voltages Available: no
Average Efficiency: 94% at half rated power
Recommended Fusing: 300A Class T or Trace DC-175
Dimensions: 8.5˝ x 7.25˝ x 21˝
Weight: 45 lb

27-245 Trace DR2412 Inverter/Charger, 12V $1,395⁰⁰

Trace DR-Series, 24-Volt Models

Trace DR1524 Inverter/Charger

Input Voltage: 21.6 to 31.0 volts DC
Output Voltage:
120 volts AC true RMS ± 5%
Continuous Output: 1,500 watts
Surge Capacity: 4,200 watts
Idle Current: 0.030A @ 24V (0.72 watt)
Charging Rate: 0 to 35 amps
World Voltages Available: yes, call for details
Average Efficiency: 94% at half rated power
Recommended Fusing: 110A Class T or Trace DC-175
Dimensions: 8.5″ x 7.25″ x 21″
Weight: 35 lb

27-238 Trace DR1524 Inverter/Charger, 24V $979⁰⁰

Trace DR2424 Inverter/Charger

Input Voltage: 21.6 to 31.0 volts DC
Output Voltage: 120 volts AC true RMS ± 5%
Continuous Output: 2,400 watts
Surge Capacity: 7,000 watts
Idle Current: 0.030A @ 24V (0.72 watt)
Charging Rate: 0 to 70 amps
World Voltages Available: yes, call for details
Average Efficiency: 95% at half rated power
Recommended Fusing: 200A Class T or Trace DC-175
Dimensions: 8.5″ x 7.25″ x 21″
Weight: 40 lb

27-242 Trace DR2424 Inverter/Charger, 24V $1,395⁰⁰

Trace DR3624 Inverter/Charger

Input Voltage: 21.6 to 31.0 volts DC
Output Voltage: 120 volts AC true RMS ± 5%
Continuous Output: 3600 watts
Surge Capacity: 12,000 watts
Idle Current: 0.030A @ 24V (0.72 watt)
Charging Rate: 0 to 70 amps
World Voltages Available: no
Average Efficiency: 95% at half rated power
Recommended Fusing: 300A Class T or Trace DC-250
Dimensions: 8.5″ x 7.25″ x 21″
Weight: 45 lb

27-246 Trace DR3624 Inverter/Charger, 24V $1,595⁰⁰

Trace Sine Wave Inverters

The Trace series of sine wave inverter/chargers does it all. Conventional remote inverter, uninterruptible power supply, generator management, or utility intertie—they can do any, or sometimes all of these functions simultaneously!

These microprocessor-controlled inverters produce a multistepped approximation of a sine wave that meets all utility requirements for line intertie and will run any AC appliance within its wattage capabilities. And they do it at up to 94% efficiency. All models in the SW series are Electrical Testing Lab approved to UL specifications.

There are three primary modes of operation for the SW-series.

Inverter Mode

In this mode the SW series acts like a typical remote power inverter with a battery charger. It has adjustable threshold watts to trigger turn-on, and adjustable search spacing to save power. It will produce rated surge power for two minutes. These inverters will automatically protect themselves from overload, high temperatures, and high or low battery voltage. Fan cooling is temperature regulated; low-voltage battery disconnect is adjustable.

The sophisticated generator management system, which works in all operation modes, provides a wide range of features, including auto start by battery voltage or time, number of start attempts, cranking time for each attempt, warm-up time before loading, automatic battery charging, maximum generator load management, auto low-voltage disconnect, auto shut-off, and more.

Standby Mode

Grid power is fed through and partially conditioned by the SW series inverters. In case of utility failure, the Trace will pick up instantly and support the load. Transfer time is 16 to 30 milliseconds. The SW series inverter will support the grid if it "browns out." If the amount of AC power demanded is greater than the breaker size you've programmed into the SW inverter, it will contribute power to the system. If the grid fails and the SW inverter is running all the loads, then all the automatic generator management and battery charging discussed above will occur. Once utility power is restored, the inverter will match phases and shift the AC loads back to the utility, then shut off the generator if it is running.

Utility Interactive Mode

This mode will allow you to sell excess power back to the utility! The setup is simple, and the unit was designed to meet all utility requirements, but regulations will vary from one utility to another, and may not be formalized at all utilities. We may only be able to offer limited assistance obtaining approval from your local utility. Talk to your local utility first; possession of the equipment and the ability to do so does not constitute a right to sell them power.

A small battery bank is still required in the Utility Interactive Mode. A larger battery bank will protect you from blackouts for as long as it lasts, or until the generator starts and takes over. The inverter will float the batteries at a programmed voltage; any additional power input will be converted to AC and fed back into the grid. If the utility fails in the Utility Interactive Mode, the SW inverter will go into the normal Inverter Mode to run your house without a hiccup.

All the SW inverters are stackable in series for 240 volts AC, three-wire splitable output at double the rated wattage continuously. Stacking only requires plugging in an inexpensive stacking cord. Designed for wall mounting with conventional 16-inch stud spacing. All Trace SW series inverters carry a two-year manufacturer's warranty. USA.

Trace SW-Series, 24-Volt Models

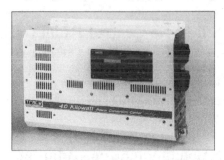

SW4024 Sine Wave Inverter/Charger

Input Voltage: 20 to 34 volts DC
Output Voltage: 120 volts AC true RMS ±2%
Continuous Output: 4 000 watts
Surge Capacity: 10,000 watts for 2 minutes
Standby Power: adjustable from 1 to 16.8 watts
Charging Rate: 0 to 120 amps
Transfer Amps: 60 amps
World Voltages Available: yes, call for details
Average Efficiency: 94% peak, 85% to 90% average
Recommended Fusing: 400A Class T or Trace DC-250
Dimensions: 22.5″ x 15.25″ x 9.0″
Weight: 105 lb

27-236 SW4024 Trace Sine Wave, 24V, 4 kW $3,495⁰⁰

**27-240 SW3024E Export Model,
various volts/Hz $3,495⁰⁰**

Trace SW Series 48-Volt Models

SW4048 Sine Wave Inverter/Charger

Input Voltage: 40 to 68 volts DC
Output Voltage: 120 volts AC true RMS ±2%
Continuous Output: 4,000 watts
Surge Capacity: 10,000 watts for 2 minutes
Standby Power: adjustable from 1 to 16.8 watts
Charging Rate: 0 to 60 amps
Transfer Amps: 60 amps
World Voltages Available: yes, call for details
Average Efficiency: 94% peak, 85% to 90% average
Recommended Fusing: 200A Class T or Trace DC-175
Dimensions: 22.5″ x 15.25″ x 9.0″
Weight: 105 lb

27-241 SW4048 Trace Sine Wave, 48V, 4 kW $3,495⁰⁰

**27-248 SW3048E Export Model,
various volts/Hz $3,495⁰⁰**

SW5548 Sine Wave Inverter/Charger

Trace also has PV and Utility versions of this inverter for utility sell-back use only, with no battery pack required. See the Intertie Inverter listings a couple of pages back.

Input Voltage: 40 to 68 volts DC
Output Voltage: 120 volts AC true RMS ±2%
Continuous Output: 5,500 watts
Surge Capacity: 10,000 watts for 2 minutes
Standby Power: Adjustable, 1 to 20 watts
Charging Rate: 0 to 75 amps
Transfer Amps: 60 amps
World Voltages Available: yes, call for details
Average Efficiency: 96% peak, 85% to 90% average
Recommended Fusing: 200A Class T or Trace DC-250
Dimensions: 22.5″ x 15.25″ x 9.0″
Weight: 130 lb

27-249 SW5548 Trace Sine Wave, 48V, 5.5kW $3,995⁰⁰

SW SERIES EFFICIENCY CURVES

Chart: EFFICIENCY (vertical axis, 0% to 100%) vs AC LOAD WATTS (horizontal axis, 0 to 5,550). Regions labeled: SW2512, SW3024E / SW3048E, SW4024 / SW4048, SW5548.

Trace PS-Series Inverters

Economical Sine Wave Quality
Without the Bells and Whistles

Trace's newest Power Station inverter family offers all the dependability and quality that we've found in the SW (Sine Wave) series for years, but with a few less bells and whistles, and for a lower price. This is basically a 2,500-watt SW-series inverter/charger in either 12V or 24V trim, with the same highly capable operating system, but in order to keep the price down there is no meter or user interface. A simple four-position switch selects Off, Search, On, or Battery Charge Only. The PS doesn't do automatic generator start, doesn't have the auxiliary relays, and only has one AC input that will accept either utility or generator power. It comes from the factory all set up to act as a conventional inverter/charger, just like the low-cost DR-series, but with the good sine wave power output. For most off-the-grid homesteads, this is perfect. As a stacked pair with conventional 240-volt 3-wire output, the PS model really shines and offers outstanding surge performance at a very good price. One word of warning, this model is noisier than the deluxe SW models and needs to live in a utility room or garage that's got a door between it and the household.

Want a little more from your PS? With the optional LCD control panel you can change the basic programming just like on the SW-series, make changes to the standard battery charging parameters, or most of the other features of the SW-series. Note: the PS-series will not do utility intertie.

PS-series inverters can be stacked in series (240VAC output), or in parallel (double the 120VAC output). Most SW options fit the PS including remote control panel, communications adapter, and stacking or paralleling kits. Only the conduit box is unique to the PS, as this inverter is slightly slimmer. European versions delivering 230V/50Hz are available by special order with the same pricing. Size is 22.5″ x 15.25″ x 6.5″, weighs 80 lb. UL approved. Two-year mfr.'s warranty. USA.

PS-Series Accessories

Remote Control or Communications Adapter is required to program either of these options.

Generator Start Module

Provides automatic generator management just like the SW-series inverters.

27-855 PS Generator Start Module $159⁰⁰

Auxiliary Load Module

Provides three voltage-controlled relays just like the SW-series inverters.

27-865 Auxiliary Load Module $159⁰⁰

PS2512 Sine Wave Inverter/Charger

Input Voltage: 12 volts DC, nominal
Output Voltage: 120 volts AC true RMS ±5%
Continuous Output: 21 Amps AC
Surge Capacity: 50 Amps AC for 100 mSec
Standby Power: 16 watts default (adjustable w/SWRC or SWCA)
Charging Rate: 0 to 130 amps
Transfer Amps: 30 or 50 amps (via selector switch)
World Voltages Available: 230VAC/50Hz
Average Efficiency: 90% peak
Recommended Fusing: 400A Class T or Trace DC-250
Dimensions: 22.5″ x 15.25″ x 6.5″
Weight: 80 lb

27-831 PS2512 Trace Sine Wave, 12V, 2.5 kW $2,165⁰⁰

PS2524 Sine Wave Inverter/Charger

Input Voltage: 24 volts DC, nominal
Output Voltage: 120 volts AC true RMS ±5%
Continuous Output: 21 Amps AC
Surge Capacity: 75 Amps AC for 100 mSec
Standby Power: 16 watts default (adjustable w/SWRC or SWCA)
Charging Rate: 0 to 65 amps
Transfer Amps: 30 or 50 amps (via selector switch)
World Voltages Available: 230VAC/50Hz
Average Efficiency: 90% peak
Recommended Fusing: 300A Class T or Trace DC-175
Dimensions: 22.5″ x 15.25″ x 6.5″
Weight: 80 lb

27-832 PS2524 Trace Sine Wave, 24V, 2.5 kW $2,165⁰⁰

Trace Inverter Options

SW/PS Remote Control Panel Option

Offers a complete LCD control panel with all controls, indicators, and display. Includes 25 feet of cable. Required for any field adjustment of control settings on PS units.

27-210 Remote Control Panel for SW/PS Series $295⁰⁰
Please specify voltage.

SW/PS Communications Adapter

This allows PC connection and monitoring of one SW- or PS-series inverter with version 4.01 or later software. Includes adapter, DOS-based software, 50-foot cable, and DB9 connector.

27-305 SW/PS Communications Adapter $175⁰⁰

SW/PS Stacking Interface

To stack a pair of SW or PS inverters, this 42″ comm-port cable is all you need. Simple phone-type plug-in connectors. 240V output from the stacked pair.

27-314 SW/PS Stacking Intertie Cable $45⁰⁰

SW/PS Paralleling Kit

This connects two SW or PS inverters in parallel to double the wattage output at 120V output. For U.S. voltage inverters only.

27-415 SW/PS Paralleling Kit $345⁰⁰

Conduit Box Options

The optional Conduit Box allows the very heavy input wiring from the batteries to be run inside conduit to comply with the NEC. Has three (one on each side) ½″, ¾″, and 2″ knockouts and plenty of elbow room inside for bending large wire. USA.

27-209 Conduit Box option-SW Series $94⁰⁰

27-833 Conduit Box option-PS Series $94⁰⁰

27-324 Conduit Box option-DR Series $69⁰⁰

DR Stacking Kit

The DR stacking kit includes the plug-in comm-port cable, like the SW, plus a pair of 2/0 battery cables to jumper from the primary to the secondary inverter. USA.

27-313 DR Stacking Intertie Kit $85⁰⁰

Battery Temperature Probe

Comes standard with the SW-series; recommended for the DR series if your batteries are subject to temperatures below 40°F or above 90°F. Will adjust the battery charger automatically to compensate for temperature. 15-ft. length, phone plug on inverter end, sticky pad on battery end. This same temp probe also works with all C-Series Trace controllers. USA.

27-319 Battery Temperature Probe, 15′ $29⁰⁰

Remote Control for DR, UX, & TS Series Inverters

A 50-foot ON/OFF remote control with LED indicator. Plugs into all Trace DR, UX, and TS series inverters. USA.

27-304 Trace RC8/50 Remote $69⁰⁰

Trace T240 Autotransformer

A very handy 2:1 bi-directional transformer with a 4,500 watt continuous rating. It can be used to step 120-volt inverter output up to run the occasional 240-volt appliance, like a water pump. It can be used to balance the load on a 240-volt generator, by delivering 120-volts to an inverter/charger while evenly loading the generator. Or it can be used as a step-down transformer, allowing a stacked pair of inverters to share the start-up and running of a large 120-volt appliance. Supplied in a powder-coated enclosure with multiple ¾″ and 1″ knock-outs and a 2-pole 25-amp circuit breaker. Weighs 39.4 lb. Measures 6.3″ x 7″ x 21″. Accepts 14 to 4 AWG wire sizes. This unit replaces the Trace T220 at the same price. USA.

27-326 T240 Autotransformer $349⁰⁰

Battery And Inverter Cables

The wire sizing between the battery and the inverter is of critical importance. Small cables limit inverter performance. We offer seriously beefy 2/0 and 4/0 cables using super flexible welding-type cable. Highly recommended to complete your inverter installation. They're available in 3-foot, 5-foot or 10-foot pairs. One-lug cables have bare wire on one end to accommodate set-screw lugs, with a crimped, soldered, and color-coded shrink-wrapped ⅜″ lug terminal on the other end. Two-lug cables have the same high-quality terminals installed at both ends. See chart below for recommended cable sizing. Extra lugs offered for the occasional installation that needs an extra lug end. USA.

15-770 2/0 3-ft. Cables, 1 lug, pair $29⁹⁵
15-771 2/0 5-ft. Cables, 1 lug, pair $49⁹⁵
15-772 2/0 10-ft. Cables, 1 lug, pair $84⁹⁵

15-773 4/0 3-ft. Cables, 1 lug, pair $39⁹⁵
15-774 4/0 5-ft. Cables, 1-lug, pair $69⁹⁵
15-775 4/0 10-ft. Cables, 1 lug, pair $115⁰⁰

27-311 4/0 5-ft. Cables, 2 lug, pair $69⁹⁵
27-312 4/0 10-ft. Cables, 2 lug, pair $125⁰⁰

27-099 2/0 5-ft. Cables, 2 lug, pair $49⁹⁵
27-098 2/0 10-ft. Cables, 2 lug pair $89⁹⁵

26-605 Copper Lug, solder-type 2/0 $4⁹⁵
26-606 Copper Lug, solder-type 4/0 $13⁹⁵
26-623 Solderless Lug, #2 to 4/0 $7⁹⁵

Wire, Adapters, and Outlets

Wires are freeways for electrical power. If we do a poor job designing and installing our wires, we get the same results as with poorly designed roads: traffic jams, accidents, and frustration. Big, wide roads can handle more traffic with ease, but costs go up with increased size, so we're looking for a reasonable compromise. The proper choices of wire and wiring methods can be confusing, but with sufficient planning and thought we can create safe and durable paths for energy flow.

The National Electrical Code (NEC) provides broad guidelines for safe electrical practices. Local codes may expand upon or supersede this code. It is important to use common sense when dealing with electricity, and this might be best done by acknowledging ignorance on a particular subject and requesting advice and help from experts when you are unsure of something. The Gaiam Real Goods technical staff is eager to help with particular wiring issues, although you should keep in mind that local inspectors will have the final say on what they consider the most appropriate means to the end of a properly installed electrical system; therefore, advice from them or from a local electrician familiar with both local code requirements and renewable energy systems might be more informative.

Wire

Wire comes in a tremendous variety of styles that differ in size, number, material, and type of conductors, as well as the type and temperature rating of insulation protecting the conductor.

One of the basic ideas behind all wiring codes is that metal conductors must have at least two layers of protection. Permanent wiring should always be run within electrical enclosures, conduit, or inside walls. Wires are prone to all sorts of threats, including but not limited to abrasion, falling objects, and children using them as a jungle gym. The plastic insulation around the metal conductor offers minimal protection, based on the assumption that the conductors will be otherwise protected. Poor installation practices that forsake the use of conduit and strain-relief fittings can lead to a breach of the insulation, which could cause a short circuit and fire or electrocution. Electricity, even in its most apparently benign manifestations, is not a force to be managed carelessly.

A particular type and gauge (thickness) of wire is rated to carry a maximum electrical current. The NEC requires that we not exceed 80% of this current rating for continuous duty applications. With the low-voltage conditions that we run across in independent energy systems, we also need to consider the proper size of wire that will move the energy efficiently enough to get the job done at the other end. As voltage decreases, amperage must increase in order to perform any particular job. This problem becomes more acute with smaller conductors over longer runs. Some voltage drop is unavoidable when moving electrical energy from one point to another, but we can limit this to reasonable levels if we choose the right size wire. Acceptable voltage drop for 120-volt circuits can be up to 5%, but for 12- or 24-volt circuits 2% to 3% is the most we want to see. There is a good all-purpose chart and formula on page 216 for figuring wire size at any voltage drop, any distance, any voltage, any current flow. Use it, or give us a call and we'll help you figure it out.

DC Adapters and Outlets

With the rising popularity and reliability of inverters that let folks run conventional AC appliances, there's less and less demand for DC fixtures and outlets, which is perhaps a good thing. The roots of the independent-energy movement grew from the automobile and recreational-vehicle industries, so plugs and outlets for low-voltage applications are often based on that ubiquitous but often somewhat flimsy creation, the cigarette lighter. This isn't the best of choices, as the limited metal to metal contact in this plug configuration seriously limits the current-carrying capacity. Cigarette lighter plugs and outlets are rarely rated for more than 15 amps surge, and 7 amps continuously is absolutely the maximum to expect. Lighter sockets are not approved by the National Electric Code. We also need to point out the obvious safety hazards of a large, open receptacle and small children. If you must use cigarette lighter outlets, be sure they're fused!

The conventional rules for DC outlets and plugs hold that the center of the receptacle (female outlet) and the tip of the plug (male adapter) are positive, and the outer shell of the receptacle and side contact of the plug are negative. This should be verified (with a handheld voltmeter) rather than assumed, because the polarity can be reversed in a variety of ways. Better to check it out and fix it rather than ruin the costly DC television you just bought.

Another convention is that the power source is presented by the female outlet. This way, it is more difficult to accidentally electrocute yourself. There are some small solar modules that have DC male adapters on them, even though they are a power source, although such a small one that they present little to no hazard. They are made this way so that you can easily charge a car battery with the solar panel through the cigarette lighter.

Some people would rather not use these cigarette lighter outlets and plugs in their houses because they look dangerous . . . and they can be. Little fingers and tools fit conveniently, and disastrously, within the supposedly "protected" live socket. The NEC recommends using an oddball AC outlet that has the prongs perpendicular, slanted, or in some other configuration which makes it impossible to plug in a standard 120-volt AC appliance. Although it is tempting to use inexpensive standard 120-volt AC outlets, we strongly advise against it. We know from personal experience that any savings are quickly lost when a wrong-voltage appliance is plugged in and destroyed. Standard outlets have AC amperage ratings, but are unlikely to have been rated for DC applications. Use common sense, and do not run a large load through light-duty hardware.

www.realgoods.com

Wire Sizing Chart/Formula

This chart is useful for finding the correct wire size for any voltage, length, or amperage flow in any AC or DC circuit. For most DC circuits, particularly between the PV modules and the batteries, we try to keep the voltage drop to 3% or less. There's no sense using your expensive PV wattage to heat wires. You want that power in your batteries!

Note that this formula doesn't directly yield a wire gauge size, but rather a "VDI" number, which is then compared to the nearest number in the VDI column, and then read across to the wire gauge size column.

1. Calculate the Voltage Drop Index (VDI) using the following formula:

VDI = AMPS x FEET ÷ (% VOLT DROP x VOLTAGE)

Amps = Watts divided by volts   Feet = One-way wire distance

% Volt Drop = Percentage of voltage drop acceptable for this circuit (typically 2% to 5%)

2. Determine the appropriate wire size from the chart below.
   A. Take the VDI number you just calculated and find the nearest number in the VDI column, then read to the left for AWG wire gauge size.
   B. Be sure that your circuit amperage does not exceed the figure in the Ampacity column for that wire size. (This is not usually a problem in low-voltage circuits.)

WIRE SIZE	COPPER WIRE		ALUMINUM WIRE	
AWG	VDI	Ampacity	VDI	Ampacity
0000	99	260	62	205
000	78	225	49	175
00	62	195	39	150
0	49	170	31	135
2	31	130	20	100
4	20	95	12	75
6	12	75	•	•
8	8	55	•	•
10	5	30	•	•
12	3	20	•	•
14	2	15	•	•
16	1	•	•	•

Chart developed by John Davey and Windy Dankoff. Used by permission.

Example: Your PV array consisting of 4 Siemens SP75 modules is 60 feet from your 12-volt battery. This is actual wiring distance, up pole mounts, around obstacles, etc. These modules are rated at 4.4 amps, times 4 modules = 17.6 amps maximum. We'll shoot for a 3% voltage drop. So our formula looks like:

$$VDI = \frac{17.6 \times 60}{3[\%] \times 12[V]} = 29.3$$

Looking at our chart, a VDI of 29 means we'd better use #2 wire in copper, or #0 wire in aluminum. Hmmm. Pretty big wire.

What if this system was 24-volt? The modules would be wired in series, so each pair of modules would produce 4.4 amps. Two pairs, times 4.4 amps = 8.8 amps max.

$$VDI = \frac{8.8 \times 60}{3[\%] \times 24[V]} = 7.3$$

Wow! What a difference! At 24-volt input you could wire your array with little ol' #8 copper wire.

Wire

Type USE Direct Burial Cable

Type "USE" (Underground Service Entrance) cable is moisture-proof and sunlight-resistant. It is recognized as underground feeder cable for direct earth burial in branch circuits and is the only wire you can install exposed for module interconnects. It is approved by the National Electrical Code and UL. It is resistant to acids, chemicals, lubricants, and ground water. Our USE cable is a single conductor with a sunlight-resistant jacket. It is much more durable than standard romex. 10-gauge is solid wire, 8-gauge is stranded.

26-521 USE 10-Gauge Wire, Price/Foot $.55

26-522 USE 8-Gauge Wire, Price/Foot $.60

Copper Lugs

We carry very heavy-duty copper lugs for connecting to the end of your large wire from 4-gauge to 4/0-gauge (#0000). The hole in the end of the lug is ⅜-inch diameter. Wire must be soldered to copper lugs.

26-601 Copper Lug, 4-Gauge $1⁹⁵

26-602 Copper Lug, 2-Gauge $3⁹⁵

26-603 Copper Lug, 1-Gauge $3⁹⁵

26-604 Copper Lug, 1/0-Gauge $3⁹⁵

26-605 Copper Lug, 2/0-Gauge $4⁹⁵

26-606 Copper Lug, 4/0-Gauge $13⁹⁵

Nylon-Coated Single Conductor Wire

This single conductor wire connects the components of your renewable energy system. We stock it in black only. We recommend you use red tape on the ends for positive and white tape on the ends for negative. Wire is stranded copper, with THHN jacketing. This wire should always be installed in conduit, and never in exposed situations.

The minimum order for 16-, 14-, and 10-gauge THHN is a 500-foot roll. Wire gauges of 8 and larger can be ordered in any length.

26-534 #8 THHN Primary Wire, Price/Foot $0.55

26-535 #6 THHN Primary Wire, Price/Foot $0.65

26-536 #4 THHN Primary Wire, Price/Foot $0.90

26-537 #2 THHN Primary Wire, Price/Foot $1³⁵

Split Bolt Kerneys

Split bolt kerneys are used to connect very large wires together or large wires to smaller wires. You must always wrap the kerney with electrical tape to prevent corrosion and the potential for short-circuiting.

26-631 Split Bolt Kerney, #6 $6⁹⁵

26-632 Split Bolt Kerney, #4 $7⁹⁵

26-633 Split Bolt Kerney, #2 $8⁹⁵

26-634 Split Bolt Kerney, #1/0 $11⁹⁵

Solderless Lugs

These solderless lugs are ideal for connecting large wire to small connections or to batteries.

26-622 Solderless Lug (#8,6,4,2) $4⁹⁵

26-623 Solderless Lug (#2 to 4/0) $7⁹⁵

Wiring 12 Volts For Ample Power

By David Smead and Ruth Ishihara. The most comprehensive book on DC wiring to date, written by the authors of the popular book *Living on 12 Volts with Ample Power*. This book presents system schematics, wiring details, and troubleshooting information not found in other publications. Leans slightly toward marine applications. Chapters cover the history of electricity, DC electricity, AC electricity, electric loads, electric sources, wiring practices, system components, tools, and troubleshooting. 240 pages, paperback. USA.

80-111 Wiring 12 Volts For Ample Power $19⁹⁵

Electrical Wiring

By the American Association of Vocational Instructional Materials (AAVIM) staff. Thoroughly covers standard electrical wiring principles and procedures. This book has become an industry standard for training students, teachers, and professionals. Includes over 350 step-by-step color illustrations, covering circuits, receptacles and switches, installing service entrance equipment, and more. Revised in 1999 to include latest changes in the National Electrical Code. 272 pages, 4-color, paperback. USA.

80-302 Electrical Wiring $32⁹⁵

Adapters

Heavy-Duty Replacement Plug

The SP-20 is designed for heavier-duty applications than the SP-6. It is supplied complete with three styles of end caps to accommodate coil cords, SJ-series jacketing, and other round wire. The plug is unbreakable and is rated at 10 amps. USA.

26-102 Heavy-Duty Plug $1⁹⁵

Fused Replacement Plug

The SP-90-F5-B fused plug has a unique polarity reversing feature. It includes a 5-amp fuse and four sizes of snap-in strain reliefs, which accommodate wire gauges from 24 to 16 AWG. It is rated at 8 amps continuous duty and can be fused to 15 amps for protecting any electric device. USA.

26-103 Fused Replacement Plug $2⁹⁵

Plug Adapter

This simple, durable, SP-70 plastic adapter provides an elegant conversion from any standard 110-volt fixture. Simply insert your 110-volt plug into the connector end of this adapter (no cutting!), insert a 12-volt bulb into the light socket of your lamp, and you're in 12-volt heaven for cheap! USA.

26-104 Plug Adapter $2⁹⁵

Triple Outlet Plug

The triple outlet plug permits the use of three cigarette lighter plugs on the same circuit simultaneously. Rated at 12 amps. Includes heavy-duty 16-gauge wire, and a self-adhesive pad. USA.

26-105 Triple Outlet Plug $9⁵⁰

Extension Cords

These low-voltage extension cords have a cigarette lighter receptacle on one end and a cigarette lighter plug on the other end. They are available in 10-foot and 25-foot lengths. The maximum recommended current is 4 amps for the 10-foot cord and 2 amps for the 25-foot cord. USA.

26-113 Extension Cord, 10-ft. $5⁹⁵

26-114 Extension Cord, 25-ft. $6⁹⁵

DC Power Converter

This converter allows you to operate a wide variety of DC powered products, including a portable stereo, cassette player, video cassette recorder, and other applications from a standard cigarette lighter receptacle. One convenient switch enables the output voltage to be set to 3, 6, 9, or 12 volts DC from a 12-volt DC input, negative ground. Output polarity may be reversed and the LED indicator shows when the adapter is in use. The complete unit, which is UL-listed, includes a 3-amp fused universal plug, 6-foot distribution cord and an assortment of four popular polarity-reversible coaxial power plugs. The fuse is replaceable. USA.

26-121 Converter $14⁹⁵

Double Outlet Adapter

The double outlet power adapter permits the use of two 12-volt products in a single outlet. Two receptacles connect to a single adapter plug with short lengths of 16-gauge wire. Rated at 10 amps. USA.

26-107 Double Outlet Adapter $5⁹⁵

Extension Cord Receptacle

This receptacle, made of all brass parts in a break-resistant plastic housing, connects with 12-volt cigarette lighter plugs. Includes 6-inch lead wire. USA.

26-108 Extension Cord Receptacle $2⁹⁵

Non-Fused Power Cord

This replacement power cord is handy when an existing power supply cord has been damaged or worn. It features a male cigarette lighter plug on one end of 8 feet of 20-gauge, polarity coded wire. The other end has the wire jacketing stripped ready for your installation. 4-amp maximum. USA.

26-109 Non-Fused Power Cord $2⁹⁵

Extension Cords And Battery Clips

These unique cords have spring-loaded clips to attach directly to battery terminals. They increase the use and value of 12-volt products like lights, TV sets, radios, or appliances. The maximum amperage that can be put through the 1-foot cord is 10 amps and the maximum that can be put through the 10-foot cord is 4 amps. USA.

26-111 Extension Cord and Clips, 1-ft. $5⁹⁵

26-112 Extension Cord and Clips, 10-ft. $6⁹⁵

12-Volt Outlets

Wall Plate Receptacle

Our most popular basic 12-volt receptacle is identical to an automobile cigarette lighter. It is made of a break-resistant plastic housing and all brass parts and fits a standard single gang junction box. Handles up to 15-amp surge, but 7 to 8 amps is the maximum continuous current from any cigarette lighter plug. USA.

For ease of installation, use 6″ of flexible lamp zip cord between receptacle terminals and incoming romex wire.

26-201 Wall Plate Receptacle $3⁹⁵

Please specify brown or ivory.

System Sizing Worksheets

CREST Solar Sizer Software

A Photovoltaic System Design Tool

This great software package from the nonprofit Center for Renewable Energy and Sustainable Technology is without a doubt the best PV sizing program ever developed. Initially created as a professional design tool, we found it incredibly easy to navigate and a blessing for anyone looking into the possibility of a PV installation. It uses a simple graphic interface to choose and customize both household loads and potential system components. Once selected, you may "install" appliances and components in the virtual house, and access detailed reports on initial, annual, and lifetime costs. Keep working until your virtual PV system meshes with your power and budget needs. Features extensive Help files, and yearly sunlight data for 250 U.S. sites and 46 sites in 27 foreign countries. USA.

System Requirements: Supplied on two 3.5˝ disks for MS Windows. Requires Windows 3.1 or later, an Intel 80486/Pentium processor, 4MB of RAM, 4MB hard drive space, and a graphics card set to display 256 colors.

80-712 CREST Solar Sizer – Windows version $125⁰⁰

For Off-The-Grid Systems

The System Sizing Worksheet (on the next few pages) provides a simple and convenient method for determining approximate total household electrical needs for off-the-grid systems. Once completed, our Gaiam Real Goods technicians can size the photovoltaic and battery system to meet your needs. Want to go with utility intertie? That's simpler, see the next paragraph. Ninety percent of the renewable energy systems we design are PV-based, so these worksheets deal primarily with PV. If you are fortunate enough to have a viable wind or hydro power source, you'll find output information for these sources in their respective chapters of the Sourcebook. Our technical staff has considerable experience with these alternate sources and will be glad to help you size a system. Give us a call.

Utility Intertie Systems

These are easy. Whatever your renewable energy system doesn't cover, your existing utility company will. So we don't need to account for every watt-hour beforehand. Either tell us how much you want to spend, how soon you want a payback, or how many kilowatt-hours of utility power you'd like to displace per average day. For direct-intertie systems without batteries, you'll spend about $1,500 for every kilowatt-hour per day your solar system delivers. For battery-based systems that can provide limited emergency back-up power you'll spend about $3,000 for every kilowatt-hour per day. These are very general ballpark figures for initial system costs. Remember, state rebates can reduce these costs by up to 50%.

General Information

Back to off-the-grid systems now. Conserve, conserve, conserve! As a rule of thumb, it will cost about $3 to $4 worth of equipment for every watt-hour you must supply. Trim your wattage to the bone! Don't use incandescent light bulbs or older standard refrigerators.

If you're intimidated by this whole process, don't feel like the Lone Ranger. Our tech staff is here to hold your hand and help you through the tough parts. We do need you to come up with an estimate of Total Household Watt-Hours per Day, which this worksheet will help you do. We can pick up the design process from there. You are in the best position to make lifestyle decisions: how late do you stay up at night, are you religious about always turning the light off when leaving a room, are you running a home business or is the house empty five days a week, do you hammer on your computer for 12 hours per day, does your pet iguana absolutely require his rock heater 24 hours a day? These are questions we can't answer for you. So figure out your watt-hours, let us know what we're shooting for.

Determine the Total Electrical Load in Watt-Hours per Day

The form following allows you to list every appliance, how much wattage it draws, how many hours per day it runs, and how many days per week. This gives us a daily average for the week, as some appliances, like a washing machine, may only be used occasionally.

Some appliances may only give the amperage and voltage on the nameplate. We need wattage. Multiply the amperage by the voltage to get wattage. Example: a blender nameplate says, "2.5A 120V 60Hz". This tells us the appliance is rated for a maximum of 2.5 amps at 120 volts/60 cycles per second. 2.5 amps times 120 volts equals 300 watts. Beware of using nameplate amperage, however. For safety reasons this must be the highest amperage the appliance is capable of drawing. Actual running amperage is often much less. This is particularly true for refrigerators and entertainment equipment. The Watt Chart for Typical Appliances may help give you a more "real" wattage use.

Line by Line Instructions

Line 1   Total all average watt-hours/day in the column above.

Line 2   For AC appliances multiply the watt-hours total by 1.1 to account for inverter inefficiency (typical by 90%). This gives the actual DC watt-hours that will be drawn from the battery.

Line 3   DC appliances are totalled directly, no correction necessary.

Line 4   Insert the total from line 2 above.

Line 5   Add the AC and DC watt-hour totals to get the total DC watt-hours/day. At this point you can fax or mail the design forms to us and after a phone consultation we'll put a system together for you. If you prefer total self-reliance, forge on.

Line 6   Insert the voltage of the battery system; 12-volt or 24-volt are the most common. Talk it over with one of our tech staff before deciding on a higher voltage as control and monitoring equipment is sometimes hard to find.

Line 7   Divide the total on line 5 by the voltage on line 6.

Line 8   This is our fudge factor that accounts for losses in wiring and batteries, and allows a small safety margin. Multiply line 7 by 1.2.

Line 9   This is the total amount of energy that needs to be supplied to the battery every day on average.

Line 10   This is where guesswork rears its ugly head. How many hours of sun per day will you see? Our Solar Insolation Map in the appendix (page 476) gives the average daily sun hours for the worst month of the year. You probably don't want to design your system for worst possible conditions. Energy conservation during stormy weather, or a back-up power source can allow use of a higher hours-per-day figure on this line and reduce the initial system cost.

Line 11   Divide line 9 by line 10, this gives the total PV current needed.

Line 12   Decide what PV module you want to use for your system. You may want to try the calculations with several different modules. It all depends on whether you need to round up or down to meet your needs.

Line 13   Insert the amps of output at rated power for your chosen module.

Line 14   Divide line 11 by line 13 to get the number of modules required in parallel. You will almost certainly get a fraction left over. Since we don't sell fractional PV modules, you'll need to round up or down to a whole number. We conservatively recommend any fraction from 0.3 and up be rounded upward.

If yours is a 12-volt nominal system, you can stop here and transfer your line 14 answer to line 19. If your nominal system voltage is something higher than 12 volts, then forge on.

Line 15   Enter the system battery voltage. Usually this will be either 12 or 24.

Line 16   Enter the module nominal voltage. This will be 12 except for unusual special-order modules.

Line 17   Divide line 15 by line 16. This will be how many modules you must wire in series to charge your batteries.

Line 18   Insert the figure from line 14 and multiply by line 17.

Line 19   This is the total number of PV modules needed to satisfy your electrical needs. Too high? Reduce your electrical consumption, or, add a secondary charging source such as wind or hydro if possible, or, a stinking, noisy, troublesome, fossil-fuel-gobbling generator.

The amount of electricity Americans use annually to illuminate digital clocks equals the total power drain of Greece, Peru, and Vietnam combined.

–Annie Berthold-Bond
in *Green Kitchen Handbook*

Battery Sizing Worksheet

Line 20 Enter your total daily amp-hours from line 9.

Line 21 Reserve battery capacity in days. We usually recommend about three to seven days of back-up capacity. Less reserve will have you cycling the battery excessively on a daily basis, which results in lower life expectancy. More than seven days capacity starts getting so expensive that a back-up power source should be considered.

Line 22 You can't use 100% of the battery capacity (unless you like buying new batteries). 80% is the maximum, and we usually recommend to size at 50% or 60%. This makes your batteries last longer, and leaves a little emergency reserve. Enter a figure from .5 to .8 on this line.

Line 23 Multiply line 20 times line 21, and divide by line 22. This is the minimum battery capacity you need.

Line 24 Select a battery type. The most common for household systems are golf carts @ 220 amp-hours, or L-16s @ 350 amp-hours. See the Battery Section for more details. Enter the amp-hour capacity of your chosen battery on this line.

Line 25 Divide line 23 by line 24; this is how many batteries you need in parallel.

Line 26 Enter your system nominal voltage from line 6.

Line 27 Enter the voltage of your chosen battery type.

Line 28 Divide line 26 by line 27; this gives you how many batteries you must wire in series for the desired system voltage.

Line 29 Enter the number of batteries in parallel from line 25.

Line 30 Multiply line 28 times line 29. This is the total number of batteries required for your system.

AC device	Device watts	X	Hours of daily use	X	Days of use per week	÷	7	=	Average watt-hours per day
		x		x		÷		=	
		x		x		÷		=	
		x		x		÷		=	
		x		x		÷		=	
		x		x		÷		=	
		x		x		÷		=	
		x		x		÷		=	
		x		x		÷		=	
		x		x		÷		=	
		x		x		÷		=	
		x		x		÷		=	
		x		x		÷		=	
		x		x		÷		=	
		x		x		÷		=	
		x		x		:		=	
		x		x		÷		=	
		x		x		÷		=	
		x		x		÷		=	
		x		x		÷		=	
		x		x		÷		=	
		x		x		÷		=	

1 Total AC watt-hours/day

2 X 1.1 = Total corrected DC watt-hours/day

DC device	Device watts	X	Hours of daily use	X	Days of use per week	÷	7	=	Average watt-hours per day
		x		x		÷		=	
		x		x		÷		=	
		x		x		÷		=	
		x		x		÷		=	
		x		x		÷		=	
		x		x		÷		=	

3 Total DC watt-hours/day

3	(from previous page)	Total DC watt-hours/day
4		Total corrected DC watt-hours/day from Line 2 +
5		Total household DC watt-hours/day =
6		System nominal voltage (usually 12 or 24) ÷
7		Total DC amp-hours/day =
8		Battery losses, wiring losses, safety factor x 1.2
9		Total daily amp-hour requirement =
10		Estimated design insolation (hours per day of sun, see map on p. 476) ÷
11		Total PV array current in amps =
12		Select a photovoltaic module for your system
13		Module rated power amps ÷
14		Number of modules required in parallel =
15		System nominal voltage (from line 6 above)
16		Module nominal voltage (usually 12) ÷
17		Number of modules required in series =
18		Number of modules required in parallel (from Line 14 above) x
19		Total modules required =

Battery Sizing

20		Total daily amp-hour requirement (from line 9)
21		Reserve time in days x
22		Percent of useable battery capacity ÷
23		Minimum battery capacity in amp-hours =
24		Select a battery for your system, enter amp-hour capacity ÷
25		Number of batteries in parallel =
26		System nominal voltage (from line 6)
27		Voltage of your chosen battery (6 or 12 usually) ÷
28		Number of batteries in series =
29		Number of batteries in parallel (from line 25 above) x
30		**Total number of batteries required**

Typical Wattage Requirements for Common Appliances

Use the manufacturer's specs if possible, but be careful of nameplate ratings that are the highest possible electrical draw for that appliance. Beware of appliances that have a "standby" mode and are really "on" 24 hours a day. If you can't find a rating, call us for advice (800-919-2400).

DESCRIPTION	WATTS
Refrigeration:	
4-yr.-old 22 cu. ft. auto defrost (approximate run time 7–9 hours per day)	500
New 22 cu. ft. auto defrost (approximate run time 7–8 hours per day)	200
12 cu. ft. Sun Frost refrigerator (approximate run time 6–9 hours per day)	58
4-yr.-old standard freezer (approximate run time 7–8 hours per day)	350
Dishwasher: cool dry	700
hot dry	1450
Trash compactor	1500
Can opener (electric)	100
Microwave (.5 cu. ft.)	900
Microwave (.8 to 1.5 cu. ft.)	1500
Exhaust hood	144
Coffeemaker	1200
Food processor	400
Toaster (2-slice)	1200
Coffee grinder	100
Blender	350
Food dehydrator	600
Mixer	120
Range, small burner	1250
Range, large burner	2100
Water Pumping:	
AC Jet Pump (⅓ hp), 300 gal per hour, 20′ well depth, 30 psi	750

DESCRIPTION	WATTS
AC submersible pump (½ hp), 40′ well depth, 30 psi	1000
DC pump for house pressure system (typical use is 1–2 hours per day)	60
DC submersible pump (typical use is 6 hours per day)	50
Shop:	
Worm drive 7 ¼″ saw	1800
AC table saw, 10″	1800
AC grinder, ½ hp	1080
Hand drill, ⅜″	400
Hand drill, ½″	600
Entertainment/Telephones:	
TV (27-inch color)	170
TV (19 inch color)	80
TV (12-inch black & white)	16
Video games (not incl. TV)	20
Satellite system, 12-ft dish/VCR	30
Laser disk/CD player	30
AC powered stereo (avg. volume)	55
AC stereo, home theater	500
DC powered stereo (avg. volume)	15
CB (receiving)	10
Cellular telephone (on standby)	5
Cordless telephone (on standby)	5
Electric piano	30
Guitar amplifier (avg. volume)	40
(Jimi Hendrix volume)	8500
General Household:	
Typical fluorescent light (60W equivalent)	15
Incandescent lights (as indicated on bulb)	
Electric clock	4
Clock radio	5
Electric blanket	400
Iron (electric)	1200

DESCRIPTION	WATTS
Clothes washer (vertical axis)	900
Clothes washer (horizontal axis)	250
Dryer (gas)	500
Dryer (electric)	5750
Vacuum cleaner, average	900
Central vacuum	1500
Furnace fan:	
¼ hp	600
⅓ hp	700
½ hp	875
Garage door opener: ¼ hp	550
Alarm/security system	6
Air conditioner: 1 ton or 10,000 BTU/hr	1500
Office/Den:	
Computer	55
17″ color monitor	100
17″ LCD "flat screen" monitor	45
Laptop computer	25
Ink jet printer	35
Dot matrix printer	200
Laser printer	900
Fax machine: (plain paper)	
standby	5
printing	50
Electric typewriter	200
Adding machine	8
Electric pencil sharpener	60
Hygiene:	
Hair dryer	1500
Waterpik	90
Whirlpool bath	750
Hair curler	750
Electric toothbrush: (charging stand)	6

CHAPTER 4

EMERGENCY PREPAREDNESS

THE WORLD CAN BE A HAZARDOUS PLACE even in the best of times. Overcrowding, growing population pressures, natural disasters, and the no-longer-deniable global warming phenomenon aren't likely to make it any friendlier in the near future. We're likely going to see bigger storms and hurricanes, stranger weather, overloaded utilities, and more electric failures in our lifetimes. And isn't it about time for that next big California earthquake?

Want Some Protection, or at Least a Little Backup?

The renewable energy industry has spent its maturing years learning how to provide highly reliable home energy systems under adverse and remote conditions. No utility power? No problem! No generator fuel? No problem! No easy access to the site? No problem! Gaiam Real Goods is ready to provide off-the-shelf solutions for short-term, long-term, or lifetime energy problems. We've collected and discussed a few of them in this chapter. Like individual lifestyles, energy needs and the necessary equipment to meet them is an individual matter. If you don't see what you need, give our experienced technical staff a call. We're masters at putting together custom systems to meet individual needs, climates, and sites. Reliable power for your peace of mind is our business.

Steps to a Practical Emergency Power System

Following are some practical solutions to utility disruption. We've arranged them in financial order with the least expensive options first. Be aware that lowest-cost options will involve some sacrifices and won't be acceptable solutions for more than short time periods—emergencies in other words. As backup systems get more robust, they impose less restrictions, and are more pleasant to live with.

1. Install a Good-Quality Backup Generator

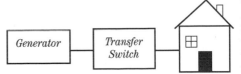

This will be useful for short power outages, or for longer ones so long as you have fuel and you, or the neighbors, are willing to endure the noise. Higher quality generators tend to be much quieter. Propane fuel is a good choice. It is delivered in bulk to your site (in all but the most remote locations), it doesn't deteriorate or evaporate with age, and it's piped to the generator, so you never need to handle it. Generators live longer on propane because there is less carbon buildup. Your electrician will need to install a transfer switch between the utility and the generator, so that your generator doesn't try to run the neighborhood, and doesn't threaten any utility workers.

Simple backup generators use manual starting, and a manual transfer switch. When power fails, somebody has to start the generator, then throw the switch. Fancier systems will do all this automatically. With any generator-based backup system, even automated ones, there will be some dead time between when the power goes out, and the backup systems come online.

Most generator-based systems will require some electrical conservation. Don't plan on running big watt-sucker appliances like air conditioning, electric room or water heaters, or electric clothes dryers, unless you're willing to support a huge generator of 12,000 watts or larger. Generators run most efficiently at 50% to 80% of full load. A big generator is going to suck almost as much fuel to run a couple lights as it is to run all the lights and the air conditioning. Generators aren't a case of "bigger is better."

2. Add Batteries and an Inverter/Charger

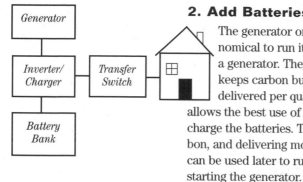

The generator only supplies power when it's running, and it isn't practical or economical to run it all the time. Running just 200 watts is particularly expensive with a generator. They're happiest running at about 80% of full rated wattage. This keeps carbon build-up to a minimum, and is the most efficient in terms of watts delivered per quantity of fuel consumed. Adding batteries and an inverter/charger allows the best use of your generator. Whenever the generator runs, it will automatically charge the batteries. This means the generator is working harder, building up less carbon, and delivering more energy for the fuel consumed. Energy stored in the batteries can be used later to run lights, entertainment equipment, and smaller loads without starting the generator.

Many medium- to larger-size inverters come with built-in, automatic battery chargers. So long as the inverter senses that utility or generator power is available, that line power is passed right through. If outside AC power fails, the inverter will automatically start using stored battery power to supply any AC demands. When outside AC power returns, the batteries will be automatically recharged. For a power failure that lasts more than a couple of days, this means you'll only need to run the generator for a few hours per day, yet you'll have AC power full time.

Putting an inverter and battery pack into the system means that any AC power interruptions will be measured in seconds, or less, depending on the equipment chosen.

3. Add Solar, Hydro, or Wind Power for Battery Charging

A generator has the lowest initial cost for power backup, but it's only cheap if you don't run it. Generators are expensive and noisy to operate, and fuel supplies need replenishment. If you anticipate power outages lasting longer than a few weeks, then solar, hydro, or wind power will save you money in the long run, and deliver a reliable long-term energy supply.

Once you have the battery pack, it can accept recharging from any source. Electrons all taste the same to your battery. Whether they come from a belching cheapo gas generator, or a clean, silent PV module, the effect is the same on the battery. Renewable energy sources have a higher initial cost, but very low costs over time for both your wallet, and the environment.

Insulation

Since warm air rises, the best place to add insulation is in the attic. This will help keep your upper floors warm. Recommended insulation levels depend on where you live and the type of heating system you use. For most climates, a minimum of R-30 will do, but colder areas, such as the northern tier and mountain states, may need as much as R-49. Two useful booklets, BOE/CE-0180 (Aug. 1997) and DOE/GO-10097-431 (Sept. 1997), provide information about insulating materials and insulation levels needed in various locations by zip code. Write the U.S. Department of Energy, Office of Technical Information, P.O. Box 62, Oak Ridge, TN 37831.

There are four types of insulating materials: batts, rolls, loose fill, and rigid foam boards. Each is suitable for various parts of your house. Batts, designed to fit between the studs in the walls, or the joist in the ceilings, are usually made of fiberglass or rock wool. Rolls are also made of fiberglass, and can be laid on the attic floor. Loose-fill insulation, which is made of cellulose, fiberglass, or rock wool, is blown into attics and walls. Rigid board insulation, designed for confined spaces such as basements, foundations, and exterior walls, provides additional structural support. It is often used on the exterior of exposed cathedral ceilings.

Emergency and Backup Power System Design

The average American home uses about 15 to 20kWh of electricity per day. While it's certainly possible to put together an emergency power system to supply this level of use, most folks would find it prohibitively expensive. Most of us are willing to modify our lifestyle in the face of an emergency.

What Do You *Really* Need in an Emergency?

Refrigeration, heat, drinking water, lights, cooking equipment, and communications are on the typical short list. Let's take a quick look at each of these necessities and see what is needed to provide for it during an emergency.

Refrigerators and Freezers

Refrigeration is a major concern for most emergency or back-up systems. The fridge will probably be your major power consumer, typically using 3 to 5 kWh of energy per day. It's going to require a 1500-watt generator, or a robust 600-watt inverter, at the very least, to start and run a home refrigerator or freezer. If your refrigerator is 1993 vintage or older, consider a new replacement. Refrigerators have made enormous advances in energy efficiency over the past few years. The average new 22 cu.ft. fridge will use half the power of one produced before '93. If you're shopping, watch those yellow EnergyGuide tags. They really level the playing field! All models are compared to the same standard, which happens to be a pessimistic 90°F ambient temperature. Most of us will see slightly lower operating costs, most of the time. Any fridge model sporting a tag for 550 kWh per year will serve you well and drop energy use to near 1.5 kWh per day. The very best mass-produced fridges will have ratings for 450 kWh/year or less. Top and bottom units have a slight energy advantage over side by side units.

If you expect power to be out for more than a few weeks, then a propane-powered fridge might be a better choice. Although small by usual American standards (about 7 to 8 cu.ft.) they'll handle the real necessities for about 1.5 gallons of propane per week. Propane freezers are also available.

Heat

If a natural disaster happens in winter, keeping warm may suddenly become overwhelmingly important, while keeping the food from spoiling is of no concern at all. Anyone with a wood stove and a supply of firewood will be sitting pretty when the next ice storm hits. Obviously, any heating solution that doesn't involve electricity is going to have an easier time keeping you warm following a natural disaster. Beyond wood stoves, other non-electric solutions may include gas-fired wall heaters, or portable gas or kerosene heaters. Wall heaters are preferable, since they're vented to the outside. Use extreme caution with portable heaters, or any other "ventless" heater that uses room oxygen and puts its waste products into the air you're breathing! Combustion waste products combined with depleted oxygen levels are a dangerous or even deadly combination. Neither brain damage or freezing are acceptable choices. Those in earthquake country shouldn't count on natural gas supply to continue following a large earthquake.

If a central furnace or boiler is your only heat choice, you'll be wanting some electricity to run it. If your generator- or inverter/battery-based emergency power system is already sized to run a fridge, it will probably be able to pick up the heating chores instead of the fridge during cold weather. (Put the food outside, Mother Nature will keep it cold.) Furnaces, which use an air blower to distribute heat, will generally use a bit more power than the average refrigerator. Boilers, which use a water pump or pumps to distribute heat, will generally use a bit less power than the fridge. Depending on the capabilities of your emergency power system, you might need to limit other uses while the heater is running.

Pellet stoves are a very poor choice for emergency heating. Although the fuel is compact and stores nicely, pellet stoves require a steady supply of power in order to operate. Pellet stoves have blowers for combustion air, other blowers for circulation of room air, and pellet feed motors. A better solution is a special stainless steel pellet basket that allows you to burn pellet fuel in any conventional wood stove. You'll find pellet baskets in Chapter 5 under Heating & Cooling appliances.

Water

If yours is a city-supplied water system, you may have a problem that an emergency power system can't solve. Most cities use a water tower or other elevated storage to provide water pressure. They can only continue to supply water for a few days, or maybe just a few hours when the power fails. Other than filling the bathtub, or draining the water heater tank, which can get you by for a few days, you may want to provide yourself with some emergency water storage. Water storage tanks are available from any farm supply store, and some larger building and home supply stores. There are "utility" and "drinking water" grade tanks. Make sure you're buying the drinking water grade with an FDA-approved nonleaching coating.

If you have your own well or other private water supply that requires a pump for water delivery, then you'll probably be wanting some kind of power system that can run your water pump. AC water pumps require a large surge current to start. The size of the surge depends on pump type and horsepower. Submersible pumps require more starting power than similar horsepower surface-mounted pumps.

For short-term backup, the easiest and cheapest solution is to use a generator to run the pump once or twice a day to fill a storage or large pressure tank. While the generator is running, you can also catch a little battery charging for later use to run more modest power needs that can be supplied by an inverter/battery system.

Longer-term solutions will either require a bigger inverter and battery bank to run the water pump on demand, or a smaller water pump that can be more easily run by solar or battery power. Our technical staff will by happy to discuss your situation and make the best recommendations for your needs and budget.

Water Quality (Is it safe to drink?)

Natural disasters often leave municipal water sources polluted and unsafe to drink, sometimes for weeks. So even though you've got water, you better think twice before drinking it. Disease, carried by polluted drinking water, does far more human harm than actual natural disasters. With just a little preparation, and at surprisingly low cost, you can be ready to treat your drinking water. And considering how truly miserable, long-lasting, and even life-threatening many water-borne diseases are, an ounce of prevention is worth many tons of cures. You can boil, which is time and energy intensive, you can treat with iodine, which is cheap and effective, but distasteful, or you can filter, which is quick and effective with the right equipment. The "right equipment" usually means either ultrafine ceramic filtration, or a reverse osmosis system. We offer several models of both types in the product section that follows.

Lighting

If you use compact fluorescent lamps instead of common incandescent lamps, you'll get four to five times as much light for your power expenditure. Because power is valuable and expensive in an emergency or backup situation, these efficient lights are a must. If these lights happen to be part of your normal household lighting in the meantime, you'll enjoy some energy savings while waiting for the big one. Since compact fluorescent lamps will outlast 10 to 13 normal incandescent lamps, and will use 75% less energy, they'll end up paying for themselves, plus putting hundreds of dollars back in your pocket over their lifetime. Compact fluorescent lamps are a great investment both for everyday living and for energy-efficient emergency backup.

Smaller, battery-powered, solar-recharged lanterns and flashlights can be extremely helpful during short-term outages, and can easily be carried to wherever light is needed. We offer several quality models that can be charged with their onboard solar cells, or, can simply have fresh batteries slipped in when time and conditions won't allow solar recharging.

Cooking

This is one of the easier problems to tackle. Let's start by looking at what you're using to cook and bake now.

Electric Stoves and Ovens. These use massive amounts of electricity. It's not going to be practical to run your electric stove top or oven with a back-up system or even with a generator. Break out the camping gear. Camp stoves running on white gas or propane are a good alternative and widely available at modest cost. Charcoal barbecues can be used as well, but *only outdoors!* They produce carbon monoxide, which is odorless and deadly if allowed to accumulate indoors. Want a longer-term solution? Consider a gas stove.

Gas Stoves and Ovens. These can be fired by either propane or natural gas. Propane, which only depends on a small, locally installed tank may be more dependable in a major emergency, particularly an earthquake. Major earthquakes break buried natural gas lines, but electrical outages don't effect natural gas systems. In either case, an electrical outage will put your modern spark-ignition burner lighters out of commission. Just use a match or camping-type stove igniter for the burners. Older stoves use a pilot light for ignition, which will work as normal with or without electricity.

Your gas oven probably won't work without electricity. Older stoves, using a standing pilot light, will be unaffected; all others will either use a "glow bulb," or a "spark igniter." Spark igniters, like those used on stovetop burners, will allow you to light the oven with a match. This is good. Glow bulbs require 200 to 300 watts of electrical power all the time the oven is on, and will not allow match lighting. This is a problem. You can tell which type of igniter you have by opening the bottom broiler door, and then turning on the oven. You'll hear or see the spark igniter, and light up will be quick. Glow bulbs take 30 seconds or longer to light up, and you'll probably see the glowing orange-red bulb once it gets warmed up.

Solar Ovens. If the weather allows, solar ovens do a great job at zero operational cost. They do require full sunlight, and are excellent for anything from a cup of rice, to a batch of muffins, to a pot full of stew. We offer several ovens, kits, and plans in the product section following.

Communications

For simple communication needs we offer several models of radios with wind-up and/or solar power sources. Many folks now consider the Internet to be basic communications. And for good reason. With no centralized control the Internet has proven to be difficult to damage, and nearly impossible to shut down or censor for content.

Internet access requires a computer, and that means some minimum power use. If you have the choice, a laptop computer will use a fraction of the power that a desktop unit requires. The average desktop with 17″ color monitor will use approximately 120 watts. The average laptop with color matrix screen only uses 20 to 30 watts. If you plan to spend more than a couple hours per day on the computer without grid support, then a good laptop is going to pay for itself in energy system savings.

Most other personal communications devices, such as CB radios, ham radios, and two-way radios are designed for, or easily adapted to 12-volt DC power use. This means the energy in your battery storage can be utilized directly.

Conclusions

There is an ancient Chinese curse, "May you live in interesting times." Like it or not, we have been thoroughly and completely cursed with interesting times. We can make the best of it with some modest preparation, or we can sit back, do nothing, and wait to see what gets dealt to us, and how much it hurts. In the Product Section following, we've put together a few items and kits that can ease the pain a bit, or even a whole lot. Choose the level of protection that feels comfortable. After you're prepared, sit back feeling secure and smug, knowing you're independent from the grid—at least for awhile. And if you're really ambitious, talk to our technicians about a full-time off-the-grid system.

—**Doug Pratt**

Emergency Preparedness Products

Super Ark Preparedness Kit

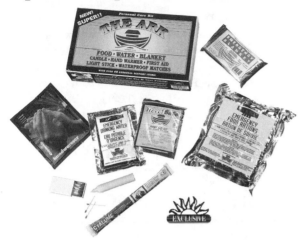

Three days waiting in the cold and hungry dark looks real different than three days with adequate food, warmth, and light. This kit assembles three days of basic necessities for one person—at $10 per day this is the best disaster insurance you can buy. Compact enough to fit in a day pack. Five-year shelf life. USA.

03-838 Super Ark Survival Kit $29⁹⁵

Kit Contains: Mainstay Food Bars (3600 calories total); 6 Water Packets; Mylar Blanket; 45 Waterproof Matches; 4-Inch Emergency Candle; 18-Hour Hand Warmer (One Use); 12-Hour Light Stick (One Use); First Aid Kit.

Multifunction Survival Tool

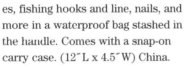

Shovel, hammer, saw, hatchet, bottle opener, nail puller, wrench. Unscrew the calibrated compass and find matches, fishing hooks and line, nails, and more in a waterproof bag stashed in the handle. Comes with a snap-on carry case. (12″ L x 4.5″ W) China.

63-114 All-Purpose Survival Tool $19⁹⁵

Emergency Lighting Products

also see Chapter Five, pages 308 to 319, for Compact Fluorescent Lamps

Guiding Light

Originally used by European hoteliers to walk their guests to bed, this slim profile oil lamp is designed to be carried safely. The sturdy carry handle is slotted for optional wall mounting. This is a clean-burning lamp with a 12-watt output, 10-hour burn time, and reliable brass alloy burner. Best results using paraffin lamp oil; it also accepts kerosene. 15″ high overall; solid brass and glass. The optional elegant 6″ etched ball shade enriches the appearance of the lamp while softening and diffusing the light. France.

35-345 Concierge Brass Lamp $79⁹⁵

35-344 Concierge Etched Shade $29⁹⁵

35-341 Chimney $9⁹⁵

35-322 Wick Replacement (Each) $3⁹⁵

35-343 Burner Replacement $29⁹⁵

Humphrey Propane Lights

Propane lighting is very bright. It is a good alternative if you don't have an electrical system. One propane lamp emits the equivalent of 50 watts of incandescent light, while burning only one quart of propane for 12 hours use. The Humphrey 9T contains a burner nose, a tie-on mantle, and a #4 Pyrex globe. The color is "pebble gray." The mantles seem to last around three months, and replacements are cheap. Note: Propane mantles emit low-level ionizing radiation due to their thorium content.

35-101 Humphrey Propane Lamp (9T) $59⁹⁵

35-102 Propane Mantle (each) $1⁹⁵

35-103 Replacement Globe $13⁹⁵

Flashlights

A Superefficient Wind-Up Flashlight

When we introduced the world's first wind-up radio, we knew a spring-wound flashlight wouldn't be far behind. This revolutionary flashlight is the ultimate choice for auto or home emergency lighting, combining a reliable clockwork power source with a built-in nicad battery pack. Run the lamp directly from the clockwork mechanism (60 turns of winding, about one minute, nets four to five minutes of run or recharge time), or from the nicads (up to two hours of continuous use). Features water-resistant construction and the familiar high-output bulb from mini-mag-type flashlights; a spare bulb is included inside the twist-off lens. Recharge by winding up, or by plugging in the included AC/DC adapter. All wind-up energy goes to recharging. A great choice for auto or home emergency lighting. Five-year mfr.'s warranty! South Africa.

37-297 Freeplay Flashlight $64⁹⁵

The Brighter, Whiter Penlight

Our new ultraefficient flashlight takes advantage of cutting-edge LED technology to bring you low power, high quality, handheld lighting. This watertight penlight produces an excellent bluish-white light with such a small power draw that the 3 AA alkaline batteries will last approximately 17 hours, or 8 hours with nicad batteries. Has wrist strap and turns on/off by tightening/unscrewing the clear plastic lens. The three LED bulbs last 60,000 hours (6.8 years of continuous use!). Brighter than typical mag lights. One-year mfr.'s warranty. USA.

37-356 White 3 LED Penlight $29⁹⁵

Create Light Without Batteries

No preparedness kit should be without the Dynamo. Continuous squeezing produces a steady beam of light. This is a reliable battery-free emergency light, with a shockproof and shatterproof lens, a spare bulb, a peg for attaching a cord, and a slide mechanism to lock the handle closed. Assorted gear colors. Canada.

37-374 Hand-Cranked Dynamo Flashlight $12⁹⁵
      3 or More $10⁹⁵ each

Solar Flashlight

Free sunlight fully recharges the built-in 600 mA nicad battery pack in about eight hours. No time for solar? Slip two AA batteries into the battery compartment, and you have light right now. These optional AA batteries will be recharged by solar too. Includes three LED warning flashers on tail with focusing lens. Flashlight and warning flashers are switched separately, so one or both can be run. An adjustable holder for bikes or whatever is included, along with two spare bulbs and wrist strap. Run time on built-in battery alone is approximately 2.5 hours for the flashlight, or six days for the warning flasher. Optional nicad batteries will increase run times by 120%; alkaline batteries by 250%. Waterproof and floats too! China.

37-295 Solar Flashlight w/Warning Flashers $24⁹⁵

Flashlight Dashlite

This is a bright, rechargeable flashlight that lives in your cigarette lighter socket, so it's always there, easy to find, and ready when you need it. Absolutely will not drain your car battery! Provides one hour of continuous use, plenty of time to fix a flat or walk to a call box. Shines up to 100 feet, and lasts up to five years. Shock-resistant ABS plastic. Mexico.

37-310 Flashlight Dashlite $12⁹⁵

Battery-Free, NightStar™ LED Flashlight—Guaranteed For Life!

With a waterproof sealed body, battery-free operation, and lifetime warranty, this could be the last flashlight you ever buy. Recharge the NightStar™ by shaking it back and forth—and it works every time, guaranteed. Shaking causes a high-strength rare earth magnet to slide back and forth through a coil of copper wire (1500 loops), generating electricity, which is then stored in a capacitor. Delivers up to 15 minutes of useable light. The NightStar™ produces a 6-foot diameter circle of soft illumination at a distance of 30 feet, about the same intensity as a full moon on a clear night. The polycarbonate/ABS plastic body is virtually indestructible, with a sealed switch and double O-ring seals for protection from submersion. Weighs 16 oz., measures 11″ L x 2″ Dia. Lifetime mfr.'s warranty. USA.

37-357 NightStar™ LED Flashlight $75⁰⁰

Dead Cell Phone Battery? Never Again

Now you can generate instant power anytime, anywhere, for almost any device powered by rechargeable batteries. This hand-squeeze-regulated generator quickly recharges cell phones, rechargeable CD players, radios, walkie-talkies, and just about anything that has compatible rechargeable batteries. It's equipped with a regulator that regulates the proper flow to the device being recharged. A universal connecting cable is included for attaching to your cell phone's cigarette lighter cord. Also includes a high-powered emergency LED light and a three-compartment leather carrying case with belt loop. 90-day limited mfr.'s warranty. 1″H x 5.3″L x 1.7″W. Weighs only 4 oz. USA.

17-326 AladinPower Hand Generator $79⁹⁵

The Real Goods Solar Lantern

EXCLUSIVE

A 3-way lantern with solar charging and great run time! Our solar lantern features a 5-watt fluorescent flood light, a powerful flashlight, and red-lens flashing warning light. The built-in nicad battery pack will run the fluorescent lamp for over 4 hours, the flashlight for 8 hours, or the warning light for 40 hours. Solar recharging takes 8–10 hours in full sun. Want more run time? Slip three optional D cells into the battery compartment. Nicad batteries will boost run times by 300%, alkaline batteries will boost by 700%! D cells will be recharged by solar. Comes with 4 spare bulbs, a shoulder strap, and a clear vinyl "raincoat" tote bag for waterproof use. Equipped with a 6-volt external charging port for an AC/DC adapter (adapter not included). A perfect lantern for camping, auto, emergency, or general use. 11.4″ x 5.6″ x 2.1″, 1.75 lb. Taiwan.

37-296 Real Goods Solar Lantern $79⁹⁵

Water Collection

Rainwater Collection for the Mechanically-Challenged

Laugh your way to a home-built rainwater collection and storage system. This delightful paperback is not only the best book on the subject of rainwater collection we've ever found, it's funny enough for recreational reading, and comprehensive enough to painlessly lead a rank amateur through the process. Technical information is presented in layman's terms, and accompanied with plenty of illustrations and witty cartoons. Topics include types of storage tanks, siting, how to collect and filter water, water purification, plumbing, sizing system components, freeze-proofing and wiring. Includes a resources list and a small catalog. Paperback, 46 pages, USA.

80-704 Rainwater Collection for the Mechanically-Challenged $17⁹⁵

Rainwater Collection For The Mechanically-Challenged Video

You say you're *really* mechanically challenged and want more than a few pictures? Here's your salvation. From the same irreverent, fun-loving crew that wrote the book above. See how all the pieces actually go together as they assemble a typical rainwater collection system, and discuss your options. This is as close as you can get to having someone else put your system together. 37 minutes and lots of laughs. USA.

82-568 Rainwater Challenged Video $19⁹⁵

Build Your Own Water Tank

By Donnie Schatzberg. An informative booklet that has recently been updated and greatly expanded with more text, drawings, and illustrations. It gives you all the details you need to build your own ferro-cement (iron-reinforced cement) water storage tank. No special tools or skills are required. The information given in this book is accurate and easy to follow, with no loose ends. The author has considerable experience building these tanks, and has gotten all the "bugs" out. 58 pages, paperback. USA.

80-204 Build Your Own Water Tank $14⁰⁰

Kolaps-a-Tank

These handy and durable nylon tanks fold into a small package or expand into a very large storage tank. They are approved for drinking water, withstand temperatures to 140° F, and fit into the beds of several truck sizes. They will hold up under very rugged conditions, are self-supported, and can be tied down with D-rings. Our most popular size is the 525-gallon model, which fits into a full-sized long-bed (5 x 8 ft.) pickup truck. USA.

47-401 73 gal. Kolaps-a-Tank, 40″ x 50″ x 12″ $300⁰⁰

47-402 275 gal. Kolaps-a-Tank, 80″ x 73″ x 16″ $400⁰⁰

47-403 525 gal. Kolaps-a-Tank, 65″ x 98″ x 18″ $500⁰⁰

47-404 800 gal. Kolaps-a-Tank, 6′ x 10′ x 2′ $600⁰⁰

47-405 1140 gal. Kolaps-a-Tank, 7′ x 12′ x 2′ $800⁰⁰

47-406 1340 gal. Kolaps-a-Tank, 7′ x 14′ x 2′ $900⁰⁰

Water Purification

also see Chapter Eight, pages 383 to 392, for longer-term water quality solutions

First Need

The First Need is the best-selling portable water filtration system in America. Will remove tastes and odors, and with 0.1 micron filtration, will safely remove waterborne pests such as bacteria, giardia, viruses, or cryptosporidium. Meets EPA Guide Standard. Sturdy plastic pump, tubes, and filter fit into included nylon tote bag. Flow is 1–2 pints/min., depending on vigor of pumper. Weighs less than 16 ounces. Great for backpacking, camping, travel, or your emergency preparedness kit. Shown with the Sediment Prefilter, which greatly increases filter life if changed regularly. Rated for approximately 100 gallons per cartridge.

42-653 First Need Filtration System $79⁰⁰

42-647 First Need Sediment Prefilter $9⁹⁵

42-658 First Need Repl. Filter Cartridge $34⁹⁵

Katadyn Pocket Filter

The Katadyn Pocket Filter is standard issue with the International Red Cross and the armed forces of many nations. Essential for any survival kit, it has been manufactured in Switzerland for over half a century. These filters are of the highest quality imaginable, reminiscent of fine Swiss watches. The Katadyn system uses an extremely fine 0.2-micron ceramic filter blocking pathological organisms from entering your drinking water. A self-contained and very easy-to-use filter about the size of a 2-cell flashlight (10″ x 2″), produces a quart of ultra-pure drinking water in 90 seconds with the simple built-in hand pump. It weighs only 23 oz. and comes with its own travel case and special brush for cleaning the ceramic filter. The replaceable ceramic filter can be cleaned up to 400 times, lasting many years with average use. An indispensable tool for campers, backpackers, fishermen, mountaineers, river runners, globetrotters, missionaries, geologists, and those who work in disaster areas.

42-608 Katadyn Pocket Filter $249⁰⁰

42-609 Replacement Filter $165⁰⁰

Healthy, Clean Drinking Water

Clean, healthy drinking water. Will you be able to access it when a natural disaster strikes? Often the first warning issued by local municipalities is concerning the safety of the water supply. With this Ceramic Water Filter you can relax because your water will be filtered to an absolute 0.5-micron level. It meets EPA standards for elimination of bacteria and protozoa, including giardia, cryptosporidium, E. coli, vibrio cholera, shigella, salmonella typhi, and klebsiella terigina. It will supply you with approximately 500 gallons of filtered water, making it perfect for camping trips too. For best life of this product, use the cleanest source possible and rinse the foam sock pre-filter often. With its easy-to-use and simply designed pump, you can filter one ounce of water with each pull of the pump. And at only 8 ounces, it's extremely lightweight. Includes clip for attachment to water carrier. Great Britain.

46-233 Ceramic Filter Pump $29⁹⁵

Lightweight, Economical Personal Filter System

Ideal for back country camping and adventure travel. The MiniWorks pocket-size, lever-style pump filter (16 oz.) has a rugged polyurethane housing that encases a ceramic filter with an activated carbon core. The Marathon Ceramic filter exceeds the EPA's Purifier Standards for bacteria and protozoa removal plus reduces some chemical contaminants and bad tastes and odors. Approximately 100-gallon filter life (with periodic cleaning). Use your own container or purchase the collapsible bag and screw the filter onto the opening for a durable water carry bag or shower. The collapsible 10-liter water bag (8.5 oz) with shower kit is sold separately. The 1000-denier cordura nylon bag (with food-grade polyurethane liner) has a threaded bag opening and comes with a shower kit that includes a 3-foot hose, spigot cap, and a shower nozzle. The black color will absorb heat for a nice warm shower and the multiple grommets laced with webbing make for easy hanging and carrying. The most important 1.75 pounds you'll ever carry. USA.

42-848 MiniWorks Pump Filter $59⁹⁵

42-849 MiniWorks Replacement Filter $29⁹⁵

42-850 Dromedary 10-Liter Bag (W/Shower) $49⁹⁵

The Katadyn Drip Filter

With no moving parts to break down, superior filtration, and a phenomenal filter life, there is simply no safer choice for potentially pathogen-contaminated water. There are no better filters than Katadyn for removing bacteria, parasites, and cysts. Three 0.2-micron ceramic filters process one gallon per hour. Clean filters by brushing the surface. Ideal for remote homes, RV, campsite, and home emergency use. Food-grade plastic canisters stack to 11″ Dia. x 25″ H. Weighs 10 lb. One-year mfr.'s warranty. Switzerland.

42-842 Katadyn Drip Filter $200⁰⁰

42-843 Replacement Filter (needs 3) $75⁰⁰ each

Water Heating & Washing

5-Gallon Solar Shower

No more cold showers when you're camping. This low-tech invention uses solar energy to heat water for all your washing needs. The large 5-gallon capacity provides ample hot water for at least four hot showers. On a 70° day the Solar Shower will heat 60° water to 108° in only three hours. Great for camping, car trips, or emergency use. This has been one of our catalog's all-time best-sellers. Taiwan.

90-416 Solar Shower $14⁹⁵

Solar Air Power Shower

This solar-heated marvel creates a fresh, invigorating spray by storing pressurized air from a two-way detachable pump. After a few pumps, you're ready to rinse gear; scour clothes; or best of all, luxuriate under a warm, rejuvenating shower. Works horizontally, or suspended from its rugged nylon carrying handle. Extra-big 7-gallon capacity. Heats 60° water to 108° in just three hours on a 70° day, and runs for two minutes without pumping when fully pressurized (only 30 seconds to pressurize) at full capacity. Features double-welded, 600-denier polyester, and flexible 7-foot hose. Hong Kong.

90-330 Solar Air Power Shower $49⁹⁵

Breadbox Solar Hot Water Plans

We've gotten lots of requests for simple plans like this over the years. This passive solar water heating system is simple to build using readily available materials; uses no pumps, controllers, or sensors, and works wonderfully as a preheater or even as a stand-alone system. Uses a pair of common 40-gallon water heater tanks, either new or recycled. The system is virtually freeze-proof with a little care during construction. The detailed, easy-to-follow plans include many drawings, a materials list, construction sequence, and options for different tank configurations, reflectors, absorber surfaces, and glazing techniques. Developed by a licensed solar plumber based on 20 years of community education workshops. 8.5″ x 11″, 8 pages (folds out). USA.

80-497 Solar Hot Water Plans $12⁰⁰

James Washers

The James hand-washing machine is made of high-grade stainless steel with a galvanized lid. It uses a pendulum agitator that sweeps in an arc around the bottom of the tub and prevents clothes from lodging in the corner or floating on the surface. This ensures that hot suds are thoroughly mixed with the clothes. The James is sturdily built. The corners are electrically spot-welded. All moving parts slide on nylon surfaces, reducing wear. The faucet at the bottom permits easy drainage. Capacity is about 17 gallons. Wringer attachment pictured with the washer is available at an additional charge.

63-411 James Washer $300⁰⁰

Hand Wringer

The hand wringer will remove 90% of the water, while automatic washers remove only 45%. It has a rustproof, all-steel frame and a very strong handle. Hard maple bearings never need oil. Pressure is balanced over the entire length of the roller by a single adjustable screw. We've sold these wringers without a problem for over 13 years.

63-412 Hand Wringer $149⁰⁰

Fifty Feet Of Drying Space

This is the most versatile clothes dryer we've seen yet. Don Reese and his family are still making them by hand, from 100% sustainably harvested New England pine and birch. Don's sturdy dryer can be positioned to create either a peaked top or a flat top ideal for sweaters. A generous 50 feet of drying space—and you can fold it up to carry back in the house (or to the next sunny spot) still loaded with laundry. Measures 53″ H x 30″ W x 33″ D in peaked position; 46″ H x 30″ W x 40″ D in flat-top position, and holds an average washer load. Folds to 4.5″ D x 58″ H. USA.

◈ **55-273 New England Clothes Dryer $47⁹⁵**

Requires UPS ground delivery.

Cooking Products

Emergency Candle Lasts 120 Hours

This candle will fit in your backpack, your car trunk, or on a shelf for emergencies. It's enclosed in a metal can so you can take it anywhere. Use it for lighting, heating, and cooking for up to 120 hours. The secret is its six long-burning movable wicks. Light just one for illumination, two or three for heating food. The specially formulated FDA approved (food grade) paraffin is nontoxic. Includes six wicks, tweezers, and matches. USA.

63-452 120-Hour Candle $12⁵⁰
  3 or More $9⁹⁵ each

Zip Stove Boils Water In 90 Seconds

Use any combustible fuel for cooking, at your campsite or in an emergency. In our field test, a handful of oak twigs boiled two cups of water in only 90 seconds. It's incredibly compact and lightweight, with a built-in fan that blows like a blast furnace to get the most heat out of found fuel, quickly. The fan uses one AA battery (not included) for approximately 6 hours of cooking time. Zip Stove measures 5″ Dia. x 6.5″ H, and weighs 17.6 oz. The optional windshield and grill, folded flat for transport, measure 9.9″ L x 9″ W x .75″ D, weigh 16.7 oz., and assure fast results. USA.

63-422 Zip Stove $59⁹⁵
63-423 Zip Windshield and Grill $24⁹⁵

Eagle Stove Is A Portable Kitchen

The Eagle Stove has all the great features of the Zip Stove plus a few new ones, and it's twice as big. Use it outside as a portable campfire for cooking, or for providing warmth on a cool evening. It burns wood, charcoal, pinecones, any solid fuel source. It's safer than gas-burning stoves and provides a contained fire where open fires are not permitted. The adjustable-speed fan circulates incoming air around the stove walls, preheating the air and creating a blast-furnace effect that delivers 35,000 BTU per hour—enough heat to boil water by the gallon. A single D cell battery (not included) is all you need to provide over 35 hours of cooking time. Made of galvanized steel, the Eagle Stove measures 12.5″ H x 9.75″ W x 12.5″ L and weighs only eight pounds. Complete instructions included. USA.

63-655 Eagle Stove $69⁹⁵

Cook With Almost Any Liquid Fuel

Optimus of Sweden has been manufacturing powerful, reliable liquid fuel stoves since 1899. The Hiker Stove has been carried on expeditions to the Himalayas and both Poles, producing excellent results with almost any liquid fuel, in any climate or altitude. Whether you are car camping in the wilds of upstate New York, trekking the Alaskan wilderness, or braving a Central American jungle, you'll never be stuck for the right fuel again. This compact stove burns kerosene, paraffin, diesel oil, methylated alcohol, white gas, and Coleman fuel (do not mix different kinds of fuels) with no need for conversion between fuels—simply refill the empty tank and cook. Includes a unique cleaning needle to unclog jets without disassembling the stove mid-meal, a multi-key air/fuel mix valve to maximize fuel efficiency, a built-in pressure pump, and an easy-switch burner jet for burning alcohol. Flame control is accurate from simmer to full boil (one liter of H2O in three minutes depending on fuel, climate, and altitude). Auto gasoline contains additives that produce toxic gases when burned, and which corrode stove seals; don't burn it in this or any stove. Kerosene or paraffin give the best performance, up to two hours on high per tank. A rugged, reliable expedition stove, it weighs in at under four pounds, 7″ Square x 4″ H. Mfr.'s lifetime warranty on workmanship. Sweden.

14-089 Optimus Multi-Fuel Stove $179⁰⁰

The World's Finest Solar Oven

Our Sun Oven is a great, portable solar cooker weighing only 21 lb. It's ruggedly built with a strong, insulated fiberglass case and tempered glass door. The reflector folds up and secures for easy portability on picnics, etc. It's completely adjustable, and comes with a built-in thermometer. The interior oven dimensions are 14″ W x 14″ D x 9″ H and temperatures range from 350°–400°F. This is a very easy oven to use and it will cook most anything! After preheating, the Sun Oven will cook one cup of rice in 35 to 45 minutes. USA.

63-421 Sun Oven $229⁰⁰

Efficient, Economical Solar Cooker
Improved Family-Sized Solar Cooker Kit

A portion of the proceeds of every sale is donated to the nonprofit Solar Cookers International group. This group provides solar cookers and information to developing countries where fuel shortages are a constant problem. Sales of Solar Cookit supports their good work.

The Solar Cookit, a hybrid of solar box and solar concentrator cookers, includes a foil-laminated foam reflector that folds flat to 13″ x 13″ x 2″. The reflector is now water- and crushproof. Cookit also includes a 3-quart, black-enameled, covered steel pot, and two high-temperature cooking bags that help hold in the heat. It's highly portable and easily stored for camping and emergency use, and is large enough to easily feed three to five people. Reaches a maximum temperature of 300°F. USA.

63-576 Solar Cookit $34⁰⁰

Cooking with the Sun
How to Build and Use Solar Cookers

By Beth and Dan Halacy. Solar cookers don't pollute, they use no wood or any other fuel resource, they operate for free, and they run when the power is out. And a simple and highly effective solar oven can be built for less than $20.

The first half of this book, generously filled with pictures and drawings, presents detailed instructions and plans for building a solar oven that will reach 400°F or a solar hot plate that will reach 600°F. The second half has 100 tested solar recipes. These simple-to-prepare dishes range from everyday Solar Stew and Texas Biscuits to exotica like Enchilada Casserole. 116 pages, paperback. USA.

80-187 Cooking with the Sun $9⁹⁵

Communications Products

Dynamo AM/FM Flashlight

No power, no batteries, no stores open late at night—no problem! The rechargeable nicad battery in this Solar Flashlight/ AM/FM Radio retains its power from solar energy, incandescent light energy, hand-crank dynamo power, or with a 6-volt AC/DC adapter (not included). It can also run on three AA batteries (not included). Complete with siren and flashing signal, this is not only a quality sounding radio, but a great survival tool as well. Plus its bright, yellow color is easy to locate in an emergency. Measures 3.5″ H x 8.5″ L x 3.75″ W. China.

37-355 Dynamo AM/FM Flashlight $39⁹⁵

Freeplay AM/FM Radio

How do you improve on wind-up power? By building in a solar panel to harvest available sunlight (even indoors through a window), then seamlessly integrating the two power sources so you don't have to bother with switching back and forth between them. The patented clockwork mechanism transforms twenty seconds of human effort into 45 minutes of AM/FM reception, even without solar input. Comes in black or a clear plastic housing that lets you see it all working. Cool. 11″ L x 8″ W x 8″ H; weighs 5.5 lb. South Africa/China.

50-237 Freeplay 2 Radio—Black $79⁹⁵

50-245 Freeplay 2 Radio—Blue/Clear $79⁹⁵

Lightweight Personal-Sized Freeplay

Millions of people bought the full-size Freeplay for emergency back-up, and were pleasantly surprised by the excellent sound quality and reception, and ease of use. Then we started noticing the weight. Here's the solution—a compact version just the right size to tote along for everyday use. Weighs in at 2.2 pounds, measures 8.2″ L x 2.9″ D x 4″ H. This model produces 40 minutes of play time with 20 seconds of winding, or by using its solar panel in direct sunlight, the radio has unlimited playtime. Kids love winding it up and watching it work through the translucent case. It's a winner! South Africa/China.

50-244 Mini Freeplay Radio (Blue) $69⁹⁵

Shortwave AM/FM Radio With LED Light

The revolutionary Freeplay radio has always been a solid, reliable choice for camping, power outage, or emergency use. With the addition of a freestanding 3-bulb white LED light, it's more practical than ever. Includes a 12-foot retractable cord, so you can move the light where you need it (under the hood? across the tent?) without schlepping the power source. The built-in clockwork generator powers the LED without a huge reduction in run time. Twenty seconds of wind-up effort yields 30 minutes of light and radio (or 45 minutes of radio only). Now that's efficient. LED light lasts about 100,000 hours. The Freeplay delivers clear, reliable AM/FM and shortwave reception, and features telescoping antenna, flip-out easy-wind handle, recessed controls, and DC and headphone jacks (headphones increase play time substantially). Measures 13.5″ L x 5.5″ W x 10″ H. South Africa/USA.

50-250 Freeplay Shortwave Radio w/Light $149⁰⁰

50-239 Antenna Booster $19⁹⁵

11-251 Solar Option $24⁹⁵

AM/FM Radio Recharges by Hand or in the Sun

Turn the Dynamo AM/FM receiver's crank for just one minute to hear ten minutes of music, news, or sports. The radio contains a miniature dynamo and a small solar panel that charges the built-in nicads. So no household current is needed, no batteries that die at the beach. The back-up battery compartment holds two AA nicads (which are recharged separately) for rainy weather. Flashlight on end can be used with a white lens cover as a steady beam, or with a red lens cover as an emergency blinker. Hong Kong.

90-319 Dynamo Radio $29⁹⁵

50-125 AA Nicad (2-pack) $4⁵⁰ each

Rugged Solar Power For Laptops And Portable Electronics

So, you want to go outside and play but have nowhere to plug in power for your laptop or cell phone—a common predicament of the '90s and Gaiam Real Goods has the solution. This portable energy system delivers charging power for just about anything; from camcorders to boom boxes. Practically unbreakable, its solar cells provide all the energy you'll need, delivering 3 to 18 volts of electricity at 9.9 watts. Prop it up to soak in the sun's energy with its adjustable foldout stand. Connection is easy: it comes with a 9-foot extension cord, two computer cables, and a multi-plug pinwheel cable. And if the supplied cables don't fit, the manufacturer will replace them with the correct ones free of charge. Comes in a zippered nylon storage bag; instructions are included and there's a one-year mfr's. warranty. USA.

11-164 Portable Energy System $449⁰⁰

Laptop Power Adapters

Run Your Laptop from PV, Cigarette Socket, or Airline Power

Laptops come in a huge variety of different input voltage and power plug configurations. Even within a single brand there is rarely a standard. Here is the gizmo to standardize them. On the input side, our adapter either plugs into a standard lighter socket, or, with a separate cord, into an airline power socket. It accepts DC input over a wide voltage range, including PV modules. On the output side, the adapter provides a matching power plug, and carefully emulates your laptop's standard charger, while providing additional high- and low-voltage protection, and spike/noise protection. Output power will always be clean and safe. The adapter is in a rugged housing, and sealed from moisture.

These adapters will accept direct PV input from a 10-watt or larger module. Some laptops will recharge with as little as 10 watts; most will require 30 to 50 watts of PV input. The module needs a female lighter socket for output.

Due to the wide variety of output voltages and power plug types, adapters are drop shipped directly from the manufacturer. We can supply adapters for most current-production laptops. Call if there's any doubt. Delivery time is two weeks or less. Size and weight may vary slightly with laptop model, but averages 4″ x 2.5″ x 1.25″, weighs 11 oz., supplied with nylon bag. Three-year mfr.'s warranty. USA.

◈ **50-251 Laptop Power Adapter (specify laptop model) $99⁹⁵**

◈ *Means shipped from manufacturer.*

Small Packaged Systems

Portable Power Any Place In The World

Rafting the Colorado? Hiking the Himalayas? Braving the wilds of Central Park? Take along our deluxe Wattz-In-A-Box portable power station, a practically indestructible travel case outfitted with a reliable solid-state DC power supply (up to 12 volts), 13W PL fluorescent gooseneck camp light, a Statpower 150-watt inverter for AC applications, and plenty of room for your laptop or video camera. The case itself is airtight, watertight, and unbreakable, with a foam-padded, custom-fit interior. The Multi Power Supply unit contains a sealed 9Ah battery with multiple mini-pin, cigarette lighter, and large terminal outputs. It will power up your laptop, cell phone, video camera, or camp lights on demand. So how do you recharge your power supply in the Serengeti? Use the cigarette lighter in your Land Rover, the AC outlet at the nearest hotel, or add the optional 10-watt PV module for solar charging. Multi-step LED indicators monitor battery and charge levels, and the unit shuts off automatically when full. The Multi Power Supply has a built-in high-powered flashlight. PV module sold separately. Taiwan and USA.

27-200 Wattz-In-A-Box $395⁰⁰

11-252 Solar Charger Option $119⁰⁰

Plug-And-Go Power For People On The Move

Weighing in at 15.5 lb., with a molded carry handle and nylon shoulder strap, this infinitely portable 12-volt power source tucks into your camper, cabin, or trunk for reliable or emergency power in any setting. It has multiple mini-pin, cigarette lighter, and large terminal outputs, and will accommodate any of the lamp, fan, or voltage supply options listed below. Can recharge from a standard AC outlet, or from an automotive cigarette lighter plug. Add a PV module for solar recharging. The built-in charge controller will handle up to 50 watts, and protect the 17Ah sealed battery from overcharging. Multi-step LED gauges monitor charging and battery capacity, and a 25-amp fuse protects everything (spare fuse included). A low-voltage circuit will shut off all outputs if the battery gets too low, and play an annoying electronic tune (sorry no choice of tune). The built-in high-powered flashlight runs over 20 hours on a full charge. Battery is designed for replacement with minor technical skill. 13″ x 8.6″ x 4.0″. Taiwan.

27-182 Double-Size Multi Power Source $149⁰⁰

15-209 Repl. 17Ah Sealed Battery $49⁰⁰

11-252 10W Solar Panel Option $119⁰⁰

Multi Power Supply

All the features, plugs, switches, lights, straps, handles, and fuses of our Double-Size Multi Power Supply above, but with half the battery capacity and weight. (Has a single 9Ah sealed battery.) It works with all Multi Power Supply accessories, and will fit inside the Wattz-In-A-Box package. 10.5″ x 7.5″ x 4.6″. 9.75 lbs. Taiwan.

27-195 Multi Power Supply $119⁰⁰

15-200 Repl. Sealed Battery $27⁹⁵

13-Watt PL Lamp Option

Included with our Wattz-In-A-Box power suite, this energy-efficient, compact fluorescent, gooseneck lamp option gives the equivalent of 60 watts incandescent light for car repairs, base-camp card games, or emergency first aid. It mounts securely to the top of the Multi Power Supply unit. Taiwan.

25-425 13W PL Lamp $29⁹⁵

8-Step Voltage Supply Adds Flexibility

Included with our Wattz-In-A-Box power suite, this reliable DC regulator clips to the Multi Power Supply. It can directly run almost any battery-operated appliance with a power input plug. Includes a 4-way adapter cord. It provides stable voltage at any power level, and cannot be damaged by short circuit. Solid-state regulated voltage at 1.5V, 2V, 3V, 4.5V, 6V, 7.5V, 9V, and 12V, with a 3-amp output limit. Taiwan.

25-366 8-Step Voltage Supply $29⁹⁵

Oscillating Fan Option

An 8″ personal fan that mounts securely in the top of the Multi Power Supply unit. Has 180-degree oscillation and, if plugged into 8-step voltage supply above, features 8-step speed control. Taiwan.

64-210 Oscillating Fan $24⁹⁵

High-Intensity Spotlight Option

Keep the hyenas at bay, set up your field camp after dark, or light your way to a midnight swimming hole. This powerful beam gives you brilliant high-intensity illumination for ¼ mile. The handheld spotlight plugs directly into the Multi Power Supply unit. Taiwan.

25-426 High-Intensity Spotlight $19⁹⁵

A Higher Capacity Portable Power System

Looking for something bigger than Wattz-In-A-Box? The Sunwize Portable Power Generator boasts ten times the capacity of the Multi Power Supplies on the previous page. The Portable Power Generator can supply lighting, or run entertainment equipment at your occasional-use cabin, or on your boat. Also ideal for emergency use and disaster relief, the PPG is housed in a rugged weathertight Rubbermaid box (21″ x 15″ x 12″). Already prewired are a 15-amp PV controller, a 150-watt Statpower Prowatt inverter for AC output, a pair of lighter plugs for DC output, an externally mounted DC voltmeter "fuel gauge," an input plug for PV module(s), and all the fusing to keep everything safe. All you need to add is the PV module(s) or other charging source of choice, and a 98Ah (or smaller) battery. All components except the battery and PV module(s) are prewired and ready to rock. The Sunwize PPG weighs 9 lb. as delivered, or 81 lb. with the maximum-capacity sealed battery listed below. One year mfr. warranty. USA.

PV Module and Battery not included.

12-107 Sunwize Portable Power Generator $599⁰⁰

15-214 98AH/12V Sealed Battery $169⁰⁰

The SolProvider

Solarize Your RV, Cabin, Tipi, Boat, or Remote Home with Our User-Friendly Real Goods SolProvider Renewable Energy Center

Our techs have already done the hard part: selecting fully compatible components and state-of-the-art safety equipment, completing the most difficult wiring, clearly labeling all landing points and circuit breakers, and pulling it all together in a comprehensive and genuinely useful manual that details every aspect of installation, use, and maintenance. Pre-wired, code-compliant, and easy to install—just add battery(ies), PV module(s), and a wall for mounting.

The system is a great jumping off point, with plenty of room for expansion and customizing. The SolProvider's 6-circuit load center delivers reliable 12-volt DC power for lighting, fans, low-voltage water pump, or entertainment equipment. Add an inverter to run your personal computer, printer, TV, or even a modest household (call our technical staff for help selecting the inverter specific to your intended use). The SolProvider/PV Package adds a Uni-Solar US64 module and rooftop RV mount to the basic package. Don't know how much PV you need? Call our techs.

SolProvider includes the Trace C-12 charge controller, charge control for up to 12 amps of PV input (up to 3 US64s), and automatic disconnect of selected loads to prevent overly discharging the battery, plus two DC lighter plug sockets. Landing points for field-installed wiring from battery. Everything is mounted on a 16″ x 18″ Meadowood™ board, a chipboard-like material made from straw. With the SolProvider and a little solar savvy, you'll get the energy self-sufficiency you've been looking for. Five-year warranty on Power Board and 20-year warranty on PV. USA.

27-180 Real Goods SolProvider $395⁰⁰

27-183 SolProvider/PV Package $795⁰⁰

Gaiam Real Goods Emergency Power Systems

Concerned about natural (or unnatural) disasters that could leave you without electric power for days, weeks, or even longer?

There's little doubt that humankind is tampering with the world's ecosystems. Just look at the environmental upsets we've been experiencing: strange weather patterns, bigger storms, and more violent and frequent hurricanes. Many people are left feeling deeply concerned about future power reliability. Increasing numbers of folks have been turning to the experienced Gaiam Real Goods techs for help. Loss of power doesn't have to mean huddling in the cold and dark, or listening to the constant nerve-jangling roar of a generator. Gaiam Real Goods can deliver clean, reliable, quiet battery-power systems, supplying anywhere from a few hours to a lifetime's worth of power. Your battery system can be recharged by solar power, by the grid (when available), or by a backup generator running just a few hours per day.

Our basic Emergency Power Kits provide an inverter with built-in battery charger, transfer switch, connection cables, and necessary fusing protection. These inverter/chargers will float charge the batteries and pass utility or generator power to their output terminals, so long as incoming AC power is available. If incoming power fails, battery power will automatically be used to supply any AC demand on the output. These kits can provide limited amounts of 120VAC power for essentials. Our estimated days of backup are based on extremely conservative energy use, and battery bank sizing, not inverter sizing. The inverter size determines how much energy can be taken at once, the battery size determines how much total energy can be taken before recharging. Want more days of storage? Add more batteries.

These basic kits do not include batteries or the assembly skills to put everything together on site. Hire a professional if you don't feel comfortable doing electrical work. None of these kits will support air conditioning or electric heater elements such as electric water heaters, electric stoves or ovens, electric clothes dryers, or room heaters. These are emergency kits only.

Emergency Power Units

The 600-Watt Emergency Power Kit

Our smallest Emergency Power Kit can provide you with emergency 120VAC power. Lights, refrigerators, TVs, and computers are some examples of appliances it will power, although you won't be able to run everything at the same time. Heavy power users like water pumps, power tools, or washing machines won't be possible with this smaller inverter. Adding more battery capacity, a back-up generator, or PV power will increase run times.

This emergency power system is based around the new Trace UX612 inverter, which will surge up to 2,500 watts to start a refrigerator or other small motor, and will deliver up to 600 watts continuously. If overloaded, it will shut off automatically without damage. It has a built-in 25-amp battery charger that will activate automatically if supplied with utility or generator power.

The recommended battery pack of 220 amp-hours will store approximately 2 kWh of electricity. With very conservative use this could provide basic lights and refrigeration for about one full day. More battery capacity will allow longer run times, or heavier use. You may choose either conventional wet-cell batteries, or maintenance-free gel-cell batteries. Our kit includes appropriate battery interconnect cables for battery choices listed below, DC fusing, and inverter supply cabling. All components made in the USA.

Batteries not included.

12-206 600-Watt Emergency Power Kit $779⁰⁰

🚚 **15-101A 12V x 220Ah**
        (wet-cell) batteries (2)* $180⁰⁰

🚚 **15-216 (1 recommended) 12V x 225Ah Gel Cell**
        (sealed) battery $385⁰⁰

The 1,500-Watt Emergency Power Kit

Our smaller middle-sized Emergency Power Kit features a 1,500 watt inverter that can provide limited-use, emergency 120VAC power. Lights, refrigerators, TVs, computers, and handheld power tools are some typical examples of appliances this system can power (but not all at once). It probably won't power water pumps, and will have trouble with some washing machines.

This emergency power system is based around the Trace DR1512 inverter, which will surge up to 3,200 watts to start motors and tools, and deliver up to 1,500 watts continuously. If overloaded it will shut off automatically without damage. It has a built-in 70-amp battery charger that will activate automatically if supplied with utility or generator power.

The recommended battery pack of 440 amp-hours will store approximately 4 kWh of electricity. With very conservative use this could provide basic lights and refrigeration for about two days. More battery capacity will allow longer run times or heavier use. You may choose either conventional wet-cell batteries, or maintenance-free gel-cell batteries. Our kit includes appropriate battery interconnect cables for battery choices listed below, DC fusing, and inverter supply cabling. All components made in the USA.

Batteries not included.

12-207 1,500-Watt Emergency Power Kit $1,449⁰⁰

15-101B 24V x 220Ah (wet-cell) batteries (4)* $360⁰⁰

15-216 (2 recommended) 12V x 225Ah gel-cell (sealed) batteries $385⁰⁰

The 2,400-Watt Emergency Power Kit

Our larger middle-sized Emergency Power Kit features a 2,400-watt inverter that can provide limited-use, emergency 120VAC power. Lights, refrigerators, TVs, computers, and handheld power tools are some typical examples of appliances this system can power (but not all at once). It may be able to power 120-volt water pumps, and is unlikely to have trouble with most washing machines.

This emergency power system is based around the Trace DR2424 inverter, which will surge up to 7,000 watts to start motors and tools, and deliver up to 2,400 watts continuously. If overloaded it will shut off automatically without damage. It has a built-in 70-amp battery charger that will activate automatically if supplied with utility or generator power.

The recommended battery pack of 880 amp-hours will store approximately 8 kWh of electricity. With very conservative use, this could provide basic lights and refrigeration for about five days. More battery capacity will allow longer run times or heavier use. You may choose either conventional wet-cell batteries, or maintenance-free gel-cell batteries. Our kit includes appropriate battery interconnect cables for battery choices listed below, DC fusing, and inverter supply cabling. All components made in the USA.

Batteries not included.

12-211 2,400-Watt Emergency Power Kit $1,895⁰⁰

15-101C 24V x 880Ah Lead Acid (wet-cell) Batteries(8) $720⁰⁰

15-216 225 Ah/12V Gel Cell (Sealed) Battery (4 recommended) $385⁰⁰

Notes on Emergency Power Kits

1. These kits contain an inverter, battery cables, and safety equipment *only.*
2. Battery(ies) are sold separately.
3. These kits require electrical and assembly skills. Most folks should hire a pro.
4. The customer must provide a way to recharge. (Generator, solar, etc.)
5. Output power (wattage) and output time (watt-hours) are limited.
6. These kits won't run large appliances such as air conditioners, electric water heaters, electric stoves or ovens, electric room heaters, dryers, etc.
7. A legal installation will require a large, expensive transfer switch to prevent backfeeding the utility.

The 3,600-Watt Emergency Power Kit

Our largest Emergency Power Kit features a larger inverter and battery bank that should be able to start and run almost any 120VAC appliance, including smaller water pumps and power tools. Of course, you still can't run everything at once. With careful energy conservation, it can provide basic lights, refrigeration, and entertainment equipment for several days.

This emergency power system is based around the Trace DR3624 inverter, which will surge up to 10,000 watts, and deliver up to 3,600 watts continuously. It has a built-in 70-amp battery charger that will activate automatically if supplied with utility or generator power.

The recommended battery pack of 880 amp-hours will store approximately 8 kWh of electricity. With very conservative use, this could provide basic lights and refrigeration for about five days. More battery capacity will allow longer run times, or heavier use. You may choose either conventional wet-cell batteries, or maintenance-free gel-cell batteries. Our kit includes appropriate battery interconnect cables for battery choices listed below, DC fusing, and inverter supply cabling. All components made in the USA.

Batteries not included.

12-208 3,600-watt Emergency Power Kit $1,995⁰⁰

15-101C 24V x 880Ah lead acid (wet-cell) (8) batteries* $720⁰⁰

15-216 (4 recommended) 225Ah x 12V gel cell (sealed) battery $385⁰⁰

*Arrangements can be made for local pick-up of wet-cell batteries in most parts of the U.S. Please call a technician to make arrangements in your location. Sealed gel-cell batteries, although heavy, can be shipped via truck to almost anyplace.

Means shipped freight collect.

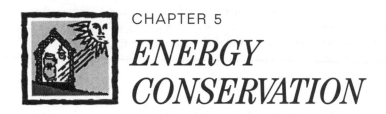

CHAPTER 5
ENERGY CONSERVATION

Living Well, but Inexpensively

Never doubt that a small group of thoughtful committed citizens can change the world. Indeed, it's the only thing that ever has.

Margaret Mead

IMAGINE A PAIR OF SIMILAR SUBURBAN HOMES on a quiet residential street. Both house a family of four living the American suburban lifestyle. The homes appear to be identical, yet one spends under $50 per month on utilities, and the other over $400. It can't be? Ah, but it can, and often is! This is the dramatic savings that careful building design, landscaping, and selection of energy-conserving appliances can make possible without affecting basic lifestyle.

Our example isn't based on some bizarre construction technique, or appliances that can be operated only by rocket scientists, but on simple, commonsense building enhancements and off-the-shelf appliances. More importantly, since average Americans spend 80% of their time inside, improvements in energy efficiency and livability through better design and retrofitting enhance our quality of life, save money, and clean up the environment. It's a win-win-win!

We will provide an overview of energy conservation techniques and ideas. Rather than exploring one single issue in depth, the chapter is designed to point out possibilities, and provide the resources you need for further research into the concepts and products best-suited to your particular lifestyle. Because heating and cooling are the single largest energy consumers for the average home, we've included those subjects here as well.

To be convinced of the need for energy conservation, one need look no further than global warming, rapid depletion of resources, toxic waste production, and burgeoning populations in developing countries. While we all care about these issues, they are distant and intellectual; too enormous for our individual contributions to have a meaningful impact. But what most of us find immediately compelling is that **energy conservation will save us money; right now, right here.** Everything else is icing on the cake. In fact, studies have shown that *no investment* pays as well as conservation. Banks, mutual funds, real estate investments . . . none of these options will bring the 100% to 300% returns that are achievable through simple, inexpensive conservation measures. Not even our own dearly beloved renewable energy systems will repay your investment as quickly as conservation.

Energy conservation is a broad and multi-faceted subject without a lot of sizzle or sex appeal. This chapter can only do justice to a small portion, and hopefully point you in the right direction for more information. At Gaiam Real Goods we are experts in energy conservation. We have to be! After 23 years designing renewable electrical systems for remote locations, we have learned to squeeze the maximum work out of every precious watt. Although our focus is on energy generation through solar modules, or wind and hydro-electric generators, we've found we often have to backtrack a bit to basic building design, or retrofitting of existing buildings before we start selecting appliances. We'll tackle the broad, multi-faceted subject of energy conservation by looking first at **Building Design,** followed by **Retrofitting an Existing Building,** and finally, **Selecting Appliances.** *Feel free to skip sections that do not apply to your current needs.*

1. It All Starts with Good Design

The single most important factor that affects energy consumption in your house is design. All the intelligent appliance selection or retrofitting in the world won't keep an Atlanta, Georgia, family room with a four-by-six-foot skylight and a west-facing eight-foot sliding glass door from overheating every summer afternoon and consuming massive amounts of air conditioning energy. Intelligent solar design is the best place to start. Passive solar design works compatibly with your local climate conditions, using the seasonal sun angles at your latitude to create an interior environment that is warm in winter and cool in summer, without the addition of large amounts of energy for heating and cooling.

Passive solar buildings cost no more to design or build than "energy-hog" buildings, yet cost only a fraction as much to live in. You don't have to build a totally solar-pow-ered house to take advantage of some passive solar savings. Just a few simple measures, mostly invisible to your neighbors, can substantially improve your home's energy performance!

Minimize Your Summer Exposure, Maximize Your Winter Exposure

In all temperate latitudes, houses perform best from a solar perspective if they are elon-gated along the east-west axis and thinner in the north-south axis. This gives maximum southern wall exposure to the low winter sun, and minimum east and west wall expo-sure to the high summer sun. A properly sized roof overhang or awning can be very effective in further protecting the southern wall in summer. Minimize north, east, and particular-ly west-facing window areas. South-facing windows can be a very good thing, *if* there is sufficient roof, awning, or other device over-hang to shade these windows from the high summer sun, but not so much as to block the low winter sun. (People south of the equator, please make appropriate direction adjustments.)

Take a look at the Solar Radiation chart on the following page. It shows heat gain in Btu per day per square foot of glass, based on which direction that glass is facing. They all peak in the summer . . . except for south-facing! South-facing windows show a big drop in Btu gain in the summer as the sun climbs higher in the sky. In the winter, as the sun drops lower in the sky, south-facing windows show a big increase in heat gain. Mother Nature is trying to tell us something very important and very simple here. We would all enjoy a home that's warm in the winter and cool in the summer with a minimum of energy input. It would be more comfortable and cost less to live in. Now, based on what you see in the Solar Radiation chart, which direction do you want the majority of your dream home's windows facing?

In the not-very-distant future, as the worldwide petroleum party starts petering out due to diminishing resources and energy costs continue skyrocketing, those folks farsighted or lucky enough to own a passive solar home are going to be feeling very, very smug.

House orientation affects heating and cooling costs.

Skylights, a Bright Idea?

Looking at the Btu/day chart again, note what happens with horizontal windows—
skylights. Hm-m-m. A huge peak in the summertime when we least want it.

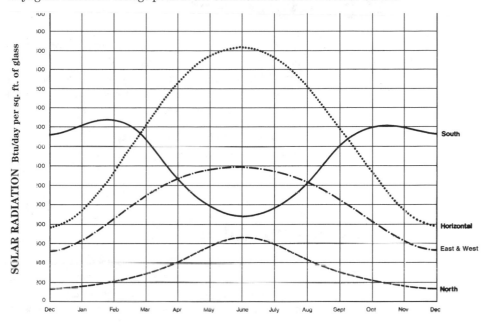

Clear-day solar radiation values, on the surfaces indicated, for 40° north latitude.

Maybe that skylight isn't such a great idea after all. The natural daylighting of skylights can transform a room magically, but some of the things that accompany skylights
we can do without. High installation costs, condensation and water staining, heat gain in
the summer, heat loss in the winter, and fading carpets and upholstery are some of the
less desirable side effects of skylights.

A newer type of skylight, using mirror-coated tubular sections, corrects these problems. Initial cost is modest, a do-it-yourselfer can install it in less than two hours, and
there are no framing cuts. These units are sealed top and bottom so there is no heat loss
or gain, no condensation, and they maintain the daylighting benefits of traditional skylights. Having delightfully retrofitted three of them, we can recommend them highly with
confidence. Gaiam Real Goods offers reflective tube skylights in several sizes; see the
product listings (page 308).

Regional Designs

Good design varies according to local climate. A house that works exceedingly well in
the high desert of New Mexico might be a flop in Florida or New Hampshire. Each cli-
mactic area has classic house designs that have evolved because they work so well. Hint:
in none of the climate zones above is a typical suburban ranch house the best choice.
This design has been created to satisfy priorities that have energy efficiency very low
on the list. Unfortunately it is a design that has been too successful for our own good.

More Passive Solar Information

If you are in the enviable position of contemplating new construction, we highly
recommend a little passive solar study. See the Shelter chapter (page 7) for discussions
on good passive solar design, appropriate building technologies, and a listing of the best
books currently in print. Intelligent shelter and passive solar design is too large a subject
to tackle here; we have co-published a complete book on the subject, *The Passive Solar
House.* **(Our item #80-372, $25.)**

2. Retrofitting: Making the Best of What You've Got

If you're like many of us, you already have a house, have no intention of building a new one, and just want to make it as comfortable and economical as possible. By retrofitting your home with energy savings in mind you can lower operating costs, boost efficiency, improve comfort levels, save money, and have a good feeling about what you've accomplished.

Weatherization & Insulation

Increasing insulation levels and plugging air leaks are favorite retrofitting pastimes, which produce well-documented, rapid paybacks in comfort and utility savings. Short of printing your own money, weatherization and insulation are the best bets for putting cash in your wallet and are a lot safer in the long run.

Weatherization

Weatherization, the plugging and sealing of air leaks, can save 25% to 40% of heating and cooling bills. The average unweatherized house in the U.S. leaks air at a rate equivalent to a four-foot-square hole in the wall. Weatherization is the first place for the average homeowner to concentrate his/her efforts. You'll get the most benefit for the least effort and expense.

Where to Weatherize

The following suggestions are adapted from Homemade Money *by Richard Heede and the Rocky Mountain Institute. Used with permission.*

Here's a basic checklist to help you get started. Weatherization points are keyed in parentheses to the illustration on the next page.

1. In the attic

• Weatherstrip and insulate the attic access door (6).

• Seal around the outside of the chimney with metal flashing and high-temperature sealant such as flue caulk or muffler cement (3).

• Seal around plumbing vents, both in the attic floor, and in the roof. Check roof flashings (where the plumbing vent pipes pass through the roof) for signs of water leakage while you're peering at the underside of the roof (8).

• Seal the top of interior walls in pre-1950s houses anywhere you can peer down into the wall cavity. Use strips of rigid insulation, and seal the edges with silicone caulk (7).

• Stuff fiberglass insulation around electrical wire penetrations at the top of interior walls and where wires enter ceiling fixtures. (But not around recessed light fixtures unless the fixtures are rated IC [for insulation contact]). Fluorescent fixtures are usually safe to insulate around; they don't produce a lot of waste heat. Incandescent fixtures should be upgraded to CF lamps, but that's another story) (2, 4).

• Seal all other holes between the heated space and the attic (9).

Each year in the U.S. about $13 billion worth of energy, in the form of heated or cooled air—or about $150 per household—escapes through holes and cracks in residential buildings.

American Council for an Energy-Efficient Economy

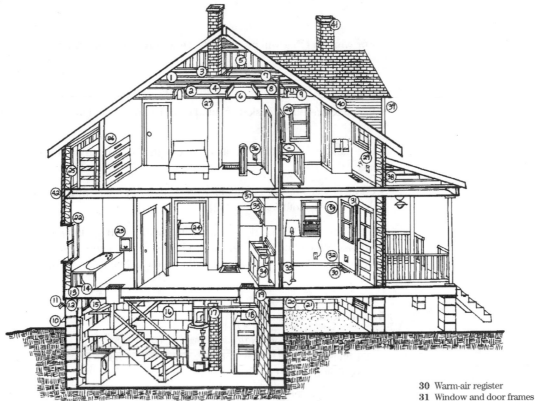

Attic:
1 Dropped ceiling
2 Recessed light
3 Chimney chase
4 Electric wires and box
5 Balloon wall
6 Attic entrance
7 Partition wall top plate
8 Plumbing vent chase
9 Exhaust fan

Basement and crawlspace:
10 Dryer vent
11 Plumbing/ utility penetrations
12 Sill plate
13 Rim joist
14 Bathtub drain penetration
15 Basement windows and doors
16 Block wall cavities
17 Water heater and furnace flues
18 Warm-air ducts

19 Plumbing chase
20 Basement/crawlspace framing
21 Floorboards

Living area:
22 Window sashes and doors
23 Laundry chute
24 Stairwell
25 Kneewall/framing intersection
26 Built-in dresser
27 Chimney penetration
28 Built-in cabinet
29 Cracks in drywall

30 Warm-air register
31 Window and door frames
32 Baseboards, coves, interior trim
33 Plumbing access panel
34 Sink drain penetration
35 Dropped soffit
36 Electrical outlets and switches
37 Light fixture

Exterior:
38 Porch framing intersection
39 Missing siding and trim
40 Additions, dormers, overhangs
41 Unused chimney
42 Floor joist

2. In the basement or crawlspace

• Seal and insulate around any accessible heating or A/C ducts. This applies to both the basement and attic (18).

• Seal any holes that allow air to rise from the basement or crawlspace directly into the living space above. Check around plumbing, chimney, and electrical penetrations (11, 14, 19, 21).

• Caulk around basement window frames (15).

• Seal holes in the foundation wall as well as gaps between the concrete foundation and the wood structure (at the sill plate and rim joist). Use caulk or foam sealant (12, 13).

3. Around windows and doors

• Replace broken glass and reputty loose panes. See the Window section about upgrading to better windows, or retrofitting yours (22).

• Install new sash locks, or adjust existing ones on double-hung and slider windows (22).

• Caulk on the inside around window and door trim, sealing where the frame meets the wall and all other window woodwork joints (31).

• Weatherstrip exterior doors, including those to garages and porches (31).

• For windows that will be opened, use weatherstripping or temporary flexible rope caulk.

4. In living areas

• Install foam-rubber gaskets behind electrical outlet and switch trim plates on exterior walls (36).

• Use paintable or colored caulk around bath and kitchen cabinets on exterior walls (26, 35).

• Caulk any cracks where the floor meets exterior walls. Such cracks are often hidden behind the edge of the carpet (32).

• Got a fireplace? If you don't use it, plug the flue with an inflatable plug, or install a rigid insulation plug. If you do use it, make sure the damper closes tightly when a fire isn't burning. See the section on Heating and Cooling for more tips on fireplace efficiency (41).

5. On the exterior

• Caulk around all penetrations where electrical, telephone, cable, gas, dryer vents, and water lines enter the house. You may want to stuff some fiberglass insulation in the larger gaps first (9, 10, 11).

• Caulk around all sides of window and door frames to keep out the rain and reduce air infiltration.

• Check your dryer exhaust vent hood. If it's missing the flapper, or it doesn't close by itself, replace it with a tight-fitting model (10).

• Remove window air conditioners in winter; or at least cover them tightly, and make rigid insulation covers for the flimsy side panels.

• Caulk cracks in overhangs of cantilevered bays and chimney chases (27, 40).

Insulation

Many existing homes are woefully underinsulated. Exterior wall cavities or underfloors are often completely ignored, and attic insulation levels are sometimes more in tune with the 1950s when heating oil was 12¢ per gallon than the turn of the century, when oil prices are more than ten times higher, and still escalating rapidly. Adequate insulation rewards you with a building that is warmer in the winter, cooler in the summer, and much less expensive to operate year-round.

Recommended insulation levels vary according to climate and geography. Your friendly, local building department can give you the locally mandated standards, but for most of North America you want a minimum of R-11 in floors, R-19 in walls, and R-30 in ceilings. These levels will be higher in more northern climes, and lower in more southern climes.

Insulate-It-Yourself?

Some retrofit insulation jobs, like hard-to-reach wall cavities, are best done by professionals who have the spiffy special tools and experience to do the job quickly, correctly, and with minimal disruption. There are insulation and weatherproofing companies that will inspect, advise, and give an estimate to bring your home up to an acceptable standard.

FORM	TYPE	R-VALUE (approx.)	RELATIVE COST (per R per sq. ft.)
Batts or Blankets	Fiberglass or Rock Wool	3½" (R-11) 5½" (R-19) 9½" (R-30)	1.8¢ to 2.0¢
Loose Fill	Fiberglass or Rock Wool	R-2.7 per inch	1.8¢ to 2.0¢
	Cellulose	R-3.7 per inch	1.6¢ to 1.8¢
Rigid Board	Expanded Polystyrene (Beadboard)	R-4 per inch	3.6¢ to 4.8¢
	Extruded Polystyrene	R-5 per inch	4.8¢ to 7.2¢
	Polyurethane or Polyisocyanurate	R-7 to R-8 per inch	4.8¢ to 6.0¢
Foamed in Place	Polyurethane	R-7 to R-8 per inch	8.4¢ to 12.0¢

Common Insulation Materials and Their Relative Cost

Illustration courtesy of New Mexico State University Cooperative Extension Service, New Mexico Home Energy Guide, p. 7, Las Cruces, NM.

R-value is a measure of the resistance of a material to heat flow. The higher the number, the greater the resistance.

Easier to reach spaces, like attics, are improvable by the do-it-yourselfer with either fiberglass batt insulation, or with easier- and quicker-to-install blown-in cellulose insulation. Many lumber and building supply centers will rent or loan the blowing equipment (and instructions!) when you buy the insulation. Some folks may choose to leave this potentially messy job to the pros.

Underfloor areas are usually easy to access, but working overhead with fiberglass insulation is no picnic. You might consider a radiant barrier product for underfloor use because of the ease of installation. A radiant barrier stapled on the bottom of floor joists performs equal to R-14 insulation in the winter. Performance will be less in the summer keeping heat out (R-8 equivalent), but this isn't a big consideration in most climates.

How Radiant Barriers Work

Heat is transmitted three ways:

Conduction *is heat flow through a solid object. A frying pan uses heat conduction.*

Convection *is heat flow in a liquid or gas. The warmer liquid or gas rises and the cooler liquid or gas comes down. Floor furnaces heat the air, which then rises in the room.*

Radiation *is the transmission of heat without the use of matter. It follows line of sight and you can feel it from the sun and a woodstove. Radiant heat flow is the most common method of heat transmission.*

Radiant barriers use the reflective property of a thin aluminum film to stop (i.e., reflect) radiant heat flow. Gold works even better, but its use is limited mainly to spacecraft, for obvious reasons. A thin aluminum foil will reflect 97% of radiant heat. If the foil is heated, it will emit only 5% of the heat a dull black object of the same temperature would emit. For the foil to work as an effective radiant barrier there must be an airspace on at least one side of the foil. A vacuum is even better and is used in glass thermos bottles, where the double glass walls are aluminized and the space between them is evacuated. Without an airspace

— continued on next page

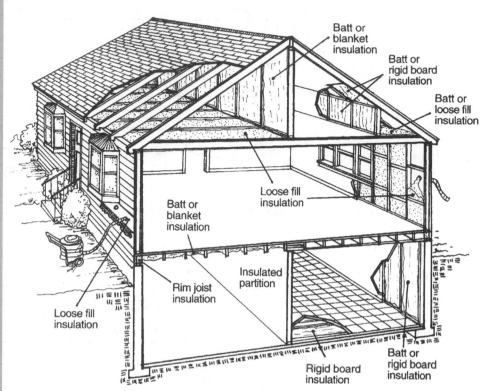

Where to Insulate

Illustration adapted form Reader's Digest (1982), Home Improvements Manual, p. 360, Pleasantville, NY.

Radiant Barriers— The Newest Wrinkle in Insulation

Over the past few years a new type of insulation called "radiant barrier" has become popular. Radiant barriers work differently from traditional "dead-air space" insulation. See sidebar.

A radiant barrier will stop about 97% of radiant heat transfer. There is no cumulative advantage to multiple barrier layers; it practically all stops at the first layer. There is no effect on conducted or convected heat transfers, so radiant barriers are usually best employed in conjunction with conventional "dead-air space" types of conventional insulation.

R-values can't be assigned to radiant barriers because they're only effective with the most common radiant transfer of heat. But for the sake of having a common understanding, the performance of a radiant barrier is often expressed in R-value equivalents. For instance, when heat is trying to transfer downwards, like from the underside of your roof into your attic in the summer, a radiant barrier tacked to the underside of your rafters will perform equivalent to R-14 insulation. When that same heat is trying to transfer back out of your attic at night, however, the radiant barrier only performs equivalent to R-8, since it has no effect on convection and convection transfer is in the upward direction. Therefore, conventional insulation is your best bet for keeping warm in the winter when all the heat is trying to go up through convection. Radiant barrier

— continued from previous page

ordinary conduction takes place and there is no blocking effect.

A radiant barrier only blocks heat flow. It has no effect on conduction or convection. But then, so long as there's a small airspace, the only practical way that heat can be transmitted across it is by radiant transfer.

See our product section for a selection of perforated radiant barriers in various widths (page 260).

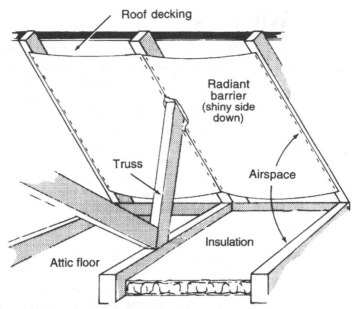

Typical Radiant Barrier Attic Installation

material is most effective when it's reflecting heat back upwards (the direction of convection). So, under rafters to keep out unwanted summer heat, and under floors to keep desirable winter heat inside your house are two of the best applications of this product.

The independent, nonprofit Florida Solar Energy Center has tested radiant barrier products extensively and has concluded that installation on the underside of the roof deck will reduce summer air conditioning costs by 10% to 20%. And that's in a climate where humidity is more of a problem than temperature. This means rapid payback for a product as inexpensive, clean, and easy to install as a radiant barrier. FSEC also paid close attention to roof temperatures, as there's a logical concern about where all that reflected heat is going. They found that shingle temperatures increased by 2° to 10°F. To a roof shingle, that's beneath notice, and will have no effect on roof lifespan.

We only offer perforated radiant barriers now, due to the moisture barrier properties of unperforated barriers causing occasional problems when misapplied.

Air-to-Air Heat Exchangers

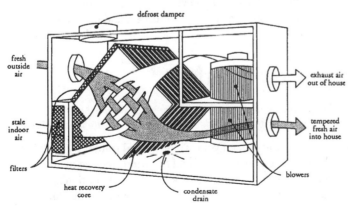

The Air-to-Air Heat Exchanger

A heat recovery ventilator's heat exchanger transfers 50–70% of the heat from the exhaust air to the intake air. Illustration courtesy of Montana Department of Natural Resources and Conservation

Building and appliance technologies have made tremendous gains in the past 20 years. Improvements in insulation, infiltration rates, and weather-stripping do much to keep our expensive warmed or cooled air inside the building. In fact we've gotten so good at making tight buildings and reducing infiltration of outside air that indoor air pollution has become a problem. A whole new class of household appliances has been developed to deal with the byproducts of normal living. Air-to-air heat exchangers take the moisture, radon, and chemically saturated indoor air and exchange it for outdoor air, while stripping up to 75% of the heating or cooling from the outbound air for energy efficiency. The power use of exchangers is minimal, and controls can be triggered by time, humidity, or usage. If you use gas for cooking, live in a moist climate, have radon or allergy problems, or simply live in a good, tight house, then air-to-air heat exchangers are worth considering.

Windows

Window technology is one of the fastest developing fields of building technology. And it's about time! Windows are the weakest link in any building's thermal barrier. Until a few years ago, a window was basically a hole in the wall that let light in and heat out. The R-value of a single-pane window is a miserable 1. Nearly half of all residential windows in the U.S. only provide this negligible insulation value. An "energy-saving" thermopane window is worth only about R-2. When you consider that the lower-cost wall around the window is insulated to at least R-12, and probably to over R-16, a window needs a solid justification for being there. In cold climates, windows are responsible for up to 25% of a home's winter heat loss. In warmer climates solar radiation, entering through improperly placed or shaded windows, can boost air-conditioning bills similarly.

Windows are the weakest link in any building's thermal barrier. Until a few years ago, a window was basically a hole in the wall that let light in and heat out.

Worth Retrofitting?

There have been major innovations in window construction recently. The most advanced superwindows now boast R-values up to 12. While superwindows only cost 20%–50% more than conventional double-pane windows, the payback period for such an upgrade in an existing house is 15–20 years due to installation costs. That's too long for most folks. But if you're planning to replace windows anyway due to remodeling or new construction, the new high-tech units are well worth the slightly higher initial cost, and will pay for themselves in just a few years.

Sneaky R-Values

When shopping for windows, be forewarned that until recently, most windows were rated on their "center of glass" R-value. Such ratings ignore the significant heat loss through the edge of the glass and the frame. Inexpensive aluminum-framed windows, which transmit large amounts of heat through the frame, benefit disproportionately from this dubious standard. A more meaningful measure is the newer "whole-unit" value. Make sure you're comparing apples to apples. Wood- or vinyl-framed units outperform similar metal-framed units, unless the metal frames are using "thermal break" construction, which insulates the outer frame from the inner frame.

Full frostbelt use of today's best windows would more than displace Alaska's entire oil output.

Amory Lovins,
The Rocky Mountain Institute

Start Cheap

With windows, as with all household energy-saving measures, start with jobs that cost the least, and yield the most. A commonsense combination of three or four of the ideas in the next section will result in substantial savings on your heating/cooling bill, enhanced comfort, minimized drafts, more constant temperatures, and more habitable areas near windows.

Cold Weather Window Solutions

The following suggestions (cold and warm weather window solutions) are adapted from Homemade Money *by Richard Heede and the Rocky Mountain Institute. Used with permission.*

First, stop the wind from blowing in and around your windows and frames by caulking and weather-stripping. After you've cut infiltration around the windows, the main challenge is to increase the insulating value of the window itself while continuing to admit solar radiation. Here are some suggestions for beefing up your existing windows in winter.

Install clear plastic barriers on the inside of windows

Such barriers work by creating an insulating dead-air space inside the window. After caulking, this is the least expensive temporary option to cut window heat loss. Such barriers can cut heat loss by 25% to 40%.

Repair and weatherize exterior storm windows

If you already own storm windows, just replace any broken glass, re-putty loose panes, install them each fall, and seal around the edges with rope caulk.

Add new exterior or interior storm windows

Storm windows are more expensive than temporary plastic options, but have the advantages of permanence, reusability, and better performance. Storm windows cost about $7.50 to $12.50 per square foot, and can reduce heat loss by 25% to 50%, depending on how well they seal around the edges. Exterior storm windows will increase the temperature of the inside window by as much as 30°F on a cold day, keeping you more comfortable.

Install tight-fitting insulating shades

These shades incorporate layers of insulating material, a radiant barrier, and a moisture-resistant layer to help prevent condensation. Several designs are available. One of my personal favorites is *Window Quilts*. This quilted-looking material consists of several layers of spun polyester and radiant barriers with a cloth outer cover. Depending on style, they fold or roll down over your windows at night providing a tight seal on all four sides, high R-value insulation, privacy, and soft quilted good looks. This allows your windows to have all the daytime advantages of daylighting and passive heat gain, while still enjoying the night time comfort of high R-values and no cold drafts. Because there is little standardization of window sizing, *Window Quilts* are generally custom cut to size, making them unsuitable for a mail-order retail operation like ours. You can order directly from the manufacturer, or they can direct you to a local retailer who can supply installation services. Call them at 800-257-4501.

Construct insulated pop-in panels or shutters

Rigid insulation can be cut to fit snugly into window openings, and a lightweight, decorative fabric can be glued to the inside. Pop-in panels aren't ideal, as they require storage whenever you want to look out the window, but they are cheap, simple, and highly effective. They are especially good for windows you wouldn't mind covering for the duration of the winter. Make sure they fit tightly so moisture doesn't enter the dead-air space and condense on the window.

Close your curtains or shades at night

The extra layers increase R-value, and you'll feel more comfortable not being exposed to the cold glass.

Open your curtains during the day

South-facing windows let in heat and light when the sun is shining. Removing outside screens for the winter on south windows can increase solar gain by 40%.

Clean solar gain windows

Keep those south-facing windows clean for better light and a lot more free heat. Be sure to keep those same windows dirty in the summer. (Just kidding!)

Low-E Films

The biggest news in window technology is "low-e" films, for low-emissivity. These thin metal coatings allow the shortwave radiation of solar energy to pass in, but block most of the long-wave thermal energy trying to get back out. A low-e treated window has less heat leakage in either direction. A low-e coating is virtually invisible from the inside, but most brands tend to give windows a semi-mirror appearance from the outside.

Low-e windows are available ready-made from the factory, where the thin plastic film with the metal coating is suspended between the glass panes; or, low-e films can be applied to existing windows.

We offer a do-it-yourself product that is permanently applied with soap and squeegee, or they can be applied professionally. These films offer the same heat reflecting performance as the factory-applied coatings, and are relatively modest in cost (especially compared to new windows!). They do require a bit of care in cleaning, as the plastic film can be scratched. Some utility companies offer rebates for after-market low-e films.

Warm Weather Window Solutions

The main source of heat gain through windows is solar gain—sunlight streaming in through single or dual glazing. Here are some tips for staying cool:

Install white window shades or miniblinds

It's a simple, old-fashioned practice. Since our grandparents didn't have air conditioners, they knew how to keep the heat out. Miniblinds can reduce solar heat gain by 40%–50%.

Close south and west-facing curtains

Do this during the day for any window that lets direct sunlight in. Keep these windows closed too.

Install awnings

Another good, old-fashioned solution. Awnings work best on south-facing windows where there's insufficient roof overhang to provide shade. Canvas awnings are more expensive than shades, but they're more pleasing to the eye, they stop the heat on the outside of your building, and they don't obstruct the view.

Hang tightly-woven screens or bamboo shades outside the window during the summer

They'll reduce your view, but are inexpensive and stop 60%–80% of the sun's heat from getting to the window.

Plant trees or build a trellis

Deciduous (leaf-bearing) trees planted to the south or particularly to the west of your building provide valuable shade. One mature shade tree can provide as much cooling as five air conditioners (although they're a bit difficult to transplant at that stage, so the sooner you plant the better). Deciduous trees block summer sun, but drop their leaves to allow half or more of the winter sun's energy to warm you on clear winter days.

Apply low-e films

Low-e films substantially reduce the amount of heat that passes through a window, with minimal effect on the amount of light passing through. The sidebar to the left explains how these films work. Don't use low-e films on, or built into, south-facing windows if you want solar gain! They can't tell the difference between winter and summer. See our product section for films that can be applied to existing windows.

Exotic infills

The other new technology commonly found in new windows is exotic infills. Instead of filling the space between panes with air, many windows are now available with argon, or krypton, exotic gas infills that have lower conductivity than air, and boost R-values. Krypton has a higher R-value, but costs more. These inert gases occur naturally in the atmosphere, and are harmless even if the window breaks. (Krypton-filled windows are also safe from Superman breakage.) All this technology adds up to windows with R-values in the 6 to 10 range.

Landscaping

When retrofitting existing buildings, most folks don't think about the impact of landscaping on energy use. Appropriate landscaping on the west and south sides of your house provides valuable summertime shading that will reduce unwanted heating as much as 50%. Those are better results than we get from more expensive projects like window and insulation upgrades! If your landscaping is deciduous, losing its leaves in winter to let the warming winter sun through, you have the best of both worlds. Landscaping can also block cold winter winds that push through the little cracks and crevices of a typical house, and make outdoor patio and yard spaces more livable and inviting.

3. Appliance Selection for Energy Conservation

Appliance selection is one area where we feel a greater sense of control. If you're renting or "financially challenged" (and who isn't?) you may not have control over your house design, window selection and orientation, or heating plant. But you *can* select the light bulb in your lamps and the showerhead in your bathroom. And you determine whether appliances get turned off (really off!) when not in use.

Many household appliances are important enough to justify entire chapters in the *Sourcebook*. Lighting, heating and cooling (with refrigeration), water pumping, water heating, composting toilets, and water purification are big topics that have significant impact on total home energy use. Please see the individual chapters or sections on the above products. Here, we are providing general information that applies to all appliances.

Don't use electricity to make heat

Avoid products that use electricity to produce heat when there's any choice. Making heat from electricity is like using bottled water for your lawn. It gets the job done just fine, but it's terribly expensive and wasteful. Heat-producing electrical appliances aren't just room heaters. This includes electric water heaters, electric ranges and ovens, hot plates and skillets, waffle irons, waterbed heaters, and the most common household electric heater . . . the incandescent lightbulb. Most of these appliances can be replaced by other appliances that cost far less to operate. Incandescent bulbs are one of the most dramatic examples. Standard lightbulbs only return 10% of the energy you feed them as visible light. The other 90% disappears as heat. Compact fluorescent lamps return better than 80% of their energy as visible light, and they last over 10 times as long per lamp.

If you have a choice about using gas or electricity for residential heating, water heating, or cooking, use gas, even if this means buying new appliances. For those lucky areas that have natural gas service, your monthly bills will be 50%–60% lower than with electricity. Even in areas that have to bring in bottled propane, gas will be 30%–40% cheaper than electric. The same goes for clothes drying. Better yet, use a zero-energy-cost clothesline, which has the side benefit of making your clothes last longer. Dryer lint is your clothes wearing out by tumbling.

Phantoms and Vampires!

Now what's that odd quote in the margin about televisions using power when "off"? Appliances that use power even when they're off create what are called phantom loads. Any device that uses a remote control is a phantom load because part of the circuitry must remain on in order to receive the "on" signal from the remote. For most TVs this power use is 15 to 25 watts. VCRs are typically 5 to 10 watts. Together that'll cost you over $20 per year. Either plug these little watt-burners into a switched outlet, or use a switched plug strip so they can really be turned off when not in use.

The other increasingly common villain to watch out for are the little transformer cubes that are showing up on the end of many small appliance power cords. Officially, in the electric industry these are known as "vampires" because they constantly suck juice out of your system, even when there's no electrical demand at the appliance. Ever noticed how those little cubes are usually warm to the touch? That's wasted wattage being converted to heat. Most vampire cubes draw a few watts continuously. Vampires need to be unplugged when not in use, or plugged into switched outlets.

All the remote control televisions in the U.S., when turned to the "off" position, still use as much energy as the output of one Chernobyl-sized plant.

Amory Lovins, The Rocky Mountain Institute

Insulation & Radiant Barrier Products

Reflectix Insulation

Reflectix is a wonderful insulating material with dozens of uses. It is lightweight, clean, and requires no gloves, respirators, or protective clothing for installation. It is a 5/16″-thick reflective insulation that comes in rolls and is made up of seven layers. Two outer layers of aluminum foil reflect most of the heat that hits them. Each layer of foil is bonded to a layer of tough polyethylene for strength. Two inner layers of bubble pack resist heat flow, and a center layer of polyethylene gives Reflectix additional strength.

See the chart below for comparable R-value performance. Multiple layers of Reflectix will not increase the R-value. Reflectix inhibits or eliminates moisture condensation and provides no nesting qualities for birds, rodents, or insects. It is Class A Class 1 Fire Rated and nontoxic.

Reflectix BP (bubble pack) is used in retrofit installations. Proper installation requires a ¾″ airspace on at least one side of any Reflectix products. The best application for Reflectix products is preventing unwanted heat gain in attics in the summertime. Reflectix is also a great add-on to already insulated walls, providing a radiant barrier, vapor barrier, and increased R-value.

R-Values for Reflectix

R-value ratings are dependent on the direction of heat flow. "Up" refers to heat escaping through the roof in the winter or heat infiltration up through the floor in the summer. "Down" refers to solar heat gain through the roof in the summer or heat loss through the floor in the winter. "Horizontal" refers to heat transfer through the walls.

UP	DOWN	HORIZONTAL
8.3	14.3	9.8

As well as its most common usage as a building insulator, Reflectix has a myriad of other uses: pipe wrap, water heater wrap, duct wrap, window coverings, garage doors, as a camping blanket or beach blanket, cooler liner, windshield cover, stadium heating pad, camper shell insulation, behind refrigerator coils, and camera-bag liner. USA.

Reflectix BP (bubble pack) comes in 16″, 24″, and 48″ widths and in lengths of 50 feet and 125 feet.

◆ 56-503-BP Reflectix 16″ x 50′ (66.66 sq. ft) $34⁹⁵

◆ 56-502-BP Reflectix 16″x 125′ (166.66 sq. ft) $85⁹⁵

◆ 56-511-BP Reflectix 24″ x 50′ (100 sq. ft) $49⁹⁵

◆ 56-512-BP Reflectix 24″ x 125′ (250 sq. ft) $125⁰⁰

◆ 56-521-BP Reflectix 48″ x 50′ (200 sq. ft) $109⁰⁰

◆ 56-522-BP Reflectix 48″ x 125′ (500 sq. ft) $259⁰⁰

We can only ship Reflectix at standard rates within the continental U.S. For air shipments to AK, HI, or foreign countries, call for freight quote before ordering.

◆ *Means shipped directly from manufacturer.*

Bubble Pak Staple-Tab Reflectix

This Reflectix product is the same bubble-pack insulation as the standard Reflectix insulation, with the addition of staple tabs, and is made to be installed between framing members as opposed to on the surface. This product makes installation far easier, as it eliminates the need to add furring strips that are needed to form an airspace with standard Reflectix. Same pricing as BP insulation above, but be sure to specify Staple-Tab by ending the product number with ST instead of BP.

◆ 56-503-ST Reflectix Staple-Tab 16″ x 50′ $34⁹⁵

◆ 56-502-ST Reflectix Staple-Tab 16″ x 125′ $84⁹⁵

◆ 56-511-ST Reflectix Staple-Tab 24″ x 50′ $49⁹⁵

◆ 56-512-ST Reflectix Staple-Tab 24″ x 125′ $125⁰⁰

◆ 56-521-ST Reflectix Staple-Tab 48″ x 50′ $109⁰⁰

◆ 56-522-ST Reflectix Staple-Tab 48″ x 125′ $259⁰⁰

We can only ship Reflectix at standard rates within the continental U.S. For air shipments to AK, HI, or foreign countries, call for freight quote before ordering.

◆ *Means shipped directly from manufacturer.*

Save Energy with Low-E Window Film

Installing this Low-E window film combines year-round comfort with energy savings. It reflects over 50% of radiant solar heat in summer, retains over 50% of radiant heat in winter. By blocking a minimum 90% of ultraviolet light, it prevents costly fading and deterioration of furniture and fabrics. Glare is cut by up to 60%, yet high light transmission remains. Use it on most single pane (except for ¼" glass) glass windows. A margin of safety is added by making your window shatter-resistant. For the average 1,600–2,000 sq. ft. home, it only takes 18–24 months to pay back your initial investment in the film, which has a lifetime of about 15 years. Installs like wallpaper, using a squeegee and other common household tools. The scratch-resistant surface allows use of normal household window cleaners. USA.

Not recommended for south-facing windows if you want solar heating.

56-496 48″ x 15′ Low-E Insulator $44⁹⁵

56-497 24″ x 15′ Low-E Insulator $29⁹⁵

56-495 36″ x 15′ Low-E Insulator $35⁹⁵

56-498 Application Tool Kit $3⁹⁵

Super Insulating Radiant Barrier

A radiant barrier is your best line of defense for keeping excess heat out of your attic in the summer and keeping your floors warm in the winter. In a barn, chicken coop, or anywhere it's vital to keep farm animals and livestock warm in winter and cool in summer, a radiant barrier will provide invaluable protection. Stapled under the rafters or joists, Super R Premium will reflect 97% of the radiant heat. The nonprofit Florida Solar Energy Center has shown that an attic radiant barrier can reduce air conditioning costs by at least 20%. Available in three widths to fit any standard stud or joist spacing, this perforated product won't become an unintentional moisture barrier; it bounces heat upwards and performs equivalent to R-14. Installation is a snap—just staple the radiant barrier to the bottom of rafters or joists. Super R Premium is a perforated, two-sided, pure aluminum foil over a lightweight, tear resistant poly fabric. Weighs 16.5 lb/1000 sq. ft. Class A fire-rated. Ships in 500 sq. ft. rolls. USA.

56-508 Super R Premium 48″ x 125′ $89⁹⁵

Cord Caulk Plugs Damaging Leaks

Cord Caulk is a great draft and weather sealing substance that comes in a 100′ roll. It stops drafts, window rattles, and moisture damage to sills and floors. A soft, fine, fiber yarn that's saturated with synthetic adhesive wax polymers, it can be applied at below freezing temperatures, and can be removed and reused numerous times without losing its sealant capabilities. Cord Caulk adheres well to wood, aluminum, steel, paint, glass, rigid vinyl, rigid plastic sheet, and nylon. Comes in woodgrain brown or white. USA.

56-501 Cord Caulk $14⁹⁵

Please specify color.

Water Conservation Products

Water Conservation Kit

An average family of four will save 30,000 gallons of water per year by simply installing one low-flow showerhead, two faucet aerators, and a set of toilet dams. We've packaged all these water-savers together with a toilet leak detection kit, an instruction card, and a 26-page booklet on saving water. This is a $45 retail value but we're offering this kit at a price that will make saving water irresistible. USA.

46-109 Water Conservation Kit $19⁹⁵

Water-Saving Showerheads

Showers typically account for 32% of home water use. A standard showerhead uses about three to five gallons of water per minute, so even a five-minute shower can consume 25 gallons. According to the U.S. Department of Energy, heating water is the second largest residential energy user. With a low-flow showerhead, energy use and costs for heating water for showers may drop as much as 50%. Add one of our instantaneous water heaters for even greater savings. Our low-flow showerheads can easily cut shower water usage by 50%. A recent study showed that changing to a low-flow showerhead saved 27¢ worth of water and 51¢ of electricity per day for a family of four. So, besides being good for the Earth, a low-flow showerhead will pay for itself in about two months!

Lowest Flow Showerhead

This is by far our best-selling and finest-design showerhead. It can save up to $250 a year for a family of four by cutting hot and cold water use by up to 70%. At 40 psi it delivers 1.8 gpm, with a maximum at any pressure of 2.4 gpm. Manufactured in the U.S. of solid brass, and chrome-plated, it exceeds California Energy Commission standards. Built-in on/off button for soaping up. Standard ½-inch threads make a wrench the only tool necessary for installation. Fully guaranteed for ten years. Specified by the City of Los Angeles. USA.

46-104 Lowest Flow Showerhead $11⁹⁵

If each member of a family of four takes a five-minute shower each day, they will use more than 700 gallons of water every week—the equivalent of a three-year supply of drinking water for one person.

—from 50 Simple Things You Can Do to Save the Earth

House Heating and Cooling

Heating and cooling a home costs an average family well over $600 a year. This is over 40% of the average family's energy bill. That's a princely sum, and any of us could find more interesting ways to spend it. Fortunately, it isn't difficult to cut our heating and cooling bills dramatically with a judicious mixture of weatherization, additional insulation, landscaping, window upgrades, improved heating and cooling systems, and careful appliance selection.

Improvements to your building envelope, including weatherization, insulation, window treatments, and landscaping for energy conservation are covered in the preceding Energy Conservation section. In this section we'll cover home heating and cooling systems, how to select the best ones for your home and improvements and fine-tuning to keep them working efficiently.

Home Heating Systems

Fine-Tuning Your Heating System

We're going to assume, perhaps foolishly, that your home is already well-insulated and weather-tight. By weatherizing and insulating your home, you've reduced the amount of heat escaping in the most cost-effective way. It now makes sense to look at how efficiently your heating system produces and delivers the heat that you need. This section gives tips for making your existing heating system run more efficiently, whether it is gas, electric, oil, wood, or solar-powered. The following suggestions are either relatively inexpensive or free. All are highly cost-effective.

For those of you in the market for a new furnace, boiler, heat pump, or woodstove, we give you some helpful ideas on which kind of heating system will keep your house warm reliably for the smallest life-cycle cash outlay. Modern heating systems have seen dramatic efficiency gains over the past few years.

It probably won't be cost-effective to replace a working heating system based on fuel savings alone. But if the system appears close to a natural death, or its maintenance costs are high, replacing it could well save you money. And if you do, buying a high-efficiency model—rather than an average one—will almost certainly be cost-effective. Also, remember that the weatherizing and insulating that you've done will greatly reduce your heat losses, allowing you to buy a significantly smaller, less expensive heating system.

For safety reasons, adjustments, tune-ups, and modifications to your heating system itself are best done by a heating system professional. Homeowners and renters can improve system performance by insulating ducts and pipes, cleaning registers, replacing filters, and installing programmable thermostats.

If a house isn't resource-efficient, it isn't beautiful.

Amory Lovins, The Rocky Mountain Institute

Furnaces and Boilers

Heating System Tune-Ups

Gas furnaces and boilers should be tuned every two years, while oil units should be tuned once a year. Your fuel supplier can recommend or provide a qualified technician. Expect to pay $60 to $150. It's money well spent. While s/he's there, have the technician do a safety test to make sure the vent does not leak combustion products into the home. You can do a simple test by extinguishing a match a couple of inches from the spill-over vent: the smoke should be drawn up the chimney.

During a furnace tune-up, the technician should clean the furnace fan and its blades, correct the drive belt tension, oil the fan and the motor bearings, clean or replace the filter (make sure you know how to perform this monthly routine maintenance), and help you seal ducts if necessary.

Efficiency Modifications

While the heating contractor is there to tune-up your system, s/he may be able to recommend some modifications such as reducing the nozzle (oil) or orifice (gas) size, installing a new burner and motorized flue damper, or replacing the pilot light with an electronic spark ignition. If you have an older oil burner, installing a flame-retention burner head (which vaporizes the fuel and allows more complete combustion) will typically pay back your investment in two to five years.

Turn Off the Pilot Light During the Summer

This will save you about $2 to $4 per month. Do this only if you can safely light it again yourself, so you don't have to pay someone to do it. Federal regulations now require that new natural gas-fired boilers and furnaces be equipped with electronic ignition, saving $30 to $40 per year in gas bills. Propane-fired units cannot be sold without a pilot light and cannot be safely retrofitted with a spark ignitor unless you install expensive propane-sniffing equipment. Propane is heavier than air, and any leaking propane can pool around the unit creating an explosion potential when the unit tries to start. Natural gas is lighter than air.

Insulate the Supply and Return Pipes on Steam and Hot-Water Boilers

Use a high-temperature pipe insulation such as fiberglass wrap for steam pipes. The lower temperatures of hydronic or hot water systems may allow you to use a foam insulation (make sure it's rated for at least 220°F).

Clean or Change Your Filter Monthly

Yes, filters are used for more than coffee and cars. A clogged furnace filter impedes air flow, makes the fan work harder, the furnace run longer, and cuts overall efficiency. Filters are designed for easy service, so the hardest part of this maneuver is turning into the hardware store parking lot with a note of the correct size in hand. While you're there, pick up several filters, since you'll want to replace the filter every month during heating season. Most full-house air conditioning systems use the same fan, ducts, and filter. So monthly changes during AC season are a must, too. They'll set you back a buck or two apiece. For five bucks you can buy a reusable filter that will need washing or vacuuming every month, but will last for a year or two.

Seal and Insulate Air Ducts

Losses from leaky, uninsulated ducts—especially those in unheated attics and basements—can reduce the efficiency of your heating system by as much as 30%. Don't blow your expensive heated air into unheated spaces! Seal ducts thoroughly with mastic, caulking, or duct tape and then insulate with fiberglass wrap.

Radiators and Heat Registers

Vacuum the Cobwebs Out of Your Registers

Anything that impedes airflow makes the fan work harder, and the furnace run longer.

Reflect the Heat from Behind Your Radiator

You can make foil reflectors by taping aluminum foil to cardboard, or use radiant barrier material (left over from your attic or basement). Place them behind any radiator on an external wall with the shiny side facing the room. Foil-faced rigid insulation also works well.

Vacuum the Fins on Baseboard Heaters

This improves airflow and efficiency. Keeping furniture and drapes out of the way also improves airflow from the radiator.

Bleed the Air out of Hot-Water Radiators

Trapped air in radiators keeps them from filling with hot water and thus reduces their heating capacity. Doing this should also quiet down clanging radiators. You can buy a radiator key at the hardware store. Hold a cup or a pan under the valve as you slowly open it with the key. Close the valve when all the air has escaped and only water comes out. If it turns out that you have to do this more than once a month, have your system inspected by a heating contractor.

Thermostats

Install a Programmable Thermostat

One-half of homeowners already turn down their heat at night, saving themselves 6% to 16% of heating energy. A programmable or clock thermostat can automatically do this for you. In addition to lowering temperatures while you sleep, it will also raise temperatures again before you get out from under the covers. It not only sounds good, it's cost-effective too. Such thermostats can be used to drop the house temperature during the day as well if the house is unoccupied. *Home Energy* magazine reports savings in excess of 20% with two eight-hour setbacks of 10°F each.

Electronic setback thermostats are readily available for $25 to $150. If you have a heat pump or central air conditioning, make sure the thermostat is designed for heat and A/C. Installation is simple, but make sure to turn off the power to the heating plant before you begin. If connecting wires makes you jittery, have an electrician or heating technician do it for you. It's well worth the investment.

Heat Pumps

Change or Clean Filters Once a Month

Dirty filters are the most common reason for failures.

Keep Leaves and Debris Away from the Outside Unit

Good air flow is essential to heat pump efficiency. Clear a two-foot radius. This includes shrubbery and landscaping.

Listen Periodically to the Compressor

If it cycles on for less than a minute, there is something wrong. Have a technician adjust it immediately.

To Burn or Not to Burn

I'm totally confused. The subject is wood, a material I have used to heat my home for the past 15 years. I know it's more work to burn wood, but I like the exercise. It makes more sense to burn calories splitting and hauling hardwood than running in front of a television set on a treadmill at the health spa. I enjoy the fresh air. I like the warmth. I like the flicker and glow of embers.

Sometimes, when I am stacking wood so that a summer's worth of sunshine can make it a cleaner, hotter fuel, I think of those supertankers carrying immense cargos of black gold from the ancient forests of the Middle East. I think of the wells that Saddam Hussein set afire in Kuwait, and the noxious roar of the jet engines of the warplanes we used to ensure our country's access to cheap oil.

My wood comes from hills that I can see. It's delivered by a guy named Paul, who cuts it with his chainsaw, then loads it in his one-ton pickup. We talk about the weather, the conditions in the woods, and the burning characteristics of different species. We finish by bantering about whether he's delivered a large cord, a medium, or a small. I can't imagine having the same conversation with the man who delivers oil or propane.

— continued on next page

Dust Off the Indoor Coils of the Heat Exchanger at Least Once a Year

This should be part of the yearly tune-up if you pay a technician for this service.

Have the Compressor Tuned Up

Every year or two a technician should check refrigerant charge, controls, filters, and oil the blower motors.

Woodstoves

Wood heat is a mixed blessing. If harvested and used in a responsible manner, firewood can be a sustainable resource. To be honest, most wood-gathering is not done sustainably. There are 27 million wood-burning stoves and fireplaces in the U.S., the majority of them older pre-EPA designs, that contribute millions of tons of pollutants to the air we breathe. In Washington state, for example, wood heating contributes 90% of the particulates in two of the counties with the worst air pollution problems.

Use a Woodstove Only If You're Willing to Make Sure It Burns Cleanly

Woodstoves aren't like furnaces or refrigerators; you can't just turn them on and forget them. "Banking" a woodstove for an overnight burn only makes the wood smolder and emit lots of pollution. I know we've all done it, but we didn't know any better. Now we do. A well-insulated house will hold enough heat that you shouldn't need to keep the stove going all night. Heavier, masonry stoves are a good idea in colder climates. Once the large mass is warmed up, it emits slow, steady heat overnight without a smoky fire.

Burn Well-Seasoned Wood and No Trash

Wood should be split at least six months before you burn it. Never burn garbage, plastics, plywood, or treated lumber.

Hotter Fires Are Cleaner

Another good case for masonry stoves, which burn hot, clean, short fires with no air restriction. The heat is soaked up by the masonry and emitted slowly over many hours.

If a Woodstove Is Your Main Source of Heat, Use an EPA-Approved Model

New EPA regulations require woodstoves to emit less than 10% of the pollution that older models produced.

Check Your Chimney Regularly

Blue or gray smoke means your wood is not burning completely.

— continued from
previous page

*Unfortunately (and this
is where I begin to get confused), there is another
side to the wood-burning
issue. Along comes evi-
dence that the emissions
from airtight stoves contain
insidious carcinogens, and
further indications that a
home's internal environ-
ment can be more adversely
affected by burning wood
than by passive cigarette
smoke. So now I burn my
wood in a "clean" stove.
But, my environmentally
active friends say that the
only good smoke is no
smoke, and that even my
supposedly high-tech stove
is smogging up their skies.*

*Theoretically, I should
be able to go back to feeling good about wood, but
life is never that simple.
On the positive side of the
ledger comes the informa-
tion that wood-burning, in
combination with responsible reforestation, actually
helps the environment by
reversing the greenhouse
effect. The oxidation of biomass, whether on the forest
floor or in your stove,
releases the same amount
of carbon into the atmos-
phere. It makes more sense
for wood to be heating my
home than contributing to
the brown skies over
Yellowstone.*

*This is obviously a sim-
plistic analysis, but wood
should be a simple subject.
In the meantime, I have to
keep warm, so I'm choos-
ing to burn wood. I keep
coming back to the fact
that humans have been
burning oil for more than
50 years and wood for
five million.*

—Stephen Morris

Fireplaces

Fireplaces offer a good light show but are not an effective source of heat. In fact, fire-places are usually a net heat loser because of the huge volume of warm interior air sucked up the chimney, which has to be replaced with cold air leaking into the house. Fireplaces average minus 10% to plus 10% efficiency. If you've done a good job of weath-erizing your house and sealing up air leaks, you may lack adequate fresh air for combustion, or the chimney won't draw properly, filling the house with smoke. The two best things to do with a fireplace are don't use it and plug the flue; or install a modern woodstove and line the chimney.

Improve the Seal of the Flue Damper

To test the damper seal, close the flue, light a small piece of paper and watch the smoke. If the smoke goes up the flue, there's an air leak. Seal around the damper assembly with refractory cement. (Don't seal the damper closed.) If the damper has warped from heat over the years, get a sheet metal shop to make a new one, or consider a woodstove con-version.

Install Tight-Fitting Glass Doors

Controlling air flow improves combustion efficiency by 10% to 20%, and may help reduce air leakage up a poorly sealed damper.

Use a C-Shaped Tubular, or Blower-Equipped Grate

Tube grates draw cooler air in the bottom and expel heated air out the top. Blower-equipped grates do the same job with a bit more strength.

Caulk Around the Fireplace and Hearth

Do this where they meet the structure of the house, using a butyl rubber caulk.

Use a Cast-Iron Fireback

They are available in a variety of patterns and sizes. Fireplace efficiency is improved by reflecting more radiant heat into the room.

Locate the Screen Slightly Away from the Opening

This allows more heated air to flow over the top of the screen. While they do prevent sparks from flying into the room, firescreens also prevent as much as 30% of the heat from entering the room.

If You Never Use Your Fireplace, Put a Plug in the Flue to Stop Heat Loss

Seal the plug to the chimney walls with good quality caulk and tell anyone who might build a fire (or leave a sign). Temporary plugs can be made with rigid board insulation, plywood with pipe insulation around the edges, or an inflatable plug.

Buying A New Heating System

If your old heating system is about to die, you're probably in the market for a new one. Or, if you're currently spending over $1,000 per year on heating, it's likely that the $800 to $4,500 you'll spend on a more efficient heating system will reduce your bills enough to pay for your investment in several years. Your system will also be more reliable, as well as increase the value of your house. As you shop, it's worth keeping in mind a few things that will save you money and increase your comfort.

If you've weatherized and insulated your home—and we'll remind you again that these are by far the most cost-effective things you can do—then you can downsize the furnace or boiler without compromising the capability of your heating system to keep you warm. An oversized heater will cost more to buy up front, more to run every year, and the frequent short-cycle on and off of an oversized system reduces efficiency. Ask your heating contractor to explain any sizing calculations and make sure s/he understands that you have a tight, well-insulated house, to verify that you don't get stuck with an oversized model (which s/he gets to charge more money for).

Natural Gas Furnaces and Boilers

It is always cost-effective to pay a little more up front in return for more efficiency. But it may not be cost-effective to pay a lot more for the most efficient unit, since your reduced heating needs mean a longer payback. The differences between various models can be significant, and it's worth your while to examine efficiency ratings before making a decision.

Fuel Switching

When available, natural gas is the lowest-cost heating fuel for most homes. If you have a choice, switching fuels may be cost-effective, but only if this does not also require substantial changes to your home's existing heat distribution system. For example, replacing the old hot water boiler with a high-efficiency forced-air furnace would require installation of all new forced-air ducts. If you are currently using electricity for heating, switching to any other fuel will be cost-effective. See the next subject.

Electric Resistance Heating

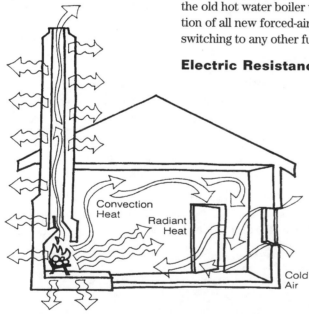

Electric baseboard heating is by far the most expensive way to warm one's home; it's cheap to buy and install, but it costs two to three times as much to heat with electricity as with gas. The life-cycle cost of electric heating is, without exception, far higher than that of gas-fired furnaces and boilers, even taking into account the higher installation cost of the latter, and even in regions with exceptionally low electricity prices. If you must heat with electricity, weatherizing and insulating your home, along with installing a programmable thermostat, are especially lucrative options. If you also use air conditioning, and live in a mild winter climate, consider switching to a heat pump.

Heat Pumps

Heat pumps are the most efficient form of electric heat. There are three types of heat pumps: air-to-air heat pumps, water-source heat pumps, and ground-source heat pumps. Heat pumps collect heat from the air, water, or ground depending on type, concentrate the warmth, and distribute it through the home. Heat pumps typically deliver three times more energy in heat than they consume in electric power. Heat pumps can also be used to cool homes by reversing the

Fireplaces can be a net heat loss due to infiltration of cold air. Weatherization and insulation help, but the best measure, though often difficult and expensive, is to provide outside combustion air to the fireplace. Illustration courtesy Public Service of Colorado (1992), Your Energy Guide to Heating, Cooling, and Home Appliances.

process—collecting indoor heat and transferring it outside the building. Some newer heat pumps are designed to also provide an inexpensive source of household hot water.

Not all heat pumps are created equal, but fortunately the Air Conditioning and Refrigeration Institute rates all heat pumps on the market. Look at the efficiency ratings and purchase a system designed for a colder climate. Some electric utilities offer rebates and other incentives to help finance the higher capital costs of these more efficient systems.

Air-to-Air Heat Pumps are the most popular, the least expensive initially, and the most expensive to run in colder weather. Air-to-air units must rely on inefficient backup heating mechanisms (usually electric resistance heat) when outside temperatures drop below a certain point. That point varies from model to model, so make sure you buy one that's designed for your climate—some models can cope with much colder weather before resorting to backup heat. A heat pump running on backup is no better than standard electric resistance heating. Some newer pumps have more efficient gas-fired backup mechanisms, but if gas is available, why not use it directly?

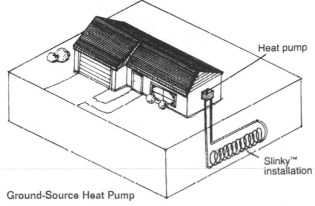

Heat pump

Slinky™ installation

Ground-Source Heat Pump

Ground-source heat pumps can be highly cost-effective in new construction over the life cycle of heating and cooling equipment compared to more conventional options. Several designs are on the market; this illustration shows the Slinky™ design. Adapted from an illustration by E SOURCE (1993), Space Heating Technology Atlas.

Ground- or Water-Source Heat Pumps are one of the most efficient heating systems, and are substantially more cost-effective in colder climates. Because of the extensive burial of plastic pipe required, these systems are best installed during new construction.

Tens of thousands of ground-source heat pumps have been installed in Canada, New England, and other frostbelt areas. See the annualized heating/cooling system cost chart at the end of this section for a comparison.

People Heaters

It often makes sense to use a gas or electric radiant or convection heater to warm only certain areas of the home, or, as in the case of radiant heaters, keeping the people warm and comfortable. Radiant systems keep you and objects around you warm just like the sun warms your skin. This allows you to set the thermostat for the main heating system lower by 6° to 8°F.

A good, long-lasting, efficient, cast iron woodstove with porcelain finish, by Vermont Castings.

Even if you heat with gas, electric connective or radiant spot heaters can save you money, depending on how the system is used. Radiant systems are designed and used more like task lights, turning them on only when and where heat is needed, rather than heating the whole house. Most central heating systems give you little control in this respect, since they are designed to heat the entire home.

As one old-timer, who was accustomed to sitting near a wood-stove, complained while pacing the floor at his son-in-law's house: "I don't know where to sit. Everything is the same temperature!"

Woodstoves and Pellet Stoves

Modern woodstoves have catalytic combustors and other features that boost their efficiencies into the 55% to 75% range while lowering emissions by two-thirds. Older woodstoves without air controls have efficiencies of only 20% to 30%. Wood burns at a higher temperature in a catalytic woodstove and the gases go through a secondary combustion at the catalyst, thereby minimizing pollution and creosote buildup.

If you're buying a new woodstove, make sure it's not too large for your heating needs. This will allow you to keep your stove stoked with the damper open, resulting in a cleaner and more efficient burn. Some stoves incorporate more cast iron or soapstone as thermal mass to moderate temperature swings and prolong fire life. If you live in a climate that requires heat continuously for several months, you definitely want the heaviest stove you can afford. The masonry heaters take the thermal mass idea one very comfortable step further.

Pellet stoves have the advantage that the fuel pellets of compressed sawdust, cardboard, or agricultural waste are automatically fed into the stove by an electric auger. The feed rate is dialed up or down with a rheostat, so they'll happily run all night if you can afford the pellets. You fill the reservoir on the stove once every day or so. Pellet stoves are cleaner burning than woodstoves as combustion air is force-fed into the burning chamber. Unlimited access to oxygen and drier fuel than firewood mean a more efficient combustion. But the pellets—which look like rabbit food—may be slightly more expensive than firewood in most parts of the country, and simply unavailable in some areas. Pellets must be kept bone dry in order to auger and burn. Pellet stoves have more mechanical parts than woodstoves, so they're subject to maintenance and breakdown, and they require electricity. No power—no fire. The fire will go out without the combustion air blower running. Pellet stoves usually cannot be used with renewable energy systems, unless you've got considerable extra power in the winter, like from a hydro system.

Fireplace Inserts

A fireplace insert fits inside the opening of a fireplace, operates like a woodstove, and offers improved heat performance. An insert is difficult to install because it must be lifted into the fireplace opening and the space around it must be covered with sheet metal and sealed with a cement grout to reduce air leaks. Such inserts are only 30% to 50% efficient, which is a good improvement over a fireplace, but less efficient than a good freestanding woodstove, which is free to radiate heat off all sides, not just the front.

A fireplace can be retrofitted with an airtight woodstove installed in, or better yet in front of, the fireplace. This is the most efficient choice, since all six heat-emitting sides sit inside the living space, on the hearth. The stove's chimney is linked into the existing flue, and good installations line the chimney all the way to the top with stainless pipe to limit creosote buildup and provide easy flue cleaning. Simply shoving a few feet of woodstove pipe into the existing chimney is begging for a chimney fire of epic proportions. Any creosote in the smoke stream will condense on the large, cool chimney walls, providing a spectacular amount of high-temperature fuel in a few years. This has been well-proven already to be an excellent way to burn houses down. Further demonstrations and experiments aren't necessary. Line your chimney when retrofitting a woodstove.

Gas Fireplaces

Gas fireplaces have combustion efficiencies up to 80%, whereas the initially less expensive gas logs are only 20% to 30% efficient. However, in case of gas leaks, many local codes require dampers to be welded open. This reduces their overall efficiency unless they have tight-fitting doors to cut down the amount of warmed interior air allowed to freely float out.

Masonry stoves are the best woodburning solution for extreme climates that need 24-hr. heating.

Masonry Stoves and Fireplaces

Masonry stoves, also called Russian stoves or fireplaces, from a people who know heating, have long been popular in Europe. They are large, free-standing, masonry fireplaces with a firebox in which the heat and smoke from the combustion process go through a long twisting labyrinth masonry chimney. The mass of the heater slowly warms up and radiates heat into the room for hours. They are best suited for extreme climates that need heat 24 hours per day. They are much more efficient than fireplaces. How many woodstoves have seats built into them so that you can snuggle in? Russian fireplaces commonly do. They have very low emissions because air flow is barely restricted during burning, giving a clean, high temperature burn. As Mark Twain observed: "One firing is enough for the day . . . the heat produced is the same all day, instead of too hot and too cold by turns."

Masonry stoves are expensive initially, but will last for the life of the home, and are one of the cleanest, most satisfying ways to heat with wood. They're easiest to install during initial construction, and are best suited to well-insulated homes with open floor plans that allow the heat to radiate freely.

Passive Solar Heat

Solar energy is the most environmentally friendly form of heating you can find. The amount of solar energy we take today in no way diminishes the amount we can take tomorrow, or tomorrow, or tomorrow. Free fuel is the cheapest fuel. Passive solar heating uses no moving parts, just sunshine through south-facing insulated windows and thermal mass in the building structure to store the heat. Those south of the equator please make the usual direction adjustments.

A carefully designed passive solar building can rely on the sun for half or more of its heating needs in virtually any climate in the U.S. Many successful designs have cut conventional heating loads by 80% or better. Some buildings, such as the Real Goods Solar Living Center in Hopland, California, or Rocky Mountain Institute's headquarters high in the Rocky Mountains in Snowmass, Colorado, have no need for a central heating system. RMI even grows banana trees indoors! By the same token, in hot climates, good design, shading, and passive cooling strategies can eliminate the need for mechanical cooling. Our Solar Living Center can survive over four weeks of daily high temperatures exceeding 100°F—and a steady flow of overheated visitors—while the interior temperature stays below 78°F without using air conditioning or mechanical cooling in any way.

Adding south-facing windows or a greenhouse is a way to retrofit your house to take advantage of free solar gain. See our chapter on Shelter and Passive Solar for more detailed information and books about this subject.

We need to point out that using passive solar requires careful design of the home. It's best done in new construction, or during major remodeling. Otherwise, put your money into weatherization and insulation, which always has a good payback.

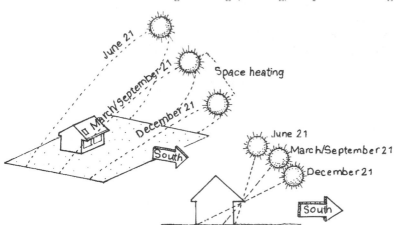

Sun Path Diagrams

Passive solar heating is practicable in every climate. This simplified illustration shows the importance of a calculated roof overhang to allow solar heating in the winter but prevent unwanted solar heat gain in the summer. Adapted from an illustration by E SOURCE (1993), Space Heating Technology Atlas.

Active Solar Heating

In many climates around the U.S., flat-plate collector systems can provide enough heat to make central heating systems unnecessary, but in most regions it is prudent to add a small backup system for cloudy periods. Active systems can be added at any time to supplement space and water heating needs. The best and most common use of active systems is for domestic water heating. Real Goods offers several solar hot water systems for differing budgets. See our Water Heating chapter. Active solar space heating has a high initial cost, is complex, and has potential for high maintenance. We don't recommend it; use passive solar! The beauty of passive systems is in their simplicity and lack of moving parts. Remember the KISS rule: Keep It Simple, Stupid.

Heating Costs Comparison

While operating costs will vary by climate and region, and fuel and electricity prices differ across the country, the following chart represents estimated installation and operating costs for selected heating and cooling systems in a typical single-family house. It assumes that heat pumps will provide air conditioning in the summer as well as space heating in winter.

	Installation Cost	Operating Cost Per Year	Annualized Cost
Advanced gas furnace + high-eff. A/C	$7,200	$746	$1,455
Standard gas furnace + standard A/C	5,775	901	1,469
Gas-fired air-source heat pump	8,333	658	1,478
Advanced ground-source heat pump	9,250	682	1,592
Advanced air-source heat pump	8,940	822	1,702
Standard air-source heat pump	5,715	1,232	1,794
Advanced oil furnace + high-eff. A/C	6,515	1,266	1,907
Electric resistance heat + standard A/C	5,515	1,769	2,312

Annualized* cost of selected heating and cooling systems (in dollars; assumed life-expectancy is 20 years; estimated national average).

Adapted from: Joan Gregerson, et al, (1993), Space Heat Technology Atlas, E SOURCE, *and* Environmental Protection Agency (1993), Space Conditioning: The Next Frontier.

*Annualized cost includes yearly operating cost plus the capital and installation cost recalculated as an annual payment (based on 20-year financing and a 10 percent discount rate to account for the time value of money).

www.realgoods.com

Appliances for Heating

Real Goods does not offer heat pumps or furnaces that require the services of a specialized technician for safe installation. We offer small, room-sized direct-vent gas heaters and wood-burning cookstoves.

Gas-Fired Space Heating

Eco-Therm Direct Vent Heaters

Direct venting makes the Eco-Therm the ideal gas heater for bedrooms, bathrooms, and all other closed rooms, as well as super airtight homes. It draws in outside air for combustion and exhausts it through the same wall penetration (pipe within a pipe). It doesn't deplete oxygen in the room and doesn't waste heated air.

The sleek, slim Eco-Therm installs easily almost anywhere on an outside wall a minimum of 6 inches thick and a maximum of 12 inches thick. It should not be located under windows that can be opened. Just cut a 4¾-inch hole in the wall, then trim the length of the direct vent tube to the thickness of the wall. Hang the heater on the wall and bring a ½-inch gas line to the valve. Since it heats so well without fans or blowers, the Eco-Therm requires no electricity. This unit is thermostat-controlled.

The special design ensures that the front panel stays at a safe temperature. All components are of the highest quality. The steel heat exchanger is enameled inside and out. The vent system is aluminum for corrosion resistance. Its strong outside terminal needs no additional protective grid, and the offset dual throat makes it waterproof. Eco-Therm heaters at altitudes above 5,000 feet must have the fuel orifice replaced. This is a no-cost conversion kit when ordered with your heater. High altitude replacement orifice kits are drop-shipped from Vermont.

Please specify either propane or natural gas. Warranty is one year, with a five-year warranty on the heat exchanger. Minimum clearances: floor 4.75"; side wall 4"; rear 0"; ceiling 36". Argentina, warranted by U.S. importer.

65-209 High Altitude Orifice Kit $0⁰⁰

(State heater model and altitude)

Eco-Therm MV-130.

Eco-Therm MV-130
Output: 10,690 Btu/hr
Dimensions: 27.125" x 24.375" x 6.625"
Net Weight: 46 lb
Shipping Weight: 55 lb
Warranty: 1 year parts, 5 years heat exchanger

65-203 Eco-Therm Heater, MV-130 $469⁰⁰

Eco-Therm MV-120
Output. 6,400 Btu/hr
Dimensions: 21.125" x 24.375" x 6.625
Net Weight: 34 lb
Shipping Weight: 42 lb
Warranty: 1 year parts, 5 years heat exchanger

65-204 Eco-Therm Heater, MV-120 $429⁰⁰

Eco-Therm MV-112
Output. 4,200 Btu/hr
Dimensions: 15.125 x 24.125 x 6.625"
Net Weight: 26 lb
Shipping Weight: 33 lb
Warranty: 1 year parts, 5 years heat exchanger

65-205 Eco-Therm Heater MV-112 $409⁰⁰

Books

Consumer Guide to Home Energy Savings

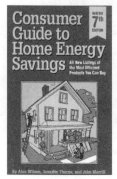

By Alex Wilson and John Morrill. The updated seventh edition of *Consumer Guide to Home Energy Savings* identifies the most energy-efficient home appliances by brand name and model number. This book is as compact and efficient as its subject matter. Its pages are crammed with money-saving information. 5″ x 7″, 296 pages, 118 illustrations, appendices, index, paperback.

82-399 7th Ed. Consumer Guide to Home Energy Savings $8⁹⁵

Build Your Own Solar Hot Air Collector

There's a free winter heat source just waiting to be taken. Do you have a more or less south-facing roof? Do you have cold sunny days in the winter? Then you could cut your winter heating costs by 50% to 80% depending on your roof area and climate. In this 52-minute video a homeowner in Massachusetts takes us carefully step-by-step through refurbishing his large hot air collector. All the details are shown, and a complete materials and resources list is included.

The Solar Hot Air Video shows how to build your own solar collector.

This is a simple, but amazingly effective homemade design that can be adapted to any size, shape, or slope of sun-facing roof. You can retrofit almost any house for inexpensive solar heating, even when no consideration was given to solar gain initially. Care is taken to ensure a maintenance-free installation for many, many years of ultralow-cost heating. Only a tiny fractional horsepower blower motor automatically controlled by a temperature switch uses power. Typically this fan draws 120 watts or less, a fraction of what a furnace fan uses. It's simple, it's homemade, it saves resources; Dr. Doug highly recommends it.

80-890 Solar Hot Air Video $29⁹⁵

Heartland Cookstoves

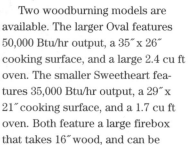

In an age when nothing seems to last, Heartland Cookstoves are a charming exception. Though modern refinements have been seamlessly incorporated into the design, the Cookstove still evokes time-honored traditions. Our black and white catalog can't do justice to the bright nickel trim, soft rounded corners, and porcelain enamel inlays in seven optional colors. We'll be happy to send the factory color brochure.

Two woodburning models are available. The larger Oval features 50,000 Btu/hr output, a 35″ x 26″ cooking surface, and a large 2.4 cu ft oven. The smaller Sweetheart features 35,000 Btu/hr output, a 29″ x 21″ cooking surface, and a 1.7 cu ft oven. Both feature a large firebox that takes 16″ wood, and can be loaded from the front or top. A summer grate position directs the heat to the oven and cooktop, while limiting heating output. Flue size for both models is common 6″, and is supplied locally.

Options include color porcelain, a stainless steel firebox water jacket for water heating, a 5-gallon copper water reservoir with spigot, a coal grate for burning coal rather than wood, and an outside combustion air kit.

Gas and electric models with all the traditional styling and color options are also available. Please call for options or a free color brochure.

61-113 Oval Cookstove, Std. White $3,675⁰⁰

61-117 Sweetheart Cookstove, Std. White $2,995⁰⁰

Options (same cost for both models):

61-106 Color Option (Almond) $50⁰⁰

**61-107 Color Option (Green, Blue, Teal, Black) $199⁰⁰
(Red) $299⁰⁰**

61-119 Stainless Steel Water Jacket (Oval) $199⁰⁰

61-122 Stainless Steel Water Jacket (Sweetheart) $199⁰⁰

61-120 Coal Grate $229⁰⁰

61-125 Outside Air Kit $75⁰⁰

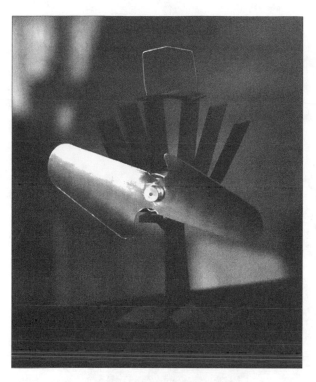

A Heat-Powered Woodstove Fan

The Ecofan was voted Real Goods Product of the Decade, as it was our best-selling item by far for the last four years!

The Ecofan uses heat from your woodstove to run a heat-distributing fan. The thermoelectric generator starts the fan automatically and adjusts speed according to stove temperature. The hotter your stove gets, the faster it runs. Peak performance occurs at surface temperatures of 400°–500°F.

The unique ultraquiet anodized brass blade is designed to silently deliver a very broad cross section of gentle air movement, increasing the heating effectiveness of the stove. Because it uses only the heat from the stove, the Ecofan costs nothing to run. It doesn't plug in, there are no batteries, and it can be used where there is no electricity. The Ecofan is anodized extruded aluminum, which won't rust or corrode, and the thermoelectric module doesn't wear out, making this a lifetime energy-saving appliance. A bi-metal strip in the base will help prevent overheating by tipping the fan slightly, breaking surface contact with the stove. Laboratory tests have shown a 30% faster increase in temperature on a wall 30 feet from a woodstove using the Ecofan. One-year mfr.'s warranty.

64-207 Ecofan Woodstove Fan $109⁰⁰

Cost-Effective Way To Increase Heater Efficiency

Install this unobtrusive room-to-room door fan and your existing heating system won't have to work as hard. Whether you heat with wood, floor or wall furnace, or space heaters, circulating heated air throughout the house is the key to comfort. Installs in any door frame, pulling heated air from the top of the room and through the doorway, without interfering with the movement of the door. This product meets our environmental criterion by reducing your existing heater's fuel consumption with minimal energy expense (30 watts). A quiet nine-blade fan moves 80 cubic feet of air per minute. Ten-foot cord. 7″ L x 7″ H. One-year limited mfr.'s warranty. ETL listed to UL standards. China.

14-010 Door Fan $30⁰⁰

www.realgoods.com

Burn Pellets in Any Woodstove

The Prometheus pellet fuel basket is a major break-through in wood-burning technology. Pellet fuel is a recycled product, produces far less creosote and ash than cordwood, is easier to handle, more readily available, and in many areas it now costs less. The only problem has been that, until now, it took a specialized stove to burn pellet fuel. The all-stainless steel Prometheus fuel basket makes pellet burning possible in any woodstove or fireplace. One year mfr.'s warranty. USA.

Model Selection

For safe and reliable heating, it's important to choose the correct size Prometheus. Just multiply the inside width by the inside depth of your stove and compare your figure to the Minimum Hearth Area column on the chart. Your hearth area must be equal to or larger than this minimum area.

Pellet fuel can be added to the basket chambers at any time for continuous burning.

Full pellet basket may burn 2 to 10 hours without refueling depending on the size of basket.

Angled base to allow viewing of the flames and easier starting.

Sloped top provides optimum fire viewing and facilitates reloading.

Ash falls through openings at the bottom.

Fuel can be added with as little as one inch of pellet fuel remaining in the basket to provide extended burn time.

Air flows through channels in the basket for an efficient burn.

Adjustable legs.

Order #	Model #	Height	Width	Depth	Woodstoves smallest hearth area allowed	Fireplaces minimum opening width Masonry	Manufactured	Price
61-127	6117	9.63″	11.63″	7.63″	192	18″	21″	**$189**⁰⁰
61-128	6147	9.63″	14.75″	7.63″	244	21″	24″	**$199**⁰⁰
61-129	6177	9.63″	17.63″	7.63″	192	24″	26″	**$219**⁰⁰
61-130	6227	9.63″	22.75″	7.63″	377	29″	37.5″	**$239**⁰⁰
61-131	5147	8.63″	14.75″	7.63″	244	21″	24″	**$199**⁰⁰
61-132	61411	9.63″	14.75″	11.13″	358	NA	NA	**$239**⁰⁰

Home Cooling Systems

There are two familiar ways to stay cool in the summertime: Install a big air conditioner, crank it up, and pay the bills; or seek the shade, lie quietly, and sweat it out. Both methods work, but there are better, more comfortable options. There are four basic strategies to evaluate: 1) reducing the cooling load; 2) exploring alternative cooling methods; 3) increasing the efficiency of your existing air conditioner; 4) installing new, efficient cooling equipment.

Reduce the Cooling Load

The best strategy for keeping a dwelling cool is to keep it from getting hot in the first place. This means preventing outside heat from getting inside, and reducing the amount of heat generated inside by inefficient appliances like incandescent lights, unwrapped water heaters, or older refrigerators.

In a hot, humid climate, 25% of the cooling load is the result of infiltration of moisture, 25% is from higher outdoor heat penetrating windows, walls, and roof, 20% from unwanted solar gain through windows, and the remaining 30% from heat and moisture generated within the home. It makes a lot of economic sense to cut these loads before getting that new monster air conditioner. You'll be able to get by with a smaller, less expensive unit that costs less to operate for the rest of its life.

Weatherizing and insulating will reduce your cooling costs substantially (funny how we keep hammering those points). But there are still a few more tricks to keep the outside heat from getting inside.

Shading

Shading that blocks summer sun on the east, south, and west sides of your house, but not summer breezes, is one of the most effective ways to keep your home cooler.

Planting shade trees, particularly on the west and south sides of your house can greatly increase comfort and coolness. Awnings, porches, or trellises on those same sides of a building will reduce solar gain through the walls as well as through the windows. A home's inside temperature can rise as much as 20°F or more if the east and west windows and walls are not shaded.

Our own Dr. Doug has greatly improved the comfort of his master bedroom by planting a jasmine-covered trellis over a west-facing wall. Without all those BTUs soaking through the wall four to eight hours after sunset, the bedroom stays 10° to 20°F cooler overnight than it used to. And the jasmine dramatically softens what used to be a glaring backyard patio in the afternoon, making both the outdoor and the indoor spaces more comfortable and pleasing.

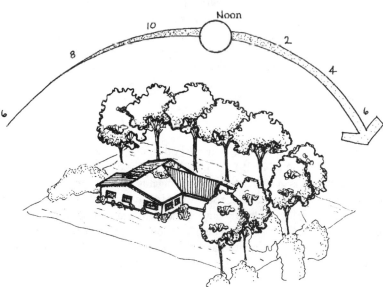

Mature trees shading your house will keep it much cooler by lowering roof and attic temperatures, blocking unwanted direct solar heat radiation from entering the windows, and providing a cooler microclimate outdoors. Adapted from an illustration by Saturn Resource Management, Helena, MT.

Windows

For too long a window has simply been a hole in the wall that lets light and heat through. Windows, even so-called "thermopane" windows, have very low R-values. Single-pane units have an R-value of 1. Thermopane, or dual-glazed windows have an R-value of 2. Both are pretty miserable. Conventional windows conduct heat through the glass and frame, permit warm, moist air to leak in around the edges, and let in lots of unwanted heat in the form of solar radiation. South-facing windows with properly sized roof overhangs or awnings won't take in the high summer sun, but will accept the lower winter sun when it's usually wanted. East- and west-facing windows cannot be protected so simply. See our chapter on Shelter and Passive Solar for a complete discussion of window treatments, including newer high-tech windows with R-values up to 8 or 10.

Radiant Barriers

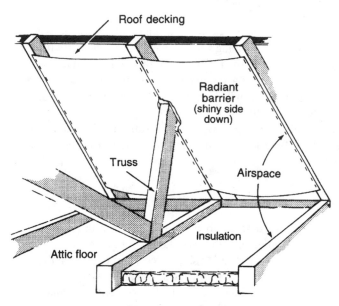

A Typical Radiant Barrier Installation in an Attic

Radiant barriers are thin metal films on a plastic or paper sheet to give them strength. They will reflect or stop approximately 95% of long-wave infrared heat radiation. Such barriers are typically stapled to the underside of attic rafters to lower summertime attic temperatures. Laying these barriers on the attic floor is also an option, but research indicates that the barriers' effectiveness is reduced by dust build-up. Lowering attic temperatures—which can easily reach 160°F on a sunny day—reduces heat penetration into the living space below and is frequently a cost-effective measure to lower air conditioning bills. A study by the Florida Solar Energy Center found that properly installed radiant barriers reduced cooling loads by 7% to 21%. The study also found that temperatures of the roofing materials were only increased by 2°F to 10°F, which will have little or no impact on the life expectancy of a shingle. A radiant barrier will reduce, but not eliminate, high attic temperatures, so insulating to at least R-19 is still advisable to slow heat penetration. The cost for radiant barrier material is very moderate at approximately 20 cents per square foot. See our product section.

Insulation

Insulation slows heat transfer from outside to inside your home, or vice versa in the winter. Attic insulation levels should be R-19 or higher. If you also need to heat your house in the winter, insulation is even more important because of the greater temperature differential between inside and outside. You'll want to have insulation levels of R-30 or higher. In hot climates, if you already have some significant amount of insulation, it will probably be more cost-effective to install a radiant barrier instead of adding more insulation. We offer a low-cost, easy-to-handle, perforated radiant barrier product for attics and for under floors. See the Energy Conservation chapter.

Renewable Energy Systems & Home Cooling

Traditional air conditioners, even small room-sized units, and renewable energy systems are generally incompatible but it can be done. The equipment and knowledge is readily available. The problem is the sheer quantity of electricity required for air conditioning, and the long hours of continuous operation, augmented by the high initial cost of renewable energy systems. As a ballpark estimate, we have found that it will require about $3.50 worth of renewable energy "stuff" for every watt-hour needed.

A 1,500-watt room-size A/C unit running for six hours consumes 9,000 watt-hours, and needs over $30,000 worth of renewable energy system to support it.

Ventilation

Ventilating your attic is important for moisture control and keeping cool. Roofs can reach temperatures of 180°F, and attics can easily exceed 160°F. This heat will eventually soak through even the best insulation unless you get rid of it. Solar-powered vent fans are a very cost-effective way to move out this unwanted heat. By using solar power they'll only run when the sun shines, and the brighter the sun, the faster they'll work. We offer complete kits in two different sizes; they include fan, temperature switch—so it will only run when attic temperature is above 80°F—photovoltaic module(s), and mounting. See the product section.

Roof Whitening

A white roof is an obvious solution. If you are replacing a roof, using a white or reflective roofing surface will reduce your heat load on the house. If you are not reroofing, coating the existing roof with a white elastomeric paint is a cost-effective measure in some climates. Be sure to get a specialty paint specifically designed for roof whitening: normal exterior latex won't do. The Florida Solar Energy Center reports savings from white roofs of 25% to 43% on air conditioning bills in two poorly insulated test homes, and 10% savings in a Florida home with good R-25 roof insulation already in place. Costs are relatively high—the coatings cost 30 to 70 cents per square foot of roof surface area—and adding insulation or a radiant barrier may be a more cost-effective measure. Light-colored asphalt or aluminum shingles are not nearly as effective as white paint in lowering roof and attic temperatures.

Weatherization

Weatherization, usually done in northern climates to keep heat in, can also effectively keep out heat and humidity in the hot, humid climates or seasons. Cutting air infiltration by half, which is easily done, can cut air conditioning bills by 15% and save $50 to $100 per year in the average southern house. Much of the work done by an air conditioner in humid climates goes into removing moisture, which increases comfort. Reducing air infiltration is more important in humid climates than in dry ones.

Alternative Cooling Methods

In this section we'll look at ways to cool your house that don't involve traditional air conditioning. These alternative methods work best in drier summer climates. Low humidity climates tend to have greater temperature swings between night time and day-time. We can take advantage of cooler nights to dump accumulated heat, and recharge the household "coolth."

Whole-House, Ceiling, and Portable Fans

A whole-house fan is an effective means of cooling and is far less expensive to run than an air conditioner. It can reduce indoor temperatures by 3° to 8° F, depending on the outside temperature. Through prudent use of a whole-house fan, you can cut air conditioner use by 15% to 55%. This is particularly true if you're lucky enough to live in a climate that reliably cools off in the evening. A high R-value house can be buttoned up during the heat of the day, heating up inside slowly like an insulated ice cube. In the evening when it's cooler outside than inside, a whole-house fan blows the hot air out of your house, while drawing cooler air through the windows. It should be centrally located so that it draws air from all rooms in the house. Be sure your attic has sufficient ventilation to get rid of the hot air. To ensure a safe installation, the fan must have an automatic shut-off in case of fire.

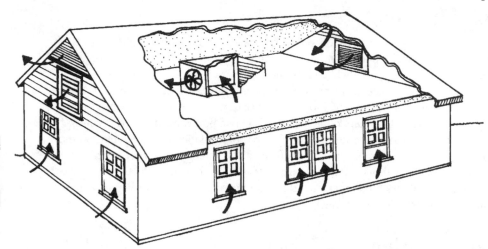

A Whole-House Fan

A whole-house fan can cool the house by bringing in cooler air—especially at night—and lowering attic temperatures. Adapted from an illustration by Illinois Department of Energy and Natural Resources (1987), More for Your Money... Home Energy Savings.

Ceiling, paddle, and portable fans produce air motion across your skin that increases evaporative cooling. A moderate breeze of one to two miles an hour can extend your comfort range by several degrees and will save energy by allowing you to set your air conditioner's thermostat higher or eliminate the need for air conditioning altogether. Less frequent use of air conditioning by setting the thermostat higher will greatly cut cooling bills. See our product section for more details.

Passive Night-Time Cooling

Opening windows at night to let cooler air inside is an effective cooling strategy in many states. This method is far more effective in hot, dry areas than in humid regions, however, because dry air tends to cool down more at night, and because less moisture is drawn into the house. The Florida Solar Energy Center found that when apartments in humid areas (and Florida certainly qualifies) opened their windows at night to let the cooler air in, the air conditioners had to work much harder the next day to remove the extra moisture that came along with the night air. Storing "coolth" works better if your house has thermal mass such as concrete, brick, tile, or adobe that will cool down at night and help prevent overheating during the day. These are the same things that make passive solar houses work better at heating in the winter.

Evaporative Cooling

If you live in a dry climate, an evaporative or "swamp cooler" is an excellent cooling choice, saving 50% of the initial cost and up to 80% of the operating cost of an air conditioner. The lower the relative humidity, the better they work. Evaporative coolers are not effective when your relative humidity is higher than about 40%.

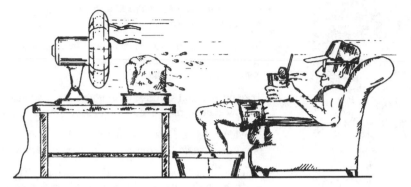

Personal Cooling

Buying a block of ice isn't cost-effective, but feeling cool is important. Cool drinks, small fans, less clothing, siestas, or sitting on a shady porch all work. Illustration courtesy of Saturn Resource Management, Helena, MT.

Personal Cooling

Cool drinks, the Gaiam Real Goods Solar Hat Fan, a minimum of lightweight clothing, siestas, going for a swim, or sitting on a shady porch all work.

Improving Air Conditioner Efficiency

Once you've reduced heat penetration and taken advantage of alternative cooling methods, if you still have a need for mechanical cooling, it makes sense to improve the efficiency of your air conditioner.

Check Ducts for Leaks and Insulate

With central A/C systems, inspect the duct system for leaks and seal with duct tape or special duct mastic where necessary. The average home loses more than 20% of its expensive cooled or heated air before it gets to the register. It's well worth your time to find and seal these leaks. Just turn the fan on and feel along the ducts. Distribution joints are the most common leaky points.

If your ducts aren't already insulated in the attic or basement, insulate them with foil-faced fiberglass duct insulation. Batts can be fastened with plastic tie wraps, wire, metal tape, or glue-on pins. Your A/C technician can do this job or recommend someone who can, if you're not up for it.

Have Your Heat Pump or Central A/C Serviced Regularly

Regular maintenance will include having the refrigerant charge level checked—both under- and overcharging compromises performance—oiling motors and blowers, removing dirt buildup, cleaning filters and coils, and checking for duct leakage. Proper maintenance will ensure maximum efficiency as well as the equipment's longevity.

- Set your A/C to the recirculate option if your system has this choice—drawing hot and humid air from outside takes a lot more energy. Many newer systems only recirculate, so if you can't find this option it's probably already built in.

- Set your A/C thermostat at 78°F—or higher if you have ceiling fans. For each degree you raise the thermostat, you'll save 3%–5% on cooling costs.

- Don't turn the A/C thermostat lower than the desired setting—cooling only happens at one speed. The house won't cool any faster, and you'll waste energy by overshooting.

- Turn off your A/C when you leave for more than an hour. It saves money.

Myth: *It's more efficient to leave the A/C running than to shut it off and re-cool the house later.*

Fact: *You begin saving as soon as you shut off your A/C. If your house is tight and well insulated, it will stay cool for many hours. This only works, of course, if your keep your house closed up. Opening windows not only heats up the house, but also allows the humidity back in that your A/C worked so hard to remove. The exception: on cool nights with low humidity, you're better off opening your windows and using ceiling or whole-house fans.*

- Close off unused rooms or, if you have central A/C, close the registers in those rooms and shut the doors.

- Install a programmable thermostat for your central A/C system to regulate cooling automatically. Such a thermostat can be programmed to ensure that your house is cool only when needed.

- Trim bushes and shrubs around outdoor condensor units so they have unimpeded airflow; a clear radius of two feet is adequate. Remove leaves and debris regularly.

- Provide shade for your room A/C, or the outside half of your central A/C if at all possible. This will increase the unit's efficiency by 5% to 10%.

- Remove your window-mounted A/C each fall. Their flimsy mounting panels and drafty cabinets offer little protection from winter winds and cold. If you are unable to remove the unit, at least close the vents and tightly cover the outside.

- Clean your A/C's air filter every month during cooling season. Normal dust buildup can reduce airflow by 1% per week.

- Clean the entire unit according to the manufacturer's instructions at least once a year. The coils and fins of the outside condenser units should be inspected regularly for dirt and debris that would reduce air flow. This is part of the yearly service if you're paying a technician for the service.

If You Live in a Humid Climate

An important part of what a conventional air conditioner does is removing moisture from the air. This makes the body's normal cooling—sweating—work better, and you feel cooler. Unfortunately, some of the highest-efficiency models don't dehumidify as well as less-efficient air conditioners. A/C units that are too big for their cooling load have this problem also. High humidity leads to indoor condensation problems and gives us a sweaty mid-August New York City feeling—minus the physical and aural assaults. People tend to set their thermostats lower to compensate for the humidity, using even more energy. Here are some better solutions.

- Reduce the fan speed. This makes the coils run cooler and increases the amount of moisture that will condense in the A/C rather than inside your house.

- Set your A/C to recirculate if this is an option and you haven't already done so.

- Choose an A/C with a "sensible heat fraction" (SHF) less than 0.8. The lower the SHF, the better the dehumidification.

- Size your A/C carefully. Contractors should calculate dehumidification as well as cooling capacity. Many air conditioning systems are oversized by 50% or more. The old contractor's rule of thumb—one ton of cooling capacity per 400 square feet of living space—results in greatly oversized A/C units when reasonable weatherization, insulation, and shading measures have been done to reduce cooling load. One ton per 800 to 1,000 square feet may be a more reasonable rule of thumb.

- Investigate using a Dinh heat pipe/pump. These devices cool the air before it runs through the A/C evaporator, and warm it as it blows out, and this without energy input. They can help your A/C dehumidify several times more effectively.

- Explore "desiccant dehumidifiers," which, coupled with an efficient A/C, may save you substantial energy. Desiccant dehumidifiers use a water-absorbing material to remove moisture from the air. This greatly reduces the amount of work the A/C has to do, thereby reducing your cooling bill.

When shopping, how can you tell which A/C units are the most efficient ones?

Room air conditioners are rated by an Energy Efficiency Ratio (EER). The higher the EER, the better. Typical new room air conditioners have EERs of about 9, but the best ones are 12 or higher. Make sure that any unit you purchase has a thermostat to shut it off when cooling is no longer necessary.

Central air conditioners and heat pumps are rated on their Seasonal Energy Efficiency Ratio (SEER). Again, the higher, the better. Older central A/Cs have SEER ratings of 7 to 8. National standards now require a minimum SEER of 10. The most efficient models have SEER values of 13 to 16. The Florida Solar Energy Center recommends a SEER of at least 12 for cost-effectiveness.

Buying a New Air Conditioner

Air conditioners have become much more efficient over the last 15 years and top-rated models are 50% to 70% better than the current average. Federal appliance standards have eliminated the least-efficient models from the market, but builders and developers have little incentive to install a model that's more efficient than required, since they won't be paying the higher electric bills.

If you have an older model, it may be cost-effective to replace it with a properly sized, efficient unit. The Office of Technology Assessment—a U.S. Congress research organization—estimates that buying the most-efficient room air conditioner, costing $70 more than a standard unit, has a payback of six and a half years and less, if the additional features the better units include are taken into account.

Air Conditioner Efficiency

Several studies have found that most central air conditioning systems are oversized by 50% or more. To avoid being sold an oversized model, or one that won't remove enough moisture, ask your contractor to show you the sizing calculations and dehumidification specifications. Many contractors may simply use the "one ton of cooling capacity per 400 square feet of living space" rule of thumb. This will be far larger than you need, especially if you have weatherized and done many of the other measures we discussed here to reduce heat gain and the need for cooling.

One ton of cooling capacity—an archaic measurement equaling approximately the cooling capacity of one ton of ice—is 12,000 Btu per hour. Room air conditioners range in size from half a ton to one and a half tons, while typical central A/Cs range from two to four tons.

Desuperheaters

It's possible to use the waste heat removed from your house to heat your domestic water. In hot climates, where A/C is used more than five months per year, it makes economic sense to install a desuperheater to use the waste heat from the A/C to heat water. Some utilities give rebates to builders or homeowners who install such equipment.

Evaporative Coolers

If you live in a dry climate, an evaporative or "swamp" cooler can save 50% of the initial cost, and up to 80% of the operating cost of normal A/C. An evaporative cooler works by evaporating water, drawing fresh outside air through wet porous pads. This is an excellent choice for most areas in the Southwestern "sunbelt." The lower the relative humidity, the better they work. At relative humidity above 30%, performance will be marginal; above 40% humidity, forget it! "Indirect" or "two-stage" models that yield cool, dry air rather than cool, moist air are also available, but the same low humidity climate restrictions apply.

Heat Pumps

If you live in a climate that requires heating in the winter and cooling in the summer, consider installing a heat pump in new construction, or if you are replacing the existing cooling systems. The ground- or water-coupled models are the most efficient; see the chapter on heating, and the comparison chart included there. The air conditioning efficiency of a heat pump is equivalent to that of a typical air conditioner, as they're basically the same thing, but the heating mode will be much more efficient than the cooling mode, outperforming everything except natural gas-fired furnaces.

Appliances for Cooling

Since our core business is renewable energy, and the high energy consumption of air conditioning equipment makes renewable energy systems very expensive, we offer a minimal selection of appliances. Apply the weatherization, insulation, shading, and alternative cooling tips we discussed, and your need for energy-intensive mechanical cooling will be eliminated in all but the most severe climates.

Real Goods sells a selection of DC-powered cooling appliances that can be run from batteries or directly from photovoltaic modules. These include small, highly efficient evaporative coolers, a selection of fans for attic venting or personal cooling, and some ceiling-mounted paddle fans.

www.realgoods.com

Cooling Products

SolarCool DC Evaporative Cooler

This is a superefficient evaporative cooler that delivers 1,000 cfm running 2.2 amps at 18 volt from a 40- to 45-watt PV module. Consumes under 1 gallon of water per hour. Features 3,456 cubic inches of evaporative pad media that is treated to resist hard water deposits; lasts up to five seasons before replacement. Quality construction is completely assembled from galvanized steel, with a tan powder-coated finish. Has a long-life 24-volt fan motor and water pump, protective fan outlet grill, a brass float valve, and a wiring junction box. Designed to provide cooling for one or two rooms or a trailer. Can be operated from a 24-volt battery using a variable fan speed control, or from a 12-volt battery using the voltage doubler below. Measures 27″W x 25″H x 20″D, weighs 35 lb dry. USA.

◈ **64-499 SolarCool Evaporative Cooler $495⁰⁰**
64-505 Fixed Voltage Doubler $59⁰⁰

◈ *Means shipped directly from manufacturer.*

Attic Ventilation

Solar-Powered Attic Vents

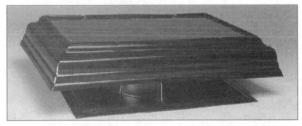

This solar-powered fan vents your attic using electricity from a built-in 10-watt PV module. The DC fan removes air at 600 to 800 cfm under full sun. Three square feet of soffit air inlets are recommended for each solar fan. Airflow is highest when the sun is hottest, at midday. Shuts off automatically at night or on cloudy days. Select from flat or pitched roof models. The pitched roof model has a flat flange base that is designed to mount under shingles; the flat roof model features a crimped flange base that mounts onto a 2″ x 4″ framed curb measuring 16″ x 19″. Venting your attic reduces cooling costs in summer, prevents moisture buildup in winter, and with this fan, never costs a cent on your electric bill. Three-year mfr.'s warranty. USA.

◈ **64-238 Solar Attic Fan, Flat Roof $359⁰⁰**
◈ **64-239 Solar Attic Fan, Pitched Roof $359⁰⁰**

12″ and 16″ DC Fans

These are the same fans used in our complete Attic Fan Kits. Both models use heavy-duty 36-volt DC motors that can be run with either 12- or 24-volt systems. Frames have three rubber grommeted mounts. A nonadjustable thermostat switch that closes at 110°F and opens at 90°F (approximate temperature) is included. Input power and output in cubic feet per minute are given in the chart above right. Specs at 14 and 28 volts simulate PV-direct applications. Specs at 12 and 24 volts are battery-driven.

12″ DC FAN

VOLTAGE INPUT	AMPS DRAW	CFM OUTPUT
12 volts	1.17 amps	740
14 volts	1.47 amps	1,050
24 volts	3.20 amps	1,335
28 volts	4.20 amps	1,430

16″ DC FAN

VOLTAGE INPUT	AMPS DRAW	CFM OUTPUT
12 volts	1.47 amps	810
14 volts	1.84 amps	1,100
24 volts	4.20 amps	2,000
28 volts	5.30 amps	2,275

64-216 12″ DC Fan $119⁰⁰
64-217 16″ DC Fan $139⁰⁰

Solar-Powered Attic Fan Kits

Moving hot, stagnant air out of your attic, greenhouse, or barn is the least energy-intensive way to keep your space cool. Using solar power to do the work makes even more sense. The brighter the sun shines, the faster the fan runs. Our Solar Fan Kits give you everything you need. Kits include a high-efficiency DC fan, matched PV modules, mounting brackets, wiring, a thermoswitch to prevent running in cold weather, and instructions. Depending on mounting locations, additional wire from PV modules to fan may be required.

The 12″ Fan Kit has a 13-watt PV module and moves 600–800 CFM in full sun. The 16″ Fan Kit uses a pair of 13-watt PV modules, and moves 800–1,000 CFM. Fan frames are rubber grommet mounted for quiet operation. Modules are 15.1″ x 14.4″ single crystal with tempered glass and aluminum frame.

64-501 12″ Attic Fan Kit $249⁰⁰

64-502 16″ Attic Fan Kit $379⁰⁰

Oscillating 8″ DC Table Fan

This 12-volt table fan features a weighted base with rubber feet, and 2-speed operation. Draws about 1.2 amps on low, and 2.0 amps on high. Oscillating feature can be switched on or off, just like a real fan! Taiwan.

64-241 12V, 8″ Table Fan $39⁹⁵

12V Personal Fans

Jet Fan

This small, personal fan will move more air with less noise and power use than anything you've ever seen before. Designed for permanent, screw-down mounting, you may want to devise your own mobile mounting base. Black, impact-resistant plastic, can mount in any direction, will swivel or tilt to any position, and is supplied with a five-foot power cord. The Jet fan is a 6″ diameter fan, 8.2″ fixture height. Has a single speed, on/off switch on the housing. 12-volt. Draws only 4 watts! Austria.

64-481 Jet Fan, 12V $39⁹⁵

Compact Desk Fan

*Powerful, Quiet, Battery-Operated Fan
Keeps Air Fresh and Cool*

Overheated offices can be unbearable. This powerful, variable-speed fan will keep you cool and comfortable. The computerized speed control allows for a complete range of fan speeds—a better option over typical three-speed fans. The last setting is conveniently remembered when the fan is turned on and off again, and the tilt of the head is adjustable. Offering an impressive air flow, it runs over 300 hours on just one set of four "D" batteries (not included). Or, use the optional 6-volt AC/DC adapter and plug in anywhere. Measures 5.75″ W x 3.625″ L x 8.75″ H. Canada.

64-255 Compact Desk Fan $24⁹⁵

64-256 AC/DC Adapter $9⁹⁵

Thermofor Ventilator Opener

The Thermofor is a compact, heat-powered device that regulates window, skylight, or greenhouse ventilation according to temperature, using no electricity and requiring no wiring. You can set the temperature at which the window starts to open between 55° and 85°F. It will open any hinged window up to 15 pounds a full 15 inches, and can be fitted in multiples on long or heavy vents. These units are ideal for greenhouses, animal houses, solar collectors, cold frames, and skylights.

64-302 Thermofor Solar Vent Opener $59⁰⁰

Our Most-Efficient DC Ceiling Fan

For keeping comfortable with a 12- or 24-volt power system, you can't beat the efficiency and quiet operation of our 42˝ diameter DC ceiling fan. It comes with both a close flush-mount for flat ceilings, or a down rod ball-mount for sloped ceilings. The fan can be assembled with either four or five blades, showing either walnut or oak finish (five blades included). The fan body is a charcoal-gray ABS injection casting that can be painted to match or left stock. Reversible and speed-adjustable using the solid-state controls sold below. Close mount hangs down 9.5˝; down rod needs 13˝.

Performance specs; at 12 volts: 60 rpm, 4.8 watts, 1800 cfm; at 24 volts: 120 rpm, 19.2 watts, 4000 cfm. Note that performance is . . . um . . . modest at 12 volts. 12-volters may be interested in the 12 to 24 voltage doubler speed control sold below. Fan weight; 11 lb. USA.

64-495 RCH DC Ceiling Fan $189⁰⁰

24V Variable Fan Speed Control With Reversing Switch

This solid-state DC speed control boasts 97% efficiency, smooth linear speed control from 0% to 100%, a reversing switch with a convenient center-off position, radio frequency filtering, a silencing capacitor for quiet motor operation at low speeds, and it can even be connected to a standard millivolt thermostat for automatic operation. Whew! That seems to cover everything. Mounted on an ivory color single-gang electric box cover. Includes mounting screws and detailed, clearly illustrated installation instructions. 24V DC only. USA.

64-497 DC Fan Control w/Reverse 24V/1.5A $59⁹⁵

12V Variable Fan Speed Control With Reversing Switch

Identical looks, performance, and features with our 24V unit above, but made for 12-volt use. 2.0 amps, 12V DC only. USA.

64-504 DC Fan Control w/Reverse 12V/2.0A $49⁹⁵

Voltage Doubler Variable Speed Control

An adaptable solid-state DC controller. With 12-volt input, the Doubler can provide smooth linear speed control up to 24 volts for our DC ceiling fan, or it can deliver a steady, regulated 12.6 to 24.0 volts for recharging laptops, cell phones, or other electronics. Output voltage is not affected by variable input voltage. Delivers up to 90% efficiency and up to 2.0 amps output. If overloaded, power output will reduce to a safe level without damage. Mounted on a double-gang ivory colored electric box cover. Includes mounting screws and detailed, clearly illustrated instruction manual. 12VDC input only. Not for use with SolarCool Evaporative Cooler. USA.

64-498 12 to 24 Voltage Doubler $129⁰⁰

A Truly Efficient AC-Powered Ceiling Fan

Here's an Emerson 42˝ diameter ceiling fan that really moves the air without burning up the power. And unlike many multi-speed ceiling fans, it runs quietly on modified sine wave inverters. This three-speed, reversible fan draws only 20, 30, and 50 watts on low, medium, and high speed. A pull chain changes speeds; a switch controls direction. The five blades are reversible, with light oak on one side, and white on the other. The sealed, permanently lubricated motor body is white. The short 4˝ stem mount can be used on flat ceilings, or sloped ceilings up to 35 degrees. 120VAC only. Manufacturer's warranty of 15 years on motor, 1 year on all else. Taiwan.

64-478 Emerson 42˝ AC Fan $119⁰⁰

Sun Tex Shadecloth And Super Solar Screen
Improved Window Screening

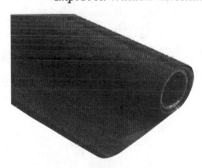

One of our best-selling energy conservation products just got better. The original Sun Tex Shadecloth blocks up to 80% of the sun's damaging glare, while still admitting plenty of natural light and air movement. The Super Solar Screen raises protection to 90%, with a finer mesh that provides for greater protection. A great way to create privacy, the cloths appear virtually opaque from outside, while allowing you to see out clearly. Made from heavy-duty coated polyester, they're mildew and fade resistant. Easy to install, easy to remove, they block the sun's glare through your windows, skylights, greenhouses, patios, and doors. USA.

63-569 Sun Tex Shadecloth 36˝ x 84˝ $15⁹⁵
63-612 Sun Tex Shadecloth 48˝ x 84˝ $21⁹⁵
63-613 Super Solar Screen 36˝ x 84˝ $19⁹⁵
63-614 Super Solar Screen 48˝ x 84˝ $24⁹⁵

Cool Your Body Temperature Way Down With Solar Hat Fans

Our best selling new product of 2000!

It's hot. It's cool. It's one heck of a nifty gadget that really cools you down on a hot day! The Solar Hat Fan will impress your friends and neighbors, and keep your head and body cool on sunny days. It clamps on to your favorite hat (stiffer-type brims and baseball caps are recommended) and gets to work harvesting enough solar energy to directly power a working fan. Features a small photovoltaic panel with an adjustable mount. We sold 40 of these solar hat fans to a roofing company and the workers love them as they cool down from the 100°+ temperature. Measures 2.75″ L x 2.125″ W x 2.25″ H. China.

90-323 Solar Hat Fan $8⁹⁵

3 or more $7⁹⁵ each

Solar Safari Hat

Not since Dr. Livingston explored the Congo has technological innovation touched the pith helmet. Now, a perennial favorite has been made much better. The Solar Safari Hat, with its open-weave construction, adjustable headband, and a fan powered by the sun or 2 AA batteries, is worn by gardeners, fishermen, boaters, and African explorers. White only. Batteries not included.

90-411 Solar Safari Hat $39⁰⁰

50-262 AA NiMH Batteries (2-pack) $4⁵⁰

www.realgoods.com

Energy-Efficient Appliances

Why Bother to Shop for Energy Efficiency?

Wasted energy translates into carbon dioxide production, air pollution, acid rain, and lots of money down the drain. The average American household spends more than $1,100 per year on appliances and heating and cooling equipment. You can easily shave off 50%–75% of this expense by putting some intelligence into your appliance choices. For example, simply replacing a 20-year-old refrigerator with a new energy-efficient model will save you about $85 per year in reduced electric bills while saving 1000 kWh of electricity, and reducing your home's CO_2 contribution by about a ton per year, contributing to global cooling. Highly efficient appliances may be slightly more expensive to buy than comparable models with lower or average efficiencies. However, the extra first cost for a more efficient appliance is paid back through reduced energy bills long before the product wears out.

What Appliances Can I Buy from Real Goods?

Most appliance manufacturers prefer to sell through established appliance stores. Your mail-order choices are limited to more specialized appliances. At present we offer the Staber horizontal-axis washing machine, the Spin-X centrifugal dryer, several gas refrigerators and freezers, high-efficiency Aquastar tankless water heaters, and solar water heaters.

We want to enable our customers to make intelligent choices on their appliance purchases. Thanks to the nonprofit American Council for an Energy Efficient Economy, we can offer the *Consumer Guide to Home Energy Savings*. This is a more-or-less annual listing of the most efficient refrigerators, clothes washers, dishwashers, tank-type water heaters, air conditioners, and home heating appliances available. Fortify yourself with this wonderful book before venturing into any major appliance purchase.

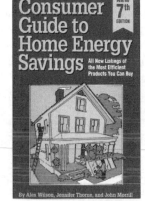

**7th Edition
Consumer Guide to
Home Energy Savings**

82-399 $8⁹⁵

For details, see page 274.

Refrigerators and Freezers

The refrigerator is likely to be the largest single power-user in your home aside from air conditioning and water heating. Refrigerator efficiency has made enormous strides in the past few years, largely due to insistent prodding from the Feds with tightening energy standards. An average new fridge with top-mounted freezer sold today uses under 650 kilowatt-hours per year, while the average model sold in 1973 used nearly 2,000 kilowatt-hours per year. These are national average figures. The most efficient models available today are under 500 kilowatt-hours per year, and still dropping. The typical refrigerator has a life span of 15 to 20 years. The cost of running over that time period will easily be two to three times the initial purchase price, so paying somewhat more initially for higher efficiency offers a solid payback. It may not be worth scrapping your 15-year-old clunker to buy a new energy-efficient model. But when it does quit, or it's time to upgrade, buy the most efficient model available. A great source for listings of highly efficient appliances is the Environmental Protection Agency's Energy Star® website, www.epa.gov/energystar/.

Be sure to have the freon removed professionally from your clunker before disposal. Most communities now have a shop or portable rig to supply this necessary service at reasonable cost.

Refrigerators and CFCs

All refrigerators use some kind of heat transfer medium, or refrigerant. For conventional electric refrigerators this has always been freon, a CFC chemical and prime bad-guy in the depletion of the ozone layer. Manufacturers are now rapidly switching to non-CFC refrigerants, and by now it should be difficult to find a new fridge using CFC-based refrigerants in an appliance showroom.

What most folks probably didn't realize is that of the 2.5 lb of CFCs in the typical 1980s fridge, 2 lb are in the foam insulation. Until very recently, most manufacturers used CFCs as blowing agents for the foamed-in-place insulation. Most have now switched to non-CFC agents, but if you're shopping for new appliances, it's worth asking.

The CFCs in older refrigerators must be recovered before disposal. For the 0.5 lb in the cooling system this is easily accomplished, and your local recycler, dump, or appliance repair shop should be able to either do it or recommend a service that will. To recover the 2.0 lb in the foam insulation requires shredding and special equipment, which sadly only happens rarely.

Refrigeration and Renewable Energy Systems

When your energy comes from a renewable energy system, which has a high initial cost, the payback for most ultraefficient appliances is immediate and handsome. You will save the price of your appliances immediately in the lower costs of your power generating system. This is particularly true for refrigerators.

Many smaller, or intermittent-use renewable energy system owners prefer gas-powered fridges and freezers, especially when the fuel is already on site anyway for water heating and cooking. Gas fridges have the clean 'n green advantage of using a mixture of ammonia and water, not freon, as a refrigerant. We offer several gas-powered freezers and refrigerator/freezer combinations.

For many years Sun Frost has had the enviable position of being the maker of "world's most efficient refrigerators," and with that has come a steady flow of renewable energy customers willing to pay the high initial cost in exchange for the savings in operating cost. Real Goods has been an unabashed supporter of this product for many years. But times are changing. The major appliance manufacturers have managed to implement huge energy savings in their mass-produced designs over the past few years. While few of the mass-produced units challenge Sun Frost for king-of-the-efficiency-hill, many are close enough, and less expensive enough, to deserve consideration. The difference between a $900 purchase and a $2,700 purchase will buy quite a lot of photovoltaic wattage. A number of mass-produced fridges have power consumption in the 1.0 to 1.5 kilowatt hours per day range, and can be supported on renewable energy systems. Any reasonably sized fridge with power use under 1.5 kilowatt hours per day is welcome in a renewable energy system; that's $45 to $50 per year on the yellow EnergyGuide tag that every new mass-produced appliance wears in the showroom.

In Sun Frost's defense, we need to point out that conventional refrigerators run on 120 VAC, which must be supplied with an inverter, and will cost an extra 10% due to inverter efficiency. Also, if your inverter should fail, you could lose refrigeration until it gets repaired. Sun Frosts (if ordered as DC-powered units) will run on direct battery power, making them immune to inverter failures.

Improving Your Existing Fridge's Performance

Here is a checklist of things that will help any fridge do its job more easily and more efficiently.

- Cover liquids and wrap foods stored in the fridge. Uncovered foods release moisture (and get dried out), which makes the compressor work harder.

- Clean the door gasket and sealing surface on the fridge. Replace the gasket if damaged. You can check to see if you are getting a good seal by closing the refrigerator door on a dollar bill. If you can pull it out without resistance, replace the gasket. On new fridges with magnetic seals, put a flashlight inside the fridge some evening, turn off the room lights, and check for light leaking through the seal.

- Unplug the extra fridge or freezer in the garage. The electricity the fridge is using—typically $130 a year or more—costs you far more than the six-pack or two you've got stashed there. Take the door off, or disable the latch so kids can't possibly get stuck inside!

- Move your fridge out from the wall and vacuum its condenser coils at least once a year. Some models have the coils under the fridge. With clean coils the waste heat is carried off faster, and the fridge runs shorter cycles. Leave a couple of inches of space between the coils and the wall for air circulation.

- Check to see if you have a power-saving switch or a summer-winter switch. Many refrigerators have a small heater (yes, a heater!) inside the walls to prevent condensation build-up on the fridge walls. If yours does, switch it to the power-saving (winter) mode.

The EPA and Refrigerator Power Use Ratings

We're all familiar with the yellow energy-use tags on appliances now. These level the playing field, and make comparisons between competing models and brands simple. With some appliances, like refrigerators, the EPA takes a "worst case" approach. Fridges are all tested and rated in 90°F ambient temperatures. The test fridge is installed in a "hot box" that maintains a very stable 90° interior temperature. This results in higher energy use ratings than will be experienced in most settings. At the mid-70° temperatures that most of us prefer inside our houses, most fridges will actually use about 20% to 30% less energy than the EPA tag indicates.

- Defrost your fridge if significant frost has built up.

- Turn off your automatic ice maker. It's more efficient to make ice in ice trays.

- If you can, move the fridge away from any stove, dishwasher, or direct sunlight.

- Set your refrigerator's temperature between 38°F and 42°F, and your freezer between 10°F and 15°F. Use a real thermometer for this, as the temperature dial on the fridge doesn't tell real temperature.

- Keep cold air in. Open the fridge door as infrequently and briefly as possible. Know what you're looking for. Label frozen leftovers.

- Keep the fridge full. An empty fridge cycles frequently without any mass to hold the cold. Beer makes excellent mass, and you probably always wanted a good excuse to put more of it in the fridge, but it tends to disappear. In all honesty, plain water in old milk jugs works just as well.

Buying a New Refrigerator

A new, more efficient refrigerator can save you $70 to $80 a year, and will completely pay for itself in nine years. It frees you from the guilt of harboring a known ozone killer. Since 1999 all new fridges in the showroom are completely CFC-free.

Shop wisely by carefully reading the yellow EPA EnergyGuide label found on all new appliances. Use it to compare models of similar size.

The American Council for an Energy-Efficient Economy is a nonprofit organization dedicated to advancing energy efficiency as a means of promoting economic development and environmental protection. Every year ACEEE publishes an extremely handy book called Consumer Guide to Home Energy Savings. This invaluable guide lists all the most efficient mass-produced appliances by manufacturer, model number, and energy use. We strongly recommend fortifying yourself with this guide before venturing into any appliance showroom. It has an amazingly calming effect on overzealous sales personnel, and allows you to compare energy use against the top of the class models. See the product section.

Refrigerator Shopping Checklist

- Smaller models will obviously use less energy than larger models. Don't buy a fridge that's larger than you need. One large refrigerator is more efficient than two smaller ones however.

- Models with top- or bottom-mounted freezers average 12% less energy use than side-by-side designs.

- Features like through-the-door ice, chilled water, or automatic ice-makers increase the profit margin for the manufacturer, and the purchase price by about $250. So ads and salespeople tend to push them. They also greatly increase energy use and are far more likely to need service and repair. Avoid these costly, troublesome options.

- Make sure that any new refrigerator you buy is CFC-free in both the refrigerant and the foam insulation.

- Be willing to pay a bit more for lower operating costs. A fridge that costs $75 more initially, but costs $20 less per year to operate due to better construction and insulation, will pay for itself in less than four years.

Energy-Efficient Home Appliances

Consumer Guide to Home Energy Savings

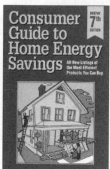

By Alex Wilson and John Morrill. The updated seventh edition of *Consumer Guide to Home Energy Savings* identifies the most energy-efficient home appliances by brand name and model number. This book is as compact and efficient as its subject matter. Its pages are crammed with money-saving information. 5″ x 7″, 296 pages, 118 illustrations, appendices, index, paperback.

82-399 7th Ed. Consumer Guide to Home Energy Savings $8⁹⁵

Refrigeration

Sun Frost Refrigerators & Freezers

For over 15 years Sun Frost has been the first choice for off-the-grid homesteads, thanks to a variety of design innovations that result in the best energy efficiency and quietest operation available. Major appliance manufacturers are getting close, but the Sun Frost is still king of the efficiency hill, particularly for off-the-grid folks who can run directly on DC power. The smaller Sun Frosts (RF-12 and RF-16) are so efficient that they can be powered by only two 75-watt PV modules.

Compressors are top-mounted for thermal efficiency. Without the usual plumbing and mechanical equipment on the bottom, a 13″ stand is recommended to raise the bottom shelf level. Sun Frost's optional stand has two drawers. Standard models are off-white Navimar (a Formica-like laminate). Colors are available for $100, wood veneers are available for $150, unfinished.

In the listings below R stands for refrigerator, F for freezer, followed by the cubic footage. All models are 34.5″ wide by 27.5″ deep. Heights vary. Allow an additional 4″ width if door will open against a wall. Doors will clear standard 24″-deep countertops. All models available in 12 or 24 volts DC or 120 VAC. Voltage is not field switchable; specify when ordering. Can be hinged on the right or left side (when facing the unit). Hinge side is not field switchable; specify when ordering. Sun Frosts are made to order, usually requires 4–6 weeks. Ships freight collect from Northern California.

62-134 Sun Frost RF-16 (DC) $2,655⁰⁰

62-134-AC Sun Frost RF-16, 120VAC $2,495⁰⁰

62-135 Sun Frost RF-19 (DC) $2,799⁰⁰

62-135-AC Sun Frost RF-19, 120VAC $2,645⁰⁰

62-125 Sun Frost R-19 (DC) $2,539⁰⁰

62-125-AC Sun Frost R-19, 120VAC $2,459⁰⁰

62-115 Sun Frost F-19 (DC) $2,855⁰⁰

62-115-AC Sun Frost F-19, 120VAC $2,699⁰⁰

62-133 Sun Frost RF-12 $1,979⁰⁰

62-133-AC Sun Frost RF-12 120VAC $1,899⁰⁰

62-122 Sun Frost R-10 (DC) $1,595⁰⁰

62-122-AC Sun Frost R-10 120VAC $1,515⁰⁰

62-112 Sun Frost F-10 (DC) $1,699⁰⁰

62-112-AC Sun Frost F-10 120VAC $1,619⁰⁰

62-131 Sun Frost RF-4 (DC) $1,385⁰⁰

62-121 Sun Frost R-4 (DC) $1,385⁰⁰

62-111 Sun Frost F-4 (DC) $1,385⁰⁰

◈ 62-116 Sun Frost Stand, White $259⁰⁰

◈ 62-117 Sun Frost Stand, Color $299⁰⁰

◈ 62-118 Sun Frost Stand, Woodgrain $325⁰⁰

◈ 62-120 Sun Frost Color Option $100⁰⁰

◈ 62-119 Sun Frost Woodgrain Option $150⁰⁰

◈ 62-190 Sun Frost Crating Charge

Shipped freight collect from Northern California.

◈ Means shipped directly from manufacturer.

SUN FROST TECHNICAL DATA

MODEL	POWER USAGE @ 70°F*	POWER USAGE @ 90°F*	WEIGHT CRATED	VOLUME CRATED	EXTERIOR DIMENSIONS
RF-19	770 W-H/Day	1030 W-H/Day	320 lb.	46 cu. ft.	66″ x 34.5″ x 27.75″
R-19	380	630	310	46	66″ x 34.5″ x 27.75″
F-19	1250	1630	320	46	66″ x 34.5″ x 27.75″
RF-16	560	810	300	44	62.5″ x 34.5″ x 27.75″
RF-12	350	560	230	36	49.3″ x 34.5″ x 27.75″
R-10	190	310	215	32	43.5″ x 34.5″ x 27.75″
F-10	690	880	215	32	43.5″ x 34.5″ x 27.75″
R-4	110	160	160	23	31.5″ x 34.5″ x 27.75″
F-4	350	450	160	23	31.5″ x 34.5″ x 27.75″
RF-4	160	240	160	23	31.5″ x 34.5″ x 27.75″
RFV-4	160	230	160	23	31.5″ x 34.5″ x 27.75″

*24-hour closed-door test with stated ambient outside temperature.

INTERIOR DIMENSIONS OF SUN FROST REFRIGERATORS AND FREEZERS

MODEL	FREEZER SECTION HEIGHT	DEPTH	WIDTH	VOLUME	REFRIGERATOR SECTION HEIGHT	DEPTH	WIDTH	VOLUME
R, F, RF-19	24″	20.75″	28″	8.1 cu. ft.	24″	20.75″	28″	8.1 cu. ft.
RF-16	13″	20″	26″	3.91 cu. ft.	31″	20.75″	28″	10.4 cu. ft.
RF-12	6.5″	21″	26″	2.05 cu. ft.	24″	20.75″	28″	8.1 cu. ft.
R-10	—	—	—	—	28″	20.5″	27.5″	9.3 cu. ft.
F-10	28″	20.5″	27.5″	9.13 cu. ft.	—	—	—	—
R-4	—	—	—	—	13″	20″	26″	3.9 cu. ft.
F-4	13″	20″	26″	3.91 cu. ft.	—	—	—	—
RF-4	2.5″	20″	26″	.68 cu. ft.	10.5″	20″	26″	3.2 cu. ft.
RFV-4	4″	20″	26″	1.2 cu. ft.	6″	20″	26″	1.8 cu. ft.

Minus 40 Refrigerators & Freezers

For Your Next Expedition or Adventure

The Minus 40 line of DC-powered refrigerators and freezers was developed for tough off-road safari conditions in Africa. These rugged units feature extra-thick polyurethane insulation, so they only need power for three or four hours per day to keep their cool. Although these robust units were designed for vehicle or trailer mounting, they work great for boats, camping, or residential service. Only needing power for a few hours per day makes them an excellent choice for emergency or disaster backup. Minus 40 coolers can be powered from 12- or 24-volt DC using the same compressor without modification or special ordering. For dependable knock-around field service, there are no exposed coils on the exterior or cooling equipment protruding on the interior. Interiors are rustproof stainless steel. Lids have positive, lockable catches, and bodies have heavy-duty carrying handles. Thermostats can be adjusted for fridge or freezer operation. All Minus 40 coolers use CFC-free R134A refrigerant. Made in South Africa, warranted for one year by the U.S. importer.

Four models are available. True interior capacity is listed. Power use is listed for 24 hours at 75°F. Higher temperatures will increase power use.

Kalahari

Safari

Minus 40 Safari

The middle-sized Minus 40 model has 2.0 cu. ft. internal capacity, measures 32.9″ L x 20.1″ W x 19.7″ H, weighs 102 pounds, and consumes about 600 watt-hours per day. This model is equipped with an external digital temperature gauge.

◈ **62-538 Minus 40 Safari $2,095⁰⁰**

Minus 40 Bushmaster

The next-to-largest Minus 40 model has 4.1 cu. ft. internal capacity, measures 36.2″ L x 22.0″ W x 27.6″ H, weighs 125 pounds, and consumes about 840 watt-hours per day.

◈ **62-537 Minus 40 Bushmaster $1,950⁰⁰**

Minus 40 Kalahari

The largest Minus 40 model has 5.4 cu. ft. internal capacity, measures 55.1″ L x 20.1″ W x 28.3″ H, weighs 165 pounds, and consumes about 960 watt-hours per day.

◈ **62-539 Minus 40 Kalahari $2,195⁰⁰**

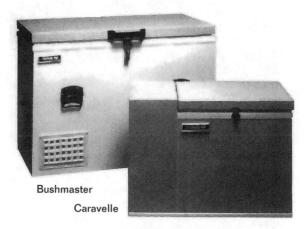

Bushmaster

Caravelle

Minus 40 Caravelle

Minus 40's smallest model has 1.4 cu. ft. internal capacity, measures 26.8″ L x 14.6″ W x 18.9″ H, weighs 77.6 pounds, and consumes about 480 watt-hours per day.

◈ **62-536 Minus 40 Caravelle $1,595⁰⁰**

An Energy-Efficient And Ozone-Friendly Refrigerator

The U.S. Dept. of Energy estimates that the typical household devotes 15% of its energy budget to the refrigeration and freezing of food. The redesigned Conserv offers high efficiency environmental design at a reasonable price. It uses newly formulated foam insulation and R-134A refrigerant that contains absolutely no CFC products. This unit also meets the strict European standards for recyclability.

Standing just over 6´ 5˝ tall, the refrigerator over freezer design allows easy access to the most commonly used items without bending over.

The Conserv has separate compressors and controls for each of the refrigerator and freezer sections. The condenser and cooling tubes are built into the walls of the unit, and are sealed so no dust collects on the working parts of the unit, which would lower the efficiency. This "sealed" design also ensures that the Conserv will operate very quietly and never require cleaning.

Internal volume is approximately 11 cu. ft. Although the power consumption is somewhat higher, this refrigerator is a reasonable alternative to Sun Frost for independent energy homes and for the environmentally-conscious homeowner. EnergyGuide tagged at $47/year. Available in off-white enamel, or brushed stainless steel exterior finish. Denmark—warranted by U.S. importer.

FEATURES:
- Built in condenser gives lower energy consumption: no dust, low noise
- Separate compressors for fridge and freezer
- Can be built in
- Left- or right-hand opening, field-switchable

FRIDGE SECTION HAS:
- Interior light
- Automatic defrost
- Flexible door shelf arrangement

FREEZER SECTION HAS:
- Defrost water outlet (fold-out) for manual defrost
- Combination of pull-out drawers and fast freeze shelves

CONTROL PANEL AT TOP OF CABINET WITH:
- Thermometer for fridge or freezer section
- Thermostats for fridge and freezer sections
- Fast Freeze button and lamp

SPECIFICATIONS:

NET CAPACITY:
Fridge: 6.9 cu. ft.
Freezer: 4.2 cu. ft.
Height: 78.7˝
Width: 23.4˝
Depth: 23.4˝
Freezing Capacity (24 hrs): 39.5 lb
Electrical consumption EPA specs @ 90°F : 1.4 KWh/day

🚚 **62-304 Conserv Refrigerator $1,039⁰⁰**
🚚 **62-544 Conserv Stainless Refrigerator $1,239⁰⁰**

Shipped freight collect from Scottsdale, AZ

Since the Conserv is taller and narrower than American-made refrigerators, it will generally not fit into the standard refrigerator spaces with cabinets above.

This is the same as the Vestfrost refrigerator we've been selling for several years. The Conserv model is a cosmetic redesign.

Super-Efficient, CFC-Free Freezer—Costs Only 5¢ a Day to Run

This 7.5 cubic foot capacity, manual defrost chest freezer uses absolutely no CFC products for insulation or coolant, is constructed of all-recyclable components, and uses a minuscule 540 watt-hours per day (at 90°F ambient temperature). That's about 50% less energy than comparable chest freezers. You can have an affordable home freezer without high energy bills or guilt about the ozone layer.

The Conserv freezer was developed three years ago to satisfy the German and Swedish "green" markets. It is manufactured by Vestfrost, a Danish manufacturer with 35 years in the refrigeration business. It is equipped with a lock and interior light on the PVC-lined, counterbalanced lid. There is an internal partition, which doubles as a drain pan during defrosting, and a pair of sliding, lift-out, wire baskets. The temperature control knob is a childproof design, plus there are external "Power On" and "Temperature Warning" lights. A fast freeze switch is provided for sudden heavy loads. Freezing capacity is 67 lb. per 24-hour period. A drainhole and plug are provided for defrosting. Operation is almost completely silent. Size is 33.5˝ H x 44.4˝ W x 25.6˝ D. Weight is approximately 150 lb. Color is white. Warranty is one year. Denmark—warranted by U.S. importer.

🚚 **62–305 Conserv Freezer $695⁰⁰**

Shipped freight collect from Scottsdale, AZ

Servel Propane Refrigerators

Servel has been a household word in gas refrigeration for over 50 years. It was first marketed by the Swedish company Electrolux in 1925. In 1956, the rights to Servel were acquired by Whirlpool, which was unsuccessful with the unit, and it disappeared for 30 years. Now Servel is back with a state-of-the-art household propane refrigerator.

We've sold more than 1,000 of the new Servels since their reintroduction in 1989. All continue to perform flawlessly and appear to be extremely well built. They average less than 1.5 gallons of propane per week! The body is all white and the door is hinged on the right, like the Model-T, with no choices available. The unit comes with rustproof racks, four in the refrigerator and two in the freezer. The spacious interior has two vegetable bins, egg and dairy racks, frozen-juice rack, ice-cube trays, and an ice bucket. An optional warranty can be purchased from the manufacturer. The total volume is 7.7 cubic feet (6 for the refrigerator, 1.7 for the freezer). Total shelf space is 11.5 square feet. The overall dimensions are 63.5″ H x 23″ W x 26.5″ D. Net weight is 167 lb. It operates on either propane or 120-volt AC (draws 325 watts almost continuously on 120 volts). Sweden/USA.

62-303 Servel RGE 400 Propane Fridge $1,295⁰⁰

Shipped freight collect from Ohio.

Frostek Propane Freezer

A freezer that works without electricity or CFCs! The Canadian-produced Frostek is the largest (and perhaps only) propane-powered freezer on the market. Its 8.5 cubic foot capacity offers reliable long-term food storage, and can accommodate an entire side of beef, or quarts upon quarts of frozen fruits and vegetables.

In carefully monitored laboratory tests, the powerful cooling unit maintained –8°F interior temperatures or less all the way up to an outside ambient temperature of 80°F. Even at 90°F it still maintains 0°F inside. Fuel use is approximately 1.8 to 2.5 gallons per week, depending on ambient temperature (or 7.5 to 11 lb. if you buy propane by the pound). An intelligent exterior thermometer shows interior conditions without opening the lid. Has easy-access bottom drain, piezo ignitor, thermostat on/off control, and runs silently.

The Frostek stands 38″ H x 44.5″ L x 31″ D. Inside dimensions are 24″ x 37″ x 15.5″. There are no awkward cooling fins jutting into the food storage area. Cabinet is rustproof steel with baked enamel finish. Color is gray and white. Approved by both the American Gas Association and Canadian Gas Association. Weight is 220 lb. Shipping weight is 275 lb. Call for availability; production quantities are limited. Canada.

62-308 Frostek Propane Freezer $1,785⁰⁰

Shipped freight collect from Ohio.

Koolatron 12V Refrigerators

Traveller

Voyager

Companion

Koolatron of Canada is the world's largest manufacturer of 12-volt coolers. The secret of their cooler is a miniature thermoelectric module that effectively replaces the bulky piping coils, compressors, and loud motors used in conventional refrigeration units. For cooling, current passing through the solid-state module draws heat from the cooler's interior and forces it to flow to the exterior, where it is fanned through the outer grill. The Koolatron is thermostatically controlled. Amp draw is 4.0 amps when on.

The coolers will maintain approximately 40°F to 50°F below outside temperature. They can be run as either a cooler or a warmer and can be used on 12-volt DC or adapted to 120-volt AC (by ordering the optional adapter). Constructed of high-impact plastic and super urethane foam insulation, the coolers include a 10-foot detachable 12-volt cord. Not recommended for remote homesteads unless you've got plenty of excess power. Average draw is 500 to 700 watt-hours per day. Canada.

Traveller

Dimensions: 10" x 12.5" x 15"; Weight: 8 lb; Capacity: 0.25 cu. ft.

◈ 62-521 Traveller 12V Cooler $99⁰⁰

Voyager

Dimensions: 16.5" x 16.5" x 20"; Weight: 14 lb; Capacity: 0.9 cu. ft.

◈ 62-526 Voyager 12V Cooler $149⁰⁰

Companion

Dimensions: 13.5" x 16" x 19.5"; Weight: 13 lb; Capacity: 0.6 cu. ft.

◈ 62-523 Companion 12V Cooler $169⁰⁰

◈ 62-520 Koolatron AC adapter $70⁰⁰

Koolatron Battery Saver

Plug this handy gizmo into your lighter plug and you'll never find the battery run down by your 12-volt fridge, cellphone, or other 12-volt appliance. It automatically shuts off power to all 12-volt appliances if battery voltage drops to 11.0 volts, saving you enough juice to start the vehicle. Resets automatically when voltage rises. Canada.

◈ 62-518 Battery Saver $39⁹⁵

◈ *Means shipped directly from manufacturer.*

Isaac Solar Ice Maker

A solar ice maker? No we haven't been out in the sun too long; this is a real product with a number of working installations. Using only the heat of the sun, an Isaac will produce 10 to 1,000 pounds of ice per day depending on model and available sunlight. The Isaac uses the Solar Ammonia Absorption Cycle that was discovered in the 1850s, and is still used on a smaller scale by all gas refrigerators. This refrigeration technology fell out of fashion in the 1930s with the arrival of cheap electricity and the discovery of the miracle refrigerant freon. The Isaac uses no electricity, gas, or freon.

During the day the Isaac stores energy in the receiving tank as high-pressure, distilled, pure ammonia. At night the user checks the sight glass to judge how much ammonia was produced, adds an appropriate amount of water to the ice compartment, and switches the valves from Day to Night positions. The ammonia is allowed to evaporate back into the collector while providing refrigeration for the ice compartment. The refrigerant cycle is sealed. In the morning the valves are switched again and the process starts all over, while the ice production from the night before is harvested.

The Isaac is constructed of stainless steel for maintenance-free outdoor installation in oceanside sites (not to mention some pure industrial beauty). Many of the current installations are for fishing villages in remote sites. Maintenance consists of re-aiming the collector every four weeks to track the sun, and an occasional bucket of water to wash away any dust.

The Isaac is built to order in a variety of models. Delivery is 30 to 90 days. An experienced technician is required on site for installation and training; this is not included in the price. Discounts available for quantity orders. Call for more information. USA.

🚚 62-532 Isaac Standard Solar Ice Maker $9,895⁰⁰

05-218 Export Crating for Standard Isaac $595⁰⁰

🚚 62-533 Isaac Double Solar Ice Maker $13,995⁰⁰

05-219 Export Crating for Double Isaac $995⁰⁰

Shipped freight collect from Maryland.

Clothes Washing & Drying

Staber Horizontal-Axis Washing Machine

The Most Advanced, Efficient, and Convenient Washer Available

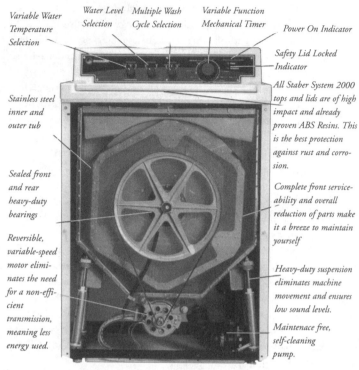

Variable Water Temperature Selection

Water Level Selection

Multiple Wash Cycle Selection

Variable Function Mechanical Timer

Power On Indicator

Safety Lid Locked Indicator

All Staber System 2000 tops and lids are of high impact and already proven ABS Resins. This is the best protection against rust and corrosion.

Complete front serviceability and overall reduction of parts make it a breeze to maintain yourself

Heavy-duty suspension eliminates machine movement and ensures low sound levels.

Maintenace free, self-cleaning pump.

Stainless steel inner and outer tub

Sealed front and rear heavy-duty bearings

Reversible, variable-speed motor eliminates the need for a non-efficient transmission, meaning less energy used.

The Staber System 2000 clothes washer combines top-loading convenience with the superior cleaning and low power use of horizontal-axis design. With the money you'll save on water, electricity, and laundry products, the Staber washer can actually pay for itself in about three years.

The Staber is a unique horizontal-axis machine that loads from the top. This is more convenient for the user, and allows stronger support on both sides of the six-sided all stainless-steel drum. Horizontal-axis machines have proven to deliver superior cleaning with less power and water use due to the tumbling action. Compared to conventional vertical-axis washers, the Staber uses 66% less water, 50% less electricity, and 75% less detergent. The well-supported drum design allows much higher spin speeds, which shortens drying time too.

Other features that set the Staber apart include power use so low that an 800-watt inverter will run it happily; no complex, power robbing transmission; a self-cleaning, maintenance-free pump; rust and dent-proof ABS resin

tops and lids with a 25-year warranty; easy front access to all parts, and a complete, fully illustrated service/owners manual for easy owner maintenance. Average power use per wash cycle is only 150 watt-hours. Average electric draw is 300 watts, peak is 550 watts. Water use: 6 gal. wash, 6 gal. first rinse, 6 gal. second rinse (6 gal. pre-wash).

Standard features include a large 18-pound capacity; variable water temperature; high/low water level selection; multiple wash cycles (normal, perm-press, delicate, pre-wash); an automatic dispensing system for detergent, pre-wash detergent, bleach, and softener; and a safety lid lock when running. EnergyGuide tagged at $8/year for natural gas water heater, or $36/year for electric water heater (hot water costs are included in washer comparisons). Measures 27″ W x 26″ D x 42″ H. Mfr.'s warranty: 25 years on outer tub module, 10 years on tub and lid, 5 years on bearings & suspension, 1 year on all else. USA.

63-645 Staber HXW-2304 Washer $1,199⁰⁰

Shipped freight collect from Ohio.

Going to run on an inverter? Tell us the make and model. Staber fits custom power boards for perfect operation with inverters.

Spin X Spin Dryer

Saves Time, Clothes, and Money!

Clothes removed from your washer still contain considerable water. A tumble dryer will take up to an hour to remove that water, while the tumbling wears out your clothes, and the heat costs you money. Line drying takes even longer. In three minutes at 3,300 rpm Spin X removes 60% of the water with centrifugal force. Clothes come out barely damp, greatly reducing or even eliminating expensive dryer run time. A three-minute cycle in the Spin X costs 15 seconds worth of electric dryer time.

The compact white enamel Spin X dryer is only 13.5″ diameter by 23.5″ tall with chrome-plated top rim and extra-durable plastic lid. Its stainless steel drum has an 8-lb. load capacity or the newest model has a 10-lb. capacity. The one-hand safety catch/control switch is simple and obvious. Start up is smooth and gradual. An electronic timer limits run time to three minutes, and the drum is braked to a stop in less than three seconds when turned off. Plugs in any 120VAC outlet, UL approved. Warranted one year by the U.S. importer. Will ship by UPS. Germany.

◈ **63-610 Spin X Spindryer 8 lb. $329⁰⁰**
◈ **63-812 Spin X Spindryer 10 lb. $359⁰⁰**

James Washers

The James hand-washing machine is made of high-grade stainless steel with a galvanized lid. It uses a pendulum agitator that sweeps in an arc around the bottom of the tub and prevents clothes from lodging in the corner or floating on the surface. This ensures that hot suds are thoroughly mixed with the clothes. The James is sturdily built. The corners are electrically spot-welded. All moving parts slide on nylon surfaces, reducing wear. The faucet at the bottom permits easy drainage. Capacity is about 17 gallons. Wringer attachment pictured with the washer is available at an additional charge.

63-411 James Washer $300⁰⁰

Hand Wringer

The hand wringer will remove 90% of the water, while automatic washers remove only 45%. It has a rustproof, all-steel frame and a very strong handle. Hard maple bearings never need oil. Pressure is balanced over the entire length of the roller by a single adjustable screw. We've sold these wringers without a problem for over 13 years.

63-412 Hand Wringer $149⁰⁰

Fifty Feet Of Drying Space

This is the most versatile clothes dryer we've seen yet. Don Reese and his family are still making them by hand from 100% sustainably harvested New England pine and birch. Don's sturdy dryer can be positioned to create either a peaked top or a flat top ideal for sweaters. A generous 50 feet of drying space—and you can fold it up to carry back in the house (or to the next sunny spot) still loaded with laundry. Measures 53″ H x 30″ W x 33″ D in peaked position; 46″ H x 30″ W x 40″ D in flat-top position, and holds an average washer load. Folds to 4.5″ D x 58″ H. USA.

◈55-273 New England Clothes Dryer $49⁰⁰

Requires UPS ground delivery.

Solar Rotary Clothes Dryer

Having beautifully fragrant clothes with the scent of the fresh outdoors is easier than ever! Manufactured in the Netherlands, this Retro Solar Clothesline offers just about 50 yards of drying space and can accommodate five loads of laundry, yet folds to only a few inches wide (from an impressive 9′6″ at the diagonal). The height conveniently adjusts from 4′ to 6′. Easy opening and closing for stowing between uses.

88-841 Umbrella Clothes Dryer $159⁰⁰

Save Space And Conserve Energy With Fold-Out Clothes Dryer

This wooden drying rack is easy to install on any wall surface. Position unit on porch for solar drying, near back door for easy access, or in laundry room for indoor drying. It's a perfect drying rack for flowers, herbs, and pasta, too! Space-saving dryer protrudes only 6″ from wall when folded. Measures 12″ x 27″. Extends to 22″ when in use. Constructed of sturdy Eastern white pine with durable birch hardwood dowels and four Shaker-style pegs. Comes fully assembled. Includes mounting hardware. USA.

◈ 54-542 Wall Shelf Drying Rack $59⁹⁵

Requires UPS ground delivery.

Super-Efficient Lighting

A revolution has taken place in commercial and residential lighting. Until recently, we were bound by the ingenious discoveries that Thomas Edison made over 100 years ago and which now appear crude.

Mr. Edison was never concerned with how much energy his incandescent bulb consumed. Today, however, economical, environmental, and resource issues force us to look at energy consumption much more critically. Incandescent literally means "to give off light as a result of being heated." With standard lightbulbs, typically only 10% of the energy consumed is converted into light, while 90% is given off as heat.

New technologies are available today that improve on Edison's invention in terms of both reliability and energy savings. Although the initial price for these new, energy-efficient products is higher than that of standard lightbulbs, their cost-effectiveness is far superior in the long run. (See the discussion of whole-life cost comparison below.) When people understand just how much money and energy can be saved with compact fluorescents, they often replace regular bulbs that haven't even burned out yet.

Light output of a lamp is measured in lumens; energy input to a lamp is measured in watts. The efficiency of a lamp is expressed as lumens per watt. In the Efficiency of Lamps chart, this number is listed for various lamps. High Pressure Sodium and Low Pressure Sodium lights are primarily used for security and street lighting.

EFFICIENCY AND LIFETIME OF LAMPS

Lamp Type And Power	Lumens Per watt	Lifetime In hours
Incandescent bulb - 25w	8.4	2500
Incandescent bulb - 60w	14	1000
Halogen bulb - 50w	14.2	2000
Incandescent bulb - 100w	16.9	750
Halogen bulb - 90w	17.6	2000
Halogen floodlight - 300w	19.8	2000
G-16, compact fluorescent - 16w	53	10,000
SLS 20, compact fluorescent - 20w	60	10,000
Metal Halide, floodlight - 150w	60	10,000
D lamp, compact fluorescent - 39w	71	10,000
T8, 4 ft fluorescent tube - 32w	86	20,000
High Pressure Sodium light - 150w	96	24,000
Low Pressure Sodium light - 135w	142	16,000

The types of efficient lighting on the market include new-design incandescents, quartz-halogens, standard fluorescents, and compact fluorescents.

"Energy-Saving" Incandescent Lights?

Inside an incandescent light is a filament that is heated, giving off light in the process. The filament is delicate and eventually burns out. Some incandescents incorporate heavy-duty filaments or introduce special gases into the bulb to increase life. While this does not increase efficiency, it increases longevity by up to four times. Most manufacturers are now offering "energy-saving" incandescent bulbs. These are nothing more than lower-wattage bulbs that are marketed as having "essentially the same light output as . . ." In fact, General Electric lost a multi-million-dollar lawsuit relating to this very issue. A 52-watt incandescent bulb does not put out the same amount of light as a 60-watt bulb. An "energy-saving" incandescent bulb will use less power, true, but it will also give you less light. This specious advertising claim is just a mega-corporation excuse to avoid retooling by selling you a wimpier lightbulb at a higher price. Don't be hoodwinked; you can do much better.

Quartz-Halogen Lights

Clean Lights Save Energy

Building owners can save up to 10% of lighting costs simply by keeping light fixtures clean, says NALMCO, the National Association of Lighting Management Companies. Over the past three years, crews working on the U.S. EPA-funded project have collected scientific field measurements at offices, schools, health facilities, and retail stores to determine the benefits—improved lighting as well as reduced costs—of simple periodic cleaning of light fixtures. Recognizing that fixtures produce less light as they get dirty, current design standards compensate by adding more fixtures to maintain the desired light level. But if building owners committed to regular cleanings, designers could recommend fewer fixtures, reducing installation costs while trimming lighting energy use by at least 10%. For more information, call NALMCO at 515-243-2360.

Tungsten-halogen (or quartz) lamps are really just "turbocharged" incandescents. They are typically only 10% to 15% more efficient than standard incandescents; a step in the right direction, but nothing to write home about. Compared to standard incandescents, halogen fixtures produce a brighter, whiter light and are more energy-efficient because they operate their tungsten filaments at higher temperatures than standard incandescents. In addition, unlike the standard incandescent lightbulb, which loses approximately 25% of its light output before it burns out, a halogen light's output depreciates very little over its life, typically less than 10%.

To make these gains, lamp manufacturers enclose the tungsten filament inside a relatively small, quartz-glass envelope filled with halogen gas. During normal operation, the particles that evaporate from the filament combine with the halogen gas and are eventually redeposited back on the filament, minimizing bulb blackening and extending filament life. Where halogen lamps are used on dimmers, they need to be occasionally operated at full output to allow this regenerative process to take place.

There are several styles of halogens available for both 12-volt and 120-volt applications. The most popular low-voltage models for residential applications include the miniature multifaceted reflector lamp (product #33-108) and the halogen version of the conventional A-type incandescent lamp, which incorporates a small halogen light capsule within a protective outer glass globe. The reflector lamp offers a light source with very precise light-beam control, allowing you to distribute light exactly where you need it without wasteful spillover. A new family of 120-volt AC, sealed-beam reflector halogen lamps is now on the market to replace standard incandescent reflectors.

The higher operating temperatures used in halogen lamps produce a whiter light, which eliminates the yellow-reddish tinge associated with standard incandescents. This makes them an excellent light source for applications where good color rendition is important or fine-detail work is performed. Because tungsten-halogen lamps are relatively expensive compared to standard incandescents, they are best suited for applications where the optical precision possible with the compact reflector models can be effectively utilized. They are a favorite of high-rent retail stores.

Never touch the quartz-glass envelope of a halogen lamp with your bare hands. The natural oils in your skin will react with the quartz glass and cause it to fail prematurely. This applies to automotive halogen headlight bulbs as well. Because of this phenomenon, and for safety reasons, many manufacturers incorporate the halogen lamp capsule (generally about the size of a large flashlight bulb) within a larger outer globe.

Fluorescent Lights

Fluorescent lights are still trying to overcome a bad reputation. For many people the term "fluorescent" connotes a long tube emitting a blue-white light with an annoying flicker, a death-warmed-over color, and headaches. However, these limitations have been completely overcome with technological improvements.

The fluorescent tube, no matter how it is shaped, contains a special gas at low pressure. When an arc is struck between the lamp's electrodes, electrons collide with atoms in the gas to produce ultraviolet light. This, in turn, excites the white phosphors (the white coating on the inside of the tube), which emit light at visible wavelengths. The quality of the light fluorescents produce depends largely on the blend of chemical ingredients used in making the phosphors; there are dozens of different phosphor blends available. The most common and least expensive are "cool white" and "warm white." These provide a light with relatively poor color-rendering capabilities, making colors appear washed out, lacking luster and richness.

The tube may be long and straight, as with standard fluorescents, or there may be a series of smaller tubes, in a configuration that can be screwed into a common light fixture. This latter configuration is called a compact fluorescent light (or CF light).

The growth of the compact fluorescent market in recent years has been enormous. There is now a good variety of lamp wattages, sizes, shapes, and aesthetic packages. This is a growing and rapidly changing marketplace, and the more-frequently published Gaiam Real Goods catalogs may feature lamp types not described here.

LIGHT, COLOR RENDERING INDEX, AND COLOR TEMP

Type Of Light	CRI	Deg. K
Incandescents	90–95	2700
Cool White Fluorescent	62	4100
Warm White Fluorescent	51	3000
Compact Fluorescent	82	2700

Color Quality

The color of the light emitted from a fluorescent bulb is determined by the phosphors that coat the inside surface. The terms "color temperature" and "color rendering" are the technical terms used to describe light. Color temperature, or the color of the light that is emitted, is measured in degrees Kelvin (°K) on the absolute temperature scale, ranging from 9000°K (which appears blue) down to 1500°K (which appears orange-red). The color rendering index (CRI) of a lamp, on a scale of 0 to 100, rates the ability of a light to render an object's true color when compared to sunlight. 100 is perfect, 0 is a cave. The accompanying table will demystify the quality of light debate.

Phosphor blends are available that not only render colors better, but also produce light more efficiently. Most notable of these are the fluorescent lamps using tri-stimulus phosphors, which have CRIs in the eighties. These incorporate relatively expensive phosphors with peak luminance in the blue, green, and red portions of the visible spectrum (those which the human eye is most sensitive to), and produce about 15% more visible light than standard phosphors. Wherever people spend much time around fluorescent lighting, specify lamps with higher (80+) CRI.

Ballast Comparisons

All fluorescent lights require a ballast to operate, in addition to the bulb. The ballast regulates the voltage and current delivered to a fluorescent lamp, and is essential for proper lamp operation. The electrical input requirements vary for each type of compact fluorescent lamp, and so each type/wattage requires a ballast specifically designed to drive it. There are two types of ballasts that operate on AC: magnetic and electronic. The magnetic ballast, the standard since fluorescent lighting was first developed, uses electromagnetic technology. The electronic ballast, only recently developed, uses solid-state technology. All DC ballasts are electronic devices.

Magnetic ballasts can last up to 50,000 hours, and often incorporate replaceable bulbs. Magnetic ballasts flicker when starting and take a few seconds to get going. They also run the lamp at 60 cycles per second, and some people are affected in a negative way by this flicker.

Electronic ballasts weigh less than magnetic ballasts, operate lamps at a higher frequency (30,000 cycles per second vs. 60 cycles), are silent, generate less heat, and are more energy-efficient. However, electronic ballasts cost more, particularly DC units. Much longer-lived electronic ballasts are possible, but costs go up dramatically.

Electronic ballasts last about 10,000 hours, the same as the bulb, and most do not have replaceable bulbs. Electronic ballasts start almost instantly with no flickering. They run the lamp at about 30,000 cycles per second. For the many people who suffer from the "60-cycle blues," this is the energy-efficient lamp of choice.

All of the compact fluorescent lamp assemblies and prewired ballasts reviewed here are equipped with standard medium, screw-in bases (like normal household incandescent lamps.) The ballast portion, however, is wider than an incandescent light bulb just above the screw-in base. Fixtures having constricted necks or deeply recessed sockets may require a socket "extender" (to extend the lamp beyond the constrictions). These are readily available either in our catalogs or at most hardware stores.

Note: There is no difference between a fluorescent tube for 120 volts and one for 12 volts. Only the ballasts are different.

Savings

The main justification for buying fluorescent lights is to save energy and money. The new compact fluorescents provide opportunities for tremendous savings without any inconveniences. Simple payback calculations prove their cost-effectiveness (see the Cost Comparison table). Fluorescent lights typically last ten to thirteen times longer and use one-fourth the energy of standard incandescent lights. The U.S. Congress Office of Technology Assessment calls compact fluorescents the best investment in America today, with a 1.2-year payback.

The next question, of course, is, "How much do compact fluorescents save?" The more a light is used, the more that can be saved by replacing it with a compact fluorescent. Installing compact fluorescents in the fixtures that are used most and have high wattage saves the most. An excellent candidate is an all-night security light. The calculations in the table below assume that the light is on for an average of six hours per day and power costs 10¢ per kWh (the approximate national average). The total cost of operation is the cost of the bulb(s) plus the cost of the electricity used. The table shows that using the compact fluorescent saves about $30 and 540 kWh of electricity.

An Energy Tale, or Real Goods Walks Its Talk

In May of '95 Real Goods headquarters moved into a new-to-us 10,000 sq. ft. office building. Built in the early '60s, the building had conventional four-foot fixtures, each holding four 40-watt cool white tubes. The dreary, glaring "office standard." Light levels were much brighter than recommended when employees are using computers a lot. Plus, a number of our employees were having headache, energy level, and "attitude" problems after moving, probably caused by the 60-cycle flicker from the old magnetic ballasts. Reducing overall light levels for less eyestrain was needed, as was giving employees a healthier working environment. And if we could save some money too, it wouldn't hurt.

— continued on next page

LIFETIMES OF LIGHTS (IN HOURS)

Standard incandescents	1,000
Long-life incandescents	3,000
Quartz-halogen	2,250
Compact fluorescents	10,000

EFFICIENCY OF LIGHTS (LUMENS PER WATT—APPROXIMATE)

Incandescents	16
Compact fluorescents	60

Health Effects

Migraine headaches, loss of concentration, and general irritation have all been blamed on fluorescent lights. These problems are caused not by the lights themselves, but by the way they are run. Common magnetic ballasts run the lamps at the same 60 cycles per second that is delivered by our electrical grid. This causes the lamps to flicker noticeably 120 times per second, every time the alternating current switches direction. Approximately one-third of the human population is sensitive to this flicker on a subliminal level.

The cure is to use electronic ballasts, which operate at around 30,000 cycles per second. This rapid cycling totally eliminates perceptible flicker and avoids the ensuing health complaints.

COST COMPARISON*

	Philips SLS20, 20-Watt	Incandescent Bulb, 75-Watt
Cost of bulb	$19.95	50¢
Product life	4.5 years	167 days
Bulbs used in 4.5 years	1	10
Energy used annually	44 kWh	164 kWh
Energy used in 4.5 years	198 kWh	739 kWh
Total cost in 4.5 years	$39.75	$78.90
Savings in 4.5 years	$39.15	
Energy saved in 4.5 years	541 kWh	

*Computed at 6 hours of lamp use a day and electricity cost of 10¢ per kWh.

Fluorescents and Remote Energy Systems

Fluorescent lights are available for both AC and DC. Most people using an inverter choose AC lights because of the wider selection, a significant quality advantage, and lower price. Some older inverters may have problems running some magnetic-ballasted lights. If you have an older Heart inverter, buy one light to try it first.

Most inverters operate compact fluorescent lamps satisfactorily. However, because all but a few specialized inverters produce an alternating current having a modified sine wave (versus a pure sinusoidal waveform), they will not drive compact fluorescents that use magnetic ballasts as efficiently or "cleanly" as possible, and some lamps may emit an annoying buzz. Electronic-ballasted compact fluorescents, on the other hand, are tolerant of the modified sine wave input, and will provide better performance, silently.

Compact Fluorescent Applications

Due to the need for a ballast, a compact fluorescent lightbulb is shaped differently from an "Edison" incandescent. This is the biggest obstacle in retrofitting light fixtures. Compact fluorescents are longer, heavier, and sometimes wider. The ballast, the widest part, is located at the base, right above the screw-in adapter.

Compact fluorescents have many household applications—table lamps, recessed cans, desk lamps, bathroom vanities, and more. As manufacturers become attuned to this relatively new market, more light fixtures suited for compact fluorescents are becoming available. We now offer a limited selection of fixtures; watch future catalogs for the latest offerings in this rapidly expanding field.

Table lamps are an excellent application for compact fluorescents. The metal "harp" that supports the lampshade may not be long or wide enough to accommodate these bulbs. We offer an inexpensive replacement harp that can solve this problem. Be aware also that heavier CFs may change a light but stable lamp into a top-heavy one.

Recessed can lights are limited by diameter, and sometimes cannot accept the wide ballast at the screw-in base. The base depth is adjustable on most recessed cans. There are a number of special bulbs available now just for recessed cans.

Desk lamps can be difficult to retrofit, but complete desk lights that incorporate compact fluorescents are readily available.

Hanging fixtures are one of the easiest applications for compact fluorescent lights. One of the best applications is directly over a kitchen or dining-room table. Usually the shade is so wide that it won't get in the way. It may even be possible to use a Y-shaped two-socket adapter (available at hardware stores) and screw in two lights if more light is desired.

Track lighting is one of the most common forms of lighting today. When selecting your track system, choose a fixture that will not interfere with the ballasts located right above the screw-in base. It is best to use one of the reflector lamps, such as the SLS/R series.

Dimming

With limited exceptions, the current generation of compact fluorescent lights should never be dimmed. Using these lights on a dimmer switch may even pose a hazard. If you have a fixture with a dimmer switch that you wish to retrofit, you can either buy a CF lamp that specifically says that it is dimmable, or replace the switch. Replacing the switch is a simple, inexpensive procedure. Remember to turn off the power first!

— continued from previous page

Our solution was to retrofit the existing fixtures with electronic ballasts for no-flicker lighting, install specular aluminum reflectors, and convert from four cool white 40-watt T-12 lamps per fixture to two warm-colored 32-watt T-8 lamps. Power use per fixture was reduced by over 60%, but desktop light levels, because of the reflectors and more efficient lamps and ballasts, were reduced by only 30% or less. The entire retrofit cost about $6,000 (not counting the $2,300 rebate from our local utility), yet our electrical savings alone are close to $5,000 per year. Plus our employees enjoy flicker-free, warm-colored, non-glaring light and greatly reduced EMF levels. The entire project paid for itself in less than a year! Improved working conditions came as a freebie!

Three-Way Sockets

Compact fluorescents can be screwed into any three-way light socket, but, just like a standard bulb in a three-way socket, they will only operate on full light output.

Start-Up Time

The start-up time for compact fluorescent lamps varies. It is normal for most magnetic compact fluorescents to flicker for up to several seconds when first turned on while they attempt to strike an arc. Most electronically ballasted units start their lamps instantly. All fluorescent lamps start at a lower light output; depending on the ambient temperature, it may take anywhere from several seconds to several minutes for the lamp to come up to full brightness.

The very brief start-up time, which is only apparent in some of the fluorescent lights, is a small price to pay for the energy savings and the subsequent good feeling about doing less harm to our fragile environment.

Cold Weather/Outdoor Applications

Fluorescent lighting systems are sensitive to temperature. Manufacturers rate the ability of lamps and ballasts to start and operate at various temperatures. Light output and system efficiency both fall off significantly when lamps are operated above or below the temperature range at which they were designed to operate. Most fluorescent lamps will reach full brightness at 50° to 70°F. In general, the lower the ambient temperature, the greater the difficulty in starting and attaining full brightness. Manufacturers are usually conservative with their temperature ratings. We have found that most lamps will start at temperatures 10° to 20°F lower than those stated by the manufacturer. Most compact fluorescent lamps are not designed for use in wet applications (for example, in showers or in open outdoor fixtures). In such environments, the lamp should be installed in a fixture rated for wet use.

Service Life

All lamp-life estimates are based on a three-hour duty cycle, meaning that the lamps are tested by turning them on and off once every three hours until half of the test batch of lamps burn out. Turning lamps on and off more frequently decreases lamp life. Since the cost of operating a light is a combination of the cost of electricity used and the replacement cost, what is the optimum no-use period before turning off a light? Generally, it is more cost-effective to turn standard fluorescent lights off whenever you leave a room for more than 15 minutes. For compact fluorescents this is 3 minutes and you should turn incandescents off whenever you leave the room. Since the rated lamp life represents an average, some lamps will last longer and some shorter. A typical compact fluorescent lamp will last as long as (or longer than) ten standard incandescent AC lightbulbs or five standard incandescent AC floodlights, saving you the cost of numerous bulbs, in addition to a lot of electricity due to the greater efficiency. See the Cost Comparison table on page 304.

Full-Spectrum Fluorescents

Literally, "full-spectrum" refers to light that contains all the colors of the rainbow. As used by several manufacturers, "full-spectrum" refers to the similarity of their lamps' light (including ultraviolet light) to the midday sun. While we do believe that the closer an electric light source matches daylight the healthier it is, we have several practical reservations. The quality of light produced by "full-spectrum" lighting available today is very "cool" in color tone and in our opinion, not flattering to people's complexions or most interior environments. Also, the light intensity found indoors is roughly 100 times less than that produced by sunlight. At these low levels, we question whether people receive the full benefits offered by full-spectrum lighting.

Ideally, we encourage you to get outdoors every day for a good dose of sunshine. Indoors, we think it is best to install lighting that is complimentary to you and the ambiance you desire to create. For most applications, people prefer a warmer toned light source. Most of the compact fluorescent lamps we offer mimic the warm rosy color of 60-watt incandescent lamps, but we do offer a small selection of full-spectrum lamps for those who prefer them.

—Real Goods Staff

Natural Lighting

SunPipe Skylights

Free Sunlight Is the Best Light

The SunPipe adds beautiful natural daylight to your house without high expense, heat loss or gain, bleaching of fabrics, or framing and drywall work. Sunlight is collected above the roof and transferred thru a mirror-finish aluminum pipe to a diffusing lens in your ceiling. The SunPipe transmits no direct sun rays, so there's no solar heat gain, bleaching, or color fading. No cutting of joists and rafters, or framing and drywall work is required. Installation takes two to three hours on average, and can be easily done by any home handyman. The light is uniformly diffused by the interior lens, and the pipe is sealed top and bottom to prevent heat loss or condensation.

Two sizes are available. Approximately speaking, the 13˝ pipe is for rooms 10´ x 10´ or larger, and the 9˝ pipe is for smaller rooms. The 13˝ pipe delivers illumination equivalent to about 900 watts of incandescent light on sunny days, or 100 to 500 watts on cloudy days. The 9˝ pipe delivers about half as much light.

The Basic Kits do not include any pipe; they supply everything for the top and bottom assemblies: the clear dome top, galvanized roof flashing, sealing materials, hardware, installation guide, and bottom diffuser. Purchase telescoping 2 ft., 3 ft., and 4 ft. sections as needed to fit your installation. Allow a minimum of 10 inches above your roofline. The optional SunScoop is attached inside the dome, facing south, to capture up to 170% more sunlight during the low sun of winter. The steep flashing option is for SunPipes in steeply pitched roofs of 6/12 to 12/12. The price is for a trade out, replacing the standard flashing with the steep flashing in the basic kit. The optional elbows adjust from 0° to 45°. USA.

◈ **63-156 Basic SunPipe Kit 9˝ $169⁰⁰**

◈ **63-000 SunPipe 9˝ Extension 1 ft. $22⁹⁵**

◈ **63-158 SunPipe 9˝ Extension 2 ft. $44⁹⁵**

◈ **63-001 SunPipe 9˝ Extension 3 ft. $69⁹⁵**

◈ **63-157 SunPipe 9˝ Extension 4 ft. $89⁹⁵**

◈ **63-159 SunPipe 9˝ SunScoop $30⁰⁰**

◈ **63-823 SunPipe 9˝ Steep Flashing Option $53⁰⁰**

◈ **63-002 SunPipe 9˝ Elbow Option $49⁹⁵**

◈ **63-180 Basic SunPipe Kit 13˝ $199⁰⁰**

◈ **63-003 SunPipe 13˝ Extension 1 ft. $34⁹⁵**

◈ **63-181 SunPipe 13˝ Extension 2 ft. $74⁹⁵**

◈ **63-155 SunPipe 13˝ Extension 3 ft. $109⁰⁰**

◈ **63-182 SunPipe 13˝ Extension 4 ft. $149⁰⁰**

◈ **63-183 SunPipe 13˝ SunScoop $39⁹⁵**

◈ **63-814 SunPipe 13˝ Steep Flashing Option $87⁰⁰**

◈ **63-004 SunPipe 13˝ Elbow Option $64⁹⁵**

◈ *Means shipped directly from manufacturer.*

Basic Kits do not include pipe.
Order pipe lengths as needed for your particular installation.

AC Lighting

LIGHTING COMPARISON CHART

Description	Item #	CFL Usage (watts)	Incandescent Equivalent (watts)	Lumens	Operating Temp.°F	Size (inches)	CRI*
SLS 15-Watt	36-115	15	62	900	–10 to 140	4.75 x 2.25	82
SLS 20-Watt	36-116	20	76	1200	–10 to 140	5.5 x 2.25	82
SLS 25-Watt	36-117	25	102	1750	–10 to 140	6.2 x 2.25	82
Spring Lamp 15-Watt	36-508	15	65	980	–20 to 160	5.0 x 2.0	84
Spring Lamp 18-Watt	36-509	18	76	1200	–20 to 160	5.0 x 2.0	84
Spring Lamp 23-Watt	36-510	23	92	1500	–20 to 160	5.0 x 2.0	84
Twist Light Bug	31-734	11	50	550	–20 to 140	5.75 x 2.5	82
Twist Light	36-187	11	50	550	–20 to 140	5.75 x 2.5	82
G-25 Globe	36-098	25	86	1370	–22 to 122	5.25 x 3.75	84
G-15 Globe	36-504	15	60	850	–22 to 122	5 x 3.75	84
T-28 Capsule	36-099	28	102	1750	–22 to 122	6.625 x 2.75	84
T-15 Capsule	36-503	15	60	850	–22 to 122	5.25 x 2.25	84
SLS R-40	36-191	20	52	625	–10 to 140	6.6 x 3.75	82
3-Way CF	36-202	low 15 med. 25 high 41	57 90 150	750 1370 2780	0 to 140	8 x 8 x 4	82
1-Way CF	31-119	41	150	2780	0 to 140	8 x 8 x 4	82
Bug-A-Way Bulb	36-192	17	75		–10 to 140	5.875 x 3.0	
Decorative 13-Watt	36-175	13	56	750	–10 to 140	6.75 x 2.0	84
Candelabra 5-Watt	31-677	5	27	250	–4 to 120	4.75 x 1.5	82

Real Goods CFL bulbs have a 10,000-hour life (SLS R-40 is 8,000 hours) and an electronic ballast.
*CRI: Color Rendering Index: a measure of color accuracy relative to sunlight. 0 is dark, 100 is noon sunlight.

Earthlight CFs For Almost Anywhere

We've been carrying the Phillips "Earthlight" series for years; it's our original high color rendering warm color, electronically ballasted compact fluorescent. When you buy a $10 CF, you get what you pay for. The 15-, 20-, and 25-watt bulbs are good for replacing 60- to 100-watt incandescents. The 15- and 20-watt versions are available with a snap-on reflector for use as a floodlight. The 9- and 11-watt bulbs are great for sconces. Their narrow ballast makes them skinnier than any other CF we sell. These bulbs should not be used in an enclosed fixture. UL listed. USA.

36-115 SLS 15-Watt $19⁹⁵
36-116 SLS 20-Watt $19⁹⁵
36-117 SLS 25-Watt $19⁹⁵
36-212 SLS 9-Watt $19⁹⁵
36-210 SLS 11-Watt $19⁹⁵
36-190 SLS R-30 15-Watt Flood $29⁹⁵
36-191 SLS R-40 20-Watt Flood $29⁹⁵

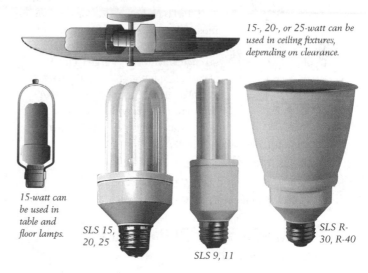

15-, 20-, or 25-watt can be used in ceiling fixtures, depending on clearance.

15-watt can be used in table and floor lamps.

SLS 15, 20, 25

SLS 9, 11

SLS R-30, R-40

Enclosed Fixture: Ceiling and wall fixtures. Exterior yard post lights, wall brackets, and hanging fixtures. (Check minimum starting temperatures.)

Shaded Lamps: Table and floor lamps.

Open Indoor: Hanging pendant, ceiling lamps, and wall fixtures.

Recessed Can: Recessed ceiling downlights or track lights.

Dimmable Spring Lamp

At Last!

The innovative, lateral twisted tube design of this CF is dimmable to 20%. Use almost anywhere standard incandescents are used. UL-listed.China.

36-508 Spring Lamp 15-Watt $29⁹⁵

36-509 Spring Lamp 18-Watt $29⁹⁵

36-510 Spring Lamp 23-Watt $29⁹⁵

11-Watt Replaceable CF

Save Up to 75% in Energy Costs!

Fits almost anywhere standard incandescent bulbs fit—lamps, hanging lamps, and ceiling fixtures. Comes with a reusable base—just replace the easy twist off bulb once every five years! Produces a soft white light. The same configuration is available in yellow as a bug light. The replacement bulbs are interchangeable. Not for use with dimmers. Ballast life is 30,000 hours. UL-listed. China.

36-187 Twist Light $17⁹⁵

36-188 Twist Replacement Bulb $7⁹⁵

31-734 Twist Light Bug Light Bulb $19⁹⁵

31-735 Twist Light Bug Light Repl. Bulb $8⁹⁵

Compact Globes And Capsules

Both this Panasonic G- (globe) and the T- (tube) type bulbs have proven our most reliable CFs in the presence of dubious quality AC power (brownouts, strange waveforms, surges, etc.). These compact globes are the perfect visual solution for open, decorative fixtures. They provide a pleasing warm-white light. UL listed. Japan.

36-504 G-16 Globe $29⁹⁵

36-098 G-25 Globe $29⁹⁵

36-503 Panasonic T-15 Capsule $29⁹⁵

36-099 Panasonic T-28 Capsule $29⁹⁵

3-Way Compact Fluorescent

Our Brightest CF

The unique D-shaped design provides an even, overall soft-white light. UL-listed. Lamp, Great Britain. Ballast, Italy.

36-202 3-Way with Adapter $39⁹⁵

31-119 1-Way with Adapter $29⁹⁵

31-124 39-Watt 2D Replacement Bulb $19⁹⁵

Decorator Fluorescent For Finer Fixtures

This compact fluorescent, in decorator styling, looks great in open fixtures and sconces. Not to be used on dimming circuits. This is not a candelabra bulb, it has a standard medium base. China.

36-175 Decorative Bulb 13W $17⁹⁵

Candelabra 5-Watt Bulb

Designed especially for candelabras, this bulb has the smaller base so hard to find in a compact fluorescent. Instant on; no flickering; noiseless ballast. UL-listed. China.

31-677 5-Watt Candelabra Bulb $10⁹⁵

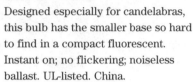

Keep The Bugs Away And Save Energy Too!

This compact fluorescent is ideal for porch and patio lights. The special yellow shatter-proof plastic cover produces a yellow light that is not visible to most insects. The warm, pleasing glow creates a welcoming ambiance on your front porch. The 17-watt Earth Light produces 800 lumens of light, which is equivalent to a 75-watt incandescent light bulb. It measures 5.875″ long x 3″ wide, fitting easily into most open porch lights. Philips Amalgam technology provides stable light over a wide range of temperatures. UL-listed for damp locations. Mexico.

• Electronic ballast
• Lamp life: 10,000 hours
• Operating temperature: −10° to 140°F

36-192 Bug-A-Way Bulb $20⁰⁰

Dimmable Fluorescent Torchiere

This metallic, nickel-finished torchiere is designed to be versatile; it blends in with a variety of room decors and provides a fresh alternative in torchiere lighting. Employs the latest technology for bright light output and safe operation. Comes with a 55-watt, 2-D™ energy-saving bulb, providing nearly the same light intensity as a 300-watt halogen bulb, but without the dangerous heat intensity. 2-D™ light-bulb will last three to five years (or 10,000 hours), depending on frequency of use. Torchiere stands 72″ high on an 11″ Dia. base. Assembly required. Electronic ballast. UL-listed. China.

35-430 Glencoe Torchiere $99⁹⁵

35-429 2-D™ Replacement Bulb $24⁹⁵

Lamp Harp Retrofits

Some compact fluorescent lamps are larger than standard incandescent bulbs and will not fit ordinary floor or table lamps unless you change the harp. The harp is the rigid piece that holds the shade over the bulb. We offer 10″ harps to satisfy most lamp conversion needs. They are very easy to install, requiring no tools.

36-403 10″ Lampshade Harp $2⁹⁵

Socket Extender

Sometimes your compact fluorescent lamps may have a neck that is too wide for your fixture. This socket extender increases the length of the base. The extender is also useful if a bulb sits too deeply inside a track lighting or can fixture, losing its lighting effectiveness. Extends lamp 13/16″. USA.

36-405 Socket Extender $2⁹⁵

Full-Spectrum Lighting

Energy-Saving 20-Watt CFL Ott Light Fits Standard Fixtures

If you want the health benefits of full spectrum lighting, but can't bear to part with your antique lamp or fixture, this is the bulb for you. Not only does the standard edison base fit almost all lamps (plus many track lighting systems and recessed fixtures), but it offers all the environmental advantages of a compact fluorescent, producing full-spectrum lighting equivalent to a 75-watt incandescent while using only 20 watts of electricity. Made by the original manufacturer of quality Full-Spectrum lighting products, the Ott 20-Watt Full-Spectrum CFL is the easiest way to boost your well-being without replacing fixtures and lamps. Standard incandescent type edison screw-in base, 5.5″ H x 2.7″ Dia. 1400 lumens with 10,000-hour bulb life. USA.

31-722 Ott 20-Watt Full-Spectrum CFL $34⁰⁰

Happylite Jr. Chases Away The Blues

An estimated 25 million people suffer needlessly from light-deprivation-related depression and low energy levels. The Happylite Jr. provides the full-spectrum light needed to boost your mood, without busting your energy budget. Two energy-efficient 36-watt fluorescent full-spectrum bulbs provide a 10,000-lux bright light system, enough to promote alertness and increase energy naturally. Flicker-free, glare-free, with UV/EMF blocking. Includes a stand for desktop use, or wall mount to save space. UL-listed. Compact and lightweight plastic light weighs only 7 lb. (12″ x 3.5″ x 17″ H). China.

31-741 Happylite Jr. $199⁰⁰

Whiter, Brighter Full-Spectrum Task Light

This Full-Spectrum halogen lamp provides heightened contrast (eases eyestrain), and superior color rendering. The bulb used in this lamp was chosen by the National Gallery of Art for "Van Gogh's Van Goghs." The bulbs were specifically chosen for their excellent color-rendering capability and low ultraviolet production (no fading!). Heavy weighted base (8.5″ Dia.) with a four-joint pivoting neck (36″ fully extended) for maximum directionality. 50-watt halogen bulb with a 3000-hour bulb life. 9′ cord. Taiwan.

**31-719 Full-Spectrum Halogen
        Task Lamp $169⁰⁰**

**31-720 Full-Spectrum Halogen
        Replacement Bulb $17⁹⁵**

Ott Lite® Floor Lamp

The Natural Feeling of Sunlight
Now in a Floor Lamp

Designed to fit over a chair, desk, or work-bench, this is the ultimate floor lamp for reading, needlework, sewing, painting, and more. The head of the lamp includes an 18-watt "Ott-Lite"® —a true color, total spectrum and radiation-shielded light source closer to natural lighting than anything yet devised. Also great for indoor plants. It measures approximately four feet high with a flexible neck, allowing you to focus light right on your project. Has a 11″ x 8″ weighted base for stability. UL-listed. Taiwan.

A 35-418 Ott Lite Floor Lamp
      $179⁰⁰

B 31-721 Ott Desk Lamp $169⁰⁰

   36-201 18-Watt True Color
      Repl. Bulb $29⁹⁵

Portable Full-Spectrum Task Light

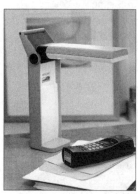

Sleek, compact design makes this task light the ideal choice for people on the move. Just 3.5″ square x 10.5″ high closed, it tucks into your brief-case, work basket, or handbag so you can enjoy all the advantages of Full-Spectrum lighting on the road. On a desk or wall mounted, it uses only 13 watts of power, 650 lumens. Flip it up for on (up to 19.5″ H), closed for off. Weighs only 3.5 lb. and comes with a carrying handle. UL-listed. Taiwan.

31-718 Ott Portable Task Light $69⁹⁵
35-366 13-Watt True Color Replacement Bulb $24⁹⁵

Efficient Full-Spectrum Plant Light Grows With Your Plant

The closest thing to natural sunlight, by the original full-spectrum lighting maker. The 10,000-hour, 13-watt fluorescent bulb lets you grow your own without a drastic increase in energy consumption. Plus, low-heat fluorescent won't dry or burn leaves. The adjustable neck flexes for optimum lighting control, the shade turns 180°, and the stand telescopes up to approximately 4′5″ high from the clamp. Clamp fits up to 1.25″ thick pots, planters, or tabletops. All-metal construction. UL-listed. Taiwan.

31-736 OTT Adjustable Plant Light
      $99⁹⁵

31-737 Replacement Bulb $14⁹⁵

Natural Spectrum Tubes

Natural Spectrum fluorescent tubes come in two sizes. The 40-watt T-12 has a 2100-lumen output and is good for existing magnetic ballast fixtures or newer electronic designs. The 32-watt T-8 has a 1750-lumen output and works with the newer electronic ballast fixtures only. Both sizes have a 26,000-hour life and are sold as a set of 4. USA.

31-683 Natural Spectrum T-8 Tubes
      (4) (48″ x 1.5″) $49⁹⁵

36-082 Natural Spectrum T-12 Tubes
      (4) (48″ x 1.5″) $49⁹⁵

DC Lighting

When designing an independent power system, it is essential to use the most efficient appliances possible. Fluorescent lighting is four to five times more efficient than incandescent lighting. This means you can produce the same quantity of light with only 20% to 25% of the power use. When the power source is expensive, like PV modules, or noisy and expensive, like a generator, this becomes terribly important. Fluorescent lighting is quite simply the best and most efficient way to light your house.

Usually we recommend against using DC lights in remote home applications except in very small systems where there is no inverter. The greater selection, lower prices, and higher quality of AC products has made it hard to justify DC lights. DC wiring and fixture costs are significantly higher than with AC lighting, and lamp choices are limited. However, you might want to consider putting in a few DC lights for load minimization and emergency back-up. A small incandescent DC night-light uses less than its AC counterpart plus inverter overhead, and a single AC device may not draw enough power to be detected by the inverter's load-seeking mode.

Since the introduction of efficient, high quality, long-lasting inverters, the low-voltage DC appliance and lighting industries have almost disappeared except for RV equipment, which tends to be of lower quality. This is because most RV equipment is designed for intermittent use, where short life or lower quality isn't as noticeable.

Tube-type fluorescents, such as the Thin-Lite series, are great for kitchen counters and larger areas. These are RV units, with lower life expectancies compared to conventional AC units. Replacement ballasts for these units are available, and the fluorescent tubes are standard hardware-store stock. We also offer incandescent and quartz halogen DC lights, but suggest caution in using these for alternative energy systems, where every watt is precious.

Solsum 12VDC Compact Fluorescent Lamps

Here's a pair of truly compact, and even attractive DC fluorescent lamps. Using instant-start solid-state electronic ballasts, they will operate reliably from 10.5 to 14.5 volts, and use a standard edison screw base. The 7-watt lamp is equivalent to a 40-watt incandescent and measures 5.1″ length, 2.25″ diameter. The 11-watt is equivalent to a 60-watt and measures 6.3″ length, 2.25″ diameter. Life expectancies are approximately 6,000 hours. China.

33-502 Solsum 7W, 12VDC Lamp $24⁹⁵
33-503 Solsum 11W, 12VDC Lamp $24⁹⁵

12-Volt Circline Lamp

This 22-watt circline lamp delivers 1100 lumens, about the equivalent of a 75-watt incandescent. The 12-volt electronic ballast screws into a standard edison base. A good choice for ceiling fixtures and table lamps. Diameter is 9.5″, height is 4″. Uses a replaceable standard bulb. USA.

36-614 12V 22W Circline Lamp $34⁹⁵

A 12-Volt, 15-Watt Compact Fluorescent Bulb

Using a highly compact and efficient tri-tube fluorescent tube, this 15-watt bulb delivers the equivalent light output of a 60-watt incandescent bulb. Fits most fixtures including recessed and track lighting. China.

33-507 12V, 15W Compact Fluorescent $32⁹⁵

Portable 12-Volt Fluorescent Worklight

This efficient, rugged 15-watt fluorescent worklight is watertight down to 50 feet, is virtually unbreakable in a Lexan polycarbonate thickwall tubing with rubber end caps, and has a 12-foot line cord. Light output is equal to a 75-watt incandescent worklight, but without the heat and burn potential. Can be safely used in wet weather, marine environments, even underwater, and it floats. The electronic ballast is enclosed in the Lexan tube, so there's no bulky transformer on the line cord. It's reverse polarity protected. Excellent for fishing, marine, utility, RV, and emergency service. Draws 600 milliamps at 12VDC. Six-month mfr.'s warranty. USA.

32-958 Portable 12V Worklight $50⁰⁰

PL Twin-Tube Fluorescent Bulbs

PL bulbs are available in 5-, 7-, 9-, and 13-watt versions. The Quad-13 consists of two 7-watt PL bulbs on a single base. All PL bulbs must use a ballast! USA.

31-101 PL-5 Bulb, Pin Base $6⁹⁵
31-102 PL-7 Bulb, Pin Base $6⁹⁵
31-103 PL-9 Bulb, Pin Base $6⁹⁵
31-104 PL-13 Bulb, Pin Base $4⁹⁵
31-122 Quad-13 Bulb, Pin Base $11⁹⁵

Replace Your Watt-Burning DC Lights with Ultra-Efficient LED Lamps

A major breakthrough in LED technology has transformed the humble Light Emitting Diode from a so-so monochromatic lamp, suitable only for gauges and indicators, to an ultraefficient, high-quality, white light source appropriate for many off-the-grid home applications. And, the power use is so minimal, you can literally turn these lights on and forget them.

Each diode produces about the same output as a small penlight, yet uses only 30 milliamps (.36 watts!) Light output is boosted by combining two or more diodes. These exciting new LED lamps produce an excellent bluish-white light, ideal for reading and task lighting where depth perception is critical.

It may be hard to believe, but the life expectancy of an LED lamp is a whopping 60,000 hours—that's 6.8 years of 24-hour-a-day use! In occasional-use cabins or RVs, your LED lamp could last longer than your motor home. Sorry, they're only available in 12-volt DC so far.

Moonglow White LED Lamp

Using a single white LED, this 4.5″ aluminum fixture with glass lens and on/off switch is powder-coated white and tolerant of protected outdoor or marine mounting. Good for low-level night-light, porch, or stairway light. Draws only 30 milliamps. 12-volt DC. One-year mfr.'s warranty. USA.

32-198 Moonglow LED Lamp $23⁰⁰

Comet White LED Lamp

Featuring three white LEDs for triple the light output, and using the same white powder-coated aluminum fixture as the Moonglow above, the Comet still only draws 30 milliamps at 12-volt DC. (The LEDs operate on 3.5 volts, so we can either use an inexpensive resistor, or three LEDs in series for 12V input.) Bright enough for reading without strain. One-year mfr.'s warranty. USA.

32-199 Comet LED Lamp $39⁰⁰

Nova White LED Lamp

Featuring six white LEDs, the Nova is our brightest LED lamp. Has a 5.5″ stainless steel fixture with glass lens and on/off switch that is okay for protected outdoor or marine installations. Draws 55 milliamps at 12-volt DC. One-year mfr.'s warranty. USA.

32-200 Nova LED Lamp $59⁹⁵

Sunspot White LED Spotlight

Featuring an outdoor spotlight fixture that fits a standard 4″ round outdoor lamp junction box, this 12-volt DC fixture is available with either 6 LEDs drawing 55 milliamps, 12 LEDs drawing 110 milliamps, or 18 LEDs drawing 165 milliamps. Fixture is black rubber with mirror reflector and clear plastic lens, stainless steel bracket, and standard 4″ round aluminum base plate. One-year mfr.'s warranty. USA.

32-202 6 LED Sunspot $59⁹⁵

32-284 12 LED Sunspot $109⁰⁰

32-285 18 LED Sunspot $139⁰⁰

LED Strip Lights

These 12-volt DC lights are great for mounting under cabinets for direct counter lighting, and for accent lighting strips. They have one LED per inch. Each ⅝″ tube has a rotating end cap for mounting. Available in 6 LED strips drawing 55 milliamps and 8″ long, 12 LED strips drawing 110 milliamps and 14″ long, or 18 LED strips drawing 165 milliamps and 20″ long. One-year mfr's. warranty. USA.

32-286 6 LED Strip $48⁰⁰

32-287 12 LED Strip $96⁰⁰

32-288 18 LED Strip $128⁰⁰

Screw-In LED Conversion

In a standard medium edison base, here's an easy LED conversion for 12-volt DC systems. Uses 12 LEDs, which gives illumination approximately similar to a 15-watt incandescent lamp, but only uses 110 milliamps. That's an incredible 1.3 watts! LEDs have a life expectancy of 60,000 hours, or just under seven years of continuous use. It's unlikely that you'll ever wear these out. For 12-volt DC ONLY!

32-289 12 LED Edison Base, 12V $59⁹⁵

Cabin Lite LED Fixture

Uses either one or three white LEDs in a wall-mounted reading light with swivel neck. Black plastic with mirror reflector and on/off switch. Both draw 30 milliamps at 12-volt DC. One-year mfr.'s warranty. USA.

32-201 Cabin Lite 1 LED $24⁹⁵

32-196 Cabin Lite 3 LED $39⁹⁵

Sunrise/Sunset Switch For LED Lamps

Want your minimal-power-use LED porch, stairway, or night-light to come on automatically at sunset and switch off at sunrise? Here's the simple, inexpensive solution. Switch A can handle 1 to 18 LEDs. Switch B can handle 18 to 30 LEDs. 12-volt DC only. Switch will fit inside most fixtures. One-year mfr.'s warranty. USA.

32-204 Sunrise/Sunset Sw. A $9⁹⁵

32-205 Sunrise/Sunset Sw. B $24⁰⁰

DC Indoor Lighting

While not always as efficient as the new compact fluorescent lamps, these standard indoor fluorescent lamps are nonetheless far more efficient than incandescents and halogens, and are reasonably priced and well built.

Warm 12VDC Fluorescent Fixtures

Thin-Lite offers a line of DC lighting fixtures using the modern, warm-colored, U-shaped compact fluorescent lamps. They all feature instant-starting, no flicker electronic ballasts with radio frequency suppression, and deliver more lumens per watt than any previous straight tube type. All have anodized aluminum housing and feature a clear acrylic diffuser lens. Measure 4.5″ wide, 1.5″ deep. Length varies by model. 12VDC only. The 13-watt fixture draws 1.1 amps, delivers 900 lumens, equivalent to a 60W incandescent bulb. Length is 8.8 inches. The 27-watt DC fixture draws 2.1 amps, delivers 1620 lumens, equivalent to a 100W incandescent bulb. Length is 14.8 inches. The 39-watt DC fixture draws 2.45 amps, delivers 2620 lumens, equivalent to a 150W incandescent bulb. Length is 18.8 inches.

32-949 Thin-Lite 13-Watt Fixture $44⁹⁵

36-505 13-Watt 4-pin Repl. Bulb $19⁹⁵

32-953 Thin-Lite 27-Watt Fixture $59⁹⁵

36-506 27-Watt 4-pin Repl. Bulb $19⁹⁵

32-957 Thin-Lite 39-Watt Fixture $69⁹⁵

36-507 39-Watt 4-pin Repl. Bulb $19⁹⁵

Thin-Lites

Thin-Lites are made by REC Specialties, Inc. Their 12-volt DC fluorescents are built to last, and are all UL-listed. Easy to install, they have one-piece metal construction, non-yellowing acrylic lenses, and computer-grade rocker switches. A baked white enamel finish, along with attractive woodgrain trim, completes the long-lasting fixtures. Two-year warranty. All use easy-to-find, standard fluorescent tubes, powered by REC's highly efficient inverter ballast. Bulbs used in DC fixtures are exactly the same as those used in AC fixtures and can be purchased at any local hardware store—only the ballast is different. USA.

30-Watt 12-Volt DC Light

Uses two F15T8/CW fluorescent tubes (included).
- 18″ x 5.5″ x 1.375″
- 1.9 amps
- 1760 lumens

32-116 Thin-Lite #116, 30-Watt $48⁰⁰

15-Watt 12-Volt DC Light

Uses one F15T8/CW fluorescent tube (included).
- 18″ x 4″ x 1.375″
- 1.26 amps
- 800 lumens

32-115 Thin-Lite #115, 15-Watt $40⁰⁰

22-Watt 12-Volt DC Circline Light

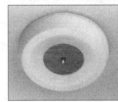

Uses one FC8T9/CW fluorescent tube (included).
- 9.5″ diameter x 1.5″
- 1.9 amps
- 1100 lumens

32-109 Thin-Lite #109C, 22-Watt $46⁰⁰

Hi-Tech Styles

Thin-Lite Hi-Tech styles were developed for both efficient and attractive lighting where maximum light is required. Anodized aluminum housings and clear acrylic diffuser lenses provide high light output on three sides. They are designed for commercial and industrial vehicles, and for use in remote area housing, schools, and medical facilities in conjunction with alternative sources of energy. Fixtures have almond-colored end caps. USA.

8-Watt 12-Volt DC Light

Uses one F8T5/CW fluorescent tube (included).
- 12.375" x 2.5" x 2.438"
- 0.9 amps
- 400 lumens

◈ **32-191 Thin-Lite #191, 8-Watt $42⁰⁰**

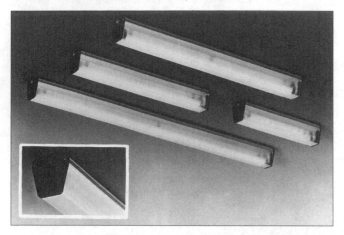

15-Watt 12-Volt DC Light

Uses one F15T8/CW fluorescent tube (included).
- 18.125" x 2.25" x 2.438"
- 1.3 amps
- 870 lumens

◈ **32-193 Thin-Lite #193, 15-Watt $39⁰⁰**

Twin Lights

A permanent fixture for boat or cabin, this 12-volt DC unit has an extra energy-saving feature: a selection switch that allows use of only one or both of the two 12-inch, 8-watt fluorescent tubes. It's equipped with screws for mounting the fixture, and connecting wire for easy installation. Lamps included. China.
- 14.3" x 1.6" x 2"
- 1.8 amps
- 800 lumens

32-107 Twin Light, 8-Watt $14⁹⁵

Tail-Light Bulb Adapters

This simple adapter has a standard medium edison base and accepts a standard automotive-type bulb. It's a very easy way to convert lamps to 12-volt with this ½-inch long adapter. Bulb pictured but not included. Get tail-light or turn-signal bulbs at your local auto supply store. USA.

33-404 Tail-Light Bulb Adapter $7⁹⁵

12-Volt Christmas Tree Lights

It just doesn't seem like Christmas without lights on the tree and these 12-volt lights will dazzle any old fir bush. These lights are actually great all year round for decorating porches, decks, your rolling art automobile, or your Harley. The light strand is 20 feet long and consists of 35 mini, colored, non-blinking lights (draw is 1.2 amps) with a 12-volt cigarette lighter socket on the end. Note: you can't connect them in series like you can with many 120-volt lights. Uses conventional 3-volt mini-lights, same as any 120-volt mini-light string, just wired for 12-volt input. Taiwan.

37-301 12-Volt Christmas Tree Lights $14⁹⁵

3 for $12⁹⁵ each

Halogen Lighting

Littlite High-Intensity Lamp

The Littlite is a great gooseneck lamp for your desk, stereo, headboard, or worktable. It's available in a 12- inch or an 18-inch length. The "A" series comes with base and dimmer, 6-foot cord, gooseneck, hood, and halogen bulb. Two pieces of snap mount are included for permanent mounting. Options include a weighted base for a movable light source, plastic snap mount with adhesive pads, and an adjustable mounting clip that adjusts from ¹⁄₁₆- to ¾-inch wide. Also available is a replacement halogen bulb. The power transformer is available for 120-volt users to convert the Littlite from 12-volt to standard house current. Available with a cigarette lighter plug (but won't work with weighted base). 12-volt DC only. USA.

33-307 18″ Littlite $47⁰⁰

33-309 18″ Littlite, 12V plug $49⁹⁵

33-401 Littlite Weighted Base $14⁰⁰

33-402 Littlite Snap Mount $3⁹⁵

33-403 Littlite Mounting Clip $12⁹⁵

33-109 Replacement 5w Bulb $7⁰⁰

33-501 120V Transformer $15⁰⁰

Quartz Halogen Lamps Inside Frosted Globes

A small cottage industry in the Northeast manufactures these incredibly ingenious 12-volt light bulbs. On the outside they appear identical to an incandescent, but they have the increased efficiency and longevity of quartz halogens. Multiply the watts on halogens by 1.5 and you'll get an idea of their equivalent light output compared to an incandescent lamp. Available in 12- or 24-volt. USA.

33-102 Quartz Lamp, 10W/0.8amp $14⁹⁵

33-103 Quartz Lamp, 20W/1.7amp $14⁹⁵

33-104 Quartz Lamp, 35W/2.9amp $14⁹⁵

33-105 Quartz Lamp, 50W/4.2amp $14⁹⁵

33-106 Quartz Lamp, 20W/24V $19⁹⁵

33-107 Quartz Lamp, 50W/24V $24⁹⁵

Halogen Flood Lamp

The MR-16 is a very bright flood lamp with a faceted reflector and a quartz halogen bulb in the center, and comes covered with a glass lens. It is ideal for situations where a bright light is desired but not overly directed to a pinpoint. The bulb is a 20-watt halogen which draws 1.7 amps but is far brighter than you'd expect. Fits into an edison base. 12-volt DC only! USA.

33-108 20-Watt Halogen Flood, 12V $24⁹⁵

12-Volt Motion Sensor

Want your porch light to come on automatically to light the way into your renewable energy palace? Here's a solution that doesn't leave your inverter sucking watts in the "on" mode all the time you're gone. This passive infrared sensor runs on 12-volt DC, and will switch up to 10 amps. Use it to switch on a DC light directly, or through a relay to switch on an AC light. Run time is adjustable from 5 seconds to 20 minutes. Standby power use is a miniscule 0.05 watts. Has standard ½″ threaded mount. Measures 9″ x 5.5″ x 1.5″. USA.

34-349 12-Volt Motion Sensor $59⁹⁵

Solar Outdoor Lighting

Solar Sensor Light

The Sensor Light eliminates the need to wastefully burn an outside light while you're away. The solar-charged detection circuit automatically turns on the light to welcome you when triggered by heat or motion at approximately 35 feet, then turns off after you leave the area. This can also make it effective in deterring prowlers. The bright 20-watt quartz-halogen bulb will last over 100 hours and lights instantly at even the coldest temperatures. There are no timers and only one switch to set (off, charge only, on). An adjustable sensitivity control reduces false triggers by your cat, a raccoon, tiny UFOs, etc. The unit mounts easily to wall, fascia, soffit, or roof eaves. No wiring or electrician required. Has a 14-foot cord on solar module for best sun exposure. Extra battery capacity allows up to two weeks of operation without sun. Comes with mounting bracket hardware and bulb included. Replaceable battery and bulb. Full one-year warranty. China.

34-310 Solar Sensor Light $119⁰⁰

15-198 6V 4Ah Repl. Battery $11⁹⁵

34-313 Solar Sensor Replacement Bulb $11⁰⁰

A Motion Sensor For Sleepy Inverters

To save power, you want your inverter to drop into sleep mode whenever there's no load (this saves 5 to 20 watts times many hours). But it sure would be nice if that motion sensor would light up to show you the way up the steps when you come home late juggling three bags of groceries, wouldn't it? Here's the answer. This motion sensor and socket screws into any standard socket. It uses a 9-volt battery to run the motion sensor, while letting the inverter sleep. When motion is sensed it closes a relay for the light, which wakes up the inverter, which runs the light while you stumble in. When the light turns off, the inverter returns to sleep mode. Too cool. USA.

25-717 AC Motion Sensor $39⁹⁵

The Latest And Most Innovative Solar-Powered Garden Light

During the day the 20% larger solar panel charges the 100% larger battery bank. Around sundown (adjustable sensor) three amber LEDs turn on and stay on all night. The LEDs provide a warm glow to mark a path while drawing minimal power from the battery. An infrared motion sensor turns on a bright 10-watt halogen light for full illumination when a warm body moves in its adjustable range. The light stays on as long as motion is present, or can be turned to 30 seconds or one minute. Two 8″ stake sections support the 8″ diameter by 8.5″ high light unit. China.

57-174 Deluxe Pagoda Sensor Light $99⁰⁰

Solar Moonlight Markers

Don't Rip Up Your Yard
Installing Electrical Wiring

Our Moonlight Markers have non-corrosive stakes, just place them in your garden and that's it—no complicated installation. Make your driveway or garden path a safer place for evening visitors. Our Moonlight Markers exude an amber glow from their built-in LED lamps; their heads measure 4″H x 4.1″Dia., while the stakes measure 7.75″L x .5″Dia. Can run continuously for eight hours at night, and they have built-in sensors to automatically turn on at dusk and shut off in the morning. Solar cells will allow for a modest charge on cloudy days. Comes as a set of two and can be wall mounted for porch marking (hardware included). Hong Kong.

34-300 Moonlight Markers (Set of 2) $39⁹⁵

2 or More Sets $36⁹⁵ each set

Flashlights

A Super Efficient Wind-Up Flashlight

When we introduced the world's first wind-up radio, we knew a spring-wound flashlight wouldn't be far behind. This revolutionary flashlight is the ultimate choice for auto or home emergency lighting, combining BayGen's reliable clockwork power source with a built-in nicad battery pack. Run the lamp directly from the clockwork mechanism (60 turns of winding, about one minute, nets 4 to 5 minutes of run or recharge time), or from the nicads (up to two hours of continuous use). Features water-resistant construction and the familiar high-output bulb from mini-Mag-type flashlights; a spare bulb is included inside the twist-off lens. Recharge by winding up, or by plugging in the included AC/DC adapter. All wind-up energy goes to recharging. A great choice for auto or home emergency lighting. Five-year mfr.'s warranty! South Africa.

37-297 Freeplay Flashlight $75⁰⁰

Depend On This Versatile Flashlight —It Doesn't Rely On Batteries

Don't lend this flashlight to anyone, because you'll have a hard time getting it back. It uses either of two independent power systems. First is a wind-up system, a battery-free source of power generated by you. Your winding energy is stored in a patented spring mechanism that releases the power on demand, converting it to electricity that energizes a cluster of three bright-white LED lights. A full wind-up (60 turns in 20 seconds) provides about 10 minutes of shine time. The second energy source is an AC power converter that recharges the NiMH battery; this powers a bright Xenon-filled bulb, for about 45 minutes of shine time from a full charge. Great technology from a very forward-thinking manufacturer. 4.75″H x 9″L x 3.5″W. South Africa/China.

31-750 Freeplay 20/20 Flashlight $64⁹⁵

Create Light Without Batteries

No preparedness kit should be without the Dynamo. Continuous squeezing produces a steady beam of light. This is a reliable, battery-free emergency light, with a shockproof and shatterproof lens, a spare bulb, a peg for attaching a cord, and a slide mechanism to lock the handle closed. Assorted gear colors. Canada.

37-374 Hand-Cranked Dynamo Flashlight $12⁹⁵

3 or More $10⁹⁵ each

The Brighter Whiter Penlight

Our new ultraefficient flashlight takes advantage of cutting-edge LED technology to bring you low power, high quality, handheld lighting. This watertight penlight produces an excellent bluish-white light with such a small power draw that one set of 3 AA alkaline batteries will last approximately 17 hours, or 8 hours with nicad batteries (included). Has wrist strap and turns on/off by tightening/unscrewing the clear plastic lens. The three LED bulbs last 60,000 hours (6.8 years of continuous use!). Brighter than typical mag lights. One year mfr.'s warranty. USA.

37-356 White 3 LED Penlight $29⁹⁵

Solar Flashlight

Free sunlight fully recharges the built-in 600 mA nicad battery pack in about eight hours. No time for solar? Slip two AA batteries into the battery compartment, and you have light right now. These optional AA batteries will be recharged by solar too. Includes 3 LED warning flashers on tail with focusing lens. Flashlight and warning flashers are switched separately, so one or both can be run. An adjustable holder for bikes or whatever is included, along with two spare bulbs and wrist strap. Run time on built-in battery alone is approximately 2.5 hours for the flashlight, or 6 days for the warning flasher. Optional nicad batteries will increase run times by 120%, alkaline batteries by 250%. Waterproof and floats too! China.

37-295 Solar Flashlight w/Warning Flashers $24⁹⁵

Flashlight Dashlite

This is a bright, rechargeable flashlight that lives in your cigarette lighter socket, so it's always there, easy to find, and ready when you need it. Absolutely will not drain your car battery! Provides one hour of continuous use, plenty of time to fix a flat or walk to a call box. Shines up to 100 feet, and lasts up to five years. Shock-resistant ABS plastic. Mexico.

37-310 Flashlight Dashlite $12⁹⁵

Flashlight Needs No Batteries Or Bulbs—Ever!

Simple shaking recharges this revolutionary flashlight. It will always work, and there's a lifetime warranty. How's it do that? The Starlight uses a high-output white LED light source and a precision lens. It delivers a 6-foot diameter illumination at a distance of 30 feet, like the glow of a full moon. Shaking the flashlight causes a high-power rare earth magnet to slide back and forth through a coil of wire, generating a small current. This electricity is stored in a capacitor, lighting the LED, and delivering up to five minutes of run time. Double O-ring seals allow use underwater or in severe weather. The casing is virtually indestructible recycled polycarbonate. This is the last emergency light you'll ever need to purchase! Weighs 16 oz., measures 11″ x 2″. Lifetime manufacturer's warranty! Made in USA.

37-357 Nightstar Flashlight $75⁰⁰

The Real Goods Solar Lantern

A 3-way lantern with solar charging and great run time! Features a 5-watt fluorescent floodlight, a powerful flashlight, and red-lens flashing warning light. The built-in nicad battery pack will run the fluorescent lamp for over 4 hours, the flashlight for 8 hours, or the warning light for 40 hours. Solar recharging takes 8–10 hours in full sun. Want more run time? Slip three optional D cells into the battery compartment. Nicad batteries will boost run times by 300%, alkaline batteries will boost by 700%! Optional D cells are not solar charged. Comes with four spare bulbs, a shoulder strap, and a clear vinyl "raincoat" tote bag for waterproof use. Equipped with a 6-volt external charging port for an AC/DC adapter (adapter not included). A perfect lantern for camping, auto, emergency, or general use. 11.4″ x 5.6″ x 2.1″, 1.75 lb. Hong Kong.

37-296 Real Goods Solar Lantern $79⁹⁵

50-123 Nicad Battery "D" (2-pack) $13⁵⁰ (not included)

CHAPTER 6

WATER SYSTEMS

WATER IS THE SINGLE MOST IMPORTANT INGREDIENT in any homestead. Without a dependable water source, we can't call anyplace home for very long. Gaiam Real Goods offers a wide variety of water pumps that are specifically made for solar-, battery-, wind-, or water-powered pumping. These pumps are designed for long hours of dependable duty in out-of-the-way places where the utility lines don't reach. For instance, ranchers are finding that small solar-powered submersible pumps are far cheaper and more dependable for moderate lift applications than the old wind pumps we're used to seeing dotting the great plains. Many state and national parks use our renewable-energy powered pumps for their backcountry campgrounds. The great majority of pumps we sell are solar- or battery-powered electric models, so we'll cover them first.

The initial cost of solar-generated electricity is high. Because of this high initial cost, most solar pumping equipment is scaled toward more modest residential needs, rather than larger commercial or industrial needs. By wringing every watt of energy for all it's worth, we keep the start-up costs reasonable. Solar-powered pumps tend to be far more efficient than their conventional AC-powered cousins.

The Three Components of Every Water System

Every rural water system has three easily identified basic components.

A Source. This can be a well, a spring, a pond, creek, or surface water.

A Storage Area. This is sometimes the same as the source, and sometimes it is an elevated or pressurized storage tank.

A Delivery System. This used to be as simple as the bucket at the end of a rope. But given a choice, most of us would prefer to have our water arrive under pressure from a faucet. Hauling water, because it's so heavy, and we use such surprising amounts of it, gets old fast.

The first two components, source and storage, you need to produce locally; however, we can offer a few pointers based on experience and some of the better, and worse, stories we've heard.

A Mercifully Brief Glossary of Pump Jargon

Flow: The measure of a pump's capacity to move liquid volume. Given in gallons per hour (gph), gallons per minute (gpm), or for you worldly types who have escaped the shackles of archaic measurement, liters per minute (lpm).

Foot Valve: A check valve (one-way valve) with a strainer. Installed at the end of the pump intake line, it prevents loss of prime, and keeps large debris from entering the pump.

Friction Loss: The loss in pressure due to friction of the water moving through a pipe. As flow rate increases and pipe diameter decreases, friction loss can result in significant flow and head loss.

Head: Two common uses. 1) The pressure or effective height a pump is capable of raising water. 2) The height a pump is actually raising the water in a particular installation.

Lift: Same as Head. Contrary to the way this term sounds, pumps do not suck water, they push it.

— continued on next page

Sources

Wells

The single most common domestic water source is the well, which can be hand-dug, driven, or drilled. Wells are less prone to pick up surface contamination from animals, etc. Hand-dug wells are usually about 3 feet or larger in diameter and rarely more than 40 feet deep. The most common sizes for driven or drilled wells is 4 inches or 6 inches. Beware of the "do-it-yourself" well drilling rigs, which can rarely insert pipe larger than 2 inches. This restricts your pump choices to only the least efficient jet-type pumps. A 4-inch casing is the smallest into which a submersible pump can be inserted.

Springs

Many folks with country property are lucky enough to have surface springs which can be developed. At the least, springs need to be fenced to keep wildlife out, or more commonly, are developed with either a backhoe and 2-ft. or 3-ft. concrete pipe sections sunk into the ground, or developed with one of the newer, lightweight horizontal drilling rigs. This helps to ensure that the water supply won't be contaminated by surface runoff or animals. Before developing a spring, you should consider that it is probably supporting a large and diverse ecosystem. If there are no other springs nearby, try to ensure that some runoff will still continue after development.

Ponds, Streams and Other Surface Water

Ponds are one of the least expensive and most pleasing methods of supplying water. They are often used to hold winter and spring runoff for summertime use. Some water systems are as simple as tossing the pump intake into the lake or stream. Surface water is usually used for livestock or agricultural needs. Don't drink or cook with it unless it's treated first (see the Sourcebook chapter on Water Purification for more help on this topic).

Storage

The purpose of water storage may be any or all of the following: to get through long dry periods, to provide pressure through elevated storage or pressurized air, to keep drinking water clean and uncontaminated, or to prevent freezing.

Surface Storage

Those with ponds, lakes, streams, or springs may not need any additional storage. However, many systems will need to pump water to a freeze-proof or elevated storage site in order to develop pressure.

Liners

Custom made polyvinyl liners are available for ponds or leaking tanks. Gaiam Real Goods carries a brand of very heavy-duty, custom sized liners. These ingenious liners are designed for installation during pond construction and are then buried with 6 inches of dirt around the edges, practically guaranteeing a leak-free pond even when working with sandy or other problem soils. When buried, the life expectancy of these liners is 50 years plus. These make reliable pond construction possible in locations that normally could not accommodate such inexpensive water storage methods.

Tanks

Covered tanks of one sort or another are the most common and longest lasting storage solutions. The cover must be screened and tight enough to keep critters like mice and squirrels from drowning in your drinking water (always an exciting discovery). The most common materials are polypropylene, fiberglass, and concrete.

— continued from previous page

Prime: *A charge of water that fills the pump and the intake line, allowing pumping action to start. Centrifugal pumps will not self-prime. Positive displacement pumps will usually self-prime if they have a free discharge no pressure on the output.*

Suction Lift: *The difference between the source water level and the pump. Theoretical limit is 33 feet; practical limit is 10–15 feet. Suction lift capability of a pump decreases 1 foot for every 1,000 feet above sea level.*

Submersible Pump: *A pump with a sealed motor assembly designed to be installed below the water surface. Most commonly used when the water level is more than 15 feet below the surface, or when the pump must be protected from freezing.*

Surface Pump: *Designed for pumping from surface water supplies such as springs, ponds, tanks or shallow wells. The pump is mounted in a dry, weather-proof location less than 10 15 feet above of the water surface. Surface pumps cannot be submerged and be expected to survive.*

Plastic Tanks. Both fiberglass and polypropylene tanks are commonly available. Both types will suffer slightly from UV degradation in sunlight. Simply painting the outside of the tank will stop the UV degradation and will probably make the tank more aesthetically pleasing. With fiberglass, make sure the tank is internally coated with an FDA approved material for drinking water. Plastic tanks can be found at your local farm supply store. Some plastic tanks can be partially or completely buried for freeze protection. Ask before you buy if this is a consideration for you.

Concrete Tanks. These are one of the best storage solutions, but expensive initially. Concrete tanks are built on site, and can be concrete block, ferrocement, or monolithic block pours. Although monolithic pours require hiring a contractor with specialized forming equipment, these tanks are usually the most troublefree in the long run. Any concrete tank will need to be coated internally with a special sealer to be watertight. Concrete tanks can be buried for freeze protection, and to keep the water cool in hot climates.

Pressure Tanks. These are used in pumped systems to store pressurized water so that the pump doesn't have to start for every glass of water. They work by squeezing a captive volume of air, since water doesn't compress. The newer, better types use a diaphragm—sort of a big heavy-duty balloon—so that the water can't absorb the air charge. This was a problem with the older plain pressure tanks. Pressure tanks are rated by their total volume. Draw down volume, the amount of water that can actually be loaded into and withdrawn from the tank under ideal conditions, is typically about 40% of total volume.

Delivery Systems

A few lucky folks are able to collect and store their water high enough above the level of intended use, so that the delivery system will simply be a pipe, and the weight of the water will supply the pressure free of charge. Most of us, though, are going to need a pump, or maybe even two, either to get the water up from underground, or to provide pressure. We'll cover electrically driven solar and battery-powered pumping first, then water-powered and wind-powered pumps.

The standard utility-powered water delivery system consists of a submersible pump in the well delivering water into a pressure tank. A pressure tank extends the time between pumping cycles by saving up some pressurized water for delivery later. This system usually solves any freezing problems by placing the pump deep inside the well, and the pressure tank indoors. The downside is that the pump must produce enough volume to keep up with any potential demand, or the pressure tank will be depleted and the pressure will drop dramatically. This requires a 1/3 hp pump minimally, and usually 1/2 hp or larger. Well drillers may sell a larger than necessary pump because it increases their profit and guarantees that no matter how many sprinklers you add in the future, the water delivery will be there. This is fine when you have large amounts of utility power available to meet heavy surge loads, but is very costly to power with a renewable energy system because of the large equipment requirements. We try to work smarter, smaller, and use less expensive resources to get the job done.

Solar-Powered Pumping

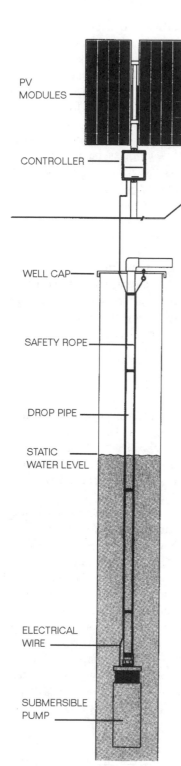

PV MODULES

CONTROLLER

WELL CAP

SAFETY ROPE

DROP PIPE

STATIC WATER LEVEL

ELECTRICAL WIRE

SUBMERSIBLE PUMP

A Simple PV-Direct Solar Pumping System

Where Efficiency Is Everything

PV modules can be quite expensive, and water is surprisingly heavy. These two facts dominate the solar-pumping industry. At 8.3 pounds per gallon, a lot of energy is needed to move water uphill. Anything we can do to wring a little more work out of every last watt of energy is going to make the system less expensive to set up. Because of these harsh economic realities, the solar-pumping industry tends to use the most efficient pumps available. For many applications that means a positive displacement pump. In this class of pumps there is no possibility of the water slipping from high-pressure areas to lower-pressure areas inside the pump. Positive displacement pumps also ensure that even when running very slowly—such as when powered by a PV module under partial light conditions—water will still be pumped.

There are some disadvantages to using positive displacement pumps. They tend to be noisier, as the water is expelled in rapid, discrete segments. They usually pump smaller volumes of water, they must start under full load, they require periodic maintenance, and some types won't tolerate running dry. These are reasons why this class of pumps isn't used extensively in the AC-powered pumping industry.

Most AC-powered pumps are centrifugal types. This class of pumps is preferred because of easy starting, low noise, smooth output, and minimal maintenance requirements. Centrifugal pumps are good for moving large volumes of water at relatively low pressure. As pressure rises however, the water inside the centrifugal pump "slips" increasingly, until finally a pressure is reached at which no water is actually leaving the pump. This is 0% efficiency.

In the solar industry centrifugal pumps are used for pool pumping, and some circulation duties in hot water systems. But in all applications where pressure exceeds 20 psi, you'll find us recommending the slightly noisier, occasional-maintenance-requiring, but vastly more-efficient positive displacement type pumps. For instance, an AC submersible pump running at 7%–10% efficiency is considered "good." Our solar submersible pump runs at 30%–35%.

Running PV-Direct for even higher efficiency.

We often design solar pumping systems to run PV-direct. That is, the pump is connected directly to the photovoltaic (PV) modules with no batteries involved in the system. The electrical to chemical conversion in a battery isn't 100% efficient. When we avoid batteries and deliver the energy directly to the pump, 20% to 25% more water gets pumped. This kind of system is ideal when the water is being pumped into a large storage tank, or is being used immediately for irrigation. It also saves the initial cost of the batteries, the maintenance and periodic replacement they require, plus charge controllers, and fusing/safety equipment that batteries demand. PV-direct pumping systems, running all day long, also help us get around the lower gallon-per-minute output of most positive displacement pumps.

However (every silver lining has its cloud), there is one piece of modern technology we like to use on PV-direct systems that isn't often found on battery-powered systems. A Linear Current Booster, or LCB for short, is a solid-state marvel that will help get a PV-direct pump running earlier in the morning, keep it running later in the evening, and sometimes make running at all a possibility on hazy or cloudy days. An LCB will convert excess PV voltage into extra amperage when the modules aren't producing quite enough current for the pump. The pump

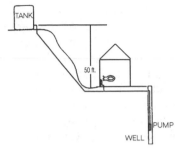

Pump Option 1

will run slower than if it had full power, but with positive displacement pumps, if it runs at all, it delivers water. LCBs will boost water delivery in most PV-direct systems by 20% or more, and we usually recommend a properly sized one with every system.

Direct Current (DC) motors for variable power.

Most of our pumps use DC electric motors. PV modules produce DC electricity, and all battery types store it. DC motors have the advantage of accepting variable voltage input without distress. Common AC motors will overheat if supplied with low voltage. DC motors simply run slower when the voltage drops. This makes them ideal to work with PV modules. Day and night, clouds and shadows; these all affect the PV output, and a DC motor simply "goes with the flow!"

Which Solar-Powered Pump Do I Want?

That depends on what you're doing with it, and what your climate is. We'll start with the most common and easiest choices, and work our way through to the less common.

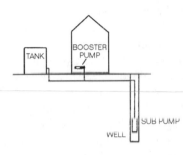

Pump Option 2

Pumping from a Well

Got a well that's cased with a 4 inch or larger pipe, and a static water level that is no more than 600 feet below the surface? Fine. We carry several brands of proven DC-powered submersible pumps with a range of prices, lift, and volume capabilities. The Shurflo Solar Sub is the lowest-cost system with lift up to 230 feet, and sufficient volume for most residential homesteads. The bigger Sunrise pumps are available in four models, with lift up to 600 feet, or volume over 6 gpm depending on the model. The latest addition to DC submersibles is a helical rotor-type pump, the TSP-1000, available in six different models with lift up to 230 feet, or volume up to 25 gpm (wow!).

Performance, prices, and PV requirement for each model are listed in the Products Section. The Shurflo and Sunrise are diaphragm-type pumps, and unlike AC-powered submersible pumps, they can tolerate running dry. The manufacturer says just don't let it run dry for more than a month or two! This feature makes these pumps ideal for many low output wells.

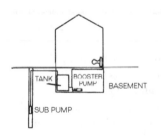

Pump Option 3

Complete submersible pumping systems—PV modules, mounting structure, LCB, and pump—range from $1,300 to $8,200 depending on lift and volume required. Options like float switches that will automatically turn the pump on and off to keep a distant storage tank full are inexpensive and easy to add when using an LCB with remote control as we recommend.

Because these are usually low output pumps, averaging 1.25 to 1.5 gpm, they won't directly keep up with average household fixtures. For household use we usually recommend, in order of cost and desirability:

Option 1. Pumping into a storage tank at least 50 feet higher than the house, if terrain and climate allow.

Option 2. Pumping into a house-level storage tank if climate allows, and using a booster pump to supply household pressure.

Option 3. Pumping into a storage tank built into the basement for hard-freeze climates, and using a booster pump to supply household pressure.

Option 4. Using battery power from your household renewable energy system to run the submersible pump, and using a big pressure tank (80 gallons min.).

Option 5. Using a conventional AC-powered submersible pump, large pressure tank(s), and your household renewable energy system with large inverter to run it. (Note that there is a great loss of efficiency running this way, but it's the standard way to get the job done in freezing climates, and your plumber won't have any problems understanding the system. Besides, if you've already got the renewable energy system . . .)

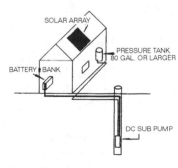

Pump Option 4

If you would like the help of our technical staff in selecting an appropriate pump, controller, power source, etc., we have a Solar Water Supply Questionnaire at the end of this chapter. It will give our staff the information we need in order to thoroughly and accurately recommend a water supply system for you. You can mail or fax your completed form. Please give us a day-time phone number if at all possible!

Many folks, for a variety of reasons, already have an AC-powered sub pump in their well when they come to us, but are real tired of having to run the generator to get water. For wells with 6-inch and larger casings it's usually possible to install both the existing AC pump and a submersible DC pump. If your AC pump is 4 inches in diameter, it's possible to install the DC pump underneath it. The cabling, safety rope, and ½-inch poly delivery pipe from the DC submersible will slip around the side of the AC pump sitting above it. Just slide both pumps down the hole together. It's often comforting to have emergency backup for those times when you need it, like when it's been cloudy for three weeks straight, or the fire is coming up the hill, and you want a lot of water fast!

Pumping from a Spring, Pond, or other Surface Source

Your choices are a bit more varied here, depending on how high you need to lift the water, and how many gallons per minute you want. Surface mounted pumps are not freeze tolerant. If you are in a freezing climate, make sure that your installation can be (and is!) completely drained before freezing weather sets in. If you need to pump through the hard freeze season, we recommend a submersible pump as described above.

Pumps don't like to pull water up from a source. Or, put simply, Dr. Doug's #1 Pumping Rule: **Pumps don't suck, pumps push**. In order to operate reliably, your surface-mounted pump must be installed as close to the source as practical, and in no case should the pump be more than 10 feet above the water level. With some positive displacement pumps a 15-foot suction lift is possible, but not recommended. If you can get the pump closer to the source, and still keep it dry and safe, do it! You'll be rewarded with more dependable service, longer pump life, and less power consumption.

For modest lifts up to 50 or 60 feet and volumes of 1.5 to 3.0 gpm we have found *diaphragm*-type pumps to be dependable and moderately priced. The SHURflo series are good examples. They can tolerate running dry for moderate amounts of time. They can be rebuilt quite easily in the field. Diaphragm life expectancy is usually one to five years depending on how hard, and how much, the pump is working. Diaphragm pumps will tolerate sand, algae, and debris without damage, but these may stick in the internal check valves reducing or stopping output, and necessitating disassembly to clean out the debris. Who needs the hassle? Filter your intake!

For higher lifts, or more volume, we usually go to a pump type called Rotary-Vane. Examples of these pumps are the Solar Slow pump and Flowlight Booster pumps. Rotary-vane pumps are capable of lifts up to 440 feet, and volumes up to 4.5 gpm, depending on model. Of all the positive displacement pumps, they are the quietest and smoothest. But they will not tolerate running dry, or abrasives of any kind in the water. It's very important to filter the input of these pumps with a 5-micron or finer filter in all applications. Rotary-vane pumps are very long-lived, but will eventually require a pump head overhaul, which must be done at the factory.

Household Water Pressurization

We have two pumps that are commonly used for this application: The Chevy and the Mercedes models if you will.

The Chevy model. SHURflo's Medium-Flow pump is our best-selling pressure pump. It comes with a built-in 20–40 psi pressure switch. With a 2.5 to 3.0 gpm flow rate it will keep up with most household fixtures, garden hoses excluded. The diaphragm pump is reliable, easy to overhaul, but somewhat noisy. We recommend 24-inch flexible connector in a loop on both sides of this pump, and a pressure tank plumbed in as close as possible to absorb most of the noisy buzz.

The Mercedes model. Flowlight's rotary vane pump is our smoothest, quietest, largest-volume pressure pump. It delivers 3 to 4.5 gpm at full pressure, and is very long-lived, but quite expensive initially. This pump will easily keep up with garden hoses,

Diaphragm-Type Pump

sprinklers, and any other normal household use. Brushes are externally replaceable, and will last five to ten years. Pump life expectancy is 15 to 20 years.

Any household pressure system requires a pressure tank. A 20-gallon tank is the minimum size we recommend for a small cabin; full-size houses usually have 40-gallon or larger tanks. Pressure tanks are big, bulky, and expensive to ship. Get one at your local hardware/building center store.

Solar Hot Water Circulation

Hot Water Circulation Pump

Most of the older solar hot water systems installed during the tax-credit heydays of the early 1980s used AC pumps with complex controllers/brains and multiple temperature sensors at the collector, tank, plumbing, ambient air, etc. This kind of complexity allows too many opportunities for Murphy's Law.

The smarter systems simply used a small PV panel wired directly to a DC pump. When the sun shines a little bit, producing a small amount of heat, the pump runs slowly. When the sun shines bright and hot, producing lots of heat, the pump runs fast. Very simple system control is achieved with an absolute minimum of "stuff." We carry several hot water circulation pumps, the best of which is the El-Sid pump. This is the first solid-state, brushless DC circulation pump that was designed from scratch for PV-direct applications. It only requires a 5-watt module to drive it, and life expectancy is three to four times longer than any other DC circulation pump. Volume and lift are sharply limited however. The El-Sid is standard equipment in some of the solar hot water systems we sell.

Swimming Pool Circulation

Yes, it's possible to live off-the-grid and still enjoy luxuries like a swimming pool. In fact pool systems dovetail nicely with household systems in many climates. Houses generally require a minimum of PV energy during the summer because of the long daylight hours, yet there is a maximum of energy available. By switching a number of PV modules to pool pumping in the summer, then back to battery charging in the winter we get better utilization of resources. We offer several DC pool pumps. DC pumps run somewhat more efficiently than AC pumps, so a slightly smaller DC pump can do the same amount of work as a larger AC pump. We also strongly recommend using a low-back pressure cartridge type pool filter. See the Product Section for specs and prices. Please consult with our technical staff regarding pool filters, PV array sizing, and switching equipment.

Water-Powered Pumps

A few lucky folks have an excess supply of falling water available. This falling water energy can be used to pump water for you. Both the High Lifter and Ram type water pumps use the energy of falling water to force a portion of that water up the hill to a storage tank.

Ram Pumps

Ram pumps have been around for many decades, providing reliable water pumping at almost no cost. They are commonly used in the Eastern U.S. where modest falls and large flow rates are the norm, but they will happily work almost anyplace their minimum flow rate can be satisfied. Rams will work with a minimum of 1.5 feet to a maximum of about 20 feet of fall feeding the pump. Minimum flow rates depend on the pump size; see the Product Section for specs. A flow is started down the drive pipe and then is shut off suddenly. The resulting pressure surge from the momentum of moving water slamming to a stop sends a little squirt of water up the hill. How much of a squirt depends on the pump size, the amount of fall, and the amount of lift. Output charts are with the pumps in the Product Section. Each ram needs to be carefully tuned for its particular site. Ram pumps are not self-starting. If they run short of water they will stop pumping and simply dump incoming water, so don't buy too big. Rams make some noise. A lot less than a gasoline-powered pump for sure, but the constant chunk-chunk-chunk is a consideration for some sites. Ram pumps deliver less than 5% of the water that passes through them, and the discharge must be into an unpressurized storage tank or pond. But they work for free and have life expectancies measured in decades.

The High Lifter Pump

This pump is unique. It works by simple mechanical advantage. A large piston at low water pressure pushes a smaller piston at higher water pressure. High Lifters recover a much greater percentage of the available water than ram pumps do, but they require greater fall into the pump generally. This makes them better suited for more mountainous territory. They are available in two ratios, 4.5 to 1 and 9 to 1. Fall to lift ratios and waste water to pumped water ratios are also either 4.5 to 1 or 9 to 1. Note, however, that as the lift ratio gets closer to theoretical maximum the pump is going to slow down and deliver fewer gallons per day. High Lifters are self-starting. If they run out of water they will simply stop and wait, or slow down to match what water is available. A very handy trait for unattended, or difficult to attend sites. The only serious disadvantage of the High Lifter is wear caused by abrasives. O-rings are used for sealing the pistons against the cylinder walls. Any abrasives in the water wear out the O-rings quickly, so filtering of the intake is strongly encouraged. High Lifter pumps can be overhauled fairly easily in the field, but the O-ring kit costs a lot more than filter cartridges. Output charts for this pump are included in the Product Section.

Wind-Powered Water Pumps

If you've been reading this far because you want to buy a nostalgic old-time jack-pump windmill, we're going to disappoint you. They are still made, but we don't sell them, and don't recommend them very strongly except in unusual cases. They are very expensive, try $6,000 and up, quite a big deal to set up and install into the well, and require routine yearly service at the top of the tower. This is technology that has largely seen its day, except for very deep remote location pumping. We do have a wind-powered pump that has seen good success.

Bowjon pumps

Bowjon pumps use compressed air. There are two models available depending on lift and volume needs. Both models use a simple pole-mounted turbine that direct-drives an air compressor. The air is piped down the well and run through a carefully engineered air injector. As it rises back up the supply tube, it carries water in between the bubbles. The lift/submergence ratio of this pump is fairly critical. Approximately 30% of the total lift is the recommended distance for the air injector to be submerged below the static level. Too little submergence and the air will separate from the water, too much and the air will not lift the water, though there is considerable latitude between these performance extremes. This pump isn't bothered by running dry. Output depends on wind speed, naturally, but the larger Rancher model is capable of up to 9 gpm at lower lifts, or can lift a maximum of 275 feet. The air compressor requires an oil level check twice a year. See the Product Section for more details and output specs.

Wind Generators running submersible pumps

The larger wind generator manufacturers, Bergey and Whisper, both offer options that allow the three-phase wind turbine output to power a three-phase submersible pump directly. These aren't residential-scale systems and are mostly used for large agricultural projects, or village pumping systems in underdeveloped countries. They are relatively expensive initially. Contact our technical staff for more information on these options.

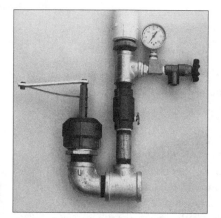

Ram Pump

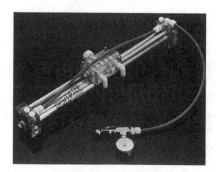

High Lifter Pump

Freeze Protection

In most areas of the country freezing is a major consideration when installing plumbing and water storage systems. For outside pipe runs the general rule is to bury the plumbing below frost level. For large storage tanks burial may not be feasible, unless you go with concrete. In moderate climatic zones, simply burying the bottom of the tank a foot or two along with the input/output piping is sufficient. In some locations, due to climate or lack of soil depth, outdoor storage tanks simply aren't feasible. In these situations we can go for a smaller storage tank built into a corner of the basement with a separate pressure boosting pump, or we can pump directly into a large pressure tank system.

Other Considerations & Common Questions

Wind-Powered Pump

Hopefully by this point you've zeroed in on a pump or pumps that seem to be applicable to your situation. If not, our technical staff will be happy to discuss your needs, and recommend an appropriate pumping system—which may not be renewable-energy powered). If you need help from the tech staff, please fill out the Solar Water Supply Questionnaire and fax or mail it to us. At this point another crop of questions usually appears, such as . . .

How Far Can I Put the Modules from the Pump?

Often the best water source will be deep in a heavily wooded ravine. It's important that your PV modules have clear, shadow-free access to the sun for as many hours as practical. Even a fist-sized shadow will effectively turn off most PV modules. The hours of 10 to 2 are usually the minimum your modules want to see, and if you can capture full sun from 9 to 3, that's more power for you. If the pump is small, running off one or two modules, then distances up to 200 feet can be handled economically. Longer distances are always possible, but consult with our tech staff or check out the wire sizing formula in the wiring chapter first, as longer distances require large (expensive!) wire. Many pumps routinely come as 24-volt units now, as the higher voltage makes long distance transmission twice as easy. Some of our pumps can be special-ordered as 120-volt units when distances over 500 feet are involved.

What Size Wire Do I Need?

This depends on the distance, and the amount of power you are trying to move. We always recommend 10-gauge copper submersible pump wire going down the well to the submersible pump, simply because #10 is the largest commonly available submersible pump wire. For topside wiring consult the wire sizing formula in the wiring chapter, or give our tech staff a call.

What Size Pipe Do I Need?

Most of the pumps we offer have modest flow rates of 4 gpm or less. This makes it okay to use smaller pipe sizes without increasing friction loss. If the pipe run is being used for pumping delivery only, then ¾-inch or 1-inch pipe sizes are usually sufficient. Please note we said for pumping delivery only. There's no reason you can't use the same pipe to take the water up the hill to the storage tank, and also bring it back down to the house or garden. Pipes don't care which way the water is flowing through them. But if you do this, you'll want a larger pipe to avoid friction loss and pressure drop when the higher flow rate of the household or garden fixtures comes into play. We usually recommend at least 1¼-inch pipe for household and garden use, and 2-inch for fire hose lines.

Water Pumping Truths & Tips

*1. Pumps prefer to **push** water, not **pull** it. In fact most pumps are limited to 10 feet or 15 feet of lift on the suction side. Mother Nature has a theoretical suction lift of 33.9 feet at sea level, but only if the pump could produce a perfect vacuum. Suction lift drops 1 foot with every 1,000 feet rise in elevation. To put it simply, **Pumps Don't Suck, They Push**.*

2. Water is heavy, 8.33 lb. per gallon. It can require tremendous amounts of energy to lift and move.

3. DC electric motors are generally more efficient than AC motors. If you have a choice, use a DC motor to pump your water. Not only can they be directly powered by solar modules, but your precious wattage will go further.

4. Positive displacement pumps are far more efficient than centrifugal pumps. Most of our pumps are positive displacement types. AC powered submersibles, jet pumps, and booster pumps are centrifugal types.

5. As much as possible we try to avoid batteries in pumping systems. When energy is run into and out of a battery, 25% is lost. It's more efficient to take energy directly from your PV modules and feed it right into the pump. At the end of the day, you'll end up with 25% more water in the tank.

6. One pound per square inch (psi) of water pressure equals 2.31 feet of lift, a handy equation.

What Size PV Modules Do I Need?

A number of the pumps listed in the Product Section have performance and wattage tables listed with them. For instance, the rotary-vane Solar Slowpump model 2503 lifting 40 feet will deliver 2.5 gpm and requires 60 watts. Hey, what's to figure here? All PV modules are rated by how many watts they put out, right? So I just need a 60-watt module. Wrong. PV modules are rated under ideal laboratory conditions, not real life. If you want your pump to work on hot or humid days (heat reduces PV output, water vapor cuts available sunlight), then you must add 30% to the pump wattage when figuring PV wattage. So in our example we actually need 78 watts. Looking at available PV modules, we don't find exactly 78-watt modules, which means we buy a 75-watt module if this is a relatively temperate climate, or an 80- to 90-watt module if this is a hot (over 80°F) climate. And, of course, an LCB is practically standard equipment with any PV-direct system.

Can I Automate the System?

Absolutely! Life's little drudgeries should be automated at every opportunity. The LCBs that we so strongly recommend with all PV-direct systems help us do this. The three models that cover most applications are all supplied with the remote control option. This option allows you to install a float switch at the holding tank and a pair of tiny 18-gauge wires can be run as far as 5,000 feet back to the pump/controller/PV modules area. Float switches can be used to either pump up or pump down a holding tank, and will automatically turn the system off when satisfied. The three common LCBs, plus float switches and other pumping accessories are presented in the Product Section. LCBs can also be used on battery-powered systems when remote sensing and control would be handy.

Where Do We Go from Here?

If you haven't found all the answers to your remote water pumping needs yet, please give our experienced technical staff a call. They'll be happy to work with you selecting the most appropriate pump, power source, and accessories for your needs. We run into situations occasionally where renewable energy sources and pumps simply may not be the best choice, and we'll let you know if that's the case. For 95% of the remote pumping scenarios there is a simple, cost-effective, long-lived renewable energy powered solution, and we can help you develop it.

—Doug Pratt

Solar Water Supply Questionnaire

In order for us to thoroughly and accurately recommend a water supply system, we need to know the following information about your system. Please fill out the form as completely as possible. If at all possible please give us a daytime phone number.

Name:_____

Address:_____

City:_____ State:_____ Zip: _____

email:_____ Phone:_____

Describe Your Water Source:

Depth of well: _____
Depth to standing water surface: _____
If level varies, how much? _____
Estimated yield of well (gallons per minute): _____
Well casing size (inside diameter): _____
Any problems? (silt, sand, corrosives, etc.)_____
_____

Water Requirements:

Is this a year-round home? _____ How many people full-time?_____
Is house already plumbed? _____ Conventional flush toilets? _____
Residential gal/day estimated: _____
Is gravity pressurization acceptable? _____
Hard freezing climate? _____
Irrigation gal/day estimated: _____ Which months? _____
If you have a general budget in mind, how much? _____
Do you have a deadline for completion? _____

Describe Your Site:

Elevation: _____
Distance from well to point of use: _____
Vertical rise or drop from top of well to point of use: _____
Can you install a storage tank higher than point of use? _____
  How much higher? _____ How far away? _____
Complex terrain or multiple usage? _____ (Please enclose map)
Do you have utility power available? _____ How far away? _____
Can well pump be connected to nearby home power system? _____
  How far? _____ Home power system power battery voltage: _____

Mail your completed questionnaire to:
Tech Staff/Water Supply, Real Goods, 13771 South Hwy. 101, Hopland, CA 95449.
Or fax to : 707-462-4807

Pumping Products

Books

Planning for an Individual Water System

By the American Association of Vocational Instructional Materials (AAVIM). The definitive book on water systems and one of our all-time favorites. It was a surprise and a complete delight when one of our customers pointed it out to us. Incredible graphics and charts. Presents a thorough discourse on all forms of water pumps, and discusses water purity, hardness, chemicals, water pressure, pipe sizing, windmills, freeze protection, and fire protection. If you have a new piece of land or are thinking of developing a water source, you need this book! 160 pages, paperback. USA.

80-201 Planning for an Individual
**   Water System $18⁹⁵**

Cottage Water Systems

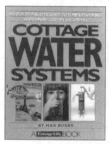

An out-of-the-city guide to water sources, pumps, plumbing, water purification, and wastewater disposal, this lavishly illustrated book covers just about everything concerning water in the country. Each of the 12 chapters tackles a specific subject such as sources, pumps, plumbing how-to, water quality, treatment devices, septic systems, outhouses, alternative toilets, graywater, freeze protection, and a good bibliography for more info. This is the best illustrated, easiest to read, most complete guide to waterworks we've seen yet. 150 pages, softcover. USA.

80-098 Cottage Water Systems $24⁹⁵

Rainwater for the Mechanically-Challenged Video

You say you're *really* mechanically challenged and want more than a few pictures? Here's your salvation. From the same irreverent, fun-loving crew that wrote the book to the right. See how all the pieces actually go together as they assemble a typical rainwater collection system, and discuss your options. This is as close as you can get to having someone else put your system together. 37 minutes and lots of laughs. USA.

82-568 Rainwater Challenged Video $19⁹⁵

Build Your Own Water Tank

By Donnie Schatzberg. An informative booklet by one of our original customers that has recently been updated and greatly expanded with more text, drawings, and illustrations. It gives you all the details you need to build your own ferro-cement (iron-reinforced cement) water storage tank. No special tools or skills are required. The information given in this book is accurate and easy to follow, with no loose ends. The author has considerable experience building these tanks, and has gotten all the "bugs" out. 58 pages, paperback. USA.

80-204 Build Your Own Water Tank $14⁰⁰

Earth Ponds Sourcebook

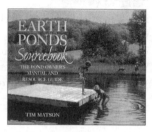

By Tim Matson. This book presents new tips and techniques, and hundreds of resources for supplies and support, in a comprehensive guide to the care and feeding of ponds. Big ponds, skating ponds, recreational fishing ponds, swimming ponds, duck ponds, hydropower, and fire protection ponds are just some of the pond types explored. Matson focuses on processes and resources, with a strong emphasis on wildlife ponds, wetlands, and biologically controlled aquatic environments. A wealth of information for anyone interested in renovating a pond or reservoir, or general wetlands preservation. 171 pages, paperback. USA.

80-557 Earth Ponds Sourcebook $19⁹⁵

Rainwater Collection for the Mechanically-Challenged

Laugh your way to a home-built rainwater collection and storage system. This delightful paperback is not only the best book on the subject of rainwater collection we've ever found, it's funny enough for recreational reading, and comprehensive enough to painlessly lead a rank amateur through the process. Technical information is presented in layman's terms, and accompanied with plenty of illustrations and witty cartoons. Topics include types of storage tanks, siting, how to collect and filter water, water purification, plumbing, sizing system components, freeze-proofing and wiring. Includes a resources list and a small catalog. Paperback, 46 pages, USA.

80-704 Rainwater Collection for the
**   Mechanically-Challenged $17⁹⁵**

Rainwater Collection Video

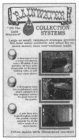

There is an untapped, inexpensive water source on your own property: rainwater. Rainwater Collection: Texas Style tells the story of three Texan hill country families who collect rainwater for all their household needs, including drinking. The half-hour video details how low-quality well water drove a physician and his wife to devise a successful rainwater collection method that inspired others in the area to follow suit. John Dromgoole, from the PBS series *The New Garden*, hosts the video. With it comes a 50-page booklet that elaborates on the systems illustrated in the video and gives information on equipment suppliers and prices. 30 minutes. USA.

80-132 Rainwater Video $29⁹⁵

The Home Water Supply

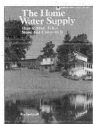

By Stu Campbell. Explains completely and in depth how to find, filter, store, and conserve one of our most precious commodities. A bit of history; finding the source; the mysteries of ponds, wells, and pumps; filtration and purification; and a good short discussion of the arcana of plumbing. After reading this book, you may still need professional help with pipes and pumps, but you will know what you're talking about. 236 pages, paperback. USA.

80-205 The Home Water Supply $18⁹⁵

A Great Water Pumping Video

Part of the *Renewable Energy with the Experts* series, this 59 minute video features solar-pumping pioneer Windy Dankoff, who has more than 15 years experience in the field. Windy demonstrates practical answers to all the most common questions asked by folks facing the need for a solar-powered pump, and offers a number of tips to avoid common pitfalls. This is far and away the best video in this series, and offers some of the best advice and knowledge available for off-the-grid water pumping. This grizzled old technician/copyeditor even learned a few new tricks about submersible pump installations. Highly recommended! USA

80-368 RE Experts Water Pumping Video $39⁹⁵

Build Your Own Ram Pump

Utilizing the simple physical laws of inertia, the hydraulic ram can pump water to a higher point using just the energy of falling water. The operation sequence is detailed with easy-to-understand drawings. Drive pipe calculations, use of a supply cistern, multiple supply pipes, and much more are explained in a clear, concise manner. The second half of the booklet is devoted to detailed plans and drawings for building your own 1″ or 2″ ram pump. Constructed out of commonly available cast iron and brass plumbing fittings, the finished ram pump will provide years of low-maintenance water pumping for a total cost of $50 to $75. No tapping, drilling, welding, special tools, or materials are needed. This pump design requires a minimum flow of three to four gallons per minute, and three to five feet of fall. It is capable of lifting as much as 200′ with sufficient volume and fall into the pump. The final section of the booklet contains a setup and operation manual for the ram pump. Paperback, 25 pages. USA.

80-501 All About Hydraulic Ram Pumps $9⁹⁵

www.realgoods.com

Storage

Kolaps-a-Tank

These handy and durable nylon tanks fold into a small package or expand into a very large storage tank. They are approved for drinking water, withstand temperatures to 140° F, and fit into the beds of several truck sizes. They will hold up under very rugged conditions, are self-supported, and can be tied down with D-rings. Our most popular size is the 525-gallon model, which fits into a full-sized long-bed (5 x 8 ft.) pickup truck. USA.

◈ **47-401 73 gal. Kolaps-a-Tank, 40″ x 50″ x 12″ $300⁰⁰**
◈ **47-402 275 gal. Kolaps-a-Tank, 80″ x 73″ x 16″ $400⁰⁰**
◈ **47-403 525 gal. Kolaps-a-Tank, 65″ x 98″ x 18″ $500⁰⁰**
◈ **47-404 800 gal. Kolaps-a-Tank, 6′ x 10′ x 2′ $600⁰⁰**
◈ **47-405 1140 gal. Kolaps-a-Tank, 7′ x 12′ x 2′ $800⁰⁰**
◈ **47-406 1340 gal. Kolaps-a-Tank, 7′ x 14′ x 2′ $900⁰⁰**

◈ *Means shipped directly from manufacturer.*

Leak-Free Pond Liners

Heavy-Duty Gauges Now Available

A pond is a major investment, requiring lots of earth work, soil blending, and compaction. It's a gamble if it will hold water or not, or for how long. When we realized that the porous gravel soil at our creekside Solar Living Center made pond liners an absolute necessity, we spent a lot of time checking out the many brands and styles on the market. The very tough, UV-stabilized polyethylene is nontoxic, chemically inert, and nonleaching—and it's FDA/USDA approved for potable water supplies. Earth work may be kept to a minimum, and your pond will not have a leak. You can establish a pond on any kind of soil without fear of having the water percolate away. Whether your plans call for a stocked aquaculture pond, an agricultural reservoir, or a garden feature, you need a liner that is absolutely leak-free to protect your investment of time and money. Our PondLiners are custom-sized to order, and allow you to establish a successful pond on practically any kind of soil with a minimum of earthwork. Factory seams are as strong, or stronger than the material itself. It resists punctures, tears, root penetration, and rodents.

Available in tough 20-mil thickness, in super-heavy-duty 30-mil, or for extreme sites, 40-mil. (We used 30-mil for our rough, gravelly site.) Per 1,000 sq. ft. of liner, the 20-mil

We used these pond liners for our three ponds at the Real Goods Solar Living Center in Hopland, California.

weighs 100 lb., 30-mil weighs 150 lb., and 40-mil weighs 200 lb. Free freight on all PondLiners under 2,000 sq. ft.! Under 500 sq. ft. there is a $1500 handling charge. USA.

Pond liners are made to order. To figure size: Double the depth of your pond, then add to the width and length. Add an additional 2 to 5 feet for the buried apron. For instance, a 50′ x 50′ pond that's 5′ deep needs a 62′ x 62′ liner at minimum. 100 sq. ft. minimum order.

PondLiner 100–200 sq. ft.
47-203   20-mil   .56/sq. ft.
47-217   30-mil   .61/sq. ft.
47-217   40-mil   .69/sq. ft.

PondLiner 201–600 sq. ft.
47-205   20-mil   .43/sq. ft.
47-218   30-mil   .51/sq. ft.
47-218   40-mil   .58/sq. ft.

PondLiner 601–2,000 sq. ft.
47-206   20-mil   .39/sq. ft.
47-219   30-mil   .45/sq. ft.
47-219   40-mil   .52/sq. ft.

PondLiner 2,001–5,000 sq. ft.
47-207   20-mil   .34/sq. ft.
47-220   30-mil   .40/sq. ft.
47-220   40-mil   .49/sq. ft.

PondLiner 5,001–8,000 sq. ft.
47-208   20-mil   .31/sq. ft.
47-221   30-mil   .37/sq. ft.
47-221   40-mil   .46/sq. ft.

PondLiner 8,001–12,000 sq. ft.
47-209   20-mil   .28/sq. ft.
47-222   30-mil   .34/sq. ft.
47-222   40-mil   .43/sq. ft.

PondLiner 12,001–20,000 sq. ft.
47-210   20-mil   .26/sq. ft.
47-223   30-mil   .32/sq. ft.
47-223   40-mil   .41/sq. ft.

Pond liners are made to order, so please have dimensions ready when ordering. 100 sq. ft. minimum order. Free freight Continental U.S. only.

47-212 PondLiner Fab Tape 2″ x 30′ $19⁹⁵
47-227 Handling Charge under 500 sq. ft. $15⁰⁰

Fountain Pumps

A Solar Fountain Pump That Lasts

We've searched for years for a reliable PV-direct fountain pump. The extremely high-quality German-made submersible Aquasolar 700 finally meets our needs. This brushless, solid-state, sealed DC pump is designed to run at voltages from 12 to 24 volts; it will happily operate continuously at the 16- to 20-volt output of a 12V PV module. Using only 8 watts at the pump, it is rated for 12 liters per minute (over three gallons) maximum flow, or one meter (3.2 feet) maximum lift. This pump will run to full output with any 10- to 15-watt PV module. Output volume is externally adjustable. The pump housing acts as a large surface area intake screen. The outlet adapter can accept ½″ or ¾″ pipe thread. The housing, pump, and rotor can be easily disassembled for cleaning. Has a 15-foot cord with quick disconnect. Three-year manufacturer's warranty. Germany.

41-914 Aquasolar 700 Fountain Pump $179⁰⁰

Basic Solar Pump Kit

EXCLUSIVE

Deluxe Fountain Kit

Deluxe Fountain Kit: High-Volume Pump And Solar Module

Combines our 13-watt PV module (below) with our best brushless, solid-state, sealed DC pump, the Aquasolar 700. Adjustable flow volume up to 3 gpm (up to 3-feet maximum lift); accepts ½″ or ¾″ fountain tubing; easy disassembly for cleaning; 15-foot cable, three-year mfr.'s warranty. Germany/China.

41-919 Deluxe Fountain Kit $299⁰⁰

Basic Fountain Kit: Compact Pump And Solar Module

Combines our 13-watt PV module with a reliable, efficient, ultracompact, German submersible pump. Maximum 20″ lift. With a respectable 2-gpm water flow volume, it accepts ⅜″ or ½″ fountain tubing for small to mid-size fountains. 20-foot cable. Expect to replace pump every two to three years. Germany/China.

41-920 Basic Solar Pump Kit $169⁰⁰

41-129 Basic Pump Only $34⁹⁵

13-Watt PV Module

This workhorse of a module directly powers both of our fountain kit pumps, and it's eminently useful for any 12-volt charging application. China.

11-562 13-Watt Module with Clips $139⁰⁰

Attwood Solar Fountain Pumps

We've searched a long time for this one. These submersible 12-volt pumps are designed as continuous-duty marine bilge pumps. We've found they make very impressive solar-powered fountain pumps. They won't lift very high, but they move an incredible volume and use a tiny amount of power. The pump motor twist-locks onto the pump base to allow easy cleaning and removal of debris. They can withstand dry running for a limited time. Three-year mfr's. warranty. USA.

Tech Note: In full-time PV-direct use, these pumps can be expected to last about one summer. Attwood is great about honoring the 3-year warranty, so we recommend buying two pumps, and being prepared to rotate your stock.

Attwood V625 Pump

Model V625 is a good match with a 30-watt module. (Solarex MSX30L, #11-511, recommended.).
• Amps: 1.3 @ 3 ft. head
• Max gpm: 7.5 @ 3 ft. head
• Max head: 5 ft.

41-157 V625 Solar Fountain Pump $25⁰⁰

Attwood V1250 Pump

Model V1250 is a good match with a 50-watt module. (Siemens SM55 or Evergreen EC-51 recommended.) 3-year warranty!
• Amps: 2.9 @ 3 ft. head
• Max gpm: 15.5 @ 3 ft. head
• Max head: 8 ft.

41-158 V1250 Solar Fountain Pump $35⁰⁰

Centrifugal Pumps

El-Sid Hot Water Circulation Pumps

These completely solid-state "motors" have no moving parts! There are no brushes, bearings, or seals to wear out. The rustproof bronze pump section is magnetically driven and is an existing off-the-shelf pump head that can be easily replaced if ever necessary. Life expectancy is many times longer than any other DC circulation pump.

Two models are offered at 1.7 and 3.0 gpm flow rates. Both are designed to be driven PV-direct for simple system control. These are circulation pumps for closed loop systems, and will not do any significant lift. See the specifications chart. The smaller pump requires a 5-watt PV module, the larger pump requires a 10-watt module. Rated for temperatures up to 240°F, one year mfr's. warranty. USA.

41-134 El-Sid 1.7 gpm Pump $197⁰⁰

41-133 El-Sid 3.0 gpm Pump $219⁰⁰

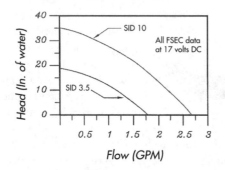

Hartell Hot Water Circulation Pumps

Hartell pumps feature magnetic drive with no rotating seals. Designed for solar hot water systems or other low-flow, modest-lift applications, they will operate directly off an 18- to 40-watt PV module. Brighter sun means faster pumping. Rated for temperatures up to 200°F, and pressure up to 150 psi. Brass pump bodies have ½″ MNPT fittings. We offer one brushless pump model, and two conventional brushed models, all with six-month mfg. warranty. USA.

Hartell MD-10-HEH

A brushless model that offers almost zero maintenance. Requires an 18- to 20-watt PV module. Measures 5.25″ diameter by 9″ length. Maximum volume, 5 gpm; maximum lift, 10 feet. See chart below.

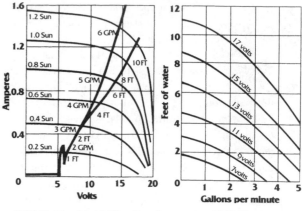

HEH Motor with 18-Watt Panel   HEH Motor

41-529 Hartell Pump MD-10-HEH $410⁰⁰

Hartell MD-3-DCL

A conventional brushed model that will need periodic maintenance. Requires an 18- to 20-watt PV module. Measures 5.25″ diameter by 7.75″ length. Maximum volume, 4.5 gpm; maximum lift, 5 feet. See chart below.

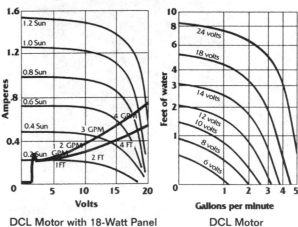

DCL Motor with 18-Watt Panel   DCL Motor

41-526 Hartell Pump MD-3-DCL $260⁰⁰

Hartell MD-10-DCH

A slightly more robust conventional brushed model. Requires a 30- to 40-watt PV module. Measures 5.25″ diameter by 7.75″ length. Maximum volume, 6 gpm; maximum lift, 11 feet. See chart below.

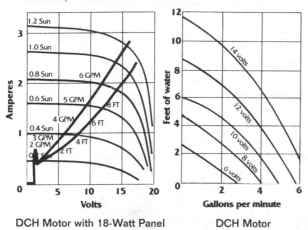

DCH Motor with 18-Watt Panel   DCH Motor

41-527 Hartell Pump MD-10-DCH $215⁰⁰

SunCentric Solar Surface Pumps

A reliable cast-iron DC pump with one moving part. Good for swimming pools, agriculture, hydronic and solar heating, pond aeration, and more. Can do lifts up to 80 feet, or volumes up to 60 gpm depending on model; see performance charts. Maximum suction lift is 10 feet with a foot valve; centrifugal pumps will not self-prime. Mount the pump as close to the water source as possible. Easy-starting on PV-direct systems; no electronic boost controls are needed. Both PV-direct and battery-powered models are available. Standard pumps use a glass-filled poycarbonate impeller with a temperature limit of 140°F. A no-cost high-temperature option uses a brass impeller with a 212°F limit. HT option reduces flow volume by 15%; power input is the same. Brush life is typically three to ten years. Two-year mfr's warranty. USA.

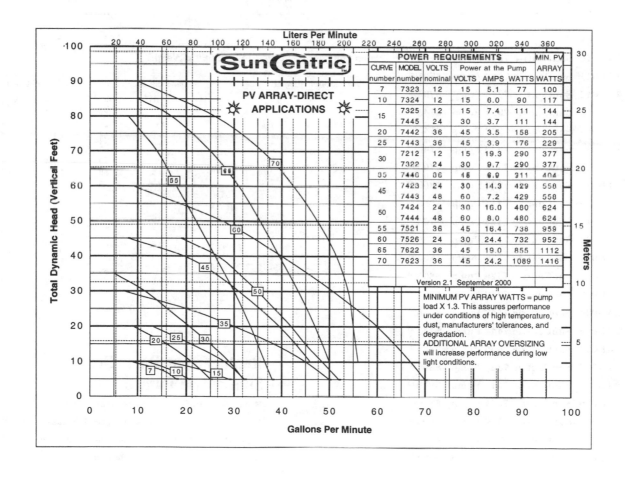

Version 2.1 September 2000

MINIMUM PV ARRAY WATTS = pump load X 1.3. This assures performance under conditions of high temperature, dust, manufacturers' tolerances, and degradation.
ADDITIONAL ARRAY OVERSIZING will increase performance during low light conditions.

CURVE number	MODEL number	VOLTS nominal	VOLTS	AMPS	WATTS	MIN. PV ARRAY WATTS
7	7323	12	15	5.1	77	100
10	7324	12	15	6.0	90	117
15	7325	12	15	7.4	111	144
	7445	24	30	3.7	111	144
20	7442	36	45	3.5	158	205
25	7443	36	45	3.9	176	229
30	7212	12	15	19.3	290	377
	7322	24	30	9.7	290	377
35	7446	36	45	6.9	311	404
45	7423	24	30	14.3	429	558
	7443	48	60	7.2	429	558
50	7424	24	30	16.0	480	624
	7444	48	60	8.0	480	624
55	7521	36	45	16.4	738	959
60	7526	24	30	24.4	732	952
65	7622	36	45	19.0	855	1112
70	7623	36	45	24.2	1089	1416

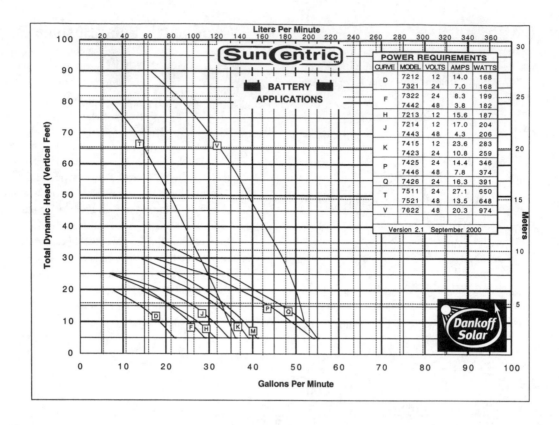

Power Requirements

CURVE	MODEL	VOLTS	AMPS	WATTS
D	7212	12	14.0	168
	7321	24	7.0	168
F	7322	24	8.3	199
	7442	48	3.8	182
H	7213	12	15.6	187
J	7214	12	17.0	204
	7443	48	4.3	206
K	7415	12	23.6	283
	7423	24	10.8	259
P	7425	24	14.4	346
	7446	48	7.8	374
Q	7426	24	16.3	391
T	7511	24	27.1	650
	7521	48	13.5	648
V	7622	48	20.3	974

Version 2.1   September 2000

◈41-010 SunCentric 7212 Pump $705⁰⁰
◈41-011 SunCentric 7213 Pump $705⁰⁰
◈41-012 SunCentric 7214 Pump $705⁰⁰
◈41-013 SunCentric 7321 Pump $745⁰⁰
◈41-014 SunCentric 7322 Pump $745⁰⁰
◈41-015 SunCentric 7323 Pump $745⁰⁰
◈41-016 SunCentric 7324 Pump $705⁰⁰
◈41-017 SunCentric 7325 Pump $860⁰⁰
◈41-018 SunCentric 7415 Pump $860⁰⁰
◈41-019 SunCentric 7423 Pump $765⁰⁰
◈41-020 SunCentric 7424 Pump $765⁰⁰
◈41-021 SunCentric 7425 Pump $860⁰⁰
◈41-022 SunCentric 7426 Pump $855⁰⁰
◈41-023 SunCentric 7442 Pump $765⁰⁰
◈41-024 SunCentric 7443 Pump $765⁰⁰
◈41-025 SunCentric 7444 Pump $765⁰⁰
◈41-026 SunCentric 7445 Pump $860⁰⁰
◈41-027 SunCentric 7446 Pump $860⁰⁰
◈41-028 SunCentric 7511 Pump $810⁰⁰
◈41-029 SunCentric 7521 Pump $810⁰⁰
◈41-030 SunCentric 7526 Pump $905⁰⁰
◈41-031 SunCentric 7622 Pump $905⁰⁰
◈41-032 SunCentric 7623 Pump $905⁰⁰

Dimensions, Pipe Fitting Sizes, Shipping Weight

PUMP MODEL NUMBER				DIMENSIONS			PIPE SIZE	
				LENGTH in.	HEIGHT in.	SHIP WT lbs.	INLET NPT	OUTLET NPT
72x1	72x2	72x3	72x4	15.5	9.1	49	1 1/4"	1"
			72x5	17	10.5	54	1 1/2"	1 1/4"
			72x6	17	10.5	54	2"	1 1/2"
73x1	73x2	73x3	73x4	17	9.1	50	1 1/4"	1"
			73x5	18	10.5	55	1 1/2"	1 1/4"
			73x6	18	10.5	55	2"	1 1/2"
74x1	74x2	74x3	74x4	17	9.1	58	1 1/4"	1"
			74x5	18.5	10.5	63	1 1/2"	1 1/4"
			74x6	18.5	10.5	63	2"	1 1/2"
75x1	75x2	75x3	75x4	18	9.1	60	1 1/4"	1"
			75x5	19.5	10.5	65	1 1/2"	1 1/4"
			75x6	19.5	10.5	65	2"	1 1/2"
76x1	76x2	76x3	76x4	19	9.1	65	1 1/4"	1"
			76x5	20.5	10.5	70	1 1/2"	1 1/4"
			76x6	20.5	10.5	70	2"	1 1/2"

Replacement Parts For SunCentric Pumps

◈ 41-033 SunCentric Brush Set $45⁰⁰

◈41-034 SunCentric 7xx1 to 4 Seal & Gasket Set $18⁰⁰

◈41-035 SunCentric 7xx5 or 6 Seal & Gasket Set $18⁰⁰

◈ *Means shipped directly from manufacturer.*

M3 Solar-Driven Agricultural Pump

A professional, long-lived floating submersible pump for low-head, medium-volume agricultural needs from surface water supplies. Just the thing to keep the stock from breaking down banks and mucking up ponds. Common uses are livestock watering, small-scale irrigation, pond aeration, landscape ponds, and water transfer systems. The pump is magnetically coupled to eliminate any shaft seal and possibility of leakage. Also provides overload protection for the motor. This easy-starting centrifugal pump doesn't need a controller or booster to start. Fifty feet of submersible cable is already attached, so no watertight splice kit is needed. Motor brushes have approximately a two-year life expectancy, and are easily field replaceable with ordinary hand tools. An input screen prevents clogging by any foreign material. Anything that can pass through the screen is easily lifted by the pump. Kit includes pump, 50′ wire, 50′ ¼″ poly safety rope to hold pump in position, and 25′ flexible output hose. Depending on lift and volume required, it needs either a single 75-watt or two 50-watt PV modules with mounting to complete a system. See output chart below. Canada.

❖**41-468 M3 Agricultural Pump $995⁰⁰**

11-165 Astropower 75W module $419⁰⁰

❖**13-399 Pole-Top Mount, 1 AP75 $90⁰⁰**

11-002 Evergreen EC-51 module $269⁰⁰

13-593 Uni-Rac Pole-Top mount 2-EC51 $139⁰⁰

M3 AGRICULTURAL PUMP PERFORMANCE CHART

Meters Lift (Feet)	1–75W Liters/Min. (GPM)	Approx. Daily Output in GPM	2–50W Liters/Min. (GPM)	Approx. Daily Output in GPM
1 (3.3)	22.7 (6.0)	1,500	31.8 (8.4)	2,100
2 (6.6)	21.5 (5.7)	1,400	28.6 (7.6)	1,900
3 (9.8)	20.0 (5.3)	1,300	25.4 (6.7)	1,700
4 (13.1)	16.7 (4.4)	1,200	22.0 (5.8)	1,450
5 (16.4)	13.6 (3.6)	1,100	18.3 (4.8)	1,250
6 (19.7)	11.4 (3.0)	900	13.9 (3.7)	950
7 (23.0)	7.6 (2.0)	600	9.9 (2.6)	650

Rotary Vane Pumps

Solar Slowpumps

Slowpumps are the original solar-powered pumps developed by Windy Dankoff in 1983. They set the standard for efficiency and reliability in solar water pumping where water demand is in the range of 50 to 3,000 gallons per day. The Slowpump is designed to draw water from shallow wells, springs, cisterns, tanks, ponds, rivers, or streams and to push it as high as 440 vertical feet for storage, pressurization, or irrigation.

These positive displacement rotary-vane type pumps feature forged brass bodies with carbon graphite vanes held in a stainless steel rotor. No plastic! Motor brushes are externally replaceable, and last 5 to 10 years. Pump life expectancy is 15 to 20 years. Rebuilt/exchange pump heads are available for under $150. Slowpumps are NSF approved for drinking water. A wide variety of pump and motor combinations are available for a variety of lifts and delivery volumes. Consult the performance charts.

Rotary-vane pumps must have absolutely clear water. They will not tolerate any abrasives. A 5-micron cartridge pre-filter is highly recommended for all installations. If your water is very dirty, improve the source or consider a diaphragm pump.

Fittings on 1300 and 1400 series pumps are ½-inch female, 2500 and 2600 series pumps have ¾-inch male fittings. Sizes and weights vary slightly with model but are approximately 6″ x 16″ and 16 lb. One year mfr's. warranty. USA.

¼ HORSEPOWER 1300 & 2500 MODELS PV OR BATTERY POWER, 12V, 24V, OR 115VAC

Total Lift In Feet	1322		1308		1303		2507	
	GPM	Watts	GPM	Watts	GPM	Watts	GPM	Watts
20	0.51	27	1.25	30	2.50	48	4.00	57
40	0.51	32	1.25	48	2.50	60	3.95	78
60	.051	36	1.20	54	2.40	78	3.90	102
80	0.49	40	1.20	60	2.30	93	3.90	120
100	0.49	45	1.20	66	2.30	105	3.85	144
120	0.48	50	1.20	70	2.25	121	3.80	165
140	0.47	56	1.20	75	2.20	138	3.65	195
160	0.47	62	1.20	84	2.20	153		
180	0.47	66	1.18	93	2.15	165		
200	0.45	74	1.16	101	2.15	180		
240	0.44	90	1.14	117	2.15	204		
280	0.41	102	1.12	135				
320	0.41	120	1.10	153				
360	.041	134	1.05	171				
400	0.40	150	1.00	198				
440	0.39	168						

Actual performance may vary ±10% from specifications. Performance listed at 15V or 30V (PV-direct). Deduct 20% from flow and watts for battery. Watts listed are power use at pump.

¼-Horsepower PV Or Battery Powered Pumps, 12 Or 24 Volts

See chart above. Specify voltage when ordering.

❖ **41-102 Slowpump 1322 pump $475⁰⁰**

❖ **41-104 Slowpump 1308 pump $475⁰⁰**

❖ **41-106 Slowpump 1303 pump $475⁰⁰**

❖ **41-108 Slowpump 2507 pump $475⁰⁰**

❖ *Means shipped directly from manufacturer.*

½-Horsepower Slowpumps

See chart below. State voltage when ordering.

❖ **41-931 Slowpump 1404 Pump $695⁰⁰**

❖ **41-932 Slowpump 1403 Pump $695⁰⁰**

❖ **41-933 Slowpump 2605 Pump $695⁰⁰**

❖ **41-934 Slowpump 2607 Pump $695⁰⁰**

AC Version Of Any Slowpump Model

❖ **41-110 Slowpump AC Option   Add $145⁰⁰**

½-HORSEPOWER 1400 & 2600 MODELS, 24 OR 48V BATTERY OR 36V PV DIRECT

Total Lift in Feet	1404		1403		2605		2607	
	GPM	Watts	GPM	Watts	GPM	Watts	GPM	Watts
160							4.30	283
180					3.35	280	4.25	305
200					3.33	296	4.20	338
240			2.55	266	3.30	331	4.05	396
280			2.50	302	3.25	373	4.00	444
320	1.66	255	2.50	338	3.20	410		
360	1.64	280	2.50	374	3.16	450		
400	1.62	312	2.50	406				
440	1.60	341	2.50	451				

Actual performance may vary ±10% from specifications. Watts listed are power use at pump.

Flowlight Booster Pumps

Sharing all the robust design features of the smaller Slowpumps, these larger rotary vane pumps are specifically designed for household water pressurization. They use one-half to one-third as much energy as a similar capacity AC pressure pump, and eliminate the starting surges that are so hard on inverters. These pumps are quieter, smoother running, and far more durable than diaphragm-type pressure pumps. For full-time off-the-grid living, this is the pressure pump you want. Life expectancy is 15 to 20 years, and then they just need a simple pump head replacement.

Two models are available. Standard models feature higher flow rates, but are noisier, require 1″ or larger intake plumbing, and have less than 10 feet of suction capacity. Low Speed models are quieter, can accept ¾″ intake plumbing and suction lifts greater than 10 feet, but have lower flow rates. Both models are available in 12 or 24 volts. Standard models also available in 48VDC and 115VAC.

A pressure tank *must* be used with this, or any, pressure pump. Forty-gallon capacity is the minimum size recommended; larger is better. The Booster Pump Installation Kit and the Inline Sediment Filter, as shown with the pump picture, are strongly recommended. Flexible hose input and output hose assemblies and pressure relief valve are included with the pump (shown below the pump in the picture). One year mfr's. warranty. USA.

BOOSTER PUMP PERFORMANCE CHART

PSI	STANDARD MODEL				LOW SPEED MODEL			
	30	40	50	65	30	40	50	65
GPM	4.5	4.3	4.3	4.1	3.4	3.3	3.1	2.7
Amps @24V	6.5	7.5	8.0	11.0	5.0	5.5	6.0	7.5
W/H per Gal	0.6	0.67	0.75	1.1	0.6	0.67	0.75	1.1

◈41-141 Standard Booster Pump, 12V $540⁰⁰

◈41-142 Standard Booster Pump , 24V $540⁰⁰

◈41-000 Standard Booster Pump, 48V $590⁰⁰

◈41-147 Standard Booster Pump, 115VAC $630⁰⁰

◈41-145 Low Speed Booster Pump, 12V $520⁰⁰

◈41-146 Low Speed Booster Pump, 24V $520⁰⁰

Accessories to Make Your Slowpump Installation Easier and More Reliable

Inline Sediment Filter

This is the recommended minimum filter for every Slowpump installation. This filter is designed for cold water only, and meets NSF standards for drinking water supplies. It features a sump head of fatigue resistant Celcon plastic, an opaque body to inhibit algae growth if installed outdoors, and is equipped with a manually operated pressure release button to simplify cartridge replacement. Rated for up to 125 psi and 100°F, it is equipped with ¾-inch female pipe thread inlet and outlet fittings, and is rated for up to 6 gpm flow rates (with clean cartridge). It accepts a standard 10-inch cartridge with 1-inch center holes. Supplied with one 5-micron fiber cartridge installed.

41-137 Inline Sediment Filter $74⁹⁵

42-632 5 Micron Sediment Cartridge $12⁹⁵

Dry Run Shutdown Switch

Rotary vane pumps will be damaged if allowed to run dry. This thermal switch clamps to the pump head of any new or older Slowpump. It will sense the temperature rise of dry running and shut off the pump before damage can occur. A manual reset is required to restart the pump. This switch will pay for itself many times over even if only used once during the lifetime of your Slowpump.

◈41-135 Dry Run Sw. for 1300 Series Pumps $49⁹⁵

◈41-144 Dry Run Sw. for all
         Larger Series Pumps $49⁹⁵

Booster Pump Installation Kit

This kit contains all the plumbing bits and pieces you'll need for installation of the Booster Pump. It includes a one-way check valve, a pressure switch, a pressure gauge, a drain valve, a shut-off valve, and the pressure tank tee all manifolded together. Shown with Booster pump picture.

◈41-143 Booster Pump Install Kit $99⁹⁵

Replacement Motor Brush Sets

The Slowpump brushes are easy to inspect and replace. Simply unscrew and withdraw the old brushes, then screw the new ones in if needed. Normal life expectancy is five to ten years. Sold in sets of two.

◈41-111 12V Slowpump Motor Brush Set $19⁹⁵

◈41-112 24V Slowpump Motor Brush Set $19⁹⁵

Rebuilt And Replacement Pump Heads

Because of the wide variety of pump heads used with Slow-pumps, and their 15- to 20-year life expectancy if the water is clean, Gaiam Real Goods recommends that you deal directly with the manufacturer when it's time for a rebuild. Information will be included with your pump purchase. So that you know what you are getting into, in late 2000 exchange heads were priced at $130, and brand new replacement heads were priced at $195 for all models.

Diaphragm Pumps

Guzzler Hand Pumps

These high-volume, low-lift pumps are driven by simple, dependable muscle power. Reinforced diaphragms and flapper valves ensure durable, dependable pumping. Both models have up to 10 ft. of suction lift, and about 12 ft. of delivery head. They make excellent emergency sump pumps, bilge pumps, or simple water delivery systems. Model 400 delivers up to 10 gpm, and uses 1″ hose. Model 500 delivers up to 15 gpm, and uses 1.5″ hose. Don't use garden hose with these pumps. It's too small, and will restrict delivery. USA.

41-151 Guzzler 400 Hand Pump $49⁰⁰

41-152 Guzzler 500 Hand Pump $59⁰⁰

Solar Surface Pump Kit

Our surface pump kit features one of our most popular pumps, the SHURflo Low Flow, along with enough PV power to lift up to 130 feet. This kit includes an Astropower 110-watt PV module, a poletop mount for ease of installation, an LCB for best delivery under marginal sun conditions, and a screened foot valve for reliable operation. You supply a 10-foot length of 2.5″ steel pipe for the PV mount, plumbing and wiring as needed, and a surface water supply. Mount your pump as close to the water supply as practical, with a maximum of 10-feet suction lift. Pumps would much rather push. Kit pricing saves $47⁰⁰.

Delivers approximately 1.5 gpm, depending on lift and sunlight availability. In a typical six-hour summer day it puts over 500 gallons into your garden or storage tank. All components U.S.-made.

41-037 Solar Surface Pump Kit $889⁰⁰

SHURflo Pumps

SHURflo produces a high-quality line of positive displacement, diaphragm pumps for RV and remote household pressurization, and for lifting surface water to holding tanks. These pumps will self-prime with 10 feet of suction and free discharge. The three-chamber design runs quieter than other diaphragm pumps and will tolerate silty or dirty water or running dry with no damage. However, if your water is known to have sand or debris, it's best to use an inexpensive inline cartridge filter on the pump intake (order #41-137 filter housing and #42-632 filter cartridge.) Sand sometimes will lodge in the check valves, necessitating disassembly for cleaning. No damage, but who needs the hassle? Put a filter on it!

Motors are slow-speed DC permanent-magnet types (except AC model) for long life and the most efficient performance. All pumps are rated at 100% duty cycle and may be run continuously. There is no shaft seal to fail; the ball bearing pump head is separate from the motor. Pumps are easily field repaired and rebuild kits are available for all models. All pumps have a built-in pressure switch to prevent over-pressurization. Some are adjustable through a limited range. Low flow model has ⅜-inch MIPT inlets and outlets. Medium and high flow models are ½-inch MIPT. All pumps carry a full one-year warranty. USA.

Note: Pumps may appear different from pictures. SHURflo has been upgrading design.

SHURflo Low Flow Pump

For modest water requirements or PV direct applications. Available in 12 volts DC only. Delivers a maximum of 60 psi or 135 feet of head. This is an excellent surface water delivery pump used with the PV Direct Pump Control (25-134RC) and PV modules as necessary for your lift. (Add 30% to pump wattage to figure necessary PV wattage.) ⅜-inch MIPT inlet and outlet.

SHURFLO LOW FLOW 12V		
PSI	**GPM**	**Watts**
0	1.75	37
10	1.66	41
20	1.57	50
30	1.48	59
40	1.38	67
50	1.30	76
60	1.23	86

41-450 SHURflo Low Flow Pump $109⁰⁰

SHURflo Medium Flow Pump

The Medium Flow is available in 12- or 24-volt DC; performance specs are similar for both (the 24V pump does slightly less volume). Maximum pressure is 40 psi or 90 feet of lift for both models. This model is a good choice for low-cost household pressurization; it will keep up with any single fixture. It has a built-in 20- to 40-psi pressure switch, perfect for most households. Buzzy output can be reduced by looping 24-inch flexible connectors on both inlet and outlet. ½-inch MIPT inlet and outlet.

MEDIUM FLOW 12V			MEDIUM FLOW 24V		
PSI	GPM	Amps	PSI	GPM	Amps
0	3.30	5.1	0	3.00	2.5
10	3.02	5.6	10	2.87	3.2
20	2.82	7.0	20	2.64	3.8
30	2.65	8.4	30	2.41	4.3
40	2.48	9.6	30	2.20	4.7

41-451 SHURflo Medium Flow, 12 Volt $165 165^{00}

41-452 SHURflo Medium Flow, 24 Volt 165^{00}

SHURflo High Flow Pump

This is SHURflo's highest output model. Good for household pressurization systems, if you need higher than normal 40 psi. It has a built-in 30- to 45-psi pressure switch. Buzzy output can be reduced by looping 24-inch flexible connectors on both inlet and outlet. Available in 12 volts only. Maximum pressure is 45 psi or 104 feet of lift. ½-inch MIPT inlet and outlet.

SHURFLO HIGH FLOW 12V		
PSI	GPM	Amps
0	3.40	6.0
10	3.00	6.9
20	2.75	7.5
30	2.50	8.6
40	2.25	9.5

41-453 SHURflo High Flow Pump 189^{00}

SHURflo AC Pump, 115 Volts

For those who need to send power more than 150–200 feet away, this pump can save you big $$ on wire costs! The motor is thermally protected and under heavy continuous use will shut off automatically to prevent overheating. This should only happen at maximum psi after approximately 90 minutes of running. Pressure switch is adjustable from 25 to 45 psi. Maximum pressure is 45 psi. Weight 5.8 lb. ½-inch MIPT inlet and outlet.

SHURFLO 115V AC PUMP		
PSI	GPM	Watts
0	3.05	59
10	2.85	65
20	2.60	82
30	2.40	96
40	2.24	108

41-454 SHURflo 115 volt AC Pump 179^{00}

SHURflo High Pressure Pumps

The High Pressure is available in 12 or 24 volts DC. This is a special pump for very high pressures up to 100 psi. It can pump up to 230 feet of vertical lift. The built-in pressure switch is factory adjusted to turn on at 85 psi and off at 100 psi. Adjustment range is 80 to 100 psi. Not recommended for continuous operation. Has ⅜-inch female NPT inlet and outlet fittings. USA.

HIGH PRESSURE 12V			HIGH PRESSURE 24V		
PSI	GPM	Amps	PSI	GPM	Amps
10	1.68	3.6	10	1.52	2.0
30	1.53	5.1	30	1.39	2.7
50	1.39	6.4	50	1.29	3.3
70	1.29	7.6	70	1.19	3.9
90	1.20	8.8	90	1.11	4.4
100	1.15	9.3	100	1.07	4.7

41-357 SHURflo High Pressure, 12 Volt 109^{00}

41-358 SHURflo High Pressure, 24 Volt 129^{00}

SHURflo Repair and Replacement Parts

REPLACEMENT PARTS KIT LIST

Pump Model	Pressure Sw. 3	Check Valve 4	Upper Housing With Sw. 3, 4, & 5	Valve Kit 6	Drive & Diaphragm Kit 7	Motor 8
Low Flow 12V 41-450	41-325 $19⁹⁵	41-327 $9⁹⁵	n/a	41-331 $19⁹⁵	41-333 $44⁹⁵	41-335 $99⁹⁵
Medium Flow 12V 41-451	41-326 $19⁹⁵	41-328 $9⁹⁵	41-329 $24⁹⁵	41-332 $14⁹⁵	41-334 $29⁹⁵	41-335 $99⁹⁵
Medium Flow 24V 41-452	41-326 $19⁹⁵	41-328 $9⁹⁵	41-329 $24⁹⁵	41-332 $14⁹⁵	41-334 $29⁹⁵	41-337 $119⁰⁰
High Flow 12V 41-453	n/a	41-328 $9⁹⁵	41-330 $49⁹⁵	41-332 $14⁹⁵	41-334 $29⁹⁵	41-339 $179⁰⁰
115VAC Pump 41-454	41-326 $19⁹⁵	41-328 $9⁹⁵	41-329 $24⁹⁵	41-332 $14⁹⁵	41-334 $29⁹⁵	41-338 $169⁰⁰

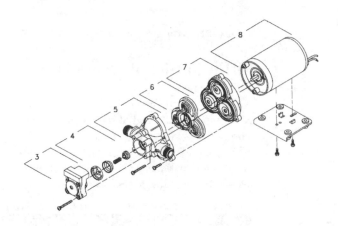

Solaram Pump

This is the highest yield, highest lift solar pump available. Using a large multi-piston industrial diaphragm pump and a permanent-magnet DC motor with positive gear belt drive, the Solaram series of pumps features models capable of lifts up to 960 feet, or over 9 gallons per minute. The performance chart shows delivery and watts required. If running PV-direct, the rated PV wattage must exceed the pump requirement by 25% or more. A pressure relief valve and flexible intake/outlet hoses are included.

In full-time daily use, yearly diaphragm replacement is recommended. Motor brushes typically last 10 years. The Diaphragm and Oil Kit provides a set of new diaphragms and the special nontoxic oil. The Long-Term Parts Kit contains three diaphragm and oil kits, plus a gearbelt and motor brush kit.

Inlet fittings are 1.5 inches, outlet is 1 inch. Dimensions are 28″ long, 16″ deep, 16.5″ high. Weight is approximately 150 lb. (varies slightly with model) and is shipped in two parcels. Mfr's. warranty is one year. USA.

Specify full 4-digit model number when ordering.

◆**41-480 Solaram 8100 series Pump $2,475⁰⁰**
◆**41-481 Solaram 8200 series Pump $2,475⁰⁰**
◆**41-482 Solaram 8300 series Pump $3,825⁰⁰**
◆**41-483 Solaram 8400 series Pump $3,825⁰⁰**
◆**41-484 Diaphragm & Oil Kit $84⁹⁵**
◆**41-485 Long Term Parts Kit $345⁰⁰**

Solaram™ Surface Pump Performance Chart

Model Numbers: First 2 digits ——————
Second 2 digits ——————

TOTAL LIFT		_ _ 21			_ _ 22			_ _ 23			_ _ 41			_ _ 42			_ _ 43			Model #
Feet	Meters	GPM	LPM	Watts	GPM	LPM	Watts	GPM	LPM	Watts	GPM	LPM	Watts	GPM	LPM	Watts	GPM	LPM	Watts	Volts
0-80	24	3.0	11.4	170	3.7	14.0	207	4.6	17.4	285	6.2	23.5	258	7.5	28.4	339	9.4	35.6	465	81__ 24V
120	37	2.9	11.0	197	3.7	14.0	238	4.5	17.1	319	6.0	22.7	305	7.3	27.7	396	9.1	34.5	539	
160	49	2.9	11.0	225	3.6	13.6	268	4.5	17.1	352	5.8	22.0	354	7.2	27.3	453	8.9	33.7	619	
200	61	2.9	11.0	247	3.6	13.6	296	4.5	17.1	388	5.7	21.6	400	7.1	26.9	513	8.9	33.7	693	
240	73	2.8	10.6	265	3.6	13.6	327	4.5	17.1	427	5.6	21.2	453	7.0	26.5	572	8.6	32.6	724	82__ 24V
280	85	2.8	10.6	286	3.6	13.6	356	4.4	16.7	466	5.5	20.8	499	6.9	26.2	628	8.4	31.8	801	
320	98	2.8	10.6	315	3.5	13.3	388	4.4	16.7	496	5.4	20.5	548	6.8	25.8	686	8.3	31.5	869	
360	110	2.8	10.6	342	3.5	13.3	416	4.4	16.7	536	5.4	20.5	592	6.6	25.0	733	8.2	31.1	927	
400	122	2.7	10.2	363	3.4	12.9	450	4.4	16.7	572	5.3	20.1	649	6.5	24.6	782	8.7	33.0	1122	83__ 180V
480	146	2.7	10.2	416	3.4	12.9	505	4.3	16.3	649	5.3	20.1	717	6.5	24.6	900	8.5	32.2	1265	
560	171	2.7	10.2	456	3.3	12.5	570	4.3	16.3	693	5.2	19.7	800	6.5	24.6	1045	8.4	31.8	1397	
640	195	2.7	10.2	502	3.3	12.5	623	4.2	15.9	774	5.1	19.3	893	6.5	24.6	1116	8.2	31.1	1540	84__ 180V
720	220	2.6	9.9	551	3.2	12.1	690	4.1	15.5	856	5.1	19.3	1031	6.4	24.3	1287	8.1	30.7	1683	
800	244	2.6	9.9	589	3.2	12.1	715	4.1	15.5	931	5.1	19.3	1114	6.4	24.3	1408	8.0	30.3	1815	
880	268	2.6	9.9	647	3.2	12.1	774	4.0	15.2	1082	5.1	19.3	1206	6.3	23.9	1529	8.0	30.3	1958	85__ 180V
960	293	2.6	9.9	705	3.1	11.7	838	4.0	15.2	1190	5.0	18.9	1289	6.1	23.1	1650	8.0	30.3	2145	

SOLAR FORCE PISTON PUMP PERFORMANCE

| Vertical Lift | Model 3110 | | Model 3020 | | Model 3040 | |
Feet	GPM	Watts	GPM	Watts	GPM	Watts
20	5.9	77	5.2	110	9.3	168
40	5.6	104	5.2	132	9.3	207
60	5.3	123	5.1	154	9.2	252
80	5.1	152	5.1	182	9.2	286
100	5.1	171	5.0	202	9.1	322
120	4.9	200	5.0	224	9.1	364
140	4.9	226	5.0	252	9.1	403
160	*This model 12 or 24V battery-driven only*		4.9	269		
180			4.9	280		
200			4.8	308		
220			4.7	314		

Actual performance may vary ± 10%. Add 20%–30% to pump wattage for PV-direct.

Piston Pumps

Solar Force Piston Pump

Built like a tank, the cast-iron Solar Force is an excellent choice for high-volume delivery of 4 to 9 gpm at lifts up to 220 feet. It can be used for water delivery to a higher storage tank, for irrigation, fire protection, or pressurization. It has a 25-foot suction lift at sea level. The durable cast-iron and brass pump is designed to last for decades, and routine maintenance is only required every two to six years. A Linear Current Booster is needed for PV-direct drive. LCB sizing depends on model and voltage. Our tech staff will advise. Solar Force uses a non-slip gear belt drive on PV-direct models, and standard V-belt drive on battery models. Intake fitting is 1.25 inches, output is 1 inch. A pressure relief valve is included. Measures is 22″ W x 16″ H by 13″ D. Weighs approximately 80 lb. maximum, shipped in two or three parcels depending on model. Two-year mfr's. warranty. USA.

The Solar Force Pump is available in three basic models:

Model 3010: available in 12 or 24 volts battery-driven only.

Model 3020: available in 12, 24, or 48 volts, battery or PV-direct, also 115VAC.

Model 3040: available in 12, 24, or 48 volts, battery or PV-direct, also 115VAC.

Please specify voltage (12, 24, 48, or 115) when ordering.

◈ **41-254 Model 3010 Solar Force**
    (12 & 24V battery only) $975⁰⁰
◈ **41-255 Model 3020 Battery Powered $1,250⁰⁰**
◈ **41-256 Model 3020 PV Powered $1,390⁰⁰**
◈ **41-257 Model 3040 Battery Powered $1,250⁰⁰**
◈ **41-258 Model 3040 PV Powered $1,420⁰⁰**
◈ **41-259 Model 3020 or 3040 115VAC Powered**
    $1,275⁰⁰

Solar Force Accessories & Parts

Heavy-Duty Pressure Switch

A DC-rated pressure switch for using the Solar Force as a pressure pump. Will handle the heavy starting surge. Good for 12-, 24-, or 48-volt systems.

◈ **41-264 Solar Force Pressure Switch $75⁰⁰**

Surge Tank

This tank is included with the PV models. It helps absorb pressure pulsations when long piping runs are required between the pump and the tank. It keeps plumbing systems from being shaken apart.

◈ **41-265 Solar Force Pulsation Tank $59⁰⁰**

Seal & Belt Kit

Contains all the repair parts needed for routine service required every two to six years. Spare gaskets, rod and valve seals, new drive belt, and two sets of piston seals are included. PV kits are higher priced due to gear belt drive instead of V-belt.

Specify pump model number when ordering.

◈ **41-267 Seal & Belt Kit for Battery Pumps $45⁰⁰**
◈ **41-268 Seal & Belt Kit for PV Pumps $75⁰⁰**

Long-Term Parts Kit

Everything included in the Seal & Belt Kit above, plus a second drive belt, a replacement cylinder sleeve, motor brushes, and two oil changes.

Specify pump model number when ordering.

◈ **41-269 Long-Term Parts Kit for**
    Battery Pumps $160⁰⁰
◈ **41-270 Long-Term Parts Kit for PV Pumps $210⁰⁰**

Submersible Pumps

SHURflo Solar Submersible Pump

This efficient, submersible, diaphragm pump will deliver up to 230 feet of vertical lift and is ideal for deep well or freeze-prone applications. Remember, you're only lifting from the water surface, not from the pump. In a PV-direct application it will yield from 300 to 1,000 gallons per day, depending on lift requirements and power supplied. Solar submersible pumps are best used for slow, steady water production into a holding tank, but may be used for direct pressurization applications as well. (See suggestions on this subject in the Editorial section a few pages back.) Flow rates range from .5 to 1.8 gpm depending on lift. The pump can be powered from either 12- or 24-volt sources. Output is best with 24-volt panel-direct. The best feature of this pump, other than proven reliability, is easy field serviceability. No special tools are needed. Repair parts sold below.

The pump is 3.75 inches in diameter (will fit standard 4-inch well casings), weighs 6 pounds, uses ½-inch poly delivery pipe, making easy one-person installation possible. Maximum pump submersion is 100 feet. One year warranty. USA.

In addition to the pump you will also need:
• PV modules and mounting appropriate for your lift (see chart)
• A Controller for panel-direct systems (PV Direct Controller #25-138RC for 24V, #25-134RC for 12V)
• ½-inch poly drop pipe or equivalent
• 10-gauge submersible pump cable (common 3-wire cable okay if flat style)
• Poly safety rope
• Sanitary well seal

Drop pipe, submersible cable, rope, and seal are all commonly available at any plumbing supply store.

Watertight splice kit for flat cable is included.

(12-volt performance will be approximately 40% of figures listed above.)

SHURFLO SUBMERSIBLE PUMP PERFORMANCE CHART AT 24 VOLTS

Total Vertical Lift	Gallons Per Hour	Minimum PV Watts	Amps
20	117	58	1.5
40	114	65	1.7
60	109	78	2.1
80	106	89	2.4
100	103	99	2.6
120	101	104	2.8
140	90	115	3.1
180	93	135	3.6
200	91	141	3.8
230	82	155	4.1

41-455 SHURflo Solar Submersible $695⁰⁰

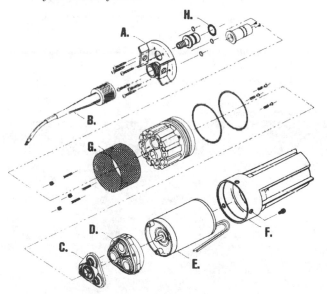

Repair and Replacement Parts for SHURflo Solar Sub Pump

REPLACEMENT PARTS KIT LIST FOR SHURFLO SOLAR SUB PUMP

Complete Tool Kit	Lift Plate Kit A	Cable Plug Kit B	Valve Kit C	Drive & Diaphragm Kit D	Motor E	Canister F	Filter Screen G	Complete O-Ring Kit H
41-340	41-341	41-342	41-343	41-344	41-345	41-346	41-437	41-348
$20⁰⁰	$80⁰⁰	$225⁰⁰	$40⁰⁰	$100⁰⁰	$225⁰⁰	$129⁰⁰	$30⁰⁰	$25⁰⁰

A Complete Solar Sub-Pump Kit

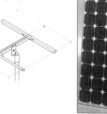

Our Sub-Pump Kit features the popular and durable Shurflo submersible pump. We've put it together with a pair of high-powered Astropower 75-watt modules for maximum output, a pole-top mount for ease of installation, and a current booster controller for good performance with marginal sun. Cost of components purchased separately is $1,772. All components made in USA except controller made in Canada.

41-128 Shurflo Sub-Pump Kit $1,695⁰⁰

SunRise Solar Submersible Pump

Lifts Up to 600 Vertical Feet, or Flow Rates Over 6 gpm!

The SunRise is a truly serious solar powered pump that picks up where smaller solar pumps give up. There are four models available. The lowest flow model is capable of lifting up to 600 feet, and still delivers 1.0 gpm. The highest flow model produces over 6.5 gpm at 25 feet of lift. See the performance charts for specific output and wattage requirements for each model. SunRise is a sealed piston pump for the highest energy efficiency, and will tolerate running dry in low-yielding wells without harm. All models will fit in a 4-inch or larger well casing. This nominal 48-volt pump uses all stainless steel construction, is built in Europe, and has been proven in

DISTANCE IN FEET FROM ARRAY TO PUMP FOR 5% VOLTAGE DROP FOR VARIOUS WIRE SIZES

Watts	#12	#10	#8	#6
100	540	900	1440	2160
150	360	600	960	1440
200	270	450	720	1080
250	215	360	575	865
300	180	300	480	720

African desert regions since 1993. Inside the pump, the oval camshaft, running in nontoxic oil, drives two horizontally opposed pistons. The stroke is very short, creating almost zero fatigue on the flexible gaskets. A permanent-magnet DC motor using replaceable brushes was chosen for reliability (brushless motors require complex electronics). Motor brushes wear at predicable rates, and will require replacement at 5- to 10-year intervals. This field service replacement is easy and takes less than 30 minutes.

Two controller choices are offered. The SunPrimer Controller is required on all PV-direct systems to provide starting surge and prevent stalling in low-sun conditions. It comes in a sealed outdoor enclosure. The Battery Controller/Converter will run the SunRise at full output from a 24-volt battery bank, or at reduced capacity from a 12-volt battery bank, while providing automatic low-voltage disconnect to protect the batteries. It comes in an indoor enclosure. Both controllers provide remote float switch sensing and status indicators. Remote float switch is reverse logic; i.e., open float switch = running pump, closed float switch = stopped pump.

The SunRise pump requires either flexible poly pipe to absorb pressure pulsations, or the installation of the Damping Line accessory if rigid drop pipe is used. The

SUNRISE SOLAR SUBMERSIBLE PUMP PERFORMANCE

Vertical Lift		Model 5218			Model 5222			Model 5226			Model 5230		
Feet	Meters	LPM	GPM	Watts	LPM	GPM	Watts	LPM	GPM	Watts	LPM	GPM	Watts
25	8	10.1	2.7	84	12.3	3.3	103	13.7	3.6	108	25.0	6.6	212
50	15	9.6	2.5	96	11.7	3.1	117	12.9	3.4	124	23.3	6.2	220
75	23	9.0	2.4	108	11.0	2.9	132	12.4	3.3	138	21.0	5.5	229
100	30	8.9	2.1	114	10.9	2.9	139	12.0	3.2	151	18.8	5.0	240
125	38	8.3	2.2	126	10.1	2.7	154	11.6	3.1	166	16.8	4.4	250
150	46	7.7	2.0	138	9.4	2.5	169	11.0	2.9	180	15.5	4.1	259
175	53	7.3	1.9	144	8.9	2.3	176	10.7	2.8	192	14.2	3.7	267
200	61	7.1	1.8	156	8.7	2.3	191	0.4	2.7	210			
250	76	6.5	1.7	168	7.9	2.1	205	9.6	2.5	229			
300	91	5.8	1.5	180	7.1	1.9	220	9.1	2.4	256			
350	107	5.4	1.4	192	6.6	1.7	235	8.3	22	276			
400	122	5.0	1.2	198	6.1	1.6	242						
450	137	4.7	1.2	204									
500	152	4.2	1.1	216*									
550	168	4.1	1.1	219*									
600	183	3.7	1.0	222*									

Wattage listed is requirement at pump. Add 20% when figuring required PV wattage.

Performance measured at 60V (typical working voltage from a 48V nominal PV array). Performance may vary ± 10%. For battery system, subtract 15% from Flow and from Watts, due to overworking voltage.

*In this range, expect reduced life of some mechanical parts.

50-foot Damping Line sections are equipped with ¾-inch threaded ends.

For installations with less than 150 feet lift, use ¾-inch 100 psi Type SDR-15 poly pipe, commonly available locally. For lifts greater than 150 feet, use ¾-inch, 200 psi Type SDR-9 poly pipe with bronze compression adapters.
• Diameter at cable guard: 3.9″ (98mm)
• Length: 30″ (76mm), including check valve assembly.
• Weight: 27 lb. (12kg)
• Outlet Fitting: ¾″ female thread
• Drop Pipe: ¾″ polyethylene (black flexible) pipe is required. Use heavy wall 200 psi pipe as required by lift. *Do not* use rigid pipe! Flexible pipe absorbs pulsation.
• Submersible Cable: 2-wire standard pump cable, with ground if required. See chart for wire sizing.
• Submergence: Minimum 3 feet (1m), Maximum 150 ft (50m) if necessary.
Warranty: Two-year mfr's. warranty. Italy.

◈41-465 **SunRise Model 5218 Sub Pump** $1,849⁰⁰

◈41-474 **SunRise Model 5222 Sub Pump** $1,849⁰⁰

◈41-466 **SunRise Model 5226 Sub Pump** $1,849⁰⁰

◈41-467 **SunRise Model 5230 Sub Pump** $2,149⁰⁰

◈41-639 **SunPrimer Controller** $429⁰⁰

◈41-349 **Battery Controller/Converter** $340⁰⁰

◈41-350 **Damping Line < 150´ lift** $29⁹⁵

◈41-351 **Damping Line > 150´ lift** $84⁹⁵

◈41-352 **SDR-9 ¾″ Poly Pipe, 100 feet** $49⁹⁵

◈41-353 **SDR-9 ¾″ Poly Pipe, 300 feet** $150⁰⁰

◈41-354 **SDR-9 ¾″ Poly Pipe, 500 feet** $250⁰⁰

◈41-355 **Bronze Adapter, ¾″ MPT to SDR-9** $16⁹⁵

◈41-356 **SDR-9, ¾″ Bronze Coupling** $19⁹⁵

◈41-471 **SunRise Brush Kit** $29⁹⁵

TSP-1000 Submersible Pump

The Most Powerful Positive Displacement Available

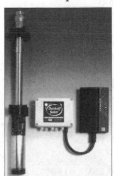

The TSP-1000 is a high-output, extremely high-efficiency pump for PV-direct use. It can deliver up to 25 gallons per minute at low lift. Even at its maximum lift of 230 feet it still manages 5 gpm. Performance is impressive, but efficiency is even better. It only needs 300 to 600 watts of PV for this incredible delivery. No other pump on the market comes close, or has such a strong four-year warranty!

The TSP-1000 is a helical-rotor type pump. It has one moving part, a hard, chrome-plated rotor that revolves inside a rubber stator. The contacting surfaces form sealed cavities that progress upward as the rotor turns. Friction is low, with almost no back leakage. This type of pump, rod-driven from above ground, has been proven for decades, and is commonly used in Africa. Using a brushless high-torque motor and a specialized above-ground control box, it fits in a 4″ well

casing. The controller will accept PV-direct arrays with 6, 7, or 8 modules in series, and is current-booster equipped to help run the pump under low light conditions. The controller accepts level sensor probes to prevent dry running, and/or a float switch to turn off when the storage tank is full. Float switch is reverse-logic; closed circuit turns pump off. The controller displays on/off, relative rotational speed, overload, low water, or full tank conditions.

The pump, controller and a junction box for easy wiring are supplied as a package. The controller and j-box are weatherproof. There are six different models, depending on lift and volume required. See the Performance/ Selection chart. Wattage listed in the performance chart is the PV factory rating. Listed wattage is a minimum. More watts at the PV array will improve low-light performance. Four-year mfr's warranty. France.

TSP 10000 PERFORMANCE CHART

PV Array Peak Watts / # of Modules x Watts Each*					
300 Watt 6 x 50W	350 Watt 7 x 50W	400 Watt 8 x 50W	450 Watt 6 x 75W	525 Watt 7 x 75W	600 Watt 8 x 75W
50 ft.** model # 11 gpm E	14 gpm E	19 gpm A	20 gpm A	22 gpm A	25 gpm A
65 ft. model # 9.5 gpm E	11 gpm E	13.5 gpm C	12 gpm B	18 gpm B	20 gpm A
80 ft. model # 7.5 gpm E	9 gpm E	11 gpm C	12 gpm B	14 gpm B	16 gpm B
100 ft. model # 7 gpm E	8 gpm E	9 gpm E	11 gpm E	12 gpm E	15 gpm C
115 ft. model # 6.5 gpm E	7 gpm E	8 gpm E	9 gpm E	10.5 gpm E	12.5 gpm C
130 ft. model # 5 gpm E	6 gpm E	7 gpm E	8 gpm E	9.5 gpm E	11 gpm E
150 ft. model # 5 gpm E	5.5 gpm E	6 gpm E	7 gpm E	8.5 gpm E	10 gpm E
165 ft. model # 4 gpm F	4.5 gpm F	5 gpm F	6 gpm E	7 gpm E	9 gpm D
180 ft. model # 3.5 gpm F	4 gpm F	4.5 gpm F	5.5 gpm E	6.5 gpm E	8 gpm D
200 ft. model # 3 gpm F	3.75 gpm F	4 gpm F	5 gpm F	5.5 gpm F	6.8 gpm D
215 ft. model # 3 gpm F	3.5 gpm F	4 gpm F	4.75 gpm F	5 gpm F	6 gpm D
230 ft. model # 2 gpm F	3 gpm F	3.5 gpm F	4 gpm F	4.5 gpm F	5 gpm F

This is a minimum. More watts at the PV array will result in improved low-light performance.

**Vertical feet.*

◈ **41-003 TSP-1000A Helical Rotor Sub Pump & Controller** $4,800⁰⁰

◈ **41-004 TSP-1000B Helical Rotor Sub Pump & Controller** $4,800⁰⁰

◈ **41-006 TSP-1000C Helical Rotor Sub Pump & Controller** $4,800⁰⁰

◈ **41-007 TSP-1000D Helical Rotor Sub Pump & Controller** $4,600⁰⁰

◈ **41-008 TSP-1000E Helical Rotor Sub Pump & Controller** $4,600⁰⁰

◈ **41-009 TSP-1000F Helical Rotor Sub Pump & Controller** $4,600⁰⁰

◈ *Means shipped directly from manufacturer.*

Submersible Pump Options

The Universal Solar Pump Controllers
The Extraordinary Controller for Ordinary Pumps

From the same incredibly innovative company that produced GM's EV1 electric vehicle, and the human-powered plane that crossed the English Channel, comes an inverter/controller that will run any standard single- or three-phase AC submersible pump directly from PV power. The advanced USPC maintains a constant volts/Hz ratio that allows standard AC pumps to run at lower speeds without distress. This allows easy soft starts, and the efficient use of lower than peak PV outputs with off-the-shelf pumps. Minimum speed is owner-selectable to avoid no-flow conditions. The USPC's power electronics are 97% efficient, and will provide automatic shutdown in case of dry running, motor lock, or wiring problems. The USPC can drive any standard 50- or 60-Hz three-wire single-phase pump or three-phase induction motor that has a rated voltage of 120, 208, or 230VAC. The model 2000 will run any 0.5 to 2.0 horsepower pump. The model 5000 is for pumps up to 5.0 horsepower, and a 10 horsepower model is available by special order. The enclosure is rugged outdoor-rated NEMA 3 steel. USPC complete systems, with PV array, controller, and pump run $5,000 to $50,000. Two-year mfr's. warranty. UL-certified. USA.

Please call our tech staff for help with PV sizing, which will be unique for your installation.

◈**41-913 USPC-2000 Controller $1,899**⁰⁰
◈**41-923 USPC-5000 Controller $3,500**⁰⁰

PV Direct Pump Controllers

These current boosters from Solar Converters will start your pump earlier in the morning, keep it going longer in the afternoon, and give you pumping under lower light conditions when the pump would otherwise stall. Features include; Maximum Power Point Tracking, to pull the maximum wattage from your modules; a switchable 12- or 24-volt design in a single package; a float or remote switch input; a user-replaceable ATC-type fuse for protection; and a weatherproof box. Like all pump controllers, a closed float switch will turn the pump off, an open float switch turns the pump on.

Four amperage sizes are available in switchable 12/24 volt. Input and output voltages are selected by simply connecting or not connecting a pair of wires. Amperage ratings are surge power. Don't exceed 70% of amp rating under normal operation. The 7-amp model is the right choice for most pumping systems, and is the controller

included in our submersible and surface pumping kits. 7- and 10-amp models are in a 4.5″ L x 2.25″ W x 2″ D plastic box. 15-amp model is a 5.5″ L x 2.9″ W x 2.9″ D metal box. 30-amp model is supplied in a NEMA 3R raintight 10″ x 8″ x 4″ metal box. One-year mfr.'s warranty. Canada.

25-002 7A, 12/24V Pump Controller $89⁹⁵
◈ **25-003 10A, 12/24V Pump Controller $125**⁰⁰
◈ **25-004 15A, 12/24V Pump Controller $199**⁰⁰
◈ **25-005 30A, 12/24V Pump Controller $349**⁰⁰

◈ *Means shipped from manufacturer.*

Float Switches

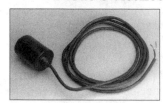

These fully encapsulated mechanical *(no mercury!)* float switches make life a little easier by providing automatic control of fluid level in the tank. You regulate the water level by merely lengthening or shortening the power cable and securing it with a band clamp or zip tie. Easily installed, the switches are rated at 15 amps at 12 volts, or 13 amps at 120 volts. The "U" model will fill a tank (closed circuit when float is down, open circuit when up) and the "D" model will drain it (closed circuit when up, open circuit when down). When used with a PV-Direct Pump Controller, the "U" and "D" functions are reversed: "D" will fill and "U" will empty.

Not sure if you want an "up" or "down" float switch? Confused by reverse logic controllers? The Goof-Proof float switch is your salvation. This 3-wire float switch can be connected either way. There's a common wire, an "up" wire, and a "down" wire. If it doesn't work right, just switch wires. 5-amp maximum current. Safe for domestic water. Two-year mfr.'s warranty. USA.

41-637 Float Switch "U" $34⁹⁵
41-638 Float Switch "D" $34⁹⁵
41-036 Goof-Proof Float Switch $44⁹⁵

12-Volt To 28-Volt Controller

Want to run your submersible pump from your household 12-volt battery bank, but worried that it will pump too slowly? This controller is your salvation. It will draw 12 volts from your batteries and convert it to 28 volts for the pump. Good for up to 10 amps input. It has the same remote control functions as the PV Direct Controller listed to the left.

41-632 Controller/Converter, 12V to 28V $180⁰⁰

Flat 2-Wire Sub Pump Cable

This flat, two-conductor, 10-gauge submersible cable is sometimes difficult to find locally. Works with the SHURflo Solar Sub splice kit. Cut to length; specify how many feet you need.
26-540 10-2 Sub Pump Cable $0.79/ft.

Wind Powered Pumps

Bowjon Wind-Powered Pumps

Bowjon utilizes a unique "air injection" system that is ideal in areas with high wind velocities. A small air compressor is direct-driven by the turbine. You can place your Bowjon where the wind blows best, as far as ¼ mile away from your well. The compressed air output is directed to an injector at the bottom of the well. Rising air bubbles in the delivery tube carry water up to a collection tank. It uses five high-torque, heavy-duty aluminum blades. It is quickly installed on a single post tower. Its minimal maintenance only requires checking the compressor oil every six months.

More than one windmill can service the same well. The hub is a massive 17 lb. of tempered aluminum that will withstand 3,200 psi. Bowjon's five blades are made of 6061-T6 tempered aluminum, each consisting of three layers for the maximum strength that's required in extremely high winds.

10-12 MPH WIND

GALLONS PER HOUR with adequate submersion*

Propeller: Tempered triple-layer reinforced blades to minimize flex in high winds. Blades have a unique varied pitch to allow low-wind, high-torque start-up. They have high rpm ability.

Compressor: Industrial strength twin or single cylinder, depending on model selected. Compressor is capable of 5 cubic feet of air per minute.

Start-Up: 5 to 8 mph.

Pump: Air injection. No moving parts, no cylinders, valves, rods, or leathers to wear out. Will run dry, accepting silt, sludge, and sand without harm. Length is 5 feet. Minimum well casing diameter is 2 inches.

Air Line: ⅜-inch polyethylene tubing (200 psi rated).

Water Line: 0- to 50-ft. lift: 1-inch tubing or PVC Schedule 40 plastic pipe. 50-to 100-ft. lift: ¾-inch tubing or PVC Schedule 40 plastic pipe. 100 ft. lift and over: ½-inch tubing or PVC Schedule 40 plastic pipe.

Submersion: At least 30% of the vertical lift distance from static water level in the well to the highest point of delivery or storage.

Submersions less than 30 feet restrict vertical lift; but, where water is only to be lifted a few feet, submersions of 5 or 10 feet with air tank and regulator are practical.

If the submergence is too low for the amount of lift, the air will separate from the water. If the submergence is too high, the air will not lift the water.

LIFT/SUBMERGENCE RATIOS

Vertical Lift	Submergence Below Water	Optimum Total Well Depth
50 ft	35 ft	90 ft
80 ft	56 ft	140 ft
120 ft	85 ft	210 ft
200 ft	100 ft (50% of lift)	350 ft

Homesteader Model Bowjon Pump

The Homesteader model comes with 6 ft. 8 inch diameter blades with a single cylinder compressor. It has a ⅜-inch air line 250 feet long and AL2 air injection pump. One-year warranty. USA.

◈41-701 Bowjon Pump, Homesteader **$1,375⁰⁰**

Shipped freight collect from Southern California.

Rancher Model Bowjon Pump

The Rancher model comes with 8 ft. 8 inch diameter blades with a twin cylinder compressor. It has a ⅜-inch air line 250 feet long and an AL3 air injection pump. One-year warranty.

◈41-711 Bowjon Pump, Rancher **$1,695⁰⁰**

Shipped freight collect from Southern California.

Water Powered Pumps

Ram Pumps

The ram pump works on a hydraulic principle using the liquid itself as a power source. The only moving parts are two valves. In operation, the water flows from a source down a "drive" pipe to the ram. Once each cycle, a valve slaps shut causing the water in the drive pipe to suddenly stop. This causes a water-hammer effect and high pressure in a one-way valve leading to the "delivery" pipe of the ram, thus forcing a small amount of water up the pipe and into a holding tank. Rams will only pump into unpressurized storage. In essence, the ram uses the energy of a large amount of water falling a short distance to lift a small amount of water a large distance. The ram itself is a highly efficient device; however, only 2% to 10% of the liquid is recoverable. Ram pumps will work on as little as 2 gpm supply flow. The maximum head or vertical lift of a ram is about 500 feet.

Selecting A Ram

Estimate amount of water available to operate the ram. This can be determined by the rate the source will fill a container. Make sure you've got more than enough water to satisfy the pump. If a ram runs short of water it will stop pumping and simply dump all incoming water.
Estimate amount of fall available. The fall is the vertical distance between the surface of the water source and the selected ram site. Be sure the ram site has suitable drainage for the tailing water. Rams splash big-time when operating! Often a small stream can be dammed to provide the 1½ feet or more head required to operate the ram.
Estimate amount of lift required. This is the vertical distance between the ram and the water storage tank or use point. The storage tank can be located on a hill or stand above the use point to provide pressurized water. Forty or fifty feet water head will provide sufficient pressure for household or garden use.
Estimate amount of water required at the storage tank. This is the water needed for your use in gallons per day. As examples, a normal two- to three-person household uses 100 to 300 gallons per day, much less with conservation. A 20- by 100-foot garden uses about 50 gallons per day. When supplying potable water, purity of the source must be considered.

Using these estimates, the ram can be selected from the following performance charts. The ram installation will also require pouring a small concrete pad, a drive pipe five to ten times as long as the vertical fall, an inlet strainer, and a delivery pipe to the storage tank or use point. These can be obtained from your local hardware or plumbing supply store. Further questions regarding suitability and selection of a ram for your application will be promptly answered by our technical staff.

Aqua Environment Rams

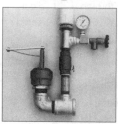

We've sold these fine rams by Aqua Environment for over 15 years now with virtually no problems. Careful attention to design has resulted in extremely reliable rams with the best efficiencies and lift-to-fall ratio available. Working component construction is of all bronze with O-ring seal valves. Air chamber is PVC pipe. The outlet gauge and valve permit easy startup. Each unit comes with complete installation and operating instructions.

41-811 Ram Pump, ¾″ $249⁰⁰

41-812 Ram Pump, 1″ $249⁰⁰

41-813 Ram Pump, 1¼″ $299⁰⁰

41-814 Ram Pump, 1½″ $299⁰⁰

Typical Performance and Specifications					
Vertical Fall (feet)	Vertical Lift (feet)	Pump Rate (Gallons/Day)			
		¾" Ram	1" Ram	1¼" Ram	1½" Ram
20	50	650	1350	2250	3200
20	100	325	670	1120	1600
20	200	150	320	530	750
10	50	300	650	1100	1600
10	100	150	320	530	750
10	150	100	220	340	460
5	30	200	430	690	960
5	50	100	220	340	460
5	100	40	90	150	210
1.5	30	40	80	130	190
1.5	50	20	40	70	100
1.5	100	6	12	18	25

Water Required to Operate Ram
¾" Ram - 2 gallons/minute   Maximum Fall - 25 feet
1" Ram - 4 gallons/minute   Minimum Fall - 1.5 feet
1¼" Ram - 6 gallons/minute   Maximum Lift - 250 feet
1½" Ram - 8 gallons/minute

Folk Ram Pumps

The Folk is the most durable and most efficient ram pump available. It is solidly built of rustproof cast aluminum alloy and uses all stainless steel hardware. With minimal care, it will probably outlast your grandchildren. It is more efficient because it uses a diaphragm in the air chamber, preventing loss of air charge, and eliminating the need for a snifter valve in the intake that cuts efficiency.

Folk rams require a minimum of 2 gpm and 3 feet of drive pipe fall to operate. The largest units can accept up to 75 gpm, and a maximum of 50 feet of drive pipe fall. Delivery height can be up to 15 times the drive fall, or a maximum of 500 feet lift. The drive pipe should be galvanized Schedule 40 steel, and the same size the entire length. Recommended drive pipe length is 24 feet for 3 feet of fall, with an additional 3 feet length for each additional one foot of fall. Be absolutely sure you have enough water to meet the minimum gpm requirements of the ram you choose. Rams will stop pumping and simply dump water if you can't supply the minimum gpm.

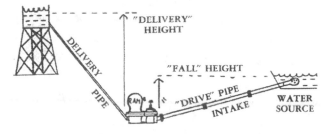

To estimate delivery volume, multiply fall in feet times supply in gpm, divide by delivery height in feet, then multiply by 0.61. Example: a ram with 5 gpm supply, a 20-foot drive pipe fall, and delivering to a tank 50 feet higher. 20ft x 5gpm ÷ 50ft x 0.61 = 2.0gpm x 0.61 = 1.22gpm, or 1,757 gallons per day delivery.

Warranted by manufacturer for one year. USA.

All Folk rams are drop-shipped via UPS from the manufacturer in Georgia.

FOLK RAM PUMP CAPACITY

Folk Model Drive Pipe Size	Capacity In GPM Min.— Max.	Delivery Pipe
1″	2—4	1″
1¼″	2—7	1″
1½″	3—15	1″
2″	6—30	1¼″
2½″	8—45	1¼″
3″	15—75	1¼″

◆41-815 Folk 1″ Ram Pump $795.00
◆41-816 Folk 1¼″ Ram Pump $795.00
◆41-817 Folk 1½″ Ram Pump $795.00
◆41-818 Folk 2″ Ram Pump $1,095.00
◆41-819 Folk 2½″ Ram Pump $1,095
◆41-820 Folk 3″ Ram Pump $1,095

High Lifter Pressure Intensifier Pump

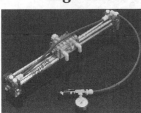

The High Lifter Water Pump offers unique advantages for the rural user. Developed expressly for mountainous terrain and low summertime water flows, this water-powered pump delivers a much greater percentage of the input water than ram pumps can. This pump is available in either 4.5:1 or 9:1 ratios of lift to fall. The High Lifter is self-starting and self-regulating. If inlet water flow slows or stops the pump will slow or stop, but will self-start when flow starts again.

The High Lifter pump has many advantages over a ram (the only other water-powered pump). Instead of using a "water hammer" effect to lift water as a ram does, the High Lifter is a positive displacement pump that uses pistons to create a hydraulic lever that converts a larger volume of low-pressure water into a smaller volume of high-pressure water. This means that the pump can operate over a broad range of flows and pressures with great mechanical efficiency. This efficiency means more recovered water. While water recovery with a ram is normally about 5% or less; the High Lifter recovers 1 part in 4.5 or 1 part in 9 depending on ratio. In addition, unlike the ram pump, no "start up tuning" or special drive lines are necessary. This pump is quiet, and will happily run unattended.

The High Lifter pressure intensifier pump is economical compared to gas and electric pumps, because no fuel

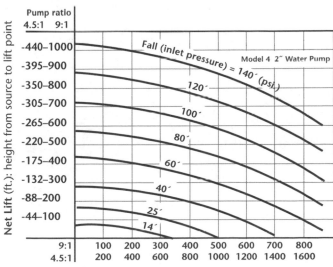

1 psi= 2.3′ **Delivery (gal./day): assuming adequate water @ source**

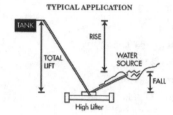

TYPICAL APPLICATION

is used and no extensive water source development is necessary. A kit to change the working ratio of either pump after purchase is available, as are maintenance kits. Maintenance consists of simply replacing a handful of O-rings. The cleaner your input water, the longer the O-rings last. Choose your model High Lifter pump from the specifications and High Lifter performance curves. One year parts and labor warranty from mfr's. USA.

◆ **41-801 High Lifter Pump, 4.5:1 Ratio $895⁰⁰**

◆ **41-802 High Lifter Pump, 9:1 Ratio $895⁰⁰**

◆ **41-803 High Lifter Ratio Conversion Kit $99⁹⁵**

◆ **41-804 High Lifter Rebuild Kit, 4.5:1 ratio $65⁹⁵**

◆ **41-805 High Lifter Rebuild Kit, 9:1 ratio $65⁹⁵**

◆ *Means shipped directly from manufacturer.*

Pump Accessories

Inline Sediment Filter

This filter is designed for cold water only, and meets NSF standards for drinking water supplies. It features a sump head of fatigue-resistant Celcon plastic, an opaque body to inhibit algae growth if installed outdoors, and is equipped with a manually operated pressure release button to simplify cartridge replacement. Rated for up to 125 psi and 100°F, it is equipped with ¾-inch female pipe thread inlet and outlet fittings, and is rated for up to 6 gpm flow rates (with clean cartridge). It accepts a standard 10-inch cartridge with 1-inch center holes. Supplied with one 5-micron fiber cartridge installed.

41-137 Inline Sediment Filter $75⁰⁰

42-632 5-Micron Sediment Cartridge $12⁹⁵

Friction Loss Charts for Water Pumping

Friction Loss- PVC Class 160 PSI Plastic Pipe
Pressure loss from friction in psi per 100 feet of pipe.

Flow GPM	1	1.25	1.5	2	2.5	3	4	5	6	8	10
1	0.02	0.01									
2	0.06	0.02	0.01								
3	0.14	0.04	0.02								
4	0.23	0.07	0.04	0.01							
5	0.35	0.11	0.05	0.02							
6	0.49	0.15	0.08	0.03	0.01						
7	0.66	0.20	0.10	0.03	0.01						
8	0.84	0.25	0.13	0.04	0.02						
9	1.05	0.31	0.16	0.05	0.02						
10	1.27	0.38	0.20	0.07	0.03	0.01					
11	1.52	0.45	0.23	0.08	0.03	0.01					
12	1.78	0.53	0.28	0.09	0.04	0.01					
14	2.37	0.71	0.37	0.12	0.05	0.02					
16	**3.04**	0.91	0.47	0.16	0.06	0.02					
18	3.78	1.13	0.58	0.20	0.08	0.03					
20	4.59	1.37	0.71	0.24	0.09	0.04	0.01				
22	5.48	1.64	0.85	0.29	0.11	0.04	0.01				
24	6.44	1.92	1.00	0.34	0.13	0.05	0.02				
26	7.47	2.23	1.15	0.39	0.15	0.06	0.02				
28	8.57	**2.56**	1.32	0.45	0.18	0.07	0.02				
30	9.74	2.91	1.50	0.51	0.20	0.08	0.02				
35		3.87	**2.00**	0.68	0.27	0.10	0.03				
40		4.95	2.56	0.86	0.34	0.13	0.04	0.01			
45		6.16	3.19	1.08	0.42	0.16	0.05	0.02			
50		7.49	3.88	1.31	0.52	0.20	0.06	0.02			
55		8.93	4.62	1.56	0.62	0.24	0.07	0.02			
60		10.49	5.43	1.83	0.72	0.28	0.08	0.03	0.01		
65			6.30	2.12	0.84	0.32	0.09	0.03	0.01		
70			7.23	2.44	0.96	0.37	0.11	0.04	0.02		
75			8.21	2.77	1.09	0.42	0.12	0.04	0.02		
80			9.25	3.12	1.23	0.47	0.14	0.05	0.02		
85			10.35	3.49	1.38	0.53	0.16	0.06	0.02		
90				3.88	1.53	0.59	0.17	0.06	0.03		
95				4.29	1.69	0.65	0.19	0.07	0.03		
100				4.72	1.86	**0.72**	0.21	0.08	0.03	0.01	
150				10.00	3.94	1.52	0.45	0.16	0.07	0.02	
200				6.72	2.59	**0.76**	0.27	0.12	0.03	0.01	
250				10.16	3.91	1.15	0.41	0.18	0.05	0.02	
300					5.49	1.61	**0.58**	0.25	0.07	0.02	
350					7.30	2.15	0.77	0.33	0.09	0.03	
400					9.35	2.75	0.98	0.42	0.12	0.04	
450						3.42	1.22	**0.52**	0.14	0.05	
500						4.15	1.48	0.63	0.18	0.06	
550						4.96	1.77	0.76	0.21	0.07	
600						5.82	2.08	0.89	0.25	0.08	
650						6.75	2.41	1.03	0.29	0.10	
700						7.75	2.77	1.18	0.33	0.11	
750						8.80	3.14	1.34	**0.37**	0.13	
800							3.54	1.51	0.42	0.14	
850							3.96	1.69	0.47	0.16	
900							4.41	1.88	0.52	0.18	
950							4.87	2.08	0.58	0.20	
1000							5.36	2.29	0.63	0.22	
1500								4.84	1.34	0.46	
2000									2.29	0.78	
2500									3.46	1.18	
3000											1.66

Friction Loss- Polyethylene (PE) SDR-Pressure Rated Pipe
Pressure loss from friction in psi per 100 feet of pipe.

Flow GPM	0.5	0.75	1	1.25	1.5	2	2.5	3
1	0.49	0.12	0.04	0.01				
2	1.76	0.45	0.14	0.04	0.02			
3	3.73	0.95	0.29	0.08	0.04	0.01		
4	**6.35**	1.62	0.50	0.13	0.06	0.02		
5	9.60	2.44	0.76	0.20	0.09	0.03		
6	13.46	3.43	1.06	0.28	0.13	0.04	0.02	
7	17.91	4.56	1.41	0.37	0.18	0.05	0.02	
8	22.93	**5.84**	1.80	0.47	0.22	0.07	0.03	
9		7.26	2.24	0.59	0.28	0.08	0.03	
10		8.82	2.73	0.72	0.34	0.10	0.04	0.01
12		12.37	**3.82**	1.01	0.48	0.14	0.06	0.02
14		16.46	5.08	1.34	0.63	0.19	0.08	0.03
16			6.51	1.71	0.81	0.24	0.10	0.04
18			8.10	2.13	1.01	0.30	0.13	0.04
20			9.84	2.59	1.22	0.36	0.15	0.05
22			11.74	**3.09**	1.46	0.43	0.18	0.06
24			13.79	3.63	1.72	0.51	0.21	0.07
26			16.00	4.21	1.99	0.59	0.25	0.09
28				4.83	2.28	0.68	0.29	0.10
30				5.10	**2.59**	0.77	0.32	0.11
35				7.31	3.45	1.02	0.43	0.15
40				9.36	4.42	1.31	0.55	0.19
45				11.64	5.50	1.63	0.69	0.24
50				14.14	6.68	**1.98**	0.83	0.29
55					7.97	2.36	0.85	0.35
60					9.36	2.78	1.17	0.41
65					10.36	3.22	1.36	0.47
70					12.46	3.69	**1.56**	0.54
75					14.16	4.20	1.77	0.61
80						4.73	1.99	0.69
85						5.29	2.23	0.77
90						5.88	2.48	0.86
95						6.50	2.74	0.95
100						7.15	3.01	**1.05**
150						15.15	6.38	2.22
200							10.87	3.78
300								8.01

www.realgoods.com

CHAPTER 7

WATER HEATING

MOST OF US TAKE HOT WATER AT THE TURN OF A TAP for granted. It makes civilized life possible, and we get seriously annoyed by cold showers. Yet most people probably do not realize how much this civilized convenience costs them. The average household spends an astonishing 20% to 40% of its energy budget on water heating. And all those energy dollars are used with an appliance that has a life expectancy of only 10 to 15 years. Any improvements you provide in efficiency or longevity for your appliances reduces your environmental impact and leaves a better world for your children. There are better, cheaper, and more durable ways to heat your home water supply. We will start with the simplest solutions and work our way up to the more complex ones. We've provided a discussion of all the common water heater types, including solar water heaters. We'll discuss the good and bad points of each heater type, and we've developed a fairly accurate Life-Cycle Cost Comparison chart at the end of the section, so you can compare the real operation costs over the lifetime of the appliance. Points to consider for each heater type are initial cost, cost of operation, recovery time (how fast does it heat water?), ease of installation or ability to retrofit, life expectancy, and finally, life-cycle cost.

Common Water Heater Types

Storage or Tank-Type Water Heaters

Storage or tank-type water heaters are by far the most common type of residential water heater used in North America. They range in size from 20 to 80 gallons, and can be fueled by electricity, natural gas, propane, or oil. Storage heaters work by heating up water inside an insulated tank. The good points of storage heaters are the modest initial cost, the ability to provide large amounts of hot water for a limited time, and they're well understood by plumbers and do-it-yourself homeowners everywhere. Bad points include constant standby losses, slow recovery times, and low life expectancy. Because heat always escapes through the walls of the tank (standby heat loss), energy is consumed even when no hot water is being used. This wastes energy and raises operation costs. Newer "high-efficiency" storage heaters use more insulation to reduce standby losses. Customer-installed water heater blankets do the same thing (although maybe not quite so well). Standby losses for gas and oil heaters are higher because air is constantly drawn through the internal flue passages, warmed, and exhausted. Although storage tanks can deliver large volumes of hot water rapidly, once they're exhausted, the limited energy input means a long recovery time. Who hasn't had to wait to get a hot shower, or suffered through a quick cold one, after someone else used all the hot water? Life expectancy for tank-type water heaters averages 10 to 15 years. Usually the tank rusts through at this age, and the entire appliance has to be replaced. Not a very efficient way to design relatively expensive appliances. Life expectancy can be prolonged significantly if the sacrificial anode rod is replaced at about five-year intervals. This sacrificial metal rod is installed by the tank manufacturer. It keeps the tank from rusting, but only has about a five-year life span.

Storage water heaters are fairly cost-effective if you have natural gas, which happens to be a real energy bargain. If you're heating with electricity or propane, then a storage water heater, though cheaper initially, is your highest-cost choice in the long run. See the Life-Cycle Comparison Chart.

An "Add-On" Heat Pump Water Heater

Heat Pump Water Heaters

If you use electricity to heat water, heat pumps are three to five times more efficient than conventional tank-type resistive heaters. Heat pump water heaters use a compressor and refrigerant fluid to transfer heat from one place to another. This is like a refrigerator in reverse; electricity is the only fuel choice. The heat source is air in the heat pump vicinity, although some better models can duct in warm air from the attic or outdoors. The warmer the air the better. Heat pumps do best in warm climates where they don't have to work as hard to extract heat from the air. The upper element of the conventional electric water heater usually remains active for backup duty. The good points of heat pump water heaters are they use only 33% to 50% as much electricity as a conventional electric tank-type water heater, they will provide a small cooling benefit to the immediate area, and life expectancy is a good 20 years. The negatives are that heat pump heaters are quite expensive initially (average installed cost is approximately $1,200) and recovery rates are modest, variable, and can be fairly low if the pump doesn't have a warm environment from which to pull heat. Like all storage tank systems, heat pumps have standby losses, and if installed in a heated room will steal heat from Peter (the furnace) to pay Paul (the water heater).

Heat pumps make better use of electricity because it's much more efficient to use electricity to *move* heat than to *create* it. Heat pump water heaters are available with built-in water tanks, called integral units, or as add-on units to existing water heaters. Add-on units may be a smarter investment, as the heat pump will probably outlive the storage tank. If you live in a warmer climate and use electricity to heat water, a heat pump is your best choice. In fact, heat pump units stack up quite favorably against everything but natural gas if you can afford the initial purchase cost.

Heat pump water heaters are a bit complex to install, particularly the better units that take their heat from a remote site, or use a separate tank. You will probably require a contractor to install one. Call your local heating and air conditioning contractor for more details on availability in your area, cost, and installation estimates.

An "Integral" Heat Pump Water Heater

Tankless Coil Water Heaters

This type of heater is more common in older oil- or gas-fired boilers. A tankless coil heater operates directly off the house boiler; it does not have a storage tank, so every time there is a demand for hot water the boiler must run. This may be fine in the winter, when the boiler is usually hot from household heating chores anyway, but during the rest of the year this results in a lot of start-and-stop boiler operation. Tankless coil boilers may consume 3 Btu of fuel for every 1 Btu of hot water they deliver. This type of water heating system is not recommended.

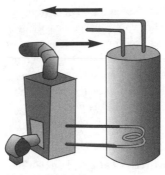

Indirect Water Heater

Indirect Water Heaters

Indirect water heaters also use the home heating system's boiler, but in a smarter setup. Hot water is stored in a separate insulated tank. Heat is transferred from the boiler using a small circulation pump and a heat exchanger. The separate insulated storage tank adds reserve capacity. This means the boiler doesn't have to turn on and off as frequently, which greatly improves fuel economy. When used with the new high-efficiency, gas-fired boilers and furnaces, indirect water heaters are usually the least expensive long-term solution to provide hot water. They will add about $700 to the heating system cost. The life expectancy of the heat exchanger and circulation pump is an exceptional 30 years, although the storage tank may need replacement at 15 to 20 years. The disadvantages of indirect water heating systems are that they are an integral part of the household boiler/heating system that is best installed during new construction by your heating and cooling contractor. They suffer the usual standby losses of storage tanks.

Demand or Tankless Water Heaters

This strategy is so obvious it's a no-brainer. Why keep 30 to 80 gallons of water at 120°F all the time? Do you leave your car idling 24 hours a day just in case you need to run an errand in the middle of the night? Of course not. You start it when you need it. Our water heaters should work the same way, starting up only when we need hot water. Otherwise we heat the water up, it sits there in a big tank and loses heat, then we heat it up again, it sits there . . . and soon, ad infinitum. In many cases this waste heat must even be removed from the home with expensive air conditioning! Due to a long honeymoon with cheap power, the United States is one of the few countries that still uses tank type water heaters. Nearly everybody else in the world figured out the virtues of demand water heating a long time ago. Tank-type heaters use a minimum of 20% more energy than demand systems, due to heat loss. If yours is a small one- or two-person household, your heat losses are even greater, because the hot water spends more time sitting around, waiting to be used.

Advantages of demand water heaters are very low standby losses, lowest operation costs, unlimited amounts of hot water delivery, a very long life expectancy, and a friendliness to do-it-yourselfers and retrofitting. Disadvantages are they only run on natural gas or propane, the standing pilot light wastes some of the energy saved from standby losses, they cost a bit more initially, and the unlimited amounts of hot water can't be taken too rapidly, because flow rates are limited.

How Demand Water Heaters Work

Demand or tankless water heaters only go to work when someone turns on the hot water. When water flow reaches a minimum flow rate the gas flame comes on, heating the water as it passes through a radiator-like heat exchanger. Tankless heaters do not store any hot water for later use, but heat water only as needed. The minimum flow rates required for turn-on prevent any possibility of overheating at very low flow, and ensure that the unit turns off when the faucet is turned off. Minimum flow rates vary from model to model, but are generally about 0.5 to 0.75 gpm for household units. Other safety devices include the standard pressure/temperature relief valve that all water heaters in North America are required to carry, plus, tankless heaters use an additional overheat sensor on the heat exchanger.

The great advantage of tankless heaters is that you run out of hot water only when either the gas or the water runs out. On the other hand, tankless heaters are limited to a fixed output in terms of gallons per minute. Excess water flow will result in lower temperature output. So most tankless heaters are limited to running just one fixture at a time. Larger heaters, like the Aquastar 125, can run multiple fixtures simultaneously, so long as one of those fixtures isn't a shower, where things get quite temperature-sensitive.

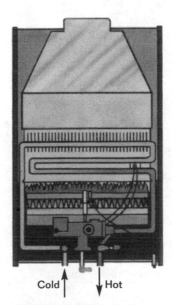

Aquastar 125—A Demand or Tankless Water Heater

Cold ↑   ↓ Hot

Basic Demand Water Heater Internal Construction

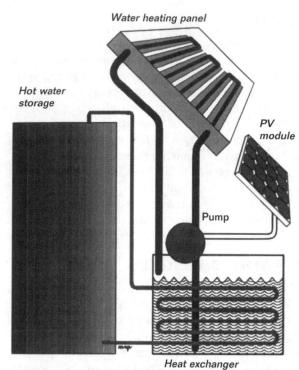

Water heating panel

Hot water storage

PV module

Pump

Heat exchanger

A flat plate solar hot water system using an antifreeze solution in the collector loop with PV-powered pumping and a heat exchanger. This is a top end "Lexus" solution.

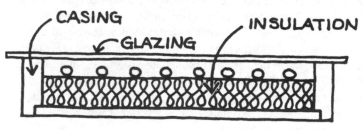

CASING

GLAZING

INSULATION

Standard "Flat Plate" Collector Construction

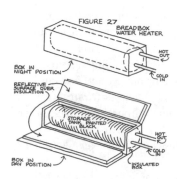

FIGURE 27
BREAD BOX WATER HEATER

BOX IN NIGHT POSITION

HOT OUT

COLD IN

REFLECTIVE SURFACE OVER INSULATION

STORAGE TANK PAINTED BLACK

HOT OUT

BOX IN DAY POSITION

COLD IN

INSULATED BOX

Very simple homemade batch heater made from a water heater tank and an insulated box. Must be manually opened and closed daily. This is the low-end "bicycle" solution.

Tankless heaters are probably the best choice for smaller homesteads of three people or less where hot water use can easily be coordinated. Larger homes with an intermittent use a long distance from the rest of the household hot water plumbing, such as a master bedroom at the end of a long wing, are also good candidates for a tankless heater just to supply that isolated area.

*See the **Best Choices** section on the next page for a complete discussion of demand water heaters if this seems like a good choice for you.*

Can I Get an Electric Demand Water Heater?

Gaiam Real Goods doesn't sell them, but there are some small point-of-use, electric demand water heaters on the market. These units are primarily for washing hands or other small tasks. They will not support a shower or other household fixtures. A household-sized electric demand heater would draw more power than the average house is wired to supply.

Solar Water Heaters

The sun drops approximately 1,000 watts worth of free energy on every square meter of the earth's tropical and temperate zones surface at midday. Solar water heaters simply collect and store some of this free thermal energy. Although expensive initially, they can save large amounts of money over the long term. The typical solar heater in North America will provide approximately two-thirds of a four-person family's hot water needs. As our Comparison Chart shows, a solar heater, working as a preheater before an electric- or propane-fueled backup heater, will save you hundreds to thousands of dollars over its lifetime. If you're lucky enough to have low-cost natural gas as your energy source, then solar heaters have a marginal return on investment.

The advantages of solar heaters are a no-cost/no-impact energy source (the amount you take today in no way diminishes the amount you can take tomorrow, and tomorrow, and tomorrow), the lowest operation cost, a retrofit-friendly technology, and a long life expectancy. The disadvantages are high initial cost, the potential of freezing with some less expensive models, and the need for a backup heater in most climates.

Types of Solar Water Heaters

Solar heaters are divided into two basic types: flat plate collectors and batch collectors. These are discussed briefly below, and covered in more detail in our "Best Choices" section.

Flat plate collectors are as simple as putting a black metal plate in the sun, then collecting the heat. Flat plate collectors circulate a small amount of fluid through little passages in a sun-exposed, blackened copper plate, which is enclosed in an insulated, glass-covered box. Heat picked up by the black plate is transferred to the fluid, which carries it to the storage tank.

Batch collectors are even simpler. Set a tankful of water out in the sun, then wait for it to get hot. Paint the tank black to help absorb heat, and put it into an insulated glass-covered box, and you've got a batch collector. Or for an even simpler setup, open and close your insulated box every day as in our illustration. The batch collector is plumbed between incoming cold water and the inlet port of the backup water heater. Any time hot water is used, the backup heater gets preheated water from the solar collector, rather than stone-cold water.

*If you think solar water heating is a good possibility for you, read about it in more detail in our **Best Choices** section ahead.*

The Best Choices for Water Heating

Demand or Tankless Water Heaters

The Best Choice for Natural Gas or Propane

Tankless heaters are moderate-cost initially, feature long (20 years or more) life expectancies, and next to solar, have the lowest operation costs, making them the best choice if you have either natural gas or propane as your fuel source. We covered operation basics above; what follows are more specifics of operation, and some installation tips.

Thermostatic Control

The Aquastar and Tagaki tankless heater units we sell are thermostatically controlled. At lower flow rates the gas flame is automatically reduced to maintain a stable output temperature. The temperature is front-panel adjustable from roughly 90°F to 140°F. (We consider 120°F to be an optimal setting.) If the water flow rate increases, the heater will respond by increasing the gas flow, up to the heater's Btu maximum. This makes a heater that performs more like the storage tank heaters we're all used to. The system isn't perfect, the sudden opening of a second tap will still cause the shower temperature to fluctuate. But this is also true of the plumbing in most homes with storage tank heaters. With all the tankless heater models we sell, we recommend only running a single fixture, if that fixture is a shower. We're all touchy about our shower temperature, and this is the best way to ensure a steady temperature. The Aquastar units have built-in flow restrictors, so that it is difficult to run more water through the unit than it can heat. If too many taps are opened simultaneously, the water pressure will fall and the water will simply run lukewarm, rather than ice-cold.

Life Expectancy of a Tankless Water Heater

Another great advantage of tankless heaters is their life expectancy. Demand heaters are designed so that all parts are repairable or replaceable. Aquastar warrants their heat exchanger for a full fifteen years. No model has any corrosive parts that touch water. The manufacturers have toll-free 800 service numbers backed up with full-time technicians and fully-stocked parts warehouses. There is no reason why these heaters should not last the rest of your home's life. Compare this to tank-type heaters, which have a 10- to 15-year life expectancy at best. On the downside, the Aquastar units have a water valve that needs service every few years with a $20 factory kit.

Where Should I Install a Tankless Heater?

Just like tank-type heaters, the shorter the hot-water delivery pipe, the less energy is lost to warming up the plumbing. However, if you are willing to throw away the heat energy, and to wait for the hot water to reach the fixture, pipe length doesn't matter. A tankless heater does not need to be installed right at the fixture any more than a storage-type heater. If you are replacing an existing tank-type heater, it is probably most convenient

Ask Dr. Doug

Is it possible to use one of the tankless heaters on my hot tub?

With some ingenuity and creativity it has been done successfully. You need to bear in mind that these heaters are designed for installation into pressurized water systems. For safety reasons, they won't turn on until a certain minimum flow rate is achieved. The flow rate is sensed by a pressure differential between the cold inlet and the hot outlet. Tankless heater flow rates are three gallons per minute and less, which is a much lower flow rate than hot tubs. A hot tub system is very low pressure, but high volume. What usually works is to tee the heater inlet into the hot tub pump outlet, and the heater outlet into the pump inlet. This diverts a portion of the pump output through the heater, and the pressure differential across the pump is usually sufficient to keep the heater happy. Aquastar has a very inexpensive "recirc kit" for its heaters that lowers the pressure differential required. This kit or one of the new "L" low-pressure Bosch models, is mandatory! In addition, you'll need an aquastat to regulate the temperature by turning the pump on and off (your tub may already have one if there was a heating system already installed). You'll also need a willingness to experiment and a good dose of ingenuity.

to install the tankless unit in the same space. Your water and gas piping are already in the vicinity and will require only minor plumbing changes. If this is new construction, just pick the most central location. As with all gas appliances, the heater requires venting to the outside. In a retrofit installation the existing flue pipe will probably need to be upsized for the tankless heater. ***Do not*** reduce the flue size of the tankless unit to fit a smaller flue already in place. *Do not* install a tankless heater outside or in an unheated space, unless it never freezes in your climate. The pilot light can be blown out easily if the unit is exposed to the wind.

Solar System Back-Up

Aquastar tankless heaters can be used with solar- or woodstove-preheated water. As we mentioned above, the standard tankless units sense the outgoing water temperature and adjust the flame accordingly to maintain a steady output temperature. If the incoming water has been preheated, the gas will be modulated down to compensate. The standard Aquastar or Tagaki units can only modulate heat input down to about 20,000 Btu. The optional Aquastar "S" units can modulate all the way down to zero Btu, and are highly recommended if there's any possibility of preheated water in the future. Standard Tagaki units are not recommended for solar backup, although a special unit for preheated water is supposed to be available in the near future. If the incoming water is already preheated to your selected output temperature, then the tankless heater will only come on briefly. If your preheated water is at 90°F, and you've got a 120°F output set, then the water heater will come on just enough to give 120°F output. We think that's pretty slick, and it gracefully slides us into the next good choice, solar water heaters.

Tankless Water Heater Installation Tips

Gas Piping

Most tankless heaters require a larger gas supply line than the average tank-type heater. Just adapting an existing ½-inch supply line to ¾-inch at the heater will choke the fuel supply and limit the water heater output. If the heater's Installation Manual says "¾-inch supply line," that means all the way back to the gas pressure regulator.

Most of these heaters have a small safety override regulator on the gas inlet. This is a back-up for, and in addition to, your standard regulator at the tank or gas entrance. This secondary regulator is also the means for adjusting gas flow at higher altitudes (fully explained in the Installation Manual). Gas inlet is bottom center for all tankless heaters.

Pressure/Temperature Relief Valve

Unlike conventional tank-type heaters, there may not be a P/T valve port built into the heater. This important safety valve must be plumbed into your hot water outlet during installation. Simply tee into the hot water outlet.

Venting

Tankless heaters must be vented to the outside. All models come with draft diverter installed. Aquastar tankless heaters use conventional double-wall Type B vent pipe. This is the same type of vent used with conventional water and space heaters. Tagaki tankless heaters use inexpensive 4-inch single-wall vent pipe, due to their power-vented exhaust. Vent pipe is not included with the heater, but is easily available at plumbing, building supply, or hardware stores. The vent piping used with most of these heaters is larger than tank-type vent piping. Do not adapt down to an existing vent size! Replace the vent and cut larger clearance holes as necessary if doing a retrofit. Venting may be run horizontally as much as 10 feet. Maintain 1-inch rise per foot of run. Aquastar recommends ten feet of vertical vent at some point in the system to promote good flow.

Energy Conversions

Btu: British Thermal Unit, an archaic measurement, the energy required to raise one pound of water one degree Fahrenheit.

1 gal. liquid propane = 4 lb. (if you buy propane by the pound)

1 gal. liquid propane = 91,500 Btu

1 gal. liquid propane = 36.3 cubic feet propane gas @ sea level

1 Therm natural gas = 100,000 Btu

1 cubic foot natural gas = 1,000 Btu

1 kW electricity = 3414.4 Btu/hr

1 horsepower = 2547 Btu/hr

Installation Location

Don't install your tankless heater outside unless you are in a location where it never freezes. Freeze damage is the most common repair on these heaters. Don't tempt fate by trying to save a few bucks on vent pipe. If the heater is installed in a vacation cabin, put tee fittings and drain valves on both the hot and cold plumbing directly below the heater to ensure full drainage. The heat exchanger that works so well to get heat into the water works just as well to remove heat.

Warning

It is illegal and dangerous to install a tankless water heater in an RV or trailer unless the heater is certified for RV service.

The following chart will help you plan your tankless heater installation before you buy. Aquastar heaters have zero clearance on the back for wall mounting. Tagaki T-K1 needs 1″ clearance on back, provided by the included wall-mount brackets.

TANKLESS WATER HEATER CLEARANCE CHART

Heater	Btu/hr	Top	Bottom	Side	Front
Aquastar 38B	40,000	12″	12″	4″	4″
Tagaki T-K1	165,000	11″	–	R-2″, L-6″	4″
Aquastar 125B	117,000	12″	12″	4″	4″

Solar Water Heaters

A Typical Heliodyne Flat Plate Collector Installation

The Best Choice for the Long Run

Although solar water heaters are expensive initially, their long life expectancy and nearly zero cost of operation gives them the lowest life-cycle cost. Solar water heaters enjoyed a huge surge of interest following the Arab oil embargo during the '70s and early '80s when they were supported by 50% federal tax credits. On the down side, those same tax credits encouraged sleazy door-to-door salesmen pushing less-than-the-best technology at astronomical prices. A lot of people got burned in the process, and the solar hot water industry still hasn't recovered from the bad image this circus left behind. All the solar heaters carried by Real Goods were designed after the solar tax credit days, and have benefited from lessons learned during this period. The solar heaters available now *cost far less* and are *much more reliable*. So the first lesson was quality. Don't sell junk. All these modern water heaters are high-quality units warranted for 5 to 10 years by solid companies that are not going to disappear the day after the solar tax credits expire. These water heaters can reasonably be expected to last as long as your house. The second lesson was simplicity. The systems we sell generally don't use controllers, sensors, air vents, or draindown valves—all the complex hardware that experience showed us was most prone to breakdown. These heaters use simple passive designs with sunlight and heat as the controls.

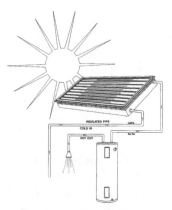

The ProgressiveTube collector preheats the water before it reaches the hot water tank.

Flat Plate Collectors

Now, let's discuss a few details. Flat plate collectors are great at collecting solar energy, and moderately easy to construct, but almost always require a pump to circulate the heated fluid to storage, and a controller to tell the pump when to run. They're also prone to freezing easily, because of the many small passages and large surface area. The better systems use a freeze-proof fluid, which then requires a heat exchanger between this fluid and the household water supply. This results in a moderately expensive system, but one that is absolutely freeze-proof, and has an excellent life expectancy in any climate. The Heliodyne solar hot water systems we carry are examples of freeze-proof flat plate collector systems. This is a high-quality solar hot water system that will work year-round, in any location no matter how cold, with no attention from the homeowner.

How about simplicity, or how do we get rid of all the sensors and controllers that make the flat plate collector work? Obviously you don't want the circulation pump(s) to run unless the collector is hotter than the storage tank. Most flat plate collector systems accomplish this by using multiple temperature sensors and a little controller "brain." Experience has shown that temperature sensors, and the wires, and splice connections leading to them, are the most trouble-prone parts of solar hot water systems, followed closely by controllers. Our Heliodyne systems eliminate controllers, wiring splices, and sensors by using a small PV module connected directly to the circulation pump. When the sun shines, the PV module starts producing electricity, and the pump starts running! The hotter and more direct the sunlight, the faster the pump runs. When the sun goes down, the pump stops working. Circulation speed is directly proportional to sunlight intensity. Simple elegance.

Batch Collectors

Batch heaters are designed to be simple, starting right out by eliminating the pumps, controllers, sensors, and wiring. The water isn't circulated from the collector to a storage tank. Batch collectors *are* the storage tank. This is simplicity itself. The primary disadvantage of this simple collector type is the potential for freezing. Batch collectors heat the domestic water directly, so we can't use an antifreeze fluid. Because of the mass of 20 to 40 gallons of water inside the collector, the chance of freezing overnight in a cold snap is minimal, although the supply and return plumbing can be vulnerable. In summary, batch collectors are less expensive and less complex than flat plate collectors, but are for climates without extended hard freeze weather, or for three-season use. Hard freeze is defined as temperatures of 20°F or colder that last longer than a day. Three-season use is actually the most viable, cost-effective alternative for much of North America. Cloud cover in the winter will prevent collecting any meaningful amount of energy in much of the Northeast, and other portions of the continent. In a hard freeze climate the batch collector is simply shut off and drained in the autumn, then refilled in the spring. The ProgressivTube collector we carry is an example of a batch collector system, and we also carry simple do-it-yourself plans for an effective batch heater using a standard water heater tank.

—**Doug Pratt**

A Homemade Batch Collector with Glass Glazing Made from Our Simple Plans

Fuel Choices for Water Heating– Electric vs. Gas

Electricity is wonderfully useful for many tasks, but using it for larger-scale heating chores, such as household water and space heating, isn't a smart choice. While inexpensive and easy to install initially, electric tank heaters cost two to three times more to operate than gas-fired heaters. Over an average 13-year lifetime this amounts to a $3,000–$8,000 difference! See our Life-Cycle Cost Comparison Chart below for details of electric, gas, tankless, and solar water heating options. If you have gas or oil available, by all means use it! Over the lifetime of the heater you'll save many times the initial purchase and installation cost. In many rural homes where natural gas is not available, the higher life-cycle cost when using electricity may be a strong argument for bringing propane fuel on site, or for investing in one of the efficient heat pump water heaters. While propane costs about twice as much as natural gas, as our Comparison Chart shows, it's still significantly less expensive than electricity.

LIFE-CYCLE COST COMPARISON FOR WATER HEATERS

Type	Cost[1]	Yearly Energy Cost[2]		Life	Cost (over 20 years)[3]
Conventional	$425	@ 8¢/kWh	$390	13 years	$8,650
Electric Tank		@ 10¢/kWh	$468		$10,610
		@ 12¢/kWh	$585		$12,550
High-Efficiency	$500	@ 8¢/kWh	$374	13 years	$8,480
Electric Tank		@ 10¢/kWh	$468		$10,360
		@ 12¢/kWh	$561		$12,220
Electric Heat Pump	$1,200	@ 8¢/kWh	$160	20 years	**$4,400**
		@ 10¢/kWh	$200		$5,200
		@ 12¢/kWh	$240		$6,000
Conventional	$425	w/ NG	$231	13 years	$5,470
Gas Tank		w/ LP	$446		$9,770
High-Efficiency	$500	w/ NG	$205	13 years	$5,100
Gas Tank		w/ LP	$396		$8,920
Tankless Gas	$650	w/ NG	$199	20 years	$4,630
		w/ LP	$319		$7,030
Solar with Electric	$2,500	@ 8¢/kWh	$125	20 years	$5,000
Tank Back-up		@ 10¢/kWh	$156		$5,620
		@ 12¢/kWh	$187		$6,240
Solar with Gas	$2,500	w/ NG	$75	20 years	**$4,000**
Tank Back-Up		w/ LP	$144		$5,380
Solar w/Tankless	$2,725	w/ NG	$64	20 years	$4,005
Gas Back-Up		w/ LP	$102		**$4,765**

[1] *Approximate, includes installation.*

[2] *Energy cost based on hot water needs for typical family of four. Energy costs of $.85/therm for natural gas, and $1.50/gal for propane (late 2000 national averages).*

[3] *Future operation costs are neither discounted nor adjusted for inflation. Includes replacement costs for systems with less than 20-year life expectancy.*

Source: Adapted from American Council for an Energy-Efficient Economy life-cycle charts.

Water Heating

Tankless Water Heaters

Aquastar Water Heaters

An Aquastar, besides being the most efficient gas water heater you can use, is the last water heater you'll ever purchase for your house. All parts are repairable or replaceable, and all parts that actually touch water are noncorrosive stainless steel, brass, or copper, with a ten-year warranty on the heat exchanger. All models are thermostatically controlled to maintain your set water temperature, making them perform like a tank-type heater, but without the wasteful standby losses. Adjustment range is approximately 90°F to 140°F. For safety, all models have a standing pilot thermocouple, an overheat fuse, and a manual shut-off valve.

Aquastar was purchased by Robert Bosch Corp. in 1996, and we're starting to see some new models as a result. The biggest changes are to the most popular 125 model. Now with a more efficient heat exchanger, it burns 6.5% less fuel, but supplies the same amount of hot water. Activation flow rate is down from 0.75 to 0.5 gpm, making this unit less prone to turning off at low flow rates. Other new standard equipment includes a piezo ignitor, a pressure relief valve to comply with U.S. plumbing code, and a drain plug for easier, complete draining to prevent freeze damage. Cabinets have also been redesigned. Happily, the one thing that stayed the same was the price! New models have a "B," for Bosch, following the model number.

The smallest model—38B—will run a low-flow shower so long as incoming water temperature is above 45°F, making it ideal for summer cabins. Model 125B is our "standard" for multi-person households with normal suburban demands.

Optional "S" models are for installations that have (or may have in the future) solar preheated water. "S" models will modulate the gas burner all the way down to zero if needed to maintain set output temperature. This option won't affect normal operation with or without preheated water, but costs a whole lot more to add as a retrofit in the future.

Aquastar heaters are manufactured in Portugal or France depending on model, and are warranted by the U.S. importer for 15 years on the heat exchanger, and for 2 years on all other components. Support includes a toll-free 800 # for parts and service advice, with real technicians, who are backed up with a fully stocked parts warehouse. AGA approved. Specify LP (propane) or NG (natural gas) when ordering. Portugal or France.

Aquastar 38B

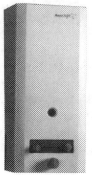

The new improved Aquastar 38B is the perfect water heater for a small cabin with a low-flow shower or anywhere the demand for hot water is low. Will run a shower so long as incoming water temperature is 45°F or above. The model 38 is equipped with a piezo igniter. Now approved for RVs and marine use. Features a 77.5% efficiency rating. Specify Propane or Natural Gas.

- Btu Input: 40,000 Btu/hr.
- 45°F Temp. Rise: 1.3 gpm
- 55°F Temp. Rise: 1.0 gpm
- 65°F Temp. Rise: 0.9 gpm
- Minimum Flow: 0.6 gpm
- Vent Size: 4″ Type B gas vent
- Water Connections: ½″ NPT
- Min. Gas Supply Line: ½″ NPT
- Min. Water Pressure: 15 psi
- Dimensions: 25.375″ H x 10.625″ W x 9.125″ D
- Shipping Weight: 25 lb

45-101LP Aquastar 38B LP $359⁰⁰

45-101NG Aquastar 38B NG $359⁰⁰

Aquastar 125B

The new improved Aquastar 125B is the perfect residential heater. It's capable of delivering a constant supply of hot water at the temperature you select. The 125B can support a single shower, or smaller multiple fixtures simultaneously. This model is the best choice for the average home. Appropriate for small-volume commercial installations also. All 125 models feature an excellent 82% efficiency rating.

Standard models will only modulate down to approximately 20,000 Btu; "S" models will modulate down to zero for preheated water. Get the "S" model if you have, or plan to have, a solar or woodstove preheating system, which will not affect normal operation (but costs lots more to add as a retrofit). The 125B model also has a new "L" option for low-flow or recirculation applications such as hot tub or slab floor heating.

Aquastar is offering a pilotless "X" version of the 125. This model does away with the standing pilot light, using electronic ignition, and significantly boosting energy efficiency. No electric plug is required, the ignition is powered by a pair of D-cells. Batteries are included! Life expectancy is about two years, and it will peep like a smoke alarm when batteries get low. All other specifications are identical to the standard 125 B model. The "X" model is not available with the "S" solar option.

- Btu Input: 117,000
- 55°F Temp. Rise: 3.3 gpm
- 75°F Temp. Rise: 2.4 gpm
- 90°F Temp. Rise: 2.0 gpm
- Minimum Flow: 0.5 gpm
- Vent Size: 5″ Type B gas vent
- Water Connections: ½″ NPT
- Min. Gas Supply Line: ¾″ NPT
- Min. Water Pressure: 15 psi
- Dimensions: 29.75″ H x 18.25″ W x 8.75″ D
- Shipping Weight: 44 lb

45-105-LP Aquastar 125B Propane $599⁰⁰

45-105-NG Aquastar 125B Natural Gas $599⁰⁰

"S" Units For Use With Preheated Water

45-107-LPS Aquastar 125B Propane $699⁰⁰

45-107-NGS Aquastar 125B Natural Gas $699⁰⁰

"L" Units For Use In Recirculation Applications

45-113-L-LP Aquastar 125B L Propane $659⁰⁰

45-113-L-NG Aquastar 125B L Natural Gas $659⁰⁰

Note: "S" option not available in low-flow models.

"X" Units with Pilotless Electronic Ignition

45-114-X-LP Aquastar 125 X Propane $699⁰⁰

45-114 X-NG Aquastar 125 X Natural Gas $699⁰⁰

Note: "S" option not available in pilotless models.

Aquastar 170

The large Aquastar 170 has been replaced by the Tagaki T-K1, see next page for details and specifications. The T-K1 is suitable for domestic and residential use, unlike the old Aquastar 170.

Aquastar 130B

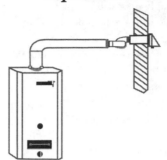

The model 130B is actually a 125X model teamed up with a Power Vent exhaust. The power vent boosts Btu input/output slightly to 125,000 Btu, and allows the use of smaller 4″ vent pipe with a horizontal thru-the-wall termination. No chimney is needed, but 120VAC power is required for the power vent. All other sizing and connection info is the same as the standard 125X listed above. As a package, the 130B saves more than $200 over purchasing the heater and power vent separately! The 130B sports a top-of-the-line 85% efficiency rating.

45-003-LP Aquastar 130B Propane $749⁰⁰

45-004-NG Aquastar 130B Natural Gas $749⁰⁰

TANKLESS WATER HEATER CLEARANCE CHART					
Heater	Btu/hr	Top	Bottom	Side	Front
Aquastar 38B	40,000	12″	12″	4″	4″
Tagaki T-K1	165,000	11″	–	R-2″, L-6″	4″
Aquastar 125B	117,000	12″	12″	4″	4″

The Most Powerful And Efficient Tankless Water Heater Available

Indoor

Outdoor for
Mild Climates

The high-performance T-K1 sets a new instant water heater standard by delivering up to 5 gallons per minute of hot water continuously. Want to shower while filling the washer? The T-K1 handles multiple uses without distress. Innovative features include computerized temperature control, electronic ignition, power venting, and automatic freeze protection. Advanced computerized sensors monitor the water temperature, and modulate the gas valves to deliver steady output. Electronic ignition means no pilot light, and no gas use till there's a demand, giving this heater an excellent Energy Factor of .84. For comparison, a 50-gallon tank heater has an Energy Factor of .53. Power venting allows a smaller 4″ single-wall vent pipe that can be run vertically or horizontally. Automatic freeze protection will protect the unit down to 5°F (with no wind). Water drains are included for seasonal installations. Outdoor installations are okay in mild Sunbelt climates using an optional vent hood. Life expectancy is 20–30 years.

The T-K1 modulates the gas input down to a 35,000 Btu minimum, so it isn't recommended for solar hot water backup. Turn-on requires 0.75 gpm flow; once lit, flow can be reduced to 0.6 gpm without turning off. Temperature output is factory set at 140°F. Temperature is adjustable only by using the remote control option. All water, gas, and electric connections are on the left side facing the heater. Clearances to combustibles are: back-1″, front-4″, right-2″, left-6″, top-11″, noncombustible floor, if sitting on the floor. Vent can run up to 21 feet vertical or horizontal; deduct 5 feet for each elbow. Stand-off wall mounting brackets are included. AGA approved. Two-year mfr.'s warranty on the complete unit, five years on the heat exchanger. Technical help and parts available from U.S. importer via toll-free 800 number. Japan.

Indoor/Outdoor
Wall Hanging

45-505 LP T-K1 Tankless Water Heater for Propane $999⁰⁰

45-505 NG T-K1 Tankless Water Heater for Natural Gas $999⁰⁰

45-506 T-K1 Outdoor Vent Hood $89⁹⁵

45-507 T-K1 Remote Control/Temperature Control Option $195⁰⁰

- Btu Input: 35,000–165,000
- 50°F Temp. Rise: 5.3 gpm
- 60°F Temp. Rise: 4.4 gpm
- 70°F Temp. Rise: 3.8 gpm
- 80°F Temp. Rise: 3.3 gpm
- 90°F Temp. Rise: 2.9 gpm
- Minimum Flow: 0.75 gpm
- Vent Size: 4″ Single Wall
- Water Connections: 3/4″ Male NPT
- Min. Gas Supply: 3/4″ (if over 25′ use 1″)
- Min. Water Pressure: 15 psi
- Dimensions: 24.5″ H x 16.5″ W x 8.3″ D
- Shipping Weight: 65 lb

Other Water Heating Products

Consumer Guide to Home Energy Savings

By Alex Wilson and John Morrill. The updated seventh edition of Consumer Guide to Home Energy Savings identifies the most energy-efficient home appliances by brand name and model number. This book is as compact and efficient as its subject matter. Its pages are crammed with money-saving information. 5″ x 7″, 296 pages, 118 illustrations, appendices, index, paperback.

82-399 7th Ed. Consumer Guide to Home Energy Savings $8⁹⁵

TANKLESS WATER HEATER SELECTION CHART

Mfg.	Model	Recommended Use
Aquastar	38B	Single fixture, small cabin with low-flow shower
Aquastar	125B	Conventional home, single shower or several smaller fixtures simultaneously
Aquastar	125BS	Same as above, but with possibility of preheated water
Tagaki	T-K1	Conventional home, two showers or several smaller fixtures simultaneously

Snorkel Hot Tub Heaters

Snorkel stoves are woodburning hot tub heaters that bring the soothing, therapeutic benefits of the hot tub experience into the price range of the average person. Snorkel stoves are simple to install in wood tubs, easy to use, extremely efficient, and heat water quite rapidly. They may be used with or without conventional pumps, filters, and chemicals. The average tub with a 450-gallon capacity heats up at the rate of 30°F or more per hour. Once the tub reaches the 100°F range, a small fire will maintain a steaming, luxurious hot bath.

Stoves are made of heavy-duty, marine-grade plate aluminum that is powder-coated for the maximum in corrosion resistance. This material is very light, corrosion resistant, and strong. Aluminum is also a great conductor of heat, three times faster than steel. Stoves are supplied with secure mounting brackets and a sturdy protective fence.

Two stove models are offered, the full-size 120,000 Btu/hr Snorkel, or the smaller 60,000 Btu/hr Scuba. The Snorkel is recommended for 6- and 7-foot tubs. The Scuba is for smaller 5 foot tubs. These stoves are for wooden hot tubs only.

We also offer Snorkel's precision-milled, long-lasting Western Red Cedar hot tub kits that assemble in about four hours. A 5-foot tub holds two to four adults, a 6-foot tub holds four to five adults. The 7-foot tub is for large families or folks who entertain frequently. USA.

🚚 **45-400 Scuba Stove $599⁰⁰**

🚚 **45-401 Snorkel Stove $799⁰⁰**

🚚 **45-002 Cedar Hot Tub 5′ x 3′ $1,449⁰⁰**

🚚 **45-207 Cedar Hot Tub 6′ x 3′ $1,749⁰⁰**

🚚 **45-208 Cedar Hot Tub 7′ x 3′ $2,295⁰⁰**

🚚 **45-504 Drain Kit, PVC 1.5″ $45⁹⁵**

🚚 **45-501 6′ Tub 3-Bench Kit $349⁰⁰**

🚚 *Shipped freight collect from Washington State.*

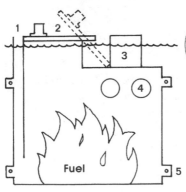

1. *Special "Snorkel" air intake.*
2. *Sliding, cast aluminum, tilt-up door that services the fire box as well as regulates the air intake.*
3. *Sunken stack to increase efficiency.*
4. *Heat exchange tubes.*
5. *Mounting brackets for securing fence and stove.*

Solar Water Heating

Domestic Water Heaters

ProgressivTube Solar Water Heater

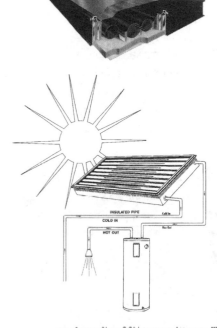

This passive batch-type heater is one of the simplest, yet most elegant and durable water heaters we've seen. Plumbed inline ahead of your conventional water heater's cold inlet, it will preheat all incoming water, reducing your conventional heater's work load to near zero. Rated at approximately 30,000 Btu/day, the 4 ft x 8 ft collector is housed in a top-quality, bronze-finished aluminum box with stainless and aluminum hardware. There are no rusting components. The dual glazing has tempered low-iron solar glass on the outside, and non-yellowing or -degrading 96% transmittance Teflon film on the inside. High temperature, non-degrading phenolic foam board insulation is used on the sides (R-12.5), between internal tubes (R-12.5), and the bottom (R-16.7).

The collector is composed of large 4-inch diameter selective-coated copper pipes aligned horizontally. They are connected in series, with cold water introduced at the bottom, and hot water taken off at the top. As the water warms, it stratifies and gradually works its way from tube to tube toward the top. The outgoing hot water is never diluted by incoming cold water. 30-, 40-, and 50-gallon models are produced by varying the number of collector pipes inside. Exterior size is the same for all models, and BTUs collected per day vary only slightly from 22,100 for the PT-30, to 28,400 for the PT-40, to 28,700 for the PT-50, according to FSEC.

The collector and mounting system have been officially tested and approved for 180 mph winds, and 250 units installed in St. Croix were unofficially tested by the 200+ mph winds of Hurricane Hugo in 1989. Only six units suffered minor damage, and that was due to flying debris. Use caution if roof-mounting a ProgressivTube collector. Filled, they weigh close to 600 lb. Rafters may need to be doubled up. Ground mounting is often a better choice.

Because of the thermal mass and highly insulated box, the collector is unlikely to freeze (the warranty will cover freeze damage in the Sunbelt), but the supply piping is vulnerable. In hard-freeze climates the ProgressivTube must be drained for the winter.

ProgressivTube has the best warranty we've ever seen: ten years. And during the first five years, there is no charge to the homeowner for any warranty claim. The warranty covers labor and transportation, in addition to parts. The second five years covers parts only. Both fixed (roof angle) and tilting mounts with 36 inches of cut-to-fit back leg tubes are offered. The Valve Kit includes hard-to-find hardware, including drain valves, a very important tempering valve, and ball valves to choose between solar preheating or solar bypass. USA.

45-085 ProgressivTube PT-30 $995⁰⁰
45-094 ProgressivTube PT-40 $1,295⁰⁰
45-089 ProgressivTube PT-50 $1,369⁰⁰
45-095 ProgressivTube Fixed Mounting $59⁹⁵
45-096 ProgressivTube Tilt Mounting $79⁹⁵
45-097 ProgressivTube Valve Kit $199⁰⁰

Shipped freight collect from Sarasota, FL. Collector pricing includes $50 crating charge.

Heliodyne Solar Hot Water Systems
PV-Powered, Freeze-Proof, and the Highest Quality

If you value quality, or live in a hard freeze climate, then Heliodyne is the solar hot water system for you. In the solar business since 1976, Heliodyne makes the best equipment, then preassembles as much as possible in their Helio-Paks, saving over half the typical plumbing assembly time on site. The Heliodyne systems we offer all feature freeze-proof closed-loop collectors with nontoxic antifreeze. Pumping and control are completely grid-independent, and are done with simple PV power. The brighter and hotter the sun, the faster the pump runs. Heat is transferred to the household water through an efficient side-mounted thermosiphon heat exchanger, so only a single pump is required, and any conventional electric heater tank can be used for storage (60 gallons or larger recommended).

Using the same great selective-coated, highly efficient Heliodyne Gobi solar collectors described nearby, these kits are supplied with collector, PV pumping station, flush mounting hardware, blind unions, and sidemount heat exchanger. Two single-collector sizes are offered below that will cover most households at about 30,000 Btu/day for the HX 3366 kit, or 44,000 Btu/day for the HX 1410 kit. The average North American uses about 10,000 to 15,000 Btu worth of hot water per day. Larger collector packages are available by special order. Shipped from San Francisco Bay Area. USA.

45-508 Heliodyne HX 3366 G PV System $2,045⁰⁰
45-509 Heliodyne HX 1410 G PV System $2,350⁰⁰

System pricing includes $75 crating charge.
Means shipped freight collect.

Racking Option
(for HX systems only)

The Heliodyne systems above come with flush-mount hardware. If you need to tilt up your collector, these 42″ cut-to-fit legs give you the angle needed.

❖45-510 HX Rack Option $45⁰⁰

Cal Code Kit

Most installations will need or want these items. Includes an air vent, dial thermometer, tempering valve, and fittings. Complies with California solar hot water plumbing code.

❖45-511 Cal Code Kit $105⁰⁰

❖ *Means shipped directly from manufacturer.*

Heliodyne Solar Hot Tub Heaters

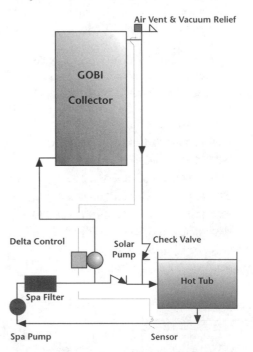

Our solar hot tub kit comes complete with a top-quality solar collector, flush-mount hardware, 120-volt AC controller, pump, temperature sensors, and air vent. You add the tub, and plumbing to suit your site. These kits are designed to mount the collector on a nearby roof and connect to existing hot tub plumbing in such a way that all water can drain from the collector and pipes when not operating. Automatic operation is provided by differential temperature sensors and the controller. The system will only run if the tub needs heat, and there's heat to be had at the collector. Tub temperature is user adjustable, with an upper limit of 105°F. The pump is not self-priming, and must be installed at or below tub water level. It will lift water up to two stories.

The smaller 3366 collector kit can deliver 18,000 to 30,000 Btu/day. With reasonable tub insulation, it will maintain daily tub temperature in North American climates, and will bring smaller to medium-size tubs up to temperature within a few days after a water change. The larger 4´ x 10´ kit can deliver 29,000 to 46,000 Btu/day. Use it for larger tubs or faster unassisted recovery times.

Prepaid shipping quote will save money. USA.

45-000 Solar Hot Tub Kit 3366 $1,320⁰⁰

45-209 Solar Hot Tub Kit 410 $1,625⁰⁰

Shipped by truck from San Francisco Bay area.

System pricing includes $75 crating charge.

Heliodyne Gobi Solar Collectors

Heliodyne has produced the absolutely best solar collectors in the business since 1976. Absorbers feature 360° wrap-around fin-to-tube bonding that will never delaminate. Absorbers have a black chrome selective coating that is better at absorbing heat than radiating it off. Headers are extra-large 1-inch diameter, with wonderful solid brass O-ringed unions at all four corners for complete adaptability. Multiple collectors will couple up without any modification. The glazing is tempered 5/32″ low-iron solar glass, with an antiglare finish. It's mounted in a full perimeter rubber gasket. Frames are exceptionally strong bronze-anodized aluminum extrusions, with a mounting flange around the full perimeter. They are certified for survival at wind speeds up to 50 lb/sq.ft.; that's a bit over 100 mph. Each solar collector system will require a pair of blind unions to block off the unused header ends, and either flush mount or tilting rack mounts.

The new 3366 collector is specifically designed to fit into trucks and standard shipping containers for reduced shipping costs. It measures 43.6″ x 88.6″, and is SRCC rated to deliver 18,000 to 30,000 Btu/day in domestic water heating service.

The 408 collector is slightly larger at 48″ x 97.6″, and is SRCC rated to deliver 23,000 to 37,000 Btu/day in domestic water heating service. Variations in collector ratings are due to low or medium temperature input to collector.

The 410 collector is the largest Heliodyne produces at 48″ x 121.6″. It is SRCC rated to deliver 29,000 to 46,000 Btu/day in domestic water heating service. Because of its large size, this collector can be expensive to ship.

The typical North American home requires about 10,000 to 15,000 Btu worth of hot water per person per day for comparison.

Shipped from San Francisco Bay Area. USA

45-512 Heliodyne 3366 Gobi Collector $620⁰⁰

45-498 Heliodyne 408 Gobi Collector $775⁰⁰

45-500 Heliodyne 410 Gobi Collector $919⁰⁰

05-226 Heliodyne Collector Crating $75⁰⁰

Crating charge must be added for each collector ordered.

Means shipped freight collect.

Heliodyne Options

Solar Collection Fluid

A propylene glycol that is food-grade safe and nonflammable. It dilutes 25% to 50% with water depending on climate. Each 4-ft x 8-ft collector needs almost one gallon. Fluid should be changed every five years. USA.

❖45-100 Dyn-O-Flo, 4 one-gallon bottles $94⁹⁵

45-515 Dyn-O-Flo, single one-gallon bottle $29⁹⁵

Blind Unions

Every system needs a pair of these to block off the two unused unions after the plumbing is assembled. (The Complete PV-Powered Systems are supplied with these already.)

❖45-495 Heliodyne Blind Unions, 1 Pair $19⁹⁵

Flush Mount Hardware

This hardware kit will mount one collector parallel to the roof, with an attractive stand-off clearance for code compliance and longest roof life. (The Complete PV-Powered Systems are already supplied with these.) See the Tilting Rack Options below if your collector needs to be tilted in relation to the roof surface.

❖45-496 Heliodyne Flush Mount Kit $44⁹⁵

Tilting Rack Options

Flush mounts are included with complete systems. If you need to tilt your collectors at something other than roof angle, here's the answer. Includes 42 inches of back leg tubing to allow cutting at needed collector angle.

❖45-109 Tilt Rack for 1 Collector $89⁹⁵

❖45-110 Tilt Rack for 2 Collectors $139⁰⁰

❖45-111 Tilt Rack for 3 Collectors $199⁰⁰

Breadbox Solar Hot Water Plans

We've gotten lots of requests for simple plans like this over the years. This passive solar water heating system is simple to build using readily available materials; uses no pumps, controllers, or sensors, and works wonderfully as a preheater or even as a stand-alone system. Uses a pair of common 40-gallon water heater tanks, either new or recycled. The system is virtually freeze-proof with a little care during construction. The detailed, easy-to-follow plans include many drawings, a materials list, construction sequence, and options for different tank configurations, reflectors, absorber surfaces, and glazing techniques. Developed by a licensed solar plumber based on 20 years of community education workshops. 8.5″ x 11″, 8 pages (folds out). USA.

80-497 Solar Hot Water Plans $11⁹⁵

Safe Water Systems

The Sol*Saver can provide safe water for an entire village, school, or health clinic.

One of the greatest health and environmental problems our world faces is providing safe drinking water. Pasteurizing water, raising it to a temperature of 174°F (79.4°C), has proven to be 99.999% effective for microbiologically contaminated drinking water. It's how the dairy industry has sterilized milk for decades. Pasteurizing requires only one-eighth the energy of boiling.

The Sol*Saver and Wood*Saver pasteurizing systems can provide safe water for villages, schools, hospitals, health clinics, orphanages, or other such uses using either solar energy, wood fuel, or both. Two simple concepts create this revolutionary technology. The temperature of water standing in a glazed solar collector can easily exceed 200°F on a sunny day. A patented thermal outlet control valve opens only when the temperature exceeds the 174°F pasteurization point. This valve is rated for more than one million cycles. In the event of malfunction, no water will flow to the sanitary water storage tank (not included) but will be discharged through a safety dump port. The second great innovation is to run the now hot, sterile water through a heat exchanger, which preheats the water that is about to be pasteurized and saves much of the input energy! A supply tank just above the pasteurizer can provide simple gravity feed.

The Sol*Saver

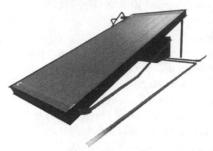

Constructed of durable copper, aluminum, brass, and tempered glass for a long life, this unit includes an approximate 4-ft x 10-ft solar collector with mounts, high-efficiency heat exchanger, and fail-safe thermal control valve. Operation is fully automatic, just add water and sun. Adjustable wrenches are the only tools needed for plumbing on site, no pipe soldering is required. An optional backup solid fuel burner can provide sterile water during prolonged cloudy weather. The solar collector can pasteurize up to 200 gallons per day. The backup burner can produce more than 25 gallons per hour and will burn wood, coal, charcoal or other solid fuels. Weight (crated): 521 pounds. Volume (crated): 63 cu. ft. USA.

42-190 Sol*Saver $1,995⁰⁰

42-191 Sol*Saver w/ Burner $2,395⁰⁰

Drop shipped freight collect from central California.

The Wood*Saver

Simply the solid wood burn chamber by itself, fail-safe thermal control valve, and heat exchanger. Weight (crated): 45 pounds. Volume (crated): 10 cu. ft. USA.

All Safe Water Systems have a life expectancy of 15 years and are backed by a three-year warranty.

42-192 Wood*Saver $1,149⁰⁰

Drop shipped freight collect from central California.

Solar Or Electrical Roof Flashing

It's often tough to find these high-quality rubber gasketed flashings in the smaller plumbing and electrical sizes. Each flashing will fit one single ½-inch, ¾-inch, or 1-inch copper pipe or electrical conduit. Standard size flashing pan for conventional asphalt shingle roofs. Not for wood shakes. Accommodates 0/12 to 12/12 pitch. Most solar hot water installations require two.

45-414 Solar Flashing $6⁹⁵

Five-Gallon Solar Shower

This incredible low-tech invention uses solar energy to heat water for all your washing needs. The large 5-gallon capacity provides ample hot water for several hot showers. On a 70°F day the Solar Shower will heat 60°F water to 108°F in only three hours for a tingling-hot shower. This unit is built of four-ply construction for greatest durability and efficiency.

90-416 Solar Shower $14⁹⁵

CHAPTER 8

WATER & AIR PURIFICATION

IF YOU ARE CONCERNED about the degradation of our environment, you are probably also conscientious about the foods you eat. If you are conscientious about the foods you eat, then you ought to be attentive to the quality of the air that you breathe and the water that you drink. Like food, air and water can carry into your body the contaminants that saturate the natural world. That seems obvious. But you may not realize the extent to which your personal source of domestic water and the air inside your house may be polluted. We will describe here the bad stuff that may be lurking in your air and water, the conditions that contribute to potentially unhealthful levels of these contaminants in your living space, and what you can do about it. And, in the best Gaiam Real Goods tradition, we offer our recommended air and water purification products and resources.

The Need for Water Purification

The Environmental Protection Agency (EPA) has recently released information stating that no matter where we live in the U.S., there is likely to be some toxic substance in our groundwater. Indeed, the agency estimates that one in five Americans, supplied by one-quarter of the nation's drinking water systems, consume tap water that violates EPA safety standards under the Clean Water Act. Even some of the substances that are added to our drinking water to protect us, such as chlorine, can form toxic compounds—such as trihalomethanes, or THMs—and have been linked to certain cancers. The EPA has established enforceable standards for more than 100 contaminants. However, credible studies have identified more than 2,110 contaminants in the nation's water supplies.

The most obvious solution to water pollution is a point-of-use water purification device. The tap is the end of the road for water consumed by our families, so this is the logical, and most efficient, place to focus water treatment. Different water purification technologies each have strengths and weaknesses, and are particularly effective against specific kinds of impurities or toxins. So the system you need depends first and foremost on the nature of the problem you have, which in turn requires testing and diagnosis. We will describe the options available with such systems in detail after we discuss the contaminants you might encounter. Most treatment systems are point-of-use and deal only with drinking and cooking water, which is less than 5% of typical home use. Full treatment of all household water is a very expensive undertaking, and is usually reserved for water sources with serious, health-threatening problems.

Contaminants in Water

Presumably your drinking water comes from a municipal system, a shallow dug well, or a deep, drilled well. If you drink bottled water, you've already taken steps to control what you consume. Bottled water drinkers should read on, too: most good purification systems provide tasty potable water for far less money than the cost of regularly drinking bottled water. Each type of water supply is more or less vulnerable to different kinds of pollutants, because surface water (rivers, lakes, reservoirs), groundwater (underground aquifers), and treated water each are exposed to environmental contaminants in different ways. Obviously, the hydrogeological characteristics of the area you live in, and localized activities and sites that create pollution, such as factories, agricultural spraying, or a landfill, will potentially impact your water quality. And some impurities, such as lead and other metals, can be introduced by the piping that delivers the water to your tap.

Reliable and inexpensive tests are available to identify the biological and chemical contaminants that may be in your water. Here's what you may find.

Biological Impurities: Bacteria, Viruses, and Parasites.

Microorganisms originating from human and animal feces, or other sources, can cause waterborne diseases. Approximately 4,000 cases of waterborne illness are reported each year in the U.S. Additionally, many of the minor illnesses and gastrointestinal disorders that go unreported can be traced to organisms found in water supplies.

Biological impurities have largely been eliminated in municipal water systems with chlorine treatment. However, such treated water can still become biologically contaminated. Residual chlorine throughout the system may not be adequate, and therefore microorganisms can grow in stagnant water sitting in storage facilities or at the ends of pipes.

Water from private wells and small public systems is more vulnerable to biological contamination. These systems generally use untreated groundwater supplies, which could be polluted due to septic tank leakage or poor construction.

Organic Impurities: Tastes, and Odors

If water has a disagreeable taste or odor, the likely cause is one or more organic substances, ranging from decaying vegetation to algae to organic chemicals (organic chemicals are compounds containing carbon.)

Inorganic Impurities: Dirt and Sediment, or Turbidity

Most water contains suspended particles of fine sand, clay, silt, and precipitated salts. This cloudiness or muddiness is called turbidity. Turbidity is unsightly, and it can be a source of food and lodging for bacteria. Turbidity can also interfere with effective disinfection and purification of water.

Total Dissolved Solids (TDS)

Total dissolved solids consist of rock and numerous other compounds from the earth. The significance of TDS in water is a point of controversy among water purveyors, but here are some facts about the consequences of higher levels of TDS:

1. *High TDS results in undesirable taste, which can be salty, bitter, or metallic.*
2. *Certain mineral salts may pose health hazards. The most problematic are nitrates, sodium, barium, copper sulfates, and fluoride.*
3. *High TDS interferes with the taste of foods and beverages.*
4. *High TDS makes ice cubes cloudy, softer, and faster melting.*
5. *High TDS causes scaling on showers, tubs, sinks, and inside pipes and water heaters.*

Toxic Metals or Heavy Metals

The presence of toxic metals in drinking water is one of the greatest threats to human health. The major culprits include lead, arsenic, cadmium, mercury, and silver. Maximum limits for each of these metals are established by the EPA's Drinking Water Regulations.

Toxic Organic Chemicals

The most pressing and widespread water contamination problem results from the organic chemicals (those containing carbon) created by industry. The American Chemical Society lists more than four million distinct chemical compounds, most of which are synthetic (man-made) organic chemicals, and industry creates new ones every week. Production of these chemicals exceeds a billion pounds per year. Synthetic organic chemicals have been detected in many water supplies throughout the country. They get into the groundwater from improper disposal of industrial waste (including discharge into waterways), poorly designed and sited industrial lagoons, wastewater discharge from sewage treatment plants, unlined landfills, and chemical spills.

Studies since the mid-1970s have linked organic chemicals in drinking water to specific adverse health effects. However, only a fraction of these compounds have been tested for such effects. Well over three-quarters of the substances identified by the EPA as priority pollutants are synthetic organic chemicals.

Volatile Organic Compounds (VOCs)

Volatile organic compounds are very lightweight organic chemicals that easily evaporate into the air. VOCs are the most prevalent chemicals found in drinking water, and they comprise a large proportion of the substances regulated as priority pollutants by the EPA.

Chlorine

Chlorine is part of the solution and part of the problem. In 1974 it was discovered that VOCs known as trihalomethanes (THMs) are formed when the chlorine added to water to kill bacteria and viruses reacts with other organic substances in the water. Chlorinated water has been linked to cancer, high blood pressure, and anemia.

The scientific research linking various synthetic organic chemicals to specific adverse health effects is not conclusive, and remains the subject of considerable debate. However, given the ubiquity of these pollutants and their known presence in water supplies, and given some demonstrated toxicity associations, it would be reasonable to assume that chronic exposure to high levels of synthetic organic chemicals in water could be harmful. If they're in your water, you want to get them out.

Pesticides and Herbicides

The increased use of pesticides and herbicides—cide means a thing that kills—in American agriculture since World War II has had a profound effect on water quality. Rain and irrigation carry these deadly chemicals into groundwater, as well as into surface waters. These are poisonous, plain and simple.

Asbestos

Asbestos exists in water as microscopic suspended fibers. Its primary source is asbestos-cement pipe, which was commonly used after World War II for city water systems. It has been estimated that some 200,000 miles of this pipe are currently in use delivering drinking water. Because pipes wear as water courses through them, asbestos shows up with increasing frequency in municipal water supplies. It has been linked to gastrointestinal cancer.

Radionuclides

The earth contains naturally occurring radioactive substances. Due to their geological characteristics, certain areas of the U.S. exhibit relatively high background levels of radioactivity. The three substances of concern to human health that show up in drinking water are uranium, radium, and radon (a gas). Various purification techniques can be effective at reducing the levels of radionuclides in water.

Testing Your Water

The first step in choosing a treatment method is to find out what contaminants are in your water. Different levels of testing are commercially available, including a comprehensive screening for nearly 100 substances. See the product section that follows.

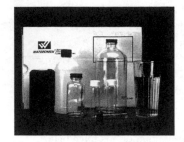

Even before you test your water, a simple comparison can help you figure out what level of treatment your domestic water supply may need to make you happy. Find some bottled water that you like. Note the normal levels of total dissolved solids (TDS), hardness, and pH, and also how high these levels range. If you don't find this information on the label, call the bottler and ask. Then get the same information about your tap water, which can be obtained by calling your municipal supplier. If your tap water has lower levels of these things than the bottled water, a good filter should satisfy your needs. If your tap water has higher levels than the bottled water you prefer, you may need more extensive purification.

If you're drinking water from a private well, you'll have to get it tested to know what's in it. Some common problems with well water, such as staining, sediment, hydrogen sulfide (rotten-egg smell), and excess iron or manganese, should be corrected before you purchase a point-of-use treatment system, because they will interfere with the effective operation of the system. These problems can easily be identified with a low-priced test; many state water quality agencies will perform such tests for a nominal fee, or we offer some tests in our product section.

If you have any reason to worry that your water may be polluted, we strongly advise you to conduct a more comprehensive test.

Water Choices: Bottled, Filtered, Purified

If you're not satisfied with your drinking water, you suspect it may be contaminated, or you've had it tested and you know it's unhealthy, you have two choices. You can buy bottled water, or you can install a point-of-use water treatment system in your home.

There are three kinds of bottled water: distilled, purified, and spring. Distillation evaporates the water, then recondenses it, thereby theoretically leaving all impurities behind (although some VOCs can pass right through this process with the water). Purified water is usually prepared by reverse osmosis, de-ionization, or a combination of both processes (see below for explanations). Spring water is usually acquired from a mountain spring or an artesian well, but it may be no more than processed tap water. Spring water will generally have higher total dissolved solids than purified water. Distilled water and purified water are better for batteries and steam irons because of their lower content of TDS.

But why pay for bottled water forever when treatment will be cheaper? Bottled water typically costs up to $2.00 per gallon or more (more than gasoline!). Many manufacturers claim that their purifiers produce clean water for just pennies a gallon; certainly these systems will pay for themselves in cost savings within a few years.

There are two basic processes used to clean water: filtration and purification. The word "filter" usually refers to a mechanical filter, which strains the water, and/or a carbon filter system, which reduces certain impurities by chemically bonding them — especially chlorine, lead and many organic molecules. Purification refers to a slower process, such as reverse osmosis, which greatly reduces dissolved solids, hardness, and certain other impurities, as well as many organics. Many systems combine both processes. Make sure that the claims of any filter or purifier you consider have been verified by independent testing.

Methods of Water Filtration & Purification

The most efficient and cost-effective solution for water purity is to treat only the water you plan to consume. A point-of-use water treatment system eliminates the middleman costs associated with bottled water, and can provide purified water for pennies per gallon. Devices for point-of-use water treatment are available in a variety of sizes, designs, and capabilities. Some systems only improve your water's taste and odor. Other systems reduce the various contaminants of health concern. Different systems work most effectively against certain contaminants. A system that utilizes more than one technology will protect you against a broader spectrum of biological pathogens and chemical impurities.

When considering a treatment device, always pay careful attention to the independent documentation of the performance of the system for a broad range of contaminants. You should carefully read the data sheets provided by the manufacturer to verify its claims. Many companies are certified with the National Sanitation Foundation (NSF), whose circular logo appears on their data sheets.

The following is a brief analysis of the strengths and weaknesses of each option.

Mechanical Filtration

Mechanical filtration can be divided into two categories:

Strainers are usually just fine mesh screens which remove only the largest of impurities.

Sediment filters (or prefilters) remove smaller particles such as suspended dirt, sand, rust, and scale—in other words, turbidity. When enough of this particulate matter has accumulated, the filter is discarded. They greatly improve the clarity and appeal of the water. They also reduce the load on any more expensive filters downstream, extending their useful life. Sediment filters will not remove the smallest particles or biological pathogens.

Activated Carbon Filtration

Carbon adsorption is the most widely sold method for home water treatment because of its ability to improve water by removing disagreeable tastes and odors, including objectionable chlorine. Activated carbon filters are a very important piece of the purification process, although they are only one piece. Activated carbon effectively removes many chemicals and gases, and in some cases it can be effective against microorganisms. However, generally it will not affect total dissolved solids, hardness, or heavy metals. Only a few carbon filter systems have been certified for the removal of lead, asbestos, VOCs, cysts, and coliform. There are two types of carbon filter systems, each with advantages and disadvantages: granular activated carbon, and solid block carbon.

Activated carbon is created from a variety of carbon-based materials in a high-temperature process that creates a matrix of millions of microscopic pores and crevices. One pound of activated carbon provides anywhere from 60 to 150 acres of surface area. The pores trap microscopic particles and large organic molecules, while the activated surface areas cling to, or adsorb, small organic molecules.

The ability of an activated carbon filter to remove certain microorganisms and certain organic chemicals, especially pesticides, THMs (the chlorine by-product), trichloroethylene (TCE), and PCBs, depends upon several factors:

1. *The type of carbon and the amount used.*

2. *The design of the filter and the rate of water flow (contact time).*

3. *How long the filter has been in use.*

4. *The types of impurities the filter has previously removed.*

5. *Water conditions (turbidity, temperature, etc.)*

Granular Activated Carbon

Any granular activated carbon filter has three inherent problems. First, it can provide a base for the growth of bacteria. When the carbon is fresh, practically all organic impurities (not organic chemicals) and even some bacteria are removed. Accumulated impurities, though, can become food for bacteria, enabling them to multiply within the filter. A high concentration of bacteria is considered by some people to be a health hazard.

Second, chemical recontamination of granular activated carbon filters can occur in a similar way. If the filter is used beyond the point at which it becomes saturated with the organic impurities it has adsorbed, the trapped organics can release from the surface and recontaminate the water, with even higher concentrations of impurities than in the untreated water. This saturation point is impossible to predict.

Third, granular carbon filters are susceptible to channeling. Because the carbon grains are held (relatively) loosely in a bed, open paths can result from the buildup of impurities in the filter and rapid water movement under pressure through the unit. In this situation contact time between the carbon and the water is reduced, and filtration is less effective.

To maximize the effectiveness of a granular activated carbon filter and avoid the possibility of biological or chemical recontamination, it must be kept scrupulously clean. That generally means routine replacement of the filter element at six- to twelve-month intervals, depending on usage.

Solid Block Carbon

These are created by compressing very fine pulverized activated carbon with a binding medium and fusing them into a solid block. The intricate maze developed within the block ensures complete contact with organic impurities and, therefore, effective removal. Solid block carbon filters avoid the problems just discussed with granular carbon filters.

Block filters can be fabricated to have such a fine porous structure that they filter out coliform and associated disease bacteria, pathogenic cysts such as giardia, and lighter-weight volatile organic compounds such as THMs. Block filters eliminate the problem of channeling. Also, they are so dense that they do not allow the growth of bacteria within the filter.

Compressed carbon filters have two primary disadvantages compared with granular carbon filters. They have smaller capacity for a given size, because some of the adsorption surface is taken up by the inert binding agent, and because they tend to plug up with particulate matter. Thus, block filters may need to be replaced more frequently. In addition, block filters are substantially more expensive than granular carbon filters.

Limitations of Carbon Filters

To summarize, a properly designed carbon filter is capable of removing many toxic organic contaminants, but it will fall short of providing protection against a wide spectrum of impurities.

1. Carbon filters are not capable of removing excess total dissolved solids (TDS). To gloss over this deficiency, many manufacturers and sellers of these systems assert that minerals in drinking water are essential for good health. Such claims are debatable. However, there is scientific evidence suggesting that minerals associated with water hardness may have some preventive effect against cardiovascular disease.

2. Only a few systems have been certified for the removal of cysts, coliform, and other bacteria.

3. Carbon filters have no effect on harmful nitrates, or on high sodium or fluoride levels.

4. For both granular and block filters, the water must pass through the carbon slowly enough to ensure that complete contact is made. A system must have an appropriate balance between a useful flow rate and adequate contact time.

Reverse Osmosis (Ultrafiltration)

Reverse osmosis (RO) is a water purification technology that utilizes normal household water pressure to force water through a selective semipermeable membrane that separates contaminants from the water. Treated water emerges from the other side of the membrane, and the accumulated impurities left behind are washed away. Eventually, sediment builds up along the membrane and it needs to be replaced.

Reverse osmosis is highly effective in removing several impurities from water: total dissolved solids (TDS), turbidity, asbestos, lead and other heavy metals, radium, and many dissolved organics. RO is less effective against other substances. The process will remove some pesticides (chlorinated ones and organophosphates, but not others), and most heavier-weight VOCs. However, RO is not effective at removing lighter-weight VOCs, such as THMs (the chlorine by-product) and TCE (trichloroethylene), and certain pesticides. These compounds are either too small, too light, or of the wrong chemical structure to be screened out by an RO membrane.

Reverse osmosis and activated carbon filtration are complementary processes. Combining them results in the most effective treatment against the broadest range of water impurities and contaminants. Many RO systems incorporate both a pre-filter of some sort, and an activated carbon post-filter.

RO systems have two major drawbacks. First, they waste a large amount of water. They'll use anywhere from 3 to 9 gallons of water per gallon of purified water produced. This could be a problem in areas where conservation is a concern, and it may be slightly expensive if you're paying for municipal water. On the other hand, this wastewater can be recovered or redirected for purposes other than drinking, such as watering the garden, washing the car, etc. Second, reverse osmosis treats water slowly: it takes about three to four hours for a residential RO unit to produce one gallon of purified water. Treated water can be removed and stored for later use.

Other Treatment Processes

There are a couple of other water treatment processes that you should know about. Distillation is a process that creates clean water by evaporation and condensation. Distillation is effective against microorganisms, sediment, particulate matter, and heavy metals; it will not treat organic chemicals. Good distillers will have a carbon filter to remove organic chemicals. Ultraviolet (UV) systems use UV light to kill microorganisms. These systems can be highly effective against bacteria and other organisms; however, they may not be effective against giardia and other cysts, so any UV system you buy should also include a 0.5-micron filter. Other than some moderately expensive solar-powered distillers on the market, any distiller or UV unit will require a power source.

A Final Word on Water

Our best advice for those of you considering a drinking water treatment system consists of three simple points.

1. Test your water so you know precisely what impurities and contaminants you're dealing with.

2. Buy a system that is designed to treat the problems you have.

3. Carefully check the data sheets provided by the manufacturer to make sure that claims about what the system treats effectively have been verified by the National Sanitation Foundation (NSF) or another third party.

These actions will enable you to purchase a system that will most effectively meet your particular needs for safe drinking water.

WATER TREATMENT METHODS—GENERAL TREATMENT CAPABILITIES

	GAC	Carbon Block	RO	Distillation	UV
Bacteria	no	maybe	no	yes	yes
Cysts	no	maybe	yes	yes	yes w/0.5 mic filter
Asbestos	no	maybe	yes	yes	
Heavy metals	lead only	maybe	yes	yes	
Turbidity	no	maybe	yes	yes	
TDS	no	no	yes	yes	
Chlorine	yes	yes	yes w/tfc	yes	
Chlorine by-products (THMs)	no	maybe	maybe	yes	
Organic matter	maybe	maybe	maybe	maybe	
Heavy organic chemicals	some	maybe	maybe	maybe	
Light organics (VOCs)	some	maybe	maybe	maybe	
Pesticides and herbicides	some	some	maybe	maybe	
Radium	no	no	yes	yes	
Radon	no	no	yes	yes	

GAC = Granular Activated Carbon

RO = Reverse Osmosis

UV = Ultraviolet Light

Water Purification Products

HOUSEHOLD WATER FILTER COMPARISON CHART

Description	Item #	Filter Life Gallons	Operating Cost Per Gal.	Filter Type	Absolute Micron Size	CYSTS	VOC	CH	THM	L	M	B	Taste Odor	Indicator	Warranty (unit only)
Countertop UV Water Purifier	42-805	1300	0.08	sed./carbon/UV	1.0	x	x	x	x	x	x	x	x	x	lifetime
Undercounter UV Water Purifier	42-806	1300	0.08	sed./carbon/UV	1.0	x	x	x	x	x	x	x	x	x	lifetime
TerraFlo Countertop	42-592	400	0.12	carbon	0.5	x	x	x	x	x	x		x		lifetime
TerraFlo Under-Counter	42-593	400	0.12	carbon	0.5	x	x	x	x	x	x		x		lifetime
Chrome Countertop	42-885	400	0.1	carbon	1.0	x	x	x	x	x	x		x		90-day
PUR Ultimate Faucet Filter	42-298	100	0.2	carbon	1.0	x	x	x	x	x	x		x	x	1-year
PUR Mini Countertop Filter	42-865	200	0.15	carbon	1.0	x	x	x	x	x	x		x	x	1-year
Terra-Cotta Water Filter Cooler	42-692	2,500	0.02	ceramic/carbon	0.5			x					x		1-year
Hydrotech Countertop R.O. Water System	42-218	200	0.08	reverse osmosis*		x		x		x	x		x	x	2-year
U.S. Pure RO System	42-223	200	0.08	reverse osmosis*		x				x	x		x	x	2-year
Seagull Undercounter	42-650	1000	0.09	carbon	0.4	x	x	x	x	x	x		x		10-year
Seagull Countertop	42-651	1000	0.09	carbon	0.4	x	x	x	x	x	x		x	x	10-year
Seagull Traveler	42-652	1000	0.09	carbon	0.4	x	x	x	x	x	x		x	x	10-year
PUR Advantage Water Filter Pitcher	42-867	40	0.33	carbon	1.0		x	x	x	x	x		x	x	90-day

Life expectancy of any filter is dependent on the concentration of the contaminants to which the system is exposed. NSF/CA certificate is not required for portable filter units.
Common pollutants: CYSTS: giardia, cryptosporidium; VOC: pesticides, herbicides; CH: chlorine; THM: trihalomethane; L: lead; M: mercury; B: bacteria
*Carbon filter

PORTABLE WATER FILTER COMPARISON CHART

Description	Item #	Filter Life Gallons	Operating Cost Per Gal.	Filter Type	Absolute Micron Size	CYSTS	VOC	CH	THM	L	M	B	Taste Odor	Indicator	Warranty (unit only)
Solar Still	42-872	N/A	$.00	evaporation	N/A	x		x		x	x	x	x	N/A	1-year
Miniworks Pump	42-848	100	0.3	ceramic/carbon	0.3	x	x	x			x	x	x	caliper	3-year
First Need Filtration	42-653	100	0.3	ceramic/carbon	0.4	x	x	x				x	x	N/A	1-year
Katadyn Pocket	42-608	13000	.02	ceramic	0.2	x						x		caliper	lifetime
Katadyn Drip	42-842	39000	.006	ceramic	0.2	x						x		N/A	1-year
Exstream Bottle Filter	42-869	26	1.15	virustat/iodine/carbon		x						x		N/A	1-year
Ceramic Filter Pump	46-233	500	0.05	ceramic	0.5	x						x		caliper	2-year

Life expectancy of any filter is dependent on the concentration of the contaminants to which the system is exposed. NSF/CA certificate is not required for portable filter units.
Common pollutants: CYSTS: giardia, cryptosporidium; VOC: pesticides, herbicides; CH: chlorine; THM: trihalomethane; L: lead; M: mercury; B: bacteria

NTL Water Check With Pesticide Option

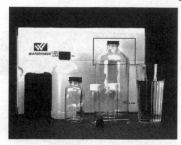

This is the best analysis of your water available in this price range. Your water will be analyzed for 73 items, plus 20 pesticides. If you have any questions about your water's integrity, this is the test to give you peace of mind. The kit comes with five water sample bottles, a blue gel refrigerant pack (to keep bacterial samples cool for accurate test results), and easy-to-follow sampling instructions. You'll receive back a two-page report showing 93 contaminant levels, together with explanations of which contaminants, if any, are above allowed values. You'll also receive a follow-up letter with a personalized explanation of your test results, plus knowledgeable, unbiased advice on what action you should take if your drinking water contains contaminants above EPA-allowed levels. NTL, located outside of Detroit, must receive the package within 30 hours of sampling in order to provide accurate results. If you live in Alaska or Hawaii, you will need express shipping. Check with your shipper. The test will check for the accompanying parameters and pesticides.

42-003 NTL Check & Pesticide Option $149⁰⁰

Affordable Water Testing Kit

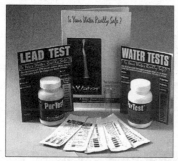

You've done the comprehensive mail-in lab tests, and taken treatment steps. Now you want to follow up, but who can afford $150 every 4 to 6 months? This EPA-based water test kit is ideal for ongoing maintenance, especially if you've had problems with coliform bacteria. It's easy, fast, accurate, convenient, and affordable, so you can take control of your water, making sure your family's water supply is safe and healthy. It reads for lead, chlorine, nitrates, and eight other potentially harmful contaminants. Results are easy to interpret at home, with toll-free phone support. Lab certified. Includes 10-page water information booklet. USA.

42-004 PurTest™ Home Water Analysis Kit $29⁹⁵

Household Point-of-Use Water Filters

Six-Step Ultraviolet Sterilization Leaves You With Clean, Healthy Water

You Choose Either Countertop or Undercounter Model

If local water supplies leave you feeling concerned about safety, this new ultraviolet water purifier utilizes a six-step system that should put your mind at ease.

Step One: Cleans the water: Large, heavy-duty sediment filters remove dirt, rust, scale, silt, and other suspended particles, including cysts such as cryptosporidium and giardia, to produce sparkling clean, clear water.

Step Two: Removes chemicals: A molded carbon block filter removes chlorine, chloroform-carbon tetrachloride-benzene, chlorinated solvents, herbicides, pesticides and other related volatile organic hydrocarbons.

Step Three: Removes heavy metals: The lead-sorbent matrix mixed with the carbon block filter irreversibly locks onto any lead, reducing the concentration well below EPA guidelines.

Step Four: Reduces fluorides and heavy metals: Activated tricalcium phosphate removes fluoride and metals such as aluminum and copper through adsorption and ion exchange processes.

Step Five: Removes chemicals, bad tastes, and odors: Granular activated carbon is used to adsorb chemical contaminants of light molecular weight such as chlorinated compounds and radon.

Step Six: Destroys microorganisms: After going through the above five steps, your water will then pass through an ultraviolet sterilization chamber to destroy disease-causing viruses, bacteria and parasites.

Both the countertop and undercounter models measure 11.5″ high x 15″ wide x 7.5″ deep, weigh 12 pounds, and have a flow rate of ½ to ¾ gallons per minute. They both deliver 1300 gallons of purified water at a cost of 5 cents per gallon. Replace the filter cartridge and ultraviolet lamp every 12 to 13 months. UL-listed. Mfr. warranty. USA.

◈**42-805 Countertop Ultraviolet**
Water Purifier $459⁰⁰

◈**42-806 Undercounter Water Purifier $459⁰⁰**

◈**42-808 Replacement Set (Filter & Lamp) $115⁰⁰**

◈ *Means shipped directly from manufacturer.*

TerraFlo Cleans Up Chemical And Organic Contaminants

The TerraTop™ CBLX provides superior protection against certain pesticides and herbicides detected in agriculture-intensive regions. This system is also the best choice for reducing trihalomethanes, a known carcinogen. It reduces lead, mercury, chlorine, turbidity, VOCs, giardia and cryptosporidium cysts. Included is a prepaid shipping label so you can send your used filter cartridge back to the manufacturer. Under Counter and Countertop systems are identical in size (11.75″ H x 5.75″ W x 5.25″ D) and use the same replacement cartridge. The Countertop model is outfitted with a built-in faucet and diverter valve for your existing fixture. The Under Counter unit requires 12.75″ clearance and a ⅜″ access hole in sink top. Both include hardware and tubing, plus detailed installation and maintenance instructions. The TerraTop CBLX Countertop system and TerraFil CBLX solid carbon block cartridge have been tested and approved by NSF; meets NSF Standards 42 and 53, and is also California certified. Filter life approx. 400 gallons. USA.

42-592 TerraFlo® Countertop $99⁹⁵

42-593 SubTerranean™ Under Counter $119⁰⁰

42-594 TerraFil™ Replacement Cartridge $44⁹⁵

Sleek Countertop Water Filter

Sleek and shiny, this chrome water filter looks right in kitchens, whether modern, minimalist, or retro, and it takes up only a little of your precious counter space. The carbon block filter reduces chlorine, lead, VOCs, mercury, and cysts such as cryptosporidium and giardia from your family's drinking water—improves the taste and odor, too. Meets NSF Standard #53 and #42, and is CA certified. The filter remains effective for up to 400 gallons (about 6–12 months). Initial installation is quick and easy—no plumbing experience required!—and replacing the cartridge is even simpler. 90-day mfr.'s warranty. 12″ H x 5.75″ D. USA.

42-885 Chrome Countertop $149⁰⁰

42-886 Replacement Cartridge $39⁹⁵

PUR Ultimate Faucet Mount: Economical, Effective, And Easy

Using this faucet-mount purifier, pure water costs only 20¢ per gallon! This PUR filter meets NSF standard #53 for lead and cysts, and #42 for chlorine, and is California certified for 5 VOCs. Swivel switches from water purifier to standard sink tap with spray fixture easily. Approximate filter life: 100 gallons (about 3 months). Light indicates when filter needs replacement. 6.19″ L x 3″ W x 8.19″ D. One-year mfr.'s warranty. USA.

42-298 PUR Ultimate Faucet Filter $59⁹⁵

42-299 PUR Ultimate Replacement Filters (2) $45⁰⁰

Mini Faucet Filter

No counter space? You can still have filtered water! Great for bathrooms, RVs, or any small space, this smart, compact system can be mounted on the wall, or set vertically or horizontally on the counter. With its diverter valve to prolong filter life and its filter monitor gauge, this is effective PUR technology. Approximate filter life: 200 gallons (4–6 months). California certified and meets NSF standards #42 and #53. 9″ L x 4″ W x 3.5″ D. USA.

42-865 PUR Mini Water Filter $69⁹⁵

42-866 Replacement Filter $29⁹⁵

www.realgoods.com

Clean, Cool Water in Two Beautiful Terra-Cotta Packages

If you and your family are big water drinkers, you need to have lots of it ready to serve, nice and cool. But who has room in the refrigerator for a large and cumbersome cooler? Not many do. So the alternative is to leave that large and cumbersome cooler out on a kitchen countertop, which is fine except they can sometimes be, well, kind of ugly. This hand-made Terra-Cotta Cooler is the solution. Not only is it lovely to look at, cooler-wise, but it naturally keeps about a gallon of water chilled and ready to drink, which is nice with warm weather approaching. A Real Goods best-seller since it first appeared, our customers swear this is the best thing since sliced bread. They say it removes bad taste and odor, and really, really keeps the water cool—but we knew that anyway. It's the natural evaporation process through the terra-cotta that actually keeps the water constantly cooled—without the need for refrigeration. It also comes with an activated carbon and ceramic filtration system built into the center of the cooler—that's where the good taste comes from. Filters six gallons a day. Filter needs replacing about once a year, or 2,500 gallons. It measures 17″ H x 10″ Dia., with a 1-gallon capacity. The Water Dispenser Cooler, measuring 12″ H x 10″ Dia., allows you to put your own 2-, 3-, or 5-gallon water jug on top of the cooler/base—the same natural evaporation process works here, dispensing chilled water, electricity-free. Brazil.

◈**42-692 Terra-Cotta Water Filter Cooler $109⁰⁰**

◈**42-693 Water Dispenser Cooler $59⁹⁵**

◈**42-685 Replacement Filter $44⁹⁵**

Hydrotech Countertop R.O. Drinking Water System

This state-of-the-art Reverse Osmosis system features the patented SMAR-TAP water quality monitor that provides instant performance verification at the touch of a button. A split-second power pulse compares the feed water vs. the product water to accurately monitor the membrane performance. Power is supplied by a 9-volt battery which is included with the system.

No installation is required, it simply snaps onto the faucet. The system operates with feed pressures from 35 to 100 psi, in temperatures from 40° to 100° F, and with pH values from 3 to 11. Maximum TDS is 2000 ppm.

The reservoir has a two-gallon capacity with a water-level viewing window. Output varies with feed pressure, temperature, and total dissolved solids, but averages from 2 to a maximum of 25 gallons per day. Output continues so long as feed water is on. When reservoir is full, it overflows into brinewater drain (no countertop floods).

The advanced four-stage treatment process allows operation with either chlorinated or unchlorinated water supplies. There are both pre-sediment and pre-carbon filters as well as a post-carbon polishing filter. All filters are simple to replace and the cabinet is easy to clean.

As with all R.O. units, there is a continual trickle of water that flushes off the TFC membrane. This brinewater is intended to drain into the sink, but can be rerouted and used for watering of gardens or landscaping. The increased total dissolved solids in "brinewater" is minuscule. The diversion rate on this R.O. unit is user-adjustable depending on the source purity, but should be in the range of 3 to 5 gallons of brinewater per 1 gallon of purified water.

Unit dimensions are 13″H x 10″D x 16.5″W, weight is 10 lb. Warranty is two years on the system, 12 months prorated on the TFC membrane. USA.

42-218 Hydrotech Countertop R.O. Unit $399⁰⁰

42-219 Replacement TFC Membrane $109⁰⁰

42-220 Sediment prefilter $14⁹⁵

42-221 Carbon prefilter $14⁹⁵

42-222 Carbon postfilter $14⁹⁵

U.S. Pure Reverse Osmosis With Water Quality Monitor

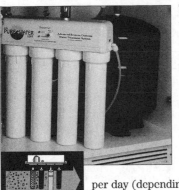

This advanced four-stage filtration system ends guesswork with a push-button water quality monitor that tells you exactly when the membrane needs changing. This is a top-quality system, utilizing the best in filtration technology to purify your family's drinking water. Flow rate is just under five gallons per day (depending on water pressure and temperature). Tested and certified to ANSI/NSF® 58 for specific contaminant claims and is California certified. Includes carbon pre- and post-filters, a sediment pre-filter, and membrane filter. Carbon and sediment filters last 6 to 12 months; maximum recommended service life for membrane filter is 36 months. The two-gallon storage tank and filters install under the sink, with a stainless steel lead-free faucet on the sink top for convenient access (system install time is about one hour). The reliable manufacturer insists on independent testing, top-quality materials, and outstanding performance. Two-year limited manufacturer's warranty. USA.

◈ **42-223 U.S. Pure Reverse Osmosis System $399⁰⁰**

◈ **42-224 Replacement Membrane $109⁰⁰**

◈ **42-225 Sediment Filter $12⁹⁵**

◈ **42-227 Carbon Pre/Post Filter (requires 2) $15⁹⁵**

◈ *Means shipped from manufacturer.*

Seagull Carbon Filters

To expand your options for the purest, safest water, we are introducing the Seagull IV systems. Seagull IV (X-1) products are among the very few qualified for continuous use aboard international airliners. They are manufactured in the USA from stainless steel and other high-grade components. Water is filtered through a unique microfine structured matrix, with a high flow rate of 1 gpm at standard pressures (30–40 psi). The replaceable cartridge should last anywhere from 9 to 15 months with ordinary use.

Here are some of Seagull IV's unique features:

1. Ultrafine filtration down to 0.1 micron—small enough to remove all visible particles, as well as giardia, cysts, harmful bacteria, and larger parasites.*

2. Molecular sieving and "broad spectrum" adsorption mechanisms remove chlorine and many organic chemicals such as pesticides, herbicides, solvents, lead, and foul tastes and bad odors.*

3. Electrokinetic attraction removes colloids and other particles even smaller than those removed by microfine filtration.*

*See performance data sheet for specific contaminants. Certain states may prohibit health claims as a matter of local or state law. Such claims not in compliance are hereby withdrawn. Please check with appropriate officials as necessary.

Seagull Undercounter

Undercounter unit, with revolutionary ceramic disc faucet for countertop and accessories for easy installation.

42-650 Seagull Undercounter Filter $389⁰⁰

42-657 Seagull Replacement Cartridge $90⁰⁰

Seagull Countertop

Countertop unit, faucet-attaching, diverter unit for apartments, cabins, etc.

42-651 Seagull Countertop Filter $359⁰⁰

42-657 Seagull Replacement Cartridge $90⁰⁰

Seagull Traveler

Same as countertop with the addition of a high-capacity hand-pump for use on non-pressurized systems.

42-652 Seagull Traveler Filter $499⁰⁰

42-657 Seagull Replacement Cartridge $90⁰⁰

Get Fresh

PUR Water: Reduces Pesticides, Herbicides, and Chemicals Linked to Cancer

PUR is an easy, convenient way to get fresh, pure water without a whole-house system. PUR conveniently eliminates chemical-based carcinogens from two gallons daily, and the .5-gallon pitcher fits neatly in a refrigerator door. Approximate filter life: 40 gallons (about 2 months); an indicator lets you know when the filter needs to be changed. Meets NSF standards #42 and #53, and is California certified. USA.

42-867 PUR Advantage Water Filter Pitcher $29⁹⁵

42-868 Replacement Filter Inserts (3) $35⁰⁰

Inline Sediment Filter

Sometimes your water isn't dirty enough to mess with fancy and expensive filtration systems and all you need is a simple filter. Our inline sediment filters accept standard 10-inch filters with 1-inch center holes. They are designed for cold-water lines only and meet National Sanitation Foundation (NSF) standards. Easily installed on any new or existing cold-water line (don't forget the shutoff valve), they feature a sump head of fatigue-resistant Celcon plastic. This head is equipped with a manually operated pressure release button to relieve internal pressure and simplify cartridge replacement. The housing is rated for 125 psi maximum and 100°F. It comes with a ¾-inch FIPT inlet and outlet and measures 14 inches high by 4 9/16 inches in diameter. It accepts a 10-inch cartridge and comes with a 5-micron high-density fiber cartridge.

41-137 Inline Sediment Filter $74⁹⁵

42-632 Rust & Dirt Cartridge $12⁹⁵

Shower and Bath Filters

Our Best Shower Chlorine Filter, More Efficient And Less Expensive

Chlorine in water is extremely harmful to hair, skin, eyes, nose and mouth membranes, and lungs. We actually can take in more chlorine from one fifteen-minute shower than from drinking eight glasses of the same water in one day. Chlorine plays havoc with our skin and hair, chemically bonding with the protein in our bodies. It makes hair brittle and dry and can make sensitive skin dry, flaky, and itchy. Why live with chlorine gas in your shower water, when the Rainshower CQ™ effectively removes it. The easy-to-install ABS plastic unit is only five inches long, and includes a low-flow shower head. Increase water savings with the off/on flow valve, so you can turn the water off to suds up, then on again to rinse without readjusting the hot/cold mix. The filter removes 90% or more of residual-free chlorine from shower water and aids in the control of fungi and mildew. Filter is good for 15,000 gallons. Savings in water and energy costs can pay for this product in less than six months for the average family. USA.

42-704 Rainshower CQ™ Filter $49⁹⁵

42-705 Replacement CQ™ Filter $19⁹⁵

Shower Filters Out Chlorine And Saves Headroom

Chlorine in shower water causes dry skin, brittle hair, and soap scum buildup. Install one of these filters to dramatically reduce chlorine and sediment, eliminate odors, and improve water clarity. The cartridge works for 30,000 gallons, up to 18 months for a family of four. Most shower filters installed above the showerhead place the showerhead too low. But this efficient backflow design doesn't protrude into the shower stall, so you don't sacrifice showerhead height. Choose the premium chrome model including a deluxe massage showerhead; or use the standard ABS white plastic model (not shown) with your own showerhead. Replacement filter fits either model. (Filter measures 3.25 D x 4" L)) USA.

42-882 Chrome Spring Shower Filter w/Massage Head $59⁹⁵

42-883 White Spring Shower Filter $39⁹⁵

42-884 Replacement Cartridge $19⁹⁵

Chlorine Protection And Shower Massage, Too

Gotta have that shower massage? You can still enjoy protection from free chlorine up to 99%. These 10″ long Handheld Shower Filters combine effective chlorine reduction, the convenience of a handheld shower, and an adjustable massaging showerhead. Easy to install, they come with mounting bracket, 72″ reinforced nylon hose, and a six-month filter cartridge. USA.

42-887 Chrome Handheld Shower Filter $49⁹⁵

42-708 White Handheld Shower Filter $39⁹⁵

42-709 Replacement Filters (2) $24⁹⁵

At Last! Filter Chlorine From Your Tub Water

While chlorine is necessary in municipal water, it does you no good in the bath, where it's readily absorbed through your skin. This Bath Ball easily hangs on your tub's faucet by its sturdy strap, and as the water runs through its KDF® copper-zinc filtering media it removes chlorine as well as hard-water minerals. Given typical water quality and family use rates, the Bath Ball should last about a year. How will you know it's time to replace? Because you'll notice that your skin and hair no longer have that silky, supple, smooth softness you'll be so accustomed to by then. Not a bath toy. About 3.5″ D. USA.

42-881 Bath Ball $15⁹⁵

Portable Water Filters

Purify Water Anywhere

This ingenious design allows the lightweight (less than 4 pounds) portable unit to operate on any water surface, from salt water lagoons to lakes, ponds, and streams. This inflatable unit allows sunlight in through the clear, conical condenser. When the light strikes the wet black surface, it changes to heat and evaporates water molecules, leaving the impurities behind. The water vapor rises and condenses on the relatively cool surface of the inside of the conical condenser. The droplets of water will run down the condenser to a collection trough that runs around the inside of the base of the unit for "harvesting." The name is somewhat of a misnomer, because the unit can be even more productive at night with relatively warm source-water combined with cool night air. Perfect for boaters and outdoor adventurers or just as emergency backup. When deflated, the Solar Still fits into its carrying case, which is only 6 inches in diameter and 21 inches long. One-year mfr.'s warranty. USA.

42-872 Solar Still $149⁰⁰

Lightweight, Economical Personal Filter System

Ideal for back country camping and adventure travel. The MiniWorks pocket-size, lever-style pump filter (16 oz.) has a rugged polyurethane housing that encases a ceramic filter with an activated carbon core. The Marathon Ceramic filter exceeds the EPA's Purifier Standards for bacteria and protozoa removal plus reduces some chemical contaminants and bad tastes and odors. Approximately 100-gallon filter life (with periodic cleaning). Use your own container or purchase the collapsible bag and screw the filter onto the opening for a durable water carry bag or shower. The collapsible 10-liter water bag (8.5 oz) with shower kit is sold separately. The 1000-denier cordura nylon bag (with food-grade polyurethane liner) has a threaded bag opening and comes with a shower kit that includes a 3-foot hose, spigot cap, and a shower nozzle. The black color will absorb heat for a nice, warm shower and the multiple grommets laced with webbing make for easy hanging and carrying. The most important 1.75 pounds you'll ever carry. USA.

42-848 MiniWorks Pump Filter $59⁹⁵

42-849 MiniWorks Replacement Filter $29⁹⁵

42-850 Dromedary 10-Liter Bag (W/Shower) $49⁹⁵

First Need

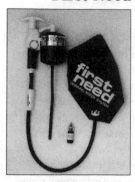

The First Need is the best-selling portable water filtration system in America. Will remove tastes, odors, and with 0.1 micron filtration, will safely remove waterborne pests such as bacteria, giardia, viruses, or cryptosporidium. Meets EPA Guide Standard. Sturdy plastic pump, tubes, and filter fit into included nylon tote bag. Flow is 1–2 pints/min., depending on vigor of pumper. Weighs less than 16 ounces. Great for backpacking, camping, travel, or your emergency preparedness kit. Shown with the included Sediment Prefilter, which greatly increases filter life if changed regularly. Rated for approximately 100 gallons per cartridge.

42-653 First Need Filtration System $79⁹⁵

42-647 First Need Sediment Prefilter $9⁹⁵

42-658 First Need Repl. Filter Cartridge $34⁹⁵

Katadyn Pocket Filter

The Katadyn Pocket Filter is standard issue with the International Red Cross and the armed forces of many nations. Essential for any survival kit, it has been manufactured in Switzerland for over half a century. These filters are of the highest quality imaginable, reminiscent of fine Swiss watches. The Katadyn system uses an extremely fine 0.2-micron ceramic filter, blocking pathological organisms from entering your drinking water. A self-contained and very easy-to-use filter about the size of a 2-cell flashlight (10″ x 2″), produces a quart of ultrapure drinking water in 90 seconds with the simple built-in hand pump. It weighs only 23 oz. and comes with its own travel case and special brush for cleaning the ceramic filter. The replaceable ceramic filter can be cleaned up to 400 times, lasting many years with average use. An indispensable tool for campers, backpackers, fishermen, mountaineers, river runners, globetrotters, missionaries, geologists, and those who work in disaster areas. Switzerland.

42-608 Katadyn Pocket Filter $249⁰⁰

42-609 Replacement Filter $165⁰⁰

The Katadyn Drip Filter

With no moving parts to break down, superior filtration, and a phenomenal filter life, there is simply no safer choice for potentially pathogen-contaminated water. There are no better filters than Katadyn for removing bacteria, parasites, and cysts. Three 0.2-micron ceramic filters process one gallon per hour. Clean filters by brushing the surface. Ideal for remote homes, RV, campsite, and home emergency use. Food-grade plastic canisters stack to 11″ Dia. x 25″ H. Weighs 10 lb. One-year mfr.'s warranty. Switzerland.

42-842 Katadyn Drip Filter $289⁰⁰

42-843 Replacement Filter $79⁰⁰ each

End Your Worries About the Water

Scoop water from any supply, squeeze, then sip. Registered with the Environmental Protection Agency, this personal water purification system has been shown to deliver on its claims. Using a three-stage purification system, it eliminates pathogenic cysts, bacteria and viruses from water, so you can travel anywhere without worry. The 26-oz. sports flask fits into any standard bottle holder and takes up minimal space in travel packs and suitcases. Each replaceable cartridge delivers 26 gallons of filtered water. USA and Taiwan.

02-0281 Extream Personal Water Filter $60⁰⁰

Healthy, Clean Drinking Water

Clean, healthy drinking water. Will you be able to access it when a natural disaster strikes? Often the first warning issued by local municipalities is concerning the safety of the water supply. With this Ceramic Water Filter you can relax because your water will be filtered to an absolute 0.5-micron level. It meets EPA standards for elimination of bacteria and protozoa, including giardia, cryptosporidium, E. coli, vibrio cholera, shigella, salmonella typhi, and klebsiella terigina. It will supply you with approximately 500 gallons of filtered water, making it perfect for camping trips too. For best life of this product, use the cleanest source possible and rinse the foam sock pre-filter often. With its easy-to-use and simply designed pump, you can filter one ounce of water with each pull of the pump. And at only 8 ounces, it's extremely lightweight. Includes clip for attachment to water carrier. Great Britain.

46-233 Ceramic Filter Pump $29⁹⁵

Pool and Spa Water Purification

Solar-Powered Pool Purifier Slashes Chlorine Use

The Floatron kills algae, bacteria, and is a safer and economical alternative to the chemical marinade of most pools. It is solid-state, with no moving parts, no batteries, and is portable and cost-effective. The electrote element delivers copper and silver ions into the water, which kill algae and bacteria cells on contact. The electrode element is gradually consumed, and needs replacement every one to three seasons, depending on pool size. The element is available directly from the manufacturer for about $35, and takes a couple of minutes to replace. No more chlorine allergies, red eyes, discolored hair, or bleached bathing suits! Floatron will typically reduce chemical expenses by an average of 80%. Comes complete including a 24-month manufacturer's warranty. Effective for pools up to approximately 40,000 gallons. USA.

42-801 Floatron $299⁰⁰

Keep Your Spa Water Sparkling Clean With This Porous Metal Foam Filter

One of the key reasons for having a spa or Jacuzzi is to relax. But what about the quality of the water? Can you truly unwind when you're worried about bacteria? Place this small metal filter in the skimmer or filter hole of your spa and in one pass it will kill 99.9% of the bacteria in your spa's water, including legionella and E. coli. Unfortunately, every spa and Jacuzzi requires that an oxidizer like chlorine or bromine be added to the water to burn off unwanted things like hair oil and sweat. The good news is that the Metal Foam Purifier is so effective it reduces oxidizer usage up to 90%—no more chlorine soaked skin! With the proper care, one two-inch copper and silver mesh "foam" disc will last up to six months. Includes disc-holder and two metal foam discs. USA.

42-804 Metal Foam Purifier $49⁹⁵

Ozone Purifiers

Clearwater Tech makes several different ozone generators for different applications. The ozone is manufactured in the generator by intaking air, which is composed of 20% oxygen (O_2), and bombarding it with a specific light frequency. This frequency causes the oxygen molecules to disassociate and reassemble as ozone (O_3). Ozone is the most powerful oxidizing agent available. When ozone is drawn into the spa or pool water, it will kill the bacteria, virus, or mold spores that come in contact with it. Ozone has a life expectancy of approximately 20 minutes. Several short cycles through the day are recommended. Each UV lamp is rated at 9,000 hours. Running the S-1200, the PR-1300, or the CS-1400 for four one-hour intervals per day equates to five-years of operation, with an energy-consumption worth of approximately 85 cents per month. USA.

S-1200 Ozone Purifier

The S-1200 features a polished stainless-steel reaction chamber, thermally protected self-starting ballast, weather-tightness (outdoor approved housing), and a 17-inch specially designed high-output ultraviolet lamp. The S-1200 is wired to the pump circuit to be on when your pump is on. The unit has convenient mounting brackets, comes with all necessary fittings, and is easy to install. If the spa is run daily, as recommended by most spa manufacturers, four separate 1-hour cycles in a 24-hour period will generate a sufficient amount of ozone to keep the spa free of biological contamination. The S-1200 will purify up to 1,000 gallons for spas and 2,000 gallons for pools. USA.

42-811 S-1200 Ozone Purifier, 110V $305⁰⁰

PR-1300 Ozone Purifier

This system comes with its own 24-hour timer and compressor. This means that the PR-1300 can run independently of the circulation pump in your spa or pool. It comes with all necessary tubing, check valve, and fittings for installation. There is also a diffuser stone that can be attached to the ozone delivery line and submerged into any vessel of water.

You can treat yourself to a lavish chlorine-free bath by using the system in your bathroom, or anyone else's bathroom since the system is totally portable. You can treat your friend's spa before you use it with this portable water treatment system. Treating your water with ozone instead of chlorine when you are dealing with bacteria, iron, or other problems in your storage tank is the solution for those who have a spring or catchment system. The PR-1300 features a GFCI (Ground Fault Circuit Interruption) circuit breaker. This gives state-of-the-art electrical protection. It has a weather-tight cabinet, coated with a baked enamel finish for years of corrosion-free service. The UV lamp is encased in a polished stainless-steel reaction chamber, and can be replaced by the homeowner in minutes. Average lamp life is 9000 hours. Power consumption is approximately 80–90 watts. The units are rated up to 1,000 gallons for spas and 2,000 gallons for pools. The size is 20″ x 9″ x 4″. Maximum pressure is 20 psi. USA.

42-821 PR-1300 Ozone Purifier, 110V $535⁰⁰

CS-1400 Ozone Purifier

The CS-1400 is a UV ozone generator designed to be used on swimming pools up to 15,000 gallons or spas up to 2,500 gallons. You can double the output and capacity by adding another CS-1400. It has a polished stainless-steel reaction chamber, a thermally protected self-starting ballast, a weather-tight enclosure (outdoor approved), and a 29-inch specially designed high-output ultraviolet lamp. With no moving parts, the CS-1400 requires virtually no maintenance and will provide years of uninterrupted service. Wire the CS-1400 to the pump circuit, so each will work together to keep your pool or spa perfectly clear. It will also work with a compressor (not included). The CS-1400 requires a venturi injector to suit your pumping system. Choose either 1.5″ or 2″, and single-speed or two-speed. USA.

42-831 CS-1400 Ozone Purifier, 110V $635⁰⁰

42-832 CS-1400 Ozone Purifier, 12V $579⁰⁰

42-837 Injector 1.5″ Single-Speed $105⁰⁰

42-838 Injector 1.5″ Two-Speed $105⁰⁰

42-836 Injector 2″ Single-Speed $125⁰⁰

42-839 Injector 2″ Two-Speed $125⁰⁰

Air Purification

Most of us are familiar with the idea that the water we drink at home might be polluted or contaminated. But did you realize that the air inside your home (or place of work) could potentially be even more hazardous to your health than the water?

The Problem with Indoor Air

Problems of indoor air quality have begun to attract attention only recently. For one thing, the sources of indoor air pollution are more mundane and subtle than the sources of outdoor air and water pollution. In addition, before highly insulated, tightly constructed buildings became common in the 1970s, most American homes were drafty enough that the build-up of harmful gases, particles, and biological irritants indoors was very unlikely. However, the combination of tighter, well-insulated buildings that allow minimal infiltration of outside air, and the continual increase of synthetic products that we bring into our homes, has added up to a possible public health problem of major proportions.

The EPA has concluded that many of us receive greater exposure to pollutants indoors than outdoors.

The preceding remark is qualified because so little is known with relative certainty about the unhealthful consequences of constant exposure to polluted indoor air. In any case, the EPA has concluded that many of us receive greater exposure to pollutants indoors than outdoors. Recent EPA studies found concentrations of a dozen common organic pollutants to be two to five times higher inside homes than outside, regardless of whether they were located in rural areas or industrialized areas. Furthermore, people who spend more of their time indoors and are thus exposed for longer periods to airborne contaminants are often the very people considered most susceptible to poor indoor air quality: very young children, the elderly, and the chronically ill.

Sources of Indoor Air Pollution

You can probably identify some of the sources of indoor air pollution: new carpeting, adhesives, cigarette smoke, and maybe you've heard about dust mites. But many things that contribute potentially irritating or harmful substances to indoor air may not be obvious.

Indoor air contaminants typically fall into one or more of the following categories:

- *Combustion products*
- *Volatile chemicals and mixtures*
- *Respirable particulates*
- *Respiratory products*
- *Biological agents*
- *Radionuclides (radon and its by-products)*
- *Odors*

What kinds of substances are we talking about? Let's start with chemicals and synthetics. Building materials and interior furnishings emit a broad array of gaseous volatile organic compounds (VOCs), especially when they are new. These materials include adhesives, carpeting, fabrics, vinyl floor tiles, some ceiling tiles, upholstery, vinyl wallpaper, particle board, drapery, caulking compounds, paints and stains, and solvents. In addition to producing emissions, many of these materials also act as "sponges", absorbing VOCs and other gases that can then be reintroduced into the air when the "sponge" is saturated. These kinds of building and furnishing materials can also be the source of respirable particulates, such as asbestos, fiberglass, and dusts.

Appliances, office equipment, and office supplies are another major source of VOCs

Building materials and interior furnishings emit a broad array of gaseous volatile organic compounds (VOCs), especially when they are new.

and particulates. Among the culprits are leaky or unvented heating and cooking appliances; computers and visual display terminals; laser printers, copiers, and other devices that use chemical supplies; common items such as preprinted paper forms, rubber cement, and typewriter correction fluid; and routine cleaning and maintenance supplies, including carpet shampoos, detergents, floor waxes, furniture polishes, and room deodorizers.

Your own routine "cleaning and maintenance supplies" are not above suspicion, either. Many of the personal care and home cleaning products we use emit various chemicals (especially from the components that create fragrance), some of which may contribute to poor indoor air quality. Clothes returning from the dry cleaner may retain solvent residues, and studies have shown that people do breathe low levels of these fumes when they wear dry-cleaned clothing. Even worse, the pesticides we use to control roaches, termites, ants, fleas, wasps, and other insects are by definition toxic. Sometimes these poisons are tracked indoors on shoes and clothing. Do you have any of this stuff stored underneath the sink? One EPA study has suggested that up to 80% of most people's exposure to airborne pesticides occurs indoors.

Other human activities can also have a dramatic impact on indoor air quality. Tobacco smoke from cigarettes, cigars, and pipes is obviously an unhealthy pollutant. Residential heating and cooking activities may introduce carbon monoxide, carbon dioxide, nitrogen oxide and dioxide, and sulfur dioxide into the air. Candles and wood fires contribute CO and CO2 as well as various particulates.

Many biological agents contribute to indoor pollution: we are surrounded by a sea of microorganisms. Bacteria, insects, molds, fungi, protozoans, viruses, plants, and pets generate a number of substances that contaminate the air. These range from whole organisms themselves to feces, dust, skin particles, pollens, spores, and even some toxins. Humans contribute too: we exhale microbes, and our sloughed skin is the primary source of food for the infamous dust mites. Furthermore, certain indoor environments, such as ducts and vents, or humid areas such as bathrooms and basements, provide ideal conditions for microbial growth.

Radon buildup can be a problem in certain parts of the U.S. where it is present in rocks and soil. A radioactive gas, radon has been implicated as a cause of lung cancer.

Radon (and its decay products, known as "radon daughters") is one type of indoor pollutant that has received widespread notice. As noted above, indoor radon buildup can be a problem in certain parts of the U.S. where it is present in rocks and soil. A radioactive gas, radon has been implicated as a cause of lung cancer. Simple and inexpensive radon test kits are available at hardware stores. Contact your regional EPA office—you can look them up in the phone book—for more information about effective radon remediation strategies.

Odors are always present. Some are pleasant, others are unpleasant; some signal the presence of irritating or harmful substances, others do not. The perception of odors is highly subjective and varies from person to person. Most odors are harmless, more of a threat to state of mind than to health, but eliminating them from an indoor environment may be a high priority to keep people content and productive.

Health Effects of Indoor Air Pollution

Before you become too alarmed about the dangers to your health that may be floating around in the air you breathe at home, remember the following proviso. As with most environmental pollutants, adverse health effects depend upon both the dose received and the duration of exposure. The combination of these factors, plus an individual's particular sensitivity to specific substances, will determine the potential health effect of the contaminant on that individual.

With that in mind, the health effects of indoor air pollution can be short-term or long-term, and can range from mildly irritating to severe, including respiratory illness, heart disease, and cancer. Most often the effects of such pollutants are acute, meaning they occur only in the presence of the substance; but in some cases, and perhaps in many cases, the effects may be cumulative over time and may result in a chronic condition or illness.

The clinical effects of indoor air pollution can take many forms. The most common clinical signs include eye irritation, sneezing or coughing, asthma attacks, ear-nose-throat infections, allergies, and migraine. Some bacteria and molds can cause diseases such as flu. More seriously, prolonged exposure to radon, tobacco smoke, and other carcinogens in the air can cause cancer. Most volatile organic compounds (VOCs) can be respiratory irritants, and many are toxic, although little applicable toxicity data exists about low levels of indoor VOCs. One VOC, formaldehyde, is now considered a probable human carcinogen. Needless to say, the best way to protect your health is to avoid or reduce your exposure to indoor air contaminants.

Nearly all observers agree that more research is needed to better understand which health effects occur after exposure to which indoor air pollutants, at what levels and for how long.

As with most environmental pollutants, adverse health effects depend upon both the dose received and the duration of exposure.

Improving Indoor Air Quality

There are three basic strategies: source control, ventilation, and air cleaning.

Source control is the simplest, cheapest, and most obvious strategy for improving indoor air quality. Remove sources of air pollution from your home. Store paint, pesticides, and the like in a garage or other outbuilding. Screen the products you bring into your home, and buy personal care, cleaning, and office products that contain fewer synthetic volatile compounds or other suspicious and potentially irritating substances.

Other sources can be modified to reduce their emissions. Pipes, ducts, and surfaces can be enclosed or sealed off. Leaky or inefficient appliances, such as a gas stove, can be fixed.

Regular house cleaning is effective: sweeping, vacuuming and dusting can keep a broad spectrum of dust, insects, and allergens under control. You should be conscientious about cleaning out ducts, and replacing furnace filters on a regular basis. Also, reduce excess moisture and humidity, which encourage the growth of microorganisms, by emptying evaporation trays of your refrigerator and dehumidifier.

Ventilation is another obvious and effective way to reduce indoor concentrations of air pollutants. It can be as simple as opening windows and doors, operating a fan or air conditioner, and making sure your attic and crawl spaces are properly ventilated. Fans in the kitchen or bathroom that exhaust outdoors remove moisture and contaminants directly from those rooms. It is especially important to provide adequate ventilation during short-term activities that generate high levels of pollutants, such as painting, sanding, or cooking. Energy-efficient air-to-air heat exchangers that increase the flow of outdoor air into a home without undesirable infiltration have been available for several years. You should especially consider installing such a system if you live in a tight, weatherized house.

Regular house cleaning is effective: sweeping, vacuuming and dusting can keep a broad spectrum of dust, insects, and allergens under control.

Unfortunately, indoor air cleaning or purification is more complicated than purifying water. Not many technologies are available for household use, most of them have limited effectiveness, and some of the more promising methods are highly controversial. Air filtration is problematic because it's only really effective if you draw all of the indoor air through the filter. Even then, if there are no leaks, such filters only reduce particulate matter. Any furnace or whole-house air conditioning system (that's air conditioning, not air conditioning) has a filter. Some, such as electrostatic filters or HEPA (High Efficiency Particulate Air) filters, can significantly reduce levels of even relatively small particulates. Portable devices incorporating these technologies have also been developed, which can be used to clean the air in one room. Follow the manufacturer's instructions about how frequently to change the filter on your furnace or stand-alone air cleaner. Air filters are not designed to deal with volatile organic compounds (VOCs) or other gaseous pollutants.

Ionization is a second method of cleaning particulates out of the air. A point-source device generates a stream of charged ions (via radio frequencies or some other method), which disperse and meet up with oppositely charged particles. As tiny particles clump together, they eventually become heavy enough to settle out of the air, and you won't breathe them in. (Ionization is not effective against VOCs.)

What About Ozone?

A somewhat controversial method of air purification uses ozone, in combination with ionization. Ozone (O_3) is a highly reactive and corrosive form of oxygen. You know about the ozone layer in the upper atmosphere, which prevents a lot of ultraviolet light from reaching the earth's surface, where it can do damage to living things. Ozone also exists naturally in small concentrations at ground level, where it has various effects. Ozone is a component of smog, produced by the reaction of sunlight upon auto exhaust and industrial pollution. But ozone is also produced by lightning, and it is partially responsible for that fresh, clean smell you notice outside after a thunderstorm. Some studies suggest that slightly greater than natural outdoor levels of ozone can reduce respiratory, allergy, and headache problems, and possibly even enhance one's ability to concentrate. However, there is no doubt among scientists that at slightly elevated levels in the air, ozone can be dangerous and unhealthful to people. In concentrations that exceed safe levels as determined by the EPA, ozone is an air pollutant.

Manufacturers of air cleaners that generate ozone make strong claims about their products. They claim that these devices kill biological agents—such as molds and bacteria—that contribute to indoor air pollution; that the ozone penetrates into drapes and furniture, and filters down to the floor, into carpets and crevices, to oxidize and burn up absorbed contaminants and those that fall out of the air; and that the ozone breaks down potentially harmful or irritating gases, including many VOCs, into carbon dioxide, oxygen, water vapor, and other harmless substances. They claim that their machines accomplish all this with levels of ozone that are safe, and that ozone production can be monitored, either automatically or manually, to ensure that ozone concentration does not become excessive.

The EPA strongly disputes many of these claims and considers them misleading. The agency's position is that "'available' scientific evidence shows that at concentrations that do not exceed public health standards, ozone has little potential to remove indoor air contaminants [and] does not effectively remove viruses, bacteria, mold, or other biological pollutants." While the EPA agrees that some evidence shows that ozone can combat VOCs, it points out that these reactions do produce potentially irritating or harmful by-products. In addition, the EPA is skeptical of manufacturer's claims that the amount of ozone produced by such devices can be effectively controlled so as not to exceed safe levels. The bottom line for the EPA is that "no" agency of the federal government has approved ozone generation devices for use in occupied spaces. The same chemical properties that allow high concentrations of ozone to react with organic material outside the body give it the ability to react with similar organic material that makes up the body, and potentially cause harmful health consequences. When inhaled, ozone will damage the lungs. Relatively low amounts can cause chest pain, coughing, shortness of breath, and throat irritation." (This information was obtained in December 1998 from a publication posted on the EPA website, at www.epa.gov/iaq/pubs/ozonegen.html.

The ozone issue is still being debated and at present, Gaiam Real Goods has limited the ozone purifiers in its product mix.

Unfortunately, indoor air cleaning or purification is more complicated than purifying water.

There is no doubt among scientists that at even slightly elevated levels in the air, ozone is dangerous and unhealthful to people.

Resources

The best source of up-to-date information about indoor air pollution and residential air cleaning systems is the federal Environmental Protection Agency. The EPA maintains several useful telephone services, and distributes, free of charge, a variety of helpful publications. You can also request these publications from the regional EPA office in your area, which you can find in your local telephone book, under U.S. Government, Environmental Protection Agency. The following resources may be particularly valuable:

The Inside Story: A Guide to Indoor Air Quality, publication #402K93007.
Indoor Air Facts, No. 7: Residential Air Cleaners, publication #20A-4001.
Residential Air Cleaning Devices: A Summary of Available Information, publication #402K96001.

National Center for Environmental Publications and Information
800-490-9198.

Indoor Air Quality Information Clearinghouse
P.O. Box 37133
Washington, D.C. 20013-7133
800-438-4318

National Radon Hotline
800-SOS-RADON; information recording operates 24 hours a day.

www.realgoods.com

Air Purification Products

AIR FILTER COMPARISON CHART

Description	Item #	Max Sq Ft	# of speeds	Size (inches) (H x W x D)	Lbs	Filter Type	CADR S /P /D	Warranty
Blueair Hepa	53-638	542	4	25.5 x 19.75 x 13.5	35	HEPA/ion	350/310/330	10-year
Bemis HEPA Air Purifier	53-873	340	3	18 x 17 x 12	15	carbon/HEPA	220/215/220	5-year
Healthmate	53-668	1500	3	23 x 14.5 x 14.5	45	carbon/HEPA		5-year
Healthmate Junior	53-657	500	3	16.5 x 11 x 11	18	carbon/HEPA		5-year
Vornado AQS-15	53-683	224	3	14.6 x 9.6 x 13.8	13	carbon/HEPA	146/142/150	5-year
Roomaid HEPA Filter	53-860	150	2	8.5 x 8.5 dia.	4	HEPA/VOC	not rated	5-year
Portable EMF Air Cleaner	53-879	200	3	23 x 12 dia.	18	enhanced media	not rated	2-year
Desk Top Air Purifier	53-667	108	2	6.5 x 6 x 6	3	carbon/electrostatic/ion	not rated	1-year
Sun Pure Air Purifier	53-681	2000	4	18 x 21.5 x 8	23	carbon/HEPA/UV/ion	265/265/265	lifetime

CADR: Clean Air Delivery Rate, S/P/D: Smoke / Pollen / Dust (effective cubic feet per minute).
Life expectancy of any filter media is dependent on the concentration of the contaminants to which the system is exposed.
Chart is based on information provided by the manufacturer.

Kills Viruses And Germs
Reduces Toxic Gases

Lifetime Warranty

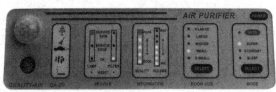

Control Panel Closeup

This unit covers all the bases, from allergens to airborne viruses. The six-stage filtering process combines gas absorption, ultraviolet technology, and ionization with the standard HEPA and carbon filters, and throws in a macro pre-filter for longer filter life—replace expensive filters every two years, instead of every six months! This is the system of choice for any environment where an active germ pool threatens family health, and it provides maximum relief for chemical sensitivities, asthma, and allergies in rooms as large as 50´ x 40´. To keep energy expenditure at a minimum and convenience at a maximum, the Ultraviolet Air Purifier utilizes an infrared motion detector that "wakes up" the system when it's needed most (when you and your family are active and awake). When there is no movement in the room, or no dust in the air, it cycles down to the lowest setting to reduce operating costs. It also features a "sleep" mode that provides super quiet operation while still purifying a room at 70% of normal capacity.

Step One: A treated pre-filter removes larger particles like dust, (.5 microns and above) before they get to your expensive HEPA filter. The pre-filter also preserves the gas-absorption media from dust fouling.

Step Two: A specially formulated gas absorption media substantially reduces exhaust fumes, organic hydrocarbons, pesticides, formaldehyde, and other noxious gases emitted from household cleaners, solvents, chlorine, and paint. Critical for those with chemical sensitivities!

Step Three: Activated carbon (from coconuts) removes unpleasant household odors and industrial pollution. This filter is pleated to provide maximum surface area and a long filter life.

Step Four: The same HEPA filter used in professional health care settings, removing airborne contaminants as small as .3 microns (pollen, mold, fungal spores, dust mites, tobacco smoke, most bacteria).

Step Five: High intensity ultraviolet light kills disease causing viruses and bacteria on contact. This is what makes the system a "purifier," and not just a cleaner. A godsend for those with suppressed immune systems (Chronic Fatigue Syndrome, AIDS).

Step Six: An ionization chamber produces negative ions and activated oxygen, so indoor air tastes and smells like the real thing. Studies suggest that negative ions may even induce and enhance sleep for your resident insomniac.

The Ultraviolet Air Purifier measures 18˝ H x 21.5˝ W x 8˝ D, and boasts a limited lifetime mfr.s warranty. Components are UL-listed. Replacement set includes all filters and lamp needed for two years. USA.

◆ **53-681 Sun Pure Air Purifier $549⁰⁰**

◆ **53-685 Sun Pure Replacement Set
(UV Lamp and Filters) $129⁰⁰**

◆ *Means shipped from manufacturer.*

The Blueair—Sweden's Top Selling Air Filter

Minimal Power Consumption; Quiet Operation; Slim, Contemporary Design

Don't let its sleek design fool you—the Blueair is not just another pretty face. At the highest speed setting, the Blueair uses just 94 watts of power to clean a 20´ x 28´ foot room (only 28 watts at the lowest of four settings). Not only is this product among the most energy-efficient filters on the market, its triple HEPA filters reduce airborne pollutants effectively and quietly. In EPA tests measuring CADR (Clean Air Delivery Rate), the Blueair performed as well or better than 150 other systems. It delivers clean air at the rate of 400 cubic feet per minute on high (47 per minute on low), and it does so quietly. Technological advances allow the Blueair to work almost silently, without the commensurate loss of efficiency common to earlier filter systems.

The unit uses a two-step filter system, incorporating three patented HEPASilent® filters (the same kind commonly used in European hospitals), and two encapsulated ion chambers. The HEPASilent® filters trap up to 99.9% of airborne particles (dander, smoke, pollen, dust) under .03 microns; the ion chamber produces negative ions that bond to particles and trap them inside the unit. The three HEPA filters should be replaced twice a year, based on 24-hour-per-day usage.

With a minimal 20˝ x 13.5˝ footprint, the award-winning Swedish design makes the Blueair an unobtrusive, even attractive, part of your household. It's easily stored behind a sofa or in a standard closet when not in use, and outfitted with casters for easy portability from room to room. Ten-year mfr.'s warranty. 25.5˝ high; 35 pounds. Sweden.

❖**53-638 Blueair HEPA Air Purifier $499⁰⁰**

❖**53-639 Blueair Replacement Filters (Set of 3) $79⁹⁵**

Bemis HEPA Air Purifier

The Bemis HEPA air purifier eliminates 99.97% of all dust, pollen, smoke, bacteria, and most viruses from the air. The advanced sensor that monitors the airflow signals when a filter needs to be changed. Filters last approximately one year. The three-caster base makes it easy to move and the triangular design allows for optimal cleaning when placed against a wall. Cleans a 17´ x 20´ room. UL-listed. USA.

53-873 Bemis Air Purifier $229⁰⁰

53-874 Replacement HEPA Filter $79⁹⁵

53-875 Replacement Pre-filter $12⁹⁵

Energy Efficient, Longer Filter Life, Adjustable To Varying Room Sizes

These units clean rooms with minimal energy outlay, and relatively quietly. Because they adjust to a variety of room sizes, the Healthmates work only as hard as the room really needs, so power consumption is kept to a minimum. The larger Healthmate Air Cleaner draws 115 watts on the highest setting (for rooms up to 10´ x 15´); the Healthmate Junior draws 80 watts (for smaller rooms up to 7´ x 10´). Both use a three-stage filtering system to effectively clear the air. The HEPA filter traps airborne particles (dander, pollen, dust, mold spores, smoke) as small as .3 microns; the pre-filter traps larger particles before they reach the HEPA, extending the HEPA filter life (and saving replacement costs). The carbon/zeolite post-filter is primarily aesthetic, eliminating unpleasant household odors so your indoor air smells and tastes better. Features a 360 degree air intake, 3-speed adjustable air flow, casters for portability, and three color choices (black, white, or sandstone). The Healthmate Air Cleaner measures 23˝H x 14.5˝ square, and weighs 45 pounds. The compact Junior model is only 16.5˝H x 11˝ square, and weighs 18 pounds. Five-year mfr. warranty. UL-listed. USA.

❖**53-668 Healthmate Air Cleaner $379⁰⁰**

❖**53-657 Healthmate Junior $249⁰⁰**

Please specify color.
Requires UPS ground delivery.

❖**53-654 Healthmate Jr. Repl. Filter $109⁰⁰**

❖**53-668A Healthmate Std. Repl. Filter $150⁰⁰**

❖*Means shipped from manufacturer.*

Powerful Air Filter Delivers Massive Volumes Of Clean Air, Quietly

This is not an ordinary air filter. It's packed with features like the top exit air outlet, circulating clean air up and out into the room, not across dirty floors. It comes with a simple, top-mounted, three-speed control and power light. It has a wide louver inlet grill, which can be easily removed for HEPA and filter replacement and cleaning. The true HEPA filter provides over 24 square feet of super-efficient filter action, removing a minimum of 99.97% of all particles. The PreSorb® pre-filter reduces odors and organic compounds including cooking odors, paint and solvent fumes, and musty bathroom odors. And with its small footprint, the Vornado doesn't take up a lot of floor space either; the flat back lets you put it against a wall or in a corner out of the way. Using an exclusive vortex action, all the air in the room is circulated so all of the air in the room is cleaned. It measures 15″ high x 13.5″ wide x 10″ deep and can filter a room as large as 16′ x 14′. UL listed. Comes with three extra PreSorb® filters. USA.

53-683 Vornado AQS-15 Air Filter $159⁰⁰

53-684 PreSorb® Filters (set of 2) $14⁹⁵

Quiet HEPA Filtration for Small Spaces

The Roomaid HEPA Filter offers the benefits of larger units in a compact, ultra-quiet design. Small enough to fit comfortably on a desk, night tableor counter, the Roomaid makes so little noise it's perfect for bedrooms, nurseries or offices. Hidden within the Roomaid's unobtrusive package is a powerful 3-stage filtration system that begins with a washable ⅛″-thick foam pre-filter, which eliminates large particulates such as dust and lint.

Next, a replaceable HEPA filter removes 99.97% of particulates larger than 0.3 microns (pollen, bacteria, dander, smoke and mold spores). Finally, a special activated-carbon VOC filter traps dangerous household gases from cleaners, paints, solvents and other chemical substances. The two-speed unit uses just 4.5 watts, yet it filters a 10′ x 10′ room 2 times an hour. 7.5″H, 8.5″diameter, 5 lbs. Canada.

01-0156 Roomaid HEPA Filter $150⁰⁰

01-0157 Replacement Filter, lasts up to 5 years $89⁰⁰

01-0158 Roomaid Filter Kit—3 voc filters,
**   2 pre-filters, each lasting 6-12 months $22⁰⁰**

Air Cleaner Filters Out Contaminants And Destroys Them

Enhanced Media Filtration (EMF) means this air cleaner actually destroys active microorganisms, as well as filtering out tiny airborne contaminants (99.97% @ .15 micron size). It destroys airborne viruses, bacteria, and mold and fungi spores, reducing them to benign substances of carbon dioxide and water vapor. The filtration system holds up to four times more pollutants than a HEPA filter of similar size, and features a pressure drop indicator showing when the main filter has reached its capacity. The sealed filtration unit guarantees there is no air leakage around the filter system. The patented airflow chamber is designed for quiet operation at any of its three speeds, making this air cleaner suitable for bedrooms and private offices as well as other public and private spaces. This unit cleans a room as large as 10′ x 20′. The main filter lasts for 14–18 months, and the post-filter for 90 days. 22″ H x 12″ D. USA.

53-879 Portable EMF Air Cleaner $329⁰⁰

53-884 Main Filter Replacement $119⁰⁰

53-883 Post-Filter Replacement $15⁹⁵

Micro Purifier—Macro Results

It's quiet, effective and really small—only 6″ square. Features dust, odors, and particulate pollutants filters, plus a separately controlled negative ion generator and a two-speed fan. Offers adjustable air louvers and easy controls. UL-listed. One-year mfr.'s warranty. China.

53-667 Desk Top Air Purifier (6″ x 6″ x 6.75″ H) $59⁹⁵

Fresh Air As You Drive

A commuter's best friend, the Auto Ionizer gets rid of the toxic soup inside your car—exhaust fumes and other airborne gaseous contaminants, tobacco smoke, mold spores, and dust. Plug it into your cigarette lighter socket, and it produces negative ions that bond with pollutants and odors, removing them from the air. It's extremely durable; with normal use and care it should last the life of your vehicle without maintenance. Emits an inaudible low frequency that can affect lower AM radio bands. China.

53-856 Auto Ionizer (4.25″ long.) $34⁹⁵

Cleaner Air—Anywhere

A lightweight, pager-size pollution solution. Tucked discreetly in a suit pocket or worn around your neck, it helps eliminate airborne pollutants, irritants, allergens, and viruses from your immediate breathing space. Runs on one 9-volt battery (included). Great for airplanes, malls, cars, offices, theatres, or anywhere you go. Comes in black. Consult your doctor if you use a pacemaker. 2.75″ x 2.25″ x .875″ D. USA.

53-872 Personal Air Purifier $129⁰⁰

Plug-In Radon Detector Continuously Monitors Household Air

Exposure to radon can be a deadly problem; it accounts for up to 20,000 lung cancer-related deaths in the U.S. each year. In a home environment, levels of radon can vary greatly during the course of a single month, fluctuating from safe to toxic and back again. Our plug-in detector is a simple, affordable way to protect your family from the dangers of this colorless, odorless, radioactive gas. Designed to continuously monitor household air, it provides easy-to-understand digital readouts displaying the exact levels of radon in your home. If radon levels become harmful, a loud alarm is immediately sounded—no waiting for expensive test results or contractor's reports. Features a long-term/short-term switch to display average readings for seven consecutive days. Comes with complete instructions. USA.

57-134 Safety Siren Radon Detector $69⁹⁵

CHAPTER 9

COMPOSTING TOILETS & GREYWATER SYSTEMS

Novel Solutions to an Age-Old Human Problem

"Actually, all pollution is simply an unused resource.
Garbage is the only raw material that we're too stupid to use."

—Arthur C. Clarke

Composting Toilets

The way humans dispose of their waste is a bellweather of civilized society. Does this society protect its members from the horrors of dysentery and even worse diseases? Over the past century sanitation in North America has gradually evolved from a system where almost every home had an outhouse in the backyard, and rivers in major cities were simply open sewers, to modern flush-and-forget-it systems, where everything seems to simply disappear, and rivers run cleaner. Making distasteful items disappear and rivers run clean is certainly going to win public approval, and we won't be so foolish

A Real Two-Story Outhouse from the Early 1900s

as to suggest that there was some mystical good in the old days. There wasn't. Outhouses smell bad in the summer, are too close to the house, and allow flies to spread filth and disease. In the winter they don't smell, but are too far away. However, in our rush to sanitize everything in sight, we end up throwing away a potentially valuable and money-saving resource (not to mention over-designing expensive and energy-intensive disposal systems that still may pollute groundwater). Think about the absurdity of mixing human waste with drinking water and giving wings to the bacteria and pathogens.

Composting toilets can close the nutrient cycle, turning a dangerous waste product into safe compost, without smell, hassle, or fly problems. They are usually less expensive than conventional septic systems and they will reduce household water consumption by at least 25%. But like the venerable outhouse, composting toilets only deal with human excreta. Unlike a modern septic system, they won't provide greywater treatment. Greywater is covered later in this section.

What Is a Composting Toilet?

A composting toilet is a treatment system for toilet wastes that does not use a conventional septic system. Composting toilets were originally developed in Scandinavia, where almost no topsoils suitable for conventional septic systems exist. A composting toilet is basically a warm, well-ventilated container with a diverse community of aerobic microbes living inside that break down the waste materials. The process creates a dry, fluffy, odorless compost, similar to what's in a well-maintained garden compost pile. Flowers and fruit trees love it. We don't recommend using this compost on kitchen gardens, as some human pathogens could possibly survive the composting process. The composting process in such a toilet does not smell. Rapid aerobic decomposition—active composting—which takes place in the presence of oxygen, is the opposite of the slow, smelly process that takes place in an outhouse, which works by anaerobic decomposition. Anaerobic microbes cannot survive in the presence of oxygen and the more energetic microbes that flourish in an oxygen-rich environment. If a composting toilet smells bad, it means something is wrong. Usually smells indicate pockets of anaerobic activity caused by lack of mixing.

A Modern Biolet Composting Toilet

How Do Composting Toilets Work?

A composting toilet has three basic elements: a place to sit, a composting chamber, and a drying tray. Most models combine all three elements in a single enclosure, although some models have separate seating, with the composting chamber installed in the basement or under the house. In either case, the drying tray is positioned under the composting chamber, and some sort of removable finishing drawer is supplied to carry off the finished and composted material.

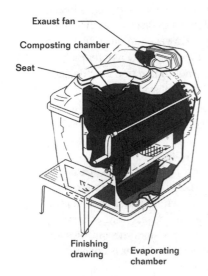

Exaust fan

Composting chamber

Seat

Finishing
drawing

Evaporating
chamber

**A Typical Sun-Mar
Composting Toilet**

Ninety percent of what goes into a composting toilet is water. Compost piles need to be damp to work well, but most composting toilets suffer from too much water. Evaporation is the primary way a composting toilet gets rid of excess water. If evaporation can't keep up, then many units have an overflow that is plumbed to the household greywater or septic system. Heat and air flowing through the unit assist the evaporation process. Every composting toilet has a vertical vent pipe to carry off moisture. Air flows across the drying trays, around and through the pile, then up the vent to the outside of the building. The low-grade heat produced by composting is supposed to provide sufficient updraft to carry vapor up the vent. However, like any passive vent with minimal heat, these are subject to downdrafts. Electric composters use vent fans and a small heating element as standard equipment. We offer an optional vent fan for nonelectric models that can be battery- or solar-driven. Adding a cup every day of high-carbon-content bulking agent, such as peat moss, wood chips, or dry-popped popcorn, helps soak up excess moisture, makes lots of little wicks to aid in evaporation, and creates air passages that prevent anaerobic pockets from forming.

The solids are treated with by a diverse microbiological community—in other words, composted. Keeping this biological community happy and working hard requires warmth and plenty of oxygen, the same things needed for speedy evaporation. Smaller composters usually employ some kind of mixing or stirring mechanism to ensure adequate oxygen to all parts of the pile, and faster composting action. Composting toilets work best at temperatures of 70°F or higher; at temperatures below 55°F the biological process slows to a crawl, and at temperatures below 45°F it comes to a stop. The composting action itself will provide some low-level heat, but not enough to keep the process going in a cold environment. It is okay to let a composting toilet freeze,

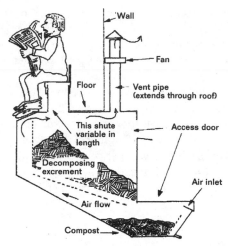

An Early Multrum Toilet—the First Generation of Composting Toilets

although it shouldn't be used when cold or frozen. Normal biological activity will resume when the temperature rises again.

The earliest composter designs, such as the Clivus Multrum system, use a lower-temperature decomposition process known as moldering, which takes place slowly over several years. These composters have air channels and fan-driven vents, but they lack supplemental heat or the capability of mixing and stirring. Therefore, these very large composters don't promote highly active composting activity. There are two other disadvantages. Liquids often have to be manually removed or pumped out of these units because the lower temperatures' slow evaporation. And, because temperatures don't get very high, there's a greater chance of pathogens or parasite eggs surviving the composting process. This type of large, slow composter is most effective in public access sites, and many are currently in successful use at state or national parks. Their large bulk lets them both absorb sudden surges of use, and weather long periods of disuse without upset.

Manufacturer Variations

Every compost toilet manufacturer will be delighted to tell you why their unit is the best. We'll attempt to take a more impartial attitude here, and give the pluses and minuses of each design. First, some general comments. Smaller composters certainly cost less, but because the pile is smaller, they are more susceptible than larger models to all the problems that can plague any compost pile, such as liquid accumulation, insect infestations, low temperatures, and an unbalanced carbon/nitrogen ratio. Smaller composters require the user to take a more active role in the day-to-day maintenance of the unit. We have found that the smaller units with electric fans and thermostatically controlled heaters have far fewer problems than the totally nonelectric units. So be aware that the less-expensive composting toilets have hidden, and long-range costs.

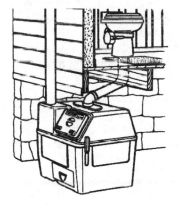

Some Sun-Mar models have the toilet separated from the composting chamber.

Sun-Mar Composting Toilets

Sun-Mar composting toilets use a rotating drum design, like a clothes dryer, that allows the entire compost chamber to be turned and mixed easily—and remotely! This ensures that all parts of the pile get enough oxygen, and that no anaerobic pockets form. Routine once- or twice-a-week mixing also tumbles the newer material in with older material, so the microorganisms can get to work on recent additions more quickly. Periodically, as the drum approaches one-half to two-thirds full, a lock is manually released and the drum is rotated backwards. This action lets a portion of the composting material drop from the drum into a finishing drawer at the bottom of the composter. When the compost is finished, the drawer is emptied on your orchard, ornamental plants, or selected parts of your garden, reinstalled, and a fresh load dumped in.

Sun-Mar has introduced a Centrex Plus line that essentially has two composting drums in series and stacked drying trays. This design doubles the solids capacity, triples the drying tray area, and doubles the time that material can spend fully composting, alleviating many of the more common complaints we've heard about Sun-Mar units.

The main disadvantage of the Sun-Mar drum system is that fresh material may be included with the material dumped into the finishing drawer, and some pathogens may survive the composting process. Also, the limited size of most Sun-Mar composters means they work best for intermittent-use cabins, or small households. With the

Cross-Section of a Typical Sun-Mar Composter

Carbon/ Nitrogen Ratios

Proper composting requires a balance of carbon and nitrogen in the organic material being composted. Human excreta are not properly balanced as they are too high in nitrogen. They require a carbon material to be added for the encouragement of rapid and thorough microbial decomposition. In the mid 1800s the concept of balancing carbon and nitrogen was not known, and the high nitrogen content of humanure in dry toilets prevented the organic material from efficiently decomposing. The result was a foul, fly-attracting stench. It was thought that this problem could be alleviated by segregating urine from feces (which thereby reduced the nitrogen content of the fecal material) and dry toilets were devised to do just that. Today, the practice of segregating urine from feces is still widespread, even though the simple addition of a carbonaceous material to the feces/urine mix will balance the nitrogen of the material and render the segregation of urine unnecessary.

from *The Humanure Handbook*

exception of the double-drum Centrex Plus line, these composters can on occasion be overwhelmed by full-time use in larger households.

Sun-Mar produces both fully self-contained composting toilets and centralized units, in which the compost chamber is located outside the bathroom and connected to either an air-flush toilet, or an ultralow-flush toilet with standard three-inch waste pipe. Electric models all feature thermostatically controlled heating elements to assist with evaporation, and a small fan to ensure fresh airflow and negative air pressure inside the compost chamber. We strongly recommend the electric models for customers who have utility power available. For those with intermittent AC power, AC/DC models are available that take advantage of AC power when it's available but can operate adequately without it as well. Finally, Sun-Mar makes nonelectric models that don't need any power, although use should be limited to one or two people or a low-use cabin with these models.

BioLet Composting Toilets

BioLet composters use a composting chamber that sits upright like a bucket. Fresh material enters the top of the chamber, works its way down as composting progresses, and eventually is allowed to sift out the bottom of the chamber into the finishing drawer. Mixing and stirring rods rotate inside the drum to ensure oxygen supply, and to sift finished material out the bottom screen. There is no possibility of fresh material passing through the composter to the finishing tray in this design, although the mixing and stirring may not result in as complete an aeration process as with the tumbling drum design.

BioLet produces two basic sizes, which, with various options, produces four different models. The Standard is the smaller size, and is available in a nonelectric manual-mixing model, an electric manual-mixing model, and an electric automatic-mixing model. The 40% larger XL size comes in one electric model with automatic mixing and NSF approval. All models except the nonelectric require utility power, or a plentiful RE source such as a strong hydro system.

BioLet designs feature a number of innovative and user-friendly elements. There are a tasteful pair of clamshell doors immediately below the seat that only open when weight is put on the seat. Heat level is user adjustable, and there is a clear sight tube to see if liquid is accumulating (which means it's time to turn up the heat). All the fans, heaters, thermostat, and mechanical components are mounted on a single easy-access cassette in the top rear.

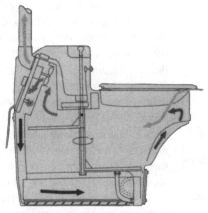

Cross-Section of a Biolet Composter Showing Air Flow

The Carousel Composter for Full-Time and Larger Family Use.

Carousel Composting Toilets

The Carousel is a large composter for full-time residential use. It features a round carousel design with four separate composting chambers. Like most composters, it has drying trays under the composting chamber and a fan-assisted venting system to ensure no-smell operations. Each of the four chambers is filled in turn, which can take from two to six months; then a new chamber is rotated under the dry toilet. This gives each of the four batches up to two years to finish composting without being disturbed. Batch composting has the advantage of not mixing new material with older, more advanced compost, allowing the natural ecological cascading of compost processes. This results in more complete composting, and greatly reduces the risk of surviving disease organisms. The Carousel can handle full-time use and large families or groups. Its only disadvantage is that it costs more and takes up more space than smaller composting toilets.

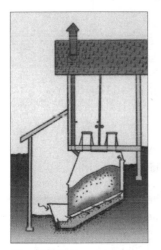

The CTS System for High-Use Public Sites

Composting Toilet Systems (CTS)

Got a campground, golf course, visitor center, or other high-use public site that needs non-smelly sanitary facilities? A CTS digestor system may be your best choice. CTS makes large digestor tanks that accept two to four air-flush toilets and can handle over 100 uses per day. These are the units we usually specify for eco-resorts in sensitive environments. Using a simple, trouble-free design, the only moving parts in a CTS digestor are the toilet seat and the fan, which ensures adequate airflow and negative air pressure inside the tank. When the seat is lifted, air is drawn down, ensuring a no-smell restroom. CTS can supply everything from a single digestor up through a completely equipped restroom building. Because the CTS uses a large digestor tank, this means a larger, more diverse biological community that is less prone to upset, or short-term overloading. On the downside, this means a more expensive composter initially. Although, considering the use levels they will accommodate, these large composters are a great bargain.

Will My Health Department Love It As Much As I Do?

Possibly. This depends greatly on the enlightenment quotient of your local health official. Some health departments will welcome composting toilets with open arms, and even encourage their use in some locales, while others will deny their very existence. We have often found that in lakeside summer cabin situations, health officials prefer to see composting toilets rather than pit privies or poorly working septic systems that leach into the lake. Several composters now carry the NSF (National Sanitation Foundation) seal of approval, which makes acceptance easier for local officials, but still doesn't mandate acceptance. It is really up to your local sanitarian. He can play God within his own district and there is nothing that you, Gaiam Real Goods, or the manufacturer can do about it. Our advice if you're trying to persuade your authorities? Be courteous. Health officials have to consider that a composting toilet might suit you to a tee, and you'll take good care of it, but what if you sell the house? Will the next homeowner be willing, or capable of taking care of it too?

Also, be aware that in all cases composting toilets only deal with toilet wastes, the blackwater. You still need a way to treat your greywater wastes—all the shower, tub,

WE DO IT EVERY DAY. BUT DO WE EVER THINK ABOUT IT!

I DEFECATE, THEREFORE I AM . . . .

sink, and washer water. Unconventional greywater systems will face the same approval problems as composting toilets. Some localities will happily allow alternatives that use the greywater for landscape watering, others may require a standard full-size septic system. In all cases greywater systems must use subsurface disposal. Once the water goes down the drain it can't see daylight again. Greywater alternatives are covered in more detail below.

Composting Toilet Installation Tips

Keep It Warm

Compost piles like warmth, because it makes all the little microbes work faster. Thus, composting toilets work best in warm environments. If your installation is in a summer-use cabin, then a composting toilet is ideal. A Montana outdoor installation in wintertime isn't going to work at all. If you plan to use the composter year-round, then install it in a heated space on an insulated floor. The small electric heater inside the compost chamber will not keep it warm in an outdoor winter environment, but it will run your electric bill up about $25 a month trying. Talk to one of Real Goods' technicians if you have any doubts about proper model selection or installation.

Plug It In

If you have utility power, by all means use one of the electrically assisted units; they have far fewer problems overall. If you have intermittent AC power from a generator or other source, use one of the AC/DC, or hybrid units. Anything that adds warmth or increases ventilation will help these units do their work. We strongly recommend adding vent-top fans to nonelectric units.

Let It Breathe

The compost chamber draws in fresh air and exhausts it to the outside. If the composter is inside a house with a woodstove and/or gas water heater which are also competing for inside air, you may have lower air pressure inside the house than outside. This can pull cold air down the composter vent and into the house. Composters shouldn't smell bad, and something is wrong if they do, but you don't want to flavor your house with their gaseous by-products either. Do the smart thing: give your woodstove outside air for combustion.

Don't Condense

Insulate the composter's vent stack when it passes through any unheated spaces. The warm, humid air passing up the stack will condense on cold vent walls, and moisture will run back down into the compost chamber. Most manufacturers include vent pipe and some insulation with their kits. Add more insulation if needed for your particular installation.

Keep Everything Afloat

With central units such as the Centrex and WCM models, where the composting chamber is separate from the ultralow-flush toilet, the slope of any horizontal waste pipe run is critical. The standard three-inch ABS pipe needs to have ⅛ to ¼ inch of drop per foot of horizontal run. More drop per foot allows the liquids to run off too quickly, leaving the solids high and dry. Vertical runs are no problem, so if you want a toilet on the second floor, go ahead. Sun-Mar recommends horizontal runs of 18 feet maximum if possible, but customers have successfully used runs of well over 20 feet. Play it safe and give yourself clean-out plugs at any elbow. If you are using an air flush unit, it must be installed directly over the composter inlet.

Greywater Systems

"The future is up for grabs. It belongs to any and all who will take the risk and accept the responsibility of consciously creating the future they want."

—Robert Anton Wilson

What Is Greywater?

Greywater is the mix of water, soap, and whatever we wash off at the shower, sink, dishwasher, or washing machine. Basically, it is any wastewater other than what comes from toilets. Most homes produce 20 to 40 gallons of greywater per person, per day. The main contaminant is phosphorus from soap and detergent, but greywater is also likely to contain bacteria, including small amounts of fecal coliform, protozoans, viruses, oil, grease, hair, food bits, and petroleum-based "whatevers."

Why Would I Want to Use Greywater?

Utilizing greywater has many advantages. It greatly reduces septic system demands, gives otherwise wasted nutrients to plants, reduces water use, increases your awareness of natural cycles, provides very effective water purification, and probably reduces your use of energy and chemicals. Used primarily for landscape irrigation, as a substitute for freshwater, greywater can make landscaping possible in dry areas. Similarly, it can keep your landscaping alive during drought periods. Sometimes a greywater system can make a house permit possible in sites that are unsuitable for a septic system.

When Would Greywater Be a Poor Choice?

Greywater recycling isn't for everybody. High-rise apartment dwellers would be seriously challenged. But let's assume we're speaking to folks with at least a suburban plot around them. Even then, you may have insufficient yard and landscaping space, your drain plumbing may be encased in concrete, your climate may be too wet or too frozen, costs may outweigh benefits, or it might just be illegal where you live.

How Does a Greywater System Work?

Although the trace contaminants in greywater are potentially useful, even highly beneficial, for plants and landscaping, their presence demands some modest caution in handling and disposal. How do you get rid of it safely? All responsible greywater designs stem from two basic principles:

- *Natural purification occurs while greywater passes slowly through healthy topsoil.*
- *No human contact should occur before purification.*

Following these principles, a greywater system should provide some minimal filtering for food bits, hair, and grease, and then send the water out to a subsurface watering system under the landscaping. In practice, there is usually a 30- to 60-gallon drum that acts as a surge tank to handle sudden rapid discharges, a simple sand filter, and a network of subsurface watering emitters, like a drip irrigation watering system, only buried. Usually a pump is required to push the greywater through the filter and out to the emitters, but some hilly sites can get by using only gravity. The filter needs

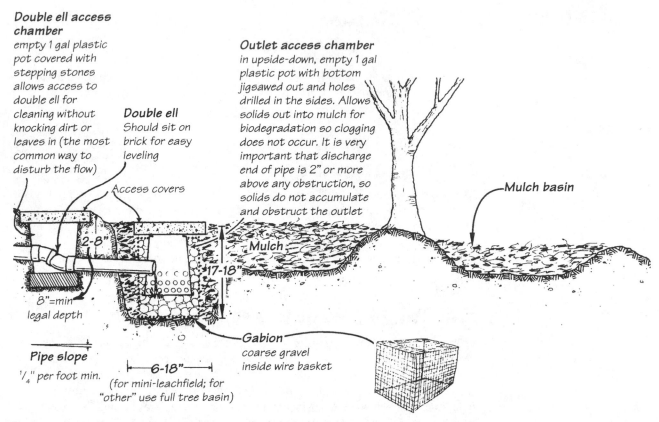

Double ell access chamber
empty 1 gal plastic pot covered with stepping stones allows access to double ell for cleaning without knocking dirt or leaves in (the most common way to disturb the flow)

Double ell
Should sit on brick for easy leveling

Access covers

Outlet access chamber
in upside-down, empty 1 gal plastic pot with bottom jigsawed out and holes drilled in the sides. Allows solids out into mulch for biodegradation so clogging does not occur. It is very important that discharge end of pipe is 2" or more above any obstruction, so solids do not accumulate and obstruct the outlet

Mulch basin

2-8"

17-18"

Mulch

8"=min legal depth

Pipe slope
1/4" per foot min.

6-18"

(for mini-leachfield; for "other" use full tree basin)

Gabion
coarse gravel inside wire basket

to be cleanable or backflushable. The subsurface emitters are highly specialized to prevent roots from growing into them and clogging. Don't use conventional above-ground drip hardware!

In almost all cases you'll still need a conventional (although drastically downsized) septic system and the ability to switch from one system to another. This allows for hard-freezing weather, washing dirty diapers, the occasional seriously sick household member, backwashing your greywater filter, or other potentially infectious problems.

Will My Health Department Love Greywater as Much as I Do?

You stand a better chance of official approval with greywater than with a composting toilet, but this might be a case of, "out of the fire, into the frying pan." Some health departments will give you a choice about greywater; some won't. Some may allow greywater systems, so long as you also install a full-size, approved septic system. Remember . . . it's the people who will buy the house from you that the health department is protecting you from. In most areas, the legality of greywater is ambiguous. Generally, official attitudes are moving toward a more accepting and realistic stance, and authorities will often turn a blind eye toward greywater use. In the late 1970s, during a drought period, the state of California published a pamphlet that explained the illegality of greywater use, while showing how to do it! Since then, California has become one of many places you can legally install a greywater system and downsize the septic system accordingly.

The best advice is to simply call your health department and ask. But ask as a hypothetical example. Health officials recognize the "hypothetical" question readily. This allows you to ask specific pointed questions, and them to answer fully and honestly, without anyone admitting that any crime or bending of the rules has occurred or is likely to occur.

How Do I Make a Greywater System?

Legal greywater systems tend toward engineering overkill, while many of the simple and economical methods that folks actually use are still technically illegal. Start with one or both of the Oasis booklets listed in our product section. These provide the best overview of the subject and detail every type of greywater system from the simple dish pan dump to fully automated systems. There are many possible ways to create a greywater system, and since the Oasis booklets cover them all with grace, style, a wealth of illustrations, and at minimal cost, we will not duplicate their work here. Aim for the best possible execution with the simplest possible system. During new home construction it's fairly easy to separate the washer, tub, sink, and shower drains from the toilet drains. Retrofits are a bit more trouble, and in some cases impossible, like when the plumbing is encased in a concrete slab floor. Kitchen sinks are often excluded from greywater systems because of the high amount of oil, grease, and food bits. But this is a personal choice, and depends on how much maintenance you're willing to do on a regular schedule.

—**Doug Pratt**

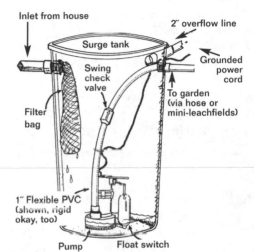

Inlet from house
Surge tank
2″ overflow line
Swing check valve
Grounded power cord
Filter bag
To garden (via hose or mini-leachfields)
1″ Flexible PVC (shown, rigid okay, too)
Pump
Float switch

www.realgoods.com

Composting Products

Books

Create an Oasis with Greywater

A Complete Guide to Managing Greywater

The concise, readable format is long enough to cover everything you need to consider, and short enough to stay interesting. Plenty of charts and drawings cover health considerations, sources, designs, what works and what doesn't, bio-compatible cleaners, maintenance, preserving soil quality, the list goes on. . . . There's also an extensive bibliography. 47 pages, paperback. USA.

82-440 Create an Oasis with Greywater $14⁹⁵

Building Professional's Greywater Guide

A Companion to
Create an Oasis with Greywater

Will help you successfully include greywater systems in new construction or remodeling. Includes reasons to install or not install a greywater system, flowcharts for choosing an appropriate system, dealing with inspectors, legal requirements checklist, design and maintenance tips, and the complete text of the new California greywater code with extensive (plain English!) annotations. This California code is very similar to the greywater appendix in the Uniform Plumbing Code, which applies to most of the U.S. 46 pages, paperback. USA.

**80-097 Building Professional's
Greywater Guide $14⁹⁵**

Branched Drain Greywater Systems

A companion volume to *Create an Oasis with Greywater*, this book gives detailed plans and specific systems for the reliable, sanitary, and low-maintenance distribution of household greywater to downhill plants without filtration or pumping. This inexpensive and easy-to-install greywater system is inherently simple and surprisingly easy to use. 52 pages, paperback. USA.

82-494 Branched Greywater $14⁹⁵

The Humanure Handbook 2nd Ed.

A Guide to Composting Human Manure

Deemed "Most Likely to Save the Planet" by the Independent Publishers in 2000. For those who wish to close the nutrient cycle, and don't mind getting a little more personal about it than our manufactured composting toilets require, here's the book for you. Provides basic and detailed information about the ways and means of recycling human excrement, without chemicals, technology, or environmental pollution. Includes detailed analysis of the potential dangers involved and how to overcome them. The author has been safely composting his family's humanure for the past 20 years, and with humor and intelligence has passed his education on to us. 302 pages, paperback. USA.

82-308 The Humanure Handbook $18⁹⁵

The Composting Toilet Book

Even though off-the-grid homeowners have been using them for decades, convincing the local inspector to permit a composting toilet system in your conventional year-round home can be a frustrating experience. This book gives you the ammunition you need to get this water-saving, eco-friendly technology installed and permitted. This is the definitive handbook we've been wanting for at least ten years. It's packed with technical details of various permittable systems, state-by-state permitting information, maintenance and operation tips the manufacturers often do not provide, profiles of long-time composting toilet users, and a bonus section on greywater applications. While costs and restrictions for conventional systems have been rising, composting toilets have dropped in price and risen in quality. There's never been a better time to switch over to a composting toilet system, and there's never been a better resource on the topic than this one. 150 pages, paperback. USA.

80-708 The Composting Toilet Book $29⁹⁵

Composting Accessories

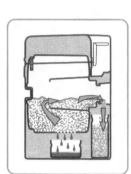

Sun-Mar Home Composter

No-Smell Composting for Kitchen Waste

Sun-Mar has been handling tough composting jobs for more than 25 years. Their practical experience shows in this new Home Composter for kitchen and garden waste. Designed for odorless indoor or outdoor use, the Home Composter employs Sun-Mar's proven mixing drum for easy and complete aeration. A well-turned compost pile is a happy, active, good-smelling compost pile, and the rotating drum on this composter makes it easy. Kitchen scraps, garden waste, peat moss, and a bit of water go into the composter upper chute. Material up to 1.5″ chunks is acceptable. Smaller chunks compost faster. The gear-driven drum is rotated by turning a handle. A drum lock stops the drum in position to accept fresh material. As the compost tumbles and aerates, it moves along the main drum, and then back through the inner drum before dropping automatically into the finished compost drawer. To prevent saturation, any excess liquid drains through a screen into the liquid drawer to be used as compost tea.

Finished compost can be used for ornamental flower beds, fruit trees, or vegetable gardens. The compost tea can be used as liquid fertilizer.

Composting works best at temperatures over 60°F. Our busy little microbes pretty much go into hibernation at lower temperatures, so an indoor installation is recommended unless you live in a warm climate or only plan on seasonal use. Basements, heated garage, mudrooms, utility rooms, or laundry rooms are all good choices. The Home Composter is supplied with a 3-foot length of flexible dryer-type 3″ vent with a through-the-wall exhaust fitting, and a low-wattage vent fan to insure no-smells operation. The fan is 12-volt DC for safe operation, with a transformer cube that plugs into any conventional AC outlet. Compost starter and microbe mix are included to ensure a well-balanced biological community from initial start up. Canada.

Dimensions: 23″ x 24″ footprint, 33″ height
Weight: 51 lb. (empty)
Shipping weight 70 lb.
Ships via: UPS directly from mfr.
Mfr.'s warranty: one year. Canada.

◆ **44-814 Sun-Mar Home Composter $360⁰⁰**

◆ *Means shipped from manufacturer.*

12-Volt Toilet Exhaust Fan

Our powerful 12-volt fan is designed to fit easily over a standard 4-inch ABS or PVC vent pipe. When used as a standard bathroom ventilation fan, it can be wired into your 12-volt lighting system via a switch. It can also be used on a continuous daytime basis by connecting it to a 10-watt solar module, such as the Solarex SX-10 (#11-513). When used for ventilating composting toilets, such as the Sun-Mar NE, it should be connected to a battery to provide round-the-clock ventilation. It is also a good fan for greenhouse ventilation. The bathroom fan uses only 9 watts (¾ amp at 12 volts), and is rated at 90 cfm. The quiet, brushless fan motor is encapsulated to protect it from corrosion. The exhaust fan is highly recommended for residential use and in locations where downdraft can occur, such as areas surrounded by mountains or high trees, it will greatly improve airflow and capacity. USA.

44-802   12-Volt Toilet Exhaust Fan $69⁰⁰
44-801   24-Volt Toilet Exhaust Fan $69⁰⁰

Compost Starter Mulch

An organic starter mulch specially formulated for composting toilets. Using this mulch introduces a healthy, well-rounded biological community for faster, hotter, more complete composting. Independently tested and recommended by compost toilet manufacturers. Highly recommended for initial and springtime startup on all composters. Can be used occasionally in lightly-used units, or can be used daily in place of the usual peat moss on heavily-used units. Highly recommended for any composter that has trouble keeping up with use. 8-gallon bag, (approx. 12 lb.) Sold by volume. USA.

44-252 SoilTech Composter Starter Mulch $29⁹⁵

Composting Toilets

Sun-Mar Composting Toilets

Made in Canada, Sun-Mar has been in the composting toilet business longer than almost anyone. Sun-Mar designs are all based on a rotating drum, like a clothes dryer, that locks into the "use" position. Turning a crank on the outside of the composter rotates the drum, which ensures complete mixing and good aeration of the pile. A well-turned compost pile works faster, and more effectively. Mixing should happen once or twice a week whenever the composter is in active use. When the drum is rotated in the normal mixing direction, the inlet flap swings shut, and all material stays in the drum. Periodically, when the drum approaches one-half to two-thirds full, a lock is released that allows the drum to rotate backwards. This lets some compost drop into the finishing tray at the bottom. It finishes composting here until the next time the drum needs to dump some material.

The fully self-contained Sun-Mar composters, the Excel, Compact, and Ecolet models, are a good choice for summer camps, sites with intermittent use, sites without electricity, or full-time residences with winter heating. If you want your composter to work in the winter, it *must* be in a 65°F or warmer space.

The Centrex series puts the composting drum outside or in the basement for those who want a more traditional-looking toilet or toilets in the bathroom. Centrex models are available for a full range of uses from intermittent summer camps to full time residential use.

If you have utility electricity available at your site, we strongly recommend using one of the standard AC models with fan and heater. They compost faster, are more tolerant of variable uses, and are much less likely to suffer any performance problems. This electrical advice applies to *all* composting toilet brands!   ·

Sun-Mar Centrex 3000-Series
Our Highest-Capacity Composting Toilet

The Sun-Mar Centrex 3000-series is designed for heavy cottage use, or medium continuous residential use. The double-length composting drum in the 3000-series has 40% more capacity than any other Sun-Mar unit. This allows more time for composting, handles more volume or surge loads, and compost now cleverly drops into the finishing drawer automatically when the drum is turned. There are two finishing drawers to complete the composting process, and by moving a filled drawer from the end of the drum to the second station under the drum, the composting cycle can be extended further. The evaporating chamber for excess liquids now has two cascading trays above the composter bottom, greatly increasing available evaporation area. Electric models have a pair of 250-watt thermostatically-controlled heaters to boost evaporation rates, and keep the compost chamber warm and friendly. Both heaters have manual override switches. A 1″ overflow security drain hose is included with all models, which should be connected to the household greywater system.

The Centrex 3000 is available in a standard AC version, a nonelectric version, or an AC/DC version. All version are further available for use with either one-pint, ultralow-flush RV toilet(s), or an air-flush (dry) toilet. Toilets are purchased separately. Suggested capacities vary slightly with model, but generally, the 3000-series is good for five to eight adults in residential use, or seven to ten adults in vacation use.

Use a standard version if you have utility power. It comes with a 2″ vent and 30-watt fan. The AC/DC version

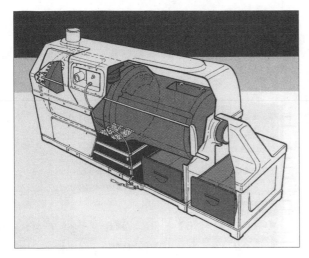

is for folks who run a generator more or less daily and has the 2″ vent and fan, plus a second 4″ vent with 2.4-watt DC fan. The nonelectric version has the 4″ vent with DC fan. Canada.

Dimensions: All versions 31.5″ H x 69.5″ W x 26.5″ D. To remove finishing drawers, 52″ of depth needed.
Weight: approximately 120 lb.
Shipping weight: 220 lb. (some models slightly less)
Ships via: Freight collect
Ships from: Buffalo, NY
Mfr.'s warranty: 3-yr. parts, 25-yr. body

44-118 Centrex 3000 Std $1,599⁰⁰
44-119 Centrex 3000 NE $1,419⁰⁰
44-120 Centrex 3000 AC/DC $1,699⁰⁰
44-121 Centrex 3000 Std A/F $1,649⁰⁰
44-122 Centrex 3000 A/F NE $1,449⁰⁰
44-123 Centrex 3000 A/F AC/DC $1,799⁰⁰
◈ 44-815 Sun-Mar 12v/3.4w 4″ Stack Fan $49⁹⁵
◈ 44-816 Sun-Mar 24v/3.4w 4″ Stack Fan $49⁹⁵

Means shipped freight collect.

◈ *Means shipped directly from manufacturer.*

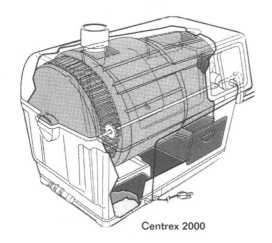

Centrex 2000

Sun-Mar Centrex 2000-Series
Newest Addition to Centrex Line

The Centrex 2000-series is new for 2001, and will probably become the most popular of the central composting units. Good for medium to heavy seasonal use, or light residential use, the 2000-series is the refined result of over 25 years of composting experience. Available in a standard AC version, a non-electric version, or an AC/DC version. In addition, each version can be ordered for either a low-flush, RV-type toilet(s), or for a dry, air-flush toilet. Multiple low-flush toilets can serve a single central composter, but only a single air-flush toilet can be used with the A/F models. Toilet options are sold separately.

Standard AC units, and AC/DC units come with a 350-watt thermostatically controlled heater, a 2″ vent, and a 30-watt fan. Nonelectric units have no heater, and use a 4″ vent. AC/DC units have both 2″ and 4″ vents. 1.4-watt DC vent fans are included with the NE and AC/DC air-flush composters only; they are optional with any other composter. A 1″ overflow drain is included, and should be connected to the household greywater system with all composters. Suggested capacities vary slightly with model, but generally, the 2000-series is good for three to six adults in residential use, or five to eight adults in vacation use. Canada.

Dimensions: 27.5″ H x 26.5″ D x 44.75″ W
(48.75″ width to turn handle)
Weight: 100 lb.,
Shipping weight: 125 lb. (approx.; varies slightly with model)
Ships via: freight collect
Ships from: Buffalo, NY
Mfr.'s warranty: 3-yr. parts, 25-yr. body

44-817 Centrex 2000 Std. $1,199⁰⁰
44-818 Centrex 2000 N.E. $949⁰⁰
44-819 Centrex 2000 AC/DC $1,299⁰⁰
44-820 Centrex 2000 Std. A/F $1,249⁰⁰
44-821 Centrex 2000 N.E. A/F $999⁰⁰
44-822 Centrex 2000 AC/DC A/F $1,349⁰⁰
◈ 44-803 Sun-Mar 12v/1.4w 4″ Stack Fan $49⁹⁵
◈ 44-804 Sun-Mar 24v/1.4w 4″ Stack Fan $49⁹⁵

Dimensions: 27.5″ H x 26.25″ D x 31.5″ W (35.5″ width to turn handle)
Weight: 50 lb.
Shipping weight: 100 lb. (approx. varies slightly with model)
Ships via: freight collect
Ships from: Buffalo, NY.
Mfr.'s warranty: 3-yr. parts, 25-yr. body.

44-105 Centrex 1000 Std. $1,049⁰⁰

44-106 Centrex 1000 N.E. $879⁰⁰

44-107 Centrex 1000 AC/DC $1,149⁰⁰

◆ 44-803 Sun-Mar 12v/1.4w 4″ Stack Fan $49⁹⁵

◆ 44-804 Sun-Mar 24v/1.4w 4″ Stack Fan $49⁹⁵

Sun-Mar Centrex 1000-Series

Our Smallest Central Composter

The Centrex 1000 is designed for light to medium seasonal use. It is not recommended for continuous residential use. The Centrex 1000 has all controls, access, and venting on the front panel for ease of service. Front venting makes cottage installation much easier, particularly when using the NE (nonelectric) model, which requires a straight vent with no bends.

The Centrex 1000 uses an RV toilet with an ultralow flush of approximately 1 pint. Toilets are sold separately. Toilets use standard 3-inch ABS waste pipe for hook-up to the composting unit. Horizontal waste pipe run should be no more than 20 feet with precisely ¼-inch drop per foot. Steeper pipe slope allows liquids to run off faster than solids. Vertical waste pipe run can be any distance. The 1″ overflow drain should be connected to the household greywater system.

Available in three models:

Centrex 1000 is the standard version for installations with utility AC power. Has a single 2-inch vent stack with a 30-watt fan and thermostatically controlled 250-watt heater. Good for five to seven people in intermittent weekend/vacation use.

Centrex 1000 NE for installations without any electricity. No heater, no fan. Has a single 4-inch vent stack. The optional 1.4-watt DC vent fan is recommended. Good for four to six people in intermittent weekend/vacation use.

Centrex 1000 AC/DC for installations that run an AC generator several hours per day. Has the heater, fan, and 2″ vent just like the standard version, plus, it has a 4-inch vent like the NE version. The 1.4-watt DC fan is optional, but recommended. Good for four to six people in intermittent weekend/vacation use. Canada.

Toilets For Use With Centrex Composters

Sun-Mar Dry Toilet   Sealand

We offer the fine china Sealand, one-pint, RV-type toilet for use with the Centrex Series composters.

The Dry Toilet includes a bowl liner that can be removed for cleaning as required. The Dry Toilet is only used with the Centrex Air Flush models.

44-823 Sealand 510 Plus White China Toilet $229⁰⁰

44-116 Sun-Mar White Dry Toilet $249⁰⁰

◆ 44-117 Dry Toilet 29″ Pipe Extension $55⁰⁰

All toilets shipped freight collect when ordered with a Sun-Mar composter; shipped UPS when ordered alone.

Sun-Mar Excel–AC

The fully self-contained, National Sanitation Foundation (NSF) approved, Excel-AC is our most popular composter for installations with 120-volt electricity. It requires no water connections. Installed in a protected, moderate-temperature indoor location, this high-capacity unit can handle three to five people full-time, or even more intermittently, making it ideal for year-round or seasonal use. The Excel has a 25-watt fan that runs continuously, and a 250-watt thermostatically-controlled heater. Normally the Excel will evaporate all liquids. A ½″ emergency drain is fitted and should be connected if heavy use, or prolonged power outages are expected. Black bowl liner can be easily removed for cleaning. The drum crank handle is recessed. The 2″ central vacuum-type vent pipe exits at the top rear of the unit, and can be installed invisibly through the wall.

The AC fan speed control is an option to install into the fan door cover. Use for composters that overdry the compost, or where slight fan noise is a customer concern. Canada.

Dimensions: 22.5″ W x 32″ H x 33″ D. Requires 48″depth to remove finishing drawer.
Weight: 60 lb.,
Ship weight: 100 lb.
Ships via: freight collect
Ships from: Buffalo, NY
Mfr.'s warranty: 3-yr. parts, 25-yr. body

🚚 **44-102 Sun-Mar Excel-AC Toilet $1,049⁰⁰**

◆ **44-824 AC Fan Speed Control $55⁰⁰**

Excel AC/DC Composting Toilet

The Excel AC/DC hybrid was co-developed by Real Goods and Sun-Mar specifically for Real Goods' off-the-grid customers who derive part of their power from generators. The AC/DC is identical to the Excel model, described above, except that it is fitted with an NE drain and with an additional 4-inch NE vent installed next to the Excel's 2″ vent stack. The AC/DC provides the increased capacity of the Excel unit when a generator is running, but operates as a nonelectric unit when 120-volt electricity is not available. For residential use, a 12-volt stack vent fan is included with this model, for use when the AC/DC is running in a nonelectric mode. Stack fans are assembled inside a length of 4″ vent pipe; they draw 1.4 watts. Canada.

Dimensions: 22.5″ W x 32″ H x 33″ D
Weight: 60 lb.
Ship weight 110 lb.
Ships via: freight collect
Ships from: Buffalo, NY
Mfr.'s warranty: 3-yr. parts, 25-yr. body

🚚 **44-103 Sun-Mar Excel AC/DC Hybrid $1,149⁰⁰**

◆ **44-803 Sun-Mar 12v/1.4w 4″ Stack Fan $49⁹⁵**

◆ **44-804 Sun-Mar 24v/1.4w 4″ Stack Fan $49⁹⁵**

◆ **44-805 Sun-Mar 110v AC 4″ Stack Fan $49⁹⁵**

Sun-Mar Excel–N.E.

The Non-Electric is our most popular composting toilet. It's perfect for many of our customers living off-the-grid and not wanting to be dependent on their inverters. The N.E. uses the same body as the standard Excel, but with a larger 4-inch vent that exits straight out the top, and no fan or heater. The aeration and mixing action of the Bio-Drum, coupled with the help of a 4-inch vent pipe and the heat from the compost, creates a "chimney" effect which helps draw air through the system in a manner similar to that of a woodstove. In full-time residential use an optional vent stack fan is recommended to improve capacity, aeration, and evaporation. A 1″ drain is supplied, and must be connected to a greywater or other disposal system. Canada.

Dimensions: 22.5″ W x 31″ H x 33″ D
Weight: 50 lb.
Ship weight 95 lb.
Ships via: freight collect
Ships from: Buffalo, NY
Mfr.'s warranty: 3-yr. parts, 25-yr. body

🚚 **44-101 Sun-Mar Excel-NE Toilet $899⁰⁰**

◆ **44-803 Sun-Mar 12v/1.4w 4″ Stack Fan $49⁹⁵**

◆ **44-804 Sun-Mar 24v/1.4w 4″ Stack Fan $49⁹⁵**

◆ **44-805 Sun-Mar 110v AC 4″Stack Fan $49⁹⁵**

Sun-Mar Compact

This is a scaled-down version of the Excel model. For those who don't need the larger capacity of the Excel and have access to 120-volt power, this is the recommended unit. The working components are the same as the Excel: a bio-drum for mixing and aeration, a thermostatically controlled base heater, and a small fan for positive air movement.

What makes the Compact model different is a smaller variable diameter bio-drum, which allows a smaller overall size; an attractive rounded design; and most importantly, no more footrest! Also, the handle for rotating the bio-drum is now hinged and folds into the body when not being used. Recommended for one person in full-time residential use, or two to four people in intermittent cottage use. Canada.

Dimensions: 22″ W x 27.5″ H x 33″ D
Installation length to remove drawer: 45″
Seat height: 21.5″
Weight: Shipping weight 90 lb., product weight 50 lb.
Ships via: freight collect
Ships from: Buffalo, NY
Vent & Drain: 2″ vent pipe & fitting (supplied with unit),
½″ emergency overload drain (optional hookup).
Electrical Power Requirements: 115-volt, 2.5-amp. 25-watt fan, 200-watt heater with replaceable thermostat.
Approximate average demand: 125 watts.
Mfr.'s warranty: 3-yr. parts, 25-yr. body

🚚 **44-104 Sun-Mar Compact $949⁰⁰**

Sun-Mar Ecolet Space Saver Composting Toilet

The smallest model from Sun-Mar, the Ecolet is 19″ wide by 22″ deep, only slightly larger than a regular-sized toilet seat. It will fit almost anywhere! Can handle two to three people in weekend use, or one to two people in residential use. Constructed of high-quality fiberglass and marine grade stainless steel. Stands 28″ high and uses a 3″ vent stack. The 1″ overflow drain should be connected to household greywater, or mobile holding tank.

Available in three models: a standard 120VAC unit and a pair of mobile units in RV or marine configurations. The Space Saver is the standard AC model with a thermostatically controlled 120-watt heater and a 20-watt fan. The mobile unit has the same thermostatically controlled 120-watt AC heater for when AC power is

available, plus a 12-volt, 4-watt stack fan. A 12-volt, 120-watt heater is optional for either model. The Marine unit is identical to the RV unit, but with a hull-shaped sloped back. It is U.S. Coast Guard certified. Both mobile units have mounting brackets, a gasketed finishing drawer, and higher air intakes to accommodate violent motion. Canada.

Weight: 45 lb.
Ship weight: 80 lb.
Ships via: freight collect
Ships from: Buffalo, NY
Mfr.'s warranty: 3-yr. parts, 25-yr. body

🚚 **44-207 Ecolet Space Saver 120VAC Toilet $999⁰⁰**
🚚 **44-208 Ecolet Marine AC/DC $999⁰⁰**
🚚 **44-111 Ecolet RV AC/DC $999⁰⁰**
◈ **44-825 Optional 12v/120w Heater $75⁰⁰**

BioLet Composting Toilets

Made in Sweden, BioLet composting toilets are some of the best-designed, and best-selling toilets in the world. They offer excellent performance and a compact design for homes that have utility power. Features include:

• Clamshell doors below the seat that hide the compost chamber until weight is put on the seat.
• A composting process that works with gravity. Fresh material enters the top; only the oldest, most completely composted material drops out the bottom, into the finishing tray.
• An adjustable thermostat to control liquid buildup. Won't use electricity unless actually needed.
• A mechanical cassette with all fans, heaters, and thermostats in the top rear makes any service quick, easy, and clean.

BioLet toilets are supplied with an installation kit consisting of 14 feet of 2″ vent pipe; 39″ of insulation; 39″ of outer pipe jacket to cover insulation; vent cap; and a flexible roof flashing. All Biolet models are made in Sweden.

BioLet produces two basic sizes, which, with various options, produces four different models. The Standard is the smaller size, and is available in a nonelectric manual-mix model, an electric manual-mix model, and an electric automatic-mixing model. The 40% larger XL size comes in one electric model with automatic mixing and NSF approval. All models except the nonelectric require utility power, or a plentiful RE source such as a strong hydro system.

BioLet XL Model

BioLet's top-of-the-line model is National Sanitation Foundation (NSF) approved for residential use. Has 40% larger composting capacity than the Manual Model. The thermostatically controlled heating element is 305 watts. The mixer motor runs automatically for 30 seconds every time the lid is lifted and replaced. The quiet 25-watt fan runs continuously, the heating elements run as required by thermostat setting and room temperature. For homes with utility power. The XL can handle four persons full-time, or six persons for intermittent vacation use.

Dimensions: 26″ H x 26″ W x 32″ D. Requires 26″ x 54″ of floor space for the finishing tray to slide out the front.
Weight: approx. 50 lb.
Ship weight: approx. 75 lb.
Ships via: FedEx surface freight
Ships from: Massachusetts, $85 to continental U.S.
Mfr.'s warranty: 3-yr.
44-109 BioLet XL Model $1,559⁰⁰

BioLet Deluxe Model

The Deluxe model, using BioLet's 40% smaller ABS plastic body, features the top-of-the-line automatic-stirring system, fans, and thermostatically controlled heaters like the XL model above. The heater for this smaller model is 250 watts, and will run as required by thermostat setting and room temperature. The quiet 25-watt fan runs continuously. The Deluxe can handle three persons full-time, or five persons for intermittent vacation use. For homes with utility power. Sweden.

Dimensions: 26″ H x 22″ W x 29″ D. Requires a footprint of 22″ x 40″ to allow the finishing tray to slide out.
Weight: approximately 50 lb.
Shipping weight: approx. 75 lb.
Ships via: FedEx surface freight
Ships from: Massachusetts, $40 to continental U.S.
Mfr.'s warranty: 3 yr.
44-124 BioLet Deluxe Model $1,495⁰⁰

BioLet Standard Model

This model, BioLet's best-seller, has a smaller composting chamber than the XL model and a thermostatically controlled 250-watt heating element. Mixing is done manually with the handle on top; instead of flushing you simply twirl the handle a few times. The mechanism is geared down ten to one, so little effort is needed. The quiet 25-watt fan runs continuously; the heater element

runs as required by thermostat setting and room temperature. For homes with utility power. The Standard model can handle three persons full-time, or five persons for intermittent vacation use.

Dimensions: 27″ H x 22″ W x 29″ D. Requires 22″ x 40″ of floor space for the finishing tray to slide out the front.
Weight: approx. 50 lb.
Ship weight: approx. 75 lb.
Ships via: FedEx surface freight
Ships from: Massachusetts, $40 to continental U.S.
Mfr.'s warranty: 3-yr.
44-110 BioLet Standard Model $1,299⁰⁰

BioLet Basic Non-Electric

The Basic Non-Electric model uses BioLet's smaller ABS body and the manual mixing system just like the Standard model. Unlike the Standard model, there is no heater, thermostat or fan. An overflow fitting is supplied with this model, which needs to be routed to the home's greywater or septic systems. It needs a nice warm environment to live in, as there's no heater to help out the microbes. Temperatures below 70°F will slow it way down. In a friendly environment, this composter can handle up to three persons full-time. An optional 12-volt vent fan is available and recommended. Sweden. Price *includes* shipping anyplace in the continental U.S.

Dimensions: 26″ H x 22″ W x 29″ D. Requires footprint of 22″ x 40″ to allow the finishing tray to slide out.
Weight: approx. 50 lb.
Ship weight: approx. 75 lb.
Ships via: Freight prepaid (within continental U.S.)
Ships from: Massachusetts
Mfr.'s warranty: 3-yr.
44-125 BioLet Basic Non-Electric Model $995⁰⁰
◆**44-126 Optional BioLet 12V Vent Fan $59⁹⁵**

BioLet Composting Toilet Mulch

BioLet strongly recommends using their own mulch for starting and ongoing maintenance, rather than commercial peat moss. We're increasingly finding that commercial peat moss has been ground almost to dust, and has very little structure or ability to open air channels. This leads to poor composting and material clumping.

BioLet's mulch is specially formulated for composting toilets. It has more tiny air channels, speeds composting by introducing oxygen, and resists clumping. Your composting toilet only needs a cup a day, or less, so it lasts a long time. 8-gallon bag, approximately 12 lbs. USA.

44-827 BioLet Composting Toilet Mulch $29⁹⁵

Carousel Composting Toilets

The Best Choice for Full-Time or Heavy Use

Suitable for full-time residential use, the Carousel BioReactor system comes in two sizes to comfortably accommodate 3 to 5 people year-round, and as many as 32 people in a seasonal-use cabin. These systems use the superior batch composting method, resulting in more complete composting and greatly reducing the risk of disease organisms surviving the composting process. It features a round carousel design with four separate composting chambers, each of which is filled in turn. When one chamber is full, usually every two to six months, a new chamber is rotated under the dry toilet. This method has the distinct advantage of not mixing new material with older, more advanced compost; each of the four batches has from six months to a couple of years of undisturbed time to finish composting, allowing a more natural ecological cascading of the process.

Like most composters, it has drying trays under the compost chamber and a venting system with optional fan-assist to insure no-smell operations. An excess liquid overflow is also provided. With the optional 120VAC heater and vent system, the Carousel is NSF-certified.

Carousel composters can either use a dry toilet with 8″ connecting pipe, as sold below, or they can use an ultralow-flush RV toilet, such as the Sealand china toilet sold as a Sun-Mar accessory. Venting can be either passive or fan-assisted, and vent size is adaptable to comply with local codes. The complete and helpful Installation Manual details the many options available during installation.

Warranty is 25 years on fiberglass components, 2 years on all other components. USA.

Carousel Large Bio-Reactor

All Carousel models are 52″ diameter, the Large is 51.2″ high. With the heater option it handles 5+ people full-time, year-round, or 32 people seasonally.
🚚 **44-401 Carousel Large Bio-Reactor w/heater & fan $4,349⁰⁰**

Carousel Dry Toilet

Carousel Medium Bio-Reactor

Handles 3 people full-time, year-round, or 15 people seasonally. 52″ diameter x 24.8″ high.
🚚 **44-402 Medium Carousel Bio-Reactor Only $2,059⁰⁰**

Carousel Options

◆ **44-404 Composting Toilet Starter Kit $40⁰⁰**
◆ **44-403 Dry Toilet with Liner $559⁰⁰**

The Storburn Incinerating Toilet

Let's face it, much as they suit our company philosophy of recycling *everything*, composting toilets just don't work everyplace you might need sanitary facilities. Say, an unheated cabin that sees use a few weekends every winter. Leave it to the Canadians to develop a cold-weather solution. The self-contained Storburn toilet uses no water, no electricity, no plumbing, no holding tank, has no moving parts beyond the lid and seat, and won't freeze. It uses propane or natural gas to incinerate wastes. It can be used 40 to 60 times between incineration cycles, which take about 4.5 hours, and leave a sterile ash. Ash needs removal every other cycle. To reduce excitement, the burner cannot be activated while the unit is in use. A packet of anti-foam is added before lighting. The lid is locked, and the pilot light is started with the built-in ignitor. The burner shuts off automatically when finished. It has a written guarantee for no foul odors inside or outside. The cabinet is fiberglass plastic, top deck is stainless steel with a heavy-duty seat and lid, storage chamber is cast nickel alloy for a long life expectancy. Maximum propane use for a capacity load will be about 10 pounds or 2.5 gallons. Smaller loads use slightly less gas, although partial loads cost almost as much due to preheating the combustion chamber. Chamber capacity is 3 gallons. Requires 17.75″ x 31.25″ footprint. Height is 53″. Unit weighs 170 lb. One-year mfr.'s warranty. Canada.

Vent kit is purchased separately, and consists of 6″ ID Type L vent pipe (don't use the cheaper Type B pipe). Vent kits consist of pipe, firestop and stack support, flashing, storm collar, and raincap. Mobile home kit is 82″ total height; standard vent kit is 115″ total height.

A packet of MK-1 Antifoam is required every burn cycle. Aerosol masking foam is optional. The foam helps ensure quick, clean, odorless operation. A sample of each is included.

Specify white (with white seat), or beige (with natural wood seat).
Choose propane or natural gas.
Shipped by truck from Ontario, Canada.

44-807 **Storburn Model 60K, propane** $2,695⁰⁰

44-806 **Storburn Model 60K, natural gas** $2,740⁰⁰

◆ 44-808 **Storburn Std. Vent Kit** $290⁰⁰

◆ 44-809 **Storburn Mobile Vent Kit** $240⁰⁰

◆ 44-811 **Storburn MK-1 Antifoam, box of 24** $20⁰⁰

◆ 44-810 **Storburn Masking Foam, 4 Can Pack** $25⁰⁰

Composting Toilet Systems

Restrooms where you need them!

CTS digestor systems with air flush toilets are widely used in parks, recreation areas, campgrounds, golf courses, visitor centers, or even homes and cabins. These large digestor tanks will accept from two to four toilets, with some models handling over 100 uses per day. Designed for odorless operation, baffle walls and air channels are molded into the fiberglass digestor tank to provide an oxygen-rich environment for aerobic decomposition. The heat produced by composting rises naturally, carrying moisture-laden air up the vent stack. Natural airflow is assisted by either an AC or DC fan to create negative air pressure inside the tank. When the seat is lifted, air is drawn down, leaving the restroom odorless. With a minimum of moving parts (the fan and the toilet seat), or bells and whistles, there's less to break or maintain. Larger digester tanks mean larger, more diverse biological communities that are less prone to upset or short-term overloading. For high use sites, a CTS composter is the ecological toilet of choice! These are the units we usually specify for eco-resorts in sensitive environments. CTS toilets are NSF listed, and carry a five-year mfr. warranty. Basic digester packages start at about $3,800 and can go up to $5,500 depending on model. Please call for more information and help with sizing: 800-919-2400, or email: techs@realgoods.com.

CHAPTER 10

OFF-THE-GRID LIVING & HOMESTEADING TOOLS

EVERY JOB NEEDS TOOLS. The better the quality of your tools, the better your craftsmanship is likely to be, plus the job will be easier and more enjoyable. We are a tool-using, tool-loving species. It's one of the primary traits that sets Homo Sapiens apart from the rest of the animal kingdom. Good tools improve the quality of our lives, and make them more fulfilling.

Tools for better lives and more efficient living is what this entire book is about. Most of these tools we've managed to neatly compartmentalize into chapters about solar electric, water pumping, composting toilets, etc. But we've always carried a selection of useful gizmos, such as our solar-cooled safari hat, a motion-detecting deer chaser, or our push lawn mower, that defy nice, neat compartmentalization. In previous editions of the Sourcebook we've tossed all these nonconformists into a catch-all chapter called Energy Conservation. Now, we're not saying that the title Off-the-Grid Living is any more descriptively accurate than Energy Conservation, but in this new edition of the Sourcebook, we were already using the title, Energy Conservation, in Chapter Five, covering insulation, weatherization, and heating/cooling. So here, in Off-the-Grid Living, you'll find our wild, diverse, and fun collection of things that defy being compartmentalized. Enjoy browsing!

Wood-Fired Stoves & Accessories

Due to shipping costs, and safe installation concerns, Gaiam Real Goods only offers a couple of very special woodburners. The Heartland stoves are nostalgic cookstoves, with all the charm of Grandma's house, and all the quality and refinement of over 100 years of production. They are also available in propane, natural gas, or electric versions, which are more popular than the wood-fired models.

The Snorkel and Scuba stoves are unique and durable wood-fired hot tub heaters.

Heartland Cookstoves

In an age when nothing seems to last, Heartland Cookstoves are a charming exception. Though modern refinements have been seamlessly incorporated into the design, the Cookstove still evokes time-honored traditions. Our black and white catalog can't do justice to the bright nickel trim, soft rounded corners, and porcelain enamel inlays in seven optional colors (but we'll be happy to send the factory color brochure).

Two woodburning models are available. The larger Oval features 50,000 Btu/hr output, a 35˝ x 26˝ cooking surface, and a large 2.4 cu. ft. oven. The smaller Sweetheart features 35,000 Btu/hr output, a 29˝ x 21˝ cooking surface, and a 1.7 cu. ft. oven. Both feature a large firebox that takes 16˝ wood, and can be loaded from the front or top. A summer grate position directs the heat to the oven and cooktop, while limiting heating output. Flue size for both models is common 6˝, and is supplied locally. Options include color porcelain, a stainless steel firebox water jacket for water heating, a 5-gallon copper water reservoir with spigot, a coal grate for burning coal rather than wood, and an outside combustion air kit.

Gas and electric models with all the traditional styling and color options are also available. Please call for options or a free color brochure.

61-113 Oval Cookstove, Std. White $3,675⁰⁰

61-117 Sweetheart Cookstove, Std. White $2,955⁰⁰

Options (same cost for both models)

61-106 Color Option (Almond) $50⁰⁰

61-107 Color Option
         (Green, Blue, Teal, Black) $199⁰⁰

61-107 Color Option (Red) $299⁰⁰

61-119 Stainless Steel Water Jacket $199⁰⁰

61-120 Coal Grate $229⁰⁰

61-125 Outside Air Kit $75⁰⁰

Snorkel Hot Tub Heaters

Snorkel stoves are woodburning hot tub heaters that bring the soothing, therapeutic benefits of the hot tub experience into the price range of the average person. Snorkel stoves are simple to install in wood tubs, easy to use, extremely efficient, and heat water quite rapidly. They may be used with or without conventional pumps, filters, and chemicals. The average tub with a 450-gallon capacity heats up at the rate of 30°F or more per hour. Once the tub reaches the 100°F range, a small fire will maintain a steaming, luxurious hot bath.

Stoves are made of heavy-duty, marine-grade plate aluminum that is powder-coated for the maximum in corrosion resistance. This material is very light, corrosion resistant, and strong. Aluminum is also a great conductor of heat, three times faster than steel. Stoves are supplied with secure mounting brackets and a sturdy protective fence.

Two stove models are offered, the full-size 120,000 Btu/hr Snorkel, or the smaller 60,000 Btu/hr Scuba. The Snorkel is recommended for 6- and 7-foot tubs. The Scuba is for smaller 5-foot tubs. These stoves are for wooden hot tubs only.

We also offer Snorkel's precision-milled, long-lasting Western Red Cedar hot tub kits that assemble in about four hours. A 5-foot tub holds two to four adults, a 6-foot tub holds four to five adults. The 7-foot tub is for large families or folks who entertain frequently. USA.

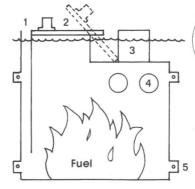

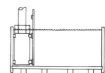

1. *Special "Snorkel" air intake.*
2. *Sliding, cast aluminum, tilt-up door that services the fire box as well as regulates the air intake.*
3. *Sunken stack to increase efficiency.*
4. *Heat exchange tubes.*
5. *Mounting brackets for securing fence and stove.*

45-400 Scuba Stove $599.00

45-401 Snorkel Stove $799.00

45-002 Cedar Hot Tub 5´ x 3´ $1,449.00

45-207 Cedar Hot Tub 6´ x 3´ $1,749.00

45-208 Cedar Hot Tub 7´ x 3´ $2,295.00

45-504 Drain Kit, PVC 1.5˝ $45.95

45-501 6´ Tub 3-Bench Kit $349.00

Shipped freight collect from Washington State.

Yard and Garden Aids

Our Gaiam Real Goods color catalogs offer our widest selection of gardening tools and gizmos, but we have a few perennial favorites that deserve year-round space here in the Sourcebook. To order one of our periodic Gaiam Real Goods catalogs call 1-800-762-7325.

Parmak Solar Fence Charger

The 6-volt Parmak Fence Charger will operate for 21 days in total darkness and will charge up to 25 miles of fence. It includes a solar panel and a 6-volt sealed, leakproof, low internal resistance gel battery. It's made of 100% solid-state construction with no moving parts. Fully weatherproof, it has a full two-year warranty.

63-128 Parmak Solar Fence Charger $195.00

63-138 Replacement Battery for Parmak, 6V $39.95

Track The Months And Days As Well As The Hours

Inscribed with the classic quotation "I Count None But Golden Hours," this heirloom sundial not only shows accurate solar time (within 15 minutes), but it has a unique 40-year perpetual calendar embedded in the dial face, revealing the day of the week and the month. Place it outdoors on any flat surface, in a sunny location, for a lifetime of accurate timekeeping. Handcast of solid recycled aluminium, with a weather-resistant verdigris finish that will not rust or corrode. Some assembly required. 12˝ diameter, optional pedestal is 15.26˝ tall. USA.

63-265 Perpetual Calendar Sundial $59.95

14-076 Roman Pedestal $79.95

14-092 Sundial & Pedestal $129.00

Manual Lawn Mower

We've found a great manual lawn mower made by the oldest lawn mower manufacturer in the U.S. This mower is safe, lightweight, and very easy to push. It's perfect for small lawns and hard-to-cut landscaping. The reel mower provides a better cut than power mowers, keeping lawns healthy and green, and it doesn't create harmful fumes or noise pollution. The short grass clippings from the mower can be left on the lawn as natural fertilizer, or you can purchase the optional grass catcher and add the grass to your compost pile. Cutting width is 16 inches. It has 10-inch adjustable wheels, five blades, and a ball bearing reel.

63-505 Lawn Mower $119.00

Add $10 for additional shipping

63-506 Grass Catcher $29.95

Put Earthworms To Work For A Healthier Garden

With our user-friendly composting system, a team of red worms digests kitchen wastes into worm castings—an organic, nutrient-rich garden amendment. The perforated stacking trays separate the worms from their castings automatically, making it very easy for you to gather the worm castings left behind. The handy spigot allows you to capture the "worm tea"—a rich liquid amendment your plants will love. Sturdy, odorless, and pest-resistant, this earth-friendly composting system is made from 100% post-consumer recycled plastic. Used indoors or out, it comes with complete instructions. Order a worm coupon, mail it to our worm supplier, and in about a week you'll receive 1,000–1,200 (1 lb.) healthy worms. Can/Australia, worms/USA.

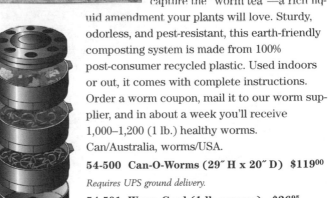

54-500 Can-O-Worms (29˝ H x 20˝ D) $119.00

Requires UPS ground delivery.

54-501 Worm Card (1 lb. worms) $36.95

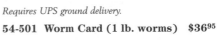

Quick, Compact, Small-Scale Composter

This is the ideal composter for urban settings; it's compact enough to store on a balcony or patio, producing up to seven cubic feet of fertile compost in as little as 30 days, without odors close neighbors might complain about. Meets all the requirements for sustainable living: It's manufactured in the U.S. of 50% post-consumer plastic, performs outstandingly, has a long useful life, and is totally recyclable. A liquid fertilizer reservoir is built into the base, designed to collect the nutritious liquid run-off for use on indoor plants, window boxes, or in the garden. It's an elegantly simple tumbler design that rolls with minimal effort on a stable base with eight built-in rollers. Secure latching door is hinged for filling, removes for emptying. USA.

65-362 Spinning Composter
(26˝L x 21˝ W x 31˝ H) $119⁰⁰

Requires UPS ground delivery.

Our Most Efficient Composter

Today's enlightened gardener recognizes composting as a simple, efficient way to reduce waste and create a nutrient-rich amendment for plants. Made of 100% recycled plastic, this unobtrusive bin traps solar heat to accelerate the production of compost. Designed for convenience, it has adjustable air vents, an easy-access hinged lid for adding materials, and a sliding bottom door for compost removal. It snaps together in minutes—no tools are required. Includes a guide to successful backyard composting, and has a five-year mfr.'s warranty. 10.2-cubic-foot capacity. Canada.

81-436 Garden Composter w/Rodent Screen
(40˝ H x 23˝ W) $69⁰⁰

Requires UPS ground delivery.

Turn And Aerate Compost More Easily

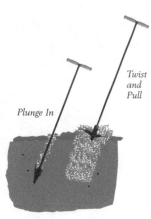

Plunge In

Twist and Pull

This perfectly simple compost turner is useful for any stationary compost pile, bin, or system, large or small, but it's especially well suited for bins with small top openings where it's difficult to manipulate a shovel or fork for the necessary turning and aerating of your compost. Just plunge the turner into the heart of the compost pile, then twist and pull. The folding wings at the tip open up to aerate and turn your compost. Won't rust or jam. Robust steel construction, with plastic handle grips. 36.5˝ long, handle is 7.75˝ across. Mexico.

81-435 Compost Turner $19⁹⁵

How Hot Is Your Compost?

Ideal composting temperatures range from about 95° to 130°F. At the high end of the range, weed seeds, disease, and insect eggs are destroyed. As decomposition is completed, the pile slowly cools. Checking the interior temperature of the compost is a good way to monitor its progress, and know when it's done. This 20˝ long, quick-read thermometer allows you to do just that. It doesn't have to be left in place, so you can monitor several piles in various stages of doneness with just one thermometer. Taiwan.

65-341 Compost Thermometer $19⁹⁵

Let It Rot
The Gardener's Guide to Composting

Stu Campbell. Gold for the garden from garbage! This is one of the basic composting texts; it details all the important methods for recycling waste, from the kitchen, the garden or the barn, for growing exceptional gardens. Sidebars and diagrams let you understand the science and explore the art. 153 pages, paperback.

82-459 Let It Rot $11⁹⁵

What To Do With Dog & Cat Poo

This is a legitimate, worm-based composting system for pet poo. Stop laughing! It really works. Most of us pick it up off the sidewalk, the grass, or the bottom of the cage already. It's a short step to dropping it in a composter instead of a trash can. Why not? You get a nutrient-rich soil conditioner great for flowers. It's odor-free and compact. Australia.

14-098 Pet Poo Composter
(9.84″ H x 22.63″ L x 14.96″ W) $99⁹⁵

Solar Dehydrator Plans

Drying is the oldest, simplest, and most natural method of preserving food. Solar food dehydration does not require added energy to presere your foods. This set of plans makes it easy to build your own professional-quality dryer. Totally enclosed, your food is protected from the direct rays of the sun and from insects. The air that dries the food is heated on the curved solar collector panel and moves through the food by natural convection. The dehydrator has 20 square feet of tray area and on an average sunny summer day maintains a temperature of 115°F. It takes two or three days for most food to dry.

80-229 Solar Dehydrator Plans $17⁹⁵

PANtrie Dries Abundant Summer Produce

Hang this PANtrie pantry in your breezeway, kitchen, motor home, or from a tree, for over five square feet of drying space for summer fruits and vegetables, then collapse to just 3″ H x 15″ square to store. Also works great for sprouting grains, beans, and seeds, plus will grow wheatgrass without any soil. Fine mesh netting walls protect food from insects and birds, while permitting free air circulation. Polypropylene drying trays are dishwasher safe. Sprouting and drying instruction booklet included. China.

63-363 Food PANtrie $49⁹⁵

The World's Most Effective Deer Deterrent

Hook the Scarecrow up to your garden hose and stake it. The smart motion sensor detects an intruder (up to 35 feet away), and sends a full-pressure blast of water right at it. It switches off immediately, using a conservative two cups of water per discharge. You pre-set the detection area (protects a 1,000-sq. ft. area) and sensitivity to prevent triggering by household pets. Sensor sensitivity is automatically dampened in windy conditions to avoid false triggers. Up to 1,000 discharges on a single 9-volt battery (not included). Reinforced nylon stake with sturdy step and hose flow-through. 24″ High overall. Canada.

54-657 Scarecrow Sprinkler $69⁹⁵

Bats Are The Ultimate House Guests

Mosquitoes don't have a chance around bats. Because they devour up to 600 night-flying insects in an hour and rarely bother humans, it makes sense to welcome bats to take up nearby residence. Made of Western Red Cedar sawmill trim, this handsome slatted bat conservatory provides shelter for approximately 40 bats. Measures 24.5″ x 16″ x 5.25″; attaches easily to your home, barn, or nearby tree. Put out the welcome mat for our native North American bats, and these fascinating creatures will reward your hospitality year after year. No more nasty mosquito repellents! Learn more about bats with Merlin Tuttle's 96-page book. USA.

54-185 Bat Conservatory $49⁹⁵
80-379 America's Neighborhood Bats $10⁹⁵

Recycled Tires Make Long-Lasting Buckets

Our buyers say these are the best buckets ever to bring compost to the garden or to get the car washed. It's the kind of tool that's irreplaceable when you need it, and having the right one on hand always makes your job easier. These long-lasting buckets are made of recycled tires and waste plastic, using a manufacturing process that is gentle to the earth. Unlike metal buckets, they won't corrode or buckle, and the handsome galvanized steel handles hold up to the hardest knocks. To suit every need, we offer them in a great value set of three: 8-quart, 12-quart, and 18-quart. A product line that delivers where the rubber meets the road. USA.

63-819 Tire Buckets (Set of 3) $19⁹⁵

Heirloom Quality Hatchet And Axe

This fine hatchet and splitting axe are heirloom quality tools crafted by Gränsfors Bruks, the historic Germany firm. The hatchet is lightweight, and handy for small jobs at the hearth or on the trail. The axe, with its grooved handle, is designed especially for splitting wood safely. Both have 3″ faces and come with custom-crafted leather sheaths; both sport time-tested hickory handles designed to fit your hand. Classically finished with linseed oil and beeswax. An informative guide to using and maintaining these tools is included with each purchase. 20-year mfr.'s guarantee. Germany.

◈ **63-817 Hatchet (1-pound head) $64⁹⁵**

◈ **63-818 Splitting Axe (3.5-pound head) $79⁹⁵**

◈ *Means shipped from manufacturer.*

The New Organic Grower

By Eliot Coleman. This is an excellent companion book to Coleman's new work, *Four-Season Harvest*. With 50,000 copies already in print, Coleman has established himself as one of the leading voices pointing a way toward solving the current ecological crisis in agriculture. Crop rotation, green manures, and garden pests are covered in depth, as are seeding, transplanting, and cultivation. Whether you garden for home consumption or for the market, this is required reading for those who want to plant and reap without pillaging and raping. 310 pages, 128 illustrations, paperback, 1993.

82-109 The New Organic Grower $24⁹⁵

Solar Gardening: A Real Goods Independent Living Book

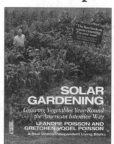

By Leandre Poisson and Gretchen V. Poisson, *Solar Gardening* shows how to increase the effects of the sun during the coldest months of the year and how to protect tender plants from the intensity of the scorching sun during the hottest months through the use of solar "mini-greenhouses." The book includes instructions for building a variety of solar appliances plus descriptions of more than 90 different crops, with charts showing when to plant and harvest each. The result is a year-round harvest even from a small garden. In *Solar Gardening* the Poissons show you how to:

• Dramatically increase the annual square-foot yield of your garden

• Extend the growing and harvest seasons for nearly every kind of vegetable

• Select crops that will thrive in the coldest and hottest months of the year, without artificial heating or cooling systems

• Build solar appliances for your own garden, including detailed instructions on how to build Solar Cones, Pods, and Pod Extenders that provide an ideal growing environment and protect plants from both extreme heat and cold

Armed with nothing but this book and a few simple tools, even novice gardeners can quickly learn to extend their growing season and increase their yields without increasing the size of their garden plot. 296 pages, paperback, 1994.

80-247 Solar Gardening $24⁹⁵

Seed to Seed

Best Book Ever on Seed Saving for Home Gardeners

Comprehensive and easy to understand, *Seed to Seed* is an outstanding reference work that enables you to save your own vegetable seeds for over 250 plants from all continents and climates. Excellent photographs teach step-by-step hand pollination. Clear instructions cover all aspects of seed production. Published by Seed Savers Exchange, the world renowned network of gardeners who relentlessly pursue and preserve our eroding genetic heritage. 222 pages, paperback.

80-255 Seed to Seed $19⁹⁵

Kitchen and Home Aids

An eclectic mix of products. Now we're really getting down to things that defy compartmentalization! The only common thread here is that these are all things you would use inside your home (probably).

This Mouse Trap Will Outlast You

If you are serious about food storage, the last thing you need is a horde of mousekins dancing in your pantry. Time to bring out the big guns. Uses no bait, simply wind the knob to catch multiple mice 24 hours a day and release. This heavy-duty galvanized steel trap, with a simple goof-proof design is the last mouse trap you'll ever need. USA.

54-493 Ketch All Mouse Trap
(5.5″ H x 9″ L x 7.25″ W) **$19⁹⁵**

www.realgoods.com

12V Blenders Make Off-The-Grid Margaritas

Make your margaritas without AC power. These two blenders come with a cigarette lighter plug for 12-volt DC operation. Grind ice, make smoothies, etc., from your boat, RV, tailgate, or 12-volt cabin. The American-made Waring unit has a long fused cord (15 feet instead of 5) and is heavier duty, with a steel base instead of plastic. It comes with a five-year warranty on the motor (one year on overall unit). If you are planning to do nonstop ice grinding, the Waring will last longer. The manufacturer of the lighter weight, portable unit recommends a 5-minute cooldown after a maximum of 3 minutes of grinding ice. For light and intermittent use, particularly where weight is an issue, it is a sensible choice. The portable unit comes with a one-year warranty and is made in China. Both units will surge to around 12 amps starting up under a heavy load. More typical consumption is under 100 watts.

63-815 Portable 12-Volt Blender $79⁹⁵

63-314 Waring 12-Volt Blender $145⁹⁵

Grind Grains, Nuts, Corn, Or Cereals—Fast

This vintage mechanical wonder lets you grind family-sized quantities quickly, making long-term storage of whole grains, beans, and nuts vastly more practical. When you do the grinding, you know exactly what's in your "whole" wheat flour. Freshly ground grains, nuts, and herb seeds retain more flavor, freshness, and nutritional value. Heavy-duty cast-iron, electroplated for rust-resistance, easy to use and clean, adjustable from fine to coarse, cranks out about one pound per minute. Clamps to your table. China.

51-325 Universal Grain Mill $39⁹⁵

Champion Juicers

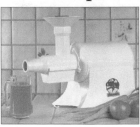

The Champion Juicer is the finest, most reliable, and most versatile juicer on the market. It works great on carrots, all vegetables, apples, and also makes nut butters. It is 120-volt AC only and runs great off of Trace inverters (812 and larger). Draw is 5.7 amps from a 120-volt source. Available in Almond or White. All units come with a one-year warranty on the motor and a five-year warranty on parts. Specify color.

◈**63-401 Champion Juicer $279⁰⁰**

◈**63-402 Champion Grain Mill Attachment $89⁹⁵**

Wheat Grass—The Most Concentrated Vegetable On Earth

Containing vitamin C, calcium, iron, and amino acids, wheat grass is a virtual pharmacopoeia of essential nutrition, sometimes called nature's medicine. This heavy-duty cast-iron, electroplated for rust-resistance, wheat grass juicer is also great for nut butters and purées. A stainless steel screen separates pulp from juice. Clamps to counter; 13″ H x 8″ W. China.

63-822 Wheat Grass Juicer $44⁹⁵

Electricity-Free Citrus Juicer

This earth-friendly mechanical device is simpler to use and easier to clean than electric models. Makes a sweeter juice, free of bitter skin oil. Features a powerful, long-lasting, rack-and-pinion gear system, leveraging up to 800 pounds of pressure with a single pull of the lever. Has stainless steel cone and filter. Good for fruit up to 4″ in diameter. Mexico.

51-250 Chrome Citrus Juicer (7.5″ H x 7″ W x 6.75″ D) $59⁹⁵

Cold Process Coffee— Beyond Compare

Bitter oils, acids, and sediments are a nasty byproduct of hot-water brewing. You won't believe the taste difference this cold process gravity fed system will make. Ideal for people with sensitive stomachs. It's easy to use, just steep the grounds in cold water overnight to produce a coffee concentrate you store in the fridge. To serve, add hot water to the strength you like—diner style for casual drinkers or full-strength for serious addicts. Great for iced coffee drinks too. No paper filters to buy. The system has everything you need except the coffee, even a handy storage carafe. USA.

54-576 Filtron Cold Water Coffee Extractor (42-Cup) $39⁹⁵

Grind Grains, Nuts, Corn, Or Cereals—Fast

This vintage mechanical wonder lets you grind family-sized quantities quickly, making long-term storage of whole grains, beans, and nuts vastly more practical. When you do the grinding, you know exactly what's in your "whole" wheat flour. Freshly ground grains, nuts, and herb seeds retain more flavor, freshness, and nutritional value. Heavy-duty cast-iron, electroplated for rust-resistance, easy to use and clean, adjustable from fine to coarse, cranks out about one pound per minute. Clamps to your table. China.

51-325 Universal Grain Mill $39⁹⁵

Updated Rotary Sweeper Does It Better Without Electricity

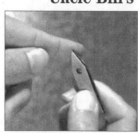

A cleaning solution that exceeds our sustainability standards—effective cleaning of floors and low-pile rugs without power, no liners to throw away, exceptional design and craftsmanship. 11″ square and only 3.5″ high for easy sweeping under furniture. Simple bottom-mount dustpan. Compact, and much lighter than the average vacuum (5 pounds). Germany.

65-361 Rotary Sweeper (11″ Square, 37″ H Overall) $69⁹⁵

Uncle Bill's Tweezers

The "Sliver Grippers" made by Uncle Bill's Tweezer Company are quite simply the finest tweezers you'll ever use. Made of spring-tempered stainless steel, the precision points are accurately ground and hand-dressed. With these tweezers it's easy to find and grip even the tiniest splinter or stinger. No pocket, purse, first-aid kit, or tool box should be without a pair! All tweezers come with a lifetime money back guarantee and a convenient holder that fits on your keychain! Our local Lyme Disease Control Center is now recommending Uncle Bill's Tweezers for removing ticks.

63-428 Uncle Bill's Tweezers (set of 3) $16⁰⁰

Rapid-Fire Can Compactor For Easy Recycling

Try as you may to limit your purchases of canned drinks, the empties somehow seem to keep piling up. Multi-Crush reduces the size of the problem. It makes recycling aluminum cans faster and easier than any other can compactor. This incredibly sturdy unit holds up to six cans at once, automatically feeding them into the crushing chamber and ejecting without interruption. Your kids will fight for the chance to use this extra fun machine. Just mount it on the wall and place your recycling bin beneath. Multi-Crush features all-steel construction and a lifetime guarantee. Measures 18″ H x 6″ W x 8″ D. USA.

51-210 Multi-Crush Can Crusher $24⁹⁵

Five Grocery Bags In One!

Our Five-in-One bag creates an efficient, eco-friendly way to shop. The roomy, sturdy bag has two pockets, each with another bag inside—just pull them out as you need them. Bottles, canned goods, and magazines fit nicely into the empty pockets. All the bags have flat bottoms and easy-carry handles. USA.

51-905 Five-in-One Bag $29⁹⁵

80% Post Consumer Toilet Tissue

Try the 12-roll sampler, then compare Seventh Generation to other "recycled" brands: Green Forest® and Quilted Northern® brands use only 10% post-consumer papers. While we applaud any effort to keep waste paper out of landfills, it only makes sense to choose the highest post-consumer content available. Also they use environmentally friendly and safe hydrogen peroxide and sodium hydrosulphite bleaching agents. Seventh Generation is 80% post-consumer, the other 20% post-industrial. 500-sheet rolls of 2-ply tissue, in the 12-roll Sampler, or by the half case (48 rolls), full case (96 rolls). USA.

52-751 Seventh Generation Toilet Tissue— Sampler $7⁹⁵

52-750 Seventh Generation Toilet Tissue— Half Case $30⁹⁵

52-786 Seventh Generation Toilet Tissue— Full Case $61⁹⁵

100% Post-Consumer Napkins

Seventh Generation recycled products saved 89,000 cubic feet of landfill last year. These napkins are 100% post-consumer, manufactured without chlorine bleach or other toxins. We offer a sampler pack, with two 500-napkin Jumbo packs. Or choose a half case (6 packages) or a full case (12 packages). USA.

52-760 Seventh Generation Napkins—Sampler $7⁹⁵

52-759 Seventh Generation Napkins—Half Case $22⁹⁵

52-758 Seventh Generation Napkins—Full Case $43⁹⁵

Ultra-Soft Facial Tissue, 100% Recycled

New Seventh Generation Facial Tissues are so soft and cushy, it's hard to believe they're not made from virgin paper pulp. No fragrances or dyes to irritate stuffy noses, no chlorine bleach, and a fresh new floral theme box. 175 tissues per box. Choose the Sampler (9 boxes), half case (18 boxes), or full case (36 boxes). USA.

52-757 Seventh Generation Facial Tissue— Sampler $17⁹⁵

52-756 Seventh Generation Facial Tissue— Half Case $32⁹⁵

52-755 Seventh Generation Facial Tissue— Full Case $65⁹⁵

Paper Towels Without Chlorine Bleach

Chlorine is not environmentally friendly. Bleaching paper with it releases dioxins and other toxins, which threaten aquatic life and poison our waterways. Seventh Generation offers an absorbent, 100% post-consumer paper towel, unbleached; a 2-ply towel that performs comparably to supermarket brands. 180 sheets per roll. Try the 4-roll Sampler, or choose a half case (8 rolls) or full case (15 rolls). USA.

52-754 Seventh Generation Paper Towels— Sampler $7⁹⁵

52-753 Seventh Generation Paper Towels— Half Case $15⁹⁵

52-752 Seventh Generation Paper Towels— Full Case $29⁹⁵

Non-Toxic Spot Remover Cleans Everything from Carpets to Clothing

We've discovered a safe, earth-conscious method for eliminating even stubborn, ground-in stains. Used for years in the world's leading hotels, this powerful spot remover works on coffee stains, wine, chocolate, ink, tomato sauce, oil, grass, and more. Just rub a small amount on the stain, then wipe off with a clean cloth. In seconds, the spot is gone, without leaving behind telltale rings or film. Use this odorless, nontoxic cream to clean any colorfast fabric, leather, or plastic—including clothing, shoes, carpets, furniture, and car interiors. Comes in a reusable 10-oz. container. USA.

53-824 White Wizard Spot Remover $7⁹⁵
      2 for $6⁹⁵ each

The Finest Organic, All-Purpose Spot Remover

Ossengal gets rid of spots and stains better than anything we've ever tried, and it's organic, odorless, chemical-free, and easy to use. It's made from the purified gall of oxen, an animal who needs tough-acting stuff to break down what it eats. Ossengal acts the same way on fabric spots and stains. Ossengal is a pure white stick of soap-like substance. Just moisten the spot, rub it well with the stick, work it in, then rinse. Most stains just disappear. Holland.

54-101 Ossengal Spot Remover $6⁹⁵
      3 for $5⁹⁵ each   6 for $4⁹⁵ each

Natural Papaya Formula Will Keep Whites Their Whitest

It's good to have options. Some detergents, especially bleach, are great for keeping your whites really white and for removing stubborn stains, but they're often too harsh for babies, allergy sufferers, or anyone with skin sensitivities. This super-concentrated and completely biodegradable bleach replacement provides an all-natural option in laundry maintenance. It's made of papaya enzymes, ginger extract, and purified water. It's excellent for brightening white fabrics, removing stains, and cleansing heavily soiled clothes. It's completely safe for all colors and fabrics, and also helps de-scale lime buildup in washing machine pipes and acts as a continual defense against further lime deposits. USA.

52-121 Papaya Enzyme Formula (16 Oz.) $17⁹⁵

Citra-Solv Natural Citrus Solvent

Citra-Solv is one of our favorite products. It will handle nearly all your cleaning needs. It dissolves grease, oil, tar, ink, gum, blood, fresh paint, and stains, to name just a few. It replaces carcinogenic solvents, such as lacquer and paint thinner, toxic drain cleaners, and caustic oven and grill cleaners. You won't believe how well it cleans your oven! Use Citra-Solv in place of soap scum removers, which use bleaching agents. Citra-Solv is composed of natural citrus extracts derived from the peels and pulp of oranges. It is highly concentrated and can be used on almost any fiber or surface in the home or workplace, except plastic. Citra-Solv is 100% biodegradable, and the packaging for Citra-Solv is 100% recyclable. USA.

54-141 Citra-Solv (quart) $16⁹⁵
54-135 Citra-Solv (gallon) $48⁹⁵

Smells Be Gone

Tackles heavy-duty odors like tobacco smoke or skunk spray; is nontoxic, odorless, nonallergenic and uses a deodorant system safe for a nursery or kitchen. USA.

54-102 (12 Oz.) $12⁹⁵
53-826 (24 Oz.) $22⁹⁵

Banish Musty Closet Odors

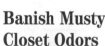

Banish musty odors from your closets with 100% natural refreshers made of absorbent volcanic minerals. Simply hang a packet of zeolite granules in air-starved closets. It actually cleans the air of powerful odors from smoke, perspiration, and perfumes—without the use of chemicals or synthetic scents. Expose the granules to the sun every eight to ten months to renew their absorbing action. Set includes two 6-oz. cloth bags for standard size closets and one 32-oz. cloth bag for walk-ins—enough for an average home. Also available, the 32-oz. Refresh-A-Basement in a mesh bag. Clears the air within 600 square feet. USA.

54-104 Refresh-A-Closet (3 Pc. Set) $17⁹⁵
54-103 Refresh-A-Basement $9⁹⁵
      2 or More $7⁹⁵ each

Renew Old Wood Finishes Safely

Restores and protects old, tired wood finishes without sanding. Ideal for cabinetry, paneling, banisters, moldings, furniture. Adds shine and UV protection; resists food stains, dust, fingerprints. 16-ounce bottle. USA.

52-793 Cabinet, Paneling & Finished Wood Restorer & Protector $12⁹⁵

Floor Restorer

Fills in scratches and eradicates dull spots, restoring vinyl, wood, marble, linoleum—almost any floor. Covers 625 square feet. USA.

52-100 Floor Restorer and Finish (32-Oz.) $15⁹⁵

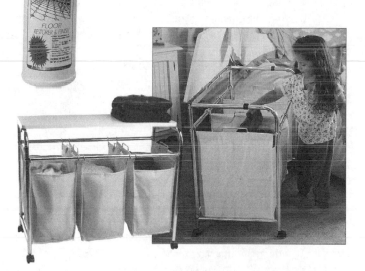

Rolling Three-Compartment Laundry Sorter

What a find! This handy organizer makes color-sorting your dirty clothes a cinch. Three roomy cotton-blend canvas bags lift out easily for wash day, and the laminate top flips down for a convenient folding table. Sturdy caster wheels let you move it wherever you need it, and lock steady while you fold. Chrome-plated frame. Taiwan.

55-279 Laundry Sorter (32˝ H x 36˝ W x 18˝ Dia.) $88⁹⁵

From the Fryer to the Fuel Tank

How to Make Cheap, Clean Fuel from Free Vegetable Oil

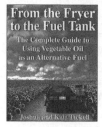

New updated and expanded version by Joshua and Kaia Tickell, who, with sponsorship from Real Goods, drove their unmodified Veggie Van over 10,000 miles around the United States during the summer of 1997, while stopping at fast food restaurants for fuel fill-ups of used deep fat fryer oil. Vegetable oil provided 50% of their fuel. They towed a small portable fuel processing lab in a trailer behind their Winnebago. This inexpensive and environmentally friendly fuel source can be easily modified to burn in diesel engines by following the simple instructions. A complete introduction to diesel engines and potential fuel sources is included. Over 130 photographs, diagrams, charts, and tables, 162 pages, paperback. USA.

80-961 From the Fryer to the Fuel Tank $24⁹⁵

Steam Power Video

For many folks steam energy may be the most easily accessible and reliable homestead power source. This one-hour video introduces the subject and gives plenty of resources if you decide to pursue the subject further. *Home Scale Steam* is a one-hour video that introduces steam physics and combustion technology. It shows all the details of a 5.5-hp steam power plant being fired up, and producing 4 kilowatts of AC power while heating the shop. This plant is demonstrated burning wood, coal, and oil. How it all goes together is carefully explained, and potential improvements are pointed out. USA.

80-042 Home Scale Steam Video $29⁹⁵

True Sight Level

A simple handheld precision instrument that allows quick establishment of levels. Simply look through the scope and line up the bubble with the cross wire. You now have a level line of sight. Will work as far as you can see. Have your partner hold a tape measure to determine rise or fall. Great for retaining walls, walks, drives, construction grades, determining total fall for hydro systems. Every homestead should have one. USA.

17-309 True Sight Level $19⁹⁵

Zip Stove

Use any combustible fuel for cooking, at your campsite or in an emergency. In our field test, a handful of oak twigs boiled two cups of water in only 90 seconds. It's incredibly compact and lightweight, with a built-in fan that blows like a blast furnace to get the most heat out of found fuel, quickly. The fan uses one AA battery (not included). Zip Stove measures 5″ dia. x 6.5″ H, and weighs 17.6-oz. The optional windshield and grill folded flat for transport measures 9.875″ L x 9″ W x .75″ D; 16.7-oz. and assures fast results. USA.

63-422 Zip Stove $59⁹⁵

63-423 Zip Windshield & Grill $24⁹⁵

Eagle Stove

The Zip Stove has been featured in the Real Goods catalog for years—it's a proven best seller and a Real Goods favorite. But some of you have told us that you need something larger to meet the needs of bigger families. Introducing, The Eagle. It has all the great features of the Zip plus a few new ones, and it's twice as big. Use it outside as a portable campfire for cooking, or providing warmth on a cool evening. It burns wood, charcoal, pinecones, any solid fuel source. It's safer than gas burning stoves and provides a "contained fire" where open fires are not permitted. The adjustable speed fan circulates incoming air around the stove walls, preheating the air and creating a blast-furnace effect that delivers 35,000 BTU per hour—enough heat to boil water by the gallon. A single D-cell battery (not included) is all you need to provide over 35 hours of cooking time. Made of galvanized steel, the Eagle Stove measures 12.5″ H x 9.75″ W x 12.5″ L and weighs only eight pounds. Complete instructions included. USA.

63-655 Eagle Stove $69⁹⁵

Solar Thermal

Besides the big job of making worldwide weather, or the small local job of heating your domestic hot water or pool, the thermal energy from sunlight can be used for heating buildings (which we fully cover in our Shelter & Passive Solar Design chapter), and for cooking food. Solar-powered cooking has enjoyed a huge surge in interest over the past few years. In many developing countries gathering fuel wood for cooking can occupy half the family members full-time, and has resulted in barren moonscapes where every available scrap of fuel gets snatched up—including any reforestation efforts. Solar cookers are quickly taking the place of the family firepit. A simple reflective box with a clear cover can easily attain 300°–400°F temperatures on an average sunny day, making an excellent oven at no cost for fuel, and with no time spent collecting the fuel. The nonprofit Solar Cookers International group provides cookers and simple cardboard and aluminum foil construction plans to developing countries. Some of our solar cooker product sales support their good efforts.

Solar cookers can be as simple as a cardboard box with aluminum foil reflectors and a dark pot with a clear turkey roasting bag wrapped over it, or they can be as fancy as our Sun Oven with insulated sides, self-leveling tray, and folding reflectors. Both will get the job done. Obviously, the Sun Oven will last longer, and has more user-friendly features, but it doesn't cook lunch any faster.

Cooking with the Sun
How to Build and Use Solar Cookers

Solar cookers don't pollute, use no wood or other fuel resources, operate for free, and still run when the power is out. A simple and highly effective solar oven can be built for less than $20 in materials.

The first half of this book, generously filled with pictures and drawings, presents detailed instructions and plans for building a solar oven that will reach 400°F or a solar hot plate that will reach 600°F. The second half has one hundred tested solar recipes. These simple-to-prepare dishes range from everyday Solar Stew and Texas Biscuits, to exotica like Enchilada Casserole. 116 pages, paperback. USA.

80-187 Cooking with the Sun $9⁹⁵

Solar Stovetop Cooker Pattern

This full-size pattern with instruction book guides you through building a reflective hotplate cooker using locally purchased supplies. Total cost will be $20–$25. Simply cut out the pattern, trace onto cardboard, and assemble. This cooker uses a circular curve rather than a parabolic curve, and a series of wedge-shaped pieces rather than trying to mold a compound surface. This makes for easier construction, and most importantly, results in a 6-inch hot spot instead of a pinpoint, for better cooking. Also includes recipes from Stovetop Pizza to Poached Salmon on Bed of Greens. USA.

80-039 Solar Stovetop Cooker Pattern $11⁹⁵

The Sun Oven

The Sun Oven is a great solar cooker. We have been using it for several years now, and love it. The Sun Oven is very portable (one piece) and weighs only 21 pounds. It is ruggedly built with a strong, insulated fiberglass case and tempered glass door. The reflector folds up and secures for easy portability on picnics, etc. It is completely adjustable, with a unique leveling device that keeps food level at all times. The interior oven dimensions are 14″ W x 14″ D x 9″ H. It ranges in temperature from 360° to 400°F. This is a very easy-to-use oven and it will cook anything! After preheating, the Sun Oven will cook one cup of rice in 35 to 45 minutes. USA.

63-421 Sun Oven $249⁰⁰

Improved Solar Cookit

A complete family-sized solar cooker kit from the nonprofit Solar Cookers International group. This helpful group provides solar cookers and information to developing countries where fuel shortages are a constant problem. Sales of Solar Cookit supports their good work.

The Solar Cookit, a hybrid of solar box and solar concentrator cookers, includes a foil-laminated cardboard reflector that folds flat to 13 inches x 13 inches x 2 inches, a 3-quart black-enameled covered steel pot, and two high-temperature cooking bags that help hold in the heat. It's highly portable and easily stored for camping, emergency use, or keeping the kitchen cool, and is large enough to easily feed three to five people. Makes a comfy seat pad when not being used too! USA.

63-576 Solar Cookit $34⁹⁵

Keep Mosquitoes Far Away— The Solar Way!

We've received countless raves from satisfied Solar Mosquito Guard owners from Hawaii to the Amazon Basin, from Alaska to Louisiana, all with the same theme: "These amazing things really work, we're mosquito free, while our friends are swatting away." The principle is they put out a high-frequency audible wave that actually repels some species of mosquitoes. Funny part is there is no scientific evidence that mosquitoes can even hear, yet they really seem to work! There is an on/off switch so you don't have to activate until the mosquitoes arrive. The battery will recharge in three hours of sunlight. China.

90-419 Solar Mosquito Guard $8⁹⁵
3 for $7⁹⁵ each   6 for $6⁹⁵ each

"When I travel upriver in the Amazon, all the natives are slapping mosquitoes at sunset, but not me. My Real Goods Solar Mosquito Guard gives me total protection from those bloodsucking predators! What a tool. I'd never leave home without it!"

Amazon Joe Easterly
Amazon Herb Company
Jupiter, Florida

Solar Safari Hat

Not since Dr. Livingston explored the Congo has technological innovation touched the pith helmet. Now, a
perennial favorite has been made much better. The Solar
Safari Hat, with its open-weave construction, adjustable
headband, and a fan powered by the sun (or 2-AA battery
back-up), is worn by gardeners, fishermen, boaters, and
African explorers. Runs directly on sunlight, or you can
switch to the backup batteries. White only. Batteries not
included.

90-411 Solar Safari Hat $49⁹⁵

50-262 AA NiMH Batteries (2-pack) $4⁵⁰

Cool Your Body Temperature
Way Down With Solar Hat Fans

Our best selling new product of 2000!

It's hot. It's cool. It's one heck of
a nifty gadget that really cools
you down on a hot day! The Solar
Hat Fan will impress your friends
and neighbors, and keep your head
and body cool on sunny days. It
clamps on to your favorite hat
(stiffer-type brims and baseball
caps are recommended) and gets to
work harvesting enough solar energy to directly power a working fan.
Features a small photovoltaic panel
with an adjustable mount. We sold
40 of these solar hat fans to a roofing company and the
workers love them as they cool down from the 100°+ temperature. Measures 2.75″ L x 2.125″ W x 2.25″ H. China.

90-323 Solar Hat Fan $8⁹⁵

Two or more $7⁹⁵ each

CHAPTER 11

MOBILITY

"Technology doesn't happen instantly; it evolves as it learns, taking along the people and stuff associated with it."

—J. Baldwin

YOUR PERSONAL CHOICE OF TRANSPORTATION TECHNOLOGY, what you choose to drive, probably has more direct ecological effect than anything else you do with your life. Most folks use more energy and nonrenewable resources in commuting, than in all other activities combined. That's an ugly little fact most of us would rather sidestep. We *love* our cars! And let's not even start on SUVs. Driving an SUV for just one year instead of a 25 mpg vehicle is equivalent to leaving your refrigerator door open 24 hours a day for six years, in terms of environmental impact. But it's a destructive love affair that's raping the world's resources to keep us all in debt. If we continue at our turn-of-the-millennium pace, all known oil reserves will be exhausted within 35 years. This is the legacy we're going to leave our kids . . . the people who will choose our nursing homes?

What Moves You?

Driving an SUV for just one year instead of a 25 mpg vehicle is equivalent to leaving your refrigerator door open 24 hours a day for six years, in terms of environmental impact.

For most folks, it's a private car or truck burning gasoline or diesel. And Lord knows, it's a cheap fuel source! People may whine like crazy when gas prices spike up above $2/gallon. But if you adjust for inflation, gas has never been cheaper. Ever. Not in the entire history of the automobile. Is this the way to treat a finite resource that's going to run out within our lifetimes? When gasoline costs less than bottled water it isn't going to be valued very highly. Prices need to rise, and eventually, either through intelligent planning, or through increasing scarcity, they will. For folks who have settled in locales, or adopted lifestyles, that require ridiculously long commutes, this will be a difficult transition period, to put it mildly. A lucky few might find public mass transit available. But all the suburban building of the past 50 years has been based on easy, cheap, individual automotive transport. What choices will be found for individual transport? In terms of a commercial availability timeline, choices will probably be hybrids, pure electric vehicles, and fuel cells. We'll look at each.

Hybrids

Hybrids will be our bridge to the future. Within 10 years almost every new vehicle will likely be some sort of hybrid drive.

A Hybrid powertrain takes the best points of internal combustion and electric drive, while eliminating the worst parts of each. Hybrid vehicles have both an internal combustion engine, an electric motor, and a highly sophisticated automatic control system that chooses what's running under what conditions. Electric motors are used at low speeds and for acceleration boost, internal combustion motors deliver cruising speed and long range, but automatically shut off at stop lights. Energy is captured during deceleration or braking by generating electricity to recharge the batteries. Hybrid vehicles are available right now. In fact, 4 of 20 Gaiam Real Goods employees at our Hopland Solar Living Center drive a Toyota Prius hybrid to work every day! Honda came out with their two-seater Insight in early 2000, Toyota started delivering their four-door Prius to North America in the summer of 2000. All major auto manufacturers are planning to offer a hybrid or two within the next two years. The ultralightweight Insight delivers 60 to 70 mpg. The more utilitarian Prius delivers 40 to 50 mpg. Both are vast improvements over the standard full-time internal combustion engine. Both these early models do continue to lean heavily on the internal combustion engine, because it's the technology we understand best. Future hybrids will rely more on electric storage and drive, and will greatly increase fuel mileage. The good news is that a Prius or an Insight can be had right now for only $20,000.

Hybrids are the technology that is most immediately accessible to manufacturers right now. Every major manufacturer has hybrid models in the works. Technology evolves. Hybrids will be our bridge to the future. Within 10 years almost every new vehicle will likely be some sort of hybrid drive.

Electric Vehicles

Pure electric vehicles thrive on cold starts, stop and go driving, and short trips. These are all the factors where internal combustion engines falter. However, it's unlikely that EVs will ever be more than one piece of our transportation puzzle. Range per charge and time required to recharge are the shortcomings of the technology. Anything more than a 50- or 60-mile one-way commute will seriously challenge most electric vehicles. The auto industry and the Department of Energy have spent substantial amounts on battery research, but nobody's come close (. . . yet . . .) to the "magic battery" that's lightweight, small, has enormous capacity, recharges quickly, is nontoxic, and has a low cost. This doesn't mean it's impossible, or that substantial gains and improvements haven't been realized; they have. But we haven't reached the performance level of jumping in your EV at a moment's notice, driving 300 miles, and recharging it in ten minutes. In other words, the kind of vehicle performance we're used to.

In the EV's defense, according to the EPA, close to 80% of all North American auto trips involve a round trip of less than 20 miles. This is very comfortably within the range of even the most modest EVs. EVs are far and away the best choice for shorter commutes and running errands in town. Another common argument is that EVs just move the pollution someplace else. This foolish argument ignores the fact that it's far easier to regulate one stationary smokestack than to regulate two million mobile tailpipes. Even if you're getting your EV juice from a coal-burning power plant, your total emissions will be about 90% less than running on gasoline.

The bottom line for EVs is that few people are willing (or financially able) to have one vehicle for around town, and a second vehicle for longer trips. We've been spoiled, we expect our vehicles to do it all.

Fuel Cell Vehicles

Functionally, a fuel cell is close kin to a gas-powered generator. A fuel cell takes a hydrogen-rich fuel source and delivers electricity, heat, and some kind of exhaust. But unlike internal combustion engines, fuel cells skip the combustion step. They do the conversion chemically. So fuel cells operate quietly, are two to three times more efficient, and produce less than half the emissions. The electrical output from the fuel cell is delivered to an electric motor, and you've got your basic fuel cell–powered vehicle. A modest battery pack is sometimes used to boost acceleration and help even out the electrical load on the fuel cell. There have been a number of fuel cell–powered research vehicles produced by a variety of manufacturers in recent years. These have ranged from trucks and buses down to little two-seater compacts. Fuel cell vehicles for retail sale to the public are at least 5 years away, and probably closer to 15 years. There is *much* research and development required yet.

In the ideal world that we're shooting for, a fuel cell running on pure hydrogen (the most common element in the universe) will deliver only pure water vapor as exhaust. On a practical level, hydrogen is found bound up with other elements to make water or hydrocarbon fuels, so pure hydrogen is expensive. Most fuel cell vehicles under development use some kind of hydrocarbon fuel, such as natural gas, alcohol, or even gasoline. The fuel is run through an onboard "fuel processor" that boosts the hydrogen content. If you run something other than perfectly pure hydrogen into your fuel cell, you'll get something other than perfectly pure water vapor for exhaust. It still beats the pants off any kind of combustion process for cleanliness, particularly the nasty smog-producing types of chemicals we get from combustion.

Since it seems apparent there won't be any significant effort to cut back on petroleum use until it's all gone, a hydrogen-based economy seems like our best bet.

Fill 'er Up With Premium Hydrogen Please

Hydrogen-enriched fuels running through a fuel processor are the easiest and least expensive fuel cell "fuel" in the short term, but most of these fuels are still petroleum-based. Since pure hydrogen runs cleaner, and isn't likely to run out, it will be the preferred fuel, if it's available for reasonable cost. In the long term hydrogen can be cracked from water by PV modules. This technology isn't well developed or very efficient yet, but it's got good possibilities. Perhaps every homestead will have a small hydrogen production plant, with the fuel being transferred to the homestead vehicle in the evenings. Or, there could be large central plants that supply whole cities. At least we won't have multi-billion-dollar cleanups when a drunk hydrogen tanker skipper goes aground in some pristine Alaskan sound. Hydrogen simply evaporates and rises.

Futurists, folks who get paid for worrying about the future, are pushing hard for a hydrogen-based economy to replace our current petroleum-based economy. Fuel cells are an important piece of this alternative future. Since it seems apparent there won't be any significant effort to cut back on petroleum use until it's *all gone*, a hydrogen-based economy seems like our best bet. Besides, if it works, it'll probably get some futurists into better rest homes. Refer to page 121 for a more thorough article on hydrogen technology.

—**Doug Pratt**

An Electric Vehicle Primer

What Is an EV?

Sparrow

An electric vehicle, or EV, is any means of transport that is powered by electricity. An EV can be a car, bus, truck, train, bike, boat, or even aircraft. An EV is driven by an electric motor, a device that converts electricity into mechanical turning power.

Though future EVs may draw energy from external sources such as power grids in the road, many EVs today need to store energy onboard and refill it when depleted. Most do this with a large battery pack, although some hybrid designs use a smaller battery that is automatically recharged by an onboard engine and generator. Toyota is doing a limited test marketing of a hybrid design in Japan.

An EV must regulate electric current flow from its storage cells to the electric motor in a way controllable by the driver. That method could be as simple as a knife-switch, but in anything larger than a bicycle, will most likely be a smooth, electronic controller.

Today's EVs

Today an EV usually means a road-going chassis with two to four wheels powered by an electric motor with a controller, and a means to store energy (usually batteries). Most batteries are recharged by an electrical battery charger, although some experimental battery types are renewed by liquid or plate replacement.

Zebra Motors

What Are the Basic Parts of an EV?

- *A purpose-built, or adapted vehicle chassis and body*
- *An electric motor mated with the driveline*
- *An electronic controller that accepts driver input*
- *Onboard storage batteries*
- *A charger, which can be onboard or offboard*
- *Driver feedback via gauges or instruments*

Make Mine Smooth

Present-day EVs use smooth, infinitely adjustable transistorized electronic switching, rather than mechanical switching. Electronic controllers drive the electric motor(s) to provide smooth, continuous, high-power output. Older-type mechanical switching required jumping between discrete power levels, which often made yesterday's EVs abrupt and jerky. Electronic controls allow today's EVs to behave just like the internal combustion engines we've all grown used to (but without the tailpipe and emissions).

Can We Continue Living Like This?

The pollution evils of internal combustion engines are well documented. We are poisoning our nest, and burning our oil reserves at a rate that will deplete all the world's oil resources in less than 40 years. The airborne toxics harm human health, causing emphysema, heart disease, and cancer. And that's the smaller problem. Of larger concern are the enormous quantities of greenhouse gases being released, causing global warming.

In some cities like Los Angeles, just breathing is equivalent to smoking a pack of cigarettes a day. And Los Angeles is clean compared to many developing nation cities!

Children are losing 15% of their lung function before adulthood. In some regions up to 70% of pollutants are traceable to "mobile sources," namely cars and trucks.

Catalytic converters don't do the job, partly because there are just too many vehicles, and partly because real life isn't a laboratory. The EPA admitted in August '96 that due to driving habits, aging, and warm-up in cold climates, cars emit far more on-road pollutants than certification tests show and have been doing so for more than 20 years.

EVs Greatly Reduce Air Pollution

Rabbit

Well, don't EVs just move the pollution source from the vehicle to a central power plant? The answer is a resounding *no!* Even if the power plant is fossil- or coal-fueled, total emissions to drive an EV will be 90% less than emissions to drive the same vehicle with an internal combustion engine. This was the finding of a very large-scale study by the Los Angeles area South Coast Air Quality Management District.

EVs have no point-of-use emissions, and the efficiency of electric generation and drive systems is far superior. EVs don't waste energy idling in stopped traffic. EVs don't become dirtier as they age. EVs can, and do, use renewable energy sources. And finally, it's far easier to monitor and control a few large stationary pollution sources than to monitor and control a million small mobile sources.

Oil Leaks

Catastrophic oil spills like the Exxon Valdez are only part of the problem. Leakage and loss occur all along the oil supply, transport, refining, distribution, and recycling system. Crude oil seeps from pipes, refineries burn off or illegally dump toxic byproducts, gasoline seeps into the water table from underground storage tanks, and every three weeks North American do-it-yourself oil changers dump an Exxon Valdez–worth of used oil into storm sewers that empty directly into lakes and rivers. Human beings are one sloppy species!

EVs use minimal petroleum products. A few lubricants and greases are all. There is no gasoline use, and no routine oil changes. There are far fewer chances for spillage, and resources are saved for future generations.

Energy Independence

The U.S. pays far more than just the huge cost of imported oil. Oil imports are the major cause of the U.S. trade deficit. A nation that depends so heavily on a single commodity import is vulnerable to disturbances that cut off the supply. Military excursions to protect oil sources take a crippling bite out of this country's budget, not to mention the human cost in death and suffering. Does anyone seriously think that the world would give a damn about Saddam Hussein if he wasn't sitting on top of huge oil reserves?

EVs use domestic and renewable energy sources, increasing energy independence.

Safety

We pay the price of car worship in auto accident casualties and injuries. A gasoline tank is a firebomb in a car, as the infamous Pinto demonstrated so well and so tragically. The explosions and fires we see in action movies primarily use gasoline, reflecting reality. Internal combustion-powered vehicles leak oil onto road surfaces, increasing the chance and severity of accidents.

EVs don't use gasoline or leak oil, they are nonflammable, and they use sophisticated electrical protection systems. The acid in batteries can present a modest threat, although modern sealed batteries are electrolyte-starved, and won't leak more than a couple of tablespoons of acid electrolyte even when completely shredded.

Lower Operating Costs

EVs cost about two to six cents per mile for electricity if you're paying regular residential rates. This is comparable to a 30 mpg vehicle running on dollar-a-gallon gas. Most utilities offer special rates for EV charging, if you're willing to do it at night, that are half of regular rates or less. Try finding a gas station that'll sell for half price if you come in off peak hours. In addition, EVs don't need tune-ups, air, fuel, or oil filters, oil changes, or new mufflers. Because they have far fewer moving parts, the amount and cost of routine maintenance is drastically reduced.

VW Pickup

To be fair, batteries will need periodic replacement. This cost adds an additional 3 to 5 cents per mile to the EV operation.

One of the most comprehensive cost comparisons was run by the largest utility company, PG&E, in the late 1980s. They ran an identical pair of service vans for 100,000 miles each while keeping careful track of *all* expenses. They charged themselves regular residential rates for recharging power. At the end of the test, the gas van had cost 22 cents per mile to operate, the EV van cost 9 cents per mile.

EV Achilles Heels

So why hasn't PG&E converted their entire fleet? Ah, there's the rub. Conversion cost is one of the EV industry's two biggest problems. It takes $5,000 to $8,000 worth of hardware to power the typical passenger vehicle, plus the time to custom-install it. Building fossil-fueled vehicles, and then converting them to electric simply isn't an efficient way to do this, although conversions are the most common type of EV in use now. Designing and building purpose-built EVs should make them cost-competitive with fossil-fueled vehicles.

The other big problem is range. The typical EV toting around a 600- to 1,200-pound battery pack can contain about as much energy as a gallon or two of gasoline. We've all been spoiled by growing up with vehicles sporting 200- to 300-mile ranges, and think we "need" this kind of performance. While this is true for a few people all the time, and for all of us occasionally, official government studies reveal a different pattern. The EPA tells us that 80% of all U.S. driving involves a round trip of less than 20 miles. This is the kind of stop-and-go, short-hop driving that gas vehicles hate and EVs excel at.

In an ideal household, Mom and Dad both have EVs that they use for all commuting, shopping and errands around town. Then the household has a single gas vehicle that is reserved for trips beyond the range of the EVs, or they use public transport, or they simply rent a car for those occasional long trips and pocket the savings in payments, insurance, and licensing.

Reliability

With fewer moving components, EVs have fewer breakdowns. They even run out of "juice" far more gracefully. Instead of the sudden shut-off of a gas-starved vehicle, EVs just gradually slow down, giving many miles of gentle warning. And even then, an EV can pull off and "rest" the batteries for 10 to 30 minutes, and then continue for a few more miles.

So, How Do I Get One?

EVs are rare; good ones are hung onto like family heirlooms. You have several choices to acquire one. You can lease, or in some cases even buy, a brand new EV from one of the major manufacturers. You can build one of your own using a sound donor car with a worn out engine, or pay someone else to build it for you. Or you can buy a used EV. These aren't common, but they do come up for sale occasionally. Used EVs that are basically sound but in need of a replacement battery pack are the most common. But like all used cars, there are good ones, and there are bad ones. Going the used EV route can be the least expensive, but you'd better fortify yourself with a wealth of knowledge about the technology first.

No matter what route you choose to EV ownership, your first stop is membership in the Electric Auto Association. This national organization only costs $39 per year to join, and includes their excellent monthly newsletter, which is full of industry news, how-to articles, product comparisons, ads from all the important EV vendors, and want ads for used EVs and parts. An EAA membership is the best way to start learning about EVs, and maybe even find a good used one.

OEM Conversions

Honda

Many of the first-generation EVs from the major automakers are electric versions of existing vehicles. These are really conversions, but are assembled on production lines as conventional cars. The factory simply substitutes an electric drivetrain. Examples include the Ford Ranger EV, the Toyota RAV4, Fiat's Panda, Peugot's 106 Electric, and Honda's Civic-based EV.

As deadlines for offering zero-emission vehicles draw closer in California, New York, and other states, most manufacturers are designing or starting to offer vehicles built from the ground up to be electric. The EV1 from GM is an example. These prototype vehicles are usually only available as a lease, since the manufacturers are looking for maximum feedback before doing nationwide releases.

Electric Conversions

An EV can be purpose-built, as a few major manufacturers are starting to do, or they can be conversions of an existing vehicle. The great majority of EVs on the road today are conversions of existing vehicles, often a vehicle that has worn out its original engine. A donor vehicle provides the rolling chassis, suspension, steering, and brakes.

During conversion, all internal combustion components such as the engine, exhaust, and fuel systems are removed. The transmission and clutch are usually retained. Brakes, springs and shocks are often upgraded to handle the additional weight of the battery pack. Many recent conversions are taking advantage of the higher-capacity, lighter-weight batteries being developed for major manufacturers' EVs. Many conversions were done using kits from small vendors. Some were done by individuals for their own use, and a few commercial suppliers have done small production runs.

Aftermarket Conversions

A number of small start-up companies have done, or currently do, conversions on complete cars by removing and replacing the driveline equipment. In the early '80s Jet Industries of Austin, Texas, did a large number of conversions on brand new vehicles that were supplied in "glider" form without engines by Ford and Chrysler. (This editor counts himself lucky to daily drive a 1980 Courier pickup conversion from Jet.) Jet went out of business in the mid '80s, due not to lack of sales or poor quality, but to fiscal mismanagement. Currently, Eco-Electric produces a conversion based on the Nissan pickup truck, and the Solectria Force is based on the Geo Metro.

EVs for Dummies:

An EV in the Hand . . .

So you are standing eyeball to headlight with an EV for the first time. An encounter with this critter can feel a bit daunting, especially if you weren't born with a wrench in your fist. Many EAA members or EV advocates are "gee, I've never even changed the oil before" types (this applies to both men and women). And, despite the title, you aren't dummies, your expertise lies in areas other than electric power systems and automotive mechanics.

Well, Silicon Valley software mavens, New England business barons, Arizona artisans, Pennsylvania poets, Texas tornadoes, and Georgia geniuses, in the words of the beer commercial, this one's for you.

I was once where you are now. Yes, I did have some engineering background, but it wasn't that applicable. High power scared me. I didn't know doodily-squat about batteries and only enough about motors to be dangerous. As far as automotive expertise went, I once put the lug nuts on backwards after my first brake job on a Datsun B210 and nearly lost the rear wheels while tootling along in San Jose (they were splayed out at the most amazing angle by the time I stopped). Anyway, such are my qualifications or lack thereof. Onward into the innards of an EV.

Most EVs come in two flavors, conversions and built-from-the-ground-up-ers. The most numerous at this point are conversions—gas cars that have had the gas guts replaced with things electric. So, you ask, what are these things and what do they do?

(At this point, I go into some really basic stuff for those who know zip about electricity. If you feel in danger of having your intelligence insulted, skip to part B.)

A. The *Really* Basic Stuff

Let's start with a mini-EV. If you've got kids, you've probably tripping over the little EVs all around your house—toy cars and trucks that run on flashlight batteries. You turn them on, put them on the floor and they run under a chair or into a corner and get stuck. No matter. Fish one out from behind the cat's litter-box or the coffee table and take a look at it.

Leonardo did it too

Baker Electric

All EVs, from the turn of the century Bakers to today's GM Impact are basically scaled-up versions of these little Radio Shack runabouts. Not that full-scale EVs are toys, oh no. The GM Impact can blow the doors off a Mazda Miata and an American EV just recently topped the 200 mph mark. We're just using the toys as models. So did Leonardo da Vinci.

Incidentally, toys run on electric drives because they are simple, cheap, easy to manufacture and safe enough for kids to use. Think about that.

OK, so you've got your son or daughter's rug-rat entertain-mobile up on the desk. It probably has a switch thingy on the bottom. Turn it on, it runs. Turn it off, it stops. Dump out the batteries. It'll probably have at least two, perhaps as many as six. You have in your hands the bare-bones basics of a full-size EV. The batteries that store this mysterious thing called electricity, the wires that hook the batteries up to the motor and a switch that starts and stops the motor. Put the batteries back and turn it on. It runs, right? (Unless you stuck a battery in backwards.) So what is going on here?

Racehorses

The show starts with the batteries. One end is marked plus, the other, minus. When you tape a wire to each end and connect up a light or motor, the bulb lights or the motor runs. If you play around with batteries, wires, lights, and motors, you soon discover that electricity (like a bunch of little racehorses) likes to run around in closed tracks from start to finish. These tracks are called circuits. For physical and chemical reasons, the track must be made of a metal such as copper, steel, or aluminum. These metals let electricity run through them easily; they are called conductors. Rubber, plastic, and wood block electricity; they are called insulators. Wires have a conductive metal center (the track for the electricity to run in) and an insulating coat (to keep the electrical racehorses on the track.)

What goes around comes around . . .

Epic

To change analogies, a battery is like a pump that pushes water out into a pipe. If you run your pipe back around to the pump's intake, you have a circuit ("cir" is a Latin root, meaning around, as in circle.) The same with the electricity in a battery. If you run a metal wire (don't actually do this, just imagine it) from the plus terminal to the minus terminal of a battery, the electricity will happily run around from plus to minus. That is a closed circuit. (The electricity will be so happy, in fact, that it will run like a bunch of demented little racehorses from plus to minus, drain the battery and burn or melt the wire. This is called a short circuit. It results in dead batteries, burned or melted wires, rapid heartbeats, and possibly the release of a certain amount of battery contents as steam or smoke. Avoid.)

If you cut the your pipe circuit, the water will all spill out on the ground. If you cut (or "open") your wire circuit, the electricity doesn't spill out, it just stops, like a line of frustrated racehorses facing an obstacle they can't jump. A switch is just something that opens and closes a circuit.

So we've got four really dumb basic things about electricity: 1) it likes to run around in closed tracks made of metal (wires), 2) it runs from the plus side of a battery to the negative side, 3) if you cut the wire or open a switch, it stops, and 4) if you connect a light or motor into the circuit, the light shines and the motor runs (*why* a motor runs is something this column will go into later—all we have to know now is that it does). Knowing those four things, you can understand what is going on in both mini- and maxi-EVs.

When you load up the batteries (with the switch off) the mini-EV won't go. The electricity in the batteries is like a bunch of racehorses ready to run, but until the starting gate releases, nothing happens. Slide the switch and you close or complete the circuit. OK, simple. The electricity runs from the battery positive terminal, out the wires, through the motor—making it run—and back to the negative terminal. The motor spins and turns the wheels.

The earliest EVs were exactly that—batteries, switches, and motors. Really easy. Falling-off-a-log easy. That's why the first EV was built much earlier than the first gas car—in the 1830s versus the 1880s. The batteries weren't rechargeable, but the thing ran. Rechargables came along in the 1850s when LaPlante developed the lead plate, dilute sulfuric acid technology that is still going strong today.

The battery lineup

Why does your mini-EV need more than one battery? To get the car moving requires a certain amount of oomph out of the motor. The motor power comes from the amount of "push" in the batteries. This "push" is called "voltage." To return to the fluid-circuit model, if you put another pump in-line with the first one, the water would flow much faster. It would also turn a much bigger and heavier paddlewheel, if you stuck one in. Batteries are like pumps in that respect—if you line them up, plus to minus, their individual "push" amounts (or voltages) add up. The more batteries you have "in series" or lined up with each other, the more push or voltage you get and the bigger or more powerful motor you can drive. Note that the amount of electrical flow ("current") stays the same, but it moves faster. If you prefer the equine analogy, it is the same bunch of racehorses, but their eagerness (voltage) to go from start to finish is increased and they run faster.

So we have all the bits of a very basic EV—batteries connected in series to produce a "push" or voltage that will run the motor, a switch to open or close the circuit, and a way to link the motor to the wheels. Close the switch and hi-ho Seabiscuit, off we go!

B. The Usual Basic Stuff

If you've jumped to this point, you feel you have a good understanding of how a mini-EV, such as a toy truck or car, works. You also are familiar with concepts such as current and voltage. OK, so what happens when you scale the little guy up into something like a Chevy Blazer? Let's look at the basic bits or components in turn.

1. Motors—Most of the maxi-EV motors are just scaled-up versions of the little guys. The most common ones are called series-wound DC motors. In mini-EVs, the motor is connected directly to the wheels, or uses a very simple small gearbox (in most cases, just to make the motor shaft turn a corner). In maxi-EVs things get a bit more complicated. Some built-from-the-ground-up lightweight EVs stay simple (and efficient!) by just connecting the motor directly to the wheel via a shaft or a belt. Most EVs are conversions from gas cars that have more complex gearboxes, or transmissions as well as clutches.

What's a tranny, granny?

For the nonautomotively minded, here's an explanation of why gas cars have clutches and transmissions. Otherwise, skip the next two paragraphs.

Toyota

Gas car engines have to keep running all the time. If they stop, they stall and have to be re-started (by an electric motor!). Gas car engines turn much faster than the car's wheels, delivering their peak power in a band from 3,000–6,000 revolutions per minute (RPMs). These engines have to run through a mechanical gearbox that translates every ten turns of the motor, for instance, to one turn of the wheels. To go at different speeds, the car has to vary the ratio of motor turns to wheel turns. That is done mechanically—that's what your gearshift lever does.

In order to keep the gas engine running fast enough that it doesn't stall and can deliver sufficient power to move the car, the gearbox also needs a clutch. A clutch is just a mechanical way to temporarily decouple the engine from the gearbox, either while the car is stopped (to avoid stalling the engine) or while changing ("shifting") the ratio of motor turns to wheel turns.

Power from the get-go

Series DC motors have high turning capability ("torque") right from the get-go. You don't have to worry that the motor will cough, choke, and die on you (though you can stall an electric motor if you force it to work hard enough). They also deliver power in a much wider band, from 0 to 7,000 RPM. So, theoretically, you don't need a clutch and gearbox. Some EVs, even conversions, dispense with all that clutch and gearbox nonsense. Some use a gearbox without a clutch.

However these motors do have their limits, so in order to achieve higher speeds, the car needs to change the ratio of motor turns to wheel turns; so, Fanny, we're back to the ol' tranny. Furthermore, as the series motor spins faster, it loses turning force or torque. Again the most common way to solve this is via the gearbox; i.e., let the motor turn more slowly so that it can develop more torque. Another is to go to motors that don't have this characteristic, or that can rev super-high while putting out sufficient torque. These exist, but are more expensive and/or require complex drive electronics. GM's Impact EV has such a system.

Motor-to-transmission adapters

So, most EV conversions utilize the clutch and tranny, because they happen to be conveniently there. The most common conversions use an adapter plate that bolts on one side to the electric motor and the other to the case of the gearbox. The clutch in a gas car is located inside the flared end of the transmission case ("the bell housing"). The adapter plate must connect the transmission to the electric motor and also allow the clutch to operate. It also must ensure that the shaft from the motor and the shaft from the gearbox stay in-line (concentric). If not, one shaft will wobble with respect to the other, creating vibrations that become destructive at high RPMs and create rapid self-disassembly (read as "the whole thing comes apart"). Adapter plates, therefore, have bushings or bearings that hold the two shafts in-line while allowing the clutch to couple and decouple them.

Motor to transmission adapter plates are the trickiest part of doing an EV conversion. It is much easier to buy one, since they are available for nearly all types of cars. Most EV conversion kits include adapters.

Batteries and wires

How does the mini-to-maxi scale-up affect batteries and connections? The ones in your countertop pocket rocket each have 1.5 volts worth of push; all lined up they deliver anywhere from 6 to 9 volts. That's plenty to spin the tiny motor, and it only needs one-fourth to half an amp of electrical current. The little wires can carry this easily. For a Chevy Blazer or Ford Escort, you are into the 96–120-volt range. That means a lineup of more and bigger batteries. That also means that small spring connectors become big rectangular or cylindrical lugs and wires become cables. Tight connections become very important here.

The larger batteries come in standard voltages, including 6, 8 and 12 volts. They have to be rechargeable, since they are too big and expensive to chuck in the trash (not a good idea, even with the small ones). The sixes are most common, although there is a growing shift to 12s and a pioneering use of 8s. Like batteries in a mini-EV, big batteries get run flat, so they have to be able to take it repeatedly. Batteries designed to recover from deep discharge are called traction or deep cycle batteries. They are made for industrial trucks and golf carts and adapt very well to use in road-going EVs. Both mini-and maxi-EV batteries are mostly the lead-acid type. Other types are also in use, but rare.

Car starting batteries are a different animal, called an SLI or Starting, Lighting and Ignition battery. They will deliver a heavy engine-cranking current for a short time, but run them flat too many times and they roll over and die. EV racers and hotrodders use SLIs, but they don't hold up in commute or other street service.

Switches

You can run a big EV the same way you run a little one: simply by closing a switch. A number of electric boat racers just use a big hefty knife-switch—slam 'er down and hang on! My friend, "Thousand-Amp Fred," swears by it. The big knife-switch way works. Certainly that's one way to spin your tires in the parking lot. In the low-speed, low-powered electric buggies of 1905, it worked quite well. But as the greed for speed lead to more powerful systems, the ol' all-or-nothing switch approach got dicey. At higher voltages, switches arced and burned. And the jerk at takeoff could whiplash the strongest of necks.

EV designers tried to solve the problem by putting in a device that controlled how much current went through the wire by making the current work harder to get through. This is called a resistor, because it resists the flow of current. (For you A-trackers: Think of your bunch of racehorses hurtling down on the track, getting squeezed down into a single lane or a tunnel. They'll jump around and bounce off each other and things will start getting pretty wild and hot. They'll also have to slow down. That is what a resistor does—slows down current by turning it into heat.)

Resistors can be made so that they can vary how much they impede current flow (just like narrowing or widening our tunnel so the racehorses speed up or slow down). Some early EVs used huge variable resistors called rheostats, others used networks of large resistors interconnected by switches, others tapped the battery pack at different points to get different amounts of "push" (voltage) to the motor. One of my first EV rides was in a Morgan replica kit car that used resistor switching and battery tapping. It was a lively, if rough-riding, little beast.

Enter the chopper controller

Spoiled by their exposure to gas cars, which gave a continuous surge of power as the accelerator was depressed, EV drivers and designers turned to alternatives. Dissipating current through a huge rheostat or a resistor network ate up a lot of energy and generated heat (though a variation of this, called "plug braking" is still used in electric bus and tram systems.)

Due to a characteristic called inductance, electric motors don't mind if you switch the current on and off while running, if you do it fairly fast. The motor sees this string of on/offs, or pulses of current and voltage, as a constant level, though lower than full-on. With long off times and small on times, the perceived voltage is low. As on times become longer and off times shorter, the perceived voltage rises. So, accordingly, does the power output of the motor. Because it varies or "modulates" motor power using the duration, or "width" in time, of the power pulses, this method is called Pulse-Width Modulation (PWM). Since it "chops" the power into pulses, it is also called chopper control.

You can implement chopper-type motor control mechanically (by just clicking a switch on and off manually—try it!) You can also do it with tubes (remember them?); however, it is easier with an electronic device called a transistor. William Schockly came up with the transistor in 1957. It literally made today's world. Simply, a transistor is an electronic switch, opened and closed by a very small current that acts like the lever or push-button of a mechanical switch. Power transistors can switch high levels of current and voltage without arcing or overheating.

Motor control using transistors was soon common in industry, but it wasn't until the late '60s that EAA Member Frank Willey applied it to EVs in his "Willey 9," the first transistorized PWM motor controller available to individual EV builders. This homebuilt box set the pattern of today's EV controllers, such as the Curtis 1221-B and others. To run an EV with the Willey 9 or a Curtis, you rig the accelerator pedal with a small variable resistor called a potbox. Electronics in the controller continuously read the potbox resistance as the driver depresses the pedal. Big power transistors switch the high DC voltages and currents. Electronic cleverness translates the potbox resistance reading into signals that tell the power transistors how fast to switch and how long to stay on.

Willey's use of power transistors in pulse-width modulated control made an EV's pedal response much more like a gas car's and brought mid-20th century homebuilt EVs from backyard projects to serious contenders in the fight for pollution-free transport.

So, don't be intimidated by that black box with channels or fins on the outside and big impressive-looking lugs. It's only a smartened-up, fancily packaged version of a simple knife-switch.

Now you should be able to look that EV right in the headlights without a qualm, since you now understand more about it than it does about you. Well, maybe.

Sources of Further Information

Electric Auto Association (EAA)
P.O. Box 6661
Concord, CA 94514
Website: www.eaaev.org

The national electric vehicle group since 1967. Has local chapters in many states. Yearly membership is $39, and includes an informative monthly newsletter with tech articles, industry news, and want ads. The website has links to practically everything and everybody EV related.

Thanks to Clare Bell, EAA Board Member & Editor, for providing the EV writing from pages 440–448.

Electrifying Times
63600 Deschutes Market Road
Bend, OR 97701
Website: www.teleport.com/~etimes

A commercial EV magazine that comes out three times yearly. Runs 60 pages or more. Much up-to-date information, and ads from suppliers and builders. Edited by Bruce Meland. $12/year.

Department of Energy
Office of Transportation Technologies
Hybrid Electric Vehicle Program
Website: www.ott.doe.gov/hev/hev.html

The DOE's well-intentioned, but not yet fully stocked info site. What hybrids are, how they work, where you can buy one. There's a lot of good info here, but also a lot of dead links.

Marshall Brain's *How Stuff Works* **website**
"How Hybrid Cars Work"
Website: www.howstuffworks.com/hybrid-car.htm

The How Stuff Works *website is great fun, highly informative and has good interactive graphics, but you'll need to put up with some commercial content. Their hybrid site is very complete.*

Mobility Products

The New Electric Vehicles

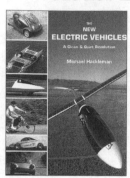

By Michael Hackelman, long-time renewable energy and EV pioneer. Distilling two decades of experience, in language you can understand, Michael covers everything, including conversions, scratchbuilts, Formula and Electrathon racers, electric motorcycles, electric bicycles, and even aircraft and watercraft. This is the most complete, and as of 1999, the most up-to-date publication on electric vehicles. 272 pages, over 500 illustrations and drawings, and over 175 sidebars. Published using solar and wind energy. USA.

80-901 The New Electric Vehicles $24^{95}

From Gasoline to Electric Power

A Conversion Saga

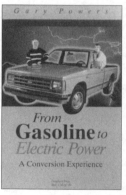

So you want to convert your gas hog to a pristine electric vehicle, but lack the automotive and electric skills to get started? Well, you won't find them here! This is a sometimes-funny, always engaging saga of one man's quest to convert his beloved pickup truck. Gary was a rank amateur when he started, but with the help of a patient and experienced supplier hundreds of miles away and a whole lot of pluck, he managed to complete a successful—and entertaining—conversion. This is definitely not a how-to manual (sometimes it reads more like a how-*not*-to manual!), but anyone considering The Big Step will not only enjoy the story, but leave it with some real insight into exactly what such a conversion really entails. Well-illustrated, with appendices, index, and references. 114 pages, paperback, 1997. USA.

80-730 From Gasoline to Electric Power $14^{95}

From the Fryer to the Fuel Tank

How to Make Cheap, Clean Fuel from Free Vegetable Oil

New updated and expanded version by Joshua and Kaia Tickell, who, with sponsorship from Real Goods, drove their unmodified Veggie Van over 10,000 miles around the United States during the summer of 1997, while stopping at fast food restaurants for fuel fill-ups of used deep fat fryer oil. Vegetable oil provided 50% of their fuel. They towed a small portable fuel processing lab in a trailer behind their Winnebago. This inexpensive and environmentally friendly fuel source can be easily modified to burn in diesel engines by following the simple instructions. A complete introduction to diesel engines and potential fuel sources is included. Over 130 photographs, diagrams, charts, and tables, 162 pages, paperback. USA.

80-961 From the Fryer to the Fuel Tank $24^{95}

Convert It

Do It Yourself

By Mike Brown. This book is a step-by-step manual for converting a gas car to an electric-powered car. It's a very readable and practical manual for the do-it-yourselfer wanting the fun and educational experience of converting a conventional automobile into an electric vehicle that will be both practical and economical to operate. Included are a generous number of illustrations of an actual conversion, along with instructions for testing and operating the completed vehicle. Brown includes many practical tips for both safety and ease of construction and does a great job of explaining the cost of the conversion components. This book has been strongly recommended by the Electric Auto Association, Electric Vehicle Progress, and Alternate Energy Transportation. 126 pages, paperback. USA.

80-404 Convert It $24^{95}

Solar Golf Cart Conversion Kit

Ideal for Both Shade and Solar Charging!

This kit is designed specifically for solar charging an electric golf cart. Golf carts are 36-volt DC devices. To charge one, you need a multiple of three 12-volt PV (solar electric) modules. You also need a special 36-volt charge controller. We have picked a PV that is hard shadow and impact tolerant. The three Uni-Solar 42-watt modules (item # 11-233) have aluminum perimeter frames, measure 29.1″ x 36.5″ each, and should be fairly easy to fit as a shade roof for the golf cart. Our kit also contains a 36-volt/8-amp charge controller (item #25-726) to provide automatic battery charge protection. You provide the mounting and the wire as needed for your particular cart, and you've got yourself a solar-powered golf cart. Twenty-year warranty on the PV modules, five-year warranty on the controller; both made in the USA.

12-110 Solar Golf Cart Conversion Kit $899⁰⁰

Fuel From Water: Energy Independence With Hydrogen

By Michael Peavey. An in-depth and practical book on hydrogen fuel technology that details specific ways to generate, store, and safely use hydrogen. Hundreds of diagrams and illustrations make it easy to understand, very informative, and practical. If you're interested in using hydrogen in any way, you will find this book indispensable. 244 pages, paperback, USA.

80-210 Fuel From Water $25⁰⁰

The ElectriCruizer SX From ZAP®

Combine Comfortable Classic Styling with a Futuristic Kick.

Use the patented ZAP® Power System to tackle nasty hills or get across town without working up a sweat. You always pedal with the motor, but now you can go up to 15 mph, depending on your weight and terrain, with much less effort. Boosting you along for 14 miles per charge, the ElectriCruizer amazingly uses less energy than a lightbulb. Recharge anywhere with the portable 120-volt battery charger. Now you have a fun excuse to leave your car at home and ZAP® into the great outdoors, pollution-free. The ElectriCruizer is a comfortable city bike with six speeds, padded seat, upright handlebars, chain guard, cantilever brakes, and big cushy 26″ balloon tires. This high-tensile bike comes fully assembled (except fenders to prevent shipping damage) in black, red, or blue. Increase the utility of your ElectriCruiser with the optional chrome wire dual baskets. Each roomy basket measures 10″ D x 14″ L x 7″ W, is joined to a central rack, and the entire unit easily attaches over the rear fender. 17 amp-hour sealed battery. USA.

◈ **91-803 ElectriCruizer SX $765⁰⁰**

◈ **91-805 Optional Rear Baskets $49⁹⁵**

Please specify color.

◈ *Means shipped from manufacturer.*

ZAP® Power System

Convert Your Own Bike

Do you already have a favorite bike and want to make it electric? The ZAP® Power System is designed to fit your bike or your money back. A powerful 300-watt permanent magnet DC motor delivers power to the rear tire through ZAP's® patented tire-transmission system. A sealed 12-volt lead-acid battery comes with a heavy-duty nylon carry case that installs at the center of the frame for a low center of gravity and optimal handling. The throttle switch is attached to the handlebar. An off-board automatic battery charge plugs into any 120-volt outlet. Includes installation manual and video. Will fit on almost all bikes with 16.5˝ frame sizes or larger. To be able to charge 12V batteries while using the bike as a stationary exercycle, add the optional charging stand and 60A Diode. USA.

◈ **91-804 ZAP® Power Kit $379**⁰⁰

◈ **63-302 Charging Stand $95**⁹⁵

25-757 60A Stottky Diode $49⁹⁵

ZAPPY™ Folding Electric Scooters

The design may remind you of your childhood, but this is a real transportation solution and a commuter's dream. The handlebar folds flat, and the scooter rolls smoothly when folded, so you can store it in your car's trunk and comfortably tote it around on public transit. Runs 8 miles on a fully-charged maintenance-free sealed dry cell battery (13 mph). Battery and charger (both included) fit neatly into a built-in, easy-access storage cubby; remove them for charging indoors, or leave them in the scooter and recharge in garage or patio. Charger plugs into any 110VAC outlet; two to three hours to full recharge. Features a gas strut to absorb sidewalk shock, adjustable handlebar height, spring-loaded on-off switch, and a rear brake. 41˝L x 11˝W x 11˝H when folded; 37 pounds. A surprisingly stable design, the ZAPPY Mobility is the standard ZAPPY scooter with a seat for easy and fun mobility! (Assembly required.) Seat is also available as an add-on kit for your existing ZAPPY. Wear your helmet! USA.

◈ **14-135 ZAPPY Mobility Scooter $649**⁰⁰

Please specify Red, Blue, or Black.

◈ **68-415 ZAPPY Seat Kit $79**⁹⁵

◈ **68-414 Zappy Electric Scooter $599**⁰⁰

APPENDIX

Gaiam Real Goods' Mission Statement

Through our products, publications, and educational demonstrations, Gaiam Real Goods promotes and inspires an environmentally healthy and sustainable future.

Part I: Who We Are & How We Got Here

Who Put the "Real" in Real Goods

As did many of his contemporaries in the 1960s and early 1970s, John Schaeffer, founder of Real Goods, experimented with an alternative lifestyle. After a protracted exposure to nearly every strand of the lunatic fringe, he graduated from U.C. Berkeley and moved to a commune outside of Boonville, California. There, in an isolated mountain community, John pursued a picturesque life of enlightened self-sufficiency.

Despite the idyllic surroundings, John soon found that certain key elements of life were missing. After several years reading bedtime stories to his children by the flickering light of a kerosene lamp, John began to squint. He grew tired of melted ice cream. He began to miss the creature comforts the family was lacking due to their "off-the-grid" lifestyle. He yearned for just a tiny amount of energy to strike a balance between the lifestyle he had grown up with and complete deprivation. In other words, John came to the realization that self-sufficiency was much more appealing as a concept than a reality.

Then he discovered power from the sun. John hooked up one humble photovoltaic module to a storage battery, with just enough juice to power lights, a radio, and the occasional television broadcast. Despite his departure from a pure lifestyle, each and every time that "Saturday Night Live" aired, John's home became the most popular place on the commune. Eventually, when the 12-hour community work days began to take their toll, John took a job as a computer operator in Ukiah, some 35 twisty miles from Boonville.

Once the word got out that John would be making the trek over the mountain to the "big city" daily, he became a one-man delivery service, picking up the fertilizer, bone meal, tools, and supplies needed for the commune. As a conscientious, thrifty person, John spent hours scrutinizing the hardware stores and home centers of Ukiah, searching for the best deals on the real goods needed for the communards' close-to-the-earth lifestyle. One day, while driving his Volvo (what else?) back to the commune after a particularly vexing shopping trip, a thought occurred to John. "Wouldn't it be great," he mused, "if there was one store that sold all the products needed for independent living, and sold them at fair prices?" The idea of Real Goods was born.

Real Goods at the Turn of the Millennium

From its humble beginnings in 1978, Real Goods has become a real business, with Real Employees serving Real Customers, and even with Real Shareowners. The company can now lay claim to the title of the Oldest and Largest catalog firm devoted to the sale and service of renewable energy products. Real Goods, now "Gaiam Real Goods" since its January 2001 merger with Gaiam, Inc., of Colorado, is still devoted to the same principles that guided its founding—quality products for fair prices, and customer service with courtesy and dignity. Early on, John managed to turn his personal commitment to right livelihood into company policy, pioneering the concept of a socially conscious and environmentally responsible business. The Company has been consistently honored and awarded for its ethical and environmental business standards. Plaudits include Corporate Conscience Awards (from the Council on Economic Priorities); inclusion in *Inc.* magazine's 1993 list of America's 500 Fastest-Growing Companies; three consecutive Robert Rodale Awards for Environmental Education; Northern California Small Business of the Year Winner for 1994; finalist for entrepreneur of the year two years running; and news coverage in *Time, Fortune, The Wall Street Journal, Mother Earth News*, numerous TV appearances, countless Japanese magazines, plus many thick scrapbooks full of press clippings. Has all this attention gone to our heads? You bet it has.

Five Principles to Live By

Gaiam Real Goods is considered so newsworthy not because our methods reflect the latest trends in corporate or business-school thinking, but because we have unwittingly helped to birth an astonishingly healthy "baby"—an ethical corporate culture based on environmental and social responsibility. Led by a certain naiveté and affection for simplicity, Gaiam Real Goods has discovered some simple principles that, by comparison to the "straight" business world, are wildly innovative. This has not, repeat not, been the work of commercial gurus or public relations mavens, but rather the result of realizing that business need not be so complicated that the average person cannot understand its workings. Our business is built around five simple principles.

Principle #1: This Is a Business.

And a business is, first and foremost, a financial institution. You can have the most noble social mission on the planet, but if you can't maintain financial viability, you cease to exist. And so does your mission. The survival instinct is very strong at Gaiam Real Goods, and that reality governs many decisions. To say it another way, you can't be truly sustainable if your business isn't economically sustainable (read profitable). To ensure the continued flourishing of our mission, we pursue profitability through our catalogs, retail stores, internet presence, design and consulting businesses, and co-publishing efforts (more about these later).

Principle #2: Know Your Stuff.

Knowledge seldom turns a profit, yet our social and environmental missions cannot be achieved without it. The independent lifestyle we advocate relies largely on technologies that often require a high degree of understanding, and a level of interaction that has been largely forgotten during our half-century binge on cheap power. We aren't interested in selling people things they aren't well-informed enough to live happily with. We want people to understand not only what we are selling and how it is used, but how a particular piece of hardware contributes to the larger goal of a sustainable lifestyle.

It goes against the grain of mainstream business to give anything away. Even a "loss leader" is designed to suck you into the store to buy other, higher-profit items. At Gaiam Real Goods, knowledge is our most important product, yet we give it away daily, through our catalogs (there's a lot in there to learn, even if you never buy a thing); through our Solar Living Center (free self-guided and group tours), now run by the non-profit Institute for Solar Living; through free workshops during the ISL's annual SolFest renewable energy celebration; and through our nonprofit Institute for Solar Living. Our website acts as a launching pad for renewable energy research, leading you to fascinating information on sustainability topics of all kinds. We believe that as our collective knowledge of sustainability principles and renewable energy technology increases, the chance of achieving the Gaiam Real Goods mission increases, too.

Principle #3: Give Folks a Way to Get Involved.

The lines between who owns the company, who works for it, and who benefits from it are deliberately fuzzy at Gaiam Real Goods. Most of our shareowners are our best customers (or is it the other way around?). Most of our employees are also our customers. Our Lifetime Membership program honors our community of customers with financial and educational benefits, and interns and volunteers are welcomed warmly by the ISL staff. Gaiam Real Goods is becoming a real community, acting in concert toward the common goal of a sustainable future. Distinctions (between who is, does, makes, buys, or sells what) become irrelevant. We are all in this together.

Principle #4: Walk the Walk.

At Gaiam Real Goods, we conduct our business in a way that is consistent with our social and environmental mission. We use the renewable energy systems we sell, and we sell what works. Our merchandising team makes absolutely sure the merchandise we sell performs as expected, is safe and nontoxic when used as directed, and is made from the highest quality sustainable materials. We challenged our customers to help us rid the atmosphere of one billion pounds of CO_2 by the year 2000, and we achieved our goal three years ahead of schedule. Real Goods has been recognized by the Rodale Award, as the business making the most positive contribution to the environment in America, for two years running. We don't just talk the talk at Gaiam Real Goods, we walk the walk.

Principle #5: Have Fun!

We throw the best party of the year for 10,000 to 15,000 of our closest friends every summer in August on the grounds of our Solar Living Center in Hopland. We look forward to our annual renewable energy celebration all year. It's a chance to give a little bit back to the faithful folks who form the lifeblood of our business. If we've learned anything since 1978, it's that all work and no inspiration makes Jack and Jill a couple of burnt out zombies. SolFest is our little reminder to take care of ourselves with some good, clean fun, so we'll be rejuvenated and reinvested in the hard work of creating a sustainable future.

Part II: Living the Dream:
The Real Goods Solar Living Center

Imagine a destination where ethical business is conducted daily amidst a diverse and bountiful landscape, where the gurgle of water flowing through its naturally revitalizing cycle heightens your perception of these ponds, these gardens, these living sculptures. You follow the sensuous curve of the hill and lazy meanders of the watercourse to a structure of sweeping beauty, where floor-to-ceiling windows and soaring architecture clearly proclaim this building's purpose—to take every advantage of the power of the sun throughout its seasonal phases. A few more steps and the spidery legs of a wind-generator fall into view, and the top of a tree that looks as though it might be planted in a Cadillac. An awesome sense of place begins to reveal itself to you. Inside the building, sunlight and rainbows play across the walls and floors of a 5,000-square-foot showroom, and you begin to understand that all of this, even the offices and cash registers, are powered by the energy of sun and wind. Welcome to the Solar Living Center in Hopland, California, the crowning achievement of the Gaiam Real Goods mission.

Our Solar Living Center began as the vision of Gaiam Real Goods founder and President, John Schaeffer. His dream was to create an oasis of biodiversity, where the company could demonstrate the culture and technology of solar living, where the grounds and structures were designed to embody the sustainable living philosophy of Gaiam Real Goods' catalogs and business. With the opening of the SLC in April 1996, John's vision is now a reality. As of mid-2001, nearly one million people have made the two-hour trek north from San Francisco to visit the center, and have left this place with an overwhelming sense of inspiration and possibility.

Form and Function United:
Designing for the Here and Now

If the "weird restrooms" sign doesn't grab them first, passersby on busy Highway 101 are bound to notice the striking appearance of the company showroom. This does not look like business as usual! The building design and the construction materials were selected with an eye toward merging efficiency of function, educational value, and stunning beauty.

The architect chosen to design the building was Sim Van der Ryn of the Ecological Design Institute of Sausalito, California. His associate, David Arkin, served as project architect, and Jeff Oldham of Real Goods managed the building of the project. Their creation is a tall and gracefully curving single-story-building which is so adept in its capture

of the varying hourly and seasonal angles of the sun that additional heat and light are virtually unnecessary. Wood-burning stoves provide backup heating for the coldest winter mornings and solar-powered fluorescent lighting is available, but is rarely used. Through a combination of overhangs and manually controlled hemp awnings, excess insolation during the hot weather months has been avoided. Solar-powered evaporative coolers provide a low energy alternative to air-conditioning, and are also used to flush the building with cool night air, storing "coolth" in the six-hundred tons of thermal mass of the building's walls, columns, and floor. Grape arbors and a central fountain with a "drip ring" for evaporative cooling are positioned along the southern exposure of the building to serve as a first line of defense against the many over-one-hundred-degree days that occur during the summer in this part of California.

Many of the materials used in the construction of the building were donated by companies and providers with a commitment similar to Gaiam Real Goods. As an example, the walls of the SLC were built with more than 600 rice straw bales donated by the California Rice Industries Association. Previously, rice straw has been disposed of by

open burning, a practice that contributes to the production of carbon dioxide, the so-called "greenhouse gas" that is the leading cause of global warming. By using this agricultural by-product as a building material, everyone benefits. The farmers receive income for their straw bales, no carbon dioxide is produced, and the builder benefits from a low cost, highly efficient building material that minimizes energy consumption.

At the SLC, visitors experience the practicality of applied solar power technology, including the generation of electricity and solar water pumping. The electrical system for the facility comprises 10 kilowatts of photovoltaic power and three kilowatts of wind-generated power. Through an intertie with the Pacific Gas and Electric Company, the SLC sells the excess power it generates to the electric company and buys it back only when necessary. In the summer of 2001, plans are to install an additional 7.5 kilowatts of PV modules, making the SLC 100% independent from the grid. Once again, like-minded companies have shared in the costs of developing the Solar Living Center as a demonstration site. Siemens Solar has donated more than 10 kilowatts of the latest state-of-the-art photovoltaic modules to the center, and intends to use the SLC as a test site for new Siemens modules in the years to come. Trace Engineering contributed four intertie inverters, which are on display behind the glass window of the SLC's "engine room" so that visitors can see the inner workings of the electrical system.

In November 1999 a partnership between GPU, Astropower, Real Goods, and the Institute for Solar Living installed a 132-kilowatt PV array, one of the largest in power-hungry Northern California. This direct-intertie array delivered 163,000 kWh of power its first full year of operation.

On tours, either self-guided or with Institute for Solar Living tour guides, visitors learn about the guiding principles of sustainable living, and are offered a chance to appreciate the beauty that lies in the details of the project. The site also provides a coherent space for presentations by guest speakers and special events, and serves as the main campus and classroom for the workshop series staged by the Institute for Solar Living, a fledgling nonprofit dedicated to education and inspiration toward sustainable living.

The Natural World Reclaimed:
The Grounds and Gardens

Learning potential is intrinsic to the (soon-to-be-award-winning) landscape, designed by Chris and Stephanie Tebbutt of Land and Place. For this project, the design of the grounds, gardens, and waterworks was the first phase of construction and contributed much to establishing the character of the site. This is a radically different approach than most commercial building projects, where the landscaping appears to be a cosmetic afterthought. At the SLC, the gardens are a synthesis of the practical and the profound. Most of the plantings produce edible and/or useful crops, and the vegetation is utilized to maximize the site's energy efficiency while portraying the dramatic aspects of the solar year. Plantings and natural stone markers follow the lines of sunrise and sunset for each equinox and solstice, emanating from a sundial at the exact center of the oasis. More sundials and unique solar calendars scattered throughout the site encourage visitors to establish a feeling for the relationship between this specific location and the sun. Those of us who work and play here daily have discovered an almost organic connection with the seasonal shifts of the solar year, and the natural rhythms of the earth.

The gardens themselves follow the Sun's journey through the seasons, with zones planted to represent the ecosystems of different latitudes. Woodland, Wetland, Grassland and Dryland zones are manifested through plantings moving from north to south, with the availability of water the definitive element. Trees are planted to indicate the four cardinal directions. The fruit garden, perennial beds, herbs and grasses reflect the abundance and fertility of a home-based garden economy. Visitors discover aesthetic statements in design and landscape tucked into nooks and crannies all over the grounds, and unexpected simple pleasures, too, like shallow water channels for cooling aching feet, and perfect hidden spots for picnics and conversation.

Unique to these gardens are the "Living Structures," which reveal their architectural nature according to the turn of the seasons. Through annual pruning, plants are coaxed into various dynamic forms, such as a willow dome, a hops tipi, and a pyramid of timber bamboo. These living structures grow, quite literally, out of the garden itself. Visitors unaccustomed to the heat of a Hopland summer find relief inside the "agave cooling tower," where the push of a button releases a gentle mist into the welcome shade of vines and agave plants. By the time visitors leave the center, they've begun to understand the subtle humor of the "memorial car grove," where the rusting hulks of '50s and '60s "gas hog" cars have been turned into planter boxes for trees. These "grow-through" cars make a fascinating juxtaposition to the famous Northern California "drive-through" redwood trees!

A Place to Play

In case this all sounds awfully serious, it should be pointed out that the SLC is a wonderful place to play! Upon entering the showroom, one is greeted by a delightful rainbow spectrum created by a large prism mounted in the roof of the building. Visitors need not understand on a conscious level that this rainbow functions throughout the year as a "solar calendar," or that the prism's bright hues mark the daily "solar noon"; this is a

deeper learning that ignites the place where inspiration happens, not a raw scientific dissertation. Outside, interactive games and play areas tempt the young at heart to forget about the theory and enjoy the pleasure of pure exploration. A bicycle "generator" is connected to a fan and to a water pump. It doesn't take long for riders to really feel how much energy is required to create the tiniest bit of energy; almost everyone takes the time to compare the aching results of muscle power to the ease with which the same amount of energy is harvested from the sun with a solar panel.

The hands-down favorite for kids is the sand and water area. A solar powered pump provides a water source that can then be channeled, diverted, dammed and flooded through whatever sandy topography emerges from the maker's imagination. A shadow across the solar panel stops the flow of water, and it doesn't take long for kids to become immersed in starting and stopping the flow at will. Without even realizing it, these little scientists are learning about engineering, hydrology, erosion, and renewable energy theory!

Where to Find the Solar Living Center

The Real Goods Solar Living Center is located 94 miles north of San Francisco on Highway 101 and is open every day except Thanksgiving, Christmas and New Year's day. There is no admission charge for regularly scheduled or self-guided tours.

Customized group tours for students, architects, gardeners, or others with special interests, are available on a fee basis by advanced reservation, through the nonprofit Institute for Solar Living. The Institute also offers a variety of structured learning opportunities, including intensive, hands-on, one-day seminars on a variety of renewable energy and sustainable living topics. Please call the Institute at 707-744-2017 for more information, or visit the website at www.solarliving.org.

Part III: Spreading the Word: The Institute for Solar Living

In April 1998, the Institute for Solar Living separated from Real Goods Trading Corporation to become a legal nonprofit (501-C-3). By severing financial ties to a for-profit corporation, the Institute is now free to develop its educational mission without the constraint of profitability, and to focus its efforts solely on environmental education.

Since Real Goods' merger with Gaiam, Inc. in January 2001, the Institute for Solar Living has received a generous donation from Gaiam, which will enable it to greatly expand its interactive displays and programs.

The ISL's Four-Fold Mission

The Institute's mission is to teach people of all ages how to live more sustainably on the earth, to teach interdependence between people and the environment, to replace fossil fuels with renewable energy, and to honor biodiversity in all its forms. To further its mission, the Institute is focusing its resources on three endeavors: 1) the Institute's Sustainable Living Workshops, 2) the Solar Living Center's Interactive Displays, Exhibits, and Educational Tours, and 3) the SolFest Summer Solstice Energy Fair & Educational Celebrations.

#1: Sustainable Living Workshops

One of the Institute's short term goals is the immediate expansion of the Sustainable Living Workshop series, offered at the Solar Living Center in Hopland, CA. Now in its tenth season, Institute workshops provide an ideal opportunity to meet people of similar interests, work beside them, and build the family up close and hands-on. The main campus (the SLC) is not only an immensely inspirational setting, but a living, breathing

model for sustainable development, restorative permaculture landscaping, and renewable technologies. With more resemblance to a museum than a store, the SLC is an invaluable resource for the Institute, the center of testing and evaluation for the Gaiam Real Goods enterprise, and a unique opportunity for students to see sustainable living technologies at work. Each Institute workshop title relates directly to technologies employed at the Solar Living Center.

Institute sessions are a fascinating blend of technological independence, rural simplicity, and community building. The secret to the success of the Institute's workshops, as near as we can tell, consists of three ingredients perfectly combined: the wonderfully real Gaiam Real Goods customers who come from far and wide, the allure and promise of living independently with energy and sustenance harvested under independently produced power, and the practical, logical skills and ideas offered by the Institute's teachers and staff. One graduate expressed it best, "I can't decide if I've just had the best short vacation of my life, or the best learning experience! Could it be both?"

These intensive one-day workshops offer a quick survey of the terrain of a sustainable lifestyle, and a thorough exploration of the energy-related aspects of the specific topic. At the end of each session, graduates have become informed consumers, able to ask the right questions and to understand the answers. They report that the experience inspires them to go further forward along the path toward independent living. Those who arrive with some technical skill leave qualified to put together a small independent energy system, to begin the design of an energy efficient home made of alternative and sustainable building materials, to know what to look for in a piece of property, or to create balance and harmony in their own garden. Energy is the heart of the Workshop series: rational, comfortable, lasting systems for homes and businesses on and off the electric grid. Most courses emphasize direct hands-on experience. The mission of each Institute workshop is to share excitement, sense of purpose, and knowledge with others interested in living sensibly and lightly on the earth.

Throughout all sessions, give and take between instructors and attendees is a topic in and of itself. Students and faculty learn to know each other, and value the efforts at independence each has contemplated, or already begun to bring to their lifestyles. If the instructors don't have an answer, often a student attending the workshop will. Product demonstrations, cost benefit analyses, and a tour of applied technologies at the SLC provide students with a practical grasp on the tools and techniques available for taking their energy lives in hand. Classes focus on hands-on instruction of popular topics on energy and sustainability. The Institute's expert faculty has decades of experience and a strong passion for passing along its expertise to hungry students. Most classes take place at the Solar Living Center. Classes are one-day, two-day, or week-long and most cost $125 per day. Accreditation is available for many classes from California State University in Sonoma and from the American Institute of Architects (AIA). Workshops are scheduled between March and October.

A Thoroughly Incomplete 2001 Class List

This is a representative list of workshops scheduled for the 2001 season. Because the Institute is strongly committed to expanding its workshop series, this cannot help but be an incomplete list. Potential students are encouraged to check with the Institute for an updated list from time to time, and to let them know which additional topics you might be interested in pursuing.

Beginning and Advanced Straw Bale Construction

Beginning and Advanced Solar Electric Systems

Passive Solar and Ecological Design

Rammed Earth Construction

Rural Living Skills

The Green Householder—3-day

Building with Bamboo

Building with Stone

Natural Building Workshop—6-day

Utility Intertie Systems

Sustainable Living Workshop—5-day

Introduction to Permaculture

Alcohol Can Be a Gas

Book Early: A Detailed Description of the Most Popular Workshops

Beginning Solar Electric Systems

Taught by the Real Goods technical staff, this one-day class covers the basics of electricity, load analysis, system sizing, and the components of various systems. The hands-on session includes wiring of representative photovoltaic systems, and a tour of the solar and wind energy systems at the Solar Living Center.

Passive Solar and Ecological Design

David Arkin, Project Architect for the Solar Living Center, shares the basic tools to design your own sustainable dwelling. Course emphasis is on passive solar design features, and includes discussion of site analysis, energy-efficient design, cost-effective alternative construction methods, recycled and nontoxic materials, and an overview of renewable energy options.

Straw Bale Construction

A brief history of the straw bale revival, fundamental design considerations, codes, and basic construction methods will be covered in this intensive class, along with hands-on construction of a small load-bearing temporary structure.

Feedback from Institute Graduates

The Institute has benefited greatly from the willingness of Institute attendees to provide feedback on the benefits and potential improvements that could be made.

It's the helpful hints and learning from mistakes that others have made that brought me to your workshop—in addition to wanting to meet the people I will someday do substantial business with (major purchases). So, less time with generalities and more specifics, hands-on or applied theory with real life examples/pictures/equipment of success and failures. That is why your people are different   real experience, not hypotheticals!

I needed to get beyond the reading stage and become more familiar with the subjects. I feel we got a very well-thought-out and balanced presentation on many subjects, and I am very appreciative that you folks are taking the lead and setting a good example in presenting the information so lovingly and enthusiastically. Thank you!

My wife is leading me toward a better lifestyle. I arrived as a skeptic and am leaving with the feeling this is possible for me.

I am so excited to get home and get started that I can't sit still. One night I was sizing my system all night. Last night I dreamt I had five different PV systems laid out around the top of a green meadow. All night I was choosing the best one—a combination—for my needs. Jeff and Ross and Nancy were there talking to other people then answering all my questions!

An honest possibility even for a conservative Republican that voted twice for Reagan. It is not as difficult as others want you to believe.

Good spiritual company, great food, wonderful setting, excellent information.

You guys are great people, and your knowledge and experience is invaluable to the rest of us—that is why we came to you. Please concentrate on conveying what is unique to you and your experience. The rest we can get from the books you sell.

Where to Find the Institute: Updated Schedules and Registration

To register for classes, or for more information on Institute sessions, content, or availability, call Gaiam Real Goods at 800-762-7325 or write to: ISL, P.O. Box 836, Hopland, CA 95449, or send an email to rgisl@isl.org or check out the website at www.solarliving.org. An Institute brochure and schedule is available upon request. Institute staff can be reached directly at 707-744-2107.

#2: Interactive Displays: Bigger, Better, Faster, More!

The Institute for Solar Living intends to expand and enrich the Solar Living Center's existing interactive and educational displays. The staff won't rest until the SLC is recognized nationwide as a major learning campus and educational tour destination. As a nonprofit organization, the ISL is free at last to pursue sources of income more in keeping with its educational goals. For example, the Institute has applied for grant monies to fund additional interactive exhibits on hydro power, solar water heating, photovoltaics, wind power and hydrogen fuel cells at the SLC. The ISL has engaged a professional designer (whose client list includes the Monterey Bay Aquarium and a smattering of other Northern California museums) to create more engaging, effective exhibits and interactive displays on the site. With Gaiam's recent generous donation, the ISL plans to install at least three new interactive displays in 2001. To complement and increase the educational value of its new displays (assuming the grants come through!) the ISL will provide age-appropriate take-home brochures. The ISL is also in the process of designing renewables curriculum materials for schools, and promoting the SLC as a destination for school groups.

#3 SolFest—Annual Summer Solstice Festival and Educational Fair

The Institute for Solar Living's premier annual event is SolFest, scheduled each year for the weekend following the Summer Solstice. The first ever SolFest took place in June 1996, celebrating the Grand Opening of the Real Goods Solar Living Center in Hopland, California. The gala three-day event was attended by 10,000 people, inaugurating the Solar Living Center as the premier destination for those interested in learning about renewable energy and other sustainable living technologies. The educational theme of the event, in

concert with the uniquely beautiful setting of the Solar Living Center, has inspired thousands of visitors. World-class speakers, unique entertainment, educational workshops, exhibitor booths with a renewable energy orientation, along with a parade and display of electric vehicles all helped to create a lively and very successful event.

Over the years SolFest has featured incredible speakers, including Amory Lovins of the Rocky Mountain Institute, Wes Jackson of the Land Institute, Ralph Nader, Jim Hightower, Ben Cohen of Ben and Jerry's Homemade, Julia Butterfly-Hill, Helen Caldicott, and many others. SolFest also features ongoing educational workshops (provided by the Institute for Solar Living) exploring topics like solar energy, straw bale construction, building with bamboo, electric vehicles, eco-design, socially responsible investing, unconventional financing, climate change, industrial hemp, community nature centers, Headwaters Forest, solar cooking, and utility deregulation.

SolFest provides a unique opportunity for renewable energy and sustainable living aficionados to gather and swap stories, and to have a great time for two days in the sun while learning and playing. Call the Institute for Solar Living at 707-744-2017 or visit the website (www.solarliving.org) to find the exact date and bill of entertainment.

The Institute's Partnership Program

It takes a lot of energy to manifest the ISL's vision, and not the kind that is measured in kilowatts. The Institute extends a warm welcome to any volunteers willing to give their time and energy to promoting sustainable living through environmental education. You don't need to live in California to help; volunteering could take the form of consultation, publicity, physical labor, day-to-day support, or . . . you tell them where your talents are and how you'd like to help. The Institute already has a well-established internship program for college students studying organic and bio-dynamic gardening, renewable energy of any kind, and general sustainable education. ISL volunteers and interns are honored and appreciated for their contributions, and come away with a lasting sense of achievement.

The other kind of energy needed is the green kind. The ISL asks each of you to seriously consider partnering with them in building the premier renewable energy education facility on the planet. They have applied for grants and are seeking large-scale donations from like-minded businesses and individuals, and expect to find some support. However, it has always been grassroots mom-and-pop generosity that got any nonprofit ball rolling. The Institute's "Partnership Program" was created to allow a tax deductible venue for its supporters. As little as $35 per year (for a Basic Partnership) will help expand the Sustainable Living Workshop Series, build more educational exhibits at the Solar Living Center, and help reach school children with environmental educational programs.

Partnership includes concrete benefits, like the occasional free magazine subscription. Donors of larger amounts receive a free copy of *Place in the Sun* (the story of the evolution of the Solar Living Center), and a personalized tour of the Solar Living Center. But the real benefits of Institute Partnership will never be counted in quantifiable units. Becoming a Partner in the Institute for Solar Living means joining a growing community of individuals whose vision and dedication have the potential to effect truly far-reaching change for our planet. To become a Partner in the Institute, to make a donation, or to volunteer, please contact the ISL at 707-744-2017 or email isl@rgisl.org.

Part IV: Getting Down to Business

Remember our first principle? We never forget it. When the profitability principle is met, we are free to pursue our educational mission effectively. We achieve profitability through our catalogs, retail stores, design and consulting groups, and co-publishing ventures. We support the principles of knowledge, involvement, credibility (and even fun!) through a variety of innovative programs, from educational opportunities to networking. When our business goals and principles work in tandem with our educational goals and principles, we sleep easier at night. These are the elements and programs which, taken together, make our business REAL.

Catalogs

The Gaiam Real Goods catalogs include products and information geared to folks who have been thinking a lot about bringing elements of sustainable living into their lives. Our focus is on high-impact, user-friendly merchandise like compact fluorescent lighting, air and water purification systems, energy-saving household implements, and mainstream products that introduce the concepts behind renewable energy without being dauntingly technical. We also include lots of information on sustainable living, environmental responsibility, and books, books, books. The full-color Gaiam Real Goods catalog is a lot like the Gideon's Bible in a hotel drawer. We don't expect it to produce any wholesale conversions, but we hope it steers a few people in the right direction.

Real Goods Renewables is our technical publication, for individuals who have made a serious commitment to reducing their impact on the planet, and need the tools to get on with it. Along with renewable energy system components (like PV modules, wind generators, hydro-electric pumps, cables, inverters, and controllers), our black-and-white Renewables catalog offers off-the-grid and DC appliances, energy-saving climate control products, solar ovens, air and water purification systems, a smattering of green building materials and even more books.

Retail Stores

As a catalog and internet business, Gaiam Real Goods can reach almost every nook and cranny in America. Even so, there will always be people who want to "kick the tires" before making a purchase. Retail stores are a two-edged sword. On the one hand, we can offer merchandise from grassroots manufacturers who can't make a thousand more by Friday, but retail stores lure more delivery trucks and personal cars out on the road. On the other hand, retail stores offer an opportunity to snag curious passers-by and show them how sustainable living can enhance their lives.

All things considered, we just can't resist the opportunity to take our show on the road. Our flagship Solar Living Center store in Hopland, California is flourishing and our Berkeley, California, store is doing equally good business. Gaiam Real Goods has opened a new store in West Los Angeles specializing in urban conservation.

Real Goods Renewables Design and Consulting Group

In response to the growing demand for large-scale commercial renewable energy systems, we launched a dynamic new division in early 1999: the Real Goods Renewables Design and Consulting Group. The Renewables D&C Group was established specifically to work on large-scale commercial renewables projects, particularly in the arena of Eco Tourism. Our experienced team of renewables experts offers a full spectrum of services for established and proposed resorts, eco-lodges, remote facilities and villages. Our D&C Group (formerly a part of Real Goods Renewables) has brought high-efficiency and renewable energy resources to thousands of clients over the last twenty years. Customers are rewarded with lower operating costs, high reliability, and diminished impact on the world's vanishing resources. We have consulted with clients ranging from the Kingdom of Tonga and the Vatican to the White House and a lone mountaineer. Each receives the same qualified, individual service.

Services range from site assessment and planning to architectural design, specializing in renewable energy system design (solar, wind, and hydroelectric), supply, installation and training. Our D&C Group will facilitate and expedite the development of site infrastructure to minimize your need for energy, water and other resources. We accomplish this through methodical assessment of the proposed site and its environment. We consult with you to develop the best design, fulfillment, and installation strategies. Our goal is to help you create an infrastructure that is economically, environmentally, and culturally sustainable.

We have assembled some of the world's leading experts in renewable energy, architecture, community planning and sustainability into an effective, productive team capable of implementing these concepts worldwide. Every member of our D&C Group has a strong and sincere commitment to the environment, local people, and cultures. Our client list includes the White House, U.S. National Park Service, U.S. Department of the Interior, U.S. Department of Defense, U.S. Department of Energy, World Health Organization, government of Brazil, Jet Propulsion Laboratory, Disney, AT&T, Pacific Bell, Los Angeles Department of Water and Power, Kingdom of Tonga, Karuk Tribe of California, The Essene Way, Women's Front of Norway, Sierra Club, the Vatican, Public Citizen, NRDC, Discovery Channel, Pan American Health Organization, Rocky Mountain Institute, NASA, City and County of San Francisco, Greenpeace, The Nature Conservancy, U.S. Agency for International Development (AID), PG&E, Sony, U.S. Forest Service, Lakota Sioux, Maho Bay Camps / Harmony / Concordia, University of California, CBS, U.S. Virgin Islands Office of Energy, Tassajara Zen Center . . . and many more. If you have a facility where utility power is unavailable or of poor quality, consider our resources:

- *Renewable Energy System Design (solar, wind, hydro)*
- *Complete Installation Services, On-Site Training, and Turnkey Systems*
- *Large-Scale Uninterruptible Backup Power including Disaster Preparedness*
- *Utility Intertie with Renewable Energy*
- *Power Quality Enhancement*
- *Site Planning and Engineering*
- *Solar Architecture*
- *Passive and Active Design*
- *Green Building Techniques*
- *Whole Systems Integration*

- *High-Efficiency Appliances*
- *Biological Waste Treatment Systems*
- *Water Quality and Management*
- *Earth Sensitive Landscaping*

The Real Goods Renewables Design & Consulting Team will help you realize your vision. Contact us at 707-468-9292; or Jeff Oldham, manager, ext. 2128 (jeff@realgoods.com).

Real Goods Renewables

The Real Goods Renewables staff are experts at configuring residential renewable energy systems. You don't want your project to be a test bed for unproved technology or experimental system design. You want information, products, and service that are of the highest quality. You want to leave the details to a company with the capability to do the job right the first time. We welcome the opportunity to design and plan entire systems. Here's how the Real Goods Renewables technical services work:

1. To assess your needs, capabilities, limitations, and working budget, we ask you to complete a specially created worksheet (see page 220). The information required includes a list of your energy needs, an inventory of desired appliances, site information, and potential for hydroelectric and wind development. Note that we only need this detailed energy use info for off-the-grid systems.

2. A member of our technical staff will determine your wattage requirements, and design an appropriate system with you. Or, for intertie systems, we will either determine how much you want to spend or how much utility power you wish to offset, and then design an appropriate system.

3. At this point, we will begin tracking the time we spend in helping you plan the details of your installation. Your personal tech rep will work with you on an unlimited time basis until your system has been completely designed and refined. He will order parts and talk you through assembly, assuring you that you get precisely what you need. He will also consult and work with your licensed contractor, if need be. The first hour is free and part of our service. Beyond this initial consultation, time will be billed in ten-minute intervals at the rate of $75 per hour. If the recommended parts and equipment are purchased from Gaiam Real Goods, however, this time will be provided at no charge.

Real Goods Renewables has been providing clean, reliable, renewable energy to people all over the planet for over twenty years. With the creation of our separate commercial division, our Renewables technicians now have a tighter focus on smaller, noncommercial residential systems; this specialization means we are better prepared than ever to design a utility intertie or independent home system that meets your unique needs.

You will leave your Real Goods Renewables consultation with the information you need to make informed decisions on which technologies are best suited for your particular application. You'll learn how to utilize proven technologies and techniques to reap the best environmental advantages for the lowest possible cost. And we won't leave you hanging once your system is installed; our technicians will help you with troubleshooting, ongoing maintenance, and future upgrades. The Real Goods commitment to custom design, proven technology, and ongoing support has resulted in more than two decades of exceptionally high customer satisfaction. Real Goods Renewables specializes in residential and commercial systems under 10kW output, including:

- *Utility Intertie with Renewable Energy*
- *Renewable Energy Systems Design and Installation (solar, wind, hydro)*
- *Large-Scale Uninterruptible Backup Power*

- *High-Efficiency Appliances*
- *Biological Waste Treatment Systems*
- *Power Quality Enhancement*
- *Whole Systems Integration*
- *Water Quality and Management*

Real Goods Renewables technical services and sales, for residential and commercial renewable energy systems are available by phone 800-919-2400, from 7:30 a.m. to 6 p.m., Pacific Time, Monday through Friday and 7:30 a.m. to 4 p.m. Saturday. We endeavor to answer technical email within 48 hours, and snail mail (USPS) within one week.

Co-Publishing with Chelsea Green

We love books. Pound for pound, and dollar for dollar, books are the best value available for entertainment and personal growth. Books are the most elegant form of information storage and transmittal ever invented. They fit all possible criteria of what is a "real good." They require no power or maintenance. They are compact and portable. When no longer needed, they can be reused by passing them along to friends, or recycled. Books entertain and instruct at a fraction of the cost of other media.

Our business was founded on the book you're currently reading, *The Solar Living Sourcebook*, now in its eleventh edition. With our publishing partner, Chelsea Green, Gaiam Real Goods proudly produces books that provide the blueprint for a sustainable future. All the books we offer show folks how to live lightly, protect the environment, and live joyful, satisfying lives. Providing the means to a sustainable end (or beginning, if you prefer) is our part in the global effort to save the planet. And our "secret weapon" is the unchallenged champion of the communications world—the good old fashioned book. You're already aware that knowledge, because it holds the potential for effecting far-reaching change, is our most important product at Real Goods. It's comforting to know that, in an age when we are awash in an ocean of information, the book still reigns supreme.

Stay tuned to our catalogs for lots more great solar living titles.

See page ii at the front of the *Sourcebook* for a current list of co-publications in our "Solar Living Series."

Cyberspace Is a Tree-Free World: The Real Goods Website

The biggest conundrum we've ever faced as a company has been trying to live our environmental mission while being in the catalog business, which by definition survives by consuming trees. We have been, and will continue to be, as environmentally responsible as possible in a print media business. Our print catalogs use maximum post-consumer content recycled paper and soy-based inks, and we mail only to customers who have the highest potential for buying from us. Still you just can't beat a virtual catalog for environmental responsibility. In fact, our website uses only post-consumer recycled electrons of the finest quality. Since we first established an Internet presence in 1995, our online business has grown by leaps and bounds. One unexpected benefit of our "Tree Free Catalog" is the almost limitless amount of space available. Our website has evolved from a simple on-line catalog to a compendium of renewables resources. You'll find links to fascinating sustainability and renewable energy sites, and you can find out just about anything about our company. It's quite possible that virtual catalogs like ours will reduce (or even eliminate) the more wasteful aspects of the mail order business in the not-too-distant future.

The free Real Goods Solar Times email newsletter is published every month. It includes tons of news on renewable energy, the environment, the latest happenings at Gaiam Real Goods, and even some great products. To subscribe, visit our home page and sign up: www.realgoods.com.

Not Your Average Open House: The Real Goods Demonstration Home Program

Pioneered in the spring of 1994, our Demonstration Home program offers a unique opportunity to see how real people live, work, and play with renewables technology. There are currently 81 Demonstration Homes throughout the country, with technologies ranging from passive solar design to energy-efficient appliances to independent renewable energy systems. Participating homeowners have generously agreed to make their homes (and themselves) available to the public, by appointment only. We will direct you to the demo homes closest to your area, or to a home using the kinds of renewables technologies you are considering for yourself. There's no better way to convince your inner skeptic of the real-life benefits of simple, sustainable living than to talk to the folks who are living with them. And, there's no better way to demonstrate a serious personal commitment to renewable energy and sustainable living than to open your home to kindred souls through participation in the Real Goods Demonstration Home Program. We extend a tasty discount on Real Goods purchases to participants. If you are interested in either touring or having your home become a Demonstration Home, write to: Demonstration Home Program, c/o Real Goods, 13771 S. Highway 101, Hopland, CA 95449, or call 800-762-7325. The Demo Home list is also now available on our website at www.realgoods.com/renew/demo.

It's Forever: The Gaiam Real Goods Hard Corps Lifetime Membership Program

Getting involved with the Gaiam Real Goods community is as easy as joining our Hard Corps Lifetime Membership Program, now 65,000 members strong. For some, it's a statement of support for the Gaiam Real Goods mission. For others, it's just enough of a discount to make that PV system affordable. For everyone, Gaiam Real Goods Lifetime Membership means you are an individual who takes energy independence seriously and who is eager to share your experience and knowledge with others who have made a similar commitment. Members receive all the news that's fit to print, a discount on all retail purchases (including sale merchandise), and about 65,000 new friends. Membership privileges are good for life, though we will stop sending mail if we don't hear from you for a long period of time; privileges are always renewable should you request to have your membership reinstated. Building community is one of the most effective ways to achieve our goals, and create a sustainable future for us all.

Real Goods Wedding Registry

Do your dreams of nuptial bliss lean toward the practical, rather than the extravagant? Does your vision of unwrapping wedding gifts reveal inverters and compact fluorescent light bulbs, rather than sterling silver tea service and a ten-slice toaster? Why not register with Real Goods, so that well-wishing friends can honor you with the gifts that won't end up in the closet. Is this a joke? No way. A number of young married couples have embarked upon their dream of living lightly on the planet with a running start, thanks to the Real Goods Wedding Registry. Then there are those reluctant folks who need a little nudge in the direction of sustainability, and those who simply prefer to let their friends and family choose for themselves. We'll happily send a gift certificate in any amount to the person of your choice, along with our latest edition of the Real Goods Catalog.

Sizing Worksheet

AC device	Device watts	X	Hours of daily use	X	Days of use per week	÷	7	=	Average watt-hours per day
		X		X		÷		=	
		X		X		÷		=	
		X		X		÷		=	
		X		X		÷		=	
		X		X		÷		=	
		X		X		÷		=	
		X		X		÷		=	
		X		X		÷		=	
		X		X		÷		=	
		X		X		÷		=	
		X		X		÷		=	
		X		X		÷		=	
		X		X		÷		=	
		X		X		÷		=	
		X		X		÷		=	
		X		X		÷		=	
		X		X		÷		=	
		X		X		÷		=	
		X		X		÷		=	
		X		X		÷		=	
		X		X		÷		=	

1 Total AC watt-hours/day

2 X 1.1 = Total corrected DC to AC watt-hours/day

DC device	Device watts	X	Hours of daily use	X	Days of use per week	÷	7	=	Average watt-hours per day
		X		X		÷		=	
		X		X		÷		=	
		X		X		÷		=	
		X		X		÷		=	
		X		X		÷		=	
		X		X		÷		=	

3 Total DC watt-hours/day

Sizing Worksheet

3	(from previous page)	Total DC watt-hours/day	
4		Total corrected DC to AC watt-hours/day from Line 2 +	
5		Total household DC watt-hours/day =	
6		System nominal voltage (usually 12, 24 or 48) ÷	
7		Total DC amp-hours/day =	
8		Battery losses, wiring losses, safety factor x	1.2
9		Total daily amp-hour requirement =	
10		Estimated insolation (hours of "noon sun" per day; see p. 476 or call) ÷	
11		Total PV array current in amps =	
12		Select a photovoltaic module for your system	
13		Module rated power amps ÷	
14		Number of modules required in parallel =	
15		System nominal voltage (from line 6 above)	
16		Module nominal voltage (usually 12) ÷	
17		Number of modules required in series =	
18		Number of modules required in parallel (from Line 14 above) x	
19		Total modules required =	

Battery Sizing

20		Total daily amp-hour requirement (from line 9)	
21		Reserve time in days x	
22		Fraction of useable battery capacity (usually .5 to .8) ÷	
23		Minimum battery capacity in amp-hours =	
24		Select a battery for your system, enter amp-hour capacity ÷	
25		Number of batteries in parallel =	
26		System nominal voltage (from line 6)	
27		Voltage of your chosen battery (6 or 12 usually) ÷	
28		Number of batteries in series =	
29		Number of batteries in parallel (from line 25 above) x	
30		**Total number of batteries required**	

Typical Wattage Requirements for Common Appliances

Use the manufacturer's specs if possible, but be careful of nameplate ratings that are the highest possible electrical draw for that appliance. Beware of appliances that have a "standby" mode and are really "on" 24 hours a day. If you can't find a rating, call us for advice 800-919-2400.

DESCRIPTION	WATTS
Refrigeration:	
4-yr.-old 22 cu. ft. auto defrost (approximate run time 7–9 hours per day)	500
New 22 cu. ft. auto defrost (approximate run time 7–8 hours per day)	200
12 cu. ft. Sun Frost refrigerator (approximate run time 6–9 hours per day)	58
4-yr.-old standard freezer (approximate run time 7–8 hours per day)	350
Dishwasher: cool dry	700
hot dry	1450
Trash compactor	1500
Can opener (electric)	100
Microwave (.5 cu. ft.)	900
Microwave (.8 to 1.5 cu. ft.)	1500
Exhaust hood	144
Coffeemaker	1200
Food processor	400
Toaster (2-slice)	1200
Coffee grinder	100
Blender	350
Food dehydrator	600
Mixer	120
Range, small burner	1250
Range, large burner	2100
Water Pumping:	
AC Jet Pump (1/3 hp), 300 gal per hour, 20' well depth, 30 psi	750

DESCRIPTION	WATTS
AC Submersible Pump (1/2 hp), 40' well depth, 30 psi	1000
DC pump for house pressure system (typical use is 1–2 hours per day)	60
DC submersible pump (typical use is 6 hours per day)	50
Shop:	
Worm drive 7 1/4" saw	1800
AC table saw, 10"	1800
AC grinder, 1/2 hp	1080
Hand drill, 3/8"	400
Hand drill, 1/2"	600
Entertainment/Telephones:	
TV (27-inch color)	170
TV (19-inch color)	80
TV (12-inch black & white)	16
Video games (not incl. TV)	20
Satellite system, 12-ft dish/VCR	30
Laser disk/CD player	30
AC powered stereo (avg. volume)	55
AC stereo, home theater	500
DC powered stereo (avg. volume)	15
CB (receiving)	10
Cellular telephone (on standby)	5
Cordless telephone (on standby)	5
Electric piano	30
Guitar amplifier (avg. volume)	40
(Jimi Hendrix volume)	8500
General Household:	
Typical fluorescent light (60W equivalent)	15
Incandescent lights (as indicated on bulb)	
Electric clock	4
Clock radio	5
Electric blanket	400
Iron (electric)	1200

DESCRIPTION	WATTS
Clothes washer (vertical axis)	900
Clothes washer (horizontal axis)	250
Dryer (gas)	500
Dryer (electric)	5750
Vacuum cleaner, average	900
Central vacuum	1500
Furnace fan:	
1/4 hp	600
1/3 hp	700
1/2 hp	875
Garage door opener: 1/4 hp	550
Alarm/security system	6
Air conditioner: 1 ton or 10,000 BTU/hr	1500
Office/Den:	
Computer	55
17" color monitor	100
17" LCD "flat screen" monitor	45
Laptop computer	25
Ink jet printer	35
Dot matrix printer	200
Laser printer	900
Fax machine: (plain paper)	
standby	5
printing	50
Electric typewriter	200
Adding machine	8
Electric pencil sharpener	60
Hygiene:	
Hair dryer	1500
Waterpik	90
Whirlpool bath	750
Hair curler	750
Electric toothbrush: (charging stand)	6

SOLAR INSOLATION MAPS

The maps below show the sun-hours per day for the U.S. Charts courtesy of D.O.E.

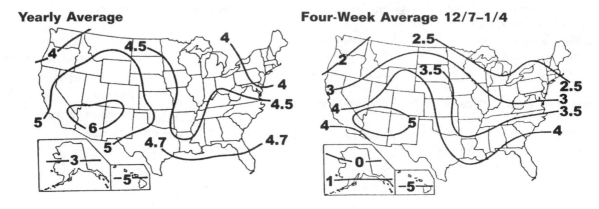

SOLAR INSOLATION BY U.S. CITY

This chart shows solar insolation in kilowatt-hours per square meter per day in many U.S. locations. For simplicity, we call this figure "Sun Hours/Day." To find average sun hours per day in your area (last column), check local weather data, look at the maps above or find a city in the table that has similar weather to your location. If you want year-round autonomy, use the lowest of the two figures. If you want only 100% autonomy in summer, use the higher figure.

State	City	High	Low	Avg	State	City	High	Low	Avg	State	City	High	Low	Avg
AK,	Fairbanks	5.87	2.12	3.99	KS,	Manhattan	5.08	3.62	4.57	NY,	Schenectady	3.92	2.53	3.55
AK,	Matanuska	5.24	1.74	3.55	KS,	Dodge City	6.5	4.2	5.6	NY,	Rochester	4.22	1.58	3.31
AL,	Montgomery	4.69	3.37	4.23	KY,	Lexington	5.97	3.6	4.94	NY,	New York City	4.97	3.03	4.08
AR,	Bethel	6.29	2.37	3.81	LA,	Lake Charles	5.73	4.29	4.93	OH,	Columbus	5.26	2.66	4.15
AR,	Little Rock	5.29	3.88	4.69	LA,	New Orleans	5.71	3.63	4.92	OH,	Cleveland	4.79	2.69	3.94
AZ,	Tucson	7.42	6.01	6.57	LA,	Shreveport	4.99	3.87	4.63	OK,	Stillwater	5.52	4.22	4.99
AZ,	Page	7.3	5.65	6.36	MA,	E. Wareham	4.48	3.06	3.99	OK,	Oklahoma City	6.26	4.98	5.59
AZ,	Phoenix	7.13	5.78	6.58	MA,	Boston	4.27	2.99	3.84	OR,	Astoria	4.76	1.99	3.72
CA,	Santa Maria	6.52	5.42	5.94	MA,	Blue Hill	4.38	3.33	4.05	OR,	Corvallis	5.71	1.9	4.03
CA,	Riverside	6.35	5.35	5.87	MA,	Natick	4.62	3.09	4.1	OR,	Medford	5.84	2.02	4.51
CA,	Davis	6.09	3.31	5.1	MA,	Lynn	4.6	2.33	3.79	PA,	Pittsburg	4.19	1.45	3.28
CA,	Fresno	6.19	3.42	5.38	MD,	Silver Hill	4.71	3.84	4.47	PA,	State College	4.44	2.79	3.91
CA,	Los Angeles	6.14	5.03	5.62	ME,	Caribou	5.62	2.57	4.19	RI,	Newport	4.69	3.58	4.23
CA,	Soda Springs	6.47	4.4	5.6	ME,	Portland	5.23	3.56	4.51	SC,	Charleston	5.72	4.23	5.06
CA,	La Jolla	5.24	4.29	4.77	MI,	Sault Ste. Marie	4.83	2.33	4.2	SD,	Rapid City	5.91	4.56	5.23
CA,	Inyokern	8.7	6.87	7.66	MI,	E. Lansing	4.71	2.7	4.0	TN,	Nashville	5.2	3.14	4.45
CO,	Grandby	7.47	5.15	5.69	MN,	St. Cloud	5.43	3.53	4.53	TN,	Oak Ridge	5.06	3.22	4.37
CO,	Grand Lake	5.86	3.56	5.08	MO,	Columbia	5.5	3.97	4.73	TX,	San Antonio	5.88	4.65	5.3
CO,	Grand Junction	6.34	5.23	5.85	MO,	St. Louis	4.87	3.24	4.38	TX,	Brownsville	5.49	4.42	4.92
CO,	Boulder	5.72	4.44	4.87	MS,	Meridian	4.86	3.64	4.43	TX,	El Paso	7.42	5.87	6.72
DC,	Washington	4.69	3.37	4.23	MT,	Glasgow	5.97	4.09	5.15	TX,	Midland	6.33	5.23	5.83
FL,	Apalachicola	5.98	4.92	5.49	MT,	Great Falls	5.7	3.66	4.93	TX,	Fort Worth	6	4.8	5.43
FL,	Belie Is.	5.31	4.58	4.99	MT,	Summit	5.17	2.36	3.99	UT,	Salt Lake City	6.09	3.78	5.26
FL,	Miami	6.26	5.05	5.62	NM,	Albuquerque	7.16	6.21	6.77	UT,	Flaming Gorge	6.63	5.48	5.83
FL,	Gainsville	5.81	4.71	5.27	NB,	Lincoln	5.4	4.38	4.79	VA,	Richmond	4.5	3.37	4.13
FL,	Tampa	6.16	5.26	5.67	NB,	N. Omaha	5.28	4.26	4.9	WA,	Seattle	4.83	1.6	3.57
GA,	Atlanta	5.16	4.09	4.74	NC,	Cape Hatteras	5.81	4.69	5.31	WA,	Richland	6.13	2.01	4.44
GA,	Griffin	5.41	4.26	4.99	NC,	Greensboro	5.05	4	4.71	WA,	Pullman	6.07	2.9	4.73
HI,	Honolulu	6.71	5.59	6.02	ND,	Bismark	5.48	3.97	5.01	WA,	Spokane	5.53	1.16	4.48
IA,	Ames	4.8	3.73	4.4	NJ,	Sea Brook	4.76	3.2	4.21	WA,	Prosser	6.21	3.06	5.03
ID,	Boise	5.83	3.33	4.92	NV,	Las Vegas	7.13	5.84	6.41	WI,	Madison	4.85	3.28	4.29
ID,	Twin Falls	5.42	3.42	4.7	NV,	Ely	6.48	5.49	5.98	WV,	Charleston	4.12	2.47	3.65
IL,	Chicago	4.08	1.47	3.14	NY,	Binghampton	3.93	1.62	3.16	WY,	Lander	6.81	5.5	6.06
IN,	Indianapolis	5.02	2.55	4.21	NY,	Ithaca	4.57	2.29	3.79					

MAGNETIC DECLINATIONS IN THE UNITED STATES

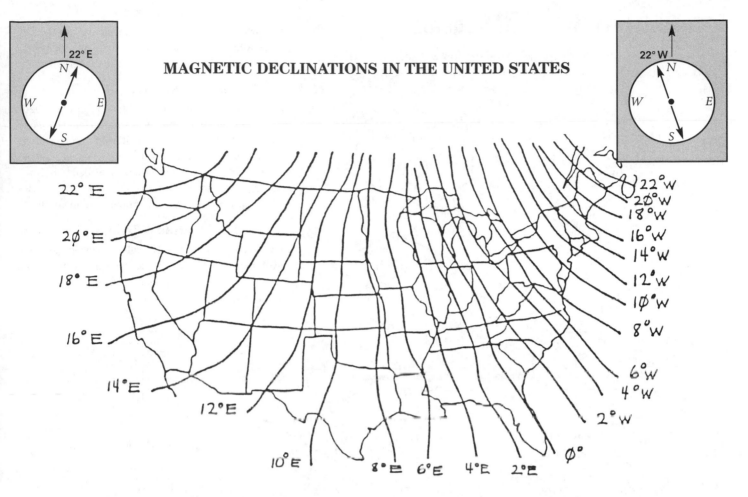

Figure indicates correction of compass reading to find true north. For example, in Washington state when your compass reads 22°E, it is pointing due north.

MAXIMUM NUMBER OF CONDUCTORS FOR A GIVEN CONDUIT SIZE

Conduit size	½"	¾"	1"	1¼"	1½"	2"
#12	10	18	29	51	70	114
#10	6	11	18	32	44	73
#8	3	5	9	16	22	36
#6	1	4	6	11	15	26
#4	1	2	4	7	9	16
#2	1	1	3	5	7	11
#1		1	1	3	5	8
#1/0		1	1	3	4	7
#2/0		1	1	2	3	6
#3/0		1	1	1	3	5
#4/0		1	1	1	2	4

Conductor size

Battery Wiring Diagrams

The following diagrams show how 2-, 6- and 12-volt batteries are connected for 12-, 24- and 48-volt operation.

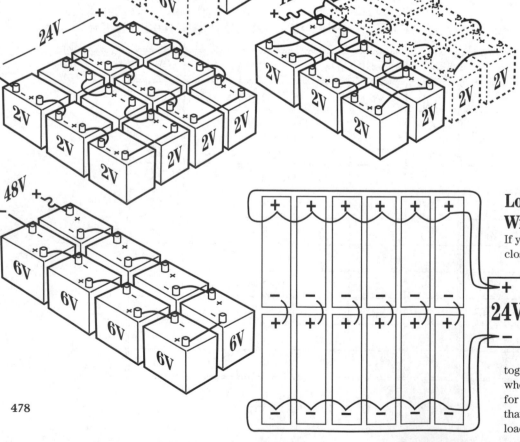

Wiring Basics

We answer a lot of basic wiring questions over the phone, which we're always happy to do, but there's nothing like a picture or two to make things apparent.

Battery Wiring

The batteries for your energy system may be supplied as 2-volt, 6-volt, or 12-volt cells. Your system voltage is probably 12, 24, or 48 volts. You'll need to series wire enough batteries to reach your system voltage, then parallel wire to another series group as needed to boost amperage capacity. See our drawings for correct series wiring. Paralleled groups are shown in dotted outline.

PV Module Wiring

PV modules are almost universally produced as nominal 12-volt modules. For smaller 12-volt systems this is fine. Most larger residential systems are configured for 24- or 48-volt input now.

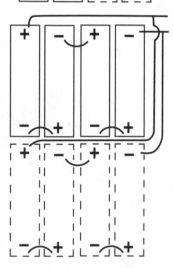

Longevity/Safety Tip for Wiring Larger PV Arrays

If you have a large PV array that produces close to, or over, 20 amps, multiple power take-off leads are a good idea. They may prevent toasted terminal boxes. Instead of only taking a single pair of positive and negative leads off some point on the array, take off one pair at one end, and another pair off the opposite end. Then join them back together at the array-mounted junction box where you're going to the larger wire needed for transmission. This divides up the routes that outgoing power can take, and eases the load on any single PV junction box.

Wire Sizing Chart/Formula

We could give you some incomprehensible voltage drop charts (like we've done in the past), but this all-purpose formula works better.

This chart is useful for finding the correct wire size for any voltage, length, or amperage flow in any AC or DC circuit. For most DC circuits, particularly between the PV modules and the batteries, we try to keep the voltage drop to 3% or less. There's no sense using your expensive PV wattage to heat wires. You want that power in your batteries!

Note that this formula doesn't directly yield a wire gauge size, but rather a "VDI" number, which is then compared to the nearest number in the VDI column, and then read across to the wire gauge size column.

1. Calculate the Voltage Drop Index (VDI) using the following formula:

VDI = AMPS x FEET ÷ (% VOLT DROP x VOLTAGE)

Amps = Watts divided by volts   Feet = One-way wire distance
% Volt Drop = Percentage of voltage drop acceptable for this circuit (typically 2% to 5%)

2. Determine the appropriate wire size from the chart below.

   A. *Take your VDI number you just calculated and find the nearest number in the VDI column, then read to the left for AWG wire gauge size.*
   B. *Be sure that your circuit amperage does not exceed the figure in the Ampacity column for that wire size. (This is not usually a problem in low-voltage circuits.)*

WIRE SIZE	COPPER WIRE		ALUMINUM WIRE	
AWG	VDI	Ampacity	VDI	Ampacity
0000	99	260	62	205
000	78	225	49	175
00	62	195	39	150
0	49	170	31	135
2	31	130	20	100
4	20	95	12	75
6	12	75	•	•
8	8	55	•	•
10	5	30	•	•
12	3	20	•	•
14	2	15	•	•
16	1	•	•	•

Chart developed by John Davey and Windy Dankoff. Used by permission.

Example: Your PV array consisting of 4 Siemens SP75 modules is 60 feet from your 12-volt battery. This is actual wiring distance, up pole mounts, around obstacles, etc. These modules are rated at 4.4 amps, times 4 modules = 17.6 amps maximum. We'll shoot for a 3% voltage drop. So our formula looks like:

$$VDI = \frac{17.6 \times 60}{3[\%] \times 12[V]} = 29.3$$

Looking at our chart, a VDI of 29 means we'd better use #2 wire in copper, or #0 wire in aluminum. Hmmm. Pretty big wire.

What if this system were 24-volt? The modules would be wired in series, so each pair of modules would produce 4.4 amps. Two pairs, times 4.4 amps = 8.8 amps max.

$$VDI = \frac{8.8 \times 60}{3[\%] \times 24[V]} = 7.3$$

Wow! What a difference! At 24-volt input you could wire your array with little ol' #8 copper wire.

Friction Loss Chart For Water Pumping

Friction Loss- PVC Class 160 PSI Plastic Pipe
Pressure loss from friction in psi per 100 feet of pipe.

NOMINAL PIPE DIAMETER IN INCHES

Flow GPM	1	1.25	1.5	2	2.5	3	4	5	6	8	10
1	0.02	0.01									
2	0.06	0.02	0.01								
3	0.14	0.04	0.02								
4	0.23	0.07	0.04	0.01							
5	0.35	0.11	0.05	0.02							
6	0.49	0.15	0.08	0.03	0.01						
7	0.66	0.20	0.10	0.03	0.01						
8	0.84	0.25	0.13	0.04	0.02						
9	1.05	0.31	0.16	0.05	0.02						
10	1.27	0.38	0.20	0.07	0.03	0.01					
11	1.52	0.45	0.23	0.08	0.03	0.01					
12	1.78	0.53	0.28	0.09	0.04	0.01					
14	2.37	0.71	0.37	0.12	0.05	0.02					
16	3.04	0.91	0.47	0.16	0.06	0.02					
18	3.78	1.13	0.58	0.20	0.08	0.03					
20	4.59	1.37	0.71	0.24	0.09	0.04	0.01				
22	5.48	1.64	0.85	0.29	0.11	0.04	0.01				
24	6.44	1.92	1.00	0.34	0.13	0.05	0.02				
26	7.47	2.23	1.15	0.39	0.15	0.06	0.02				
28	8.57	**2.56**	1.32	0.45	0.18	0.07	0.02				
30	9.74	2.91	1.50	0.51	0.20	0.08	0.02				
35		3.87	**2.00**	0.68	0.27	0.10	0.03				
40		4.95	2.56	0.86	0.34	0.13	0.04	0.01			
45		6.16	3.19	1.08	0.42	0.16	0.05	0.02			
50		7.49	3.88	1.31	0.52	0.20	0.06	0.02			
55		8.93	4.62	**1.56**	0.62	0.24	0.07	0.02			
60		10.49	5.43	1.83	0.72	0.28	0.08	0.03	0.01		
65			6.30	2.12	0.84	0.32	0.09	0.03	0.01		
70			7.23	2.44	0.96	0.37	0.11	0.04	0.02		
75			8.21	2.77	1.09	0.42	0.12	0.04	0.02		
80			9.25	3.12	1.23	0.47	0.14	0.05	0.02		
85			10.35	3.49	**1.38**	0.53	0.16	0.06	0.02		
90				3.88	1.53	0.59	0.17	0.06	0.03		
95				4.29	1.69	0.65	0.19	0.07	0.03		
100				4.72	1.86	**0.72**	0.21	0.08	0.03	0.01	
150				10.00	3.94	1.52	0.45	0.16	0.07	0.02	
200					6.72	2.59	**0.76**	0.27	0.12	0.03	0.01
250					10.16	3.91	1.15	0.41	0.18	0.05	0.02
300						5.49	1.61	**0.58**	0.25	0.07	0.02
350						7.30	2.15	0.77	0.33	0.09	0.03
400						9.35	2.75	0.98	0.42	0.12	0.04
450							3.42	1.22	**0.52**	0.14	0.05
500							4.15	1.48	0.63	0.18	0.06
550							4.96	1.77	0.76	0.21	0.07
600							5.82	2.08	0.89	0.25	0.08
650							6.75	2.41	1.03	0.29	0.10
700							7.75	2.77	1.18	0.33	0.11
750							8.80	3.14	1.34	**0.37**	0.13
800								3.54	1.51	0.42	0.14
850								3.96	1.69	0.47	0.16
900								4.41	1.88	0.52	0.18
950								4.87	2.08	0.58	0.20
1000								5.36	2.29	0.63	0.22
1500									4.84	1.34	0.46
2000										2.29	0.78
2500										3.46	1.18
3000											1.66

Friction Loss- Polyethylene (PE) SDR-Pressure Rated Pipe
Pressure loss from friction in psi per 100 feet of pipe.

NOMINAL PIPE DIAMETER IN INCHES

Flow GPM	0.5	0.75	1	1.25	1.5	2	2.5	3
1	0.49	0.12	0.04	0.01				
2	1.76	0.45	0.14	0.04	0.02			
3	3.73	0.95	0.29	0.08	0.04	0.01		
4	**6.35**	1.62	0.50	0.13	0.06	0.02		
5	9.60	2.44	0.76	0.20	0.09	0.03		
6	13.46	3.43	1.06	0.28	0.13	0.04	0.02	
7	17.91	4.56	1.41	0.37	0.18	0.05	0.02	
8	22.93	**5.84**	1.80	0.47	0.22	0.07	0.03	
9		7.26	2.24	0.59	0.28	0.08	0.03	
10		8.82	2.73	0.72	0.34	0.10	0.04	0.01
12		12.37	**3.82**	1.01	0.48	0.14	0.06	0.02
14		16.46	5.08	1.34	0.63	0.19	0.08	0.03
16			6.51	1.71	0.81	0.24	0.10	0.04
18			8.10	2.13	1.01	0.30	0.13	0.04
20			9.84	2.59	1.22	0.36	0.15	0.05
22			11.74	**3.09**	1.46	0.43	0.18	0.06
24			13.79	3.63	1.72	0.51	0.21	0.07
26			16.00	4.21	1.99	0.59	0.25	0.09
28				4.83	2.28	0.68	0.29	0.10
30				5.49	**2.59**	0.77	0.32	0.11
35				7.31	3.45	1.02	0.43	0.15
40				9.36	4.42	1.31	0.55	0.19
45				11.64	5.50	1.63	0.69	0.24
50				14.14	6.68	**1.98**	0.83	0.29
55					7.97	2.36	0.85	0.35
60					9.36	2.78	1.17	0.41
65					10.36	3.22	1.36	0.47
70					12.46	3.69	**1.56**	0.54
75					14.16	4.20	1.77	0.61
80						4.73	1.99	0.69
85						5.29	2.23	0.77
90						5.88	2.48	0.86
95						6.50	2.74	0.95
100						7.15	3.01	**1.05**
150						15.15	6.38	2.22
200							10.87	3.78
300								8.01

Nominal Pipe Size vs. Actual Outside Diameter for Steel and Plastic Pipe

Nominal Size	Actual Size
½″	0.840″
¾″	1.050″
1″	1.315″
1¼″	1.660″
1½″	1.900″
2″	2.375″
2½″	2.875″
3″	3.500″
3½″	4.000″
4″	4.500″
5″	5.563″
6″	6.625″

Temperature Conversions

°C = Degrees Celsius. 1 degree is ¹⁄₁₀₀ of the difference between the temperature of melting ice and boiling water.

°F = Degrees Fahrenheit. 1 degree is ¹⁄₁₈₀ of difference between the temperature of melting ice and boiling water.

°K = Degrees Kelvin (Absolute Temperature). This scale is based on the average kinetic energy per molecule of a perfect gas. It uses the same size units as the Celsius scale, but the degree symbol (°) is not used. Zero Kelvin (0K) is the temperature at which matter loses all energy. 0K = –273.16°C, or –459.69°F. You don't want to go there.

TEMPERATURE CONVERSION CHART

°C	°F	°C	°F	°C	°F	°C	°F
200	392	140	284	80	176	15	59
195	383	135	275	75	167	10	50
190	374	130	266	70	158	5	41
185	365	125	257	65	149	0	32
180	356	120	248	60	140	–5	23
175	347	115	239	50	122	–10	14
170	338	110	230	45	113	–15	5
165	329	105	221	40	104	–20	–4
160	320	100	212	35	95	–25	–13
155	311	95	203	30	86	–30	–22
150	302	90	194	25	77	–35	–31
145	293	85	185	20	68	–40	–40

The Real Goods Resource List

Although this Sourcebook is a complete source of renewable energy and environmental products, we can't be everything to everyone. Here is our current list of trusted outside resources. We do our best to pick the best organizations for this listing, and since we have worked directly with many of them, we are giving you the benefit of our experience. Do we need to say that we do not necessarily endorse all the actions of each group listed, nor are we responsible for what they say or do? Consider it said.

This list can never be complete. We apologize for resources that we have overlooked, and welcome suggestions. Please mail a brief description of the organization and access info to Gaiam Real Goods, Attention: Resource List.

American Council for an Energy-Efficient Economy (ACEEE)
1001 Connecticut Avenue, Suite 801
Washington, DC 20036
Website: www.aceee.org
Email: info@aceee.org
Phone (Research and Conferences):
202-429-8873
Publications: 202-429-0063
Publishes books, papers, yearly guides, and comparisons of appliances and vehicles based on energy efficiency. Their website is an excellent source of efficient appliance info.

The American Hydrogen Association
7939 West 7th Avenue
Mesa, AZ 85202
Website: www.clean-air.org/
Email: answerguy@clean-air.org
Phone: 480-827-7915
A nonprofit organization that promotes the use of hydrogen for fuel and energy storage. Publishes "Hydrogen Today," a bi-monthly newsletter.

American Society of Landscape Architects (ASLA)
636 Eye Street, NW
Washington, DC 20001-3736
Website: www.asla.org
Phone: 202-898-2444
Fax: 202-898-1185

American Solar Energy Society (ASES)
2400 Central Avenue, Unit G-1
Boulder, CO 80301
Website: www.ases.org/
Email: ases@ases.org
Phone: 303-443-3130
Fax: 303-443-3212
ASES is the United States Section of the International Solar Energy Society, a national organization dedicated to advancing the use of solar energy for the benefit of U.S. citizens and the global environment. Publishers of the bi-monthly Solar Today *magazine for members, and sponsors of the yearly National Tour of Solar Homes.*

The American Wind Energy Association.
122 C Street, NW, Suite 380
Washington, DC 20001
Website: www.awea.org/
Email: windmail@awea.org
Phone: 202-383-2500
Fax: 202-383-2505
A national trade association that represents wind power plant developers, wind turbine manufacturers, utilities, consultants, insurers, financiers, researchers, and others involved in the wind industry. These folks primarily work with utility-level wind systems. Not a good source for residential info.

American BioEnergy Association
314 Massachusetts Ave., N.E.
Suite 200
Washington, D.C. 20002
Website: www.biomass.org
Is the leading voice in the United States for the bioenergy industry. Works to build support through tax incentives, increased biomass research and development budgets, regulations and other policy initiatives.

Battery Recycling (for Nicads)
Rechargeable Battery Recycling Corp.
1000 Parkwood Circle, Suite 450
Atlanta, GA 30339
Website: www.rbrc.org
Email: consumer@rbrc.com
Phone: 678-419-9990
Fax: 678-419-9986
A nonprofit public service organization to promote the recycling of nicad batteries. Just type your zip code in the website and get a list of local stores that will accept your old nicads for recycling.

California Energy Commission
1516 Ninth Street
Sacramento, CA 95814-2950
Website: www.energy.ca.gov
Email: energia@energy.ca.gov
Phone: 800-555-7794 (inside Calif.)
916-654-4058 (outside Calif.)
California state's primary energy policy and planning agency. Strongly supports energy efficiency and small-scale utility intertie projects. A good source of honest intertie information.

CalStart
Northern California Office:
Hangar 20, Naval Air Station
1889 First Avenue
Alameda, CA 94501
Phone: 510-864-3000
Southern California Office:
3360 E. Foothill Boulevard
Pasadena, CA 91107
Phone: 626-744-5600
Website: www.calstart.org
The latest information on electric, natural gas, hybrid electric vehicles and intelligent transportation technologies.

Center for Renewable Energy & Sustainable Technology (CREST)
Mary Kathryn Campbell, Director: Marketing and Publications
1612 K St, NW, Suite 202
Washington DC 20006
Phone: 202-293-2898
Fax: 202-293-5857
Website: http://solstice.crest.org
An excellent on-line resource for sustainable energy information. Provides Internet services, software, databases, resource lists, and discussion lists on a wide variety of topics.

Chelsea-Green Publishing Co.
205 Gates-Briggs Building
P.O. Box 428
White River Junction, VT 05001
Website: www.chelseagreen.com
Email: seaton@chelseagreen.com
Phone: 802-295-6300
Fax: 802-295-6444
Gaiam Real Goods' publishing partner, offering books on sustainable living and energy conservation.

Electric Auto Association
P.O. Box 6661
Concord, CA 94514
Website: www.eaaev.org
The national Electric Vehicle association with local chapters in most states and Canada. Dues are $39/year with an excellent monthly newsletter. The website has links to practically everything in the EV biz.

Energy Information Administration
1000 Independence Ave. S.W.
Washington, DC 20585
Website: www.eia.doe.gov
Email: infoctr@eia.doe.gov
Phone: 202-586-8800
Official energy statistics from the U.S. government. Gas, oil, electricity, you name it, there's more supply, pricing, and use info here than you can shake a stick at.

Energy Star®
Environmental Protection Agency (EPA)
401 M Street, SW (6202J)
Washington, DC 20460
Website: www.energystar.gov/
Email: info@energystar.gov
Phone: 888-STAR-YES (888-782-7937)
*Good, up-to-date listings of the most effi-
cient appliances, lights, windows, home
and office electronics, and more.*

**Environmental Protection Agency
(EPA)**
Ariel Rios Building
1200 Pennsylvania Avenue N.W.
Washington, DC 20460
Website: www.epa.gov
Email: public-access@epamail.epa.gov
Phone: 202-260-2090

Florida Solar Energy Center (FSEC)
1679 Clearlake Road
Cocoa, FL 32922-5703
Website: www.fsec.ucf.edu
Email: webmaster@fsec.ucf.edu
Phone: 321-638-1000
Fax: 321-638-1010
*Provides highly regarded independent
third-party testing and certification of
solar hot water systems and other solar or
energy-efficiency goods.*

**Gas Appliance Manufacturers
Association**
2107 Wilson Boulevard, Suite 600
Arlington, Virginia 22201
Website: www.gamanet.org
Email: information@gamanet.org
Phone: 703-525-7060
Fax: 703-525-6790

Geothermal Resources Council
P.O. Box 1350
Davis, CA 95617
Website: www.geothermal.org
Email: grclib@geothermal.org
Phone: 530-758-2360
Fax: 530-758-2839
*A nonprofit organization dedicated to
geothermal research and development.*

The Green Power Network
Net Metering and Green Power info
Website: www.eren.doe.gov/greenpower
*Information on the electric power indus-
try's green power efforts. Includes
up-to-date info on green power availability
and pricing. Also the best state-by-state net
metering details on the Net.*

**The Heartwood School for the
Homebuilding Crafts**
Johnson Hill Road
Washington, MA 01235
Website: www.heartwoodschool.com
Email: info@hearwoodschool.com
Phone: 413-623-6677
Fax: 413-623-0277
*Offers a variety of hands-on building
workshops. Specializes primarily in tim-
ber frames.*

Home Power Magazine
P.O. Box 520
Ashland, OR 97520
Website: www.homepower.com
Email: hp@homepower.com
Phone: 800-707-6585
Fax: 541-512-0343
*The journal of the renewable energy indus-
try. Published bi-monthly. U.S.
subscription is $22.50/yr.*

The Institute for Solar Living
P.O. Box 836
Hopland, CA 95449
Website: www.solarliving.org
Email: isl@rgisl.org
Phone: 707-744-2017
Fax: 707-744-1682
*A nonprofit educational organization that
offers hands-on classes on a wide variety
of renewable energy, energy conservation,
and building technologies. Hosts annual
SolFest at Real Goods Solar Living Center.*

The Last Straw Online
HC 66, Box 119
Hillsboro, NM 88042.
Website: www.strawhomes.com
Email: thelaststraw@strawhomes.com
Phone: 505 895-5400
Fax: 505 895-3326
An excellent hub site with links to many other straw bale and sustainable building sites. This is also the online home of "The Last Straw," an excellent straw bale quarterly journal.

National Association of Home Builders (NAHB)
1201 15th Street, NW
Washington, DC 20005
Website: www.nahb.com
Email: info@nahb.com
Phone: 800-368-5242

National Association of State Energy Officials (NASEO)
1414 Prince Street, Suite 200
Alexandria, VA 22314
Website: www.naseo.org
Email: info @naseo.org
Phone: 703-299-8800
Fax: 703-299-6208
An excellent portal site with access info for all state energy offices.

National Center for Appropriate Technology
3040 Continental Drive
Butte, MT 59702
Phone: 800-275-6228
Website: www.ncat.org
Email: info@ncat.org
A nonprofit organization, information clearinghouse for sustainable energy, low income energy, resource efficient housing, and sustainable agriculture.

National Renewable Energies Laboratory (NREL)
1617 Cole Boulevard
Golden, CO 80401
Website: www.nrel.gov
Email: webmaster@nrel.gov
Phone: 303-275-3000
The nation's leading center for renewable energy research. Tests products, conducts experiments, and provides information on renewable energy.

Rocky Mountain Institute
1739 Snowmass Creek Road
Snowmass, CO 81654-9199
Website: www.rmi.org
Email: outreach@rmi.org
Phone: 970-927-3851
Fax: 970-927-3420
A terrific source of books, papers, and research on energy-efficient building design and components.

Solar Cookers International
1724 11th Street
Sacramento, CA 95814
Website: http://solarcooking.org/
Email: sbci@igc.org
Phone: 916-455-4499
Fax: 916-455-4498
A nonprofit organization that promotes and distributes simple solar cookers in developing countries to relieve the strain of firewood collection. Newsletter and small products catalog available.

Solar Electric Light Fund
1775 K Street, NW, Suite 595
Washington, D.C. 20006
Phone: 202-234-7265
Email: solarlectric@igc.org
Website: www.self.org/
A nonprofit organization promoting PV rural electrification in developing countries.

Solar Energy Industries Association (SEIA)
1616 H Street, NW 8th Floor;
Washington, DC 20006
Website: www.seia.org
Email: plowenth@seia.org
Phone: 202-628-7979 or 301-951-3231
Fax: 202-628-7779

Solar Energy International
P.O. Box 715
Carbondale, CO 81623-0715
Website: www.solarenergy.org
Email: sei@solarenergy.org
Phone: 970-963-8855
Fax: 970-963-8866
Offers hands-on, and online classes on renewable technology.

Solar Energy Info for Planet Earth
Website: http://eosweb.larc.nasa.gov/sse/
Complete solar energy data for anyplace on the planet! Thanks to NASA's Earth Science Enterprise program, there's more solar data here than even the Gaiam Real Goods techies can find a use for. Just point to your location on a world map.

Sustainable Architecture
Roy Prince, Architect
P.O. Box 30085
Santa Barbara, CA 93130
Website: www.SustainableABC.com
Email: royprince@sustainableabc.com
Phone: 805-898-9660
Fax: 805-898-9199
A unique compendium of links and content oriented to the global community of ecological and natural building proponents.

The Union of Concerned Scientists
2 Brattle Square
Cambridge, MA 02238
Website: www.ucsusa.org
Email: usc@ucsusa.org
Phone: 617-547-5552
Fax: 617-864-9405
Organization of scientists and citizens concerned with the impact of advanced technology on society. Programs focus on energy policy and national security.

U.S. Department of Energy's
Energy Efficiency and Renewable Energy
Clearinghouse (EREC)
P.O. Box 3048
Merrifield, VA 22116
Website: www.eren.doe.gov/
Email: doe.erec@nciinc.com
Phone: 800-DOE-EREC (800-363-3732)
Fax: 703-893-0400
The best source of information about renewable energy technologies and energy efficiency. These folks seem to be able to find good information about anything energy related. One of the best energy-related web sites.

Wind Energy Maps for U.S.
Website: http://rredc.nrel.gov/wind/pubs/
atlas/maps.html#3-54
The complete Wind Energy Resource Atlas for the United States in downloadable format. Offers averages, regions, states, seasonal; more slicing and dicing than you can imagine. Includes Alaska, Hawaii, Puerto Rico, and Virgin Islands.

Worldwatch Institute
1776 Massachusetts Avenue, N.W.
Washington, DC 20036-1904
Website: www.worldwatch.org
Email: worldwatch@worldwatch.org
Phone: 202-452-1999
Fax: 202-296-7365
A nonprofit public policy research organization dedicated to informing policymakers and the public about emerging global problems and trends. Offers a magazine and publishes a variety of books.

GLOSSARY

A

AC: alternating current, electricity that changes voltage periodically, typically sixty times a second (or fifty in Europe). This kind of electricity is easier to move.

activated stand life: the period of time, at a specified temperature, that a battery can be left stored in the charged condition before its capacity fails

active solar: any solar scheme employing pumps and controls that use power while harvesting solar energy

A-frame: a building that looks like the capital letter A in cross-section

air lock: two doors with space between, like a mud room, to keep the weather outside

alternating current: AC electricity that changes voltage periodically, typically sixty times a second

alternative energy. "voodoo" energy not purchased from a power company, usually coming from photovoltaic, micro-hydro, or wind

ambient: the prevailing temperature, usually outdoors

amorphous silicon: a type of PV cell manufactured without a crystalline structure. Compare with single-crystal and multi- (or poly-) crystalline silicon.

ampere: an instantaneous measure of the flow of electric current; abbreviated and more commonly spoken of as an "amp"

amp-hour: a one-ampere flow of electrical current for one hour; a measure of electrical quantity; two 60-watt 120-volt bulbs burning for one hour consume one amp-hour.

angle of incidence: the angle at which a ray of light (usually sunlight) strikes a planar surface (usually of a PV module). Angles of incidence close to perpendicularity (90°) are desirable.

anode: the positive electrode in an electrochemical cell (battery) toward which current flows; the earth ground in a cathodic protection system

antifreeze: a chemical, usually liquid and often toxic, that keeps things from freezing

array: an orderly collection, usually of photovoltaic modules connected electrically and mechanically secure; array current: the amperage produced by an array in full sun

avoided cost: the amount utilities must pay for independently produced power; in theory, this was to be the whole cost, including capital share to produce peak demand power, but over the years supply-side weaseling redefined it to be something more like the cost of the fuel the utility avoided burning.

azimuth: horizontal angle measured clockwise from true north; the equator is at 90°.

B

backup: a secondary source of energy to pick up the slack when the primary source is inadequate. In alternatively powered homes, fossil fuel generators are often used as "backups" when extra power is required to run power tools or when the primary sources—sun, wind, water—are not providing sufficient energy.

balance of system: (BOS) equipment that controls the flow of electricity during generation and storage baseline: a statistical term for a starting point; the "before" in a before-and-after energy conservation analysis

baseload: the smallest amount of electricity required to keep utility customers operating at the time of lowest demand; a utility's minimum load

battery: a collection of cells that store electrical energy; each cell converts chemical energy into electricity or vice versa, and is interconnected with other cells to form a battery for storing useful quantities of electricity.

battery capacity: the total number of ampere-hours that can be withdrawn from a fully charged battery, usually over a standard period

battery cycle life: the number of cycles that a battery can sustain before failing

berm: earth mounded in an artificial hill

bioregion: an area, usually fairly large, with generally homogeneous flora and fauna

biosphere: the thin layer of water, soil, and air that supports all known life on earth

black water: what gets flushed down the toilet

blocking diode: a diode which prevents loss of energy to an inactive PV array (rarely used with modern charge controllers)

boneyard: a peculiar location at Gaiam Real Goods where "experienced" products may be had for ridiculously low prices

Btu: British thermal unit, the amount of heat required to raise the temperature of one pound of water one degree Fahrenheit. 3,411 Btus equals one kilowatt-hour.

bus bar: the point where all energy sources and loads connect to each other; often a metal bar with connections on it

bus bar cost: the average cost of electricity delivered to the customer's distribution point

buy-back agreement or contract: an agreement between the utility and a customer that any excess electricity generated by the customer will be bought back for an agreed-upon amount

C

cathode: the negative electrode in an electrochemical

cell: a unit for storing or harvesting energy. In a battery, a cell is a single chemical storage unit consisting of trodes and electrolyte, typically producing 1.5 volts; several cells are usually arranged inside a single container called a battery. Flashlight batteries are really flashlight cells. A photovoltaic cell is a single assembly of doped silicon and electrical contacts that allow it to take advantage of the photovoltaic effect, typically producing .5 volts; several PV cells are usually connected together and packaged as a module.

CF: compact fluorescent, a modern form of lightbulb using an integral ballast

CFCs: chlorinated fluorocarbons, an industrial solvent and material widely used until implicated as a cause of ozone depletion in the atmosphere

charge controller: device for managing the charging rate and state of charge of a battery bank

controller terminology:

adjustable set point: allows adjustment of voltage disconnect levels

high-voltage disconnect: the battery voltage at which the charge controller disconnects the batteries from the array to prevent overcharging

low-voltage disconnect: the voltage at which the controller disconnects the batteries to prevent overdischarging

low-voltage warning: a buzzer or light that indicates low battery voltage

maximum power tracking: a circuit that maintains array voltage for maximal current

multistage controller: a unit that allows multilevel control of battery charging or loading

reverse current protection: prevents current flow from battery to array

single-stage controller: a unit with only one level of control for charging or load control

temperature compensation: a circuit that adjusts setpoints to ambient temperature in order to optimize charge

charge rate: the rate at which a battery is recharged, expressed as a ratio of battery capacity to charging current flow

clear-cutting: a forestry practice, cutting all trees in a relatively large plot

cloud enhancement: the increase in sunlight due to direct rays plus refracted or reflected sunlight from partial cloud cover

compact fluorescent: a modern form of light bulb using an integral ballast

compost: the process by which organic materials break down, or the materials in the process of being broken down

concentrator: mirror or lens-like additions to a PV array that focus sunlight on smaller cells; a very promising way to improve PV yield

conductance: a material's ability to allow electricity to flow through it; gold has very high conductance.

conversion efficiency: the ratio of energy input to energy output across the conversion boundary. For example, batteries typically are able to store and provide 90% of the charging energy applied, and are said to have a 90% energy efficiency.

cookie-cutter houses: houses all alike and all-in-a-row. Daly City, south of San Francisco, is a particularly depressing example. Runs in direct contradiction to our third guiding principle, "encourage diversity."

core/coil-ballasted: the materials-rich device required to drive some fluorescent lights; usually contains Americium (see electronic ballasts)

cost-effectiveness: an economic measure of the worthiness of an investment; if an innovative solution costs less than a conventional alternative, it is more cost-effective.

cross section: a "view" or drawing of a slice through a structure

cross-ventilation: an arrangement of openings allowing wind to pass through a structure

crystalline silicon: the material from which most photovoltaic cells are made; in a single crystal cell, the entire cell is a slice of a single crystal of silicon, while a multicrystalline cell is cut from a block of smaller (centimeter sized) crystals. The larger the crystal, the more exacting and expensive the manufacturing process.

cut-in: the condition at which a control connects its device

cutoff voltage: in a charge controller, the voltage at which the array is disconnected from the battery to prevent overcharging

cut-out: the condition at which a control interrupts the action

cycle: in a battery, from a state of complete charge through discharge and recharge back to a fully charged state

D

days of autonomy: the length of time (in days) that a system's storage can supply normal requirements without replenishment from a source; also called days of storage

DC: direct current, the complement of AC, or alternating current, presents one unvarying voltage to a load

deep cycle: a battery type manufactured to sustain many cycles of deep discharge that are in excess of 50% of total capacity

degree-days: a term used to calculate heating and cooling loads; the sum, taken over an average year, of the lowest (for heating) or highest (for cooling) ambient daily temperatures. Example, if the target is 68° and the ambient on a given day is 58°, this would account for ten degree-days.

depth of discharge: the percent of rated capacity that has been withdrawn from the battery; also called DOD

design month: the month in which the combination of insolation and loading require the maximum array output

diode: an electrical component that permits current to pass in only one direction

direct current: the complement of AC, or alternating current, presents one unvarying voltage to a load

discharge: electrical term for withdrawing energy from a storage system

discharge rate: the rate at which current is withdrawn from a battery expressed as a ratio to the battery's capacity; also known as C rate

disconnect: a switch or other control used to complete or interrupt an electrical flow between components

doping, dopant: small, minutely controlled amounts of specific chemicals introduced into the semiconductor matrix to control the density and probabilistic movement of free electrons

downhole: a piece of equipment, usually a pump, that is lowered down the hole (the well or shaft) to do its work

drip irrigation: a technique that precisely delivers measured amounts of water through small tubes; an exceedingly efficient way to water plants

dry cell: a cell with captive electrolyte

duty cycle: the ratio between active time and total time; used to describe the operating regime of an appliance in an electrical system

duty rating: the amount of time that an appliance can be run at its full rated output before failure can be expected

E

earthship: a rammed-earth structure based on tires filled with tamped earth; the term was coined by Michael Reynolds.

Eco Desk: an ecology information resource maintained by many ecologically-minded companies including Gaiam Real Goods

edison base: a bulb base designed by (a) Enrico Fermi, (b) John Schaeffer, (c) Thomas Edison, (d) none of the above. The familiar standard residential lightbulb base.

efficiency: a mathematical measure of actual as a percentage of the theoretical best. See conversion efficiency.

electrolyte: the chemical medium, usually liquid, in a battery that conveys charge between the positive and negative electrodes; in a lead-acid battery, the electrodes are lead and the electrolyte is acid.

electromagnetic radiation: EMR, the invisible field around an electric device. Not much is known about the effects of EMR, but it makes many of us nervous.

electronic ballasts: an improvement over core/coil ballasts, used to drive compact fluorescent lamps; contains no radioactivity

embodied: of energy, meaning literally the amount of energy required to produce an object in its present form. Example: an inflated balloon's embodied energy includes the energy required to blow it up.

EMR: electromagnetic radiation, the invisible field around an electric device. Not much is known about the effects of EMR, but it makes many of us nervous.

energy density: the ratio of stored energy to storage volume or weight

energy-efficient: one of the best ways to use energy to accomplish a task; for example, heating with electricity is never energy efficient, while lighting with compact fluorescents is.

equalizing: periodic overcharging of batteries to make sure that all cells are reaching a good state of charge

externalities: considerations, often subtle or remote, that should be accounted for when evaluating a process or product, but usually are not. For example, externalities for a power plant may include downwind particulate fallout and acid rain, damage to lifeforms in the cooling water intake and effluent streams, and many other factors.

F

fail-safe: a system designed in such a way that it will always fail into a safe condition

feng shui: an Asian system of placement that pays special attention to wind, water, and the cardinal directions

ferro-cement: a construction technique; an armature of iron contained in a cement body, often a wall, slab, or tank

fill factor: of a photovoltaic module's I-V (current/voltage) curve, this number expresses the product of the open circuit voltage and the short-circuit current, and is therefore a measure of the "squareness" of the I-V curve's shape.

firebox: the structure within which combustion takes place

fixed-tilt array: a PV array set in a fixed position

flat-plate array: a PV array consisting of nonconcentrating modules

float charge: a charge applied to a battery equal to or slightly larger than the battery's natural tendency to self-discharge

FNC: fiber-nickel-cadmium, a new battery technology

frequency: of a wave, the number of peaks in a period. For example, alternating current presents sixty peaks per second, so its frequency is sixty hertz. Hertz is the standard unit for frequency when the period in question is one second.

G

gassifier: a heating device which burns so hotly that the fuel sublimes directly from its solid to its gaseous state and burns very cleanly

gassing: when a battery is charged, gasses are often given out; also called outgassing

golf cart batteries: industrial batteries tolerant of deep cycling, often used in mobile vehicles gotcha!s: an unexpected outcome or effect, or the points at which, no matter how hard you wriggle, you can't escape

gravity-fed: water storage far enough above the point of use (usually 50 feet) so that the weight of the water provides sufficient pressure

greywater: all other household effluents besides black water (toilet water); greywater may be reused with much less processing than black water

grid: a utility term for the network of transmission lines that distribute electricity from a variety of sources across a large area

grid-connected system: a house, office, or other electrical system that can draw its energy from the grid; although usually grid-power-consumers, grid-connected systems can provide power to the grid.

groundwater: as distinct from water pumped up from the depths, groundwater is run off from precipitation, agriculture, or other sources.

H

heat exchanger: device that passes heat from one substance to another; in a solar hot water heater, for example, the heat exchanger takes heat harvested by a fluid circulating through the solar panel and transfers it to domestic hot water.

high-tech glass: window constructions made of two sheets of glass, sometimes treated with a metallic deposition, sealed together hermetically, with the cavity filled by an inert gas and, often, a further plastic membrane. High-tech glass can have an R-value as high as 10.

homeschooling: educating children at home instead of entrusting them to public or private schools; a growing trend quite often linked to alternatively powered homes

homestead: the house and surrounding lands

homesteaders: people who consciously and intentionally develop their homestead

house current: in the United States, 117 volts root mean square of alternating current, plus or minus 7 volts; nominally 110-volt power; what comes out of most wall outlets

hot tub: a quasi-religious object in California; a large bathtub for several people at once; an energy hog of serious proportions

HVAC: heating, ventilation, and air-conditioning; space conditioning

hydro turbine: a device that converts a stream of water into rotational energy

hydrometer: tool used to measure the specific gravity of a liquid

hydronic: contraction of hydro and electronic, usually applied to radiant in-floor heating systems and their associated sensors and pumps

hysteresis: the lag between cause and effect, between stimulus and response

I

incandescent bulb: a light source that produces light by heating a filament until it emits photons—quite an energy-intensive task

incident solar radiation: or insolation, the amount of sunlight falling on a place

indigenous plantings: gardening with plants native to the bioregion

Inductive transformer/rectifier: the little transformer device that powers many household appliances; an "energy criminal" that takes an unreasonably large amount of alternating electricity (house current) and converts it into a much smaller amount of current with different properties; for example, much lower voltage direct current

infiltration: air, at ambient temperature, blowing through cracks and holes in a house wall and spoiling the space conditioning

infrared: light just outside the visible spectrum, usually associated with heat radiation

infrastructure: a buzz word for the underpinnings of civilization; roads, water mains, power and phone lines, fire suppression, ambulance, education, and governmental services are all infrastructure. "Infra," is Latin for beneath. In a more technical sense, the repair infrastructure is local existence of repair personnel and parts for a given technology.

insolation: a word coined from incident solar radiation, the amount of sunlight falling on a place

insulation: a material that keeps energy from crossing from one place to another. On electrical wire, it is the plastic or rubber that covers the conductor. In a building, insulation makes the walls, floor, and roof more resistant to the outside (ambient) temperature.

Integrated Resource Planning: an effort by the utility industry to consider all resources and requirements in order to produce electricity as efficiently as possible

interconnect: to connect two systems, often an independent power producer and the grid; see also intertie

interface: the point where two different flows or energies interact; for example, a power system's interface with the human world is manifested as meters, which show system status, and controls, with which that status can be manipulated.

internal combustion engines: gasoline engines, typically in automobiles, small stand-alone devices like chainsaws, lawnmowers, and generators

intertie: the electrical connection between an independent power producer—for example, a PV-powered household —and the utility's distribution lines, in such a way that each can supply or draw from the other.

inverter: the electrical device that changes direct current into alternating current irradiance: the instantaneous solar radiation incident on a surface; usually expressed in Langleys (a small amount) or in kilowatts per square meter. The definition of "one sun" of irradiance is one kilowatt per square meter.

irreverence: the measure, difficult to quantify, of the seriousness of Gaiam Real Goods techs when talking with utility suits

IRP: Integrated Resource Planning: an effort by the utility industry to consider all resources and requirements in order to produce electricity as efficiently as possible

I-V curve: a plot of current against voltage to show the operating characteristics of a photovoltaic cell, module, or array

K

kilowatt: one thousand watts, a measure of instantaneous work. Ten one-hundred-watt bulbs require a kilowatt of energy to light up

kilowatt-hour: the standard measure of household electrical energy. If the ten bulbs left unfrugally burning in the preceding example are on for an hour, they consume one kilowatt-hour of electricity.

L

landfill: another word for dump

Langley: the unit of solar irradiance; one gram-calorie per square centimeter

lead-acid: the standard type of battery for use in home energy systems and automobiles

LED: light-emitting diode. A very efficient source of electrical lighting, typically lasting 50,000 to 100,000 hours

life: You expect an answer here? Well, when speaking of electrical systems, this term is used to quantify the time the system can be expected to function at or above a specified performance level.

life-cycle cost: the estimated cost of owning and operating a system over its useful life

line extensions: what the power company does to bring their power lines to the consumer

line-tied system: an electrical system connected to the powerlines, usually having domestic power generating capacity and the ability to draw power from the grid or return power to the grid, depending on load and generator status

load: an electrical device, or the amount of energy consumed by such a device

load circuit: the wiring that provides the path for the current that powers the load

load current: expressed in amps, the current required by the device to operate

low-emissivity: applied to high-tech windows, meaning that infrared or heat energy will not pass back out through the glass

low-flush: a toilet using a smaller amount (usually about 6 quarts) of water to accomplish its function

low pressure: usually of water, meaning that the head, or pressurization, is relatively small

low-voltage: usually another term for 12- or 24-volt direct current

M

maintenance-free battery: a battery to which water cannot be added to maintain electrolyte volume. All batteries require routine inspection and maintenance.

maximum power point: the point at which a power conditioner continuously controls PV source voltage in order to hold it at its maximum output current

ME: mechanical engineer; the engineers who usually work with heating and cooling, elevators, and the other mechanical devices in a large building

meteorological: pertaining to weather; meteorology is the study of weather

micro-climate: the climate in a small area, sometimes as small as a garden or the interior of a house. Climate is distinct from weather in that it speaks for trends taken over a period of at least a year, while weather describes immediate conditions.

micro-hydro: small hydro (falling water) generation

millennia: one thousand years milliamp: one thousandth of an ampere

module: a manufactured panel of photovoltaic cells. A module typically houses 36 cells in an aluminum frame covered with a glass or acrylic cover, organizes their wiring, and provides a junction box for connection between itself, other modules in the array, and the system.

N

naturopathic: a form of medicine devoted to natural remedies and procedures

net metering: a desirable form of buy-back agreement in which the line-tied house's electric meter turns in the utility's favor when grid power is being drawn, and in the system owner's favor when the house generation exceeds its needs and electricity is flowing into the grid. At the end of the payment period, when the meter is read, the system owner pays (or is paid by) the utility depending on the net metering.

NEC: the National Electrical Code, guidelines for all types of electrical installations including (since 1984) PV systems

nicad: slang for nickel-cadmium, a form of chemical storage often used in rechargeable batteries

nominal voltage: the terminal voltage of a cell or battery discharging at a specified rate and at a specified temperature; in normal systems, this is usually a multiple of 12

nontoxic: having no known poisonous qualities

normal operating cell temperature: defined as the standard operating temperature of a PV module at 800 W/m°, 20° C ambient, 1 meter per second wind speed; used to estimate the nominal operating temperature of a module in its working environment

N-type silicon: silicon doped (containing impurities), which gives the lattice a net negative charge, repelling electrons

O

off-peak energy: electricity during the baseload period, which is usually cheaper. Utilities often must keep generators turning, and are eager to find users during these periods, and so sell off-peak energy for less.

off-peak kilowatt: a kilowatt hour of off-peak energy

off-the-grid: not connected to the powerlines, energy self-sufficient

ohm: the basic unit of electrical resistance; I=RV, or Current (amperes) equals Resistance (ohms) times Voltage (volts).

on-line: connected to the system, ready for work

on-the-grid: where most of America lives and works, connected to a continent-spanning web of electrical distribution lines

open-circuit voltage: the maximum voltage measurable at the terminals of a photovoltaic cell, module, or array with no load applied

operating point: the current and voltage that a module or array produces under load, as determined by the I-V curve

order of magnitude: multiplied or divided by ten. One hundred is an order of magnitude smaller than one thousand, and an order of magnitude larger than ten.

orientation: placement with respect to the cardinal directions north, east, south, and west; azimuth is the measure of orientation.

outgas: of any material, the production of gasses; batteries outgas during charging; new synthetic rugs outgas when struck by sunlight, or when warm, or whenever they feel like it.

overcharge: forcing current into a fully charged battery; a bad idea except during equalization

overcurrent: too much current for the wiring; overcurrent protection, in the form of fuses and circuit breakers, guards against this.

owner-builder: one of the few printable things building inspectors call people who build their own homes

P

panel: any flat modular structure; solar panels may collect solar energy by many means; a number of photovoltaic modules may be assembled into a panel using a mechanical frame, but this should more properly be called an array or subarray.

parallel: connecting the like poles to like in an electrical circuit, so plus connects to plus and minus to minus; this arrangement increases current without affecting voltage.

particulates: particles that are so small that they persist in suspension in air or water

passive solar: a shelter which maintains a comfortable inside temperature simply by accepting, storing, and preserving the heat from sunlight

passively heated: a shelter that has its space heated by the sun without using any other energy

patch-cutting: clear-cutting (cutting all trees) on a small scale usually less than an acre

pathetic fallacy: attributing human motivations to inanimate objects or lower animals

payback: the time it takes to recoup the cost of improved technology as compared to the conventional solution. Payback on a compact fluorescent bulb (as compared to an incandescent bulb) may take a year or two, but over the whole life of the CF the savings will probably exceed the original cost of the bulb, and payback will take place several times over.

peak demand: the largest amount of electricity demanded by a utility's customers; typically, peak demand happens in early afternoon on the hottest weekday of the year.

peak kilowatt: a kilowatt hour of electricity take during peak demand, usually the most expensive electricity money can buy

peak load: the same as peak demand but on a smaller scale, the maximum load demanded of a single system

peak power current: the amperage produced by a photovoltaic module operating at the "knee" of its I-V curve

peak sun hours: the equivalent number of hours per day when solar irradiance averages one sun (1 kW/m°). "Six peak sun hours" means that the energy received during total daylight hours equals the energy that would have been received if the sun had shone for six hours at a rate of 1000 watts per square meter.

peak watt: the manufacturer's measure of the best possible output of a module under ideal laboratory conditions

Pelton wheel: a special turbine, designed by someone named Pelton, for converting flowing water into rotational energy

periodic table of elements: a chart showing the chemical elements organized by the number of protons in their nuclei and the number of electrons in their outer, or valence, band

PG&E: Pacific Gas and Electric, the local and sometimes beloved utility for much of Northern California

phantom loads: "energy criminals" that are on even when you turn them off: instant-on TVs, microwaves with clocks; symptomatic of impatience and our sloppy preference for immediacy over efficiency

photon: the theoretical particle used to explain light

photophobic: fear of light (or preference for darkness), usually used of insects and animals. The opposite, phototropic, means light-seeking.

photovolactics: an indicator of ignorance on the order of "nuke-ular"; makes us giggle photovoltaic cell: the proper name for a device manufactured to pump electricity when light falls on it

photovoltaics: PVs or modules that utilize the photovoltaic effect to generate useable amounts of electricity

photovoltaic system: the modules, controls, storage, and other components that constitute a stand-alone solar energy system

plates: the thin pieces of metal or other material used to collect electrical energy in a battery

plug-loads: the appliances and other devices plugged into a power system

plutonium: a particularly nasty radioactive material used in nuclear generation of electricity. One atom is enough to kill you.

pn-junction: the plane within a photovoltaic cell where the positively and negatively doped silicon layers meet

pocket plate: a plate for a battery in which active materials are held in a perforated metal pocket on a support strip

pollution: any dumping of toxic or unpleasant materials into air or water

polyurethane: a long-chain carbon molecule, a good basis for sealants, paints, and plastics

power: kinetic, or moving energy, actually performing work; in an electrical system, power is measured in watts.

power-conditioning equipment: electrical devices which change electrical forms (an inverter is an example) or assure that the electricity is of the correct form and reliability for the equipment for which it provides; a surge protector is another example.

power density: ratio of a battery's rated power available to its volume (in liters) or weight (in kilograms)

PUC: Public Utilities Commission; many states call it something else, but this is the agency responsible for regulating utility rates and practices.

PURPA: this 1978 legislation, the Public Utility Regulatory Policy Act, requires utilities to purchase power from anyone at the utility's avoided cost.

PVs: photovoltaic modules

R

radioactive material: a substance which, left to itself, sheds tiny, highly energetic pieces that put anyone nearby at great risk. Plutonium is one of these. Radioactive materials remain active indefinitely, but the time over which they are active is measured in terms of half-life, the time it takes them to become half as active as they are now; plutonium's half-life is a little over 22,000 years.

ram pump: a water-pumping machine that uses a water-hammer effect (based on the inertia of flowing water) to lift water

rated battery capacity: manufacturer's term indicating the maximum energy that can be withdrawn from a battery at a standard specified rate

rated module current: manufacturer's indication of module current under standard laboratory test conditions

renewable energy: an energy source that renews itself without effort; fossil fuels, once consumed, are gone forever, while solar energy is renewable in that the sun we harvest today has no effect on the sun we can harvest tomorrow.

renewables: shorthand term for renewable energy or materials sources

resistance: the ability of a substance to resist electrical flow; in electricity, resistance is measured in ohms.

retrofit: install new equipment to a structure which was not originally designed for it. For example, we may retrofit a lamp with a compact fluorescent bulb, but the new bulb's shape may not fit well with the lamp's design.

romex: an electrician's term for common two-conductor-with-ground wire, the kind houses are wired with

root mean square: RMS, the effective voltage of alternating current, usually about 70 percent (the square root of two over two) of the peak voltage. House current typically has an RMS of 117 volts and a peak voltage of 167 volts.

RPM: rotations per minute

R-value: Resistance value, used specifically of materials used for insulating structures. Fiberglass insulation three inches thick has an R value of 13.

S

seasonal depth of discharge: an adjustment for long-term seasonal battery discharge resulting in a smaller array and a battery bank matched to low-insolation season needs

secondary battery: a battery that can be repeatedly discharged and fully recharged

self-discharge: the tendency of a battery to lose its charge through internal chemical activity

semiconductor: the chief ingredient in a photovoltaic cell, a normal insulating substance that conducts electricity under certain circumstances

series connection: wiring devices with alternating poles, plus to minus, plus to minus; this arrangement increases voltage (potential) without increasing current.

set-back thermostat: combines a clock and a thermostat so that a zone (like a bedroom) may be kept comfortable only when in use

setpoint: electrical condition, usually voltage, at which controls are adjusted to change their action

shallow-cycle battery: like an automotive battery, designed to be kept nearly fully charged; such batteries perform poorly when discharged by more than 25% of their capacity.

shelf life: the period of time a device can be expected to be stored and still perform to specifications

short circuit: an electrical path that connects opposite sides of a source without any appreciable load, thereby allowing maximum (and possibly disastrous) current flow

short circuit current: current produced by a photovoltaic cell, module, or array when its output terminals are connected to each other, or "short-circuited"

showerhead: in common usage, a device for wasting energy by using too much hot water; in the Gaiam Real Goods home, low-flow showerheads prevent this undesirable result.

silicon: one of the most abundant elements on the planet, commonly found as sand and used to make photovoltaic cells

single-crystal silicon: silicon carefully melted and grown in large boules, then sliced and treated to become the most efficient photovoltaic cells

slow-blow: a fuse that tolerates a degree of overcurrent momentarily; a good choice for motors and other devices that require initial power surges to get rolling

slow paced: a description of the life of a Gaiam Real Goods employee . . . not!

solar aperture: the opening to the south of a site (in the northern hemisphere) across which the sun passes; trees, mountains, and buildings may narrow the aperture, which also changes with the season.

solar cell: see photovoltaic cell

solar fraction: the fraction of electricity that may be reasonable harvested from sun falling on a site. The solar fraction will be less in a foggy or cloudy site, or one with a narrower solar aperture, than an open, sunny site.

solar hot water heating: direct or indirect use of heat taken from the sun to heat domestic hot water

solar oven: simply a box with a glass front and, optionally, reflectors and reflector coated walls, which heats up in the sun sufficiently to cook food

solar panels: any kind of flat devices placed in the sun to harvest solar energy

solar resource: the amount of insolation a site receives, normally expressed in kilowatt-hours per square meter per day

specific gravity: the relative density of a substance compared to water (for liquids and solids) or air (for gases) Water is defined as 1.0; a fully charged sulfuric acid electrolyte might be as dense as 1.30, or 30 percent denser than water. Specific gravity is measured with a hydrometer.

stand-alone: a system, sometimes a home, that requires no imported energy

stand-by: a device kept for times when the primary device is unable to perform; a stand-by generator is the same as a back-up generator.

starved electrolyte cell: a battery cell containing little or no free fluid electrolyte

state of charge: the real amount of energy stored in a battery as a percentage of its total rated capacity

state-of-the-art: a term beloved by technoids to express that this is the hottest thing since sliced bread

stratification: in a battery, when the electrolyte acid concentration varies in layers from top to bottom; seldom a problem in vehicle batteries, due to vehicular motion and vibration, this can be a problem with static batteries, and can be corrected by periodic equalization.

stepwise: a little at a time, incrementally

sub-array: part of an array, usually photovoltaic, wired to be controlled separately

sulfating: formation of lead-sulfate crystals on the plates of a lead-acid battery, which can cause permanent damage to the battery

super-insulated: using as much insulation as possible, usually R-50 and above

surge capacity: the ability of a battery or inverter to sustain a short current surge in excess of rated capacity in order to start a device that requires an initial surge current

sustainable: material or energy sources that, if managed carefully, will provide at current levels indefinitely. A theoretical example: redwood is sustainable if it is harvested sparingly (large takings and exportation to Japan not allowed) and if every tree taken is replaced with another redwood. Sustainability can be, and usually is, abused for profit by playing it like a shell game; by planting, for example, a fast-growing fir in place of the harvested redwood.

system availability: the probability or percentage of time that an energy storage system will be able to meet load demand fully

T

temperature compensation: an allowance made by a charge controller to match battery charging to battery temperature

temperature correction: applied to derive true storage capacity using battery nameplate capacity and temperature; batteries are rated at 20°C.

therm: a quantity of natural gas, 100 cubic feet; roughly 100,000 Btus of potential heat

thermal mass: solid, usually masonry volumes inside a structure that absorb heat, then radiate it slowly when the surrounding air falls below their temperature

thermoelectric: producing heat using electricity; a bad idea

thermography: photography of heat loss, usually with a special video camera sensitive to the far end of the infrared spectrum

thermosiphon: a circulation system that takes advantage of the fact that warmer substances rise. By placing the solar collector of a solar hot water system below the tank, thermosiphoning takes care of circulating the hot water, and pumping is not required.

thin-film module: an inexpensive way of manufacturing photovoltaic modules; thin-film modules typically are less efficient than single-crystal or multi-crystal devices; also called amorphous silicon modules.

tilt angle: measures the angle of a panel from the horizontal

tracker: a device that follows the sun and keeps the panel perpendicular to it

transformers: a simple electrical device that changes the voltage of alternating current; most transformers are inductive, which means they set up a field around themselves, which is often a costly thing to do.

transparent energy system: a system that looks and acts like a conventional grid-connected home system, but is independent

trickle charge: a small current intended to maintain an inactive battery in a fully charged condition

troubleshoot: a form of recreation not unlike riding to hounds in which the technician attempts to find, catch, and eliminate the trouble in a system

tungsten filament: the small coil in a light bulb that glows hotly and brightly when electricity passes through it

turbine: a vaned wheel over which a rapidly moving liquid or gas is passed, causing the wheel to spin; a device for converting flow to rotational energy

turnkey: the jail warden; more commonly in our context, a system that is ready for the owner-occupant from the first time he or she turns the key in the front door lock

TV: a device for wasting time and scrambling the brain

two-by-fours: standard building members, originally two by four inches, now 1.5" by 3.5"; often referred to as "sticks"

U

ultrafilter: of water, to remove all particulates and impurities down to the sub-micron range, about the size of giardia, and larger viruses

uninterruptible power supply: an energy system providing ultrareliable power; essential for computers, aircraft guidance, medical, and other systems; also known as a UPS

V

Varistor: a voltage-dependent variable resistor, normally used to protect sensitive equipment from spikes (like lightning strike) by diverting the energy to ground

VCR: videocassette recorder, a device for making TVs slightly more responsive and useful

VDTs: video display terminals, like televisions and computer screens

vented cell: a battery cell designed with a vent mechanism for expelling gasses during charging

volt: measure of electrical potential: 110-volt house electricity has more potential to do work than an equal flow of 12-volt electricity.

voltage drop: lost potential due to wire resistance over distance

W

watt: the standard unit of electrical power; one ampere of current flowing with one volt of potential; 746 watts make one horsepower

watt-hours: one watt for one hour. A 15-watt compact fluorescent consumes 15 of these in 60 minutes.

waveform: the characteristic trace of voltage over time of an alternating current when viewed on an oscilloscope; typically a smooth sine wave, although primitive inverters supply square or modified square waveforms.

wet shelf life: the time that an electrolyte-filled battery can remain unused in the charged condition before dropping below its nominal performance level

wheatgrass: a singularly delicious potation made by squeezing young wheat sprouts; said to promote purity in the digestive tract

whole-life cost analysis: an economic procedure for evaluating all the costs of an activity from cradle to grave, that is, from extraction or culture through manufacture and use, then back to the natural state; a very difficult thing to accomplish with great accuracy, but a very instructive reckoning nonetheless

wind-chill: a factor calculated based on temperature and wind speed that expresses the fact that a given ambient feels colder when the wind is blowing

wind-spinners: fond name for wind machines, devices that turn wind into usable energy

ABBREVIATIONS:

ABS:	acrylonitrile butadiene styrene
AC:	alternating current
AGA:	American Gas Association
Ah:	amp-hour
AIC:	ampere interrupting capacity
AWG:	American Wire Gauge
Btu:	British thermal units
CFC:	chlorofluorocarbon
cfm:	cubic feet per minute
cu. ft.:	cubic feet
DC:	direct current
DPDT:	double pole double throw
EPA:	U.S. Environmental Protection Agency
FDA:	U.S. Food and Drug Administration
FIPT:	female iron pipe thread
GFCI:	ground fault circuit interruption
gpm:	gallons per minute
hp:	horsepower
hr:	hour
Hz:	Hertz (formerly "cycles per second")
ID:	inside diameter
kW:	kilowatt
kWh:	kilowatt-hour
lb:	pounds
LED:	light-emitting diode
LPG:	liquified propane gas
mA:	milliamp or milliampere
mAh:	milliamp-hour
MCM:	thousandths of circular mils
MIPT:	male iron pipe thread
NEC:	National Electrical Code
NSF:	National Sanitation Foundation
NTL:	National Testing Laboratories
OD:	outside diameter
oz:	ounce
psi:	pounds per square inch
PV:	photovoltaic
PVC:	polyvinyl chloride
RF:	radio frequency
RO:	reverse osmosis
RV:	recreational vehicle
SASE:	self-addressed stamped envelope
TDS:	total dissolved solids
UL:	Underwriters Laboratories
U.S.:	United States
UV:	ultraviolet
VAC:	volts AC
VDC:	volts DC
W-hr:	watt-hour

Unlimited Release
Revised November 2000

PHOTOVOLTAIC POWER SYSTEMS
and the
NATIONAL ELECTRICAL CODE:
SUGGESTED PRACTICES
—updated to the 1999 NEC—

A publication of
The Photovoltaic Systems Assistance Center
Sandia National Laboratories

by John Wiles
Southwest Technology Development Institute
New Mexico State University
Las Cruces, NM

ABSTRACT

This suggested practices manual examines the requirements of the *National Electrical Code (NEC)* as they apply to photovoltaic (PV) power systems. The design requirements for the balance of systems components in a PV system are addressed including conductor selection and sizing, overcurrent protection ratings and location, and disconnect ratings and location. PV array, battery, charge controller, and inverter sizing and selection are not covered as these items are the responsibility of the system designer, and they in turn determine the items in this manual. Both stand-alone, hybrid, and utility-interactive PV systems are covered. References are made to applicable sections of the *NEC*.
National Electrical Code® and *NEC®* are registered trademarks of the National Fire Protection Association, Inc., Quincy, Massachusetts 02269

ACKNOWLEDGMENTS

Numerous persons throughout the photovoltaic industry reviewed the drafts of this manual and provided comments which are incorporated in this version. Particular thanks go to Joel Davidson, Mike McGoey and Tim Ball, George Peroni, Bob Nicholson, Mark Ralph and Ward Bower, Steve Willey, Tom Lundtveit, and all those who provided useful information at seminars on the subject. Appendix E is dedicated to John Stevens and Mike Thomas at Sandia National Laboratories. Document editing and layout by Ronald Donaghe, Southwest Technology Development Institute.

TECHNICAL COMMENTS TO:

John C. Wiles
Southwest Technology Development Institute
New Mexico State University
P.O. Box 30001/Dept. 3 SOLAR
1505 Payne Street
Las Cruces, New Mexico 88003-0001
Request for copies to Photovoltaic Systems Assistance Center at Sandia National Laboratories
505-844-4383

PURPOSE

This document is intended to contribute to the wide-spread installation of safe, reliable PV systems that meet the requirements of the *National Electrical Code.*

DISCLAIMER

This guide provides information on how the *National Electrical Code (NEC)* applies to photovoltaic systems. The guide is not intended to supplant or replace the *NEC*; it paraphrases the *NEC* where it pertains to photovoltaic systems and should be used with the full text of the *NEC*. Users of this guide should be thoroughly familiar with the *NEC* and know the engineering principles and hazards associated with electrical and photovoltaic power systems. The information in this guide is the best available at the time of publication and is believed to be technically accurate. Application of this information and results obtained are the responsibility of the user.

In most locations, all electrical wiring including photovoltaic power systems must be accomplished by a licensed electrician and then inspected by a designated local authority. Some municipalities have additional codes that supplement or replace the *NEC*. The local inspector has the final say on what is acceptable. In some areas, compliance with codes is not required.

NATIONAL FIRE PROTECTION ASSOCIATION (NFPA) STATEMENT

The *National Electrical Code* including the 1999 *National Electrical Code* is published and updated every three years by the National Fire Protection Association (NFPA), Batterymarch Park, Quincy, Massachusetts 02269. The *National Electrical Code* and the term *NEC* are registered trademarks of the National Fire Protection Association and may not be used without their permission. Copies of the current edition of the *National Electrical Code* are available from the NFPA at the above address, most electrical supply distributors, and many bookstores.

In most locations, all electrical wiring including photovoltaic power systems must be accomplished by a licensed electrician and then inspected by a designated local authority. Some municipalities have additional codes that supplement or replace the *NEC*. The local inspector has the final say on what is acceptable. In some areas, compliance with codes is not required.

TABLE OF CONTENTS

LIST OF FIGURES

APPLICABLE ARTICLES FROM THE *NATIONAL ELECTRICAL CODE*

Although most portions of the *National Electrical Code* apply to all electrical power systems, including photovoltaic power systems, those listed below are of particular significance.

Article	Contents
90	Introduction
100	Definitions
110	Requirements
200	Grounded Conductors
210	Branch Circuits
240	Overcurrent Protection
250	Grounding
300	Wiring Methods
310	Conductors for General Wiring
336	Nonmetallic Sheathed Cable
338	Service Entrance Cable
339	Underground Feeders
340	Power and Control Tray Cable: Type TC
347	Rigid Non-Metallic Conduit
351	Liquid Tight Flexible Metal Conduit and Liquid Tight Flexible Non-Metallic Conduit
374	Auxiliary Gutters
384	Switchboards and Panel Boards
445	Generators
480	Storage Batteries
490	Equipment, Over 600 Volts
690	Solar Photovoltaic Systems
705	Interconnected Electric Power Production Sources
720	Low-Voltage Systems

PHOTOVOLTAIC POWER SYSTEMS
and the
NATIONAL ELECTRICAL CODE

SUGGESTED PRACTICES

OBJECTIVE
- SAFE, RELIABLE, DURABLE PHOTOVOLTAIC POWER SYSTEMS
- KNOWLEDGEABLE MANUFACTURERS, DEALERS, INSTALLERS, CONSUMERS, AND INSPECTORS

METHOD
- WIDE DISSEMINATION OF THESE SUGGESTED PRACTICES
- TECHNICAL INTERCHANGE BETWEEN INTERESTED PARTIES

INTRODUCTION

The National Fire Protection Association has acted as sponsor of the *National Electrical Code (NEC)* since 1911. The original Code document was developed in 1897. With few exceptions, electrical power systems installed in the United States in this century have had to comply with the *NEC* . This includes many photovoltaic (PV) power systems. In 1984, Article 690, which addresses safety standards for installation of PV systems, was added to the Code. This article has been revised and expanded in the 1987, 1990, 1993, 1996, and 1999 editions.

Many of the PV systems in use and being installed today may not be in compliance with the *NEC* and other local codes. There are several contributing factors to this situation:

- The PV industry has a strong "grass roots," do-it-yourself faction that is not fully aware of the dangers associated with low-voltage and high-voltage, direct-current (dc), PV-power systems.
- Some people in the PV community may believe that PV systems below 50 volts are not covered by the *NEC* .
- Electricians and electrical inspectors have not had significant experience with direct-current portions of the Code or PV power systems.
- The electrical equipment industries do not advertise or widely distribute equipment suitable for dc use that meets *NEC* requirements.
- Popular publications are presenting information to the public that implies that PV systems are easily installed, modified, and maintained by untrained personnel.
- Photovoltaic equipment manufacturers have, in some cases, been unable to afford the costs associated with testing and listing by approved testing laboratories like Underwriters Laboratories or ETL.
- Photovoltaic installers and dealers in many cases have not had significant experience installing ac residential and/or commercial power systems.

Some PV installers in the United States are licensed electricians or use licensed electrical contractors and are familiar with all sections of the *NEC* . These installer/contractors are trained to install reliable PV systems that meet the *National Electrical Code* and minimize the hazards associated with electrical power systems. Some PV installations have numerous defects that typically stem from unfamiliarity with electrical power system codes or unfamiliarity with dc currents and power systems. They often do not meet the requirements of the *NEC*. Some of the more prominent problems are listed below.

- Improper ampacity of conductors
- Improper type of conductors
- Unsafe wiring methods
- Lack of overcurrent protection on conductors
- Inadequate number and placement of disconnects
- Improper application of listed equipment
- No, or underrated, short-circuit or overcurrent protection on battery systems
- Use of non-approved components when approved components are available
- Improper system grounding
- Lack of, or improper, equipment grounding

- Use of underrated components
- Use of ac components (fuses and switches) in dc applications

The *NEC* may apply to any PV systems regardless of size or location. A single, small PV module may not present a significant hazard, and a small system in a remote location may present few safety hazards because people are seldom in the area. On the other hand, two or three modules connected to a battery can be lethal if not installed and operated properly. A single deep-cycle storage battery (6 volts, 220 amp-hours) can discharge about 8,000 amps into a short-circuit. Systems with voltages of 50 volts or higher present shock hazards. Short circuits on lower voltage systems present fire and equipment hazards. Storage batteries can be dangerous; hydrogen gas and acid residue from lead-acid batteries **must** be dealt with safely.

The problems are compounded because, unlike ac systems, there are few *UL*-Listed components that can be easily "plugged" together to make a PV system. Connectors and devices do not have mating inputs or outputs, and the knowledge and understanding of "what works with what" is not second nature to the installer. The dc "cookbook" of knowledge does not yet exist.

METHODS OF ACHIEVING OBJECTIVES

To meet the objective of safe, reliable, durable photovoltaic power systems, the following suggestions are presented:

- Dealer-installers of PV systems should become familiar with the *NEC* methods of wiring residential and commercial ac power systems.
- All PV installations should be inspected, where required, by the local inspection authority in the same manner as other equivalent electrical systems.
- Photovoltaic equipment manufacturers should build equipment to meet *UL* or other recognized standards and have equipment tested and listed.
- Listed subcomponents should be used in assembled equipment where formal testing and listing is not possible.
- Electrical equipment manufacturers should produce, distribute, and advertise, listed, reasonably priced, dc-rated components.
- Electrical inspectors should become familiar with dc and PV systems.
- The PV industry should educate the public, modify advertising, and encourage all installers to comply with the *NEC*.
- Existing PV installations should be upgraded to comply with the *NEC* or modified to meet minimum safety standards.

SCOPE AND PURPOSE OF THE *NEC*

Some local inspection authorities use regional electrical codes, but most jurisdictions use the *National Electrical Code*—sometimes with slight modifications. The *NEC* states that adherence to the recommendations made will reduce the hazards associated with electrical installations. The *NEC* also says these recommendations may not lead to improvements in efficiency, convenience, or adequacy for good service or future expansion of electrical use [90-1]. (Numbers in brackets refer to sections in the 1999 *NEC*.)

The *National Electrical Code* addresses nearly all PV power installations, even those with voltages less than 50 volts [720]. It covers stand-alone and grid-connected systems. It covers billboards, other remote applications, floating buildings, and recreational vehicles (RV) [90-2(a), 690]. The Code deals with any PV system that produces power and has external wiring or electrical components or contacts accessible to the untrained and unqualified person.

There are some exceptions. The *National Electrical Code* does not cover PV installations in automobiles, railway cars, boats, or on utility company properties used for power generation [90-2(b)]. It also does not cover micropower systems used in watches, calculators, or self-contained electronic equipment that have no external electrical wiring or contacts.

Article 690 of the *NEC* specifically deals with PV systems, but many other sections of the *NEC* contain requirements for any electrical system including PV systems [90-2, 720]. When there is a conflict between Article 690 of the *NEC* and any other article, Article 690 takes precedence [690-3].

The *NEC* suggests, and most inspection officials require, that equipment identified, listed, labeled, or tested by an approved testing laboratory be used when available [90-7,100,110-3]. Two of the several national testing organizations commonly acceptable to most jurisdictions are the *Underwriters Laboratories (UL)*, and ETL Testing Laboratories, Inc. *Underwriters Laboratories* and *UL* are registered trademarks of Underwriters Laboratories Inc., 333 Pfingsten Road, Northbrook, IL 60062.

Most building and electrical inspectors expect to see *UL* on electrical products used in electrical systems in the United States. This presents a problem for some in the PV industry, because low production rates do not yet justify the costs of testing and listing by *UL* or other laboratory. Some manufacturers claim their product specifications exceed those required by the testing organizations, but inspectors readily admit to not having the expertise, time, or funding to validate these unlabeled items.

THIS GUIDE

The recommended installation practices contained in this guide progress from the photovoltaic modules to the electrical outlets. For each component, *NEC* requirements are addressed, and the appropriate Code sections are referenced in brackets. A sentence, phrase, or paragraph followed by a *NEC* reference refers to a requirement established by the *NEC* . The words "**will**," "**shall**," or "**must**" also refer to *NEC* requirements. Suggestions based on field experience with PV systems are worded as such and will use the word "should." The recommendations apply to the use of listed products. The word "Code" in this document refers to the *NEC*.

Appendix A provides a limited list of sources for dc-rated and identified, listed, or approved products, and reference to the products is made as they are discussed.

Other appendices address details and issues associated with implementing the *NEC* in PV installations. Examples are included.

PHOTOVOLTAIC MODULES

Figure 1. Strain Reliefs

Numerous PV module manufacturers offer listed modules at the present time. Other manufacturers are considering having their PV modules listed by an acceptable testing laboratory.

Methods of connecting wiring to the modules vary from manufacturer to manufacturer. The *NEC* does not require conduit, but local jurisdictions, particularly in commercial installations, may require conduit. The Code requires that strain relief be provided for connecting wires. If the module has a closed weatherproof junction box, strain relief and moisture-tight clamps should be used in any knockouts provided for field wiring. Where the weather-resistant gaskets are a part of the junction box, the manufacturer's instructions **must** be followed to ensure proper strain relief and weather-proofing [110-3(b), *UL Standard 1703*]. Figure 1 shows various types of strain relief clamps. The one on the left is a basic cable clamp for interior use with nonmetallic sheathed cable (Romex®) that cannot be used for module wiring. The clamps in the center (T&B) and on the right (Heyco) are watertight and can be used with either single or multiconductor cable-depending on the insert.

Module Marking

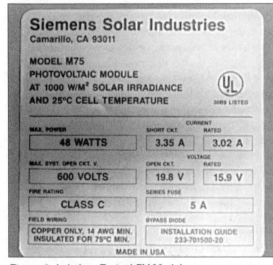

Figure 2. Label on Typical PV Module

Certain electrical information **must** appear on each module. The information on the factory-installed label **will** include the following items:

- Polarity of output terminals or leads
- Maximum series fuse for module protection
- Rated open-circuit voltage
- Rated operating voltage
- Rated operating current
- Rated short-circuit current
- Rated maximum power
- Maximum permissible system voltage [690-51]

Although not required by the *NEC* , the temperature rating of the module terminals and conductors are given to determine the temperature rating of the insulation of the conductors and how the ampacity of those conductors **must** be derated for temperature [110-14(c)]. Figure 2 shows a typical label that appears on the back of a module.

Note: Other critical information, such as mechanical installation instructions, tolerances of indicated values of Isc, Voc and Pmax, and statements on artificially concentrated sunlight are contained in the installation and assembly instructions for the module.

Module Interconnections

Copper conductors are recommended for almost all photovoltaic system wiring [110-5]. Copper conductors have lower voltage drops and good resistance to corrosion. Aluminum or copper-clad aluminum wires can be used in certain applications, but the use of such cables is not recommended—particularly in dwellings. All wire sizes presented in this guide refer to copper conductors.

The *NEC* requires 12 American Wire Gauge (AWG) or larger conductors to be used with systems under 50 volts [720-4]. Article 690 ampacity calculations yielding a smaller conductor size might override Article 720 considerations, but some inspectors are using the Article 720 requirement for dc circuits. The Code has little information for conductor sizes smaller than 14 AWG, but Section 690-31d provides some guidance.

Single-conductor, Type UF (Underground Feeder—Identified (marked) as Sunlight Resistant), Type SE (Service Entrance), or Type USE (Underground Service Entrance) cables are permitted for module interconnect wiring [690-31(b)]. Stranded wire is suggested to ease servicing of the modules after installation and for durability [690-34]. Unfortunately, single-conductor, stranded, UF sunlight-resistant cable is not readily available and may have only a 60°C temperature rating. This insulation is not suitable for long-term exposure to direct sunlight at temperatures likely to occur near PV modules. Such wire has shown signs of deterioration after four years of exposure. Temperatures exceeding 60°C normally occur in the vicinity of the modules; therefore, conductors with 60°C insulation cannot be used.

The widely available Underground Service Entrance Cable (USE-2) is suggested as the best cable to use for module interconnects. When manufactured to the *UL* Standards, it has a 90°C temperature rating and is sunlight resistant even though not commonly marked as such. The "-2" marking indicates at wet-rated 90°C insulation, the preferred rating. Additional markings indicating XLP or XLPE (cross-linked polyethylene) and RHW-2 (90°C insulation when wet) ensure that the highest quality cable is being used [Tables 310-13,16, and 17]. USE-2 is acceptable to most electrical inspectors. The RHH and RHW-2 designations frequently found on USE-2 cable allow its use in conduit inside buildings. USE or USE-2 cable, without the other markings, do not have the fire-retardant additives that SE cable has and cannot be used inside buildings.

The temperature derated ampacity of conductors at any point **must** be at least 156% of the module (or array of parallel modules) rated short-circuit current at that point [690-8(a), (b)]. If flexible, two-conductor cable is needed, electrical tray cable (Type TC) is available but **must** be supported in a specific manner as outlined in the *NEC* [318 and 340]. TC is sunlight resistant and is generally marked as such. Although frequently used for module interconnections, SO, SOJ, and similar flexible, portable cables and cordage may not be sunlight resistant and are not approved for fixed (non-portable) installations [400-7, 8].

Tracking Modules

Where there are moving parts of an array, such as a flat-plate tracker or concentrating modules, the *NEC* does allow the use of flexible cords and cables [400-7(a), 690-31(c)]. When these types of cables are used, they should be selected for extra-hard usage with full outdoor ratings (marked "WA" on the cable). They should not be used in conduit. Temperature derating information is provided by Table 690-31c. A derating factor in the range of 0.33 to 0.58 should be used for flexible cables used as module interconnects.

Trackers in PV systems operate at relatively slow angular rates and with limited motion. Normal stranded wire (USE-2 or THWN-2 in flexible conduit) has demonstrated good performance without deterioration due to flexing.

Another possibility is the use of extra flexible (400+ strands) building cable type USE-RHH-RHW or THW. This cable is available from the major wire distributors (Appendix A). Cable types, such as THW or RHW that are not sunlight resistant should be installed in flexible liquidtight conduit.

Terminals

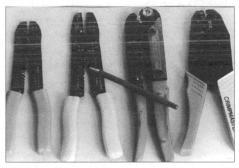

Figure 3. Terminal Crimpers

Module junction boxes have various types of terminals or pigtail leads (with and without connectors). The instructions furnished with each module **will** state the acceptable size and type of wires for use with the terminals. Some modules may require the use of crimp-on terminals when stranded conductors are used.

Light-duty crimping tools designed for crimping smaller wires used in electronic components usually do not provide sufficient force to make long-lasting crimps on connectors for PV installations even though they may be sized for 12-10 AWG. Insulated terminals crimped with these light-duty crimpers frequently develop high-resistance connections in a short time and may even fail as the wire pulls out of the terminal. It is strongly suggested that only listed, heavy-duty industrial-type crimpers be used for PV system wiring where crimp-on terminals are required. Figure 3 shows four styles of crimpers. On the far left is a stripper/crimper used for electronics work that will crimp only insulated terminals. Second from the left is a stripper/crimper that can make crimps on both insulated and uninsulated terminals. The pen points to the dies used for uninsulated terminals. With some care, this crimper can be used to crimp uninsulated terminals on PV systems. The two crimpers on the right are listed, heavy-duty industrial designs with ratcheting jaws and interchangeable dies that will provide the highest quality connections. They are usually available from electrical supply houses.

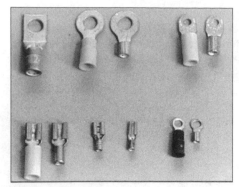

Figure 4 shows some examples of insulated and uninsulated terminals. In general, uninsulated terminals are preferred (with insulation applied later if required), but care **must** be exercised to obtain the heavier, more reliable *UL*-Listed terminals and not unlisted electronic or automotive grades. Again, an electrical supply house rather than an electronic or automotive parts store is the place to find the required items. Listed terminals are listed only when applied using the instructions supplied with the terminals and when used with the related crimping tool (usually manufactured by the same manufacturer as the terminals. If the junction box provides box-style pressure terminals, it is not necessary to use crimped terminals.

Figure 4. Insulated and Uninsulated Terminals

Transition Wiring

Figure 5. Listed Rain-proof Junction Box with Listed, Custom Terminal Strip

Because of the relatively high cost of USE-2 and TC cables and wire, they are usually connected to less expensive cable at the first junction box leading to an interior location. All PV system wiring **must** be made using one of the methods included in the *NEC* [690-31, Chapter 3]. Single-conductor, exposed wiring is not permitted except for module wiring or with special permission [Chapter 3]. The most common methods used for PV systems are individual conductors in electrical metallic tubing (EMT), nonmetallic conduit, or nonmetallic sheathed cable.

Where individual conductors are used in conduit, they should be conductors with at least 90°C insulation such as RHW-2, THW-2, THWN-2 or XHHW-2. Conduits installed in exposed locations are considered to be installed in wet locations [100-Locations]. These conduits may have water entrapped in low spots and therefore only conductors with wet ratings are acceptable in conduits that are located in exposed or buried locations. The conduit can be either thick-wall (rigid, galvanized-steel, RGS or intermediate, metal-conduit, IMC) or thin-wall electrical metallic tubing (EMT) [348], and if rigid, nonmetallic conduit is used, electrical (gray) PVC (Schedule 40 or Schedule 80) rather than plumbing (white) PVC tubing **must** be used [347].

Two-conductor (with ground) UF cable (a jacketed or sheathed cable) that is marked sunlight resistant is frequently used between the module interconnect wiring and the PV disconnect device. Splices from the conductors from the modules to this wire when located outside **must** be protected in rain-proof junction boxes such as NEMA type 3R. Cable clamps **must** also be used. Figure 5 shows a rain-proof box with a pressure connector terminal strip installed for module wiring connections. Cable clamps used with this box **must** be listed for outdoor use.

Interior exposed cable runs can also be made with sheathed, multi-conductor cable types such as NM, NMB, and UF. The cable should not be subjected to physical abuse. If abuse is possible, physical protection **must** be provided [300-4, 336 B, 339]. Exposed, single-conductor cable (commonly used between batteries and inverters) **shall not** be used-except as module interconnect conductors [300-3(a)]. Battery-to-inverter cables are normally single-conductor cables installed in conduit.

WIRING

Module Connectors

Module connectors that are concealed at the time of installation **must** be able to resist the environment, be polarized, and be able to handle the short-circuit current. They **shall** also be of a latching design with the terminals guarded. The equipment-grounding member, if used, **shall** make first and break last [690-32, 33]. The *UL* standard also requires that the connectors for positive and negative conductors should not be interchangeable.

Module Connection Access

All junction boxes and other locations where module wiring connections are made **shall** be accessible. Removable modules and stranded wiring may allow accessibility [690-34]. The modules should not be permanently fixed (welded) to mounting frames, and solid wire that could break when modules are moved to service the junction boxes should not be used. Open spaces behind the modules would allow access to the junction boxes.

Splices

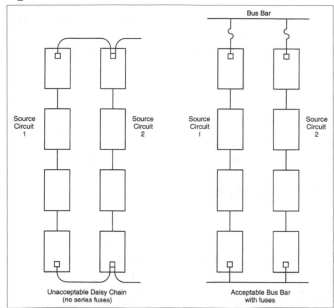

Figure 6. Module Interconnect Methods

Figure 7. Power Splicing Blocks and Terminal Strips

All splices (other than the connectors mentioned above) **must** be made in approved junction boxes with an approved splicing method. Conductors **must** be twisted firmly to make a good electrical and mechanical connection, then brazed, welded, or soldered, and then taped [110-14(b)]. Mechanical splicing devices such as split-bolt connectors or terminal strips are also acceptable. Crimped splicing connectors may also be used if listed splicing devices and listed, heavy-duty crimpers are used.

Properly used box-type pressure connectors (Figure 7) give high reliability. Fuse blocks, fused disconnects, and circuit breakers are available with these pressure connectors.

Twist-on wire connectors (approved for splicing wires), when listed for the environment (dry, damp, wet, or direct burial) are acceptable splicing devices. In most cases, they **must** be used inside enclosures, except when used in direct-burial applications.

Where several modules are connected in series and parallel, a terminal block or bus bar arrangement **must** be used so that one source circuit can be disconnected without disconnecting the grounded (on grounded systems) conductor of other source circuits [690-4(c)]. On grounded systems, this indicates that the popular "Daisy Chain" method of connecting modules may not always be acceptable, because removing one module in the chain may disconnect the grounded conductor for all of those modules in other parallel chains or source circuits. This becomes more critical on larger systems where paralleled sets of long series strings of modules are used. Figure 6 shows unacceptable and acceptable methods. Generally, 12- and 24-volt systems can be daisy chained, but higher voltage systems should not be.

Several different types of terminal blocks and strips are shown in Figure 7. The larger blocks are made by Marathon (Appendix A). Any terminal block used **must** be listed for "field-installed wiring." Many are only "Recognized" by *UL* for use inside factory assembled, listed devices.

Conductor Color Codes

The *NEC* established color codes for electrical power systems many years before either the automobile or electronics industries had standardized color codes. PV systems are being installed in an arena covered by the *NEC* and, therefore, **must** comply with *NEC* standards that apply to both ac and dc power systems. In a system where one conductor is grounded, the insulation on all grounded conductors **must** be white or natural gray or be any color except green if marked with white plastic tape or paint at each termination (marking allowed only on conductors larger than 6 AWG). Conductors used for module frame grounding and other exposed metal equipment grounding **must** be bare (no insulation) or have green or green with yellow-striped insulation or identification [200-6, 7; 210-5; 250-119].

The *NEC* requirements specify that the grounded conductor be white. In most PV-powered systems that are grounded, the grounded conductor is the negative conductor. Telephone systems that use positive grounds require special circuits when powered by PV systems that have negative grounds. In a PV system where the array is center tapped, the center tap or neutral **must** be grounded [690-41], and this becomes the white conductor. There is no *NEC* requirement designating the color of the ungrounded conductor, but the convention in power wiring is that the first two ungrounded conductors are colored black and red. This suggests that in two-wire, negative-grounded PV systems, the positive conductor could be red or any color with a red marking except green or white, and the negative grounded conductor **must** be white. In a three-wire, center-tapped system, the positive conductor could be red, the grounded, center tap conductor **must** be white and the negative conductor could be black.

The *NEC* allows grounded PV array conductors, such as non-white USE or SE that are smaller than 6 AWG, to be marked with a white marker [200-6].

Battery Cables

Battery cables, even though they can be 2/0 AWG and larger, **must** be of standard building-wire type conductor [Chapter 3]. Welding and automobile "battery" cables are not allowed. Flexible, highly-stranded, building-wire type cables (USE/RHW and THW) are available for this use. Flexible cables, identified in Section 400 of the *NEC* are allowed from the battery terminals to a nearby junction box and between battery cells. These cables **shall** be listed for hard service use and moisture resistance [690-74].

GROUND-FAULT PROTECTION AND ARRAY DISABLEMENT

Ground-Faults

Article 690-5 of the *NEC* requires a ground-fault detection, interruption, and array disconnect (GFID) device for fire protection if the PV arrays are mounted on roofs of dwellings. Ground-mounted arrays are not required to have this device. Several devices or inverters are available that meet this requirement. These particular devices may require that the system grounding conductor be routed through the device. These devices include the following functions:
- Ground-fault detection
- Ground-fault current interruption
- Array disconnect
- Array wiring overcurrent protection

Ground-fault detection, interruption, and array disablement devices might, depending on the particular design, accomplish the following actions automatically:
- Sense ground-fault currents exceeding a specified value
- Interrupt or significantly reduce the fault currents
- Open the circuit between the array and the load
- Indicate the presence of the ground fault

Ground-fault devices have been developed for some grid-tied inverters and stand-alone systems, and others are under development.

PV Array Installation and Service

Article 690-18 requires that a mechanism be provided to allow safe installation or servicing of portions of the array or the entire array. The term "disable" has several meanings, and the *NEC* is not clear on what is intended. The *NEC* Handbook does elaborate. Disable can be defined several ways:
- Prevent the PV system from producing any output
- Reduce the output voltage to zero
- Reduce the output current to zero
- Divide the array into non-hazardous segments

The output could be measured at either the PV source terminals or at the load terminals.

Firefighters are reluctant to fight a fire in a high-voltage battery room because there is no way to turn off a battery bank unless the electrolyte can somehow be removed. In a similar manner, the only way a PV system can have zero output at the array terminals is by preventing light from illuminating the modules. The output voltage may be reduced to zero by shorting the PV module or array terminals. When this is done, short-circuit current will flow through the shorting conductor which, in a properly wired system, does no harm. The output current may be reduced to zero by disconnecting the PV array from the rest of the system. The PV disconnect switch would accomplish this action, but open-circuit voltages would still be present on the array wiring and in the disconnect box. In a large system, 100 amps of short-circuit current (with a shorted array) can be as difficult to handle as an open-circuit voltage of 600 volts.

During PV module installations, the individual PV modules can be covered to disable them. For a system in use, the PV disconnect switch is opened during maintenance, and the array is either short circuited or left open circuited depending on the circumstances. In practical terms, for a large array, some provision (switch or bolted connection) should be made to disconnect portions of the array from other sections for servicing. As individual modules or sets of modules are serviced, they may be covered and/or isolated and shorted to reduce the potential for electrical shock. Aside from measuring short-circuit current, there is little that can be serviced on a module or array when it is shorted. The circuit is usually open circuited for repairs.

GROUNDING

The subject of grounding is one of the most confusing issues in electrical installations. Definitions from Article 100 of the *NEC* will clarify the situation.

Grounded: Connected to the earth or to a metallic conductor or surface that serves as earth.

Grounded Conductor: (white or natural gray) A system conductor that normally carries current and is intentionally grounded. In PV systems, one conductor (normally the negative) of a two-conductor system or the center-tapped wire of a bipolar system is grounded.

Equipment Grounding Conductor: (bare, green, or green with yellow stripe) A conductor not normally carrying current used to connect the exposed metal portions of equipment to the grounding electrode system or the grounded conductor.

Grounding Electrode Conductor: A conductor not normally carrying current used to connect the grounded conductor to the grounding electrode or grounding electrode system.

Grounding—System

For a two-wire PV system over 50 volts (125% of open-circuit PV-output voltage), one dc conductor **shall** be grounded. In a three-wire system, the neutral or center tap of the dc system **shall** be grounded [690-41]. These requirements apply to both stand-alone and grid-tied systems. Such system grounding will enhance personnel safety and minimize the effects of lightning and other induced surges on equipment. Also, the grounding of all PV systems (even 12-volt systems) will reduce radio frequency noise from dc-operated fluorescent lights and inverters.

SIZE OF GROUNDING ELECTRODE CONDUCTOR

The direct-current system-grounding electrode conductor **shall not** be smaller than 6 AWG or the largest conductor supplied by the system [250-166]. If the conductors between the battery and inverter are 4/0 AWG (for example) then the grounding electrode conductor from the negative conductor (assuming that this is the grounded conductor) to the grounding electrode **must** be 4/0 AWG. The *NEC* allows exceptions to this large grounding conductor requirement. Many PV systems can use a 6 AWG grounding electrode conductor if that is the *only connection* to the grounding electrode [250-166(c)].

POINT OF CONNECTION

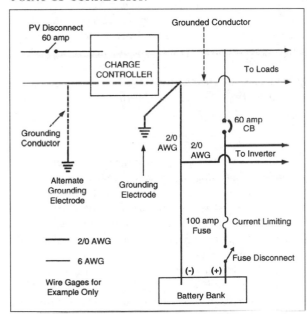

Figure 8. Typical System: Possible Grounding Conductor Locations

The system grounding electrode conductor for the direct-current portion of a PV system **shall** be connected to the PV-output circuits [690-42] at a single point. When this connection is made close to the modules, added protection from surges is afforded. Disconnect switches **must not** open grounded conductors [690-13]. In stand-alone PV systems, the charge controller may be considered a part of the PV-output circuit, and the point of connecting the grounding electrode conductor could be before or after the charge controller. But this grounding conductor may be a very large conductor (e.g., 4/0 AWG) while the conductors to and from the charge controller may be 10 AWG or smaller. Connecting the 4/0 AWG grounding conductor on the array side of the charge controller, while providing some degree of enhanced surge suppression from lightning induced surges, may not meet the full intent of the grounding requirements. Connecting the grounding conductor to the system on the battery side of the charge controller at a point where the system conductors are the largest size will provide better system grounding at the expense of less lightning protection. Since the *NEC* allows smaller grounding electrode conductors in certain circumstances, either grounding conductor point of connection may be acceptable [250-166]. Figure 8 shows two possible locations for the grounding conductor.

The *NEC* does not specifically define where the PV-output circuits end. Circuits from the battery toward the load are definitely load circuits. Since the heaviest conductors are from the battery to the inverter, and either end of these conductors is at the same potential, then either end could be considered a point for connecting the grounding conductor. The negative dc input to the inverter is connected to the metal case in some stand-alone inverter designs, but this is not an appropriate place to connect the grounding electrode conductor and other equipment-grounding conductors, since this circuit is a dc-branch circuit and not a PV-output circuit. Connection of the grounding electrode conductor to or near the negative battery terminal would avoid the "large-wire/small-wire" problem outlined above.

It is imperative that there be no more than one grounding connection to the grounded conductor of a PV system. Failure to limit the connections to one (1) will allow objectionable currents to flow in uninsulated conductors and will create unintentional ground faults in the grounded conductor [250-6]. Future ground-fault protection systems may require that this single grounding connection be made at a specific location. There are exceptions to this rule when PV arrays or generators are some distance from the main loads [250-32].

UNUSUAL GROUNDING SITUATIONS

Some inverter designs use the entire chassis as part of the negative circuit. Also, the same situation exists in certain radios-automobile and shortwave. These designs will not pass the current *UL* standards for consumer electrical equipment or PV systems and will probably require modification in the future since they do not provide electrical isolation between the exterior metal surfaces and the current-carrying conductors. They also create the very real potential for multiple grounding-conductor connections to ground.

Since the case of these non-listed inverters is connected to the negative conductor and that case **must** be grounded as part of the equipment ground described below, the user has no choice whether or not the *system* is to be grounded. The system is grounded even if the voltage is less than 50 volts and the point of system ground is the negative input terminal on the inverter. It is strongly suggested that these unlisted inverters not be used and, in fact, to use them or any unlisted component is a violation of the *NEC*.

Some telephone systems ground the positive conductor, and this may cause problems for PV-powered telephone systems with negative grounds. An isolated-ground, dc-to-dc converter may be used to power subsystems that have different grounding polarities from the main system. In the ac realm, an isolation transformer will serve the same purpose.

In larger utility-tied systems and some stand-alone systems, high impedance grounding systems or other methods that accomplish equivalent system protection and that use equipment listed and identified for the use might be used in lieu of, or in addition to, the required hard ground. The discussion and design of these systems are beyond the scope of this guide. Grounding of grid-tied systems will be discussed in Appendix C.

CHARGE CONTROLLERS—SYSTEM GROUNDING

In a grounded system, it is important that the charge controller does not have devices in the grounded conductor. Charge controllers listed to *UL Standard 1741* meet this requirement. Relays or transistors in the grounded conductor create a situation where the grounded conductor is not at ground potential at times when the charge controller is operating. This condition violates provisions of the *NEC* that require all conductors identified as grounded conductors always be at the same potential (i.e. grounded). A shunt in the grounded conductor is equivalent to a wire if properly sized, but the user of such a charge controller runs the risk of having the shunt bypassed when inadvertent grounds occur in the system. The best charge controller design has only a straight-through conductor between the input and output terminals for the grounded current-carrying conductor (usually the negative conductor).

Grounding—Equipment

All noncurrent-carrying exposed metal parts of junction boxes, equipment, and appliances in the entire electrical system **shall** be grounded [690-43; 250 D; 720-1, -10]. All PV systems, regardless of voltage, **must** have an equipment-grounding system for exposed metal surfaces (e.g., module frames and inverter cases) [690-43]. The equipment-grounding conductor **shall** be sized as required by Article 690-45 or 250-122. Generally, this will mean an equipment-grounding conductor size based on the size of the overcurrent device protecting each conductor. Table 250-122 in the *NEC* gives the sizes. For example, if the inverter to battery conductors are protected by a 400-amp fuse or circuit breaker, then at least 3 AWG conductor **must** be used for the equipment ground for that circuit [Table 250-122]. If the current-carrying conductors have been oversized to lower voltage drop, then the size of the equipment-grounding conductor **must** also be proportionately adjusted [250-122(b)]. In the PV source circuits, if the array can provide short-circuit currents that are less than twice the rating of a particular overcurrent device for the array circuits, then equipment-grounding conductors **must** be used that are sized the same as the array current-carrying conductors [690-45]. In general, the overcurrent protection is sized at 1.56 Isc, but the available short-circuit currents are much greater from the batteries and other sources. This means that Table 250-122 of the *NEC* usually applies.

Inverter AC Outputs

The inverter output (120 or 240 volts) **must** be connected to the ac distribution system in a manner that does not create parallel grounding paths. The *NEC* requires that both the green or bare equipment-grounding conductor and the white neutral conductor be grounded. The Code also requires that current not normally flow in the equipment-grounding conductors. If the inverter has ac grounding receptacles as outputs, the grounding and neutral conductors are most likely connected to the chassis and, hence, to chassis ground inside the inverter. This configuration allows plug-in devices to be used safely. However, if the outlets on the inverter are plug and cord connected (not recommended) to an ac load center used as a distribution device, then problems can occur.

The ac load center usually has the neutral and equipment-grounding conductors connected to the same bus bar which is connected to the case where they are grounded. Parallel current paths are created with neutral currents flowing in the equipment-grounding conductors. This problem can be avoided by using a load center with an isolated/insulated neutral bus bar which is separated from the equipment-grounding bus bar.

Inverters with hard-wired outputs may or may not have internal connections. Some inverters with ground-fault circuit interrupters (GFCIs) for outputs **must** be connected in a manner that allows proper functioning of the GFCI. A case-by-case analysis will be required.

Backup Generators

Backup generators used for battery charging pose problems similar to using inverters and load centers. These small generators usually have ac outlets which may have the neutral and grounding conductors bonded to the generator frame. When the generator is connected to the system through a load center, to a standby inverter with battery charger, or to an external battery charger, parallel ground paths are likely. These problems need to be addressed on a case-by-case basis. A PV system, in any operating mode (inverting or battery charging), **must not** have currents in the equipment-grounding conductors [250-6].

Suggested AC Grounding

Auxiliary ac generators and inverters should be hard-wired to the ac-load center. Neither should have an internal bond between the neutral and grounding conductors. Neither should have any receptacle outlets that can be used when the generator or inverter is operated when disconnected from the load center. The single bond between the neutral and ground will be made in the load center. If receptacle outlets are desired on the generator or the inverter, they should be ground-fault-circuit-interrupting devices (GFCI).

Section 250-32 of the *NEC* presents alternate methods of achieving a safe grounding system.

Grounding Electrode

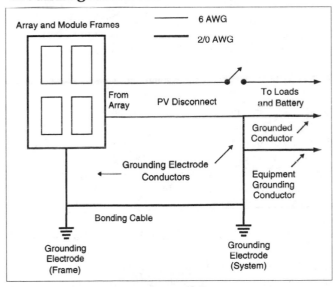

Figure 9. Example Grounding Electrode System

The dc system grounding electrode **shall** be common with, or bonded to, the ac-grounding electrode (if any) [690-47, 250-C]. The system-grounding conductor and the equipment-grounding conductor **shall** be tied to the same grounding electrode or grounding electrode system. Even if the PV *system* is ungrounded (optional at less than 50 volts [typically 125% of Voc]), equipment-grounding conductors **must** be used and **must** be connected to a grounding electrode [250-110]. Metal water pipes and other metallic structures as well as concrete encased electrodes are to be used in some circumstances [250-50]. When a grounding electrode is used, it **shall** be a corrosion resistant rod, a minimum of ⅝ inch (16mm) in diameter (½" (13mm) if stainless steel) with at least 8 feet (2.4m) driven into the soil at an angle no greater than 45 degrees from the vertical [250-52]. Listed connectors **must** be used to connect the grounding conductor to the ground rod.

A bare-metal well casing makes a good grounding electrode. It should be part of a grounding electrode system. The central pipe to the well should not be used for grounding, because it is sometimes removed for servicing.

For maximum protection against lightning-induced surges, it is suggested that a grounding electrode **system** be used with at least two grounding electrodes bonded together. One electrode would be the main system grounding electrode as described above. The other would be a supplemental grounding electrode located as close to the PV array as practical. The module frames and array frames would be connected directly to this grounding electrode to provide as short a path as possible for lightning-induced surges to reach the earth. This electrode is usually bonded with a conductor to the main system grounding electrode [250-54]. The size of the bonding or jumper cable **must** be related to the ampacity of the overcurrent device protecting the PV source circuits. This bonding jumper is an auxiliary to the module frame grounding that is required to be grounded with an equipment-grounding conductor. *NEC* Table 250-122 gives the requirements. Equipment-grounding conductors are allowed to be smaller than circuit conductors when the circuit conductors become very large. Article 250 of the *NEC* elaborates on these requirements.

Do not connect the negative current-carrying conductor to the grounding electrode, to the equipment-grounding conductor, or to the frame *at the modules.* **There should be one and only one point in the system where the grounding electrode conductor is attached to the system-grounded conductor.** See Figure 9 for clarification. The wire sizes shown are for illustration only and will vary depending on system size. Chapter 3 of the *NEC* specifies the ampacity of various types and sizes of conductors. As is common throughout the *NEC*, there are exceptions to this guidance. See *NEC* Section 250-32b.

CONDUCTOR AMPACITY

NEC Tables 310-16 and 310-17 give the ampacity (current-carrying capacity in amps) of various sized conductors at temperatures of 30°C (86°F). There are many adjustments that **must** be made to these numbers before a conductor size can be selected.

The installation method **must** be considered. Are the conductors in free air [Table 310-17] or are they bundled together or placed in conduit [Table 310-16]?

What is the ambient air temperature, if not 30°C (86°F)?

How many current-carrying conductors are grouped together?

These adjustments are made using factors presented in Chapter 2 of the *NEC*.

Additionally, most conductors used in electrical power systems are restricted from operating on a continuous basis at more than 80% of their rated ampacity [210-19, 215-2]. This 80% factor also applies to overcurrent devices and switchgear unless listed for operation at 100% of rating [210-20]. PV conductors are also restricted by this factor (0.8) [690-8(b)].

Conductors carrying PV module currents are further restricted by an additional derating factor of 80% because of the manner in which PV modules generate electrical energy in response to sunlight [690-8(a)].

It should be noted that these various ampacity adjustment factors may be applied to the basic conductor ampacities (e.g. multiply them by 0.80) or they may be applied to the anticipated current in the circuit (e.g. multiply the current by 1.25, the reciprocal of 0.8).

Photovoltaic modules are limited in their ability to deliver current. The short-circuit current capability of a module is nominally 10 to 15% higher than the operating current. Normal, daily values of solar irradiance may exceed the standard test condition of 1000W/m^2. These increased currents are considered by using the 1.25 adjustment in the ampacity calculations. Another problem for PV systems is that the conductors may operate at temperatures as high as 65-75°C when the modules are mounted close to a structure, there are no winds, and the ambient temperatures are high. Temperatures in module junction boxes frequently occur within this range. This **will** require that the ampacity of the conductors be derated or corrected with factors given in *NEC* Table 310-16 or 310-17. For example, a 10 AWG USE-2/RHW-2 single-conductor cable used for module interconnections in conduit has a 90°C insulation and an ampacity of 40 amps in an ambient temperature of 26-30°C. When it is used in ambient temperatures of 61-70°C, the ampacity of this cable is reduced to 23.2 amps.

It should be noted that the ampacity values associated with conductors having 90°C insulation can only be used if the terminals of the module and connected overcurrent devices are rated at 90°C [110-14(c)]. If the terminals are rated at only 75°C, then the ampacity values associated with 75°C insulation **must** be used, even when conductors with 90°C insulation are being used. Of course, if the 90°C insulation wire is used, the temperature derating may start with the 90°C ampacity values.

The ampacity of conductors in PV source circuits **shall** be at least 125% of the rated module or parallel-connected modules *short-circuit current* [690-8]. The ampacity of the PV-output circuit conductors **shall** be at least 125% of the short-circuit output current [690-8a]. The ampacity of conductors to and from an inverter or power conditioning system **shall** be 125% of the rated operating current for that device [690-8a]. In a similar manner, other conductors in the system should have an ampacity of 125% of the rated operating current to allow for long duration operation at full power [215-2]. These *NEC* requirements are to ensure that the connected overcurrent devices or panel boards operate at no more than 80% of their ampacity. Operation when snow or cloud enhancement increases the PV output above normal may require additional ampacity. Daily expected values of solar irradiance will exceed the standard test condition of 1000W/m^2 at many locations.

A 1989 revision to the *UL Standard 1703* for PV modules requires that module installation instructions include an *additional* 25% of the 25°C ratings for short-circuit current and open-circuit voltage to allow for expected daily peak irradiance and colder temperatures. This 1.25 factor, while still in *UL Standard 1703*, is also now contained in Section 690-8(a) of the *NEC* as mentioned above. There are only two 1.25 factors applied to PV module currents and the combined factor is 1.56 (1.25x1.25). Correct design practices require correctly determining wire size and the rating of overcurrent devices on PV source and output circuits. However, the rating of the overcurrent device should always be less than, or equal to, the ampacity of the cable. The *NEC* makes only infrequent exceptions to this rule. [240-3].

The ampacity of conductors and the sizing of overcurrent devices is an area that demands careful attention by the PV system designer/installer. Temperatures and wiring methods **must** be addressed for each site. Start with the 125% of Isc value to comply with the *UL* requirements [now in Section 690-8(a)]. Then use an additional 125% for code compliance. Finally, derate the cable ampacity for temperature. See Appendix E for additional examples.

Overcurrent devices may have terminals rated for connection to 60°C conductors necessitating a reduction in the cable ampacity when using 75°C or 90°C conductors.

When the battery bank is tapped to provide multiple voltages (i.e., 12 and 24 volts from a 24-volt battery bank), the common negative conductor will carry the sum of all of the simultaneous load currents. The negative conductor **must** have an *ampacity at least equal to the sum* of all the amp ratings of the overcurrent devices protecting the positive conductors or have an ampacity equal to the sum of the ampacities of the positive conductors [690-8(c)].

The *NEC* does not allow paralleling conductors for added ampacity, except that cables 1/0 AWG or larger may be paralleled under certain conditions [310-4]. DC-rated switchgear, overcurrent devices, and conductors cost significantly more when rated to carry more than 100 amps. It is suggested that large PV arrays be broken down into subarrays, each having a short-circuit output of less than 64 amps. This will allow use of 100-amp-rated equipment (156% of 64 amps) on each source circuit.

Stand-Alone Systems—Inverters

In stand-alone systems, inverters are used to change the direct current (dc) from a battery bank to 120-volt or 240-volt, 60-Hertz (Hz) alternating current (ac). The conductors between the inverter and the battery **must** have properly rated overcurrent protection and disconnect mechanisms [240, 690-8, 690-9]. These inverters frequently have short-duration (seconds) surge capabilities that are four to six times the rated output. For example, a 2,500-watt inverter might be required to surge to 10,000 volt-amps for 5 seconds when a motor load **must** be started. The *NEC* requires the ampacity of the conductors between the battery and the inverter to be sized by the rated 2,500-watt continuous output of the inverter. For example, in a 24-volt system, a 2,500-watt inverter would draw 134 amps at full load (85% efficiency at 22 volts) and 420 amps for motor-starting surges. The ampacity of the conductors between the battery **must** be 125% of the 134 amps or 167 amps.

To minimize steady-state voltage drops, account for surge-induced voltage drops, and to increase system efficiency, some well-designed systems have conductors that are larger than required by the *NEC*. When the current-carrying conductors are oversized, the equipment-grounding conductor **must** also be oversized proportionately [250-122].

See Appendix F for additional considerations on conductor ampacity.

OVERCURRENT PROTECTION

The *NEC* requires that every ungrounded conductor be protected by an overcurrent device [240-20]. In a PV system with multiple sources of power (PV modules, batteries, battery chargers, generators, power conditioning systems, etc.), the overcurrent device **must** protect the conductor from overcurrent from any source connected to that conductor [690-9]. Blocking diodes, charge controllers, and inverters are not considered as overcurrent devices and **must** be considered as zero-resistance wires when assessing overcurrent sources (690-9 FPN). If the PV system is directly connected to the load without battery storage or other source of power, then no overcurrent protection is required if the conductors are sized at 156% of the short-circuit current [690-8b-Ex].

When circuits are opened in dc systems, arcs are sustained much longer than they are in ac systems. This presents additional burdens on overcurrent-protection devices rated for dc operation. Such devices **must** carry the rated load current and sense overcurrent situations as well as be able to safely interrupt dc currents. AC overcurrent devices have the same requirements, but the interrupt function is considerably easier.

Ampere Rating

The PV source circuits **shall** have overcurrent devices rated at least 156 (1.25 x 1.25)% of the parallel module short-circuit current. The PV-output circuit overcurrent devices **shall** be rated at least 156% of the short-circuit PV currents [690-8]. Time-delay fuses or circuit breakers would minimize nuisance tripping or blowing. In all cases, dc-rated devices having the appropriate dc-voltage rating **must** be used.

All ungrounded conductors from the PV array **shall** be protected with overcurrent devices [Article 240, Diagram 690-1]. Grounded conductors (not shown in Diagram 690-1) **must not** have overcurrent devices since the independent opening of such a device might unground the system. Since PV module outputs are current limited, these overcurrent devices are actually protecting the array wiring from backfeed from parallel-connected modules, the battery, or the inverter.

Because the conductors and overcurrent devices are sized to deal with 156% of the short-circuit current for that particular PV circuit, overcurrents from those modules or PV sources, which are limited to the short-circuit current, cannot trip the overcurrent device in this circuit. The overcurrent devices in these circuits protect the conductors from overcurrents from parallel connected sets of modules or overcurrents from the battery bank. In stand-alone systems or grid-connected systems, these array overcurrent devices protect the array wiring from overcurrents from parallel strings of modules, the battery, or from the generator or ac utility power.

Often, PV modules or series strings of modules are connected in parallel. As the conductor size used in the array wiring increases to accommodate the higher short-circuit currents of paralleled modules, each conductor size **must** be protected by an appropriately sized overcurrent device. These devices **must** be placed nearest all sources of potential overcurrent for that conductor [240-21]. Figure 10 shows an example of array conductor overcurrent protection for a medium-size array broken into subarrays. The cable sizes and types shown are examples only. The actual sizes will depend on the ampacity needed.

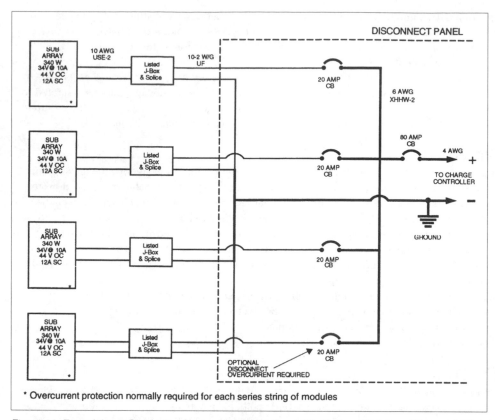

Figure 10. Typical Array Conductor Overcurrent Protection (with Optional Subarray Disconnects)

Figure 11. Listed Branch-Circuit Rated Breakers

Either fuses or circuit breakers are acceptable for overcurrent devices provided they are rated for their intended uses—i.e., they have dc ratings when used in dc circuits, the ampacity is correct, and they can interrupt the necessary currents when short circuits occur [240 E, F, G]. Figure 11 shows dc-rated, *UL*-Listed circuit breakers being used in a PV power center for overcurrent protection and disconnects. The circuit breakers in this system are manufactured by Heinemann (Appendix A). The *NEC* allows the use of listed (recognized) supplemental overcurrent devices only for PV source circuit protection.

Some overcurrent devices rated at less than 100 amps may have terminals that are rated for use with 60°C conductors. The ampacity calculations of the connected cables may have to be adjusted.

Testing and Approval

The *NEC* requires that listed devices be used for overcurrent protection. A *listed* device by *UL* or other approved testing laboratory is tested against an appropriate *UL* standard. A *recognized* device is tested by *UL* or other approved testing laboratory to standards established by the device manufacturer. In most cases, the standards established by the manufacturer are less rigorous than those established by *UL*. Few inspectors will accept recognized devices, particularly where they are required for overcurrent protection. *Recognized* devices are generally intended for use in a factory assembly or equipment that will be *listed* in its entirety.

Since PV systems may have transients—lightning and motor starting as well as others—inverse-time circuit breakers (the standard type) or time-delay fuses should be used in most cases. In circuits where no transients are anticipated, fast-acting fuses can be used. They should be used if relays and other switchgear in dc systems are to be protected. Time-delay fuses that can also respond very quickly to short-circuit currents may also be used for system protection.

Branch Circuits

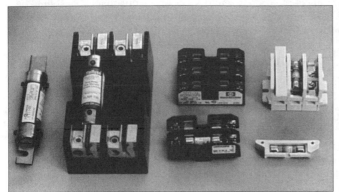

Figure 12. Listed Branch Circuit and Supplemental Fuses

DC branch circuits in stand-alone systems start at the battery and go to the receptacles supplying the dc loads or to the dc loads that are hard wired, such as inverters. In direct-connected systems (no battery), the PV output circuits go to the power controller or master power switch and a branch circuit goes from this location to the load. In utility-intertie systems, the circuit between the inverter and the ac-load center may be considered a feeder circuit.

Fuses used to protect dc or ac branch (load) and feeder circuits **must** be listed for that use. They **must** also be of different sizes and markings for each amperage and voltage group to prevent unintentional interchange [240F]. These particular requirements eliminate the use of glass fuses and plastic automotive fuses as branch-circuit overcurrent devices because they are neither tested nor rated for this application. DC-rated fuses that meet the requirements of the *NEC* are becoming more prevalent. Figure 12 shows *UL*-Listed, dc-rated, time-delay fuses on the left that are acceptable for branch circuit use, which would include the battery fuse. Acceptable dc-rated, *UL*-Listed fast-acting *supplementary* fuses are shown on the right and can be used in the PV source circuits. The fuses shown are made by Littelfuse (Appendix A) and others, and the fuse holders are made by Marathon (Appendix A). Other manufacturers, such as Bussman and Ferraz, also have *UL*-Listed dc ratings on the types of fuses that are needed in PV systems.

Automotive fuses have no dc rating by the fuse industry or the testing laboratories and **should not be used in PV systems.** When rated by the manufacturer, they have only a 32-volt maximum rating, which is less than the open-circuit voltage from a 24-volt PV array. Furthermore, these fuses have no rating for interrupt current, nor are they generally marked with all of the information required for branch circuit fuses. They are not considered supplemental fuses under the *UL* listing or component recognition programs. Figure 13 shows unacceptable automotive fuses on the left and listed ac supplemental fuses on the right. Unfortunately, even the supplemental fuses are intended for ac use and frequently have no dc ratings.

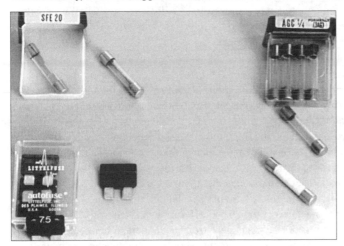

Figure 13. Unacceptable Fuses (left) and Acceptable Fuses (right) when DC-rated

Figure 14. UL-Recognized and Listed Circuit Breakers

Circuit breakers also have specific requirements when used in branch circuits, but they are generally available with the needed dc ratings [240 G]. Figure 14 shows examples of dc-rated, *UL-Recognized* circuit breakers (supplemental) on the left. When listed, they may be used in the PV source circuits for disconnects and overcurrent protection, but most are only recognized devices and may not be allowed by the inspector. The larger units are dc-rated, *UL-Listed* branch-circuit rated circuit breakers that can be used in dc-load centers for branch-circuit protection. The breakers shown are produced by Square D and Heinemann. Airpax also produces dc *UL*-Listed circuit breakers, and Potter Brumfield and others produce dc-rated, *UL*-Recognized, supplemental breakers.

To provide maximum protection and performance (lowest voltage drop) on branch circuits (particularly on 12- and 24-volt systems), the ampacity of the conductors might be increased, but the rating of the overcurrent devices protecting that cable should be as low as possible consistent with load currents. A general formula for cable ampacity and overcurrent device rating is 100% of the noncontinuous loads and 125% of the continuous loads anticipated [215-2].

Amperes Interrupting Rating (AIR)—Short-Circuit Conditions

Overcurrent devices—both fuses and circuit breakers—**must** be able to safely open circuits with short-circuit currents flowing in them. Since PV arrays are inherently current limited, high short-circuit currents from the PV array are normally not a problem when the conductors are sized as outlined above. In stand-alone systems with storage batteries, however, the short-circuit condition is very severe. A single 220 amp-hour, 6-volt, deep-discharge, lead-acid battery may produce short-circuit currents as high as 8,000 amps for a fraction of a second and as much as 6,000 amps for a few seconds in a direct terminal-to-terminal short circuit. Such high currents can generate excessive thermal and magnetic forces that can cause an underrated overcurrent device to burn or blow apart. Two paralleled batteries could generate nearly twice as much current, and larger capacity batteries would be able to deliver proportionately more current under a short-circuit condition. In dc systems, particularly stand-alone systems with batteries, the interrupt capability of every overcurrent device is important. This interrupt capability is specified as Amperes Interrupting Rating (AIR) and sometimes Amperes Interrupting Capability (AIC).

Most dc-rated, *UL*-Listed, branch circuit breakers that can be used in PV systems have an AIR of 5,000 amps. However, Heinemann Electric and AirPax make numerous circuit breakers with AIRs of 25,000 amps (Appendix A). Some dc-rated, *UL*-Recognized supplemental circuit breakers have an AIR of only 3,000 amps. Listed, dc-rated class-type fuses normally have an AIR of up to 20,000 amps if they are of the current-limiting variety.

Fuses or circuit breakers **shall never** be paralleled or ganged to increase current-carrying capability unless done so by the factory and listed for such use [240-8].

Fusing of PV Source Circuits

The *NEC* allows supplementary fuses to be used in PV source circuits [690-9(c)]. A supplementary fuse is one that is designed for use inside a piece of listed equipment. These fuses supplement the main branch-circuit fuse and do not have to comply with all of the requirements of branch fuses. They **shall**, however, be dc rated, listed, and able to handle the short-circuit currents they may be subjected to [690-9d]. Unfortunately, many supplemental fuses are not dc rated, and if they are, the AIR (when available) is usually less than 5,000 amps. The use of ac-only-rated supplementary fuses **is not** recommended for the dc circuits of PV systems [110-3(b)].

Current-Limiting Fuses—Stand-Alone Systems

A current-limiting fuse **must** be used in each ungrounded conductor from the battery to limit the current that a battery bank can supply to a short-circuit and to reduce the short-circuit currents to levels that are within the capabilities of downstream equipment [690-71(c)]. These fuses are available with *UL* ratings of 125, 300, and 600 volts dc, currents of 0.1 to 600 amps, and a dc AIR of 20,000 amps. They are classified as RK5 or RK1 current-limiting fuses and should be mounted in Class-R rejecting fuse holders or dc-rated, fused disconnects. Class J or T fuses with dc ratings might also be used. For reasons mentioned previously, time-delay fuses should be specified, although some designers are getting good results with Class T fast-acting fuses. One of these fuses and the associated disconnect switch should be used in **each** bank of batteries with a paralleled amp-hour capacity up to 1,000 amp-hours. Batteries with single cell amp-hour capacities higher than 1,000 amp-hours will require special design considerations, because these batteries may be able to generate short-circuit currents in excess of the 20,000 AIR rating of the current-limiting fuses. When calculating short-circuit currents, the resistances of all connections, terminals, wire, fuse holders, circuit breakers, and switches **must** be considered. These resistances serve to reduce the magnitude of the available short-circuit currents at any particular point. The suggestion of one fuse per 1,000 amp-hours of battery size is only a general estimate, and the calculations are site specific. The listed branch-circuit fuses shown in Figure 12 are current limiting.

For systems less than 65 volts (open circuit), Heinemann Electric 25,000 AIR circuit breakers may be used (Appendix A). These circuit breakers are not current limiting, even with the high interrupt rating, so they cannot be used to protect other types of fuses or circuit breakers. An appropriate use would be in the conductor between the battery bank and the inverter. This single device would minimize voltage drop and provide the necessary disconnect and overcurrent features. When high AIC circuit breakers are used throughout a PV system, there may be NO requirement for a current-limiting fuse.

The equipment and conductors **must** be braced for the available short-circuit current, or current-limiting devices **must** be used.

Current-Limiting Fuses—Grid-Connected Systems

Normal electrical installation practice requires that utility service entrance equipment have fault-current protection devices that can interrupt the available short-circuit currents [110-9]. This requirement applies to the utility side of any power conditioning system in a PV installation. If the service is capable of delivering fault currents in excess of the AIR rating of the overcurrent devices used to connect the inverter to the system, then current-limiting overcurrent devices **must** be used [110-9].

The equipment and conductors **must** be braced for the available short-circuit current, or current-limiting devices **must** be used.

Fuse Servicing

Whenever a fuse is used as an overcurrent device and is accessible to unqualified persons, it **must** be installed in such a manner that all power can be removed from both ends of the fuse for servicing. It is not sufficient to reduce the current to zero before changing the fuse. There **must** be no voltage present on either end of the fuse prior to service. This may require the addition of switches on both sides of the fuse location—a complication that increases the voltage drop and reduces the reliability of the system [690-16]. Because of this requirement, the use of a fusible pullout-style disconnect or circuit breaker is recommended. For the charging and dc-load circuits, it is recommended that a current-limiting fuse be used at the battery with a switch located between the battery and the current-limiting fuse. Circuit breakers can then be used for all other overcurrent devices in circuits toward the PV module where the available fault currents do not exceed their AIR or where they are protected by the current-limiting fuse.

Ungrounded 12 or 24-volt systems require an overcurrent device in both of the ungrounded conductors of each circuit. Since an equipment grounding system is required on all systems, costs may be reduced by grounding the system and using overcurrent devices only in the remaining ungrounded conductors.

DISCONNECTING MEANS

There are many considerations in configuring the disconnect switches for a PV system. The *National Electrical Code* deals with safety first and other requirements last—if at all. The PV designer should also consider equipment damage from over voltage, performance options, equipment limitations, and cost.

A photovoltaic system is a power generation system, and a specific minimum number of disconnects are necessary to deal with that power. Untrained personnel will be operating the systems; therefore, the disconnect system **must** be designed to provide safe, reliable, and understandable operation.

Disconnects may range from nonexistent in a self-contained PV-powered light for a sidewalk to those found in the space-shuttle-like control room in a large, multi-megawatt, utility-tied PV power station. Generally, local inspectors will not require disconnects on totally enclosed, self-contained PV systems like a PV-powered, solar, hot-water circulating system. This would be particularly true if the entire assembly were listed as a unit and there were no external contacts or user serviceable parts. However, the situation changes as the complexity of the device increases and separate modules, batteries, and charge controllers having external contacts are wired together and possibly operated and serviced by unqualified personnel.

Photovoltaic Array Disconnects

Article 690 requires all current-carrying conductors from the PV power source or other power source to have *disconnect* provisions. This includes the grounded conductor, if any [690-C]. *Ungrounded* conductors **must** have a switch or circuit breaker disconnect [690-13, -15, -17]. *Grounded* conductors which normally remain connected at all times may have a bolted disconnect (terminal or lug) that can be used for service operations and for meeting the *NEC* requirements. Grounded conductors of faulted source circuits in roof-mounted dc PV arrays on dwellings are allowed to be automatically interrupted as part of ground-fault protection requirements in 690-5.

In an ungrounded 12- or 24-volt PV system, both positive and negative conductors **must** be switched, since both are ungrounded. Since all systems **must** have an equipment-grounding system, costs may be reduced and performance improved by grounding 12- or 24-volt systems and using one-pole disconnects on the remaining ungrounded conductor.

Equipment Disconnects

Each piece of equipment in the PV system **shall** have disconnect switches to disconnect it from all sources of power. The disconnects **shall** be circuit breakers or switches and **shall** comply with all of the provisions of Section 690-17. DC-rated switches are expensive; therefore, the ready availability of moderately priced dc-rated circuit breakers with ratings up to 125 volts and 110 amps would seem to encourage their use in all 12-, 24-, and 48-volt systems. When properly located and used within their approved ratings, circuit breakers can serve as both the disconnect and overcurrent device. In simple systems, one switch or circuit breaker disconnecting the PV array and another disconnecting the battery may be all that is required.

A 2,000-watt inverter on a 12-volt system can draw more than 235 amps at full load. Disconnect switches **must** be rated to carry this load and have appropriate interrupt ratings. Again, a dc-rated, *UL*-Listed circuit breaker may prove less costly and more compact than a switch and fuse with the same ratings.

Battery Disconnect

When the battery is disconnected from the stand-alone system, either manually or through the action of a fuse or circuit breaker, care should be taken that the PV system not be allowed to remain connected to the load. Small loads will allow the PV array voltage to increase from the normal battery charging levels to the open-circuit voltage, which will shorten lamp life and possibly damage electronic components.

This potential problem can be avoided by using ganged multipole circuit breakers or ganged fused disconnects as shown in Figure 15. This figure shows two ways of making the connection. Separate circuits, including disconnects and fuses between the charge controller and the battery and the battery and the load, as shown in Figure 16, may be used if it is desired to operate the loads without the PV array being connected. If the design requires that the entire system be shut down with a minimum number of switch actions, the switches and circuit breakers could be ganged multipole units.

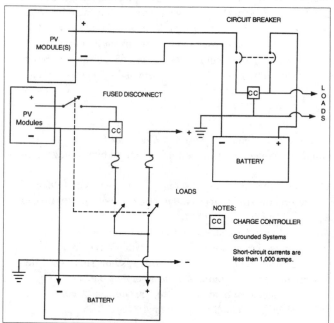

Figure 15. Small System Disconnects

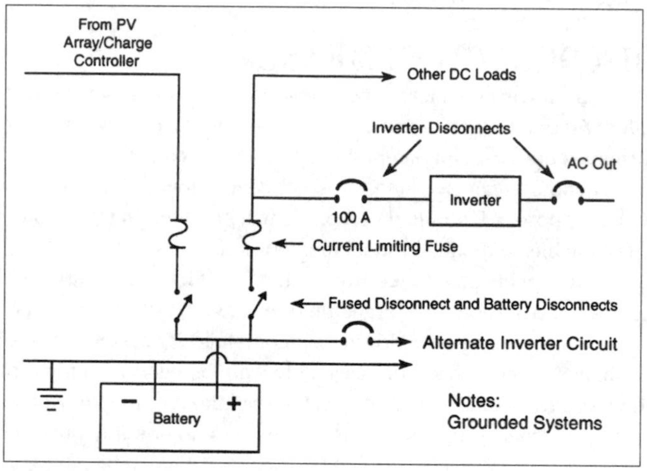

Figure 16. Separate Battery Disconnects

Charge Controller Disconnects

Some charge controllers are fussy about the sequence in which they are connected and disconnected from the system. Most charge controllers do not respond well to being connected to the PV array and not being connected to the battery. The sensed battery voltage (or lack thereof) would tend to rapidly cycle between the array open-circuit voltage and zero as the controller tried to regulate the nonexistent charge process. This problem will be particularly acute in self-contained charge controllers with no external battery sensing. The use of charge controllers listed to *UL Standard 1741* will minimize this problem.

Again, the multipole switch or circuit breaker can be used to disconnect not only the battery from the charge controller, but the charge controller from the array. Probably the safest method for self-contained charge controllers is to have the PV disconnect switch disconnect both the input and the output of the charge controller from the system. Larger systems with separate charge control electronics and switching elements will require a case-by-case analysis—at least until the controller manufacturers standardize their products. Figure 17 shows two methods of disconnecting the charge controller.

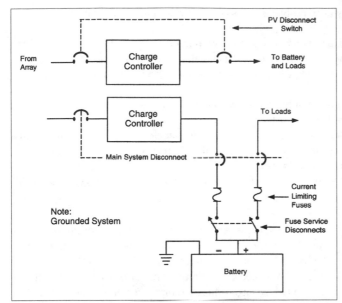

Figure 17. Charge Controller Disconnects

Non-Grounded Systems

Systems that do not have one of the current-carrying conductors grounded **must** have disconnects *and* overcurrent devices in all of the ungrounded conductors. This means two-pole devices for the PV, battery, and inverter disconnects and overcurrent devices. The additional cost is considerable.

Multiple Power Sources

When multiple sources of power are involved, the disconnect switches **shall** be grouped and identified [230-72, 690-13]. No more than six motions of the hand **will** be required to operate all of the disconnect switches required to remove all power from the system [230-71]. These power sources include PV output, the battery system, any generator, and any other source of power. Multipole disconnects or handle ties should be used to keep the number of motions of the hand to six or fewer.

PANEL BOARDS, ENCLOSURES, AND BOXES

Disconnect and overcurrent devices **shall** be mounted in listed enclosures, panel boards, or boxes [240-30]. Wiring between these enclosures **must** use a *NEC*-approved method [110-8]. Appropriate cable clamps, strain-relief methods, or conduit **shall** be used. All openings not used **shall** be closed with the same or similar material to that of the enclosure [370-18]. Metal enclosures **must** be bonded to the equipment-grounding conductor [370-4]. The use of wood or other combustible materials is discouraged. Conductors from different systems such as utility power, gas generator, hydro, or wind **shall not** be placed in the same enclosure, box, conduit, etc., as PV source conductors unless the enclosure is partitioned [690-4b]. This requirement stems from the need to keep "always live" PV source conductors separate from those that can be turned off.

When designing a PV distribution system or panel board, a listed NEMA type box and listed disconnect devices and overcurrent devices should be used. The requirements for the internal configuration of these devices are established by *NEC* Articles 370, 373, and 384 and **must** be followed. Dead front-panel boards with no exposed current-carrying conductors, terminals, or contacts are generally required. Underwriters Laboratories also establishes the standards for the internal construction of panel boards and enclosures.

BATTERIES

In general, *NEC* Articles 480 and 690-71, 72, 73 should be followed for installations having storage batteries. Battery storage in PV systems poses several safety hazards:

- Hydrogen gas generation from charging batteries
- High short circuit current
- Acid or caustic electrolyte
- Electric shock potential

Hydrogen Gas

When flooded, non-sealed, lead-acid batteries are charged at high rates, or when the terminal voltage reaches 2.3 - 2.4 volts per cell, the batteries produce hydrogen gas. Even sealed batteries may vent hydrogen gas under certain conditions. This gas, if confined and not properly vented, poses an explosive hazard. The amount of gas generated is a function of the battery temperature, the voltage, the charging current, and the battery-bank size. Hydrogen is a light, small-molecule gas that is easily dissipated and is very difficult to contain. Small battery banks (i.e., up to 20, 220-amp-hour, 6-volt batteries) placed in a large room or a well-ventilated (drafty) area may not pose a significant hazard. Larger numbers of batteries in smaller or tightly enclosed areas require venting. Venting manifolds attached to each cell and routed to an exterior location are not recommended because flames in one section of the manifold may be easily transmitted to other areas in the system. The instructions provided by the battery manufacturer should be followed.

Closed battery boxes with vents to outside-the-house air may pose problems unless carefully designed. Wind may force hydrogen back down the vent.

A catalytic recombiner cap (Hydrocap® Appendix A) may be attached to each cell to recombine some of the evolved hydrogen and oxygen to produce water. If these combiner caps are used, they will require occasional maintenance. It is rarely necessary to use power venting [*NEC* Handbook 480-8]. Flame arrestors are required by *NEC* Section 480-9, and battery manufacturers can provide special vent caps with flame-arresting properties.

Certain charge controllers are designed to minimize the generation of hydrogen gas, but lead-acid batteries need some overcharging to fully charge the cells. This produces gassing that should be dissipated.

In *no case* should charge controllers, switches, relays, or other devices capable of producing an electric spark be mounted in a battery enclosure or directly over a battery bank. Care **must** be exercised when routing conduit from a sealed battery box to a disconnect. Hydrogen gas may travel in the conduit to the arcing contacts of the switch.

It is suggested that any conduit openings in battery boxes be made below the tops of the batteries, since hydrogen rises in air.

Battery Rooms and Containers

Battery systems are capable of generating thousands of amps of current when shorted. A short circuit in a conductor not protected by overcurrent devices can melt wrenches or other tools, battery terminals and cables, and spray molten metal around the room. Exposed battery terminals and cable connections **must** be protected. Live parts of batteries **must** be guarded [690-71]. This generally means that the batteries should be accessible only to a qualified person. A locked room, battery box, or other container and some method to prevent access by the untrained person should reduce the hazards from short circuits and electric shock. The danger may be reduced if insulated caps or tape are placed on each terminal and an insulated wrench is used for servicing, but in these circumstances, corrosion may go unnoticed on the terminals. The *NEC* requires certain spacings around battery enclosures and boxes to allow for unrestricted servicing—generally about three feet [110-16]. Batteries should not be installed in living areas, nor should they be installed below any enclosures, panel boards, or load centers [110-26].

One of the more suitable, readily available battery containers is the lockable, heavy-duty black polyethylene tool box. Such a box can hold up to four L-16 size batteries and is easily cut for ventilation holes in the lid and for conduit entrances.

Acid or Caustic Electrolyte

A thin film of electrolyte can accumulate on the tops of the battery and on nearby surfaces. This material can cause flesh burns. It is also a conductor and, in high-voltage battery banks, poses a shock hazard, as well as a potential ground-fault path. The film of electrolyte should be removed periodically with an appropriate neutralizing solution. For lead-acid batteries, a dilute solution of baking soda and water works well. Commercial neutralizers are available at auto-supply stores.

Charge controllers are available that minimize the dispersion of the electrolyte and water usage because they minimize battery gassing. They do this by keeping the battery voltage from climbing into the *vigorous* gassing region where the high volume of gas causes electrolyte to mist out of the cells. A moderate amount of gassing is necessary for proper battery charging and de-stratification of the electrolyte in flooded cells.

Battery servicing hazards can be minimized by using protective clothing including face masks, gloves, and rubber aprons. Self-contained eyewash stations and neutralizing solution would be beneficial additions to any battery room. Water should be used to wash acid or alkaline electrolyte from the skin and eyes.

Anti-corrosion sprays and greases are available from automotive and battery supply stores and they generally reduce the need to service the battery bank. Hydrocap® Vents also reduce the need for servicing by reducing the need for watering.

Electric Shock Potential

Storage batteries in dwellings **must** operate at less than 50 volts (48-volt nominal battery bank) unless live parts are protected during routine servicing [690-71b(2)]. It is recommended that live parts of any battery bank should be guarded [690-71b(2)].

GENERATORS

Other electrical power generators such as wind, hydro, and gasoline/propane/diesel **must** comply with the requirements of the *NEC*. These requirements are specified in the following *NEC* articles:

Article 230	Services
Article 250	Grounding
Article 445	Generators
Article 700	Emergency Systems
Article 701	Legally Required Standby Systems
Article 702	Optional Standby Systems
Article 705	Interconnected Power Production Sources

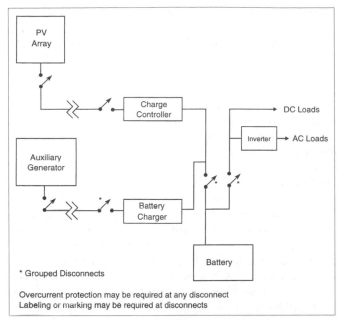

Figure 18. Disconnects for Remotely Located Power Sources

* Grouped Disconnects

Overcurrent protection may be required at any disconnect
Labeling or marking may be required at disconnects

When multiple sources of ac power are to be connected to the PV system, they **must** be connected with an appropriately rated and approved transfer switch. AC generators frequently are rated to supply larger amounts of power than that supplied by the PV/battery/inverter. The transfer switches (external to the inverter or a relay built into listed inverters) **must** be able to safely accommodate either power source.

Grounding, both equipment and system, **must** be carefully considered when a generator is connected to an existing system. There **must** be no currents flowing in the equipment-grounding conductor under any operating mode of the system. Bonds (connections) between the ac neutral and ground in generators are common and do cause problems.

The circuit breakers or fuses that are built into the generator are not sufficient to provide *NEC* required protection for the conductors from the generator to the PV system. An external (branch circuit rated) overcurrent device (and possibly a disconnect) **must** be mounted close to the generator. The conductors from the generator to this overcurrent device **must** have an ampacity of 115% of the name plate current rating of the generator [445-5]. Figure 18 shows a typical one-line diagram for a system with a backup generator.

CHARGE CONTROLLERS

A charge controller or self-regulating system **shall** be used in a stand-alone system with battery storage. The mechanism for adjusting state of charge **shall** be accessible only to qualified persons [690-72].

There are several charge controllers on the market that have been tested and listed by *UL* or other recognized testing organizations.

Surface mounting of devices with external terminals readily accessible to the unqualified person will not be accepted by the inspection authority. These charge controllers should be mounted in a listed enclosure with provisions for ventilation. Dead-front

Figure 19. Typical Charge Controller

panels with no exposed contacts are generally required for safety. A typical charge controller such as shown in Figure 19 should be mounted in a *UL*-Listed enclosure so that none of the terminals are exposed. Enclosures containing charge controllers should have knockouts for cable entry and some method of attaching conduit where required. Internal space **must** be allocated to provide room for wire bending [373,384].

Electrically, charge controllers should be designed with a "straight" conductor between the negative input and output terminals. No shunts or other signal processing should be placed in that conductor. This design will allow the controller to be used in a grounded system with the grounded conductor running through the controller. The design of the charge controller **must** be reviewed to ensure proper system grounding.

INVERTERS

Inverters can have stand-alone, utility-interactive, or combined capabilities.

The ac output wiring is not significantly different than the ac wiring in residential and commercial construction, and the same general requirements of the code apply. In the case of utility-interactive systems and combined systems, ac power may flow through circuits in both directions. This two-way current flow will normally require overcurrent devices at both ends of the circuit.

The dc input wiring associated with stand-alone or hybrid inverters is the same as the wiring described for batteries. Most of the same rules apply; however, the calculation of the dc input current needs special consideration since the *NEC* does not take into consideration some of the finer points required to achieve the utmost in reliability. Appendix F discusses these special requirements in greater detail.

The dc input wiring associated with utility-interactive inverters is similar, in most cases, to the wiring in PV source and output circuits.

Inverters with combined capabilities will have both types of dc wiring: connections to the batteries and connections to the PV modules.

DISTRIBUTION SYSTEMS

The *National Electrical Code* has evolved to accommodate supplies of relatively cheap energy. As the Code was expanded to include other power systems such as PV, many sections were not modified to reflect the recent push toward more efficient use of electricity in the home. Stand-alone PV systems *may* be required to have dc services with 60- to 100-amp capacities to meet the Code [230-79]. DC receptacles and lighting circuits *may* have to be as numerous as their ac counterparts [220, 422]. In a small one-to four-module system on a remote cabin where no utility extensions or local grids are possible, these requirements may be excessive, since the power source may be able to supply only a few hundred watts of power.

Changes in the 1999 *NEC* in Section 690-10 have clarified some of the code requirements for stand-alone PV systems.

The local inspection authority has the final say on what is, or is not, required and what is, or is not, safe. Reasoned conversations may result in a liberal interpretation of the Code. For a new dwelling, it seems appropriate to install a complete ac electrical system as required by the *NEC* . This will meet the requirements of the inspection authority, the mortgage company, and the insurance industry. Then the PV system and its dc distribution system can be added. If an inverter is used, it can be connected to the ac service entrance. *NEC* Section 690 elaborates on these requirements and allowances. DC branch circuits and outlets can be added where needed, and everyone will be happy. If or when grid power becomes available, it can be integrated into the system with minimum difficulty. If the building is sold at a later date, it will comply with the *NEC* if it has to be inspected. The use of a *UL*-Listed dc power center will facilitate the installation and the inspection.

Square D has received a direct current (dc), *UL* listing for its standard QO residential **branch** circuit breakers. They can be used up to 48 volts (125% PV open-circuit voltage) and 70 amps dc. This would limit their use to a 12-volt nominal system and a few 24-volt systems in hot climates [Table 690-7]. The AIR is 5,000 amps, so a current-limiting fuse (RK5 or RK1 type) **must** be used when they are connected on a battery system. The Square D QOM **main** breakers (used at the top of the load center) **do not** have this listing, so the load center **must** be obtained with main lugs and no main breakers (Appendix A).

In a small PV system (less than 5000 amps of available short-circuit current), a two-pole Square D QO breaker could be used as the PV disconnect (one pole) and the battery disconnect (one pole). Also, a fused disconnect or fusible pullout could be used in this configuration. This would give a little more flexibility since the fuses can have different current ratings. Figure 15 on page 28 shows both systems with only a single branch circuit.

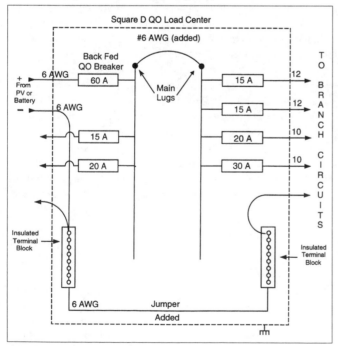

Figure 20. 12-Volt DC Load Center

In a system with several dc branch circuits, the Square D QO load center can be used. A standard, off-the-shelf Square D QO residential load center without a main breaker can be used for a dc distribution panel in 12-volt dc systems and some hot-climate 24-volt systems. The main disconnect would have to be a "back fed" QO breaker, and it would have to be connected in one of the normal branch circuit locations. Back-fed circuit breakers **must** be identified for such use [690-64b(5)]. Since the load center has two separate circuits (one for each line), the bus bars will have to be tied together to use the entire load center. Figure 20 illustrates this use of the Square D load center.

Another possibility is to use one of the line circuits to combine separate PV source circuits, then go out of the load center through a breaker acting as the PV disconnect switch to the charge controller. Finally, the conductors would have to be routed back to the other line circuit in the load center for branch-circuit distribution. Several options exist in using one and two-pole breakers for disconnects. Figure 21 presents an example.

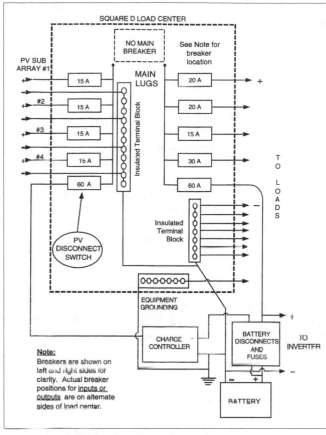

Figure 21. 12-Volt DC Combining Box and Load Center

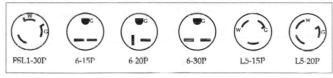

Figure 22. NEMA Plug Configurations

Interior Wiring and Receptacles

Any dc interior wiring used in PV systems **must** comply with the *NEC* [300]. Nonmetallic sheathed cable (type NM—"Romex") may be used, and it **must** be installed in the same manner as cable for ac branch circuits [336, 690-31a]. The bare grounding conductor in such a cable **must not** be used to carry current and cannot be used as a common negative conductor for combination 12/24-volt systems [336-30(b)]. Exposed, single-conductor cables are not permitted—they **must** be installed in conduit [300-3(a)]. Wires carrying the same current (i.e., positive and negative battery currents) **must** be installed in the same conduit or cable to prevent increased circuit inductances that would pose additional electrical stresses on disconnect and overcurrent devices [300-3(b)].

Equipment-grounding conductors for dc circuits only may be run apart from the current-carrying conductors [250-134(b) Ex2]. However, separating the equipment grounding conductor from the circuit conductors may increase the fault circuit time constant and impair the operation of overcurrent devices. The effects of transient pulses are also enhanced when equipment grounding conductors are separate. It is suggested that equipment-grounding conductors be run in the same conduit or cable as the circuit conductors.

The receptacles used for dc **must** be different than those used for any other service in the system [210-7f, 551-20f]. The receptacles should have a rating of not less than 15 amps and **must** be of the three-prong grounding type [210-7a, 720-6]. Numerous different styles of listed receptacles are available that meet this requirement. These requirements can be met in most locations by using the three-conductor 15-, 20-, or 30-amp 240-volt NEMA style 6-15, 6-20, 6-30 receptacles for the 12-volt dc outlets. If 24-volt dc is also used, the NEMA 125-volt locking connectors, style L5-15 or L5-20, are commonly available. The NEMA FSL-1 is a locking 30 amp 28-volt dc connector, but its availability is limited. Figure 22 shows some of the available configurations. Cigarette lighter sockets and plugs frequently found on "PV" and "RV" appliances *do not* meet the requirements of the *National Electrical Code* and should not be used.

It is not permissible to use the third or grounding conductor of a three-conductor plug or receptacle to carry common negative return currents on a combined 12/24-volt system. This terminal **must** be used for equipment grounding and may not carry current except in fault conditions [210-7].

A 30-amp fuse or circuit breaker protecting a branch circuit (with 10 AWG conductors) **must** use receptacles rated at 30 amps. Receptacles rated at 15 and 20 amps **must not** be used on this 30-amp circuit [Table 210-24].

Smoke Detectors

Many building codes require that smoke and fire detectors be wired directly into the ac power wiring of the dwelling. With a system that has no inverter, two solutions might be offered to the inspector. The first is to use the 9-volt or other primary-cell, battery-powered detector. The second is to use a voltage regulator to drop the PV system voltage to the 9-volt or other level required by the detector.

The regulator **must** be able to withstand the PV open-circuit voltage and supply the current required by the detector alarm.

On inverter systems, the detector on some units may trigger the inverter into an "on" state, unnecessarily wasting power. In other units, the alarm may not draw enough current to turn the inverter on and thereby produce a reduced volume alarm or, in some cases, no alarm at all. Small, dedicated inverters might be used, but this would waste power and decrease reliability when dc detectors are available.

Many states require detectors to be connected to the power line and have a battery backup. Units satisfying this requirement might also be powered by dc from the PV system battery and by a primary cell.

Ground-Fault Circuit Interrupters

Some ac ground-fault circuit interrupters (GFCI) do not operate reliably on the output of some non-sine-wave inverters. If the GFCI does not function when tested, it should be verified that the neutral (white-grounded) conductor of the inverter output is solidly grounded and bonded to the grounding (green or bare) conductor of the inverter in the required manner. If this bond is present and does not result in the GFCI testing properly, other options are possible. Changing the brand of GFCI may rectify the solution. A direct measurement of an intentional ground fault may indicate that slightly more than the 5 milliamp internal test current is required to trip the GFCI. The inspector may accept this. Some inverters will work with a ferro-resonant transformer to produce a wave form more satisfactory for use with GFCIs, but the no-load power consumption may be high enough to warrant a manual demand switch. A sine-wave inverter should be used to power those circuits requiring GFCI protection.

The 1999 *NEC* added a Section 690-6(d) *permitting* (not requiring) the use of a device (undefined) on the ac branch circuit being fed by an ac PV module to detect ground-faults in the ac wiring. There are no commercially available devices as of early 2001 that can meet this permissive requirement. Standard 5-milliamp anti-shock receptacle GFCIs or 30-milliamp equipment protection circuit breakers should not be used for this application. They may be destroyed if used for this application.

Interior Switches

Switches rated for ac only **shall not** be used in dc circuits [380-14]. AC-DC general-use snap switches are available by special order from most electrical supply houses, and they are similar in appearance to normal "quiet switches" [380-14(b)].

Note: There have been some failures of dc-rated snap switches when used as PV array and battery disconnect switches. If these switches are used on 12- and 24-volt systems and are not activated frequently, they may build up internal oxidation or corrosion and not function properly. Periodically activating the switches under load will keep them clean.

MULTIWIRE BRANCH CIRCUITS

Stand-alone PV and PV/Hybrid systems are frequently connected to a building/structure/house that has been previously completely wired for 120/240-volts ac and has a standard service entrance and load center.

These structures may employ one or more circuits that the *National Electrical Code* (*NEC*) defines as a multiwire branch circuit. See Section 100 in the *NEC*, "Branch Circuit, Multiwire." These circuits take a three-conductor plus ground feeder from the 120/240-volt load center and run it some distance to a location in the structure where two separate 120-volt branch circuits are split out. Each branch circuit uses one of the 120-volt hot, ungrounded conductors from the 120/240-volt feeder and the common neutral conductor. See Figure 23.

In a utility-connected system or a stand-alone system with a 120/240-volt stacked pair of inverters, where the 120/240-volt power consists of two 120-volt lines that are 180 degrees out of phase, the currents in the common neutral in the multiwire branch circuit are limited to the difference currents from any unbalanced load. If the loads on each of the separate branch circuits were equal, then the currents in the common neutral would be zero.

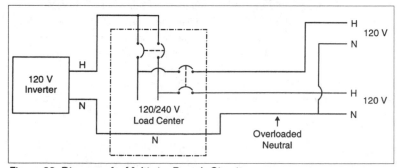

Figure 23. Diagram of a Multiwire Branch Circuit

A neutral conductor overload may arise when a single 120-volt inverter is tied to both of the hot input conductors on the 120/240-volt load center as shown in Figure 23. This is a common practice for stand-alone PV homes. At this point the two hot 120-volt conductors are being delivered voltage from the single 120-volt inverter and that voltage is in phase on both conductors. In multiwire branch circuits, the return currents from each of the separate branch circuits in the common neutral *add* together. A sketch of the multiwire branch circuit is presented in Figure 23.

Each branch circuit is protected by a circuit breaker in the ungrounded conductor in the load center. The neutral conductor is usually the same size as the ungrounded conductors and can be overloaded with the in-phase return currents. The circuit breakers will pass current up to the ampacity of the protected conductors, but when both branch circuits are loaded at more than 50%, the unprotected, common neutral conductor is *overloaded and may be carrying up to twice the currents that it was rated for.*

A definite fire and safety hazard exists. All existing stand-alone PV installations using single inverters tied to both ungrounded conductors at the service entrance should be examined for multiwire branch circuits.

The *NEC* requires that multiwire branch circuits *in some, but not all, cases* use a two-pole circuit breaker so that both circuits are dead at the same time under fault conditions and for servicing. This two-pole, side-by-side circuit breaker rated at 15 or 20 amps may be one indication that multiwire branch circuits have been used. Common handle circuit breakers rated at 30 amps and higher are usually dedicated to 240-volt circuits for ranges, hot water heaters, dryers, and the like.

Examination of the wiring in the load center may show a three-wire cable (14 or 12 AWG red, black, and white conductors) with bare ground leaving the load center. This may be connected to a multiwire branch circuit. The circuit breakers connected to this cable and the outputs of this cable should be traced to determine the presence or absence of a multiwire branch circuit.

The following options are suggested for dealing with this problem:

- **Disconnect or rewire the multiwire circuits as separate circuits ("home runs") from the load center.**
- **Connect both "hot" (ungrounded) conductors of the multiwire branch circuit to a single circuit breaker rated for the ampacity of the neutral conductor.**
- **Limit the output of the inverter with a circuit breaker rated at the ampacity of the neutral conductor (usually 15 amps). Place a warning near this circuit breaker stating that the rating *must not* be increased.**
- **Install a transformer to provide a 120/240-volt output from a 120-volt inverter.**
- **Install a stacked pair of inverters to provide 120/240V ac.**

In systems where multiwire branch circuits are used with stacked (120/240-volt) inverters, a sign should be placed near the inverters warning that single inverter use (one inverter removed for repair and the system is rewired to accommodate all branch circuits) may cause overloaded circuits. The maximum current from the single inverter should be limited to the ampacity of the common neutral conductor.

In all systems (multiwire or not), the neutral busbar of the load center **must** be rated at a higher current than the output of the inverter. In other words, do not connect an inverter with a 33-amp output to a load center rated at 20 or 30 amps.

Additional information is found in the *NEC* in sections 100, 210-4, 240-20(b), and 300-13(b), and in the *NEC* Handbook. Section 690-10 provides requirements and allowances on connecting a single inverter to a code-compliant ac wiring system.

AC PV MODULES

An AC PV module is a photovoltaic device that has an alternating current output (usually 120 volts at 60 Hz in the U.S.). It is listed (by *UL*) as a unified device and is actually a standard dc PV module with an attached (non-removable) utility-interactive inverter. The ac output is only available when the ac PV module is connected to a utility grid circuit where there is a stable 120 volts at 60 Hz present. With no utility power, there will be NO energy flow from the ac PV modules.

A number of ac PV modules may be connected on the same circuit (according to ampacity limitations), but that circuit **must** be dedicated to the ac PV module(s) and **must** terminate in a dedicated circuit breaker.

There are no external dc circuits in the ac PV module and none of the dc code requirements apply. Unlisted combinations of small listed inverters mated to listed dc PV modules do not qualify as an ac PV module and will have to have all code-required dc switchgear, overcurrent, and ground-fault equipment added.

AC PV modules **shall** be marked with the following:
- Nominal AC Voltage
- Nominal AC Frequency
- Maximum AC Power
- Maximum AC Current
- Maximum Overcurrent Device Rating for AC Module Protection [690-52]

SYSTEM LABELS AND WARNINGS

Photovoltaic Power Sources

A permanent label **shall** be applied near the PV disconnect switch that contains the following information: [690-53]
- Operating Current (System maximum-power current)
- Operating Voltage (System maximum-power voltage)
- Maximum System Voltage
- Short-Circuit Current

This data will allow the inspector to verify proper conductor ampacity and overcurrent device rating. It will also allow the user to compare system performance with the specifications.

Multiple Power Systems

Systems with multiple sources of power such as PV, gas generator, wind, hydro, etc., **shall** have a permanent plaque or directory showing the interconnections [705-10]. Diagrams are not required, but may be useful and should be placed near the system disconnects.

Interactive System Point of Interconnection

All interactive system(s) points of interconnection with other sources **shall** be marked at an accessible location at the disconnecting means as a power source with the maximum ac output operating current and the operating ac voltage [690-54].

Switch or Circuit Breaker

If a switch or circuit breaker has all of the terminals energized when in the open position, a label should be placed near it indicating: [690-17]

- WARNING – ELECTRIC SHOCK HAZARD – DO NOT TOUCH TERMINALS. TERMINALS ON BOTH THE LINE AND LOAD SIDES MAY BE ENERGIZED IN THE OPEN POSITION

General

Each piece of equipment that might be opened by unqualified persons should be marked with warning signs:

- WARNING – ELECTRIC SHOCK HAZARD – DANGEROUS VOLTAGES AND CURRENTS – NO USER SERVICEABLE PARTS INSIDE – CONTACT QUALIFIED SERVICE PERSONNEL FOR ASSISTANCE

Each battery container, box, or room should also have warning signs:

- WARNING – ELECTRIC SHOCK HAZARD – DANGEROUS VOLTAGES AND CURRENTS – EXPLOSIVE GAS – NO SPARKS OR FLAMES – NO SMOKING – ACID BURNS – WEAR PROTECTIVE CLOTHING WHEN SERVICING

INSPECTIONS

Involving the inspector as early as possible in the planning stages of the system will begin a process that should provide the best chance of obtaining a safe, durable system. The following steps are suggested.

- Establish a working relationship with a local electrical contractor or electrician to determine the requirements for permits and inspections.
- Contact the inspector and review the system plans. Solicit advice and suggestions from the inspector.
- Obtain the necessary permits.
- Involve the inspector in the design and installation process. Provide information as needed. Have one-line diagrams and complete descriptions of all equipment available.

INSURANCE

Most insurance companies are not familiar with photovoltaic power systems. They are, however, willing to add the cost of the system to the homeowner's policy if they understand that it represents no additional liability risk. A system description may be required. Evidence that the array is firmly attached to the roof or ground is usually necessary. The system **must** be permitted and inspected if those requirements exist for other electrical power systems in the locale.

Some companies will not insure homes that are not grid connected because there is no source of power for a high-volume water pump for fighting fires. In these instances, it may be necessary to install a fire-fighting system and water supply that meets their requirements. A high-volume dc pump and a pond might suffice.

As with the electrical inspector, education and a full system description emphasizing the safety features and code compliance will go a long way toward obtaining appropriate insurance.

NEC APPENDIX A

SOURCES OF EQUIPMENT MEETING THE REQUIREMENTS OF THE
NATIONAL ELECTRICAL CODE

A number of PV distributors and dealers are stocking the equipment needed to meet the *NEC* requirements. Some sources are presented here for specialized equipment, but this list is not intended to be all-inclusive or to promote any of the products.

Conductors

Standard multiconductor cable such as 10-2 with ground Nonmetallic Sheathed Cable (NM and NMC), Underground Feeder (UF), Service Entrance (SE), Underground Service Entrance (USE and USE-2), larger sizes (8 AWG) single-conductor cable, uninsulated grounding conductors, and numerous styles of building wire such as THHN can be obtained from electrical supply distributors and building supply stores.

The highest quality USE-2 cable will be listed by *UL* and will also have RHW-2, and 600V markings. Flexible USE, RHW, and THW cables in large sizes (1/0 - 250 kcmil) and stranded 8-, 10-, and 12-gauge USE single conductor cable can be obtained from some electrical supply houses and wire distributors, including:

> Anixter Bros.
> 2201 Main Street
> Evanston, Illinois 60202
> 800-323-8166 for the nearest distributor
> 847-677-2600

> Cobra Wire and Cable, Inc.
> PO Box 790
> 2930 Turnpike Drive
> Hatboro, PA 19040
> 215-674-8773

For grid-connected systems requiring cables with voltage rating higher than 600V, cable is normally special-ordered. Rubber-insulated cables are available with up to 2,000-volt insulation. They should be marked RHW-2 and be sunlight resistant when used for exposed module interconnects. The following manufacturers can supply such cable.

> American Insulated Wire
> 36 Freeman Street
> P.O. Box 880
> Pawtucket, RI 02862
> 401-726-0700

> The Okonite Company
> PO Box 340
> Ramsey, NJ 07446
> 201-825-9026

Miscellaneous Hardware

Stainless steel nuts, bolts and screws, and other hardware, insulated and uninsulated crimp-on terminals, battery terminals, copper lugs for heavy cable, battery cable, weather-resistant cable ties, heat shrink tubing and more may be obtained from the following source:

> Chesapeake Marine Fasteners
> 10 Willow Street
> P.O. Box 6521
> Annapolis, Maryland 21401
> 800-526-0658
> 410-266-9332
> Dealer's price sheet is available

The company listed below makes plastic strain reliefs that fit the standard ½" electrical knockout (⅞" diameter). These water-tight strain reliefs are needed for older ARCO PV modules, some BP Solar modules, as well as others. The single-conductor versions are hard to find, and the metal types are very expensive. A catalog and information on product 3224 (for AWG 10) or 3231 (for larger wire) can be requested. The company also makes UV-resistant black cable ties and copper, heavy-duty lugs, as well as other products that might be useful.

Heyco Molded Products, Inc.
Box 160
Kenilworth, New Jersey 07033
800-526-4182 or 908-245-0033
Quantity purchases only; call for distributor locations

DC-Rated Fuses

15, 20, 30 amps and higher rated fuses can be used for branch-circuit overcurrent protection depending on conductor ampacity and load. Larger sizes (100 amp and up) are used for current-limiting and overcurrent protection on battery outputs. DC rated, *UL*-Listed fuses are manufactured by the following companies, among others:

Bussmann
P.O. Box 14460
St. Louis, MO 63178-4460
314-527-3877
314-527-1270 (Technical Questions)

Gould/Ferraz Inc.
374 Merrimac Street
Newburyport, MA 01950
508-462-6662

Littelfuse
Power Fuse Division
800 E. Northwest Highway
Des Plaines, Illinois 60016
(708) 824-1188
800-TEC FUSE (Technical Questions)
800-227-0029 (Customer Service)

The following fuses may be used for battery circuit and branch circuit overcurrent protection and current limiting. If transients are anticipated in PV circuits, these fuses can also be used in those locations.

Fuse Description	Size	Manufacturer	Mfg #
125-volt dc, RK5 Time delay, current-limiting fuse	0.1-600 amp	Bussmann	FRN-R
125-volt dc, RK5 Time delay, current-limiting fuse	0.1-600 amp	Littelfuse	FLNR
300-volt dc, RK5 Time delay, current-limiting fuse	0.1-600-amp	Bussmann	FRS-R
300-volt dc, RK5 Time delay, current-limiting fuse	0.1-600 amp	Gould	TRS-R
300-volt dc, RK5 Time delay, current-limiting fuse	0.1-600 amp	Littelfuse	FLSR
600-volt dc, RK5 Time delay, current-limiting fuse	0.1-600 amp	Littelfuse	IDSR
600-volt dc, RK5 Time delay, current-limiting fuse	70-600 amp	Gould	TRS70R-600R

The following fuses should be used for PV source-circuit protection if problems are not anticipated with transients. They may also be used inside control panels to protect relays and other equipment.

Fuse Description	Size	Manufacturer	Mfg #
Fast-acting, current-limiting midget fuse	0.1-30 amp	Gould	ATM
Fast-acting, current-limiting midget fuse	0.1-30 amp	Littelfuse	KLK-D

FUSE HOLDERS *(ALSO SEE FUSED DISCONNECTS)*

Each fuse manufacturer makes fuse blocks matching the voltage rating and current rating of the selected fuse.

Marathon Special Projects also makes suitable fuse holders. Information and the names of distributors of Class R and Class M (midget fuse holders) should be requested. The company also makes power-distribution blocks for control panels.

> Marathon Special Products
> P.O. Box 468
> Bowling Green, Ohio 43402
> 419-352-8441

FUSED DISCONNECTS *(ALSO SEE CIRCUIT BREAKERS)*

Since fuses **must not** have power applied to either end when servicing, a combination switch and fuse can be mounted in a single enclosure to meet some, if not all, of the requirements.

> Siemens I-T-E
> Siemens Energy & Automation, Inc.
> 3333 State Bridge Rd.
> Alpharetta, Georgia 30202
> 404-751-2000

> Indoor fused switches, 250-volt dc-JN and JF series
> Outdoor fused switches, 250-volt dc-JR and FR series

> Call for nearest regional sales office that can direct you to a stocking distributor

> Square D Company
> 800-034-2003 for the nearest
> Square D electrical supply distributor

> Indoor fused switches
> 250-volt-dc-H22x, H32x, and H42x series
> 600-volt-dc-H26xx and H36xx series

> Outdoor fused switches
> 250-volt-dc-H22xR, H32xR, and H42xR series
> 600-volt-dc-H26xR and H36xR series

Boltswitch, Inc., makes pull-out fused disconnects that are dc rated for higher current applications. Contact factory for applications.

> boltswitch®, inc.
> 6107 West Lou Avenue
> Crystal Lake, IL 60014
> 815-459-6900

Circuit Breakers

> Square D QO circuit breakers (common ac residential breakers).
> UL-Listed at 5000 AIC at 48 volts dc; 1 and 2 pole, 10-70 amps; 3 pole, 10-60 amps

> Square D FA circuit breakers; 125- and 250-volt dc ratings, multiple currents

> Enclosures for QO breakers
> 2 and 3 pole units
> Indoor QO21xxBN, QO3100BN
> Rainproof QO21xxBNRB, QO3100BNRB

Any of the load centers for Square D QO breakers without main breakers may be used—main lugs should be requested instead.

> Square D Company
> 800-634-2003 for the nearest
> Square D electrical distributor

Heinemann makes a full line of dc-rated, *UL*-Listed and recognized supplemental circuit breakers, but they **must** be mounted in custom-built enclosures. (The metal is punched by the installer).

CD-CE-CF 5000 AIC at 125-volt dc, 15-110 amp

25,000 AIC available on special order. Polyester case, spun rivets, and *UL*-Listed units should be requested.

GH 10,000 AIC at 250-volts dc, 15-100 amp

GJ 10,000 AIC at 125-volts dc, 100-250 amps

GJ 25,000 AIC at 65-volts dc, 100-250 amps

GJ1P 10,000 AIC at 160-volts, 25,000 AIC at 65-volts dc, 100-700 amps

Rainshadow Solar
PO Box 541
Guthrie Cove Road
Orcas, WA 98280
360-376-5336

AIRPAX also makes a full line of dc-rated, *UL*-Listed and recognized supplemental circuit breakers, but they **must** be mounted in custom-built enclosures.

AIRPAX Corporation
P.O. Box 520
Cambridge, MD 21613-0520
410-228-4600

Call for nearest source and catalog.
Applications engineering available.

Enclosures and Junction Boxes

Indoor and outdoor (rainproof) general-purpose enclosures and junction boxes are available at most electrical supply houses. These devices usually have knockouts for cable entrances, and the distributor will stock the necessary bushings and/or cable clamps. Interior component mounting panels are available for some enclosures, as are enclosures with hinged doors. If used outdoors, all enclosures, clamps, and accessories **must** be listed for outdoor use. For visual access to the interior, NEMA 4x enclosures are available that are made of clear, transparent plastic.

Hydrocaps

Hydrocap® Vents are available from Hydrocap Corp. and some PV distributors on a custom-manufactured basis. Flame arrestors are an option.

Hydrocap
975 NW 95 St.
Miami, FL 33150
305-696-2504

Surge Arrestors

Delta makes a full line of large, silicon-oxide surge arrestors starting at 300 volts and up that are usable on low-voltage systems to clip the tops of large surges. Low-voltage versions are also available.

Delta Lightning Arrestors Inc.
P.O. Box 1084
Big Spring, TX 79721
915-267-1000

NEC APPENDIX B

PV MODULE OPERATING CHARACTERISTICS DRIVE NEC *REQUIREMENTS*

Introduction

As the photovoltaic (PV) power industry moves into a mainstream position in the generation of electrical power, some people question the seemingly conservative requirements established by Underwriters Laboratories (*UL*) and the *National Electrical Code* (*NEC*) for system and installation safety. This short discourse will address those concerns and highlight the unique characteristics of PV systems that dictate the requirements.

The *National Electrical Code* (*NEC*) is written with the requirement that all equipment and installations are approved for safety by the authority having jurisdiction (AHJ) to enforce the *NEC* requirements in a particular location. The AHJ readily admits to not having the resources to verify the safety of the required equipment and relies exclusively on the testing and listing of the equipment by independent testing laboratories such as Underwriters Laboratories (*UL*). The AHJ also relies on the installation requirements for field wiring specified in the *NEC* to ensure safe installations and use of the listed equipment.

The standards published by *UL* and the material in the *NEC* are closely harmonized by engineers and technicians throughout the electrical equipment industry, the electrical construction trades, the national laboratories, the scientific community, and the electrical inspector associations. The *UL* Standards are technical in nature with very specific requirements on the construction and testing of equipment for safety. They in turn are coordinated with the construction standards published by the National Electrical Manufacturers Association (NEMA). The *NEC*, however, is deliberately written in a manner to allow uniform application by electricians, electrical contractors, and electrical inspectors in the field.

The use of listed equipment (by *UL* or other laboratory) ensures that the equipment meets well-established safety standards. The application of the requirements in the *NEC* ensures that the listed equipment is properly connected with field wiring and is used in a manner that will result in an essentially hazard-free system. The use of listed equipment and installing that equipment according to the requirements in the *NEC* will contribute greatly not only to safety, but also the durability, performance, and longevity of the system.

Unspecified Details

The *NEC* does not present many highly detailed technical specifications. For example, the term "rated output" is used in several cases with respect to PV equipment. The conditions under which the rating is determined are not specified. The definitions of the rating conditions (such as Standard Test Conditions (STC) for PV modules) are made in the *UL* Standards that establish the rated output. This procedure is appropriate because of the *NEC* level of writing and the lack of appropriate test equipment available to the *NEC* user or inspector.

NEC Requirements Based on Module Performance

VOLTAGE

Section 690-7 of the *NEC* establishes a temperature-dependent voltage correction factor that is to be applied to the open-circuit voltage (Voc) in order to establish the system voltage. This factor on the open-circuit voltage is needed because, as the operating temperature of the module decreases, Voc increases. The rated Voc is measured at a temperature of 25°C and while the normal operating temperature is 40-50°C when ambient temperatures are around 20°C, there is nothing to prevent sub-zero ambient temperatures from yielding operating temperatures significantly below the 25°C standard test condition.

A typical crystalline silicon module will have a voltage coefficient of -0.38%/°C. A system with a rated open-circuit voltage of 595 volts at 25°C might be exposed to ambient temperatures of -30°C. This voltage (595) could be handled by the common 600-volt rated conductors and switchgear. At dawn and dusk conditions, the module will be at the ambient temperature of -30°C, will not experience any significant heating, but can generate open-circuit voltages of 719 volts (595 x (1 + (25 + 30) x 0.0038)). High wind speeds can also cause modules to operate at or near ambient temperatures, even in the presence of moderate levels of sunlight. This voltage substantially exceeds the capability of 600-volt rated conductors, fuses, switchgear, and other equipment. The very real possibility of this type of condition substantiates the *NEC* requirement for the temperature dependent factor on the rated open-circuit voltage.

Thin film PV technologies may have other voltage-temperature relationships, and the manufacturers of modules employing such technologies should be consulted for the appropriate data.

CURRENT

NEC Section 690-8(a) requires that the rated (at STC) short-circuit current of the PV module be multiplied by 125% before any other factors are applied, such as continuous currents and conduit fill. This factor is to provide a safe margin for wire sizes and overcurrent devices when the irradiance exceeds the standard 1000 W/m^2. Depending on season, local weather conditions, and atmospheric dust and humidity, irradiance exceeds 1000 W/m^2 every day around solar noon. The time period can be as long as four hours with irradiance values approaching 1200 W/m^2 again depending on the aforementioned conditions and the type of tracking being used. These daily irradiance values can increase short-circuit currents 20% over the 1000 W/m^2 value. Since these increased currents can be present for three hours or more, they are considered continuous currents. By multiplying the short-circuit current by 125%, the PV output currents are adjusted in a manner that puts them on the same basis as other continuous currents in the *NEC*

Enhanced irradiance due to reflective surfaces such as sand, snow, or white roofs, and even nearby bodies of water can increase short-circuit currents by substantial amounts and for significant periods of time. Cumulus clouds also can increase irradiance by as much as 50%. These transient factors are not considered continuous and are not addressed by either *UL* or the *NEC*

Another factor that **must** be addressed is that PV modules typically operate at 30-40°C above the ambient temperatures when not exposed to cooling breezes. In crystalline silicon PV modules, the short-circuit current increases as the temperature increases. A typical factor might be 0.1%/°C. If the module operating temperature was 60°C (35°C over the STC of 25°C), the short-circuit current would be 3.5% greater than the rated value. PV modules have been measured operating over 75°C. The combination of increased operating temperatures, irradiances over 1000 W/m^2 around solar noon, and the possibility of enhanced irradiance certainly justify the *NEC* requirement [690-8(a)] of 125% on the rated short-circuit current.

Additional *NEC* Requirements

The *NEC* requires that the continuous current of any circuit be multiplied by 125% before calculating the ampacity of any cable or the rating of any overcurrent device used in these circuits [690-8(b) and 240]. This factor is in addition to the required 125% discussed above and is needed to ensure that overcurrent devices and conductors are not operated above 80% of rating.

Since short-circuit currents in excess of the rated value are possible from the discussion of the Section 690-8(a) requirements above, and these currents are independent of the requirements established by Section 690-8(b), the *NEC* dictate that both factors will be used at the same time. This yields a multiplier on short-circuit current of 1.56 (125% x 125%).

The *NEC* also requires that the ampacity of conductors be derated for the operating temperature of the conductor. This is a requirement because the ampacity of cables is given for cables operating in an ambient temperature of 30°C. In PV systems, cables are operated in an outdoor environment and should be subjected at least to a temperature derating due to an ambient temperature of 40°C to 45°C. PV modules operate at high temperatures and, in some installations over 75°C, concentrating modules operate at even higher temperatures. The temperatures in module junction boxes approach these temperatures. Conductors in free air that lie against the back of these modules are also exposed to these temperatures. Temperatures this high require that the ampacity of cables be derated by factors of 0.33 to 0.58 depending on cable type, installation method (free air or conduit), and the temperature rating of the insulation [310-16,-17].

Cables in conduit where the conduit is exposed to the direct rays of the sun are also exposed to elevated operating temperatures.

Cables with insulation rated at 60°C have no ampacity at all when operated in environments with ambient temperatures over 55°C. This precludes their use in most PV systems. Cables with 75°C insulation have no ampacity when operated in ambient temperatures above 70°C. Because PV modules may operate at temperatures in the 45-75°C range, it is strongly suggested that only cables with an insulation rated at 90°C be used.

Summary

The conditions under which PV modules operate (high and low ambient temperatures, high and low winds, high and low levels of sunlight) and the electrical characteristics of those modules dictate that all of the requirements in the *NEC* be fully considered and applied.

There appears to be little question that the temperature-dependent correction factor on voltage is necessary in any location where the ambient temperature drops below 25°C. Even though the PV system can provide little current under open-circuit voltage conditions, these high voltages can damage electronic equipment and stress conductors and other equipment by exceeding their voltage breakdown ratings.

In ambient temperatures from 25 to 40°C and above, module short-circuit currents are increased at the same time conductors are being subjected to higher operating temperatures. Irradiance values over the standard rating condition may occur every day. Therefore the *NEC* requirements for adjusting the short-circuit current are necessary to ensure a safe and long-lived system.

NEC APPENDIX C

GRID-CONNECTED SYSTEMS

Grid-connected systems present some unique problems for the PV designer and installer in meeting the *NEC*.

Inverters

Some of the grid-tied inverters that are available do not currently meet *UL Standard 1741* for inverters. Some of the inverters cannot have both the dc PV circuits and the ac output circuits grounded without causing parallel ground current paths. Newer versions of these inverters may have solutions for this problem.

Other inverters have the internal circuitry tied to the case and force the central grounding point to be at the inverter input terminals. In some installations, this design is not compatible with ground-fault equipment and does not provide the flexibility needed for maximum surge suppression.

PV Source-Circuit Conductors

Some older grid-tied inverters operate with PV arrays that are center tapped and have cold-temperature open-circuit voltages of ±325 volts and above. The system voltage of 650 volts or greater exceeds the insulation rating of the commonly available 600-volt insulated conductors. Each disconnect and overcurrent device and the insulation of the wiring **must** have a voltage rating exceeding the system voltage rating. Type G and W cables are available with the higher voltage ratings, but are flexible cords and do not meet *NEC* requirements for fixed installations. Cables suitable for *NEC* installations requiring insulation greater than 600 volts are available (Appendix A).

Other older inverters have been designed to operate on systems with open-circuit voltages exceeding ±540 volts requiring conductors with 2000-volt or higher insulation. See Appendix D for a full discussion of this area.

Overcurrent Devices

When *UL* tests and lists fuses for dc operation, the voltage rating is frequently one-half the ac voltage rating. This results in a 600-volt ac fuse rated for 300-volt dc. Finding fuses with high enough dc ratings for grid systems operating at ±300 volts (600-volt system voltage) and above will pose problems. There are a limited number of listed, dc-rated 600 volt fuses available. See Appendix A.

Although not *UL*-Listed, Heinemann Electric Company (Appendix A) can series connect poles of dc-rated circuit breakers to obtain 750-volt ratings. Square D and others have similar products.

Circuit breakers that are "back fed" for any application (but particularly for utility interactive inverter connection to the grid) **must** be identified (in the listing) for such use and **must** be fastened in place with a screw or other additional clamp [690-64b(5), 384-16(f)].

Disconnects

In addition to the Heinemann circuit breaker mentioned above, manufacturers such as GE, Siemens, and Square D may certify their switches for higher voltage when the poles are connected in series.

Blocking Diodes

Blocking diodes are not required by the *NEC* and their use is rapidly declining.

Blocking diodes are not overcurrent devices. They block reverse currents in direct-current circuits and help to control circulating ground-fault currents if used in both ends of high-voltage strings. Lightning induced surges are tough on diodes. If isolated case diodes are used, at least 3500 volts of insulation is provided between the active elements and the normally grounded heat sink. Choosing a peak reverse voltage as high as is available, but at least twice the PV open-circuit voltage, will result in longer diode life. Substantial amounts of surge suppression will also improve diode longevity.

Blocking diodes may not be substituted for the *UL-1703* requirement for module protective fuses in each series-connected string of modules.

Surge Suppression

Surge suppression is covered only lightly in the *NEC* because it affects performance more than safety and is mainly a utility problem at the transmission line level in ac systems [280]. PV arrays mounted in the open, on the tops of buildings, act like lightning rods. The PV designer and installer **must** provide appropriate means to deal with lightning-induced surges coming into the system.

Array frame grounding conductors should be routed directly to supplemental ground rods located as near as possible to the arrays.

Metal conduit will add inductance to the array-to-building conductors and slow down any induced surges as well as provide some electromagnetic shielding.

Metal oxide varistors (MOV) commonly used as surge suppression devices on electronic equipment have several deficiencies. They draw a small amount of current continually. The clamping voltage lowers as they age and may reach the open-circuit voltage of the system. When they fail, they fail in the shorted mode, heat up, and frequently catch fire. In many installations, the MOVs are protected with fast acting fuses to prevent further damage when they fail, but this may limit their effectiveness as surge suppression devices. Other electronic devices are available that do not have these problems.

Silicon Oxide surge arrestors do not draw current when they are off. They fail open circuited when overloaded and, while they may split open on overloads, they rarely catch fire. They are not normally protected by fuses and are rated for surge currents up to 100,000 amps. They are rated at voltages of 300 volts and higher and are available from electrical supply houses or Delta Lightning Arrestors, Inc. (Appendix A).

Several companies specialize in lightning protection equipment, but much of it is for ac systems. Electronic product directories, such as the *Electronic Engineers Master Catalog* should be consulted.

NEC APPENDIX D

CABLE AND DEVICE RATINGS AT HIGH VOLTAGES

There is a concern in designing PV systems that have system open-circuit voltages above 600 volts. The concern has two main issues—device ratings and *NEC* limitations.

Equipment Ratings

Some utility-intertie inverters operate with a grounded, bipolar (three-wire) PV array. In a bipolar PV system, where each of the monopoles is operated in the 220-235-volt peak-power range, the open-circuit voltage can be anywhere from 290 to 380 volts, and above, depending on the module characteristics such as fill factor. Such a bipolar system can be described as a 350/700-volt system (for example) in the same manner that a 120/240-volt ac system is described. This method of describing the system voltage is consistent throughout the electrical codes used not only in residential and commercial power systems, but also in utility practice.

In all systems, the voltage ratings of the cable, switchgear, and overcurrent devices are based on the higher number of the pair (i.e., 700 volts in a 350/700-volt system). That is why 250-volt switchgear and overcurrent devices are used in 120/240-volt ac systems and 600-volt switchgear is used in systems such as the 277/480-volt ac system. Note that it is not the voltage to ground, but the higher line-to-line voltage that defines the equipment voltage requirements.

The *National Electrical Code* (*NEC*) defines a nominal voltage for ac systems (120, 240, etc.) and acknowledges that some variation can be expected around that nominal voltage. Such a variation around a nominal voltage is not considered in dc PV systems, and the *NEC* requires that a temperature-related connection factor on the open-circuit array voltage **must** be used. The open-circuit voltage is defined at Standard Test Conditions (STC) because of the relationship between the *UL* Standards and the way the *NEC* is written. The *NEC* Handbook elaborates on the definition of "circuit voltage," but this definition may not apply to current-limited dc systems. Section 690-7(a) of the *NEC* requires that the voltage used for establishing dc circuit requirements in PV systems be the computed open-circuit voltage for crystalline PV technologies. In new thin-film PV technologies, open-circuit voltages are determined from manufacturers' specifications for temperature coefficients.

The 1999 *NEC* specifically defines the PV system voltage as the product of a temperature-dependent factor (that may reach 1.25 at -40°C) and the STC open-circuit voltage [690-7]. The systems voltage is also defined as the highest voltage between any two wires in a 3-wire (bipolar) PV system [690-2].

The comparison to ac systems can be carried too far; there are differences. For example, the typical wall switch in a 120/240-volt ac residential or commercial system is rated at only 120 volts, but such a switch in a 120/240-volt dc PV system would have to be rated at 240 volts. The inherent differences between a dc current source (PV modules) and a voltage source (ac grid) bear on this issue. Even the definitions of circuit voltage in the *NEC* and *NEC* Handbook refer to ac and dc systems, but do not take into account the design of the balance of systems required in current-limited PV systems. In a PV system, all wiring, disconnects, and overcurrent devices have current ratings that exceed the short-circuit currents by at least 25%. In the case of bolted faults or ground faults involving currents from the PV array, the overcurrent devices do not trip because they are rated to withstand continuous operation at levels above the fault levels. In an ac system, bolted faults and ground faults generally cause the overcurrent devices to trip or blow removing the source of voltage from the fault. Therefore, the faults that pose high-voltage problems in PV, dc systems cause the voltage to be removed in ac, grid-supply systems. For these reasons, a switch rated at 120 volts can be used in an ac system with voltages up to 240 volts, but in a dc, PV system, the switch would have to be rated at 240 volts.

Another consideration that we are dealing with is the analogy of dc supply circuit and ac load circuits. An analysis of ac supply circuits would be similar to dc supply circuits.

Underwriters Laboratories (UL) Standard 1703 requires that manufacturers of modules listed to the standard include, in the installation instructions, a statement that the open-circuit voltage should be multiplied by 125% (crystalline cells), further increasing the voltage requirement of the Balance of Systems (BOS) equipment. This requirement is now in the 1999 *NEC* Section 690-7 as a temperature-dependent constant.

Current PV modules that are listed to the *UL Standard 1703* are listed with a maximum system voltage of 600 volts. Engineers caution all installers, factory and otherwise, to not exceed this voltage. This restriction is not modified by the fact that the modules undergo high-pot tests at higher voltages. *UL Standard 1703* allows modules to be listed up to 1000 volts.

Although not explicitly stated by the *NEC*, it is evident that the intent of the Code and the *UL* Standards is that all cable, switches, fuses, circuit breakers, and modules in a PV system be rated for the maximum system voltage. This is clarified in the 1999 *NEC* [690-7(a)].

While reducing the potential for line-to-line faults, the practice of wiring each monopole (one of two electrical source circuits) in a separate conduit to the inverter does not eliminate the problem. Consider the bipolar system presented in Figure D-1 with a bolted fault (or deliberate short) from the negative to the positive array conductor at the input of the inverter. With the switches closed, array short-circuit current flows, and neither fuse opens.

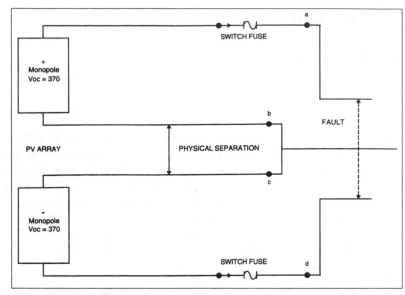

Figure D-1. Typical Bipolar System with Fault

Now consider what happens in any of the following cases.

1. A switch is opened
2. A fuse opens
3. A wire comes loose in a module junction box
4. An intercell connection opens or develops high resistance
5. A conductor fails at any point

In any of these cases, the entire array voltage (740 volts) stresses the device where the circuit opens. This voltage (somewhere between zero at short-circuit and the array open-circuit voltage) will appear at the device or cable. As the device starts to fail, the current through it goes from Isc to zero as the voltage across the device goes from zero to Voc. This process is very conducive to sustained arcs and heating damage.

Separating the monopoles does not avoid the high-voltage stress on any component, but it does help to minimize the potential for some faults.

There are other possibilities for faults that will also place the same total voltage on various components in the system. An improperly installed grounding conductor coupled with a module ground fault could result in similar problems.

Section 690-5 of the *NEC* requires a ground-fault device on PV systems that are installed on the roofs of dwellings. This device, used for fire protection, **must** detect the fault, interrupt the fault current, indicate the fault, and disconnect the array.

Some large (100 kW) grid-connected PV systems like the one at Juana Diaz, Puerto Rico, have inverters that, when shut down, crowbar the array. The array remains crowbarred until the ac power is shut off and creates a similar fault to the one pictured in Figure D-1.

NEC Voltage Limitation

The second issue associated with this concern is that the *NEC* in Section 690-7(c) only allows PV installations up to 600 volts in one and two-family dwellings. Inverter and system design issues may favor higher system voltage levels.

Voltage Remedies

System designers can select inverters with lower operating and open-circuit voltages. Utility-intertie inverters are available with voltages as low as 24 volts. The system designer also can work with the manufacturers of higher voltage inverters to reduce the number of modules in each series string to the point where the cold-temperature open-circuit voltage is less than 600 volts. The peak-power voltage would also be lowered. Transformers may be needed to raise the ac voltage to the required level. At least one inverter manufacturer has pursued this option and is offering inverters which can operate with arrays that have open-circuit voltages of less than 600 volts.

Cable manufacturers produce *UL*-Listed, cross-linked polyethylene, single-conductor cable. It is marked USE-2/RHW-2, Sunlight Resistant and is rated at 2000 volts. This cable could be used for module interconnections in conduit after all of the other *NEC* requirements are met for installations above 600 volts.

Several manufacturers issue factory certified rating on their three-pole disconnects to allow higher voltage, non-load break operation with series-connected poles. The *NEC* will require an acceptable method of obtaining non-load break operation.

Some OEM circuit breaker manufacturers will factory certify series-connected poles on their circuit breakers. Units have been used at 750 volts and 100 amps with 10,000 amps of interrupt rating. Higher voltages may be available.

High-voltage industrial fuses are available, but dc ratings are unknown at this time.

Individual 600-volt terminal blocks can be used with the proper spacing for higher voltages.

Module manufacturers can have their modules listed for higher system voltages. Most are currently limited to 600 volts.

Power diodes may be connected across each monopole. When a bolted line-to-line fault occurs, one of the diodes will be forward biased when a switch or fuse opens, thereby preventing the voltage from one monopole from adding to that of the other monopole. The diodes are mounted across points a-b and c-d in Figure D-1. Each diode should be rated for at least the system open-circuit voltage and the full short-circuit current from one monopole. Since diodes are not listed as over-voltage protection devices, this solution is not recognized in the *NEC*.

The *NEC* allows PV installations over 600 volts in non-residential applications, which will cover the voltage range being used in most current designs.

It should be noted that there are numerous requirements throughout the *NEC* that apply specifically to installations over 600 volts:

- All equipment **must** be listed for the maximum voltage.
- Clearance distances and mechanisms for achieving that clearance are significantly more stringent as voltages increase above 600 volts.

NEC APPENDIX E

EXAMPLE SYSTEMS

The systems described in this appendix and the calculations shown are presented as examples only. The calculations for conductor sizes and the ratings of overcurrent devices are based on the requirements of the 1999 *National Electrical Code* (*NEC*) and on *UL Standard 1703* which provides instructions for the installation of *UL*-Listed PV modules. Local codes and site-specific variations in irradiance, temperature, and module mounting, as well as other installation particularities, dictate that these examples should not be used without further refinement. Tables 310-16 and 310-17 from the *NEC* provide the ampacity data and temperature derating factors.

Cable Sizing and Overcurrent Protection

The procedure presented below for cable sizing and overcurrent protection of that cable is based on *NEC* requirements in Sections 690-9, 690-8, 210-20(a), 215-2, 215-3, 220-10, 240-3(b), and 240-6(a).

1. Circuit Current. For circuits carrying currents from PV modules, multiply the short-circuit current by 125% and use this value for all further calculations. For dc and ac inverter circuits in PV systems, use the rated continuous currents. AC and dc load circuits should follow the requirements of Sections 210, 220, and 215. For PV circuits in the following examples, this is called the PV 125% calculation. In the 1999 *NEC*, this requirement has been included in Section 690-8, but may also remain in *UL* 1703. Do not apply this multiplier twice.

2. Overcurrent Device Rating. The overcurrent device **must** be rated at 125% of the current determined in Step 1. This is to prevent overcurrent devices from being operated at more than 80% of rating. This calculation, in the following examples, is called the *NEC* 125%.

3. Cable Sizing. Cables **shall** have a 30°C ampacity of 125% of the current determined in Step 1 to ensure proper operation of connected overcurrent devices. There are no additional deratings applied with this calculation.

4. Cable Derating. Based on the determination of Step 3 and the location of the cable (raceway or free-air), a cable size and insulation temperature rating (60, 75, or 90°C) are selected from the *NEC* Ampacity Tables 310-16 or 310-17. Use the 75°C cable ampacities to get the size, then use the ampacity from the 90°C column—if needed—for the deratings. This cable is then derated for temperature, conduit fill, and other requirements. The resulting derated ampacity **must** be greater than the value found in Step 1. If not greater, then a larger cable size or higher insulation temperature **must** be selected. The current in Step 3 *is not* used at this point to preclude oversizing the cables.

5. Ampacity vs. Overcurrent Device. The derated ampacity of the cable selected in Step 4, **must** be equal to or greater than the overcurrent device rating determined in Step 2. If the derated ampacity of the cable is less than the rating of the overcurrent device, then a larger cable **must** be selected. The next larger standard size overcurrent device may be used if the derated cable ampacity falls between the standard overcurrent device sizes found in *NEC* Section 240-6.

Note: This step may result in a larger conductor size than that determined in Step 4.

6. Device Terminal Compatibility. Since most overcurrent devices have terminals rated for use with 75°C (or 60°C) cables, compatibility **must** be verified. If a 90°C-insulated cable was selected in the above process, the 30°C ampacity of the same size cable with a 75°C (or 60°C) insulation **must** be greater than or equal to the current found in Step 1. This ensures that the cable will operate at temperatures below the temperature rating of the terminals of the overcurrent device. If the overcurrent device is located in an area with ambient temperature higher than 30°C, then the 75°C (or 60°C) ampacity **must** also be derated.

Here is an example of how the procedure is used:

The task is to size and protect two PV source circuits in conduit, each with an Isc = 40 amps. Four current-carrying conductors are in the conduit and are operating in a 45°C ambient temperature.

Step 1: 1.25 x 40 = 50 amps. (PV 125%)

Step 2: The required fuse (with 75°C terminals) is 1.25 x 50 = 62.5 amps. The next standard fuse size is 70 amps. (*NEC* 125%)

Step 3: Same calculation as Step 2. Cable ampacity without deratings **must** be 62.5 amps.

Step 4: From Table 310-16, cables with 75°C insulation: A 6 AWG conductor at 65 amps is needed. This meets Step 3 requirements. Plan on installing a 6 AWG XHHW-2 cable with 90°C insulation and a 30°C ampacity of 75 amps. Conduit fill derate is 0.8 and temperature derate is 0.87. Derated ampacity is 52.2 amps (75 x 0.8 x 0.87). This is greater than the required 50 amps in Step 1 and meets the requirement.

Step 5: It is acceptable to protect a cable with a derated ampacity of 52.2 amps with a 60-amp overcurrent device since this is the next larger standard size. However, this circuit requires at least a 62.5 amp device (Step 2). Therefore, the conductor **must** be increased to a 4 AWG conductor with a derated ampacity of 66 amps (95 x 0.87 x0.8). A 70-amp fuse is acceptable to protect this cable since it is the next larger standard size.

Step 6. The ampacity of a 4 AWG cable with 75°C insulation (because the fuse has 75°C terminals) is 85 amps, and is higher than the calculated circuit current of 50 amps found in Step 1. Using the 75°C column in Table 310-16 or 310-17 for starting Step 4 usually ensures that this check will be passed.

EXAMPLE 1   Direct-Connected Water Pumping System

Array Size: 4, 12-volt, 60-watt modules   Isc = 3.8 amps, Voc = 21.1 volts
Load: 12-volt, 10-amp motor

Description

The modules are mounted on a tracker and connected in parallel. The modules are wired as shown in Figure E-1 with 10 AWG USE-2 single-conductor cable. A large loop is placed in the cable to allow for tracker motion without straining the rather stiff building cable. The USE-2 cable is run to a disconnect switch in an enclosure mounted on the pole. From this disconnect enclosure, 8 AWG XHHW-2 cable in electrical non-metallic conduit is routed to the well head. The conduit is buried 18 inches deep. The 8 AWG cable is used to minimize voltage drop.

The *NEC* requires the disconnect switch. Because the PV modules are current limited and all conductors have an ampacity greater than the maximum output of the PV modules, no overcurrent device is required, although some inspectors might require it and it might serve to provide some degree of lightning protection. A dc-rated disconnect switch or a dc-rated fused disconnect **must** be used. Since the system is ungrounded, a two-pole switch **must** be used. All module frames, the disconnect enclosure, and the pump housing **must** be grounded, whether the system is grounded or not.

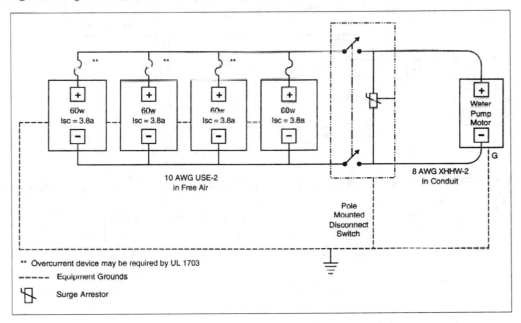

Figure E-1. Direct-Connected System

Calculations

The array short-circuit current is 15.2 amps (4 x 3.8).
PV 125%: 1.25 x 15.2 = 19 amps (Step 1)
No fuse, no Step 2
NEC 125%: 1.25 x 19 = 23.75 amps (Step 3)
The ampacity of 10 AWG USE-2 at 30°C is 55 amps.
The ampacity at 61-70°C is 31.9 amps (0.58 x 55) which is more than the 19 amp requirement. (Step 4)
The equipment grounding conductors should be 10 AWG.
The minimum voltage rating of all components is 26 volts (1.25 x 21.1).

EXAMPLE 2 Water Pumping System with Current Booster

Array Size: 10, 12-volt, 53-watt modules I_{sc} = 3.4 amps, V_{oc} = 21.7 volts
Current Booster Output: 90 amps
Load: 12-volt, 40-amp motor

Description

This system has a current booster before the water pump and has more modules than in Example 1. Initially, 8 AWG USE-2 cable was chosen for the array connections, but a smaller cable was chosen to attach to the module terminals. As the calculations below show, the array was split into two subarrays. There is potential for malfunction in the current booster, but it does not seem possible that excess current can be fed back into the array wiring, since there is no other source of energy in the system. Therefore, these conductors do not need overcurrent devices if they are sized for the entire array current. If smaller conductors are used, then overcurrent devices will be needed.

Since the array is broken into two subarrays, the maximum short-circuit current available in either subarray wiring is equal to the subarray short-circuit current under fault conditions. Overcurrent devices are needed to protect the subarray conductors under these conditions.

A grounded system is selected, and only one-pole disconnects are required. Equipment grounding and system grounding conductors are shown in Figure E-2

If the current booster output conductors are sized to carry the maximum current (3-hour) of the booster, then overcurrent devices are not necessary, but again, some inspectors may require them.

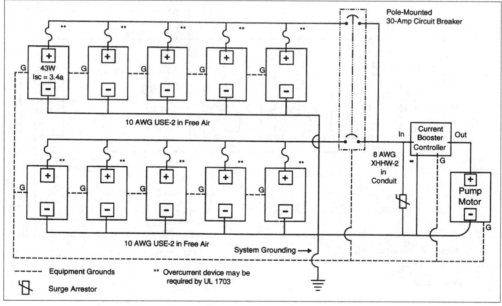

Figure E-2. Direct-Connected PV System with Current Booster

Calculations

The array short-circuit current is 34 amps (10 x 3.4).

PV 125%: 1.25 x 34 = 42.5 amps

NEC 125%: 1.25 x 42.5 = 53.1 amps

The ampacity of 8 AWG USE-2 cable at 30°C in free air is 80 amps.

The ampacity at 61-70°C is 46.4 amps (0.58 x 80), which is more than the 42.5 amp requirement, and a single array could be used. However, the array is split into two subarrays for serviceability. Each is wired with 10 AWG USE-2 conductors.

The subarray short-circuit current is 17 amps (5 x 3.4).

PV 125%: 1.25 x 17 = 21.3 amps

NEC 125%: 1.25 x 21.25 = 26.6 amps

The ampacity of 10 AWG USE-2 at 30°C in free air is 55 amps.

The ampacity at 61-70°C is 31.9 amps (0.58 x 55), which is more than the 21.3 amp requirement. Since this cable is to be connected to an overcurrent device with terminals rated at 60°C or 75°C, the ampacity of the cable **must** be evaluated with 60°C or 75°C insulation. Overcurrent devices rated at 100 amps or less may have terminals rated at only 60°C. 10 AWG 75°C cable operating at 30°C has an ampacity of 35 amps, which is more than the 21.3 amps requirement. Therefore, there are no problems with the terminals on a 75°C overcurrent device.

Thirty-amp circuit breakers are used to protect the 10 AWG subarray conductors. The required rating is 1.25 x 21.25 = 26.6 amps, and the next largest size is 30 amps.

The current booster maximum current is 90 amps.

The current booster average long-term (3 hours or longer) current is 40 amps.

NEC 125%: 1.25 x 40 = 50 amps

The ampacity of 8 AWG XHHW-2 at 30°C in conduit is 55 amps.

The ampacity at 36-40°C is 50 amps (0.91 x 55), which meets the requirements but may not meet the overcurrent device connection requirements when used.

The 8 AWG conductors are connected to the output of the circuit breakers, and there is a possibility that heating of the breaker may occur. It is therefore good practice to make the calculation for terminal overheating. The ampacity of a 8 AWG conductor evaluated with 75°C insulation (the maximum temperature of the terminals on the overcurrent device) is 50 amps, which is greater than the 40-amp requirement. This means that the overcurrent device will not be subjected to overheating when the 8 AWG conductor carries 40 amps.

All equipment-grounding conductors should be 10 AWG. The grounding electrode conductor should be 8 AWG or larger.

Minimum voltage rating of all components: 1.25 x 21.7 = 27 volts

EXAMPLE 3 Stand-Alone Lighting System

Array Size: 4, 12-volt, 64-watt modules Isc = 4.0 amps, Voc = 21.3 volts

Batteries: 200-amp-hours at 24 volts

Load: 60 watts at 24 volts

Description

The modules are mounted at the top of a 20-foot pole with the metal-halide lamp. The modules are connected in series and parallel to achieve the 24-volt system rating. The lamp, with an electronic ballast and timer/controller, draws 60 watts at 24 volts. The batteries, disconnect switches, charge controller, and overcurrent devices are mounted in a box at the bottom of the pole. The system is grounded as shown in Figure E-3.

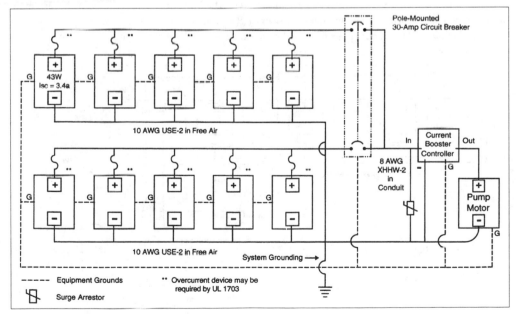

Figure E-3. Stand-Alone Lighting System

Calculations

The array short-circuit current is 8 amps (2 x 4).

PV 125%: 1.25 x 8 = 10 amps

NEC 125%: 1.25 x 10 = 12.5 amps

Load Current: 60/24 = 2.5 amps

NEC 125%: 1.25 x 2.5 = 3.1 amps

Cable size 10 AWG USE-2/RHW-2 is selected for module interconnections and is placed in conduit at the modules and then run down the inside of the pole.

The modules operate at 61-70°C, which requires that the module cables be temperature derated. Cable 10 AWG USE-2/RHW-2 has an ampacity of 40 amps at 30°C in conduit. The derating factor is 0.58. The temperature derated ampacity is 23.2 amps (40 x 0.58), which exceeds the 10-amp requirement. Checking the cable with a 75°C insulation, the ampacity at the fuse end is 35 amps, which exceeds the 10-amp requirement. This cable can be protected by a 15-amp fuse or circuit breaker (125% of 10 is 12.5). An overcurrent device rated at 100 amps or less may only have terminals rated for 60°C, not the 75°C used in this example. Lower temperature calculations may be necessary.

The same USE-2/RHW-2, 10 AWG cable is selected for all other system wiring, because it has the necessary ampacity for each circuit.

A three-pole fused disconnect is selected to provide the PV and load disconnect functions and the necessary overcurrent protection. The fuse selected is a RK-5 type, providing current-limiting in the battery circuits. A pull-out fuse holder with either Class RK-5 or Class T fuses could also be used for a more compact installation. Fifteen amp fuses are selected to provide overcurrent protection for the 10 AWG cables. They are used in the load circuit and will not blow on any starting surges drawn by the lamp or controller. The 15-amp fuse before the charge controller could be eliminated since that circuit is protected by the fuse on the battery side of the charge controller. The disconnect switch at this location is required.

The equipment grounding conductors should be 10 AWG conductors. An 8 AWG conductor would be needed to for the ground rod.

The dc voltage ratings for all components used in this system should be at least 53 volts (2 x 21.3 x 1.25).

EXAMPLE 4 Remote Cabin DC-Only System

Array Size: 6, 12-volt, 75-watt modules   Isc = 4.8 amps, Voc = 22 volts

Batteries: 700 amp hours at 12 volts

Load: 75 watts peak at 12-volts dc

Description

The modules are mounted on a rack on a hill behind the house. Non-metallic conduit is used to run the cables from the module rack to the control panel. A disconnect and control panel are mounted on the back porch, and the batteries are in an insulated box under the porch. All the loads are dc with a peak combined power of 75 watts at 12 volts due, primarily, to a pressure pump on the gravity-fed water supply. The battery bank consists of four 350-amp-hour, 6-volt, deep-cycle batteries wired in series and parallel. Figure E-4 shows the system schematic.

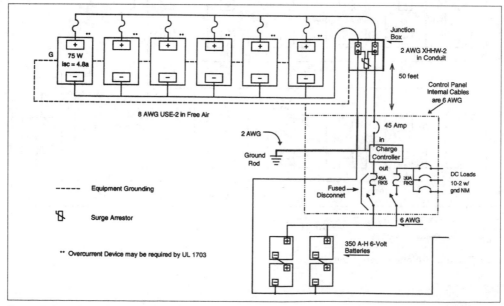

Figure E-4. Remote Cabin DC-Only System

Calculations

The array short-circuit current is 28.8 amps (6 x 4.8).

PV 125%: 1.25 x 28.8 = 36 amps

NEC 125%: 1.25 x 36.0 = 45 amps

The module interconnect wiring and the wiring to a rack-mounted junction box will operate at 65°C. If USE-2 cable with 90°C insulation is chosen, then the temperature derating factor will be 0.58. The required ampacity of the cable at 30°C is 62 amps (36/0.58), which can be handled by 8 AWG cable with an ampacity of 80 amps in free air at 30°C. Conversely, the ampacity of the 8 AWG cable is 46.4 amps (80 x 0.58) at 65°C which exceeds the 36 amp requirement.

From the rack-mounted junction box to the control panel, the conductors will be in conduit and exposed to 40°C temperatures. If XHHW-2 cable with a 90°C insulation is selected, the temperature derating factor is 0.91. The required ampacity of the cable at 30°C would be 36/0.91 = 39.6 amps in conduit. Cable size 8 AWG has an ampacity of 55 amps at 30°C in conduit. Conversely, the 8 AWG conductor has an ampacity of 50 amps (55 x 0.91) at 40°C in conduit which exceeds the 39.6 amp requirement at this temperature.

The 8 AWG cable, evaluated with a 75°C insulation, has an ampacity at 30°C of 50 amps, which is greater than the 36 amps that might flow through it on a daily basis.

The array is mounted 200 feet from the house, and the round trip cable length is 400 feet. A calculation of the voltage drop in 400 feet of 8 AWG cable operating at 36 amps (125% Isc) is 0.778 ohms per 1000 feet x 400 / 1000 x 36 = 11.2 volts. This represents an excessive voltage drop on a 12-volt system, and the batteries cannot be effectively charged. Conductor size 2 AWG (with a voltage drop of 2.8 volts) was substituted; this substitution is acceptable for this installation. The conductor resistances are taken from Table 8 in Chapter 9 of the *NEC* and are given for conductors at 75°C.

The PV conductors are protected with a 45-amp (1.25 x 36) single-pole circuit breaker on this grounded system. The circuit breaker should be rated to accept 2 AWG conductors rated at 75°C.

Cable size 6 AWG THHN cable is used in the control center and has an ampacity of 95 amps at 30°C when evaluated with 75°C insulation. Wire size 2 AWG from the negative dc input is used to the point where the grounding electrode conductor is attached instead of the 6 AWG conductor used elsewhere to comply with grounding requirements.

The 75 watt peak load draws about 6.25 amps and 10-2 with ground (w/gnd) nonmetallic sheathed cable was used to wire the cabin for the pump and a few lights. DC-rated circuit breakers rated at 20 amps were used to protect the load wiring, which is in excess of the peak load current of 7.8 amps (1.25 x 6.25) and less than the cable ampacity of 30 amps.

Current-limiting fuses in a fused disconnect are used to protect the dc-rated circuit breakers, which do not have an interrupt rating sufficient to withstand the short-circuit currents from the battery under fault conditions. RK-5 fuses were chosen with a 45-amp rating in the charge circuit and a 30-amp rating in the load circuit. The fused disconnect also provides a disconnect for the battery from the charge controller and the dc load center.

The equipment grounding conductors should be 10 AWG and the grounding electrode conductor should be 2 AWG.

All components should have a voltage rating of at least 1.25 x 22 = 27.5 volts.

EXAMPLE 5 Small Residential Stand-Alone System

Array Size: 10, 12-volt, 51-watt modules   Isc = 3.25 amps, Voc = 20.7 volts

Batteries: 800 amp-hours at 12 volts

Loads: 5 amps dc and 500-watt inverter with 90% efficiency

Description

The PV modules are mounted on the roof. Single conductor cables are used to connect the modules to a roof-mounted junction box. UF two-conductor sheathed cable is used from the roof to the control center. Physical protection (wood barriers or conduit) for the UF cable is used where required. The control center, diagrammed in Figure E-5, contains disconnect and overcurrent devices for the PV array, the batteries, the inverter, and the charge-controller.

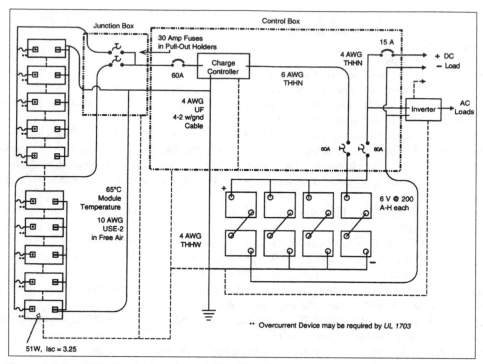

Figure E-5. Small Residential Stand-Alone System

Calculations

The module short-circuit current is 3.25 amps.

PV 125%: 1.25 x 3.25 = 4.06 amps

NEC 125%: 1.25 x 4.06 = 5.08 amps per module

The module operating temperature is 68°C.

The derating factor for USE-2 cable is 0.58 at 61-70°C.

Cable 14 AWG has an ampacity at 68°C of 20.3 amps (0.58 x 35) (max fuse is 15 amps—see notes at bottom of Tables 310-16 & 17).

Cable 12 AWG has an ampacity at 68°C of 23.2 amps (0.58 x 40) (max fuse is 20 amps).

Cable 10 AWG has an ampacity at 68°C of 31.9 amps (0.58 x 55) (max fuse is 30 amps).

Cable 8 AWG has an ampacity at 68°C of 46.4 amps (0.58 x 80).

The array is divided into two five-module subarrays. The modules in each subarray are wired from module junction box to module junction box and then to the array junction box. Cable size 10 AWG USE-2 is selected for this wiring, because it has an ampacity of 31.9 amps under these conditions, and the requirement for each subarray is 5 x 4.06 = 20.3 amps. Evaluated with 75°C insulation, a 10 AWG cable has an ampacity of 35 amps, which is greater than the actual requirement of 20.3 amps (5 x 4.06). In the array junction box on the roof, two 30-amp fuses in pull-out holders are used to provide overcurrent protection for the 10 AWG conductors. These fuses meet the requirement of 25.4 amps (125% of 20.3) and have a rating less than the derated cable ampacity.

In this junction box, the two subarrays are combined into an array output. The ampacity requirement is 40.6 amps (10 x 4.06). A 4 AWG UF cable (4-2 w/gnd) is selected for the run to the control box. It operates in an ambient temperature of 40°C and has an ampacity of 57.4 amps (70 x 0.82). This is a 60°C cable with 90°C conductors. Care **must** be used when connecting to fuses that are rated for use only with 75°C conductors.

A 60-amp circuit breaker in the control box serves as the PV disconnect switch and overcurrent protection for the UF cable. The *NEC* allows the next larger size; in this case, 60 amps, which is over the 57 amps ampacity of the cable. Two single-pole, pull-out fuse holders are used for the battery disconnect. The charge circuit fuse is a 60-amp RK-5 type.

The inverter has a continuous rating of 500 watts at the lowest operating voltage of 10.75 volts and an efficiency of 90% at this power level. The ampacity requirement of the input circuit is 64.6 amps (500 / 10.75 / 0.90) x 1.25).

The cables from the battery to the control center **must** meet the inverter requirements of 64.6 amps plus the dc load requirements of 6.25 amps (1.25 x 5). A 4 AWG THHN has an ampacity of 85 amps when placed in conduit and evaluated with 75°C insulation. This exceeds the requirements of 71 amps (64.6 + 6.25). This cable can be used in the custom power center and be run from the batteries to the inverter.

The discharge-circuit fuse **must** be rated at least 71 amps. An 80-amp fuse should be used, which is less than the cable ampacity.

The dc-load circuit is wired with 10 AWG NM cable (ampacity of 30 amps) and protected with a 15-amp circuit breaker.

The grounding electrode conductor is 4 AWG and is sized to match the largest conductor in the system, which is the array-to-control center wiring.

Equipment grounding conductors for the array and the charge circuit can be 10 AWG based on the 60-amp overcurrent devices [Table 250-95]. The equipment ground for the inverter **must** be an 8 AWG conductor.

All components should have at least a dc voltage rating of 1.25 x 20.7 = 26 volts.

EXAMPLE 6 Medium Sized Residential Hybrid System

Array Size: 40, 12-volt, 53-watt modules   Isc = 3.4 amps, Voc = 21.7 volts
Batteries: 1000 amp-hours at 24 volts
Generator: 6 kW, 240-volt ac
Loads: 15 amps dc and 4000-watt inverter, efficiency = .85

Description

The 40 modules (2120 watts STC rating) are mounted on the roof in subarrays consisting of eight modules mounted on a single-axis tracker. The eight modules are wired in series and parallel for this 24-volt system. Five source circuits are routed to a custom power center. Single-conductor cables are used from the modules to roof-mounted junction boxes for each source circuit. From the junction boxes, UF sheathed cable is run to the main power center.

Blocking diodes are not required or used to minimize voltage drops in the system.

A prototype array ground-fault detector provides experimental compliance with the requirements of *NEC* Section 690-5.

The charge controller is a relay type.

DC loads consist of a refrigerator, a freezer, several telephone devices, and two fluorescent lamps. The maximum combined current is 15 amps.

The 4000-watt sine-wave inverter supplies the rest of the house.

The 6-kW natural gas fueled, engine-driven generator provides back-up power and battery charging through the inverter. The 240-volt output of the generator is fed through a 5 kVA transformer to step it down to 120 volts for use in the inverter and the house. The transformer is protected on the primary winding by a 30-amp circuit breaker [450-3(b)(1) Ex-1]. Figure E-6 presents the details.

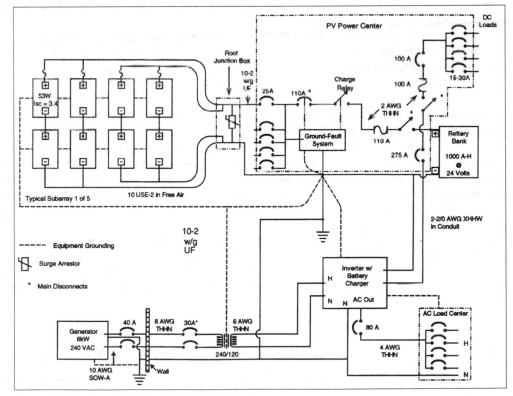

Figure E-6. Medium Sized Residential Hybrid System

Calculations

The subarray short-circuit current is 13.6 amps (4 x 3.4).

PV 125%: 1.25 x 13.6 = 17 amps

NEC 125%: 1.25 x 17 = 21.25 amps

The temperature derating factor for USE-2 cable at 61-70°C is 0.58.

The ampacity of 10 AWG USE-2 cable at 70°C is 31.9 amps (55 x 0.58).

The temperature derating factor for UF cable at 36-40°C is 0.82.

The ampacity of 10-2 w/gnd UF cable at 40°C is 24.6 amps (30 x 0.82). Since the UF cable insulation is rated at 60°C, no further temperature calculations are required when this cable is connected to circuit breakers rated for use with 60 or 75°C conductors.

The source-circuit circuit breakers are rated at 25 amps (requirement is 125% of 17 amps = 21.25).

The PV array short-circuit current is 68 amps (5 x 13.6).

PV 125%: 1.25 x 68 = 85 amps

NEC 125%: 1.25 x 85 = 106 amps

A 110-amp circuit breaker is used for the main PV disconnect after the five source circuits are combined.

A 110-amp RK5 current-limiting fuse is used in the charge circuit of the power center, which is wired with 2 AWG THHN conductors (170 amps with 75°C insulation).

The dc-load circuits are wired with 10-2 w/gnd NM cable (30 amps) and are protected with 20- or 30-amp circuit breakers. A 100-amp RK-5 fuse protects these discharge circuits from excess current from the batteries.

Inverter

The inverter can produce 4000 watts ac at 22 volts with an efficiency of 85%.

The inverter input current ampacity requirements are 267 amps ((4000 / 22 / 0.85) x 1.25). See Appendix F for more details.

Two 2/0 AWG USE-2 cables are paralleled in conduit between the inverter and the batteries. The ampacity of this cable (rated with 75°C insulation) at 30°C is 280 amps (175 x 2 x 0.80). The 0.80 derating factor is required because there are four current-carrying cables in the conduit.

A 275-amp circuit breaker with a 25,000-amp interrupt rating is used between the battery and the inverter. Current-limiting fusing is not required in this circuit.

The output of the inverter can deliver 4000 watts ac (33 amps) in the inverting mode. It can also pass up to 60 amps through the inverter from the generator while in the battery charging mode.

Ampacity requirements, ac output: 60 x 1.25 = 75 amps. This reflects the *NEC* requirement that circuits are not to be operated continuously at more than 80% of rating.

The inverter is connected to the ac load center with 4 AWG THHN conductors in conduit, which have an ampacity of 85 amps when used at 30°C with 75°C overcurrent devices. An 80-amp circuit breaker is used near the inverter to provide a disconnect function and the overcurrent protection for this cable.

Generator

The 6-kW, 120/240-volt generator has internal circuit breakers rated at 27 amps (6500-watt peak rating). The *NEC* requires that the output conductors between the generator and the first field-installed overcurrent device be rated at least 115% of the nameplate rating (6000 / 240) x 1.15 = 28.75 amps). Since the generator is connected through a receptacle outlet, a 10-4 AWG SOW-A portable cord (30 amps) is run to a NEMA 3R exterior circuit breaker housing. This circuit breaker is rated at 40 amps and provides overcurrent protection for the 8 AWG THHN conductors to the transformer. These conductors have an ampacity of 44 amps (50 x 0.88) at 40°C (75°C insulation rating). The circuit breaker also provides an exterior disconnect for the generator. Since the transformer isolates the generator conductors from the system electrical ground, the neutral of the generator is grounded at the exterior disconnect.

A 30-amp circuit breaker is mounted near the PV Power Center in the ac line between the generator and the transformer. This circuit breaker serves as the interior ac disconnect for the generator and is grouped with the other disconnects in the system.

The output of the transformer is 120 volts. Using the rating of the generator, the ampacity of this cable **must** be 62.5 amps (6000 / 120) x 1.25). A 6 AWG THHN conductor was used, which has an ampacity of 65 amps at 30°C (75°C insulation rating).

Grounding

The module and dc-load equipment grounds **must** be 10 AWG conductors. Additional lightning protection will be afforded if a 6 AWG or larger conductor is run from the array frames to ground. The inverter equipment ground **must** be a 4 AWG conductor based on the size of the overcurrent device for this circuit. The grounding electrode conductor **must** be 2-2/0 AWG or a 500 kcmil conductor, unless there are no other conductors connected to the grounding electrode; then this conductor may be reduced to 6 AWG [250-122 Exceptions].

DC Voltage Rating

All dc circuits should have a voltage rating of at least 55 volts (1.25 x 2 x 22).

EXAMPLE 7 Roof-Top Grid-Connected System

Array Size: 24, 50-volt, 240-watt modules

Isc = 5.6

Voc = 62

Inverter: 200-volt nominal dc input

240-volt ac output at 5000 watts with an efficiency of 0.95.

Description

The roof-top array consists of six parallel-connected strings of four modules each. A junction box is mounted at the end of each string which contains a surge arrestor, a blocking diode, and a fuse. All wiring is RHW-2 in conduit. The inverter is located adjacent to the service entrance load center where PV power is fed to the grid through a back-fed circuit breaker. Figure E-7 shows the system diagram.

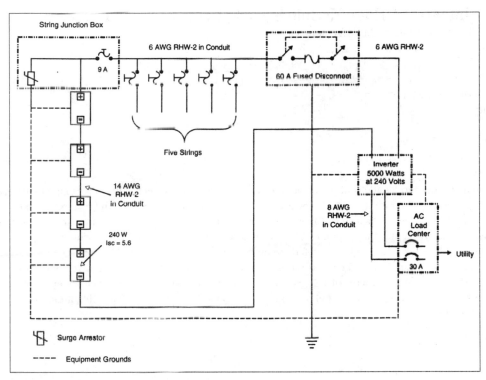

Figure E-7. Roof-Top Grid-Connected System

Calculations

The string short-circuit current is 5.6 amps.

PV 125%: 1.25 x 5.6 = 7 amps

NEC 125%: 1.27 x 7 = 8.75 amps

The array short-circuit current is 33.6 amps (6 x 5.6).

PV 125%: 1.25 x 33.6 = 42 amps

NEC 125%: 1.25 x 42 = 52.5 amps

The modules in each string are connected in series. The conductors operate at 63°C. The temperature derating factor for RHW-2 at this temperature is 0.58. The required 30°C ampacity for this cable is 15 amps (8.75 / 0.58). RHW-2 14 AWG cable has an ampacity of 25 amps with 90°C insulation and 20 amps with 75°C insulation so there is no problem with the end of the cable connected to the fuse since the 7 amps is below either ampacity.

This cable is protected with a 9-amp fuse.

The cable from the string J-Boxes to the main PV disconnect operates at 40°C. The temperature derating factor for RHW-2 with 90°C insulation is 0.91. This yields a 30°C ampacity requirement of 58 amps (52.5 / 0.91). RHW-2 6 AWG meets this requirement with an ampacity of 75 amps (90°C insulation), and a number 6 AWG cable with 75°C insulation has an ampacity of 65 amps, which also exceeds the 42 amp requirement.

Overcurrent protection is provided with a 60-amp fused disconnect. Since the negative dc conductor of the array is grounded, only a single-pole disconnect is needed.

The inverter output current is 21 amps (5000 / 240).

NEC 125%: 1.25 x 21 = 26 amps.

The cable from the inverter to the load center operates at 30°C. Cable size 8 AWG RHW-2 (evaluated with 75°C insulation) has an ampacity of 50 amps.

A back-fed 30-amp, two-pole circuit breaker provides an ac disconnect and overcurrent protection in the load center.

The equipment grounding conductors for this system should be at least 10 AWG conductors. The system grounding electrode conductor should be a 6 AWG conductor.

All dc circuits should have a voltage rating of at least 310 volts (1.25 x 4 x 62).

EXAMPLE 8 Integrated Roof Module System, Grid Connected

Array Size: 192, 12-volt, 22-watt thin-film modules

Isc = 1.8 amps

Vmp = 15.6 volts

Voc = 22 volts

Inverter: ±180-volt dc input

120-volt ac output

4000 watts

95% efficiency

Description

The array is integrated into the roof as the roofing membrane. The modules are connected in center-tapped strings of 24 modules each. Eight strings are connected in parallel to form the array. A blocking diode (required by the manufacturer) is placed in series with each string. Strings are grouped in two sets of four and a series fuse protects the module and string wiring as shown in Figure E-8. The bipolar inverter has the center tap dc input and the ac neutral output grounded. The 120-volt ac output is fed to the service entrance load center (fifty feet away) through a back-fed circuit breaker.

The manufacturer of these thin-film modules has furnished data that show that the maximum Voc under worst-case low temperatures is 24 volts. The multiplication factor of 1.25 on Voc does not apply [690-7(a)]. The design voltage will be 24 x 24 = 576 volts. The module manufacturer has specified (label on module) 5-amp module protective fuses that **must** be installed in each (+ and -) series string of modules.

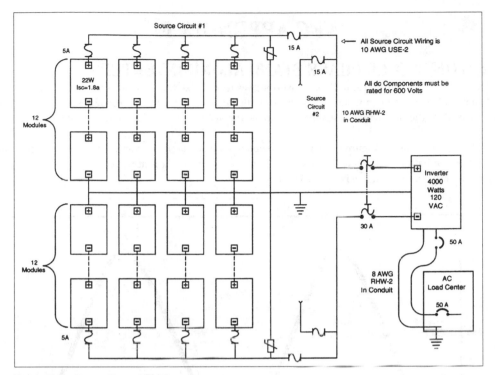

Figure E-8. Center-Tapped PV System

Calculations

Each string short-circuit current is 1.8 amps.

PV 125% (estimated for thin-film modules): 1.25 x 1.8 = 2.25 amps

NEC 125%: 1.25 x 2.25 = 2.8 amps

Each source circuit (4 strings) short-circuit current is 7.2 amps (4 x 1.8).

PV 125%: 1.25 x 7.2 = 9 amps

NEC 125%: 1.25 x 9 = 11.25 amps

The array (two source circuits) short-circuit current is 14.4 amps (2 x 7.2).

PV 125%: 1.25 x 14.4 = 18 amps

NEC 125%: 1.25 x 18 = 22.5 amps

USE-2 cable is used for the module cables and operates at 75°C when connected to the roof-integrated modules. The temperature derating factor in the wiring raceway is 0.41. For the strings, the 30°C ampacity requirement is 5.5 amps (2.25 / 0.41).

Each source circuit conductor is also exposed to temperatures of 75°C. The required ampacity for this cable (at 30°C) is 22.0 amps (9 / 0.41).

Wire size 10 AWG USE-2 is selected for moisture and heat resistance. It has an ampacity of 40 amps at 30°C (90°C insulation) and can carry 35 amps when limited to a 75°C insulation rating. This cable is used for both string and source-circuit wiring. Fifteen-amp fuses are used to protect the string and source-circuit conductors.

The array wiring is inside the building and RHW-2 is used in conduit. It is operated at 50°C when passing through the attic. The temperature derating factor is 0.82, which yields a 30°C ampacity requirement of 22 amps (18 / 0.82). Cable size 10 AWG has an ampacity of 40 amps (90°C insulation) or 35 amps (evaluated with 75°C insulation). Both of these ampacities exceed the 22-amp requirement. Twenty-five-amp fuses are required to protect these cables, but 30-amp fuses are selected for better resistance to surges. Since the inverter has high voltages on the dc-input terminals (charged from the ac utility connection), a pull-out fuse holder is used.

The inverter is rated at 4000 watts at 120 volts and has a 33-amp output current. The ampacity requirement for the cable between the inverter and the load center is 42 amps (4000 / 120) x 1.25) at 30°C. Wire size 8 AWG RHW-2 in conduit connects the inverter to the ac-load center, which is fifty feet away and, when evaluated at with 75°C insulation, has an ampacity of 50 amps at 30°C. A 50-amp circuit breaker in a small circuit-breaker enclosure is mounted next to the inverter to provide an ac disconnect for the inverter that can be grouped with the dc disconnect. Another 50-amp circuit breaker is back-fed in the service entrance load center to provide the connection to the utility.

The modules have no frames and, therefore, no equipment grounding requirements. The inverter and switchgear should have 10 AWG equipment grounding conductors. The system grounding electrode conductor should be a 8 AWG conductor.

All dc components in the system should have a minimum voltage rating of 600 volts (24 x 24 = 576).

NEC APPENDIX F

DC CURRENTS ON SINGLE-PHASE STAND-ALONE INVERTERS

When the sinusoidal ac output current of a stand-alone inverter goes to zero 120 times per second, the input dc current also goes nearly to zero. With a resistive ac load connected to the inverter, the dc current waveform resembles a sinusoidal wave with a frequency of 120 Hz. The peak of the dc current is significantly above the average value of the current, and the lowest value of dc current is near zero.

An example of this is shown in the Figure F1. This is an example of a single-phase stand-alone inverter operating with a 4000-watt resistive load. The input battery voltage is 22 volts. The figure shows the dc current waveform. The measured average dc current is 254 amps. The RMS value of this current is 311 amps.

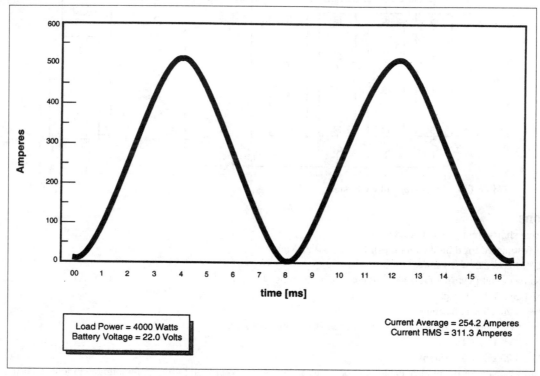

Figure F-1. Inverter Current Waveform (dc side)

The calculated dc current for this inverter (as was done in Example 6 in Appendix E) is 214 amps (4000/22/0.85) when using the manufacturer's specified efficiency of 85%.

The RMS value of current is the value that causes heating in conductors and is the value of current that causes overcurrent devices to trip. In this case, if the inverter were operated at 100% of rated power and at a low battery voltage, the conductors and overcurrent devices would have to be rated to carry 311 amps, not the calculated 214 amps. Code requirements would increase the cable ampacity requirements and overcurrent device ratings to 388 amps (1.25 x 311).

Loads that have inductive components may result in even higher RMS values of dc currents.

The systems designer should contact the inverter manufacturer in cases where it is expected that the inverter may operate at loads approaching the full power rating of the inverter. The inverter manufacturer should provide an appropriate value for the dc input current under the expected load conditions

PRODUCT INDEX

Every product carried in the Sourcebook is listed in this Index by item number. Page number on which the item is found follows the description.

SUBJECT INDEX

Information about products available from Real Goods is indicated either under the sub-entry "listings," or under specific product names. Page numbers follow the description.

NOTES

NOTES

NOTES

Real Goods Renewables: We Are Experienced!

Real Goods has provided reliable solar power systems to over 25,000 homes and villages worldwide since 1978. You don't want your project to be a test bed for unproven technology or system design. You want to go to a source of information, products, and service that are of the highest quality. You want to leave the details to a company that has the experience and capabilities to do the job right the first time. You want a company with experience!

Real Goods has been providing clean, reliable, renewable energy to people all over the planet for almost twenty years! If you have a facility where utility power is unavailable or of poor quality, consider our resources:

◆ *Renewable Energy Systems Design and Installation (solar, wind, and hydro)*

◆ *Large-Scale Uninterruptible Backup Power*

◆ *Solar Architecture*

◆ *High-Efficiency Appliances*

◆ *Biological Waste Treatment Systems*

◆ *Green Building Techniques*

◆ *Power Quality Enhancement*

◆ *Whole Systems Integration*

◆ *Utility Intertie with Renewable Energy*

◆ *Water Quality and Management*

Our eight full-time technicians with decades of combined experience in renewable energy can work with you to make your project a success. Our consultation and experience allow you to make informed decisions on which technologies are best suited for your application. Our solar, wind, and micro-hydro electric products can be coupled with diesel generators to make a hybrid electrical system scaled to your needs, offering the best environmental advantages and cost.

Gaiam Real Goods is the most experienced provider of energy conservation, renewable energy, and other sustain-

Yucatan Peninsula, Mexico: The solar power system for this resort was designed and provided by Real Goods.

able products in the world. Our commitment to excellence has resulted in our being chosen for numerous special projects. Our commitment to each unique situation results in exceptionally high levels of customer satisfaction.

We have helped millions of people become aware of the positive actions they can take to minimize environmental impact. We can introduce you to technologies and techniques that can save you thousands of dollars and free you from utility company power forever. Our experience comes from dealing with people's unusual needs every day.

"I'm writing to say thank you to both you and the staff at Real Goods who made our dream of an independent energy resort a reality. I realize that our time constraint deadline taxed everyone to the maximum. Yet even under the most adverse conditions, including working with a foreign government and regulations, and a limited skilled labor pool, Real Goods never missed a beat.

Your thoroughness, attention to detail and ability to keep all the players "rowing in the same direction" is certainly commendable. I was also very impressed by your on-going support and commitment to being there for us during our precarious start-up period.

It's not every day that an organization invests $2 million dollars in an energy system. It's nice to know that we made the right choice of a supplier.

Thank you from everyone here at The Essene Way, we look forward to seeing you on the beach again."

**Sincerely, Tom Ciola,
Director of the
Essene Way Resort**

101-Kilowatt Photovoltaic Array, The Essene Way Resort, Belize, Central America

Real Goods Renewables provides solar electric power for vital communications on Mt. Everest expeditions and many other "outback" applications.

REAL GOODS®

360 Interlocken Blvd., #300, Broomfield, CO 80021-3492

SLSB11

For credit card orders, call toll-free:
1-800-919-2400

Regular business hours: Monday–Friday: 7:30 am–6 pm (PST)
Saturday: 7:30 am–4 pm (PST)
Order Fax toll-free: 1-800-508-2342   Tech Fax: 707-462-4807
International Fax: 707-468-9486   Overseas Orders: 707-468-9214
Customer Service: 1-800-919-2400   Email: techs@realgoods.com

ORDERED BY:

A Name _____

Address _____

City _____ State _____ Zip _____

DAYTIME PHONE: _____
(In case we have questions about your order)

Email Address _____

ALTERNATIVE SHIPPING OR GIFT ADDRESS

B Name _____

Address _____

City _____ State _____ Zip _____

Message _____

ITEM #	SIZE/COLOR	QTY.	PAGE	DESCRIPTION	CIRCLE ADDRESS	PRICE	MEMBER PRICE (5% off)	TOTAL
00-100				LIFETIME MEMBERSHIP	A B	$50	✕	
					A B			
					A B			
					A B			
					A B			
					A B			
					A B			
					A B			
					A B			
					A B			
					A B			
					A B			
					A B			
					A B			
					A B			
					A B			
					A B			
					A B			
					A B			
					A B			
					A B			
					A B			
					A B			

SHIPPING AND HANDLING CHARGES

Under $25........$5.25	$100–$149.99.......$13.25
$25–$34.99......$7.25	$150–$199.99.......$16.25
$35–$49.99......$8.25	$200–$999.99.......add 8%
$50–$74.99......$9.25	$1000 and up.......add 6%
$75–$99.99....$10.95	

Does not include freight collect items. For Canadian shipments, please call for rates.

***EXPRESS DELIVERY**–Order by 12:30 pm Eastern Standard Time (EST) and for only $6.00 additional, your order will be delivered within 2 business days of the time you placed your order. For next business day delivery, add $14.50 to your shipping charges. These rates apply to all packages up to 10 lbs. and to in-stock merchandise only. Not available on products shipped directly from the manufacturer (◈). Call for rates on heavy or bulky packages, international shipments or deliveries to Alaska, Hawaii, Virgin Islands, or Puerto Rico.

PAYMENT METHOD:

❏ Check #_____

❏ Credit Card (MasterCard, Visa, Discover, or Amex)
    Account Number (*please include all numbers*)

Expiration Date: ☐☐ – ☐☐

Signature _____

WOULD YOU LIKE TO SUPPORT OUR EDUCATION EFFORTS?

Feel free to add a tax-deductible donation to your order and support our nonprofit Institute for Solar Living.

TOTAL OF GOODS _____

SALES TAX
(CA, CO & OH ONLY)
(CA 7%, CO 3%, OH 5.5%) _____

SHIPPING
(each address—
see chart at left) _____

EXPRESS
or Additional
Shipping Charges _____

YES I'd like to make a donation to the Institute _____

TOTAL ENCLOSED
(U.S. dollars only) _____

THANK YOU FOR YOUR ORDER
Prices subject to change.

Ordering Information

Order Toll-Free 800-919-2400
Monday–Friday 7:30 am–6:00 pm (Pacific Time)
Saturday 7:30 am–4:00 pm (Pacific Time)

Fax Toll-Free 800-508-2342
Please include return phone number.

Order Through Our Secure Website
http://www.realgoods.com/renew

Order by Mail
Use attached order form and envelope.

International Orders
Phone Orders: 707-468-9292 ext. 8700
Fax: 707-462-4807
(Please call or fax for shipping quote. Funds must be in U.S. dollars, drawn from a U.S. bank.) We charge $5.00 for catalogs mailed outside the U.S.

Toll-Free Customer Service
If you have a problem with your order, our Customer Service Representatives are available between 8:30 am and 5:00 pm (Pacific Time) Monday through Friday. Call toll-free at 800-919-2400.

Product Information
Our representatives can provide you with detailed information about any of our products. If you need to know more, please call us at 800-919-2400.

Technical Assistance
For technical assistance with a renewable energy system, please call 800-919-2400 or 707-468-9292 ext. 8700, Monday through Friday, 7:30 am–6:00 pm and Saturday, 7:30 am–4:00 pm Pacific Time, or fax 707-462-4807.

Shipping
Allow 7 to 10 working days for delivery, depending on your distance from our Distribution Center in West Chester, Ohio. Overweight or bulky items may require additional shipping charges.

Shipping *(continued)*
There are a few items that ship directly from the manufacturer (they are indicated by a ◆ next to the item number in the catalog). These items are designated on your invoice and are billed at the time of order. Some large items (like refrigerators) are shipped **freight collect** from the manufacturer (they are indicated by a 🚚 next to the item number in the catalog). We will be happy to provide you with a shipping estimate. Allow extra time for delivery of these items.

Price Change
Prices, specifications, and availability are subject to change.

Freight Damage, Returns, and Adjustments
Inspect all shipments upon arrival. If you discover damage, please notify the carrier immediately. Be sure to save all shipping cartons in case you have to return merchandise. If you return merchandise to us, complete the return form on the back of our invoice. Ship to us via UPS or USPS insured. We charge a stocking fee of 10% to 25% for merchandise not in original condition, damaged, or incomplete. Please return items to: Real Goods, 9107 Meridian Way, West Chester, OH 45069-6534. *Please note—if the item you ordered is shipped directly from manufacturer (◈), there may be additional instructions; please call prior to returning.*

Join Our Lifetime Membership Program—Only $50

The Real Goods Lifetime Membership Program is now 65,000 members strong and continues to grow each month. The Lifetime Membership is for people who place high priorities on knowledge about renewable energy, protecting the environment, creating a sustainable future, and getting the best deals on the best equipment. Members are serious about conservation, energy independence, and getting the best value for their dollar!

With the Real Goods Lifetime Membership Program you'll receive:
- A 5% discount on all purchases—including sale items
- All Gaiam Real Goods catalog mailings
- A current copy of the *Solar Living Sourcebook* or other offers

Lifetime Membership $50

REAL GOODS®

360 Interlocken Blvd., #300, Broomfield, CO 80021-3492

SLSB11

For credit card orders, call toll-free:
1-800-919-2400

Regular business hours: Monday–Friday: 7:30 am–6 pm (PST)
Saturday: 7:30 am–4 pm (PST)
Order Fax toll-free: 1-800-508-2342   Tech Fax: 707-462-4807
International Fax: 707-468-9486   Overseas Orders: 707-468-9214
Customer Service: 1-800-919-2400   Email: techs@realgoods.com

ORDERED BY:

A Name _____

Address _____

City _____ State ___ Zip ___

DAYTIME PHONE: _____
(In case we have questions about your order)

Email Address _____

ALTERNATIVE SHIPPING OR GIFT ADDRESS

B Name _____

Address _____

City _____ State ___ Zip ___

Message _____

ITEM #	SIZE/COLOR	QTY.	PAGE	DESCRIPTION	CIRCLE ADDRESS	PRICE	MEMBER PRICE (5% off)	TOTAL
00-100				LIFETIME MEMBERSHIP	A B	$50	✕	
					A B			
					A B			
					A B			
					A B			
					A B			
					A B			
					A B			
					A B			
					A B			
					A B			
					A B			
					A B			
					A B			
					A B			
					A B			
					A B			
					A B			
					A B			
					A B			
					A B			
					A B			
					A B			

SHIPPING AND HANDLING CHARGES

Under $25........$5.25	$100–$149.99.......$13.25
$25–$34.99......$7.25	$150–$199.99.......$16.25
$35–$49.99......$8.25	$200–$999.99.......add 8%
$50–$74.99......$9.25	$1000 and up.......add 6%
$75–$99.99....$10.95	

Does not include freight collect items. For Canadian shipments, please call for rates.

EXPRESS DELIVERY–Order by 12:30 pm Eastern Standard Time (EST) and for only $6.00 additional, your order will be delivered within 2 business days of the time you placed your order. For next business day delivery, add $14.50 to your shipping charges. These rates apply to all packages up to 10 lbs. and to in-stock merchandise only. Not available on products shipped directly from the manufacturer (◈). Call for rates on heavy or bulky packages, international shipments or deliveries to Alaska, Hawaii, Virgin Islands, or Puerto Rico.

PAYMENT METHOD:

❑ Check #_____

❑ Credit Card (MasterCard, Visa, Discover, or Amex)
   Account Number (*please include all numbers*)

Expiration Date: [ ][ ]–[ ][ ]

Signature _____

WOULD YOU LIKE TO SUPPORT OUR EDUCATION EFFORTS?
Feel free to add a tax-deductible donation to your order and support our nonprofit Institute for Solar Living.

TOTAL OF GOODS	
SALES TAX (CA, CO & OH ONLY) (CA 7%, CO 3%, OH 5.5%)	
SHIPPING (each address — see chart at left)	
EXPRESS or Additional Shipping Charges	
YES I'd like to make a donation to the Institute	
TOTAL ENCLOSED (U.S. dollars only)	

THANK YOU FOR YOUR OR
Prices subject to

Ordering Information

Order Toll-Free 800-919-2400
Monday–Friday 7:30 am–6:00 pm (Pacific Time)
Saturday 7:30 am–4:00 pm (Pacific Time)

Fax Toll-Free 800-508-2342
Please include return phone number.

Order Through Our Secure Website
http://www.realgoods.com/renew

Order by Mail
Use attached order form and envelope.

International Orders
Phone Orders: 707-468-9292 ext. 8700
Fax: 707-462-4807
(Please call or fax for shipping quote. Funds must be in U.S. dollars, drawn from a U.S. bank.) We charge $5.00 for catalogs mailed outside the U.S.

Toll-Free Customer Service
If you have a problem with your order, our Customer Service Representatives are available between 8:30 am and 5:00 pm (Pacific Time) Monday through Friday. Call toll-free at 800-919-2400.

Product Information
Our representatives can provide you with detailed information about any of our products. If you need to know more, please call us at 800-919-2400.

Technical Assistance
For technical assistance with a renewable energy system, please call 800-919-2400 or 707-468-9292 ext. 8700, Monday through Friday, 7:30 am–6:00 pm and Saturday, 7:30 am–4:00 pm Pacific Time, or fax 707-462-4807.

Shipping
Allow 7 to 10 working days for delivery, depending on your distance from our Distribution Center in West Chester, Ohio. Overweight or bulky items may require additional shipping charges.

Shipping *(continued)*
There are a few items that ship directly from the manufacturer (they are indicated by a ◈ next to the item number in the catalog). These items are designated on your invoice and are billed at the time of order. Some large items (like refrigerators) are shipped **freight collect** from the manufacturer (they are indicated by a 🚚 next to the item number in the catalog). We will be happy to provide you with a shipping estimate. Allow extra time for delivery of these items.

Price Change
Prices, specifications, and availability are subject to change.

Freight Damage, Returns, and Adjustments
Inspect all shipments upon arrival. If you discover damage, please notify the carrier immediately. Be sure to save all shipping cartons in case you have to return merchandise. If you return merchandise to us, complete the return form on the back of our invoice. Ship to us via UPS or USPS insured. We charge a stocking fee of 10% to 25% for merchandise not in original condition, damaged, or incomplete. Please return items to: Real Goods, 9107 Meridian Way, West Chester, OH 45069-6534. *Please note—if the item you ordered is shipped directly from manufacturer (◈), there may be additional instructions; please call prior to returning.*

Join Our Lifetime Membership Program—Only $50

The Real Goods Lifetime Membership Program is now 65,000 members strong and continues to grow each month. The Lifetime Membership is for people who place high priorities on knowledge about renewable energy, protecting the environment, creating a sustainable future, and getting the best deals on the best equipment. Members are serious about conservation, energy independence, and getting the best value for their dollar!

With the Real Goods Lifetime Membership Program you'll receive:
- A 5% discount on all purchases—including sale items
- All Gaiam Real Goods catalog mailings
- A current copy of the *Solar Living Sourcebook* or other offers

Lifetime Membership $50